T0403147

Conservation and Restoration of Historic Mortars and Masonry Structures

RILEM Bookseries

Volume 42

RILEM, The International Union of Laboratories and Experts in Construction Materials, Systems and Structures, founded in 1947, is a non-governmental scientific association whose goal is to contribute to progress in the construction sciences, techniques and industries, essentially by means of the communication it fosters between research and practice. RILEM's focus is on construction materials and their use in building and civil engineering structures, covering all phases of the building process from manufacture to use and recycling of materials. More information on RILEM and its previous publications can be found on www.RILEM.net.

Indexed in SCOPUS, Google Scholar and SpringerLink.

Violeta Bokan Bosiljkov · Andreja Padovnik ·
Tilen Turk

Editors

Conservation and Restoration of Historic Mortars and Masonry Structures

HMC 2022

 Springer

Editors
Violeta Bokan Bosiljkov
Faculty of Civil and Geodetic Engineering
University of Ljubljana
Ljubljana, Slovenia

Andreja Padovnik
Faculty of Civil and Geodetic Engineering
University of Ljubljana
Ljubljana, Slovenia

Tilen Turk
Faculty of Civil and Geodetic Engineering
University of Ljubljana
Ljubljana, Slovenia

ISSN 2211-0844   ISSN 2211-0852 (electronic)
RILEM Bookseries
ISBN 978-3-031-31471-1   ISBN 978-3-031-31472-8 (eBook)
https://doi.org/10.1007/978-3-031-31472-8

This Springer imprint is published by the registered company Springer Nature Switzerland AG
The registered company address is: Gewerbestrasse 11, 6330 Cham, Switzerland

Preface

It is my great pleasure to write this preface related to the 6th Historic Mortars Conference (HMC 2022) hosted by the Faculty of Civil and Geodetic Engineering of the University of Ljubljana, Slovenia, from September 21 to September 23, 2022. The HMC 2022 Conference was held for the first time as a hybrid conference, with seventy live and ten online participants. The participants came from all five continents, despite the restrictions due to the COVID-19 disease.

Today, the HMC conferences have become an internationally recognized series of conferences designed to inform the scientific and professional communities about the progress of materials and technologies in the field of preservation, conservation and strengthening of historic structures. As with HMC 2022, the implementation of these conferences is supported by the members of the technical committees of RILEM in the fields of historic and repair mortars (TC 167-COM and TC 203-RHM), grouts (TC 243-SGM) and test methods for evaluating their properties (TC 277-LHS). Ljubljana was the sixth European city in a row to host one of the HMC conferences, after Lisbon (2008), Prague (2010), Glasgow (2013), Fira, Santorini (2016) and Pamplona (2019). I would like to take this opportunity to thank my predecessors who have presided over these conferences: Rosario Veiga, Jan Valek, John Hughes, Ioanna Papayianni and Jose Ignacio Alvarez, without you there would be no HMC 2022.

Following the tradition of previous HMC conferences, the topics of HMC 2022 included the characterization of historic mortars and masonry structures—sampling and test methods; historic production, processing and application of mortars, renders and grouts; assessment of historic renders and plasters; conservation and preventing conservation case studies; repair mortars and grouts—requirements and design, compatibility issues, durability and effectiveness and adequacy of testing procedures.

This time, however, special attention is given to historic mortars in which one of the binders or the only binder is Portland cement and to the structures in which these materials are used. This is in conjunction with the aim to highlight the architectural works of the Slovenian architect Jožef Plečnik, who used Portland cement and concrete in abundance in Ljubljana, but also Vienna, Prague and Belgrade. In 2021, selected works by Plečnik in Ljubljana were included in the UNESCO List of World Cultural and Natural Heritage, and the Government of the Republic of Slovenia declared 2022, when the 150th anniversary of the architect's birth was celebrated, as Plečnik Year.

Therefore, it is no coincidence that both invited speakers at the HMC 2022 focused on Plečnik's legacy. Prof. Johannes Weber presented the advantages of image-related microanalysis using samples from Plečnik's buildings, while Ana Porok, Curator of the Plečnik House, presented Plečnik's Ljubljana to the conference audience. The opening session took place in the Church of St. John the Baptist, also known as the Trnovo Church. This church played a special role in Plečnik's life and work. At that time, the pastor of the Trnovo church and Slovenian writer Fran Saleški Finžgar was Plečnik's close friend. Plečnik's house is behind the church, and the friends never had a fence between the architect's house and the church. We are very grateful to the present pastor of the church, Tone Kompare, for allowing us to use the church as our opening venue.

The conference participants also met in other important historic buildings in Ljubljana: in the Ljubljana Castle and Cukrarna, a former sugar refinery that now serves as an exhibition space for contemporary art, which was made possible by the Mayor of Ljubljana, Mr. Zoran Janković. A heartfelt thank you to the City of Ljubljana and the Mayor.

I would like to thank many who made HMC 2022 possible: management and support services of the Faculty of Civil and Geodetic Engineering, the sponsors for their invaluable support, my colleagues at the Chair of Testing in Materials and Structures, as well as Robert Klinc and Romana Hudin.

My special thanks go to the members of the Scientific Committee of the conference for their dedication and responsiveness in the process of reviewing the papers and for many constructive comments and suggestions for improvement. Thanks to you, these proceedings contain selection of high-quality full papers submitted to the conference, arranged in topics that also reflect oral presentation sessions.

Last but not least, I would like to thank all the speakers, chairpersons and conference participants.

Together you made the HMC 2022 a success.

Violeta Bokan Bosiljkov
HMC 2022 Chair

Organization

Organizing Committee

Conference Chair

Violeta Bokan Bosiljkov   University of Ljubljana, Slovenia

Members

Andreja Padovnik	University of Ljubljana, Slovenia
Petra Štukovnik	University of Ljubljana, Slovenia
Vlatko Bosiljkov	University of Ljubljana, Slovenia
David Antolinc	University of Ljubljana, Slovenia
Marjan Marinšek	University of Ljubljana, Slovenia
Tilen Turk	University of Ljubljana, Slovenia

Scientific Committee

Alvarez, José Ignacio	University of Navarra, Spain
Biçer-Şimşir, Beril	The Getty Conservation Institute, USA
Bokan Bosiljkov, Violeta	University of Ljubljana, Slovenia
Faria, Paulina	Universidade Nova de Lisboa, Portugal
Groot, Caspar	Delft University of Technology, Netherlands
Gulotta, Davide	The Getty Conservation Institute, USA
Hughes, John	University of the West of Scotland, UK
Ioannou, Ioannis	University of Cyprus, Cyprus
Maravelaki-Kalaitzaki, Pagona-Noni	Technical University of Crete, Greece
Martínez Ramírez, Sagrario	Consejo Superior de Investigaciones Científicas, IEM-CSIC, Spain
Pachta, Vasiliki	Aristotle University of Thessaloniki, Greece
Papayianni, Ioanna	Aristotle University of Thessaloniki, Greece
Pasian, Chiara	University of Malta, Malta
Pavia, Sara	Trinity College Dublin, Ireland
Peter, Ulrike	Lhoist Group, Belgium
Secco, Michele	University of Padua, Italy

Contents

**Historic Production, Processing and Application of Mortars, Renders
and Grouts. Lime Technologies**

**Mortars in Archaeological Sites. Construction History. Archaeometry.
Dating of Historic Mortars**

**Historic Renders and Plasters. Gypsum-Based Plasters and Mortars.
Adobe and Mud Mortars. Rammed Earth Constructions. Natural
and Roman Cement**

Historic Portland Cement-Air Lime Mortars. Historic Portland Cement Mortars

Conservation Issues Concerning Mortars, Plasters, Renders and Grouts. Diagnosis. Decay and Damage Mechanisms. Case Studies

**Preservation. Consolidation Materials and Techniques. Development
of New Products. Preventive Conservation**

**Repair Mortars and Grouts. Requirements and Design. Compatibility
Issues. Durability and Effectiveness. Adequacy of Testing Procedures**

RILEM Publications

The following list is presenting the global offer of RILEM Publications, sorted by series. Each publication is available in printed version and/or in online version.

RILEM Proceedings (PRO)

PRO 1: Durability of High Performance Concrete (ISBN: 2-912143-03-9; e-ISBN: 2-351580-12-5; e-ISBN: 2351580125); *Ed. H. Sommer*

PRO 2: Chloride Penetration into Concrete (ISBN: 2-912143-00-04; e-ISBN: 2912143454); *Eds. L.-O. Nilsson and J.-P. Ollivier*

PRO 3: Evaluation and Strengthening of Existing Masonry Structures (ISBN: 2-912143-02-0; e-ISBN: 2351580141); *Eds. L. Binda and C. Modena*

PRO 4: Concrete: From Material to Structure (ISBN: 2-912143-04-7; e-ISBN: 2351580206); *Eds. J.-P. Bournazel and Y. Malier*

PRO 5: The Role of Admixtures in High Performance Concrete (ISBN: 2-912143-05-5; e-ISBN: 2351580214); *Eds. J. G. Cabrera and R. Rivera-Villarreal*

PRO 6: High Performance Fiber Reinforced Cement Composites - HPFRCC 3 (ISBN: 2-912143-06-3; e-ISBN: 2351580222); *Eds. H. W. Reinhardt and A. E. Naaman*

PRO 7: 1st International RILEM Symposium on Self-Compacting Concrete (ISBN: 2-912143-09-8; e-ISBN: 2912143721); *Eds. Å. Skarendahl and Ö. Petersson*

PRO 8: International RILEM Symposium on Timber Engineering (ISBN: 2-912143-10-1; e-ISBN: 2351580230); *Ed. L. Boström*

PRO 9: 2nd International RILEM Symposium on Adhesion between Polymers and Concrete ISAP '99 (ISBN: 2-912143-11-X; e-ISBN: 2351580249); *Eds. Y. Ohama and M. Puterman*

PRO 10: 3rd International RILEM Symposium on Durability of Building and Construction Sealants (ISBN: 2-912143-13-6; e-ISBN: 2351580257); *Ed. A. T. Wolf*

PRO 11: 4th International RILEM Conference on Reflective Cracking in Pavements (ISBN: 2-912143-14-4; e-ISBN: 2351580265); *Eds. A. O. Abd El Halim, D. A. Taylor and El H. H. Mohamed*

PRO 12: International RILEM Workshop on Historic Mortars: Characteristics and Tests (ISBN: 2-912143-15-2; e-ISBN: 2351580273); *Eds. P. Bartos, C. Groot and J. J. Hughes*

PRO 13: 2nd International RILEM Symposium on Hydration and Setting (ISBN: 2-912143-16-0; e-ISBN: 2351580281); *Ed. A. Nonat*

PRO 14: Integrated Life-Cycle Design of Materials and Structures - ILCDES 2000 (ISBN: 951-758-408-3; e-ISBN: 235158029X); (ISSN: 0356-9403); *Ed. S. Sarja*

PRO 15: Fifth RILEM Symposium on Fibre-Reinforced Concretes (FRC) - BEFIB'2000 (ISBN: 2-912143-18-7; e-ISBN: 291214373X); *Eds. P. Rossi and G. Chanvillard*

PRO 16: Life Prediction and Management of Concrete Structures (ISBN: 2-912143-19-5; e-ISBN: 2351580303); *Ed. D. Naus*

PRO 17: Shrinkage of Concrete – Shrinkage 2000 (ISBN: 2-912143-20-9; e-ISBN: 2351580311); *Eds. V. Baroghel-Bouny and P.-C. Aïtcin*

PRO 18: Measurement and Interpretation of the On-Site Corrosion Rate (ISBN: 2-912143-21-7; e-ISBN: 235158032X); *Eds. C. Andrade, C. Alonso, J. Fullea, J. Polimon and J. Rodriguez*

PRO 19: Testing and Modelling the Chloride Ingress into Concrete (ISBN: 2-912143-22-5; e-ISBN: 2351580338); *Eds. C. Andrade and J. Kropp*

PRO 20: 1st International RILEM Workshop on Microbial Impacts on Building Materials (CD 02) (e-ISBN 978-2-35158-013-4); *Ed. M. Ribas Silva*

PRO 21: International RILEM Symposium on Connections between Steel and Concrete (ISBN: 2-912143-25-X; e-ISBN: 2351580346); *Ed. R. Eligehausen*

PRO 22: International RILEM Symposium on Joints in Timber Structures (ISBN: 2-912143-28-4; e-ISBN: 2351580354); *Eds. S. Aicher and H.-W. Reinhardt*

PRO 23: International RILEM Conference on Early Age Cracking in Cementitious Systems (ISBN: 2-912143-29-2; e-ISBN: 2351580362); *Eds. K. Kovler and A. Bentur*

PRO 24: 2nd International RILEM Workshop on Frost Resistance of Concrete (ISBN: 2-912143-30-6; e-ISBN: 2351580370); *Eds. M. J. Setzer, R. Auberg and H.-J. Keck*

PRO 25: International RILEM Workshop on Frost Damage in Concrete (ISBN: 2-912143-31-4; e-ISBN: 2351580389); *Eds. D. J. Janssen, M. J. Setzer and M. B. Snyder*

PRO 26: International RILEM Workshop on On-Site Control and Evaluation of Masonry Structures (ISBN: 2-912143-34-9; e-ISBN: 2351580141); *Eds. L. Binda and R. C. de Vekey*

PRO 27: International RILEM Symposium on Building Joint Sealants (CD03; e-ISBN: 235158015X); *Ed. A. T. Wolf*

PRO 28: 6th International RILEM Symposium on Performance Testing and Evaluation of Bituminous Materials - PTEBM'03 (ISBN: 2-912143-35-7; e-ISBN: 978-2-912143-77-8); *Ed. M. N. Partl*

PRO 29: 2nd International RILEM Workshop on Life Prediction and Ageing Management of Concrete Structures (ISBN: 2-912143-36-5; e-ISBN: 2912143780); *Ed. D. J. Naus*

PRO 30: 4th International RILEM Workshop on High Performance Fiber Reinforced Cement Composites - HPFRCC 4 (ISBN: 2-912143-37-3; e-ISBN: 2912143799); *Eds. A. E. Naaman and H. W. Reinhardt*

PRO 31: International RILEM Workshop on Test and Design Methods for Steel Fibre Reinforced Concrete: Background and Experiences (ISBN: 2-912143-38-1; e-ISBN: 2351580168); *Eds. B. Schnütgen and L. Vandewalle*

PRO 32: International Conference on Advances in Concrete and Structures 2 vol. (ISBN (set): 2-912143-41-1; e-ISBN: 2351580176); *Eds. Ying-shu Yuan, Surendra P. Shah and Heng-lin Lü*

PRO 33: 3rd International Symposium on Self-Compacting Concrete (ISBN: 2-912143-42-X; e-ISBN: 2912143713); *Eds. Ó. Wallevik and I. Níelsson*

PRO 34: International RILEM Conference on Microbial Impact on Building Materials (ISBN: 2-912143-43-8; e-ISBN: 2351580184); *Ed. M. Ribas Silva*

PRO 35: International RILEM TC 186-ISA on Internal Sulfate Attack and Delayed Ettringite Formation (ISBN: 2-912143-44-6; e-ISBN: 2912143802); *Eds. K. Scrivener and J. Skalny*

PRO 36: International RILEM Symposium on Concrete Science and Engineering – A Tribute to Arnon Bentur (ISBN: 2-912143-46-2; e-ISBN: 2912143586); *Eds. K. Kovler, J. Marchand, S. Mindess and J. Weiss*

PRO 37: 5th International RILEM Conference on Cracking in Pavements – Mitigation, Risk Assessment and Prevention (ISBN: 2-912143-47-0; e-ISBN: 2912143764); *Eds. C. Petit, I. Al-Qadi and A. Millien*

PRO 38: 3rd International RILEM Workshop on Testing and Modelling the Chloride Ingress into Concrete (ISBN: 2-912143-48-9; e-ISBN: 2912143578); *Eds. C. Andrade and J. Kropp*

PRO 39: 6th International RILEM Symposium on Fibre-Reinforced Concretes - BEFIB 2004 (ISBN: 2-912143-51-9; e-ISBN: 2912143748); *Eds. M. Di Prisco, R. Felicetti and G. A. Plizzari*

PRO 40: International RILEM Conference on the Use of Recycled Materials in Buildings and Structures (ISBN: 2-912143-52-7; e-ISBN: 2912143756); *Eds. E. Vázquez, Ch. F. Hendriks and G. M. T. Janssen*

PRO 41: RILEM International Symposium on Environment-Conscious Materials and Systems for Sustainable Development (ISBN: 2-912143-55-1; e-ISBN: 2912143640); *Eds. N. Kashino and Y. Ohama*

PRO 42: SCC'2005 - China: 1st International Symposium on Design, Performance and Use of Self-Consolidating Concrete (ISBN: 2-912143-61-6; e-ISBN: 2912143624); *Eds. Zhiwu Yu, Caijun Shi, Kamal Henri Khayat and Youjun Xie*

PRO 43: International RILEM Workshop on Bonded Concrete Overlays (e-ISBN: 2-912143-83-7); *Eds. J. L. Granju and J. Silfwerbrand*

PRO 44: 2nd International RILEM Workshop on Microbial Impacts on Building Materials (CD11) (e-ISBN: 2-912143-84-5); *Ed. M. Ribas Silva*

PRO 45: 2nd International Symposium on Nanotechnology in Construction, Bilbao (ISBN: 2-912143-87-X; e-ISBN: 2912143888); *Eds. Peter J. M. Bartos, Yolanda de Miguel and Antonio Porro*

PRO 46: ConcreteLife'06 - International RILEM-JCI Seminar on Concrete Durability and Service Life Planning: Curing, Crack Control, Performance in Harsh Environments (ISBN: 2-912143-89-6; e-ISBN: 291214390X); *Ed. K. Kovler*

PRO 47: International RILEM Workshop on Performance Based Evaluation and Indicators for Concrete Durability (ISBN: 978-2-912143-95-2; e-ISBN: 9782912143969); *Eds. V. Baroghel-Bouny, C. Andrade, R. Torrent and K. Scrivener*

PRO 48: 1st International RILEM Symposium on Advances in Concrete through Science and Engineering (e-ISBN: 2-912143-92-6); *Eds. J. Weiss, K. Kovler, J. Marchand and S. Mindess*

PRO 49: International RILEM Workshop on High Performance Fiber Reinforced Cementitious Composites in Structural Applications (ISBN: 2-912143-93-4; e-ISBN: 2912143942); *Eds. G. Fischer and V. C. Li*

PRO 50: 1st International RILEM Symposium on Textile Reinforced Concrete (ISBN: 2-912143-97-7; e-ISBN: 2351580087); *Eds. Josef Hegger, Wolfgang Brameshuber and Norbert Will*

PRO 51: 2nd International Symposium on Advances in Concrete through Science and Engineering (ISBN: 2-35158-003-6; e-ISBN: 2-35158-002-8); *Eds. J. Marchand, B. Bissonnette, R. Gagné, M. Jolin and F. Paradis*

PRO 52: Volume Changes of Hardening Concrete: Testing and Mitigation (ISBN: 2-35158-004-4; e-ISBN: 2-35158-005-2); *Eds. O. M. Jensen, P. Lura and K. Kovler*

PRO 53: High Performance Fiber Reinforced Cement Composites - HPFRCC5 (ISBN: 978-2-35158-046-2; e-ISBN: 978-2-35158-089-9); *Eds. H. W. Reinhardt and A. E. Naaman*

PRO 54: 5th International RILEM Symposium on Self-Compacting Concrete (ISBN: 978-2-35158-047-9; e-ISBN: 978-2-35158-088-2); *Eds. G. De Schutter and V. Boel*

PRO 55: International RILEM Symposium Photocatalysis, Environment and Construction Materials (ISBN: 978-2-35158-056-1; e-ISBN: 978-2-35158-057-8); *Eds. P. Baglioni and L. Cassar*

PRO 56: International RILEM Workshop on Integral Service Life Modelling of Concrete Structures (ISBN 978-2-35158-058-5; e-ISBN: 978-2-35158-090-5); *Eds. R. M. Ferreira, J. Gulikers and C. Andrade*

PRO 57: RILEM Workshop on Performance of cement-based materials in aggressive aqueous environments (e-ISBN: 978-2-35158-059-2); *Ed. N. De Belie*

PRO 58: International RILEM Symposium on Concrete Modelling - CONMOD'08 (ISBN: 978-2-35158-060-8; e-ISBN: 978-2-35158-076-9); *Eds. E. Schlangen and G. De Schutter*

PRO 59: International RILEM Conference on On Site Assessment of Concrete, Masonry and Timber Structures - SACoMaTiS 2008 (ISBN set: 978-2-35158-061-5; e-ISBN: 978-2-35158-075-2); *Eds. L. Binda, M. di Prisco and R. Felicetti*

PRO 60: Seventh RILEM International Symposium on Fibre Reinforced Concrete: Design and Applications - BEFIB 2008 (ISBN: 978-2-35158-064-6; e-ISBN: 978-2-35158-086-8); *Ed. R. Gettu*

PRO 61: 1st International Conference on Microstructure Related Durability of Cementitious Composites 2 vol., (ISBN: 978-2-35158-065-3; e-ISBN: 978-2-35158-084-4); *Eds. W. Sun, K. van Breugel, C. Miao, G. Ye and H. Chen*

PRO 62: NSF/ RILEM Workshop: In-situ Evaluation of Historic Wood and Masonry Structures (e-ISBN: 978-2-35158-068-4); *Eds. B. Kasal, R. Anthony and M. Drdácký*

PRO 63: Concrete in Aggressive Aqueous Environments: Performance, Testing and Modelling, 2 vol., (ISBN: 978-2-35158-071-4; e-ISBN: 978-2-35158-082-0); *Eds. M. G. Alexander and A. Bertron*

PRO 64: Long Term Performance of Cementitious Barriers and Re inforced Concrete in Nuclear Power Plants and Waste Management - NUCPERF 2009 (ISBN: 978-2-35158-072-1; e-ISBN: 978-2-35158-087-5); *Eds. V. L'Hostis, R. Gens and C. Gallé*

PRO 65: Design Performance and Use of Self-consolidating Concrete - SCC'2009 (ISBN: 978-2-35158-073-8; e-ISBN: 978-2-35158-093-6); *Eds. C. Shi, Z. Yu, K. H. Khayat and P. Yan*

PRO 66: 2nd International RILEM Workshop on Concrete Durability and Service Life Planning - ConcreteLife'09 (ISBN: 978-2-35158-074-5; ISBN: 978-2-35158-074-5); *Ed. K. Kovler*

PRO 67: Repairs Mortars for Historic Masonry (e-ISBN: 978-2-35158-083-7); *Ed. C. Groot*

PRO 68: Proceedings of the 3rd International RILEM Symposium on 'Rheology of Cement Suspensions such as Fresh Concrete (ISBN 978-2-35158-091-2; e-ISBN: 978-2-35158-092-9); *Eds. O. H. Wallevik, S. Kubens and S. Oesterheld*

PRO 69: 3rd International PhD Student Workshop on 'Modelling the Durability of Reinforced Concrete (ISBN: 978-2-35158-095-0); *Eds. R. M. Ferreira, J. Gulikers and C. Andrade*

PRO 70: 2nd International Conference on 'Service Life Design for Infrastructure' (ISBN set: 978-2-35158-096-7, e-ISBN: 978-2-35158-097-4); *Eds. K. van Breugel, G. Ye and Y. Yuan*

PRO 71: Advances in Civil Engineering Materials - The 50-year Teaching Anniversary of Prof. Sun Wei' (ISBN: 978-2-35158-098-1; e-ISBN: 978-2-35158-099-8); *Eds. C. Miao, G. Ye and H. Chen*

PRO 72: First International Conference on 'Advances in Chemically-Activated Materials – CAM'2010' (2010), 264 pp, ISBN: 978-2-35158-101-8; e-ISBN: 978-2-35158-115-5, *Eds. Caijun Shi and Xiaodong Shen*

PRO 73: 2nd International Conference on 'Waste Engineering and Management - ICWEM 2010' (2010), 894 pp, ISBN: 978-2-35158-102-5; e-ISBN: 978-2-35158-103-2, *Eds. J. Zh. Xiao, Y. Zhang, M. S. Cheung and R. Chu*

PRO 74: International RILEM Conference on 'Use of Superabsorbent Polymers and Other New Addditives in Concrete' (2010) 374 pp., ISBN: 978-2-35158-104-9; e-ISBN: 978-2-35158-105-6; *Eds. O. M. Jensen, M. T. Hasholt and S. Laustsen*

PRO 75: International Conference on 'Material Science - 2nd ICTRC - Textile Reinforced Concrete - Theme 1' (2010) 436 pp., ISBN: 978-2-35158-106-3; e-ISBN: 978-2-35158-107-0; *Ed. W. Brameshuber*

PRO 76: International Conference on 'Material Science - HetMat - Modelling of Heterogeneous Materials - Theme 2' (2010) 255 pp., ISBN: 978-2-35158-108-7; e-ISBN: 978-2-35158-109-4; *Ed. W. Brameshuber*

PRO 77: International Conference on 'Material Science - AdIPoC - Additions Improving Properties of Concrete - Theme 3' (2010) 459 pp., ISBN: 978-2-35158-110-0; e-ISBN: 978-2-35158-111-7; *Ed. W. Brameshuber*

PRO 78: 2nd Historic Mortars Conference and RILEM TC 203-RHM Final Workshop – HMC2010 (2010) 1416 pp., e-ISBN: 978-2-35158-112-4; *Eds. J. Válek, C. Groot and J. J. Hughes*

PRO 79: International RILEM Conference on Advances in Construction Materials Through Science and Engineering (2011) 213 pp., ISBN: 978-2-35158-116-2, e-ISBN: 978-2-35158-117-9; *Eds. Christopher Leung and K. T. Wan*

PRO 80: 2nd International RILEM Conference on Concrete Spalling due to Fire Exposure (2011) 453 pp., ISBN: 978-2-35158-118-6, e-ISBN: 978-2-35158-119-3; *Eds. E. A. B. Koenders and F. Dehn*

PRO 81: 2nd International RILEM Conference on Strain Hardening Cementitious Composites (SHCC2-Rio) (2011) 451 pp., ISBN: 978-2-35158-120-9, e-ISBN: 978-2-35158-121-6; *Eds. R. D. Toledo Filho, F. A. Silva, E. A. B. Koenders and E. M. R. Fairbairn*

PRO 82: 2nd International RILEM Conference on Progress of Recycling in the Built Environment (2011) 507 pp., e-ISBN: 978-2-35158-122-3; *Eds. V. M. John, E. Vazquez, S. C. Angulo and C. Ulsen*

PRO 83: 2nd International Conference on Microstructural-related Durability of Cementitious Composites (2012) 250 pp., ISBN: 978-2-35158-129-2; e-ISBN: 978-2-35158-123-0; *Eds. G. Ye, K. van Breugel, W. Sun and C. Miao*

PRO 84: CONSEC13 - Seventh International Conference on Concrete under Severe Conditions – Environment and Loading (2013) 1930 pp., ISBN: 978-2-35158-124-7; e-ISBN: 978-2-35158-134-6; *Eds. Z. J. Li, W. Sun, C. W. Miao, K. Sakai, O. E. Gjorv and N. Banthia*

PRO 85: RILEM-JCI International Workshop on Crack Control of Mass Concrete and Related issues concerning Early-Age of Concrete Structures – ConCrack 3 – Control of Cracking in Concrete Structures 3 (2012) 237 pp., ISBN: 978-2-35158-125-4; e-ISBN: 978-2-35158-126-1; *Eds. F. Toutlemonde and J.-M. Torrenti*

PRO 86: International Symposium on Life Cycle Assessment and Construction (2012) 414 pp., ISBN: 978-2-35158-127-8, e-ISBN: 978-2-35158-128-5; *Eds. A. Ventura and C. de la Roche*

PRO 87: UHPFRC 2013 – RILEM-fib-AFGC International Symposium on Ultra-High Performance Fibre-Reinforced Concrete (2013), ISBN: 978-2-35158-130-8, e-ISBN: 978-2-35158-131-5; *Eds. F. Toutlemonde*

PRO 88: 8th RILEM International Symposium on Fibre Reinforced Concrete (2012) 344 pp., ISBN: 978-2-35158-132-2, e-ISBN: 978-2-35158-133-9; *Eds. Joaquim A. O. Barros*

PRO 89: RILEM International workshop on performance-based specification and control of concrete durability (2014) 678 pp, ISBN: 978-2-35158-135-3, e-ISBN: 978-2-35158-136-0; *Eds. D. Bjegović, H. Beushausen and M. Serdar*

PRO 90: 7th RILEM International Conference on Self-Compacting Concrete and of the 1st RILEM International Conference on Rheology and Processing of Construction Materials (2013) 396 pp, ISBN: 978-2-35158-137-7, e-ISBN: 978-2-35158-138-4; *Eds. Nicolas Roussel and Hela Bessaies-Bey*

PRO 91: CONMOD 2014 - RILEM International Symposium on Concrete Modelling (2014), ISBN: 978-2-35158-139-1; e-ISBN: 978-2-35158-140-7; *Eds. Kefei Li, Peiyu Yan and Rongwei Yang*

PRO 92: CAM 2014 - 2nd International Conference on advances in chemically-activated materials (2014) 392 pp., ISBN: 978-2-35158-141-4; e-ISBN: 978-2-35158-142-1; *Eds. Caijun Shi and Xiadong Shen*

PRO 93: SCC 2014 - 3rd International Symposium on Design, Performance and Use of Self-Consolidating Concrete (2014) 438 pp., ISBN: 978-2-35158-143-8; e-ISBN: 978-2-35158-144-5; *Eds. Caijun Shi, Zhihua Ou and Kamal H. Khayat*

PRO 94 (online version): HPFRCC-7 - 7th RILEM conference on High performance fiber reinforced cement composites (2015), e-ISBN: 978-2-35158-146-9; *Eds. H. W. Reinhardt, G. J. Parra-Montesinos and H. Garrecht*

PRO 109 (2 volumes): MSSCE 2016 - Service Life of Cement-Based Materials and Structures (2016), ISBN Vol. 1: 978-2-35158-170-4, Vol. 2: 978-2-35158-171-4, Set Vol. 1&2: 978-2-35158-172-8, e-ISBN : 978-2-35158-173-5; *Eds. Miguel Azenha, Ivan Gabrijel, Dirk Schlicke, Terje Kanstad and Ole Mejlhede Jensen*

PRO 110: MSSCE 2016 - Historical Masonry (2016), ISBN: 978-2-35158-178-0, e-ISBN: 978-2-35158-179-7; *Eds. Inge Rörig-Dalgaard and Ioannis Ioannou*

PRO 111: MSSCE 2016 - Electrochemistry in Civil Engineering (2016), ISBN: 978-2-35158-176-6, e-ISBN: 978-2-35158-177-3; *Ed. Lisbeth M. Ottosen*

PRO 112: MSSCE 2016 - Moisture in Materials and Structures (2016), ISBN: 978-2-35158-178-0, e-ISBN: 978-2-35158-179-7; *Eds. Kurt Kielsgaard Hansen, Carsten Rode and Lars-Olof Nilsson*

PRO 113: MSSCE 2016 - Concrete with Supplementary Cementitious Materials (2016), ISBN: 978-2-35158-178-0, e-ISBN: 978-2-35158-179-7; *Eds. Ole Mejlhede Jensen, Konstantin Kovler and Nele De Belie*

PRO 114: MSSCE 2016 - Frost Action in Concrete (2016), ISBN: 978-2-35158-182-7, e-ISBN: 978-2-35158-183-4; *Eds. Marianne Tange Hasholt, Katja Fridh and R. Doug Hooton*

PRO 115: MSSCE 2016 - Fresh Concrete (2016), ISBN: 978-2-35158-184-1, e-ISBN: 978-2-35158-185-8; *Eds. Lars N. Thrane, Claus Pade, Oldrich Svec and Nicolas Roussel*

PRO 116: BEFIB 2016 – 9th RILEM International Symposium on Fiber Reinforced Concrete (2016), ISBN: 978-2-35158-187-2, e-ISBN: 978-2-35158-186-5; *Eds. N. Banthia, M. di Prisco and S. Soleimani-Dashtaki*

PRO 117: 3rd International RILEM Conference on Microstructure Related Durability of Cementitious Composites (2016), ISBN: 978-2-35158-188-9, e-ISBN: 978-2-35158-189-6; *Eds. Changwen Miao, Wei Sun, Jiaping Liu, Huisu Chen, Guang Ye and Klaas van Breugel*

PRO 118 (4 volumes): International Conference on Advances in Construction Materials and Systems (2017), ISBN Set: 978-2-35158-190-2, Vol. 1: 978-2-35158-193-3, Vol. 2: 978-2-35158-194-0, Vol. 3: ISBN:978-2-35158-195-7, Vol. 4: ISBN:978-2-35158-196-4, e-ISBN: 978-2-35158-191-9; *Eds. Manu Santhanam, Ravindra Gettu, Radhakrishna G. Pillai and Sunitha K. Nayar*

PRO 119 (online version): ICBBM 2017 - Second International RILEM Conference on Bio-based Building Materials, (2017), e-ISBN: 978-2-35158-192-6; *Eds. Sofiane Amziane*

PRO 120 (2 volumes): EAC-02 - 2nd International RILEM/COST Conference on Early Age Cracking and Serviceability in Cement-based Materials and Structures, (2017), Vol. 1: 978-2-35158-199-5, Vol. 2: 978-2-35158-200-8, Set: 978-2-35158-197-1, e-ISBN: 978-2-35158-198-8; *Eds. Stéphanie Staquet and Dimitrios Aggelis*

PRO 121 (2 volumes): SynerCrete18: Interdisciplinary Approaches for Cement-based Materials and Structural Concrete: Synergizing Expertise and Bridging Scales of Space and Time, (2018), Set: 978-2-35158-202-2, Vol.1: 978-2-35158-211-4, Vol. 2: 978-2-35158-212-1, e-ISBN: 978-2-35158-203-9; *Eds. Miguel Azenha, Dirk Schlicke, Farid Benboudjema and Agnieszka Knoppik*

PRO 122: SCC'2018 China - Fourth International Symposium on Design, Performance and Use of Self-Consolidating Concrete, (2018), ISBN: 978-2-35158-204-6, e-ISBN: 978-2-35158-205-3; *Eds. C. Shi, Z. Zhang and K. H. Khayat*

PRO 123: Final Conference of RILEM TC 253-MCI: Microorganisms-Cementitious Materials Interactions (2018), Set: 978-2-35158-207-7, Vol.1: 978-2-35158-209-1, Vol.2: 978-2-35158-210-7, e-ISBN: 978-2-35158-206-0; *Ed. Alexandra Bertron*

PRO 124 (online version): Fourth International Conference Progress of Recycling in the Built Environment (2018), e-ISBN: 978-2-35158-208-4; *Eds. Isabel M. Martins, Carina Ulsen and Yury Villagran*

PRO 125 (online version): SLD4 - 4th International Conference on Service Life Design for Infrastructures (2018), e-ISBN: 978-2-35158-213-8; *Eds. Guang Ye, Yong Yuan, Claudia Romero Rodriguez, Hongzhi Zhang and Branko Savija*

PRO 126: Workshop on Concrete Modelling and Material Behaviour in honor of Professor Klaas van Breugel (2018), ISBN: 978-2-35158-214-5, e-ISBN: 978-2-35158-215-2; *Ed. Guang Ye*

PRO 127 (online version): CONMOD2018 - Symposium on Concrete Modelling (2018), e-ISBN: 978-2-35158-216-9; *Eds. Erik Schlangen, Geert de Schutter, Branko Savija, Hongzhi Zhang and Claudia Romero Rodriguez*

PRO 128: SMSS2019 - International Conference on Sustainable Materials, Systems and Structures (2019), ISBN: 978-2-35158-217-6, e-ISBN: 978-2-35158-218-3

PRO 129: 2nd International Conference on UHPC Materials and Structures (UHPC2018-China), ISBN: 978-2-35158-219-0, e-ISBN: 978-2-35158-220-6;

PRO 130: 5th Historic Mortars Conference (2019), ISBN: 978-2-35158-221-3, e-ISBN: 978-2-35158-222-0; *Eds. José Ignacio Álvarez, José María Fernández, Íñigo Navarro, Adrián Durán and Rafael Sirera*

PRO 131 (online version): 3rd International Conference on Bio-Based Building Materials (ICBBM2019), e-ISBN: 978-2-35158-229-9; *Eds. Mohammed Sonebi, Sofiane Amziane and Jonathan Page*

PRO 132: IRWRMC'18 - International RILEM Workshop on Rheological Measurements of Cement-based Materials (2018), ISBN: 978-2-35158-230-5, e-ISBN: 978-2-35158-231-2; *Eds. Chafika Djelal and Yannick Vanhove*

PRO 133 (online version): CO2STO2019 - International Workshop CO2 Storage in Concrete (2019), e-ISBN: 978-2-35158-232-9; *Eds. Assia Djerbi, Othman Omikrine-Metalssi and Teddy Fen-Chong*

PRO 134: 3rd ACF/HNU International Conference on UHPC Materials and Structures - UHPC'2020, ISBN: 978-2-35158-233-6, e-ISBN: 978-2-35158-234-3; *Eds. Caijun Shi and Jiaping Liu*

PRO 135: Fourth International Conference on Chemically Activated Materials (CAM2021), ISBN: 978-2-35158-235-0, e-ISBN: 978-2-35158-236-7; *Eds. Caijun Shi and Xiang Hu*

RILEM Reports (REP)

Report 19: Considerations for Use in Managing the Aging of Nuclear Power Plant Concrete Structures (ISBN: 2-912143-07-1); *Ed. D. J. Naus*

Report 20: Engineering and Transport Properties of the Interfacial Transition Zone in Cementitious Composites (ISBN: 2-912143-08-X); *Eds. M. G. Alexander, G. Arliguie, G. Ballivy, A. Bentur and J. Marchand*

Report 21: Durability of Building Sealants (ISBN: 2-912143-12-8); *Ed. A. T. Wolf*

Report 22: Sustainable Raw Materials - Construction and Demolition Waste (ISBN: 2-912143-17-9); *Eds. C. F. Hendriks and H. S. Pietersen*

Report 23: Self-Compacting Concrete state-of-the-art report (ISBN: 2-912143-23-3); *Eds. Å. Skarendahl and Ö. Petersson*

Report 24: Workability and Rheology of Fresh Concrete: Compendium of Tests (ISBN: 2-912143-32-2); *Eds. P. J. M. Bartos, M. Sonebi and A. K. Tamimi*

Report 25: Early Age Cracking in Cementitious Systems (ISBN: 2-912143-33-0); *Ed. A. Bentur*

Report 26: Towards Sustainable Roofing (Joint Committee CIB/RILEM) (CD 07) (e-ISBN 978-2-912143-65-5); *Eds. Thomas W. Hutchinson and Keith Roberts*

Report 27: Condition Assessment of Roofs (Joint Committee CIB/RILEM) (CD 08) (e-ISBN 978-2-912143-66-2); *Ed. CIB W 83/RILEM TC166-RMS*

Report 28: Final report of RILEM TC 167-COM 'Characterisation of Old Mortars with Respect to Their Repair (ISBN: 978-2-912143-56-3); *Eds. C. Groot, G. Ashall and J. Hughes*

Report 29: Pavement Performance Prediction and Evaluation (PPPE): Interlaboratory Tests (e-ISBN: 2-912143-68-3); *Eds. M. Partl and H. Piber*

Report 30: Final Report of RILEM TC 198-URM 'Use of Recycled Materials' (ISBN: 2-912143-82-9; e-ISBN: 2-912143-69-1); *Eds. Ch. F. Hendriks, G. M. T. Janssen and E. Vázquez*

Report 31: Final Report of RILEM TC 185-ATC 'Advanced testing of cement-based materials during setting and hardening' (ISBN: 2-912143-81-0; e-ISBN: 2-912143-70-5); *Eds. H. W. Reinhardt and C. U. Grosse*

Report 32: Probabilistic Assessment of Existing Structures. A JCSS publication (ISBN 2-912143-24-1); *Ed. D. Diamantidis*

Report 33: State-of-the-Art Report of RILEM Technical Committee TC 184-IFE 'Industrial Floors' (ISBN 2-35158-006-0); *Ed. P. Seidler*

Report 34: Report of RILEM Technical Committee TC 147-FMB 'Fracture mechanics applications to anchorage and bond' Tension of Reinforced Concrete Prisms – Round Robin Analysis and Tests on Bond (e-ISBN 2-912143-91-8); *Eds. L. Elfgren and K. Noghabai*

Report 35: Final Report of RILEM Technical Committee TC 188-CSC 'Casting of Self Compacting Concrete' (ISBN 2-35158-001-X; e-ISBN: 2-912143-98-5); *Eds. Å. Skarendahl and P. Billberg*

Report 36: State-of-the-Art Report of RILEM Technical Committee TC 201-TRC 'Textile Reinforced Concrete' (ISBN 2-912143-99-3); *Ed. W. Brameshuber*

Report 37: State-of-the-Art Report of RILEM Technical Committee TC 192-ECM 'Environment-conscious construction materials and systems' (ISBN: 978-2-35158-053-0); *Eds. N. Kashino, D. Van Gemert and K. Imamoto*

Report 38: State-of-the-Art Report of RILEM Technical Committee TC 205-DSC 'Durability of Self-Compacting Concrete' (ISBN: 978-2-35158-048-6); *Eds. G. De Schutter and K. Audenaert*

Report 39: Final Report of RILEM Technical Committee TC 187-SOC 'Experimental determination of the stress-crack opening curve for concrete in tension' (ISBN 978-2-35158-049-3); *Ed. J. Planas*

Report 40: State-of-the-Art Report of RILEM Technical Committee TC 189-NEC 'Non-Destructive Evaluation of the Penetrability and Thickness of the Concrete Cover' (ISBN 978-2-35158-054-7); *Eds. R. Torrent and L. Fernández Luco*

Report 41: State-of-the-Art Report of RILEM Technical Committee TC 196-ICC 'Internal Curing of Concrete' (ISBN 978-2-35158-009-7); *Eds. K. Kovler and O. M. Jensen*

Report 42: 'Acoustic Emission and Related Non-destructive Evaluation Techniques for Crack Detection and Damage Evaluation in Concrete' - Final Report of RILEM Technical Committee 212-ACD (e-ISBN: 978-2-35158-100-1); *Ed. M. Ohtsu*

Report 45: Repair Mortars for Historic Masonry - State-of-the-Art Report of RILEM Technical Committee TC 203-RHM (e-ISBN: 978-2-35158-163-6); *Eds. Paul Maurenbrecher and Caspar Groot*

Report 46: Surface delamination of concrete industrial floors and other durability related aspects guide - Report of RILEM Technical Committee TC 268-SIF (e-ISBN: 978-2-35158-201-5); *Ed. Valerie Pollet*

Characterization of Historic Mortars and Masonry Structures. Sampling and Test Methods

Imperial Styles, Frontier Solutions: Roman Wall Painting Technology in the Province of Noricum

Anthony J. Baragona[1]([✉]), Pavla Bauerová[2,3], and Alexandra S. Rodler[4,5]

[1] The University of Applied Arts, Vienna Salzgries 14/1, 1013 Vienna, Austria
tonybaragona@gmail.com, tony.baragona@alumni.uni-ak.ac.at

[2] Institute of Theoretical and Applied Mechanics of the Czech Academy of Sciences, Prosecká 809/76, 190 00 Prague 9, Czech Republic

[3] Faculty of Civil Engineering, Czech Technical University in Prague, Thákurova 7/2077, 166 29 Prague 6, Czech Republic

[4] Research Group Object Itineraries, Department of Historical Archaeology, Austrian Archaeological Institute, Austrian Academy of Sciences, Franz Klein-Gasse 1, 1190 Vienna, Austria

[5] Department of Lithospheric Research, University of Vienna, USZ 2, Althanstrasse 14, 1090 Vienna, Austria

Abstract. Most of today's Austria was part of the alpine province of Noricum, formally incorporated into the Roman Empire in the first century C.E. As trade flourished the area was quickly Romanized and this is reflected by surviving wall paintings exhibiting high proficiency in painting and plastering technique and utilizing precious and rare pigments.

This contribution examines the differences that can be found in roughly contemporaneous Roman wall paintings from Noricum. In the context of an ongoing study of Roman pigments, the chemical profile of the top paint layers of plaster fragments in museum collections that displayed monochrome and large-scale application of commonly available Egyptian Blue and expensive Cinnabar/Vermillion were analysed semi-quantitatively by portable XRF. Then stratigraphic cross sections of wall painting samples were made from a selection of plaster fragments that included every plaster preparation layer down to the *arriccio*. These were examined by light microscopy, SEM/EDX and digital image analysis. Through this process, this study intended to determine if there is a correlation between changes in pigment production and painting and plastering technique.

These methods were able to reveal the technical differences in how wall paintings were prepared and how pigments were used in different ways at several Roman sites of Noricum. The sites closer to Italia province showed artisanship more closely resembling that used in the central Empire, while those further north evolved a unique style. This finding reflects trade routes and the development of regional techniques in the Alpine area.

1 Introduction

During the 1st century C.E., much of the alpine region of today's Austria was part of the Roman province of Noricum, where the local Celtic cultures were in the process of rapidly becoming Romanised. As part of an ongoing geochemical study by the authors

V. Bokan Bosiljkov et al. (Eds.): HMC 2022, RILEM Bookseries 42, pp. 3–17, 2023.
https://doi.org/10.1007/978-3-031-31472-8_1

focusing on the provenance/ production of the pigments Egyptian Blue (EB) and cinnabar used in the Roman wall paintings of Noricum (as well as elsewhere in the Roman Empire), a diversity of painting and plastering techniques were revealed. Portable XRF was used to aid in the selection of fragments bearing EB or cinnabar pigments with similar chemical profiles. From these, eleven samples were selected and processed into polished cross sections to expand the study beyond the surface analysis of pigments into one that considered the entire wall painting package from paint layer to *arriccio*. This was done to reveal similarities or differences in materials and processing techniques used for wall paintings at Italian and Noric sites (ca. 1st century C.E.) as well as diachronic changes between important sites in the Roman peripheral province of Noricum (1st to 3rd century C.E.). Samples from diverse locations throughout the Empire were selected to further understand trade, production and application of Roman paint and pigment for use in decorating plastered walls, and how practices differed between the central areas of the Empire and Noricum; these sites are detailed in the following Sect. 1.1.

This work focuses on plaster bearing paint containing either Egyptian blue (EB) or cinnabar. EB is widely recognized as the first artificial pigment; it is a synthetic form of the mineral cuprorivaite, made from relatively common materials available locally [1–3]. Cinnabar (mercury sulfide), on the other hand is a rare ore found in only a few places in the Mediterranean basin. It was ground up for use as a prestigious red pigment and often mixed with more locally available (and cheaper) ochre (iron oxides) pigments. While ancient sources name the mine in Almadén, Spain as the main source of cinnabar, it is known that the Romans uses sources from Italy and Slovenia as well [4–7]. Although rare, cinnabar paint was used throughout the Roman empire and therefore is believed to have been widely traded [8–10]. Thus, the study of these two pigments and how they were used helped to form a narrative describing both local artisanship and long-distance trade in the province of Noricum and how these are reflected in wall painting technique.

1.1 Archaeological Context of the Sample Material

To discover evidence of trade in materials and (potential) knowledge transfer of wall painting techniques, samples from diverse archaeological sites were analysed, including four from the province of Noricum and two from more central and established regions of the Empire. These are detailed below, with the "central" sites listed first, and flowed by those in Noricum listed by increasing distance from Rome.

The city of Pompeii was located near Naples in the Campania region of Italy, close to the Egyptian blue production sites in the Bay of Naples [11, 12]. Due to the eruption of Mount Vesuvius in 79 C.E., wall paintings buried by ash are exceptionally well preserved and are the foundation for comparative analysis of wall painting styles across the Roman world. Two Pompeiian wall painting fragments with Egyptian blue top paint layers were used for this current work. They were collected from the frigidarium of the Sarno Baths (Regio VIII, Insula 2, modern house #17–21) in the southern part of Pompeii. This building was probably constructed in the Late Republican age and underwent several architectural changes until the destruction of Pompeii in 79 C.E. [13, 14].

Ephesus, located in today's Turkey, was an ancient port city and trading hub on the Aegean Sea [15] that became part of the Roman Empire as the capital of Asia Minor in 129 B.C.E. Wall painting fragments with cinnabar paint were collected from storage boxes at the Excavation House at Ephesus coming from two of the building complexes at Ephesus, Terrace House 1, which covers an area of app. 3000 m^2 and consists of five living units with multiple rooms each and Terrace House 2, which covers an area of app. 4000 m^2; it consists of seven living units. Both Terrace House 1 and 2 were built in the first century C.E. and destroyed during an earthquake in 262/263 C.E.

Celeia (Celje, Slovenia) received municipal rights as Municipium Claudium Celeia in 45 C.E. and a road connected Aquileia through Celeia to Pannonia. Two samples (with EB) were collected and analysed for this work. They are from the Museum square (Muzejski trg) at Celje, where a Roman domus with four different rooms and dated to the 1st to 2nd century C.E. as well as parts of a small public thermae from the 3rd century C.E. were excavated. The samples are from a stratigraphic layer which was interpreted as a levelling for the 3rd century walking surface and consists only of debris of the 1st and 2nd century C.E. domus. This domus was systematically demolished and all usable material was recycled. The walls of the domus were left up to 1.3 m high and space between these walls was filled with the levelling made from debris (up to 10 cm diameter and consisting of plaster, pieces of mortar, wall painting and stucco fragments, small pieces of brick and small stones). The wall paintings from this layer are all from the same building phase and dated to just after 80 C.E.

The Municipium Flavia Solva (today's Wagna) was founded in today's south-eastern Styria, Austria, in 70 C.E. This site had an important relevant role in connecting settlements of the early imperial period within and beyond the south-eastern Alps as well as the cultural and economic integration of Noricum into the Roman Empire [15]. Even though the Forum of Flavia Solva has not been located and there was no water transportation system or canalization, Flavia Solva was part of a far-reaching trade network and various commodities were also locally produced [16–18]. This site flourished in the 2nd century C.E. and again in the mid-3rd century C.E. until it was abandoned in the early 5th century C.E. [19]. Three samples of wall painting fragments (one EB, two with cinnabar) were taken from the excavation collection onsite and analysed for this work.

The Municipium Claudium Teurnia (today's St. Peter in Holz) was established in today's western Carinthia, Austria, in 48 C.E. and counts as one of the oldest Roman cities in the province of Noricum. The first living units of the terrace houses were most likely finalized in 66 C.E. and there was another important building phase between 141 – 161 C.E. after a fire [20]. The buildings are assumed to have been abandoned latest by the 3rd century C.E. [21, 22]. Three samples (2 EB 1, cinnabar) were taken from the excavation collection onsite and analysed for this work.

The Municipium Claudium Iuvavum (Salzburg, Austria) was founded in the 1st century C.E. and had its apogee in the mid-2nd to -3rd century C.E. A sample with a red top layer was taken from a fragment collection originating from an excavation in the city centre in 1999–2001. The Roman structures were found underneath significant urban development - the 1st and 2nd inner courtyard and middle wing of the Neue Residenz – and are dated to the 2nd-3rd century C.E. (or perhaps even 1st-4th century C.E.). These structures are interpreted as workshops as well as prestigious houses with wall paintings (up to six layers) but no mosaic floors [23, 24]. One sample of a wall painting fragment was collected and analysed for this work based on the cinnabar top layer.

2 Methods

The archaeological samples were collected during fieldwork at Ephesus in 2019 (Ephesus samples) and by collaboration with the relevant fragment collections in Austria in 2020 and 2021 (Noricum samples). The polychrome surface of wall painting fragments was first visually examined and then analysed using the handheld energy dispersive X-ray fluorescence (HH-EDXRF, XRF hereafter) spectrometer Olympus InnovX Delta Premium 6000 (Rh anode, 8–40 keV, Si-drift detector, 4W X-ray tube, current of 5–200 μA). Based on this semi-quantitative analysis of major and trace elements, wall painting fragments were selected for further analysis.

Select samples of the wall painting fragments were prepared as stratigraphic polished cross sections and examined by light microscopy followed by scanning electron microscopy coupled with energy-dispersive x-ray spectroscopy (SEM/EDX) analysis, performed using a Quanta FEG 250 (FEI, U.S.A.) scanning electron microscope coupled to the Octane Elect Plus EDX detector (Ametek/EDAX, U.S.A.) equipped with the Genesis EDX Quant software. The investigations were performed under high vacuum at an accelerating voltage of 15–20 kV in back scattered electron (BSE) detection mode[1]. Quantitative results of the size and distribution of pigment grains in the SEM images in the depth of the paint layers of the cross sections were attained by analysing the images with Image-J freeware. By this method, the SEM images of the paint film are over-exposed so that only the pigment grains are visible, which allows image-analysis software to count and measure the grains and the grain-to-grain distance of over a predetermined area. This data was then processed through custom Microsoft Excel macros.

2.1 Sample Selection by XRF

Seven wall painting fragments with Egyptian blue top layers and six wall painting fragments with a red paint surface were used for this work. The samples from Iuvavum, Pompeii and Celeia were selected by visual inspection, while all others by the detection of high mercury or copper concentrations by XRF. The semi-quantitative data obtained through XRF measurements are described below and shown in Table 1.

Blue Wall Painting Fragments
Interpretation of the semi-quantitative XRF results shows that there are significantly higher Si concentrations for the two Teurnia samples compared to the Flavia Solva sample (Table 1). The iron and titanium concentrations are similar for all three Egyptian blue samples; calcium, copper, aluminium, and sulfur concentrations vary slightly between the three samples, while only the Flavia Solva sample contains tin and zinc concentrations that can be detected by XRF analysis. Moreover, the ratio of calcium to copper to silica (4 following the stoichiometry of cruprorivaite: $CaCuSi_4O_{10}$) is similar for the two Teurnia

---

[1] Samples from Iuvavum and Flavia Solva were analysed using a Tescam MIRA II LMU SEM coupled with a Bruker EDX detector using the same settings at ITAM, Prague 9 (see co-authors above).

samples, and different between Flavia Solva and Teurnia, implying a local production for both.

Red Wall Painting Fragments

The XRF results show variable silica, calcium, aluminium, and titanium concentrations. The iron concentrations are significantly higher for one Flavia Solva and one Teurnia sample; however, these two samples have lower mercury concentrations compared to the low-iron Flavia Solva sample, while the sulfur concentrations differ more between Flavia Solva and Teurnia. The ratio of Hg to S is 1.38 for the low-iron Flavia Solva sample, while it is 0.7 for the high-iron Flavia Solva and Teurnia samples. The two Ephesian samples have iron concentrations and Hg to S ratios like those of the 1st phase at Flavia Solva, while the cinnabar in the 2nd phase at Flavia Solva is more like that at Teurnia. The interpretation of these data, as shown in Table 1, supports the hypothesis of changes in cinnabar source over time with stable trade routes throughout the Empire; this led to the further study of paint and plaster to search for similar patterns in the artisanship of the entire wall painting.

Table 1. Pigment Elemental Ratios as determined by XRF

Site/Elements	Cinnabar				Egyptian Blue	
	Flavia Solva 1st phase	Flavia Solva 2nd phase	Teurnia 1st phase	Ephesus	Flavia Solva 1st phase	Teurnia 1st phase
Ca/Cu/4*Si	-	-	-	-	0.19	0.46
Hg/S	1.38	0.73	0.70	1.52	-	-
Fe %	0.55	2.44	2.71	0.92	0.70	0.75

3 Results

Further study of the wall painting fragments was performed on polished stratigraphic cross sections of plaster bearing either Egyptian blue or cinnabar paint (or in the case of Iuvavum, both). Optical microscopic, SEM/EDX and image-analysis results are given below grouped by color to understand more clearly the similarities or differences in materials and techniques used to process pigment and apply it as a paint to the plastered walls of the diverse Roman sites described herein.

3.1 Egyptian Blue – Plaster and Paint Stratigraphy

Examples of plasters that are painted with Egyptian blue paint from 4 different sites are given below in Fig. 1. The top two images of plaster from Pompeii and Celeia display the classic recommended Vitruvian artisanship of *intonaco/ intonachino* plaster layers that combine to be 4–6 mm thick. They consist of well cured lime putty and crushed marble, providing a white, iridescent substrate to be painted on. In the case of Pompeii, a thin underpainting layer of mixed carbon black and EB is directly beneath the final EB pain layer (see inset), while Celeia lacks such underpainting. While it can be presumed that the appearance of the painted surface of the Pompeii was intended to be darker than that at Celeia, the carbon black underpainting allowed for an economizing of the use of EB, detailed further in the SEM results in Sect. 3.2 and Fig. 3 that follow.

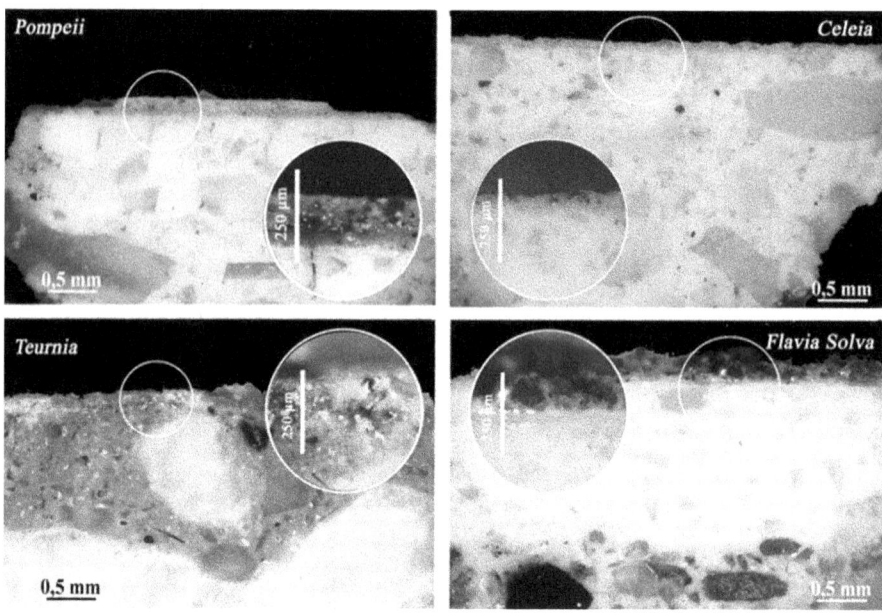

Fig. 1. 4 examples of cross sections of plaster bearing Egyptian Blue paint showing paint layer and *intonachino/* plaster stratigraphy. The three sites from Noricum (Celeia, Teurnia and Flavia Solva) were founded around the same time as the destruction of Pompeii, with abandonment or re-use occurring by the 3rd century at latest. (Color figure online)

The two bottom samples in Fig. 1 above from the more remote, mountainous regions of Noricum (Teurnia and Flavia Solva), on the other hand, display a fundamentally different approach to achieving similar chromatic results (as Pompeii). This may be partially due to the differing quality of EB pigment available at these sites (discussed in 3.2 below) which was apparently also used more liberally. The grains of EB used at Flavia Solva are larger and rougher in appearance (see inset) and there is only a single, 2 mm thick *intonachino* layer above an *arriccio* containing rounded, fine aggregate. The example from Teurnia contains 2 *intonachino/intonaco* layers, but rather than having a carbon black underpainting layer, has fine soot mixed into the final plaster layer, while in Flavia Solva, carbon black (soot) has been mixed into the paint. While the application techniques used in Pompeii and Celeia are similar, they achieved a different chromatic effect due to the underpainting, while in the Noric examples, the same final color is achieved (as Pompeii – a dark blue shown in Fig. 1 insets), but by different means (perhaps different interpretations of the same technique). These results point to artisans with similar knowledge attempting to achieve the same chromatic result.

Egyptian Blue – Pigment Grain Chemistry and Morphology, Chemistry

SEM/EDX results highlight the differences – and a few similarities - of the EB pigment used at Pompeii and the 3 sites in Noricum shown in this study. Cuprorivaite (EB) was formed in ancient times by mixing (roughly) 4 parts quartz sand with 1 part calcium carbonate and 1 part (any) copper source with an alkaline flux and firing at temperatures readily achievable in a wood-burning kiln (850 – 1050 °C) [1–3, 25–27]. The EDX results given in Table 2, which show the average values of measurements of both the crystalline (EB) and glassy phases of 5 pigment grains per sample, generally support this stoichiometry. Of particular interest are the nearly identical high content of sodium and low content of potassium remaining in the glassy phases, which would seem to indicate the use of natron and not wood ash at all sites as well as the lack of tin or zinc, indicating the use of pure copper (as opposed to bronze).

Table 2. Elemental Ratios for the crystalline and glassy phases of EB as determined by EDX

	Atm % omitting C and O								
	Si	Ca	Cu	Fe	Sn	Zn	Na	K	Al
Pompeii									
crystalline	28.00	4.50	5.75	-	-	-	-	-	-
glassy	26.40	0.47	3.22	0.45	-	-	5.25	1.00	2.50
Celeia									
crystalline	27.00	6.75	6.50	-	-	-	-	-	-
glassy	25.50	1.25	2.25	1.00	-	-	5.00	0.75	3.50
Flavia Solva									
crystalline	26.50	7.00	6.25	-	-	-	-	-	-
glassy	26.25	0.50	2.00	1.00	-	-	5.50	1.00	2.25
Teurnia									
crystalline	19.00	3.00	8.00	-	-	-	-	-	-
glassy	20.00	1.00	2.00	1.00	-	-	5.00	0.50	2.25

However, the morphology of the EB grains is quite distinct from site to site, with the alpine Noricum sites displaying the greatest differences from those found at Pompeii. Figure 2 below shows typical EB pigment grains from each of the 4 sites. While cuprorivaite (labelled "CuR") is present in all samples, the shape of the crystals and the relationship with ancillary minerals displays differences likely related to firing conditions or processing of the pigment. While the example from Pompeii shows a large, well-formed CuR crystal, what is important is not its size (which as the following section shows is atypical) but rather it's cohesiveness and close integration with interstitial glassy and quartz phases. The sample from Celeia is similar, with large single grains although the glassy phases are minimally present, perhaps due more to grain milling than the sintering process. Small out-gassing pores are present in the Celeia CuR. In the Teurnia and Flavia Solva samples, the crystals are typically smaller, but clustered and bound by relatively small glassy phases. Quartz is more prevalent and in the case of Flavia Solva there is even vestigial lime (CaO). This evidence points to rather a lower temperature of formation, perhaps caused by a firing environment that encouraged outgassing, pores of which are much more prevalent in these samples.

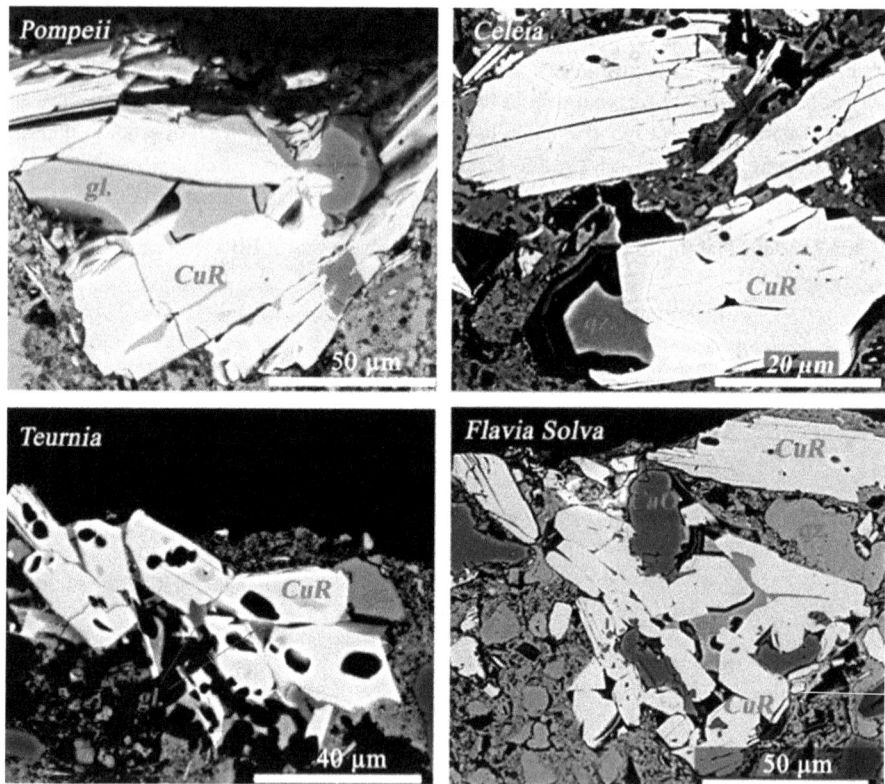

Fig. 2. SEM-BSE photomicrographs of typical Egyptian blue pigment grains from the top paint layer of the plaster fragments shown in Fig. 1 above. Key: CuR – crystalline cuprorivaite, gl. – glassy phases, qz. – quartz, CaO - lime

While the chemistry of these grains appears similar, grain morphology indicates distinct manufacturing processes. In short, the differences between the Pompeii and Celeia samples, and the differences between Teurnia and Flavia Solva samples are likely due to post-production milling, while the differences between the Pompeii samples and the Teurnia and Flavia Solva samples are likely due to both different production *and* milling.

Egyptian Blue – Use in the Paint Layer

As mentioned in Sect. 3.1, the intended chromatic effect of the artists who prepared these plaster wall paintings was likely similar. But how the final blue color was achieved with EB of differing qualities relied on the skill of the artisan applying the plaster and underpainting layers, as shown in Sect. 3.1 above, as well as how the paint was mixed, as shown here. By using the SEM to create images at high contrast, causing the metallic EB to appear very bright, it is possible to easily visualize the density and dispersion of EB within the paint film shown in Fig. 3, while also allowing quantitative analysis of the pigment grains shown in Fig. 4.

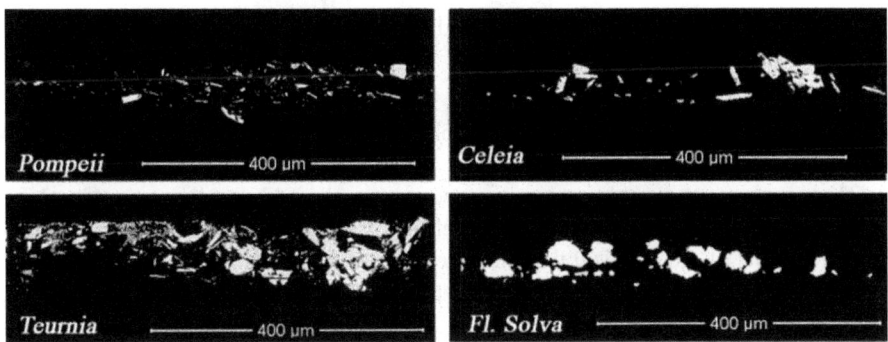

Fig. 3. SEM-BSE photomicrographs of EB grain dispersal in the paint layer of the cross sections shown in Fig. 1. Above. The larger size and greater density of the EB grains for the later Noricum sites is clearly visible.

These figures show that the EB paint used in Pompeii had both a lower grain particle density (Fig. 3) and that the grains were on the average finer (Fig. 4). Both Pompeii and Celeia use EB rather sparingly, while the Teurnia paint film is simply loaded with a high percentage of larger EB grains, in addition to (or perhaps despite) having a dark intonachino layer. In the case of Flavia Solva, the larger grains are often the forementioned "cluster" types, where it is not entirely clear if the grain morphology is due to production or milling processes. Overall, these results imply workers with similar goals and knowledge but different access to resources and means of production, rather than long distance trade in EB. They could be used to support the hypothesis that traveling groups of artisans in fact produced the wall paintings of Noricum, who adapted to local resources and styles as need be.

3.2 Cinnabar – Plaster and Paint Stratigraphy

The samples of wall painting plaster bearing cinnabar are shown in Fig. 5 below. The images highlight the differences in the paint and plaster stratigraphy of 3 sites in Noricum

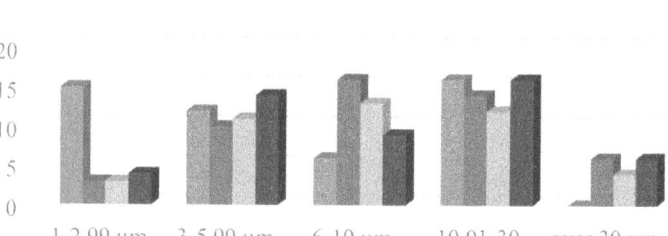

Egyptian Blue Grain Size Distribution

Fig. 4. Grain Size Distribution of EB grains found in paint fragments from Pompeii and 3 Noricum sites.

(Iuvavum, Teurnia and Flavia Solva), with Ephesus standing in as an example of practices from the central Roman Empire.

These represent 4 roughly contemporaneous examples of red wall painting, in which the final chromatic result is the same, with 4 clearly different craft practices. The Ephesus sample consists of the classical thick double layer of *intonachino* roughly 6 mm thick which was then prepared for the final cinnabar layer with 2 layers of ochre underpainting, a red one below a yellow one. The Flavia Solva example shows underpainting with ochre as well, with ochre mixed into the final cinnabar layer, perhaps reflective of a more humble, provincial building. The artists of Teurnia and Iuvavum have incorporated ochre pigment into the *intonachino* (in the case of Iuvavum up to 4 mm in depth(!)) and a yellow ochre underpainting layer beneath the cinnabar layer, which in both cases are mixed with red ochre.

Uniquely, the Iuvavum fragment contains an additional underpainting layer of Egyptian blue, with a grain morphology most like that of Pompeii in Sect. 3.1.1 above. This would point to the location being particularly prestigious as described in the Introduction. This evidence all points to artisans adapting to the unique conditions of each site, with an expensive paint material, which as the following sections show, likely had a common, distant source.

Cinnabar – Pigment Grain Morphology, Chemistry and Use
Whereas the plaster and paint stratigraphy of these examples points to diverse methods of cinnabar utilization, SEM/EDX results appear to indicate a common source for the pigment itself. In ancient times, cinnabar was a rare ore that could simply be ground up before use as a pigment and was traded throughout the Roman empire [4–10].

Starting with the EDX results shown in Table 3 below, it can be seen that apart from the low calcium content of the Flavia Solva samples, the EDX results are strikingly similar. Thus, the origin of the cinnabar pigment is the subject of ongoing lead isotope and trace mineral analysis by the authors, which may provide insight into the provenance of this widely traded mineral [28, 29].

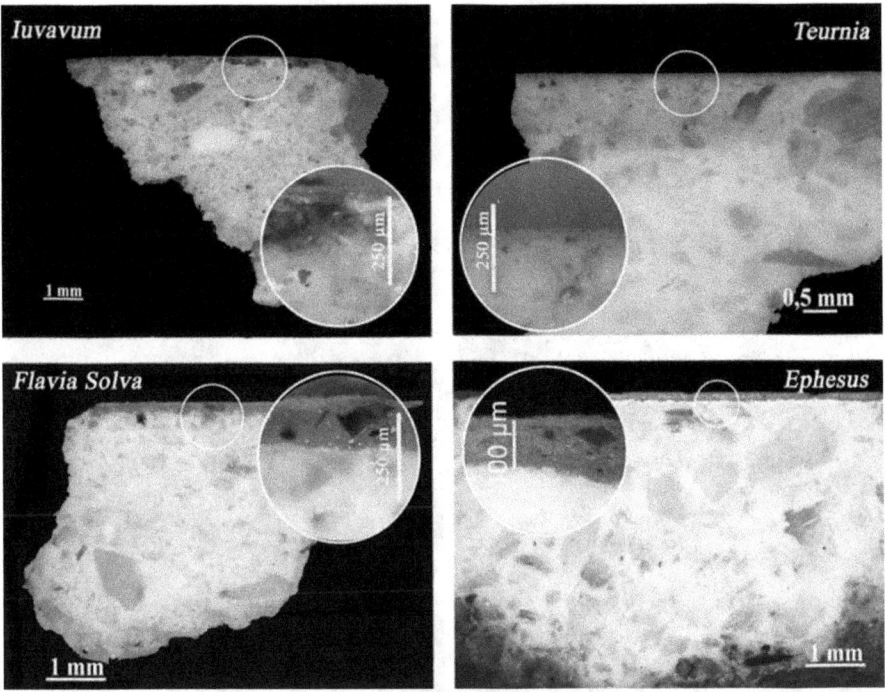

Fig. 5. 4 examples of cross sections of plaster bearing cinnabar paint showing paint layer and *intonachino*/ plaster stratigraphy. The three sites from Noricum (Iuvavum, Teurnia and Flavia Solva) had their apogee roughly contemporaneously with the Ephesus buildings discussed herein.

Table 3. Elemental Ratios of cinnabar pigments as determined by EDX Atm % omitting C and O

	Hg	S	Si	Mg	Fe	Al	Ca
Ephesus	37.50	40.00	7.00	2.50	1.00	1.75	10.00
Flavia Solva	37.50	35.50	7.00	1.75	1.00	1.25	1.00
Teurnia	37.67	36.67	5.00	1.00	2.25	2.00	12.75
Iuvavum	37.25	36.75	5.50	1.00	1.75	2.25	10.25

Figure 6 above shows SEM images of the cinnabar grains from the paint layers of the wall painting fragments shown in Fig. 5. In them it is evident that the size and shape of the cinnabar grains used is roughly the same. They were however used differently in the paint film. Using image-editing to remove the ochre[2] from the image reveals to which extent the cinnabar was "diluted" with less expensive ochre. This is additionally shown

---

[2] The iron content is "less bright" than the mercury in SEM due to mercury's additional shell of valence electrons that give a stronger signal when excited by the electron beam. This allows for their easy selection and deletion in Photoshop®.

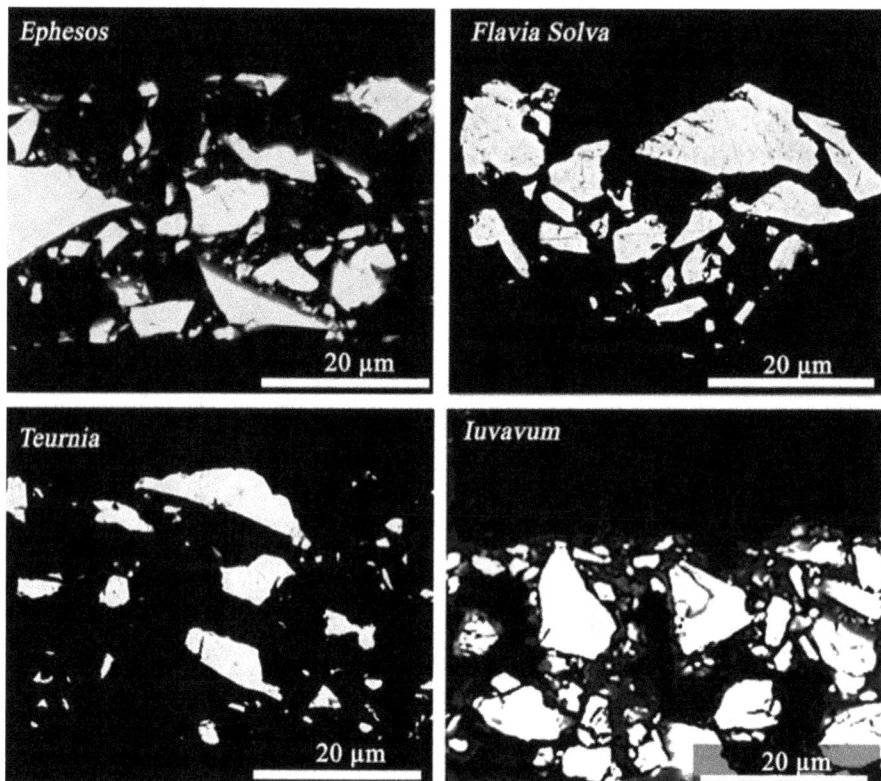

Fig. 6. SEM-BSE photomicrographs of typical cinnabar pigment grains from the top paint layer of the plaster fragments shown in Fig. 5 above. By adjusting the contrast of the images, the ochre is no longer visible, highlighting the spatial distribution of the cinnabar grains.

in Fig. 7 below, which displays this across the entirety of the samples' paint layers. This dilution is evidence that the site of Flavia Solva was perhaps a "less prestigious" site with fewer resources to spend on expensive, imported cinnabar. Image analysis can then be applied to the images shown in Fig. 7 to a large area of the paint film to produce the grain size distribution chart shown in Fig. 8., again produced by counting of the size of the grains and their spacing with image analysis software and analysing the data in Excel. The grain size distribution chart once again supports the working theory that the cinnabar used at all four sites was prepared in the same manner, and perhaps originated from a single source, then used differently as conditions allowed/ demanded at the four sites.

4   Discussion and Conclusions

This work discussed the analysis of plaster wall painting samples from several sites in the Roman Empire bearing either Egyptian blue or cinnabar-based paint as the final decorative layer. The primary difference between these two pigments is that while cinnabar

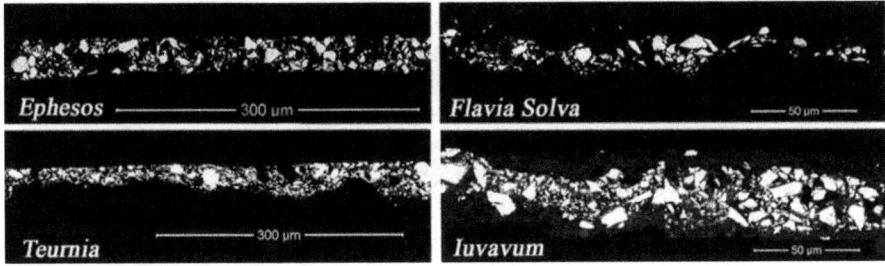

Fig. 7. SEM-BSE photomicrographs of the entire cinnabar paint layer of the plaster fragments in Fig. 5. While the pigment was used differently in each case, the grains themselves are a close match (See Figs. 6, 8 and Table 3).

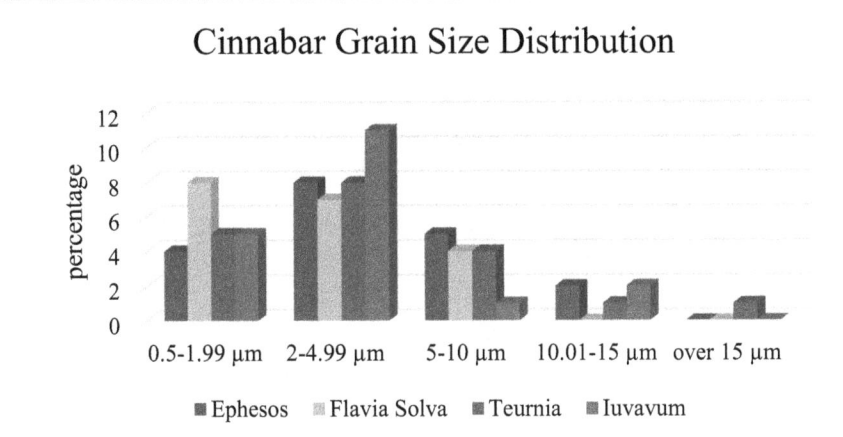

Fig. 8. Grain size distribution of the cinnabar used at Ephesus, Flavia Solva, Teurnia and Iuvavum. The size distribution of the pigment grains for each site is within a standard deviation for each site, supporting the hypothesis that the pigment was milled according to a common practice.

occurs naturally and was widely traded, Egyptian blue is synthetic and reliant on local resources and craftspeople for its production. The latter is clearly shown in the similar chemistry, but varying morphology of the pigment grains as revealed by SEM/EDX. Beyond the pigments themselves, by studying the plaster cross sections it is apparent that the solutions that artisans employed varied from site to site throughout the depth of the wall painting, i.e., from *arriccio* through the *intonaco* and *intonachino* up to under-painting and final paint layer. This is likely reflective of a group of people with similar knowledge adjusting to different locations and resources.

Overall, the sites closer to Italia Province, Celeia (Celje, Slovenia) and Flavia Solva (southern Styria, Austria), showed artisanship that more closely resembles the Pompeiian techniques, while those further north, Iuvavum (Salzburg, Austria) and Teurnia (western Carinthia, Austria) evolved a unique style. Even in this relatively small section of the Empire, regional differences can be distinguished between the materials and techniques used in the high alpine passes (and northwards) and in the areas closer to Italy, and that in each region, artisans had to use their knowledge and which materials were generally available. These findings also reflect trade routes and the development of regional techniques in the Alpine area and supports the hypothesis that apart from very high value items, such as cinnabar pigment, knowledge transfer and perhaps traveling artisans was what was drove the development of Roman wall paintings.

Acknowledgments. ASR's project on wall painting fragments from Noricum received funding from the European Union's Horizon 2020 research and innovation programme under the Marie Skłodowska-Curie grant agreement No 845075. ASR's project on wall painting fragments from Ephesus has received OeAI-internal funding for travel expenses and XRF-instrument use during fieldwork at Ephesus. The work of AJB and PB was funded in part through the above, and in collaboration with Johannes Weber at the University of Applied Arts, Vienna.

References

1. Delamare, F.: Egyptian Blue, the Blue Pigment of Mediterranean Antiquity: From Egyptian Hsbd Iryt to Roman Caeruleum. Blue Pigments: 5000 Years of Art and Industry. Archetype Publications, London, pp. 1−36 (2013)
2. Jaksch, H., Seipel, W., Weiner, K.L., Goresy, A.E.: Naturwissenschaften **70**, 525–535 (1983)
3. Tite, M.S., Bimson, M., Cowell, M.R.: Technological Examination of Egyptian Blue. Archaeological Chemistry III; Advances in Chemistry, vol. 205, pp. 215−242. American Chemical Society (1984)
4. Tsantini, E., Minami, T., Takahashi, K., Ontiveros, M.C.: Analysis of sulphur isotopes to identity the origin of cinnabar in the Roman wall paintings from Barcelona (Spain). Journal of Archaeological Science (2018). Reports https://doi.org/10.1016/J.JASREP.2018.01.032
5. Mlakar, I.: Idrijski Razgledi **3–4**(XIX), 1 (1974)
6. Mlakar, I., Drovenik, M.: Geologija. Ljubljana **14**, 67 (1971)
7. Spangenberg, J., Lavric, J., Serneels, V.: Sulfur isotope analysis of cinnabar from Roman wall paintings by elemental analysis/isotope ratio mass spectrometry – tracking the origin of archaeological red pigments and their authenticity. In: RCM Rapid Communications in Mass Spectroscopy (2010). https://doi.org/10.1002/rcm.4705
8. Mazzocchin, G.A., Baraldi, P., Barbante, C.: Isotopic analysis of lead present in the cinnabar of Roman wall paintings from the Xth Regio "(Venetia et Histria)" by ICP-MS. Talanta **74**, 690–693 (2008). https://doi.org/10.1016/j.talanta.2007.06.048
9. Minami, T., et al.: Sources of vermilion collected from ancient japanese tombs determined by the measurements of lead isotopes. Bunseki Kagaku **62**, 825–833 (2013). https://doi.org/10.2116/bunsekikagaku.62.825
10. Rodríguez, J., Montero-Ruiz, I., Hunt-Ortiz, M., García-Pavón, E.: Cinnabar provenance of Chalcolithic red pigments in the Iberian Peninsula: A lead isotope study. Geoarchaeology, 1–12 (2020). https://doi.org/10.1002/gea.21810
11. Cavassa, L.: La production du bleu égyptien durant l'époque hellenistique et l'Empire romain (III s.av. J.-C.-I s. apr. J.-C.). In: P. Jockey (ed.), Bulletin de Correspondance Hellénique. Supplément. Les arts de la couleur en Grèce ancienne et ailleurs. Approches interdisciplinaires **56**, 13–34 (2018)
12. Bernardi, L., Busana, M.S.: The sarno baths pompeii: context and state of the art. J. Cult. Herit. **40**, 231–239 (2019). https://doi.org/10.1016/j.culher.2019.04.012

13. Angelini, I., Asscher, Y., Secco, M., Parisatto, M., Artioli, G.: The pigments of the frigidarium in the Sarno Baths, Pompeii: Identification, stratigraphy and weathering. J. Cult. Herit. **40**, 231–239 (2019). https://doi.org/10.1016/j.culher.2019.04.021

14. Delile, H., et al.: Demise of a harbor: a geochemical chronicle from Ephesus. J. Archaeol. Sci. **53**, 202–213 (2015)

15. Hinker, C.: Solva vor den Flaviern. Zur Gründung von Flavia Solva. In: Porod, B. (ed.) Flavia Solva. Ein Lesebuch, pp. 8–15. Universalmuseum Joanneum, Graz (2010). ISBN 2078-0168

16. Radbauer, S.: Die römerzeitliche Keramik von Flavia Solva. In: Porod, B. (ed.) Flavia Solva. Ein Lesebuch, pp. 40–47. Universalmuseum Joanneum, Graz (2010). ISBN 2078-0168

17. Glöckner, G.: Glas in Flavia Solva. In: Porod, B. (ed.) Flavia Solva. Ein Lesebuch, pp. 48–55. Universalmuseum Joanneum, Graz (2010). ISBN 2078-0168

18. Gostenčnik, K.: Antikes Wirtschaftsleben in Flavia Solva. In: Porod, B. (ed.) Flavia Solva. Ein Lesebuch, pp. 56–65. Universalmuseum Joanneum, Graz (2010). ISBN 2078-0168

19. Csapláros, A., Sosztarits, O.: Colonia Claudia Savaria - Die Topografie der ältesten Stadt von Pannonie. In: Porod, B. (ed.) Flavia Solva. Ein Lesebuch, pp. 72–79. Universalmuseum Joanneum, Graz (2010). ISBN 2078-0168

20. Gugl, C.: Archäologische Forschungen in Teurnia. Die Ausgrabungen in den Wohnterrassen 1971–1978. Die latènezeitlichen Funde vom Holzer Berg. Österreichisches Archäologisches Institut Sonderschriften Band 33. Wien (2000). ISBN 3-900305-30-7

21. Glaser, F.: Teurnia. In: Šašel Kos, M., Scherrer, P. (eds.) The autonomous towns of Noricum and Pannonia – Die autonomen Städte in Noricum und Pannonien. Situla, Ljubljana, vol. 40, pp. 135–147 (2002). ISBN 961-6169-21-1

22. Dörfler, I.: Die römischen Wandmalereien der Wohnterrassen von Teurnia. In: Scherrer, P. (ed.) Römisches Österreich. Jahresschrift der österreichischen Gesellschaft für Archäologie Band **32**, 17–77. Wien (2009)

23. Kovacsovics, W.K.: Fundberichte aus Österreich **40**, 672 (2001). Bundesdenkmalamt 2002, ISBN 3-85028-337-2

24. Kovacsovics, W.K.: Die archäologischen untersuchungen in ersten innenhof der neuen residenz. In: Marx, E., Laub, P. (eds.) Die Neue Residenz in Salzburg. Vom "Palazzo Nuovo" zum Salzburg Museum. Jahresschrift des Salzburger Museums Carolino Augusteum 47–48/2001–2002, 113ff (2003). ISBN 3-901014-96-9

25. Pradell, T., Salvado, N., Hatton, G.D.: Physical processes involved in production of the ancient pigment, Egyptian blue, M. S. J. Am. Ceram. Soc. **89**, 1426−1431 (2006)

26. Mazzi, F., Pabst, A.: Cuprorivaite: Mineral information, data and localities. Am. Mineral. **47**, 409−411 (1962)

27. Bayer, G., Wiedemann, H.G.: Naturwissenschaften **62**, 181–182 (1975)

28. Rodler, A.S., Baragona A.J., Verbeemen, E., Goderis, S., Sørensen, L.: Connecting places: the material quality and provenance of ancient pigments 27th EAA Ann. Meeting, EAA2021 (online) Kiel, DE 10th Sep 2021 (2021)

29. Rodler, A.S., Baragona, A.J.: Red walls of Ephesus: Local or distant cinnabar sources? In: 27th The 3rd International Symposium of the Society for Near Eastern Landscape Archaeology (online) Istanbul, TR 4th Mar 2022 (2022)

The Decorative Plaster Relief in the Baroque Villa of the Argotti Botanic Gardens, Floriana, Malta: Characterisation of Original Materials and Techniques

Stephanie Parisi[1] ⓘ, Gianni Miani[2], and Chiara Pasian[1(✉)] ⓘ

[1] Department of Conservation and Built Heritage, University of Malta, Msida MSD2080, Malta
{stephanie.parisi,chiara.pasian}@um.edu.mt
[2] Pro Arte s.n.c., Via Asiago 32/9, 36025 Noventa Vicentina, VI, Italy

Abstract. On the Maltese Islands, architectural decoration in historic buildings is widely made of the local Lower Globigerina Limestone, a soft porous stone easy to carve. Little is known, on the other hand, about plaster reliefs (or 'stuccos'), particularly from a technological point of view. The present study focuses on a plaster relief found in the Baroque Argotti Villa, located in the historic Argotti Botanic Gardens (Floriana, Malta), the layout of which dates back to the period of the Knights of St John in Malta (1530–1798). The relief, Baroque in style and featuring a rope motif and scallop shells, shows the presence of three main plaster layers, and several overlapping paint layers. Remains of painted plaster above the relief and on the dado suggest the possible presence of a past wider decorative scheme. Examination on-site was complemented with analyses of samples taken from the relief. Cross- and thin sections were analysed under a Polarized Light Microscope (PLM) and a Scanning Electron Microscope (SEM) combined with Energy Dispersive X-ray Analysis (EDX). The plasters are composed of carbonated lime as the binder. Mainly carbonatic aggregates are present (possibly from local Maltese limestone, showing fossils), but also gypsum aggregates and to a much lower extent silicate-aluminate aggregates.

Keywords: Plaster relief · Baroque · plaster characterisation · PLM · SEM-EDX

1 Introduction: The Argotti Botanic Gardens and the Plaster Relief in the Argotti Villa

The Maltese Lower Globigerina Limestone is a soft and easy-to-carve stone, which was and is continuously used on the Maltese Islands for building, from the Neolithic Temples until the modern day [1]. Locally, architectural decoration in historic buildings, from low to high reliefs to sculptures, is widely made of such limestone. On the other hand, little is known about materials and techniques of plaster reliefs (or 'stuccos'). Studies have focused so far on plaster reliefs in Malta from the XIX and XX c. [2–4], but they concentrate on art historical and stylistic aspects, and and no analysis of the original materials has been carried out. Living tradition reports the use of gypsum as the binder for XX c. local reliefs (no references are found on the matter).

© The Author(s), under exclusive license to Springer Nature Switzerland AG 2023
V. Bokan Bosiljkov et al. (Eds.): HMC 2022, RILEM Bookseries 42, pp. 18–30, 2023.
https://doi.org/10.1007/978-3-031-31472-8_2

The Argotti Gardens are located in Floriana, Malta, which is a town adjacent to the capital city, Valletta. The Argotti Gardens, the significance and history of which can be traced back to the Knights of St John's period in Malta (1530–1798), are currently split into two sections: a public section, and a private section managed by the University of Malta and also referred to as the Argotti Botanic Gardens, withholding a collection of rare and exotic flora. The Argotti Gardens were first occupied by Bailiff of Acri, Don Emmanuel Pinto de Fonseca [5]. In 1741, when Pinto was elected Grand Master of the Order of St John, the Knight Argote et Guzman (Argotti) took over the garden [6] (p. 109), and in 1774, he built a palace (no longer existing) and grottos [7] (p. 29). A Nymphaeum is indeed present in the Gardens, a domed stone structure containing a unique and significant mosaic. It is reported that the Knight also erected a summerhouse (the Argotti Villa), and overall magnificently enriched the garden itself [8] (p. 110), but so far just a *terminus post quem* (after 1715) and *ante quem* (before 1827) for the Villa construction can be given (drawn from archival research and comparison of historic maps). The plaster relief focus of this paper is located in the Baroque Argotti Villa (Fig. 1).

Fig. 1. The Argotti Villa within the Argotti Botanic Garden (Parisi©, 2021)

The decorative plaster relief (Fig. 2) is located in the left room of the Argotti Villa and is largely made up of a repetitive rope motif which runs along the four walls, at the top half of the room. There is a smaller rope motif which runs directly below and parallel to the main rope motif, and is difficult to distinguish due to the multiple paint layers on top of it. At each corner of the room, a large scallop shell is present (Fig. 2). The whole relief is decorated with multiple paint layers. Remainings of an off-white painted plaster above the relief (which appears to extend behind the relief itself) and a painted reddish plaster in the lower register of the room (dado) (Fig. 2) suggests the presence of a past larger decorative scheme.

Fig. 2. SE wall. The plaster relief is located at the top and is constituted of: (Red box) Main rope motif; (Yellow box) Small rope motif; and (Blue box) Scallop shells in the corner of the room. The reddish plaster can be seen in the dado (Parisi©, 2021). (Color figure online)

2 Methodology

2.1 On-Site Examinations of the Relief

Visual examination was the first step to understand the plaster relief and the other decorative elements in the room. Observation took place using both incident and raking light and was carried out stratigraphically, by firstly observing the support, then the plaster layers and finally the paint layers of the relief, trying to understand the sequence and application technique. Images were taken using a Canon PowerShot G5 X and a Canon PowerShot SX60 HS. Close observation was undertaken using a portable light microscope, Dino-Lite AM4515ZT-EDGE.

2.2 Sampling and Analysis

The initial visual examination guided the sampling for the study of the original materials. The plaster and paint layers making up the elements of the relief (rope and scallop shell) were sampled to understand their composition. Other decorative elements in the room were sampled, including the off-white plaster topped with paint layers, which is observed right above the relief on the wall, and the painted reddish plaster, observed on the lower register of the walls. The sampling locations were chosen close to pre-existing losses.

Preliminary observation of the unmounted fragments was performed under a Nikon Stereoscopic Zoom Microscope SMZ800 at 10x-60x magnification. Images of back and front of the samples were captured with a Nikon DS-Vi1 camera using the software NISElements BR 5.21.03. The plaster samples were embedded in polyester resin (Resina Impregnante Mixer Chim-Italia Group) and prepared as thin sections with a Remet Micromet M mitre saw and a Remet LS1 grinder and polisher; petrographic analysis of the thin sections was conducted using an Olympus BX 40 polarized light microscope at 40-400X magnification. Images were captured with a Pentax K-70 camera and Adobe Photoshop CS5 Extended software.

Cross-sections of both plaster and paint layers were prepared, after examination under the stereomicroscope, embedding them in polyester resin (Tiranti Clear Casting Resin). Once cured, the cross sections were mounted in a MOPAS-XS Polisher and were firstly wet ground with a Buehler Metaserv Grinder-Polisher with 400 grit silicon carbide paper and then polished using Micro-mesh™ papers 3200–10000 grades. The cross sections were then examined under a Nikon Polarising Microscope Eclipse Ci POL in visible light and UV-induced luminescence at 100X-500X magnification. Images were captured using a Nikon DS-Vi2 and the NISElements BR 5.21.03 software.

Cross-sections and thin sections were investigated under a Scanning Electron Microscope ZEISS EVO 15. Samples were analysed using a high-definition backscattered electron detector (HDBSD) at a working distance (WD) of 10.00 mm and at magnification 57X-1070X. SmartSEM® software was used to operate the instrument and acquire the images. An energy dispersive X-rays fluorescence spectrometer was used for chemical microanalysis. Point analysis and elemental mapping were performed at 16kV for cross-sections and 20kV for thin sections, with a 3.84μs amp time and 133.7eV resolution, using the ZEISS SmartEDX® software.

3   Results and Discussion

3.1   Plaster Relief Technology: Stratigraphy and Application

The plaster relief is built layer by layer *in situ* on the stone primary support (Globigerina Limestone ashlars). Areas of loss allow to observe that the Globigerina Limestone support is keyed to provide good adhesion for the upcoming layers. The application of the plaster relief most likely began by placing wooden dowels within the mortar joints of the stone wall. The wooden dowels extend behind the relief and in some areas are visible at the top of the relief (Fig. 3a). Square and round holes were carved into the stone primary support, and were filled with plaster to anchor the relief to the wall (Fig. 3b).

On top of the stone support, there is a thin whitewash layer and a smooth white plaster layer (few mm thick), which can be observed in areas where there is loss of the relief. No analyses were performed on these layers. The smooth (not keyed) white plaster reveals the setting-out and is topped with grey drawn lines and red snapped lines. The horizontal snapped line and grey drawn line most probably guided the artist to build the relief straight across the wall, while the two vertical snapped lines are in correspondence of wooden dowels (Fig. 4). A sketch is also present on this plaster (Fig. 4), which is going to be discussed later. The smooth plaster layer appears to be related to the off-white plaster which can be found right above the relief and which is decorated with paint layers.

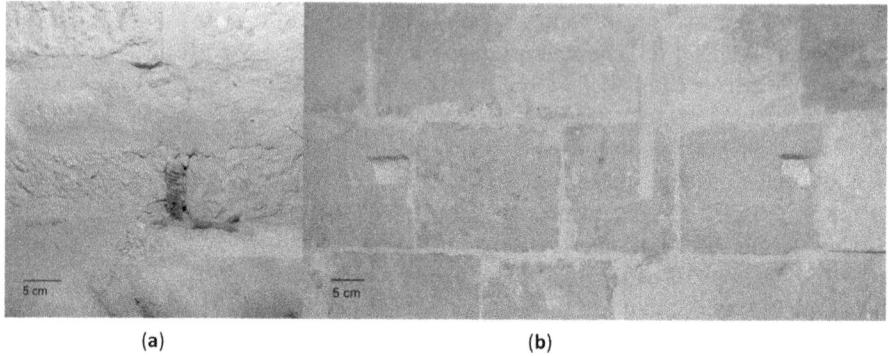

(a)   (b)

Fig. 3. (a) Wooden dowel embedded into a mortar joint; (b) Square holes filled with plaster (Parisi©, 2021).

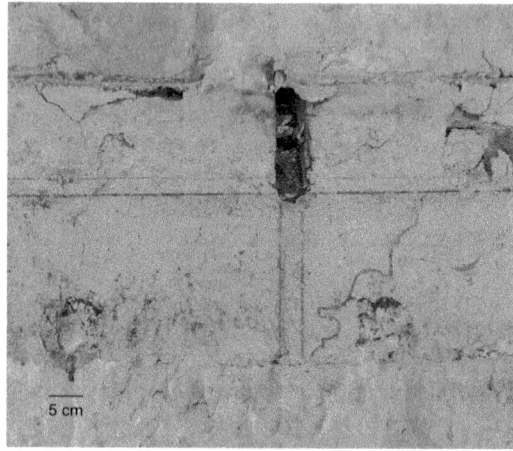

Fig. 4. The smooth plaster layer with snapped and drawn lines, and the sketch of a running mould. Remainings of a wooden dowel are also visible, and holes filled with plaster (Parisi©, 2021).

Plaster was then applied in the square and round holes (Fig. 3b and 4). The plaster in the holes appears to extend beyond them, becoming an integral part of the relief itself and anchoring it to the wall. The plaster material within these holes may be similar to the upcoming bulk beige plaster (Fig. 5), however analysis of this plaster material was not performed. In general, the bulk beige plaster (inner layer) can be considered as the first plaster layer constructing the decorative part of the relief. The term 'bulk' is commonly used to describe the first plaster layer applied in a plaster relief, giving the rough shape to it [9, 10]. A second layer of beige plaster (intermediate layer), as shown in Fig. 5, was then applied over the bulk beige plaster (inner layer), and finally a white plaster (outer layer). The rough application of the bulking beige layers indicates a quick execution by the artist.

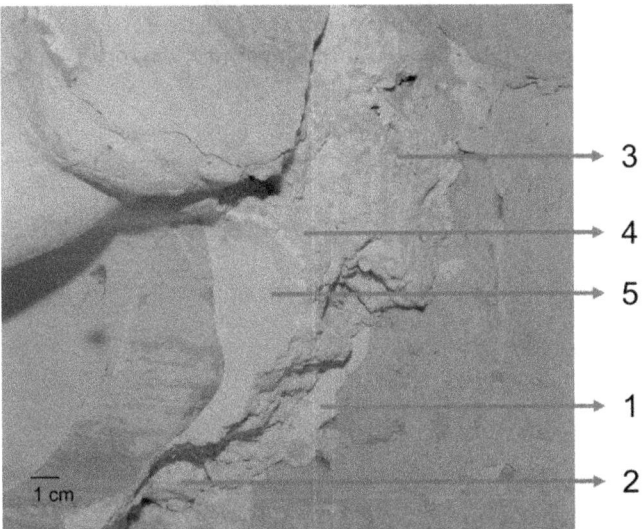

Fig. 5. The stratigraphy of the plaster layers: (1) White wash layer (in this cross section the preparation smooth white plaster is not visible), (2) Plaster in round and square holes, (3) Bulk beige plaster (inner layer), (4) Beige plaster (intermediate layer), (5) White plaster (outer layer) (Parisi©, 2021).

A single line of wooden canes was placed on top of the bulk beige plaster (inner layer) embedded in it (no other canes can be seen in cross section where loss occurred). In areas of loss of wooden canes, the indentation within the plaster indicates that the wooden cane was applied while the plaster was still fresh. The wooden canes have a similar appearance to the Maltese *qasab* which can be found locally [11]. Living tradition on the island (mid-XX c. until now) sees the use of wooden canes (*qasab*) for plaster reliefs, typically placed between the first and second plaster layers to provide strength and support. Similarly, the wooden canes in the plaster relief at the Villa appear to be placed to give structure, strength, and support to the relief.

Overtop the beige plaster layer (in the upper part of the relief, where the big rope is located) is a smooth curved band (Fig. 6). The interior of the smooth curved band is composed of beige plaster and stone fragments, which appear to act as bulking elements, while the exterior is made up of the white plaster, which is the top-most plaster layer. The band was most probably obtained using a type of running mould. Running moulds were known in the XVII c. as *horses* [12] (p. 29), and they are commonly used for architectural features such as cornices [13] (p. 28); the running mould would have a reverse shape of the intended design [13] (p. 30). The shape of a running mould is sketched on the exposed smooth preparation plaster (Fig. 4), and its profile seems to correspond to what could have been used to obtain the present relief.

The individual rope elements were then placed on top of the smooth curved band. In an area of loss of one of the rope elements on the SE wall, shallow key marks can be found, cut in the white plaster (Fig. 6), which were most likely created to provide better adhesion for the attachment of the rope elements on the band. In addition, incisions can be observed on either side of the shallow key marks. These incisions appear to mark the area where each rope element and the strips in between the rope elements were to be placed (Fig. 6).

Fig. 6. Exterior of smooth curved band formed using a running mould. Incisions were created to make the elements of the rope adhere (Parisi©, 2021).

The shape of the rope elements varies, indicating that a mould may not have been used to create exact identical shapes. Therefore, each rope element would have been modelled by hand. The rope elements appear to be made up of multiple layers of the white plaster (outer layer).

Over ten paint layers are applied on the relief, suggesting that the room was most likely frequently used over time and therefore required maintenance. The application of paint layers may have not been part of the plaster relief originally and could be a later addition: this is suggested by the fine purposely white-looking plaster which is the final plaster of the relief. However, this is not confirmed.

3.2 Analyses of Samples: Original Materials Composition

The plasters forming part of the rope motif of the relief and the off-white plaster above the relief showed similarities in their raw materials. Petrographic analysis (Table 1) showed in all the samples a colloform to micritic air lime binder, with well-sorted aggregates with a fine and very fine grain size. Both carbonatic aggregates and ground crystalline gypsum (selenite) were used for the bulk beige plaster (ARG_NE-PL-05, Fig. 7a), for the beige and white plaster (ARG_NE- PL-07, Fig. 7b for the white layer), and for the off-white plaster above the relief (ARG_NW-PL-04) (Table 1). The finishing white plaster contains much more gypsum aggregates compared to the other plasters (up to 60%, Table 1), most likely to impart a white colour to the plaster layer. The reddish plaster, still lime-based, contains mainly carbonatic aggregates, ca. 6% fragments of silicatic shale rich in iron oxides (giving the overall plaster a reddish colour), and lower amounts of carbonaceous particles and celadonite (Fig. 7c).

Table 1. Plaster Samples: Petrographic Analysis Overview

Sample #	Location	Porosity (%)	B/A ratio	Carbonate aggregate (%)	Carbonate aggregate minerals	Other aggregate (%)	Other aggregate minerals	Aggregate size
ARG_NE-PL-05 (bulk beige plaster, inner layer) & ARG_NE-PL-07[1] (beige plaster, intermediate layer)	NE (rope, relief)	17	1:2	83	Calcarenitescalcite	16	Selenite (Fig. 7a)	Well-sorted, 0.35–0.04 mm
ARG_NE-PL-07[1] (white plaster, outer layer)	NE (rope, relief)	22	1:3	40	Calcarenitescalcite	60	Selenite (Fig. 7b)	Well-sorted, 0.7–0.04 mm
ARG_NW-PL-04 (off-white plaster layer, above relief)	NW (above relief)	18	1:3	80	Calcite	20	Selenite	Very well sorted, 0.1–0.04 mm
ARG_SW-PL-06 (reddish plaster layer, dado)	SW (dado)	18	1:2	91	Calcarenitescalcite, marly limestones	9	Silicate shale (rich in iron oxides), carbonaceous particles, celadonite (green earth) (Fig. 7c)	Well-sorted, 0.3–0.04 mm

[1] The sample includes both beige and white plaster

The SEM-EDX mapping also confirmed that the bulk beige (ARG_NE-PL-05), beige and white (ARG_NE-PL-07) plasters layers have a high presence of calcium throughout the entire sample, both in the matrix (carbonated lime) and in the aggregates. Sulfur is only present within the gypsum aggregates of the plasters (Fig. 8). SEM-EDX analysis also showed silicon and aluminium-based aggregates in varying proportions in all plasters, though to a much lower extent compared to carbonatic and gypsum aggregates. It is possible that these aggregates contribute to some extent to hydraulic reaction with the lime binder.

The SEM baskscattered images (BSI) showed the presence of different fossils in all the plasters of the relief. The bulk beige and beige plasters (carbonatic aggregates 83%, Table 1) showed a higher amount of fossils present compared to the white plaster (carbonatic aggregates 40%, Table 1). Fossils are commonly found in local carbonatic rocks [14]. In Malta, there are two types of limestones which have been historically used as building materials: Globigerina Limestone and Coralline Limestone [1]. Globigerina Limestone is a pure limestone (calcite > 92%) including small quantities of quartz (up to 2%) as well as feldspars, apatite, glauconite and clay minerals, and containing minute fossils [15]. A study by Turi et al. [16] lists the main fossil components found in the Maltese Islands' sedimentary rocks, including Upper Coralline Limestone, Globigerina Limestone and Lower Coralline Limestone: Planktonic foraminifers (Globigerinacea) are the marker for Globigerina Limestone [16].

The bulk beige (ARG_NE-PL-05) and beige (ARG_NE-PL-07) plaster samples showed the presence of microfossils, which could be observed under the polarized light

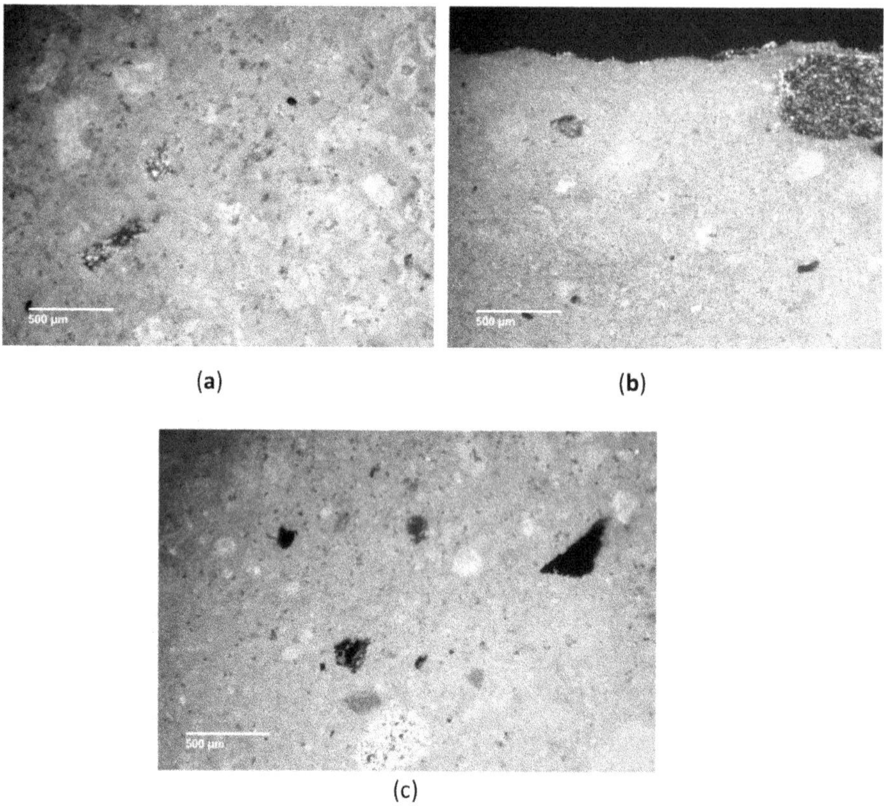

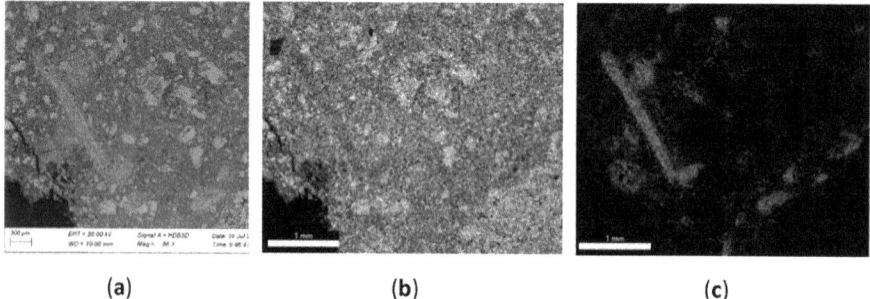

Fig. 7. Photomicrographs of thin sections, transmitted light and crossed polars: (a) Bulk beige plaster of the relief (ARG_NE-PL-05); (b) Finishing white plaster of the relief (ARG_NE-PL-07); (c) Reddish plaster of the dado (ARG_SW-PL-06) (Miani©, 2021).

(a)   **(b)**   **(c)**

Fig. 8. SEM-EDX analysis of the bulk beige plaster (ARG_NE-PL-05): (a) Backscattered electron image (BSI); (b) SEM-EDX mapping of calcium; (c) SEM-EDX mapping of sulfur (Parisi©, 2021).

microscope and at the SEM (Fig. 9). Some of these microfossils showed a strong similarity to the planktonic foraminifers (Globigerinacea) observed in thin sections of Lower

Globigerina Limestone with PLM [17] and with SEM [15]. Also the finishing white plaster and the off-white plaster above the relief contain very fine planktonic foraminifers (Globigerinacea). Therefore, crushed Globigerina Limestone is probably present in the aggregates of the bulk beige, beige, and white plaster of the relief, and in the off-white plaster. However, this does not necessarily exclude the presence of Coralline Limestone, since its typical fossils [16] may be present as well. This would require further investigations.

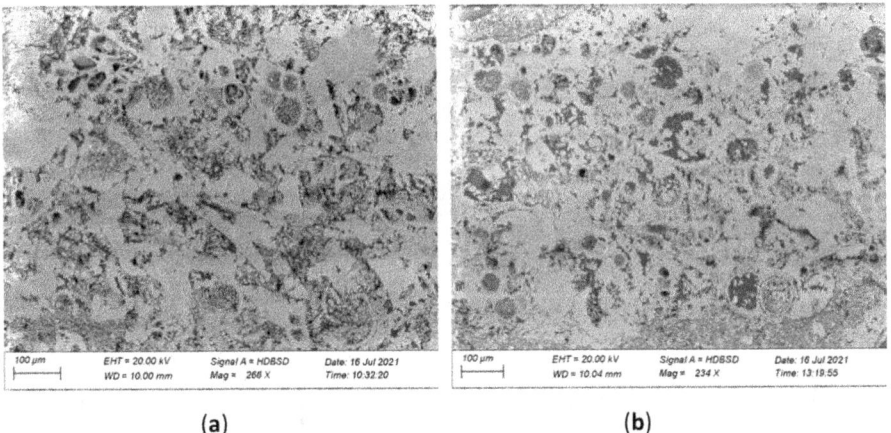

(a)   (b)

Fig. 9. SEM backscattered images (BSI) of plaster thin sections showing the presence of planktonic foraminifers (Globigerinacea): (a) Bulk beige plaster (ARG_NE-PL-05); (b) Beige plaster (ARG_NE-PL-07) (Parisi©, 2021).

Analysis of cross sections (PLM and SEM-EDX mapping) of the paint layers − on the relief, on the off-white plaster above the relief and on the reddish plaster − showed the presence of lime as the binder. Iron-based pigments were identified on samples from the relief (Fig. 10a), the off-white plaster and the reddish plaster (Fig. 10b). Paint layers on the reddish plaster showed the presence of ultramarine pigment particles (Fig. 10b): they appear opaque, irregular in size and sub-rounded, suggesting it is artificial ultramarine [18] (p. 42) [19]. The process to obtain artificial ultramarine was firstly published in 1831 [18] (p. 55), therefore the paint layers on the dado were probably applied post-1831.

3.3 Overall Discussion

Even if the addition of gypsum to the beige plasters (inner and intermediate layers) looks intentional (16%, Table 1), the reason for its use in such layers is still unclear, since the beige plasters are not visible, and mineral gypsum is not an aggregate commonly used locally (carbonatic local stones are typically preferred). Gypsum is found in the Blue Clay formation in Malta [16], therefore it is possible that the mineral gypsum in the plasters is local and originated from that formation. However, Blue Clay is not widely used for plasters, and gypsum could have been imported from abroad instead. On the other hand, as seen, the white plaster (outer layer) has a significantly higher amount of

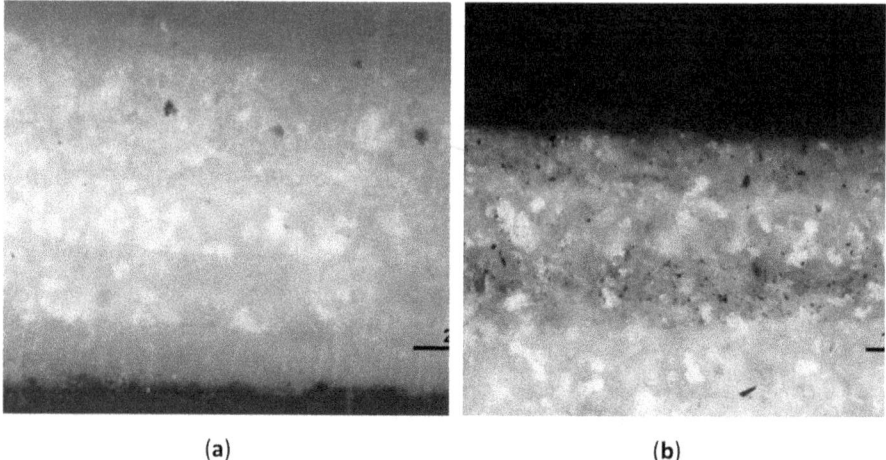

(a)   (b)

Fig. 10. Photomicrographs of cross sections: (a) Some of the paint layers on the plaster relief (ARG_SE-P-08); (b) Reddish plaster with paint layers (ARG_SW-PPL-09) (Parisi©, 2021).

gypsum aggregates (60%, Table 1) compared to the beige plasters (inner and intermediate layers), and this was most likely added to impart a white colour to the smooth finishing plaster. It is possible that the off-white plaster above the relief (20% gypsum aggregates, Table 1) was covering a larger portion of the wall (now just remainings are found), being part of a larger decorative scheme. Both the white finishing plaster of the relief and the off-white plaster above may have been left initially exposed (hence the white colour), and maybe painted at a later stage; however this cannot be confirmed.

On the other hand, although the reddish plaster in the dado appears dissimilar from the other plasters from a composition point of view, and does not have a direct physical relationship with the plasters of the relief, its presence also suggests a wider decorative scheme beyond the relief. It cannot be certain, though, if relief and dado were built together and/or when. The fact the paint layer of the dado was probably applied in the XIX c. or later does not provide information on the date of the plaster itself, which could have been applied also before.

In general, the multiple paint layers on the plaster relief and surrounding decorative elements suggest that the room was frequently used and required maintenance to possibly keep up the changing styles.

Since in Malta architectural decoration in historic buildings is typically and largely made of the local Lower Globigerina Limestone, the fact that the relief was built up *in situ* with plaster suggests that the relief may have been added at a later stage to the room, maybe as a way to embellish or enrich it.

4   Conclusions

Observations on-site allowed to comprehend the technology used to build the plaster relief, while analysis of samples gave an insight into original materials and helped to build hypothesis on the overall decorative scheme as well. An *in situ* layer-by-layer system was utilized to build the relief and all plasters proved to be lime-based: XX c. plaster reliefs living tradition in Malta typically does not include such a complex layer-by-layer technique and reliefs are rather gypsum-based. This may suggest that the relief is pre-XX c.

The analyses performed on the plaster layers making up the plaster relief (i.e. bulk beige, beige and white plasters) and the off-white plaster above the relief show that similar raw materials were used. Lime was indeed used as the binder for all the plasters, and various aggregates were employed, including carbonatic local aggregates, with presence of planktonic foraminifers (Globigerinacea) fossils, which are a marker for Globigerina Limestone [16]. However, this does not exclude the presence of crushed local Coralline Limestone, which has also been historically widely used as a building material [1]. Gypsum aggregates (therefore gypsum in its mineral form) are present in all the mentioned plasters. In the white plaster (outer layer), gypsum aggregates are present in a much higher amount compared to the other plasters, and were most likely added for optical reasons, i.e. to make the plaster appear white: this may suggest (but does not confirm) that the finishing white plaster (and possibly the off-white plaster on the wall) may have been left initially exposed, and not painted at first. However, in the bulk beige and beige plaster (inner and intermediate layers), the reason for the presence of the gypsum aggregates is not clear; they could have been added as a filler. The reddish plaster in the dado, different from the others from a composition point of view, cannot be physically related to the other plasters, but, being painted, looks nonetheless part of a decorative scheme. In fact, overall, the multiple painted plasters observed within the room (including the relief, the off-white plaster above it and the reddish plaster in the dado) suggest the presence of a larger decorative scheme compared to what is visible today. Additionally, they suggest that the room was frequently used and the plasters repainted to follow the changing styles.

As mentioned, traditionally and historically, architectural decoration in Maltese historic buildings is widely made of the local Lower Globigerina Limestone, and little is known about plaster reliefs pre-XIX-XX c. The fact that the relief object of this paper is not made of stone, and was built up *in situ* with plaster, supports the hypothesis that the relief may have been a later addition to the room. While the composition of the plasters may suggest that the relief is pre-XX c., and archival research on the origins of the Villa shows that the building was constructed after 1715, the dating of the relief is still unknown and further research is required to gather more information on this.

Ongoing studies on the decoration of the Nymphaeum at the Argotti Botanic Gardens (plasters and mosaic, including the coat of arms of Argotti), just a few meters away from the Argotti Villa, are undertaken by the Department of Conservation and Bult Heritage, University of Malta. Similarities and differences in original materials and technology will be drawn, and this will help to put the decorative schemes and their buildings in context, considering the overall highly significant historic site of the Argotti Gardens and its links with the Maltese history and history of art.

Acknowledgements. We would like to thank Mr Tony Meli, curator of the Argotti Botanic Gardens, University of Malta, for having granted access to the Villa and for the logistical support. Ms Jennifer Porter (University of Malta) is acnowledged for having generously shared her findings on archival research related to the construction of the Villa. Thanks to Prof JoAnn Cassar (University of Malta), Ms Jennifer Porter, Arch. Mark Azzopardi (Restoration Directorate, Malta) and Dr Marta Caroselli (SUPSI) for their useful comments and suggestions. Many thanks to Mr Joseph Sagona and Dr Mark Sagona (University of Malta) for having shared their knowledge on XX c. gypsum-reliefs in Malta and their living tradition. Thank you to Rosangela Fajeta (University of Malta) for her help with the SEM-EDX analysis. Finally, thanks to the employees of the Argotti Botanic Gardens.

References

1. Cassar, J.: The use of limestone in a historic context the experience of Malta. Geological Society, London, Special Publications **331**(1), 13–25 (2010)
2. Agius, F.: The Eclectic Project for Palazzo Parisio (1898–1907): The Industry of the Art and Its Protagonists. M.A. Dissertation. University of Malta (2014)
3. Attard, C.: An Italian Artistic Presence in 19th and 20th Century Malta, 1850–1960. M.A. Dissertation. University of Malta (1999)
4. Spiteri, C.: The Maltese Villa in the Eighteenth-century: A Comparative Analysis of Villa Gourgion, Villa Francia and Villa Bologna. M.A. Dissertation. University of Malta (2015)
5. Tonna, E.: The argotti gardens. Civilization **22**, 614–616 (1985)
6. Lanfranco, G.: L-lstorja tal- Ġnien ta' l-Argotti. Pronostku Malti 109–116 (1983)
7. Crosthwait, A., Ellul, M.: The Gardens of Sir Alexander Ball. Treasures of Malta **IV**(1), 27–32 (1997)
8. Mahoney, L.: 5000 years of architecture in Malta. Valletta Publishing (1996)
9. Rampazzi, L., et al.: The stucco technique of the magistri comacini: the case study of santa maria del ghirli in campione d'italia (Como, Italy). Archaeometry **630**(1), 91–100 (2012)
10. Salavessa, E., Jalali, S., Sousa, M.O.L., Fernandes, L., Duarte, A.M.: Historical plasterwork techniques inspire new formulation. Constr. Build. Mater. **48**, 858–867 (2013)
11. Portelli, P.: The giant reed. Times of Malta (October 1, 2015). https://timesofmalta.com/articles/view/the-giant-reed.586563, last accessed 2021
12. Bostwick, D.: Decorative Plasterwork: Materials and Methods. In: Forsyth, M, White, L. (eds) Interior Finished & Fittings for Historic Building Conservation, pp. 25-39. Blackwell Publishing Ltd (2015)
13. Vadstrup, S.: Conservation of Plaster Architecture on Facades - Working Techniques and Repair Methods. Centre for Building Preservation RAADVAD, Denmark (2008)
14. Cassar, J.: Deterioration of the Globigerina Limestone of the Maltese Islands. In: Siegesmund, S., Weiss, T., Villabrecht, A. (eds.) Natural stone, weathering phenomena conservation strategies and case studies, pp. 17–43. Geological Society, London (2002)
15. Cassar, J., Vannucci, S.: Petrographical and chemical research on the stone of the megalithic temples, malta. Archeological Review **5**, 40–45 (2001)
16. Turi, A., Picollo, M., Valleri, G.: Mineralogy and Origin of the Carbonate Beach Sediments of Malta and Gozo, Maltese Islands. Boll. Soc. Geol. It. 367–374 (1990)
17. Bianco, L.: Geochemistry, mineralogy and textural properties of the lower globigerina limestone used in the built heritage. Minerals **11**(7), 740 (2021). https://doi.org/10.3390/min11070740
18. Plesters, J., Roy, A.: Artist's Pigments: A Handbook of Their History and Characteristics. Ed.: A. Roy ed. National Gallery of Art: Washington 37 (1993)
19. Mactaggart, P., Mactaggart, A.: Blue Pigments (2007). https://academicprojects.co.uk/blue-pigments/, last accessed 29 March 2022

A Study on Historic Mortars for Restorative Applications in Persepolis World Heritage Site: Curing in Laboratory vs Site

Parsa Pahlavan[1(✉)], Stefania Manzi[2], and Maria Chiara Bignozzi[2]

[1] Faculty of Architecture and Urbanism, Ferdowsi University of Mashhad, Mashhad, Iran
parsa.pahlavan@um.ac.ir
[2] Department of Civil, Chemical, Environmental and Materials Engineering (DICAM), University of Bologna, Bologna, Italy

Abstract. Two types of air lime mortars with inclusion of sesame cooking oil were synthesized. The behavior of mortars in the site conditions and the laboratory can be distinct. Hence, the mortars were cured in two laboratory and natural climatic conditions of and Persepolis World Heritage Site. The mortars were monitored for two years under both conditions and the results demonstrated distinctions in characteristics of mortars, emanating from curing conditions. The air lime mortars cured in the site conditions exhibited increment in durability and hydric properties. In the natural outdoor conditions, some effects of addition of organics to mortars, such as retarding their setting time were less highlighted compared to laboratory curing mortars.

Keywords: Historic mortars · Air lime · Restoration · In-situ application

1 Introduction

Compatibility is a key requirement for restorative materials. Lime mortars are generally known compatible materials for historical masonries restorative actions; however, inclusion of various additives have been carried out to improve the function of these materials toward water and water vapour. Hydric improvements of lime mortars were usually concluded when various fatty organics were included in the mixes in modern and studies [1–3]. A study demonstrates that the unsaturation level of additive fatty acids is a key parameter for hydrophobization of oiled lime mortars [4]. Nevertheless, most of the characterizations and studies have been occurred when the mortars were made and cured in laboratory conditions, whilst the complexities of climatic parameters in real climatic conditions could occur substantial alterations in the tests results.

This study has investigated restorative mortars for actions in Persepolis world heritage site (Iran) where a great need for provision of economic, compatible and sustainable restorative mortars was reflected. Conservation actions in the historical developing is highly dependent on economic part of plan of conservation [5] as economic values of restorative materials can increase the conservation domain potentials [6–8].

V. Bokan Bosiljkov et al. (Eds.): HMC 2022, RILEM Bookseries 42, pp. 31–35, 2023.
https://doi.org/10.1007/978-3-031-31472-8_3

The organic additions has been demonstrated enhancements in a recent study under laboratory conditions, [4] with a similar mix-design used in this study. This research explores the lime mortars with local components in inclusion of sesame cooking oil in laboratory and the natural site conditions of the designated historical site (Persepolis), simultaneously.

2 Materials and Methods

As commercial lime putty production in the country of the in-situ application (Iran) is not valid, high calcium content lime putty (Calcium hydroxide) as the main component for the mortars mixes was produced in the laboratory. The putty was classified as CL 90S according to EN 459–1 [9] composed of micronized high calcium powder as a commercial product ($90\% < Ca(OH)2 < 93\%$) was slaked for 3 months with distilled water to produce 49 W.T.% water content lime putty. The non-reactive part of the mortars was composed by three different stone powders and a type of sand (particle size of 1–2 mm), from a local (Pulvar) river.

The characteristics of the additive sesame oil used in this research and the methods for mixture creation and casting is explained in detail in a previous study [10] (Table 1).

Table 1. Mix-design of the investigated samples.

Sample	Mix	Curing Condition	$Ca(OH)_2$ [wt. %]	Non-reactive part [wt. %]	Kneading water [wt. %]	Oil [wt. %]
LAB0	A0	Laboratory	27	66	7	0
LAB1	A1	Laboratory	25	66	7	2
SITE0	A0	In-situ	27	66	7	0
SITE1	A1	In-situ	25	66	7	2

The mortars were molded in disc-shaped plastic molds (diameter $= 60$ mm, thickness $= 20$ mm) over a glass support and were demolded after 24 h. The laboratory series of the mortar samples (LAB) were kept at lab conditions (Temperature $= 22 \pm 2\,°C$, RH $= 50 \pm 5\%$) for the rest of their curing period. The outdoor curing series (SITE) of mortar samples were kept in the site (temperature of the mortar production week $= 8–23\,°C$, RH $= 23 \pm 5\%$). In order to enhance the comparability of the characterizations with the recent studies, the samples were characterized after 180 days of curing.

3 Characterizations

The mortar samples were characterized in terms of

- Calcium carbonate formations by a Dietrich Fruhling calcimeter
- Open porosity by mercury intrusion porosimeter (MIP)

- Water absorption (WA) in 24 h by the formula: WA24h = [(mssd - mdry)/mdry] × 100
- Vapour permeability according to EN 1015–19 [11].
- Superficial durability by destructive freeze-thaw life cycles

The detailed process and methods of the characterizations in explained a previous paper [10].

4 Results and Discussions

The results of the open porosity values, calcium carbonate formation, impermeability (water vapour resistance) coefficients, and water absorption values are reported in Table 2.

Table 2. Results of various characterizations of the mortar samples

Samples	Open porosity [%]	Calcium carbonate formation [%]	Impermeability coeficient	WA$_{24h}$ [%]	Surface soundness after 15 destructive freaze-thaw cycles
LAB0	40.5	72	4.2	16.5 (± 0.3)	NO
LAB1	33.2	68	4.5	4.5 (± 0.5)	YES
SITE0	35.0	78	4.9	18.1 (± 0.3)	NO
SITE1	34.5	77	3.9	1.2 (± 0.4)	YES

As it can be found in the results, addition of sesame cooking oils in the air lime mortars led to substantial hydrophobic effects such as considerable reductions in water absorption values: over 70% water gain reduction when the samples were cured in laboratory and about 90% reduction when the mortars were cured in the site conditions. The oiled mortars demonstrated higher hydrophobicity when cured in the natural climatic conditions.

The open porosity values, did not demonstrated a considerable alteration for the oiled mortars, in various curing conditions. However, open porosity of the mortars cured in the laboratory, showed a 7% reduction in inclusion of additive oils.

Oil additions in the air lime mortar samples adversely affected carbonation and permeability values for the mortars cured in laboratory. This had been reported in previous studies. Nevertheless, when the mortars were cured at the outdoor conditions no negative effect on carbonation process and permeability of the samples was observed. This can be due to existence of numerous climatic parameters in the outdoor condition of natural site, compared to the laboratory such as air flow and thermal variations. These parameters affected the condition for the site cured mortars: the carbon dioxide gains and microstructural alterations consequently demonstrates 78–80% of calcium carbonate after 180 days of curing.

The mortars were furthermore applied in the in-situ applications and monitored for two years. The oiled mortars manifested a sound physical appearance, however, the non-oiled samples demonstrated various detachments due to their water gain and extended volume of the frozen water in their porous hydrophilic structure (Figs. 1 and 2).

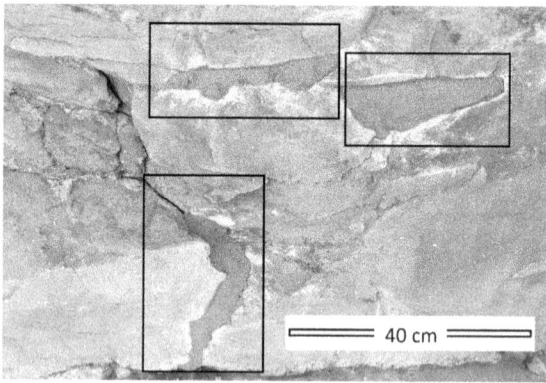

Fig. 1. Repointing joints of a real structure in Persepolis site for durability monitoring

Fig. 2. Repointing the staged condition of stones in the site

5   Conclusions

Climate change and considerable reduction of relative humidity in the recent years of the region of this application (10% reduction of RH since 2010) urges the application of air lime mortars rather than hydraulic lime for conservation actions. Addition of sesame oil to air lime mortars have considerably enhanced their hydric properties. Furthermore, no biological attack or durability issue was reported in the two years of the in-situ monitoring of the applied mortars. An enhancement of hydric properties of air lime mortars in presence of sesame oil occurred in the in-situ conditions with more complicated climatic parameters compared to the laboratory. Effect of some climatic parameters such as air flow and daily thermal variations did not manifest adverse effects on the characteristics of the studied mortars.

References

1. Čechová, E., Papayianni, I., Stefanidou, M.: Properties of lime-based restoration mortars modified by the addition of linseed oil. In: 2nd Historic Mortars Conf, pp. 937–945. RILEM, Prague (2010)
2. Nunes, C., Slížková, Z.: Hydrophobic lime based mortars with linseed oil: Characterization and durability assessment. Cem Concr Res **61–62**, 28–39 (2014)
3. Nunes, C., Slížková, Z.: Freezing and thawing resistance of aerial lime mortar with metakaolin and a traditional water-repellent admixture. Constr Build Mater **114**, 896–905 (2016)
4. Pahlavan, P., Manzi, S., Rodriguez, M.T., Bignozzi, M.C.: Valorization of spent cooking oils in hydrophobic waste-based lime mortars for restorative rendering applications. Constr Build Mater **146**, 199–209 (2017)
5. Pahlavan, P., Manzi, S., Sansonetti, A., Bignozzi, M.C.: Sustainable materials for architectural restoration in developing countries: from the new historic mixes for the future. In: International Congress on Avant-garde Contemporary Architecture and Urbanism in Islamic Countries, pp. 6–10. ACIAU, Mashhad (2018)
6. Pahlavan, P., Manzi, S., Sansonetti, A., Bignozzi, M.C.: Valorization of organic additions in restorative lime mortars: Spent cooking oil and albumen. Constr Build Mater **181**, 650–658 (2018)
7. Wei, G., Zhang, H., Wang, H., Fang, S., Zhang, B., Yang, F.: An experimental study on application of sticky rice-lime mor-tar in conservation of the stone tower in the Xiangji Temple. Constr Build Mater **28**, 624–632 (2012)
8. Barreca, F., Fichera, C.R.: Use of olive stone as an additive in cement lime mortar to improve thermal insulation. Ener Build **62**, 507–513 (2013)
9. EN 459-1 Building lime. Part 1: Definitions, specifications and conformity criteria (2010)
10. Pahlavan, P., Manzi, S., Shariatmadar, H., Bignozzi, M.C.: Preliminary valorization of climatic conditions effects on curing of air lime-based mortars for restorative applications in the pasargadae and persepolis world heritage sites. Appl. Sci. **11**(17), 7925 (2021)
11. EN 1015–19 Methods of test for mortar for masonry, Determination of water vapour permeability of hardened rendering and plastering mortars (1998)

Making Ancient Mortars Hydraulic. How to Parametrize Type and Crystallinity of Reaction Products in Different Recipes

Simone Dilaria[1,2(✉)], Michele Secco[1,2], Jacopo Bonetto[1,2], Giulia Ricci[2,3], and Gilberto Artioli[2,3]

[1] Department of Cultural Heritage (DBC), University of Padova, Padua, Italy
simone.dilaria@unipd.it

[2] Inter-Departmental Research Center for the Study of Cement Materials and Hydraulic Binders (CIRCe), University of Padova, Padua, Italy

[3] Department of Geosciences, University of Padova, Padua, Italy

Abstract. The hydraulic features of lime-based mortars are primarily determined by the occurrence in the raw materials of variable amounts of reactive silica (SiO_2) and alumina (Al_2O_3) that, in presence of water, interact with lime (CaO) to form different hydrated products (C-A-H; C-A-S-H; C-S-H). Under certain conditions, other hydrated products, based on the interaction of magnesium with silica (and sometimes also with alumina) can occur (M-S-H/M-A-S-H). The target of this research is the analysis and comparison of the characteristics and structure of calcium-based silico/aluminate hydrates and magnesium-based silico/aluminate hydrates in ancient mortars and concretes. We adopted a multianalytical approach, involving Optical Microscopy (OM), X-Ray powder diffraction (XRPD), scanning electron microscopy and energy-dispersive microanalysis (SEM-EDS) and magic-angle spinning solid-state nuclear magnetic resonance (MAS-SS NMR), for characterizing samples with different composition and structural function from the sites of Aquileia (Northern Italy), Nora (Sardinia) and Pompeii (Naples). The results we obtained demonstrate the occurrence of different hydraulic phases in the mortar-based materials, whose development primarily depends on the array of raw materials used in the manufacturing of the compounds. However, the observed hydraulic phases present a prevalent gel-like structure, that was not adequately parametrized and quantified adopting the analytical approach traditionally employed for the characterization of ancient mortars (i.e. OM, XRPD, SEM-EDS). In this perspective, the utilization of MAS-SS NMR resulted crucial in characterizing the chemical environment of hydraulic phases, and C-S-H in particular, by discriminating the degree of polymerization of the silicate tetrahedra even in disordered structures.

Keywords: Ancient mortars · pozzolanic aggregates · natural hydraulic lime · X-Ray Powder Diffraction · M-S-H · C-S-H · Nuclear Magnetic Resonance (MAS-SS NMR)

V. Bokan Bosiljkov et al. (Eds.): HMC 2022, RILEM Bookseries 42, pp. 36–52, 2023.
https://doi.org/10.1007/978-3-031-31472-8_4

1   Introduction

The hydraulic features of lime-based mortars are primarily determined by the occurrence in the raw materials of a variable amount of reactive silica (SiO_2) and alumina (Al_2O_3) that, in presence of water, interact with lime (CaO) to form different reaction products [1]. Under certain conditions, other hydrate products, based on the interaction of magnesium with silica (and sometimes also with alumina), can occur. These usually develop into M-S-H/M-A-S-H phyllosilicates having a prevalent gel-like structure [2, 3].

The target of this research is the analysis and comparison of the structures of calcium-based and magnesium-based silico/aluminate hydrates in ancient mortar-based materials, optimizing the choice of the analytical techniques adopted for the description and parametrization of these hydraulic products.

We studied by optical microscopy (OM), X-Ray powder diffraction and Rietveld quantitative phase analysis (QPA-XRPD), scanning electron microscopy and energy-dispersive microanalysis (SEM-EDS) and magic-angle spinning solid-state nuclear magnetic resonance spectroscopy (MAS-SS NMR) several mortar samples with different composition and structural function (i.e. revetments of water-tanks, foundations, floor beddings) from the sites of Aquileia (Northern Italy), Nora (Sardinia) and Pompeii (Campania). We examined "aerial" lime mortars rich in natural and artificial pozzolanic aggregates (volcanic glass, organic ashes, *terracotta* fragments) and natural hydraulic lime mortars.

Through OM and SEM-EDS investigations, we detected reaction edges around the pozzolanic aggregates and we investigated the chemistry of reacted areas. By QPA-XRPD, we were able to describe the type of crystalline hydraulic phases.

We finally performed MAS-SS NMR investigations to parameterize the actual hydraulic character of the mortars, where the hydraulic phases had maintained a prevalent gel-like structure, by observing the degree of polymerization of silica.

The hydraulic products we detected are calcium-based silico-aluminate hydrates in the form of C-A-H, C-A-S-H or pure C-S-H having a variable degree in crystallinity and arranged in different mineralogical phases, as their structure and chemistry is primarily influenced by the raw materials used in the recipe. Under certain conditions, the development of M-S-H and M-A-S-H products in some mortars has been also observed and described.

2   Case Studies and Samples

2.1   Pompeii

Pompeii is built on a volcanic plateau on the southern slopes of the Vesuvius volcano (Fig. 1, a). In the Samnite period (5th – 4th century BC), the city was already structured into a fortified urban centre. From the 3rd century BC, it entered in the Roman sphere of influence and it was monumentalized in the following decades with rich private houses and public buildings. After the 62 AD earthquake, Pompeii was interested by a renovation phase, when it was suddenly destroyed in 79 AD by the famous eruption of Vesuvius, which caused its demise but, at the same time, ensured its conservation until today.

The so-called Sarno Baths are a terraced large Roman spa located on the south-western edge of the Pompeian plateau (Fig. 1, b). The building was excavated for the very first time between 1887 and 1890, while its façade was brought to light in the 1950's. Several modern renovations reshaped the construction in its external and internal layout [4]. More than 80 mortars have been analysed from the Sarno Baths.

On the basis of the compositional features, samples were reunited in some groups according to the ancient construction phases and modern restorations. In the mortars dated to the Imperial Age phase, clasts of volcanic glass (pumices and tuff fragments) were identified (Fig. 2). Besides, vesicular tephrite lavas were detected in older mortars as well as in Imperial age ones [5, 6].

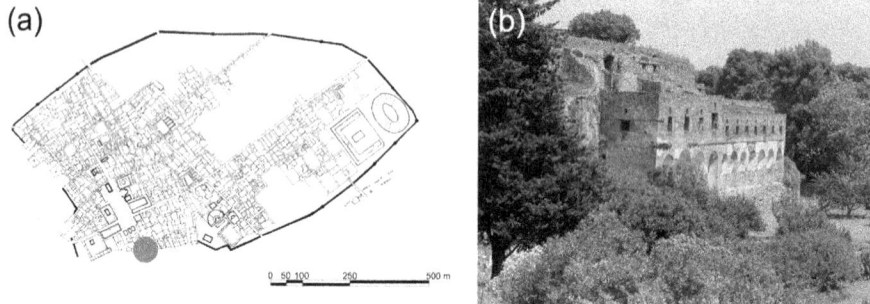

Fig. 1. The site of Pompeii and the Sarno Baths. (a) Plan of the ancient city and localization of the Sarno Baths; (b) The terraced façade of the Sarno Baths.

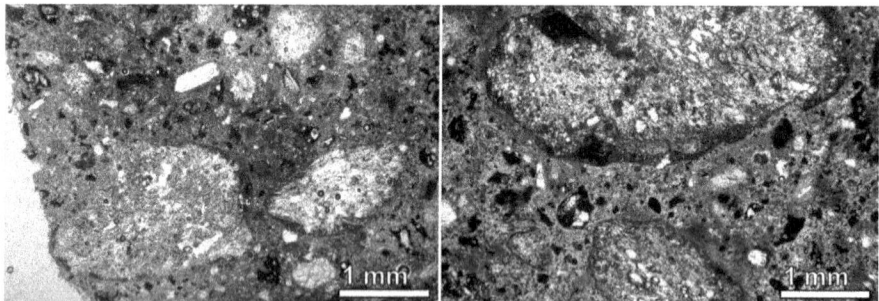

Fig. 2. Clasts of reactive volcanic glass (mainly pumice) in the wall-joint mortars from the Imperial Age phase of the Sarno Baths in Pompeii. Reaction edges around pozzolans are clearly detectable.

2.2 Nora

The ancient city of Nora is a Punic-Roman settlement located on a peninsula in the southern edge of the Gulf of Cagliari in Sardinia (Fig. 3, a). From the 8th century BC the site was occupied by the Phoenicians, that established an emporium for trading with inland Nuragic populations. By the end of the 6th century BC, Nora became a Carthaginian town. Punic inhabitants built a quartered urban system, provided with streets, stone buildings and two chambered necropolises. In 227 BC, after the First Punic War, Sardinia was conquered by Rome and Nora was gradually refurbished by Romans with new buildings such as wide forum, temples, baths and a theatre [7].

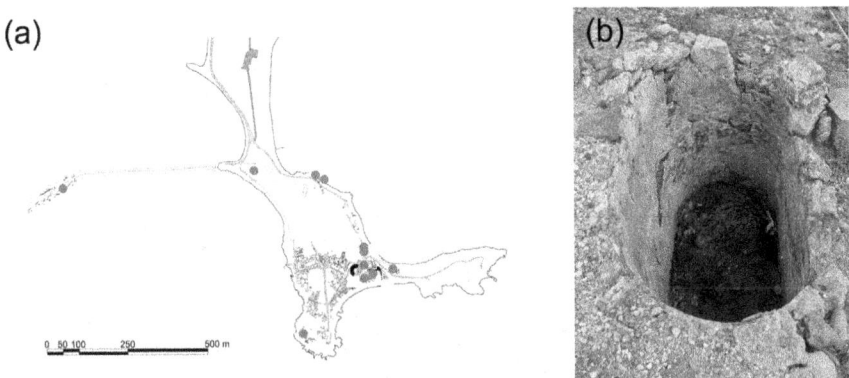

Fig. 3. The site of Nora. (a) Plan of the ancient city indicating the analysed cisterns; (b) Traditional bathtub-type cistern in the city of Nora.

Due to infrequent rainfall, water was an important good for ancient Sardinians. More than 90 cisterns for the collection of rainwater have been documented in Nora and in its suburbs (Fig. 3, b) [8]. They were plastered with hydraulic renders to provide adequate impermeabilization. Hydraulic capabilities were conferred by the addition of organic ashes, *terracotta* fragments (*cocciopesto*) or volcanic-glass (mainly pumices) to aerial lime-based mortars. These pozzolanic materials were often combined together in mixed compounds [9, 10, 31]. The cisterns of Nora were re-coated many times for restorations. We documented a clear shift in the formulations: in older coatings, mortars are mixed with organic ash (ASH type) or with organic ash and *terracotta* fragments (ASH-CP type). On the other hand, the outer restoration revetments are usually made with mortars enriched with volcanic glass (pumice) and *terracotta* fragments (PUM-CP). In the latter type, organic ash is absent or very scarce (Fig. 4). Render types are difficult to be ascribed to exact chronologies, but it seems that ASH and ASH-CP types strictly resemble ancient Phoenician and Punic plastering techniques [11], while PUM-CP one is typically Roman, and it was probably produced from the Imperial period onwards.

2.3 Aquileia

Aquileia was one of the most prosperous Roman towns in Northern Italy. After its foundation in 181 BC, the city became, during the Imperial age, a flourishing urban centre, and it was enriched by monumental buildings and prestigious private houses (Fig. 5, a-c). In the 4th century AD, Aquileia was considered by Ausonius (XI, 9, 4) one of the nine most important and extended cities of the ancient world, but after the

ASH   **ASH-CP**

CP   **PUM-CP**

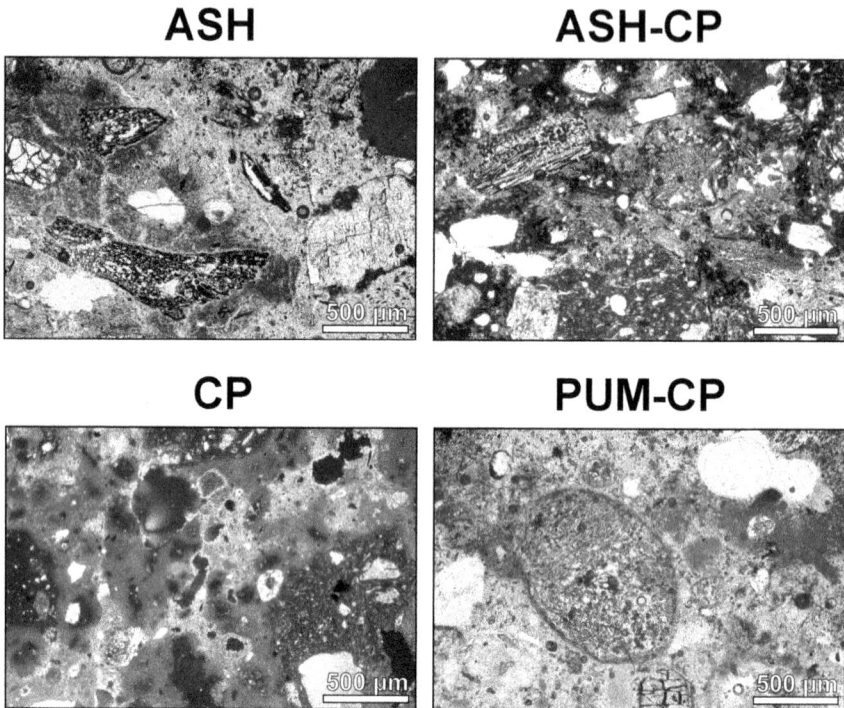

Fig. 4. The four types of cisterns' coating mortars from Nora (from [10]).

425 AD it collapsed under Attila's invasion. After this episode, the town was not totally abandoned but it faced a relentless decline in the following centuries [12].

We analysed around 300 mortars from structures of Aquileia having different function (i.e. wall joints, wall-paintings, floor beddings, foundations) and datings. Sample chronologies cover a timespan of over six centuries (2nd century BC – 6th century AD) [13–17, 32]. In this site, hydraulic mortars were usually produced by adding *terracotta* fragments to the lime-based binders. Mortars enriched in volcanic glass are extremely rare (Fig. 6, a-b), since this type of raw material was not available locally, and it was imported from far-off suppliers.

3   Describing Hydraulic Phases in Ancient Mortars

Adopting a multi-analitical approach involving OM, XRPD, SEM-EDS (for details about the analytical equipments and protocols see paragraph 6), we were able to describe in detail the mineralogical and chemical characteristics of the newformed hydraulic phases in the analysed mortars and concretes and to recognize the factors that have triggered their development.

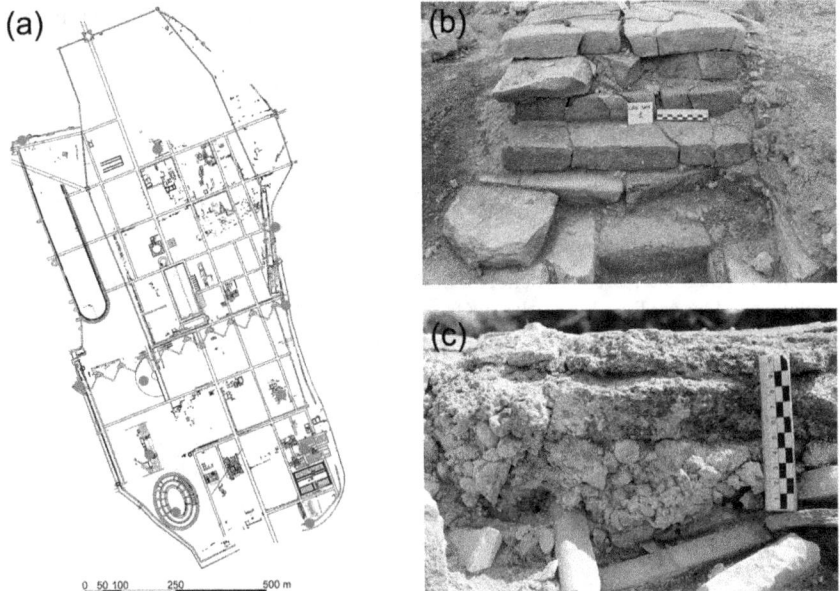

Fig. 5. The site of Aquileia. (a) Plan of the ancient city indicating the main buildings and structures we analysed; (b) joint mortar of a brick-made masonry; (c) mortar-based screed of a mosaic.

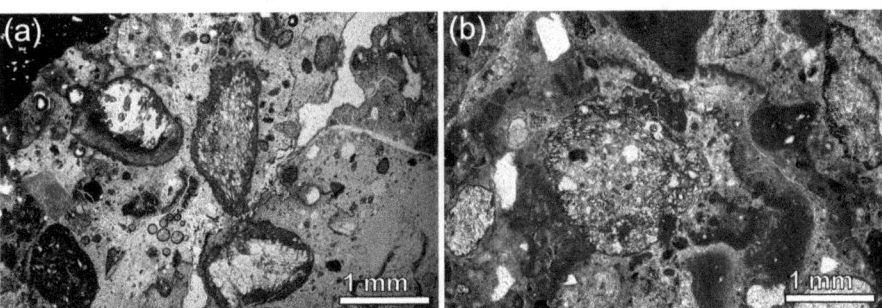

Fig. 6. Mortar-based materials enriched in *terracotta* fragments and volcanic glass (mainly pumice) from Aquileia. (a) Volcanic-rich mortar samples from the *opus caementicium* vaults of the Late Antique baths (from [16]); (b) pumice-rich mortar sample from the screed of the *orchestra* of the Roman theatre (from [32]).

3.1   AFm and C-A-S-H

C-A-H in the form of crystalline AFm [18], having hydrocalumite/hydrotalcite-like structures, are the most common reaction products we detected in alumina-rich mortars, as the cocciopesto ones.

Not all the analysed samples show the same characteristics: the degree of comminution of the *terracotta* fraction strongly affects the development of reaction products in the materials. In fact, mortars enriched in coarse *terracotta* fragments, usually employed as preparation layers of mosaics and other floors in Aquileia, present feeble hydraulic properties. In these samples, the crystalline AFm is almost absent while most of the amorphous phase should be attributed to the dehydroxylated clays rather than C-A-H or C-A-S-H products (Table 1).

Table 1. QPA-XRPD results of the bulk analysis of representative *terracotta* rich mortar-based materials from Aquileia (COS_PREP_4 CBF_PREP_29N data from [13], CBF_CM_5 and CBF_CM_6 data from [17]; GTR_CM_1 data from [16]). Abbreviations, according to [19]: Cal = Calcite; Arg = Aragonite; Brc = Brucite; AFm = AFm phases; Al-Tb = Al-Tobermorite; Dol = Dolomite; Qz = Quartz; Pl = Plagioclase; Kfs = K-feldspar; Anl = Analcime; Ms = Muscovite; Di = Diopside; Ghl = Gehlenite; Ilm = Ilmenite; Hem = Hematite; Amr = Amorphous; - = below detection limit.

Sample	COS_PREP_4	CBF_PREP_29N	CBF_CM_5	CBF_CM_6	GTR_CM_1
Function	Floor bed	Floor bed	Vat render	Vat render	Vat render
Cal	43.1	62.3	39.7	35.2	42.4
Arg	5.1	-	-	-	0.5
Brc	-	-	-	-	1.0
AFm	-	-	1.0	0.4	2.3
Al-Tb	-	-	-	-	1.9
Dol	3.7	2.1	1.1	0.6	-
Arg	-	-	-	3.0	-
Qz	8.5	3.7	7.7	6.2	5.0
Pl	3.3	1.8	1.1	1.2	0.5
Kfs	1.8	-	2.1	3.0	-
Anl	-	-	0.9	-	-
Ms	0.8	-	1.0	0.9	0.2
Di	8.9	7.8	7.7	14.0	3.5
Ghl	-	-	-	1.8	0.4
Ilm	-	-	0.7	0.3	-
Hem	0.3	0.3	0.5	0.4	0.8
Amr	24.2	22.0	36.5	45.2	32.9

Stronger peaks of crystalline AFm have been detected in cocciopesto mortars enriched in finely-ground *terracotta*. These mortars were used for coating cisterns and vats in Aquileia, demonstrating the real intent of ancient crafts to waterproof these structures. Nevertheless, also in these cases, the production technique (homogeneous mixing, size of *terracotta* fraction, lime to aggregate proportions) as well as the type of clay [11], can sensibly affect the reactivity and, as a consequence, the performances of the resulting mortars.

Al-tobermorite, in the form of crystalline C-A-S-H [20], was detected only in some mortars from Aquileia having abundant *terracotta* dust (grain size < 75 μm). The perfect blend of these mortars, which were used to waterproof the vats of a Late Imperial spa, was highlighted also by SEM-EDS analyses, which gave a hydraulic index (HI) and cementation index (CI) of the matrix of 0.7 and 1.7, respectively, indicative of extremely hydraulic materials [21]. These data provide important information about the know-how of the skilled crafts that worked in this Late Imperial spa, probably made under the sponsorship of the Imperial family.

In the renders of the cisterns of Nora, AFm was detected also in cocciopesto mortars enriched with organic ashes (ASH-CP type) and in pure ash-enriched mortars (ASH type): both *terracotta* and ashes reacted with an aerial lime binder determining the development of C-A-H/C-A-S-H. With respect to pure cocciopesto mortars (CP), ASH-CP and even ASH mortars of Nora appear to have better hydraulic properties, as suggested by the sharper AFm peaks (Fig. 7). Therefore, organic ashes can sensibly contribute to the waterproofing of aerial lime-based mortars. Again, it seems that the mixing method can influence the characteristics of the outcoming materials.

Volcanic glass-rich mortars display features similar to cocciopesto ones. However, in the volcanic glass-rich mortars from Pompeii, crystalline hydraulic phases were not detected: the development of AFm is probably extremely limited, while C-A-H/C-A-S-H phases probably maintained a prevalent gel-like structure (Table 2). Being Pompeian mortars very weak in binder to aggregate proportions, the low amount of lime probably influenced the availability of calcium in the system to develop into calcium-based silico-aluminate hydrates. In fact, most of the volcanic-glass rich mortars employed in the masonries of the Sarno Baths of Pompeii report a limited formation of hydraulic products. Therefore, the little degree of comminution of the volcanic clasts probably affected the average available surface area to react.

In some mortars from Nora, mixtures were enriched of both volcanic glass and *terracotta* fragments (PUM-CP type). As reported in Fig. 7, also in this case, the development of crystalline AFm in this type of mortars is almost absent, as most of C-A-S-H and C-A-H likely maintained a gel-like structure. In a few samples from Aquileia pumices mixed with *terracotta* fragments were also detected, but, again, the presence of crystalline AFm was just feebly detected by XRPD.

These outcomes demonstrate that the characteristics of the raw materials are important elements to be considered, but there are also other factors influencing the reactivity of the aggregates, such as the environmental pH. In Aquileia, in fact, mortars enriched with fine and coarse sized volcanic glass appear deeply reacted. SEM-EDS investigations of the binder matrix highlighted a relevant occurrence of silicon and aluminium, probably released by the reaction of volcanic aggregates that, by interacting with the calcium of the lime, determined the precipitation of C-A-S-H like phases. Volcanic clasts, in fact, resulted deeply altered, displaying a strong diffusion of Ca-enriched fluids even in the inner cores of the clast [16]. The alkalinity of the environment possibly fostered the latent reactivity of the pozzolanic materials: in the mortars of Aquileia, the alkaline activator was probably saltwater used for the production of the compounds, but there are also other factors to be considered (i.e. availability of calcium) for a full argumentation of this evidence.

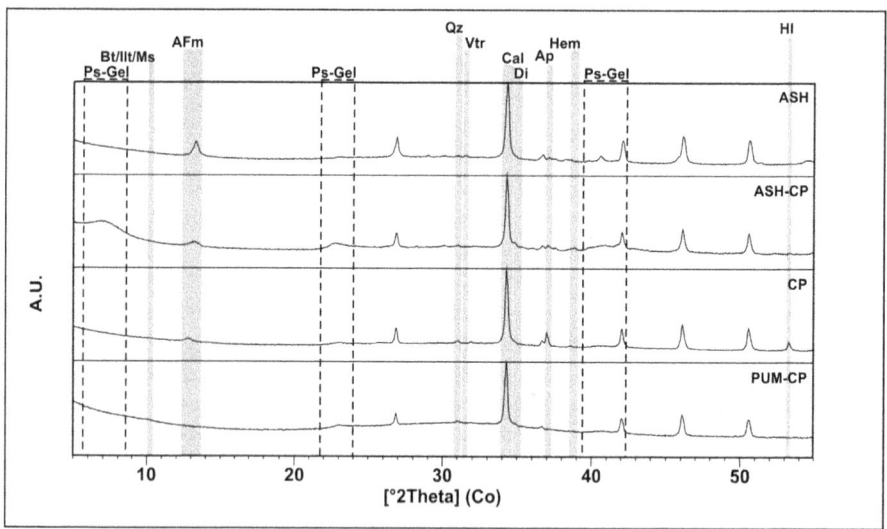

Fig. 7. XRPD diffraction patterns of the sole binder-concentrated fraction of cisterns' renders of Nora (from [9]). The main diffraction peaks of the mineral phases are highlighted. Abbreviations (according to [19]): Ps-gel = phyllosilicate gels (M-S-H/M-A-S-H); Bt/Ilt/Ms = biotite/illite/muscovite; AFm = AFm phases; Qz = quartz; Vtr = vaterite; Cal = calcite; Di = diopside; Ap = hydroxylapatite; Hem = hematite; Hl = halite.

3.2 C-S-H

In the analysed samples, C-S-H was detected only when abundant silica and low alumina were available to react in the system. In some samples of Aquileia we detected the development of pure C-S-H, structured in the form of tobermorite 14 Å, a Si-rich C-S-H [1, 22]. The formation of this phase was not triggered by a pozzolanic reaction, but it relies on the intrinsic characteristics of the binder itself. In fact, a progressive increase in the sharpness of tobermorite 14 Å peaks was observed proceeding from the diffraction pattern of the bulk sample, to that of the binder-concentrated fraction and, in particular, in the spectrum of the lime lumps, mechanically detached from the samples and analysed as they are. This evidence highlights the hydraulic behavior of the binder (Fig. 8), that can be considered as a prototype of a natural hydraulic lime, produced by the calcination of impure cherty limestones (probably local gravels) at high temperatures.

In these samples, little AFm was detected, as low alumina was available in pure binary calcium-silica reactive systems.

3.3 M-S-H and M-A-S-H

Several studies demonstrated that, in alkaline conditions, ASR (alkali-silica reaction) and ACR (alkali-carbonate reaction) can occur [23], influencing the latent reactivity of certain non-pozzolanic components of the mortars, such as dolostones and chert.

In the samples from Aquileia, the reaction in alkaline solutions of chert clasts and the de-dolomitization of dolostones caused the development of M-S-H gels, or M-A-S-H when free reactive alumina was available (Fig. 9, a-e2).

Table 2. QPA-XRPD results of bulk analysis on a selection of volcanic glass-rich mortar-based materials from Pompeii (M267, M298, M302, data from [6]), Nora (C91.I1.b, data from [9]) and Aquileia (data from [16, 17]). Abbreviations (according to [19]): Cal = calcite; Arg = Aragonite; Vtr = Vaterite; AFm = AFm phases; Dol = Dolomite; Qz = Quartz; Pl = Plagioclase; Kfs = K-feldspar; Lct = Leucite; Cpx = Clinopyroxene; Ol = Olivine; Cbz = Chabazite; Php = Phillipsite; Ms/Bi = Muscovite/Biotite; Di = Diopside; Gh= Gehlenite; Anl = Analcime; Clc = Clinochlore; Hem = Hematite; Amr = Amorphous; - = below detection limit.

Sample	M267	M298	M302	C91.I1.b	GTR_VM_4	COM_CM_25
Function	Wall joint	Wall joint	Wall joint	Cistern render	Vault opus caem.	Floor bedding
Cal	28.8	14.3	7.4	42.3	22.9	36.8
Arg	1.8	0.4	5.7	–	5.2	0.5
Vtr	0.4	–	–	1.3	3.4	–
AFm	–	–	–	–	2.2	0.6
Dol	–	–	–	–	–	0.1
Qz	0.8	1.2	0.8	6.4	1.3	1.9
Pl	5.8	10.7	15.1	2.2	–	0.8
Kfs	9.8	12.8	14.8	8.6	3.6	3.6
Lct	0.8	0.6	1.4	–	–	–
Cpx	7.3	10	14.2	–	–	3.7
Ol	–	–	1.0	–	–	–
Cbz	1.2	–	0.5	–	–	–
Php	–	–	–	–	1.0	–
Ms/Bt	0.8	0.9	–	3.1	–	0.3
Di	–	–	–	–	7.0	4.4
Gh	–	–	–	–	–	3.4
Anl	3.1	3.6	6.7	0.5	–	0.4
Clc	–	–	–	0.8	–	–
Hem	0.5	1.2	1.8	–	0.3	–
Amr	39.1	44.3	29.6	34.9	53.5	43.1

The same phases were documented in some mortars from Nora, where the magnesium component was provided by the lisciviation of Mg-rich organic ashes [9].

In both these cases, it is probable that the use of saltwater in the making of the mixtures acted as an alkaline booster for the system [24], determining the precipitation of M-S-H/M-A-S-H even in ordinary "aerial" lime-based mortars.

In those mortars having pozzolanic aggregates, the coupled development of C-S-H and M-S-H or C-A-S-H/C-A-H and M-A-S-H was suggested. These are extremely

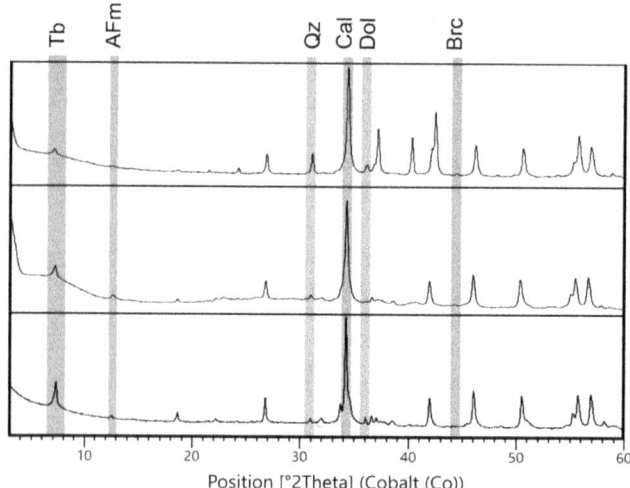

Fig. 8. XRPD diffraction patterns of a sample of mortar (M1_PREF_1) from the foundations of the Republican walls of Aquileia (from [17]). The main diffraction peaks of the mineral phases are highlighted. Upper pattern: bulk sample; middle pattern: binder-concentrated fraction; low pattern: lime lump (mechanically separated). Abbreviations (according to [19]): Tb = Tobermorite 14Å; AFm = AFm phases; Qz = Quartz; Cal = Calcite; Dol = Dolomite; Brc = Brucite.

complex situations, in which it was difficult to correctly parametrize and quantify the effective kinetics of the reactions and their possible cross-relationship.

4   Parametrizing the Effective Hydraulicity

In the mortar-based materials from Pompei, Nora and Aquileia, we observed the development of hydraulic phases, whose structure is primarily related to the sorting of the raw materials constituting the compounds. The milling rate of the aggregates, the proportions of binder and aggregates and the amalgamation of the blend are other important factors influencing the resulting hydraulic behaviour. Finally, also the environmental pH can trigger the precipitation of reaction products.

We observed that C-A-H and C-A-S-H mainly occur in *terracotta*-rich, organic ash-rich and volcanic glass-rich mortar-based materials, as well as in mixed compounds. Pure C-S-H only occurs in binary calcium-silica based systems, and we detected these phases only in mortar-based materials produced with natural hydraulic lime.

Mg-based hydrates appear to be an independent reaction product from Ca-based ones, triggered by the availability in the system of free magnesium and silica that precipitate into M-S-H/M-A-S-H compounds. Moreover, in alkali-rich (K, Na) conditions, the latent reactivity of certain non-pozzolanic aggregates (i.e. dolostones) can be induced.

In this perspective, the major issue we faced with the ordinary techniques we adopted for the characterization of mortar-based materials (OM, XRPD, SEM-EDS) was quantifying the effective hydraulic behaviour of the samples and parametrizing how the elements involved in the reaction are structured, especially in those circumstances where the

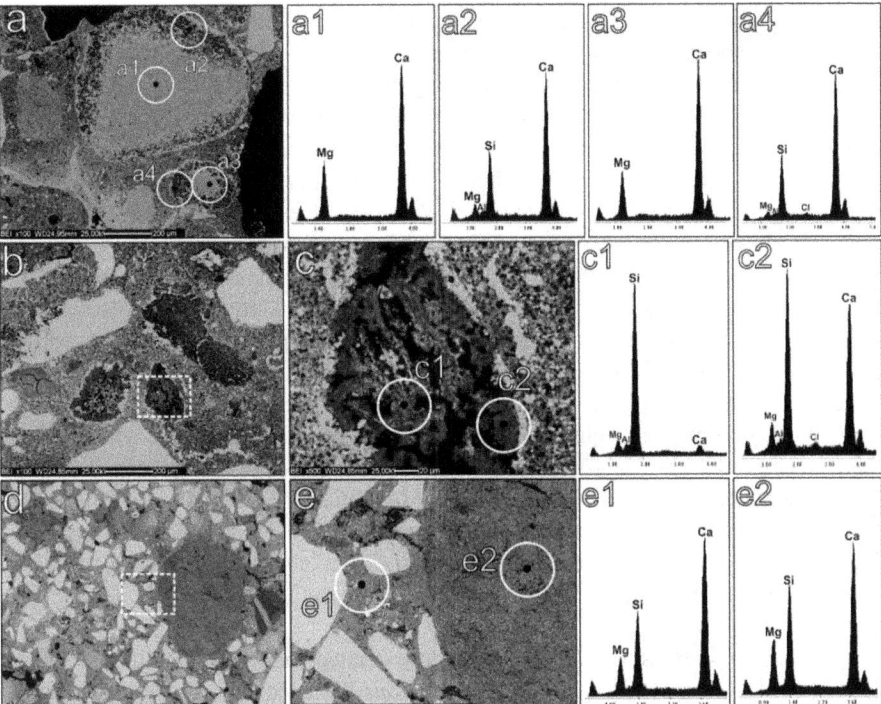

Fig. 9. SEM-EDS analysis of a selection of representative samples from Aquileia (BSE acquisition). (a) Altered dolostone; (a1) EDS microanalysis of an unaltered dolostone relict; (a2) EDS microanalysis of an altered zone of a dolostone grain, with local development of M-(A)-S-H; (a3) EDS of an unaltered dolostone relict; (a4) EDS microanalysis of an altered zone of a dolostone grain, with local enrichment of M-(A)-S-H; (b) altered chert clasts; (c) enlargement of the hatched area in fig. (b); (c1) EDS microanalysis of a weakly altered zone of a chert clast; (c2) EDS microanalysis of an altered zone of a chert clast, with local enrichment in M-(A)-S-H and growth of C-(A)-S-H phases by precipitation from calcium-enriched fluids; (d) Lime matrix on the left and a lime lump on the right; (e) magnification of the dashed area in fig. (d); (e) EDS microanalysis of a M-S-H enriched area of the binder matrix; e2) EDS microanalysis of an M-S-H enriched area in the lime lump from ([32]).

hydraulic products do not exhibit a crystalline habit. In particular, determining the real contribution in the development of hydraulic capabilities was extremely difficult when we approached "complex" formulations, where a mixed array of pozzolanic aggregates and hydraulic phases was detected.

To overcome this limitation, we adopted ^{29}Si MAS-SS NMR (for details about the analytical equipments and protocols see paragraph 6) to determine the real incidence of the hydraulic reaction amongst the analysed mortar samples (Fig. 10). More in detail, we focused on the interpretation of the chemical shifts in the −60 to −120 ppm range. This interval is crucial for the determination of the degree of condensation of SiO_4 tetrahedra [25]. By means of this analysis, we therefore attempted to parameterize the hydraulicity of the mortars by observing the structural configuration of silica in the samples. In the analysed systems, the paracrystalline nature of constituents influenced both the resulting chemical shift and the degree of broadening and overlapping of the individual ^{29}Si NMR signals related to the various siloxane building units [25], as testified by the broad ^{29}Si MAS NMR spectra with few characteristic features observed in most of the analyzed samples (Fig. 10, a-c). Nevertheless, the extent of the pozzolanic reaction was still qualitatively evaluated in the cisterns' renders of Nora by considering the shift of the fitted center of the peak from very negative chemical shift values (completely condensed Q^4 siloxane building units of the unreacted pozzolanic additives at about $-100/-105$ ppm) to less negative chemical shift values (Q^1 and Q^2 siloxane dimers and chains of the C-S-H/C-A-S-H phases at about -85 ppm). According to this analytical interpretation, ASH renders are the ones showing the lower extent of silicates pozzolanic reaction, while a progressive increase in the reaction rate is observed in PUM-CP renders, CP renders and ASH-CP renders (Fig. 10, a). Furthermore, in most of the analysed samples from the sites of Nora, Pompeii and Aquileia (Fig. 10, a-c), the overall silicate peak in the obtained NMR spectra is always characterized by ppm values in-between Q^2 siloxane chains and Q^3 siloxane sheets and wide Full Width at Half Maximum (FWHM) indicative of reduced structural order [26]. This indicates that the main hydrated calcium silicate constituting the pozzolanic network of these binding composites is likely a poorly crystalline C–(A)-S–H phase with a hybrid ino- and phyllosilicate structure of the okenite/nekoite type [27, 28].

The sole samples showing spectral features related to clearly identifiable silicate reaction products are those from the concrete foundations of the Republican walls of Aquileia characterized by an abundant occurrence of pure C-S-H, as detected by XRPD: in these materials, ^{29}Si NMR spectra are characterized by sharp peaks within chemical shift intervals of Q^1 and Q^2 siloxane dimers and chains, typical for inosilicate C-S-H phases such as tobermorite (Fig. 10, d). Nevertheless, the occurrence of broad spectra features associated to poorly crystalline C–(A)-S–H phases with a hybrid ino- and phyllosilicate structure are still detectable, indicating that the formation of such compounds is ubiquitous.

5   Conclusive Remarks

By combining OM observations, QPA-XRPD data, HI/CI of samples calculated by SEM-EDS and, in particular MAS-SS NMR, we were able to parametrize and to effectively compare the hydraulic capabilities of the investigated mortars. The best waterproofing mortars are usually those produced for the impermeabilization of structures or in foundational environments. On the other hand, mortars employed in mosaic beddings and wall joints usually display a feeble hydraulic character. This demonstrates a deep awareness by the ancient crafts of the properties of raw materials and their treatment for the production of ancient binders. The variable degree in the "quality" of the compounds probably was also influenced by the specialization of crafts, as the best hydraulic mortars are those collected from structures of public buildings, i.e. thermae or urban fortifications, where the most in-depth engineering expertise of the time was invested.

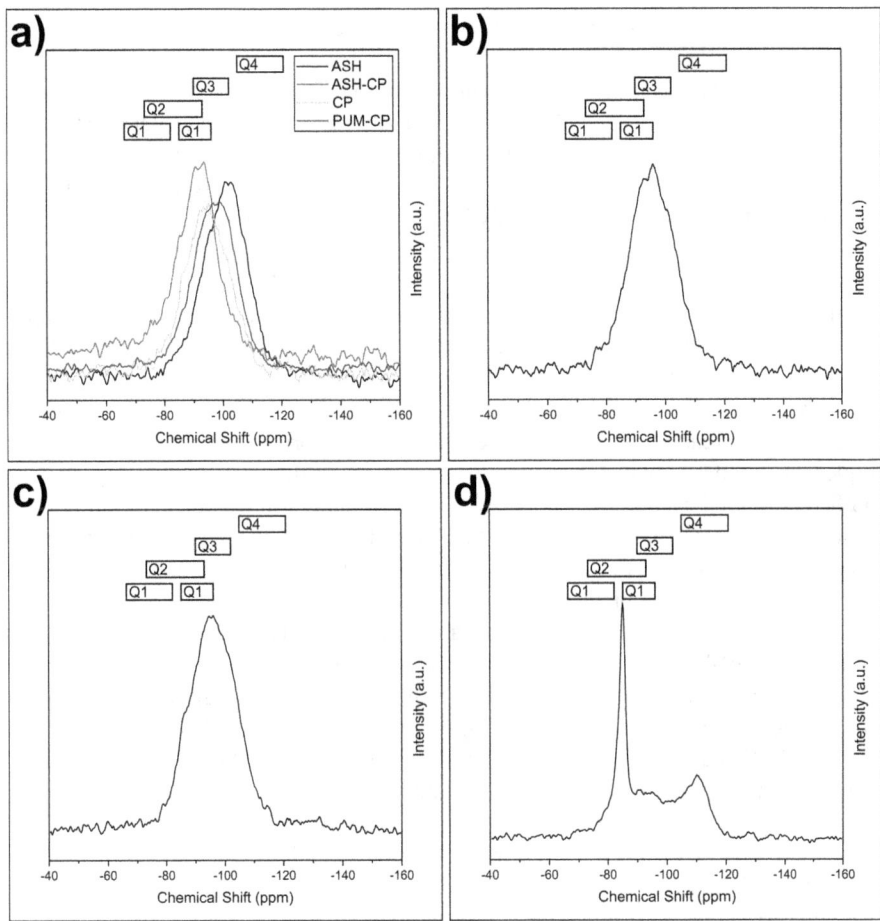

Fig. 10 a) ^{29}Si MAS-SS NMR analyses of representative binder-concentrated samples for each render typology applied in the Nora cisterns (ASH; ASH-CP; CP; PUM-CP) (edited from [9]); b) ^{29}Si MAS-SS NMR analysis of the binder-concentrated sample from a mortar representative of the main ancient construction phase of the Sarno Baths in Pompeii (sample M253); c) ^{29}Si MAS-SS NMR analysis of the binder-concentrated sample from the mortar screed of the orchestra of the Roman theatre of Aquileia (sample COM_CM_25); d) ^{29}Si MAS-SSNMR analysis of a lime lump within the concrete foundations of the Republican walls of Aquileia. Typical intervals for Q^1, Q^2, Q^3 and Q^4 SiO$_4$ tetrahedra [25] are reported.

6   Analytical Techniques

The petrographic study was performed in agreement with the macroscopic and microstratigraphic analytical procedures for the study of mortars and concretes described in Standard UNI 11176:2006 "Cultural heritage – Petrographic description of a mortar". The study was performed by TL-OM on 30 μm thin sections, obtained by vacuum impregnating portions of the materials with epoxy resin and sectioning them transversally. The microscopic study was performed on a Nikon Eclipse ME600 microscope.

Mineralogical analyses of the bulk samples were performed by XRPD-QPA. Quantification was determined by adding 20 wt% of zincite to mortars as internal standard. Data were collected using a Bragg–Brentano θ-θ diffractometer (PANalytical X'Pert PRO, Co Kα radiation, 40 kV and 40 mA) equipped with a real-time multiple strip (RTMS) detector (X'Celerator by Panalytical). Data acquisition was performed by operating a continuous scan in the range 3–85 [°2θ], with a virtual step scan of 0.02 [°2θ]. Diffraction patterns were interpreted using the X'Pert HighScore Plus 3.0 software by PANalytical, qualitatively reconstructing mineral profiles of the compounds by comparison with PDF databases from the International Centre for Diffraction Data (ICDD). Then, QPAs were performed using the Rietveld method [29].

The binder-concentrated fraction of the mortars was extracted to determine the extent of the hydraulic reaction processes. The study was performed on binder-concentrated samples obtained following the Cryo2Sonic 2.0 separation procedure [30], custom-modified through the addition of a chelating agent to the sedimentation solution (sodium hexametaphosphate 0.5% wt.) to favour the suspension of the finer, non-carbonate phases such as clay minerals and pozzolanic products, prone to flocculation due to their surface charges, according to the methodology defined in [9].

Thin sections of selected samples were also micro-structurally and micro-chemically characterized by SEM-EDS, in order to punctually investigate the chemistry of the binder and lime lumps and the extent of the hydraulic products. A CamScan MX2500 scanning electron microscope has been used, equipped with a LaB6 cathode and a four-quadrant solid state BSE detector for imaging. Furthermore, an EDAX-EDS energy dispersive X-ray fluorescence spectrometer was used for chemical microanalysis, mounting a Sapphire Detector composed by a LEAP + Si(Li) crystal and a Super Ultra-Thin Window. Qualitative interpretation of spectra and semiquantitative chemical analysis were performed through SEM Quant Phizaf software.

Finally, a selection of samples were also studied through ^{29}Si MAS-SS NMR, in order to define coordination and degree of polymerization of silicates in the binder fractions of the render samples. ^{29}Si spectra were collected on a Bruker AVANCE III 300 spectrometer, operating with a magnetic field of 7.0 T corresponding to ^{29}Si Larmor frequencies of 59.623 MHz and equipped for solid-state analysis in 4 mm diameter zirconia rotors with Kel-F caps. The magic angle was accurately adjusted prior to data acquisition using KBr. ^{29}Si chemical shifts were externally referenced to solid tetrakis(trimetylsilyl)silane at –9.0 ppm (in relation to TMS). The quantitative ^{29}Si single-pulse experiments were collected at a spinning frequency of 6 kHz, a recycling delay of 100 s and 2000 transients. The signal patterns of the spectra were deconvoluted with the Peak Analyzer routine of Origin Pro 2018 software, using combined Gaussian-Lorentzian peak functions commonly employed for the fit of MAS-SS NMR data.

Author Contributions. SD, MS and JB designed the research; GA and JB supervised the research project; SD, JB and MS performed the samplings; SD, MS and GR performed the samples preparation; SD, MS, GR analyzed the samples; SD, JB, MS, GR and GA interpreted the archeometric results; SD drafted Sect. 1, 3, 5, 6; MS drafted Sect. 4; JB drafted Sect. 2. All authors collaborated to the revision of the manuscript.

References

1. Arizzi, A., Cultrone, G.: Mortars and plasters – how to characterize hydraulic mortars. Mortar, plasters and pigments. Archaeol. Anthropol. Sci. **13**(9), 144 (2021)
2. Roosz, C., et al.: Crystal structure of magnesium silicate hydrates (M-S-H): the relation with 2:1 Mg–Si phyllosilicates. Cem. Concr. Res. **73**, 228–237 (2015)
3. Bernard, E., Lothenbach, B., Rentsch, D., Dauzères, A.: Formation of magnesium silicate hydrates (M-S-H). Phys. Chem. Earth **99**, 142–157 (2017)
4. Bernardi, L., Busana, M.S.: The Sarno Baths in Pompeii: context and state of the art. J. Cult. Herit. **40**, 231–239 (2019)
5. Secco, M., et al.: Mineralogical clustering of the structural mortars from the Sarno Baths, Pompeii: a tool to interpret construction techniques and relative chronologies. J. Cult. Herit. **40**, 265–273 (2019)
6. Dilaria, S., et al.: Phasing the history of ancient buildings through PCA on mortars' mineralogical profiles: the example of the Sarno Baths (Pompeii). Archaeometry **64**(4), 866–882 (2022)
7. Bonetto, J., et al.: Nora, Sardegna archeologica. Guide e Itinerari, 1. Delfino Editore, Sassari (2018)
8. Cespa, S.: Nora. I sistemi di approvvigionamento idrico. Scavi di Nora, VII. Quasar, Roma (2018)
9. Secco, M., et al.: Technological transfers in the Mediterranean on the verge of the Romanization: insights from the waterproofing renders of Nora (Sardinia, Italy). J. Cult. Herit. **44**, 63–82 (2020)
10. Bonetto, J., Dilaria, S.: Circolazione di maestranze e saperi costruttivi nel Mediterraneo antico. Il caso dei rivestimenti in malta delle cisterne punico-romane di Nora (Cagliari, Sardegna). ATTA, Atlante Tematico di Topografia Antica **31**, 495–520 (2021)
11. Lancaster, L.C.: Pozzolans in mortar in the roman empire: an overview and thoughts on future work. In: Ortega, F., Bouffier, S. (eds.) Mortiers et hydraulique en Méditerranée antique, Archéologies Méditerranéennes 6, pp. 31–39. Presses universitaires de Provence, Aix-en-Provence (2019)
12. Ghedini, F., Bueno, M., Novello, M.: Moenibus et portu celeberrima. Aquileia, storia di una città. Poligrafo di Stato, Roma (eds.) (2009)
13. Secco, M., et al.: Evolution of the Vitruvian recipes over 500 years of floor making techniques: the case studies of Domus delle Bestie Ferite and Domus di Tito Macro (Aquileia, Italy). Archaeometry **60**(2), 185–206 (2018)
14. Dilaria, S., Secco, M., Bonetto, J., Artioli, G.: Technical analysis on materials and characteristics of mortar-based compounds in Roman and Late antique Aquileia (Udine, Italy). A preliminary report of the results. In: Álvarez, J.I., Fernández, J.M., Navarro, Í., Durán, A., Sirera, R. (eds.). Proceedings of the 5th Historic Mortars Conference, pp. 665–679. RILEM Publications, Paris (2019)
15. Dilaria, S., et al.: Caratteristiche dei pigmenti e dei tectoria ad Aquileia: un approccio archeometrico per lo studio di frammenti di intonaco provenienti da scavi di contesti residenziali aquileiesi (II sec. a.C. – V sec. d.C.). In: Cavalieri, M., Tomassini, P. (eds.), La peinture murale antique: méthodes et apports d'une approche technique. Atti del colloquio AIRPA, pp. 125–148. Quasar, Roma (2021)
16. Dilaria, S., et al.: High-performing mortar-based materials from the late imperial baths of Aquileia: an outstanding example of Roman building tradition in Northern Italy. Geoarchaeology **37**(4), 637–657 (2022)
17. Dilaria, S.: Archeologia e archeometria dei materiali cementizi di Aquileia romana e tardoantica, Costruire nel mondo antico 8. Quasar, Roma (in press)

18. Matschei, T., Lothenbach, B., Glasser, F.P.: The AFm phase in Portland cement. Cem. Concr. Res. **37**, 118–130 (2007)

19. Whitney, D.L., Evans, B.W.: Abbreviations for names of rock forming minerals. Am. Miner. **95**, 185–187 (2010)

20. Jackson, M.D., et al.: Unlocking the secrets of Al-tobermorite in Roman seawater concrete. Am. Miner. **98**(10), 1669–1687 (2013)

21. Boynton, R.S.: Chemistry and Technology of Lime and Limestone. John Wiley & Sons, New York (1966)

22. Bonaccorsi, E., Merlino, S., Kampf, A.R.: The crystal structure of tobermorite 14 Å, a C-S–H phase. J. Am. Ceramic Soc. **88**(3), 505–512 (2005)

23. Fournier, B., Bérubé, M.A.: Alkali-aggregate reaction in concrete: a review of basic concepts and engineering implications. Can. J. Civ. Eng. **27**(2), 167–191 (2000)

24. Bechor, B., et al.: Salt pans as a new archaeological sea-level proxy: a test case from Dalmatia. Croatia. Quat. Sci. Rev. **250**, 106680 (2020)

25. Magi, M., et al.: Solid-state high-resolution silicon-29 chemical shifts in silicates. J. Phys. Chem. **88**(8), 1518–1522 (1984)

26. MacKenzie, K.J.D., Smith, M.E.: Multinuclear Solid-State NMR of Inorganic Materials. Pergamon, Oxford (2002)

27. Alberti, A., Galli, E.: The structure of nekoite, $Ca_3Si_6O_{15} \cdot 7H_2O$, a new type of sheet silicate. Am. Miner. **65**, 1270–1276 (1980)

28. Merlino, S.: Okenite, $Ca_{10}Si_{18}O_{46} \cdot 18H_2O$: the first example of a chain and sheet silicate. Am. Miner. **68**, 614–622 (1983)

29. Rietveld, H.M.: Line profiles of neutron powder-diffraction peaks for structure refinement. Acta Crystallogr. **22**, 151–152 (1967)

30. Addis, A., et al.: Selecting the most reliable ^{14}C dating material inside mortars: the origin of the Padua cathedral. Radiocarbon **61**(2), 375–393 (2019)

31. Dilaria, S., et al.: Volcanic pozzolan from the Phlegraean Fields in the structural mortars of the Roman Temple of Nora (Sardinia). Heritage **6**, 567–587 (2023)

32. Dilaria, S., et al.: Early exploitation of Neapolitan pozzolan (pulvis puteolana) in the Roman theatre of Aquileia, Northern Italy. Sci. Rep. **13**, 4110 (2023)

Physico-Chemical Characterization of Historic Mortars from the Venetian Arsenals of Chania (Greece)

Pagona-Noni Maravelaki[(✉)] [iD], Kali Kapetanaki[iD],
and Nikolaos Kallithrakas-Kontos[iD]

Technical University of Crete, 73100 Chania, Greece
`pmaravelaki@tuc.gr`

Abstract. The Venetian Arsenals of Chania in southern Greece constitute an important monument of cultural and industrial heritage, as well as a key landmark for the town. In the framework of the holistic study for the restoration of the building, mortars, plasters, and stones were sampled and analyzed in order to identify the construction phases and evaluate the decay state of the materials. Open porosity, bulk density and soluble salts were measured. Additionally, the mortars were chemically and mineralogically characterized through X-ray fluorescence analysis (XRF), infrared spectroscopy (FTIR), differential thermogravimetric analysis (DTA/TG) and X-Ray diffraction (XRD) to identify the mortars' technology and the composition of the raw materials. Furthermore, the granulometric analysis and the optical microscopy provided information about the aggregates and the consistency of the mortars. The original Venetian mortars exceebited good consistency and excellent coherence with the building stones. The Ottoman mortars, on the contrary, due to their poor consistency, have suffered the greatest decay. This research constitutes the base for the design of compatible restoration materials, thus ensuring the sustainable performance of the monument.

Keywords: mortars · characterization · binder · aggregates · construction phases

1 Introduction

The construction of the Chania Arsenals began in 1497 and gradually the addition of two impressive complexes were eventually formed. The first, at the eastern end of the harbor, comprised five Arsenals, while the other, further east, alongside today's dock, which had not been constructed at that time, consisted of seventeen adjacent Arsenals.

The dome sheds of the Arsenals, which had finial pediment facades open to the sea for the approaching ships, were almost 50 m long, 9 m wide, with an average height of 10 m. Arched passageways completed the transverse walls. The Arsenals were constructed of ashlar sandstone in the perimeters of the arches and the walls consisted of rubble stones. By the end of the 19th century the seventeen-arsenals complex had only seven remaining domes (see Fig. 1). In the beginning of the 20th century, the open sea front arches were sealed with walls, the dock was constructed, and the area changed radically. Over the

V. Bokan Bosiljkov et al. (Eds.): HMC 2022, RILEM Bookseries 42, pp. 53–65, 2023.
https://doi.org/10.1007/978-3-031-31472-8_5

years, the Arsenals had various uses such as storage, laboratories of Archaeological services, venue for events and exhibitions. Nowadays, due to the plethora of decay issues they are only used as storage [1].

In the framework of a research project of the Technical University of Crete entitled "Investigation of the historical and technological documentation and the proposal of new uses – promotion of the Arsenals in Chania", historical materials such as, mortars, plasters and stones were sampled and analyzed. More than 35 samples of mortars and plasters were extracted and analyzed in order to identify their composition, consistency, as well as their current condition.

The analysis and characterization of the original materials of historical structures constitutes a helpful tool for the determination of the construction phases, the current condition, and the design of restoration mortars [2–4]. Therefore, the physico-chemical and mineralogical characterization of the raw materials and the derived historical mortars provided information about the Arsenals' construction phases, namely the initial Venetian, the Ottoman, and modern interventions. According to the analyses carried out three main groups of original historical mortars and plasters were identified; the 1st group consists of air lime mortars, the 2nd one of moderately hydraulic mortars and the 3rd one of hydraulic mortars.

Fig. 1. Panoramic view of the seven Arsenals of Chania.

2   Experimental Part

2.1   Sampling

The samples were extracted in a gentle way without causing any destruction to the monument, trying to cover various parameters such as the interior or exterior exposure, orientation, and origin, i.e. Venetian, Ottoman or newer. The sampling positions are illustrated in Fig. 2. The samples labeled according to the following coding system:

$$N\,[1]\,[2]\,[3]\,[4] - I \tag{1}$$

where, N: First letter, used for every sample; 1: labeled as (a), (b) or (c) depending on the construction phase of Arsenal, Venetian, Ottoman, newer, respectively; 2: Filled from 1 to 7, depending on the serial number of Arsenal, starting from west to east; 3: labeled as (i) for samples from internal and as (e) for samples from external exposure sides; 4: labeled as j: for joint mortars, as r: for rubble masonry mortars and as pl: for plasters; I: the serial number of the sample.

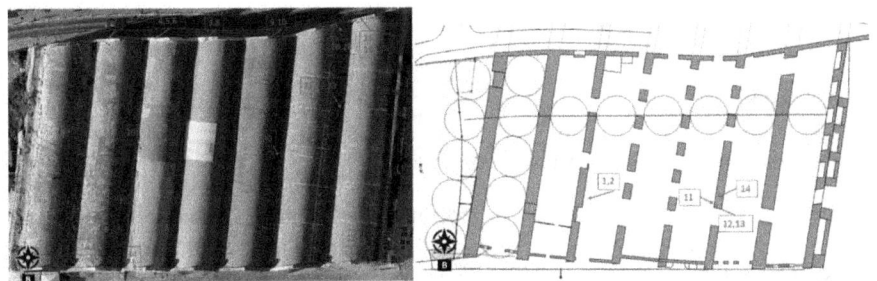

Fig. 2. Sampling location: view of vaults of the Arsenals and external sampling points (left) and plan of the ground floor with indication of the internal sampling points (right).

2.2   Methods for the Characterization of the Samples

Firstly, the original mortar samples were macroscopically observed and divided into 3 groups according to their color; the 1st group consists of the white ones, the 2nd of the grey ones and the 3rd contains the red samples.

Afterwards, the microstructure of the samples was observed using the digital optical microscope USB Dino-Lite AM4515T5 Edge with a color CMOS sensor. The apparent density, the absorbed humidity and the open porosity were also measured after immersion of desiccated samples into water for 2 days at least. Following, the grain size distribution was carried out by mechanical sieving (ASTM E 11-70 series of sieves, meshes of 4.5, 2, 1, 0.5, 0.25, 0.125 and 0.063 mm of diameter) in order to separate the binder from the aggregates and implement the grain size distribution analysis. This destructive method required a large quantity of samples and for that reason only 3 samples of each group were analyzed.

The chemical and mineralogical characterization were carried out through FTIR, XRF and X-ray diffraction (XRD) analyses, respectively. The XRF instrumentation included 109Cd and 55Fe radioactive sources and a Si(Li) semiconductor detector, and the FTIR analysis was carried out with a Thermo Scientific NicoletTM iS50 FT-IR device. Mineralogical composition was assessed with XRD on a Bruker D8 Advance Diffractometer, using Ni-filtered Cu K radiation (35 kV 35 mA) and a Bruker Lynx Eye strip silicon detector, in 40–70° 2-theta range, with 0.02° step size and 0.2 s per step. Additionally, DTA-TG analyses were carried out to further evaluate the existence of hydraulic compounds, using a Setaram LabSys Evo 1600 °C thermal analyzer up to 1000 °C under nitrogen atmosphere with a heating rate of 10 °C/min at a flow rate of 30 mL/min in aluminum crucible.

Finally, the amount of salts present in the historical mortars was estimated through conductivity measurements [5], using a GLP31 Crison conductometer. More specifically, 0.1 g of the binder is dissolved in 100 mL water and remains under magnetic stirring for 2 days. Afterwards, the conductivity of the sol is measured (ES). The salt content (TDS) is calculated according to the Eq. 2:

$$TDS(mg/L) = 0.068 * ES(\mu S/cm) \qquad (2)$$

3 Results

3.1 Macroscopic and Stereoscopic Analysis

The macroscopic and stereoscopic observation of the 1st group of white samples, illustrated in Fig. 3, revealed a moderate [sample (b) Nb3ep-8] to good [sample (a) Nb4er-4] consistency and proper mixing. However, in some cases salt deposits as well as small cracks were also observed [sample (d) Nb6er-19]. The size of aggregates varies, but it does not exceed the diameter of 4 mm.

The 2nd group of grey samples exhibited a poor consistency macroscopically as they were brittle, in accordance with the optical microscopy observations that revealed small cracks and large pores on their surface (Fig. 4). The size of aggregates was varied, exhibiting a diameter lower than 4 mm.

The sample Na1er-33-3 is a typical example of the layered mortar that protected the external part of vaults and consisted of the internal rubble mortar layer (b and d) the intermediate layer (c) and the external thin plaster. All three layers have a good consistency, and their different color could be attributed to the addition of soil material to the rubble mortar and external plaster (Fig. 5 d). Additionally, the rubble mortar also consisted of large aggregates (Fig. 5 b), whilst agglomerations of lime are observed in the internal and intermediate layers even with the naked eye.

The red group of mortars has Ottoman origin, since the samples have been selected from areas attributed to interventions during the Ottoman period (Fig. 6). The samples consisted of soil material and lime and small aggregates. Small cracks and pores are observed along with agglomerations of lime. The consistency is poor, and the mortars are too brittle. All the aggregates have a rounded shape, denoting the use of river sand.

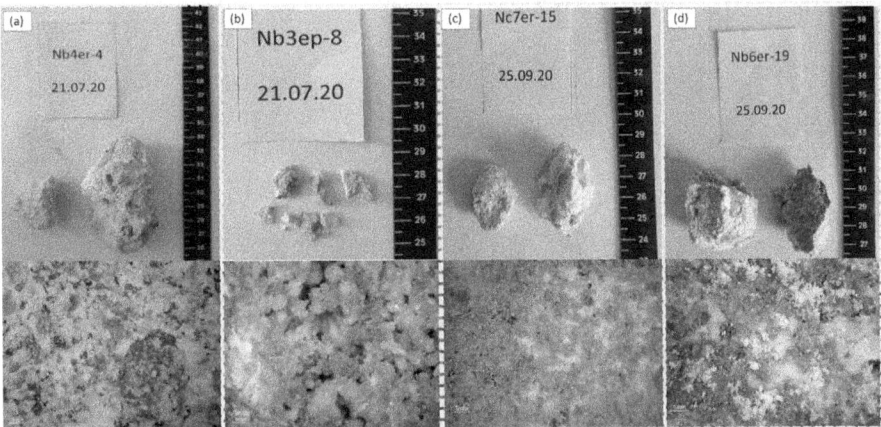

Fig. 3. Macroscopic and optical microscopic inspection of the 1st group of white mortar samples: (a) Sample Nb4er-4; (b) Sample Nb3ep-8; (c) Sample Nc7er-15; (d) Sample Nb6er-19.

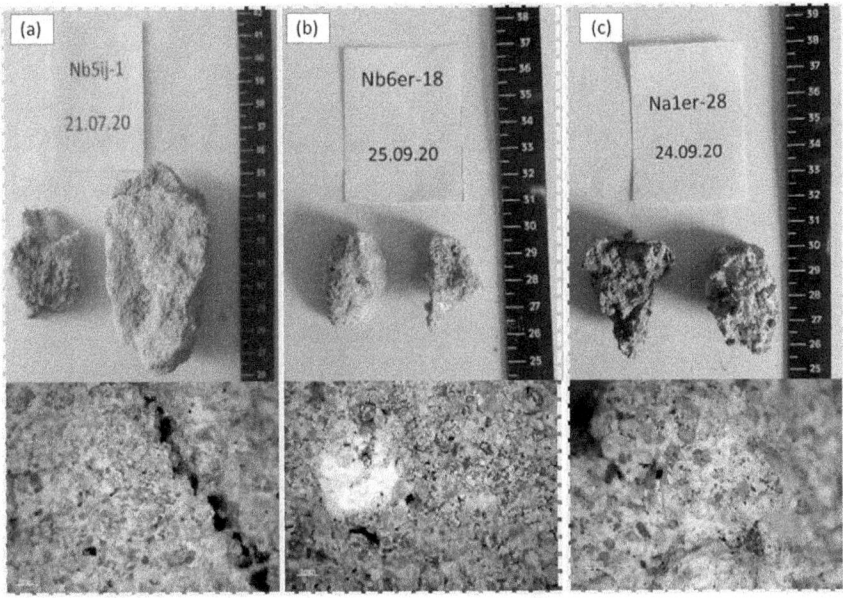

Fig. 4. Macroscopic and optical microscopic inspection of the 2nd group of grey mortar samples: (a) Sample Nb5ij-1; (b) Sample Nb6er-18–8; (c) Sample Na1er-28.

3.2 Grain Size Distribution and Physical Properties of Samples

The apparent density, the wt. % absorbed humidity, the open porosity (%) as well as the wt. % of soluble salts of the tested samples are listed in Table 1. All samples seem to have similar values concerning the accessible porosity and the apparent density. The sample Nb3er-10, mostly consisting of soil, exhibited the highest wt. % of humidity along with

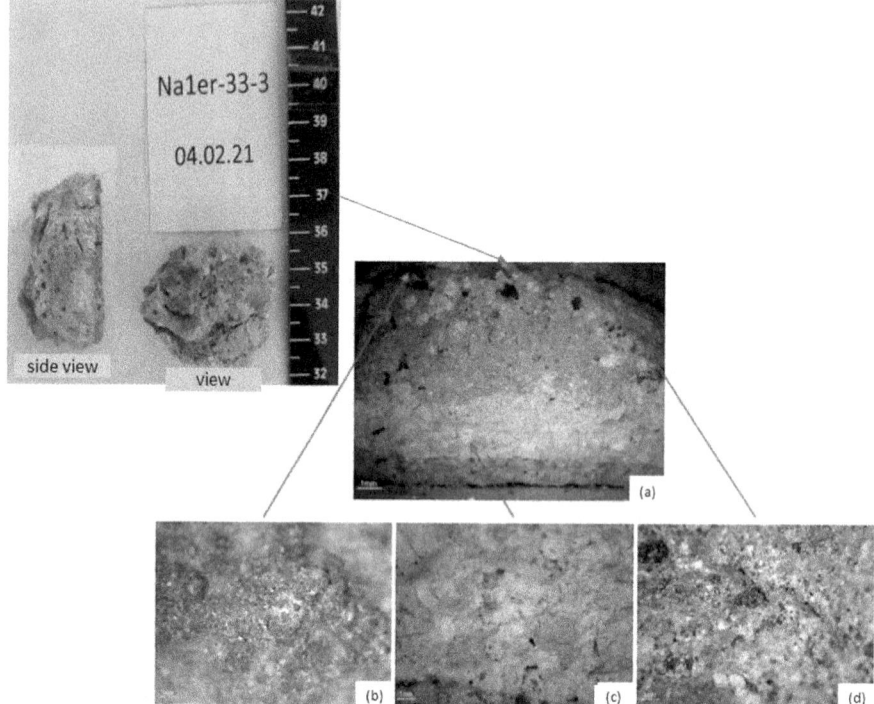

Fig. 5. Macroscopic and optical microscopic inspection of the 3-layered mortar Na1er-33-3 on the external part of vaults.

a large porosity and density, whereas the mortar Nb4er-4 has the lowest porosity, a fact that could be attributed to the good consistency and lack of cracks.

The presence of salts was quite evident in the macroscopic and microscopic observation of the samples, especially in brittle samples, where salts deposits were noted. The wt. % content of salts is higher in the rubble mortars, as they are located inside the masonry suffering from trapped moisture. The high salt content is expected, due to the proximity of the monument to the sea.

The grain size distribution of samples Nb5ij-1 (grey), Nb4er-4 (white) and Nc7er-16 (red-ottoman) was compared to the curve of standard sand in Fig. 7. It was revealed that all three tested samples have coarser aggregates compared to the grain size distribution of the standard sand's. It should be emphasized that the granulometric curve of the Nb4er-4 sample, which has a Venetian origin, is close to the standard sand's curve exhibiting a proper sigmoid grain size distribution along with good consistency, as mentioned above. On the contrary, the sample Nc7er-16 has the most heterogeneous distribution as most of the aggregates range in the zone of 15–25 mm in diameter.

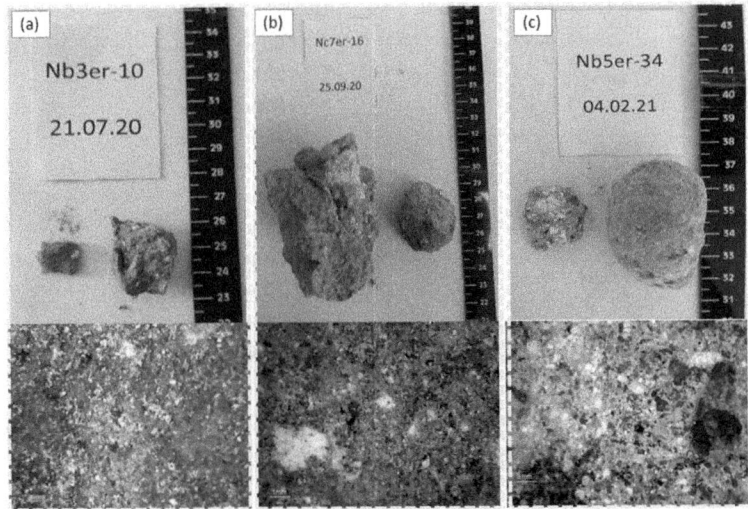

Fig. 6. Macroscopic and optical microscopic inspection of the 3rd group of red mortar samples: (a) Sample Nb3er-10; (b) Sample Nc7er-16; (c) Sample Nb5er-34.

Table 1. Characteristics of samples

Sample	Humidity (wt. %)	Open Porosity (%)	Apparent Density (g/mL)	Soluble Salts (wt. %)
Nb5ij-1	0.31	27.4	2.43	1.98
Nb4er-4	0.99	20.7	2.53	3.89
Nc7er-15	2.64	29.6	2.13	4.48
Nb3er-10	5.05	31.0	2.33	9.40

3.3   FTIR, XRF and DTA Analyses

The XRF results illustrated in Fig. 8, revealed calcium as the predominant element in the white samples (Nb4er-4 and Nb4er-15), along with silicon, iron, magnesium, sodium, and aluminum in lower proportions concentrations. The grey sample Nb5ij-1 differs slightly from the white samples, as silicon's concentration is higher, a fact that could be related to the presence of hydraulic compounds. Concerning the Ottoman sample Nb3er-10, although it has a high amount of calcium, it also contains a high percentage of silicon, aluminum, iron and magnesium. This finding could be attributed to the presence of soil material and/or the deliberate addition of ceramic crust, a methodology which was very common to Ottomans. Finally, all rubble mortars contain considerable amounts of sodium and chloride due to the presence of sea salts deposits.

The presence of hydraulic compound in the mortar samples was further evaluated though FTIR and DTA/TG analyses.

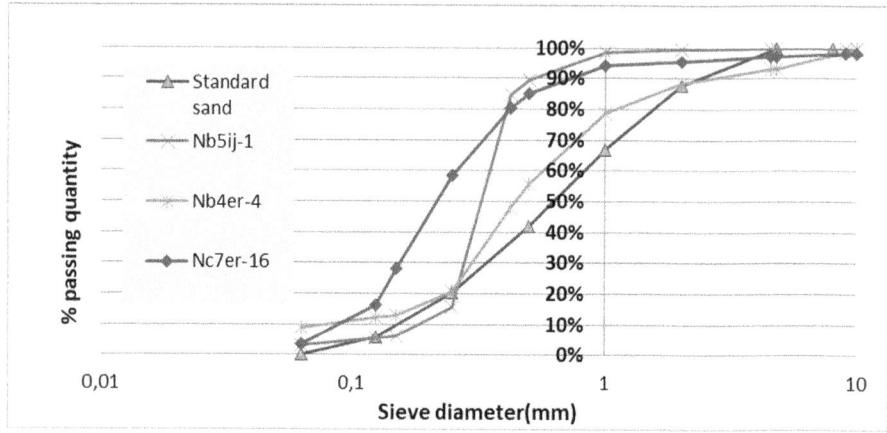

Fig. 7. Granulometric curves of the mortars' aggregates and comparison to the standard sand's.

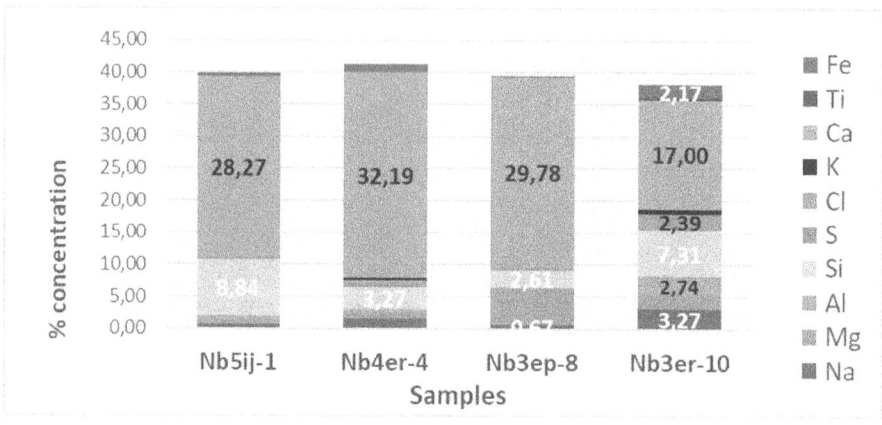

Fig. 8. Elemental composition of samples according to the XRF analysis.

Three samples of each group were analysed through DTA/TG (see Fig. 9). The endothermic reaction between 600–900 °C indicates the decomposition of calcium carbonate (Cc). Moreover, the endothermic reaction along with the corresponding mass loss around 100 °C, expresses the dehydration of the bound water of each, whilst the reaction around 400 °C shows the dehydroxylation of the hydraulic components [6, 7]. As revealed in Fig. 9, the samples of Venetian origin Nb4er-4 (a) and Nb6er-18 (b) exhibit a high percentage of calcium carbonate, whereas the sample Nc7er-16 (c) belonging to the Ottoman period shows a considerable amount of aluminosilicate components, originating from the addition of ceramic crust and soil material.

Calculating the mass losses during the endothermic reaction between 600–900 °C (ML) and considering the molar mass (MM) of calcium carbonate and carbon dioxide, the percentage of calcium carbonate (Cc) can be determined. The Venetian origin samples

Nb4er-4 and Nb6er-18 show a Cc content (wt. %) of 65.51 and 60.02, respectively, whereas in the Ottoman origin sample Nc7er-16 the Cc wt. % content is equal to 29.71.

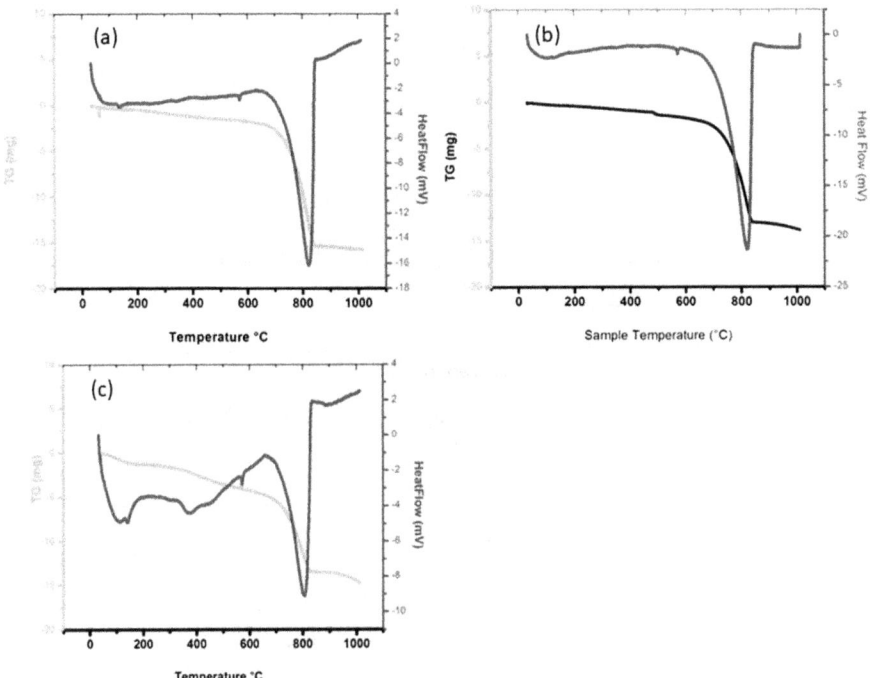

Fig. 9. DTA/TG analysis for samples of Venetian origin Nb4er-4 (a) and Nb6er-18 (b), and the Ottoman origin sample Nc7er-16 (c).

The FTIR spectra of samples are illustrated in Figs. 10–12. The predominant peaks are attributed to calcium carbonate (1440, 870 and 712 cm^{-1}). Additionally, the absorptions around 1100–1030, 780 and 460 cm^{-1} are associated with silicon oxygen bond vibrations (Si-O-Si and Si-O) [8]. More specifically, the peaks in the range 900–1000 cm^{-1} are attributed to hydraulic components, with the peak at 900 cm^{-1} to be indicative of the hydraulic lime, whereas the peak at 780 cm^{-1} is attributed to quartz [8–10]. Moreover, the intense peak at 1385 cm^{-1} of sample Nb4er-4 (Fig. 10, a) is assigned to nitrogen-oxygen (N-O) bond vibrations, due to the presence of nitrate ions and in particular KNO_3. The presence of nitrates is associated with agricultural activities through which they penetrate the aquifer. Finally, the FTIR spectra of aggregates reported in Fig. 13 showed that both carbonaceous and siliceous sand were used, depending on the origin of the river.

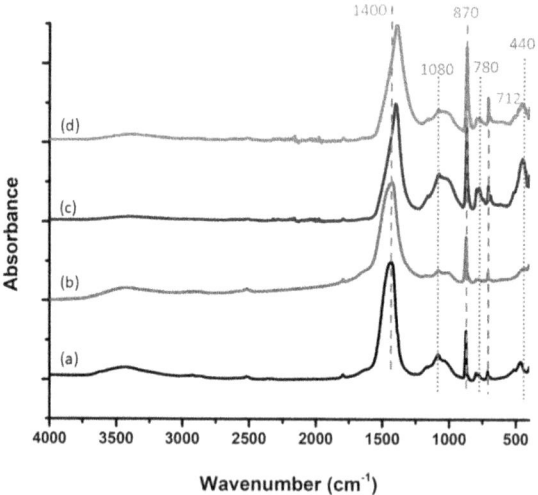

Fig. 10. FTIR spectra of the 1st group white samples; (a) Nb4er-4; (b) Nb3ep-8; (c) Nc7er15; (d) Nb6er-19.

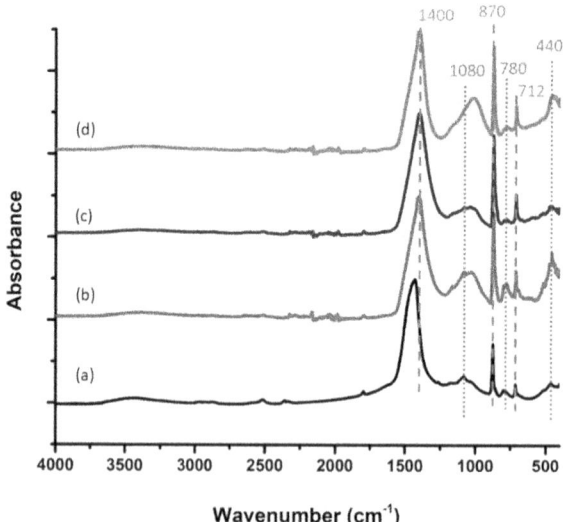

Fig. 11. FTIR spectra of the 2nd group of grey samples; (a) Nb5ij-1; (b) Nb6er-18; (c) Na1er-28; (d) Na1er-33–2.

3.4 Mineralogical Analysis

The results of the XRD analysis depicted in the x-ray diffractograms on Fig. 14, showed that the predominant mineral in the mortars was calcite (Calcite, Cc) followed by quartz (Quartz, Qz). The red Ottoman period Nb3er-10 sample, has a high content of quartz

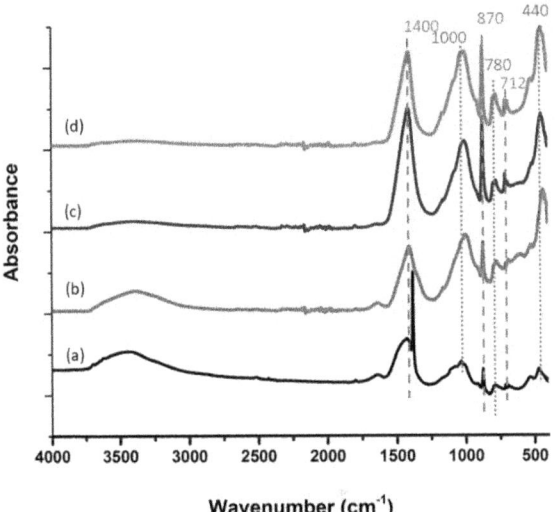

Fig. 12. FTIR spectra of the 3rd group of the red samples; (a) Nb3er-10; (b) Nc7er-16; (c) Nb5er-34; (d) Na1er-33–3 (inside layer).

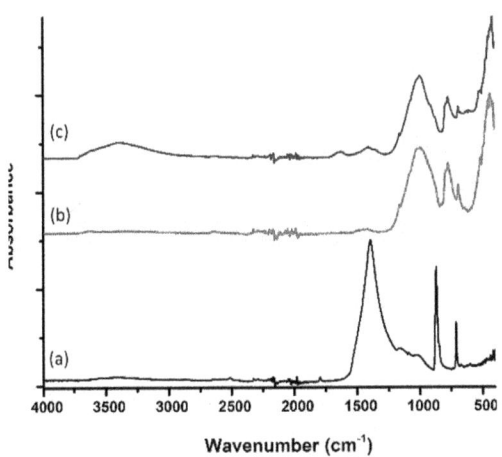

Fig. 13. FTIR spectra of the aggregates of samples: grey, Nb6er-18 (a); white, Nc7er15 (b); red, Nc7er-16 (c).

due to added ceramic material. In addition, Halite (Hl) was found in low quantity, due to sea salts [11]. Finally, the low addition of soil material was ascertained by the existence of Kaolinite (Kl).

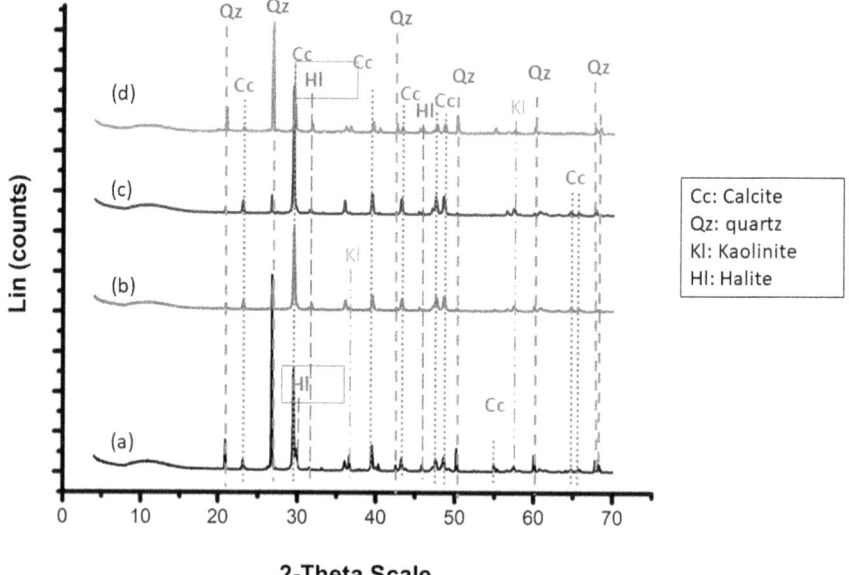

Fig. 14. XRD pattern of the samples: grey, (a) Nb5ij-1; white, (b) Nb4er-4; white, (c) Nc7er-15; red, (d) Nb3er-10.

4 Conclusions

Summarizing the above observations, it could be concluded that:

- The examined samples are quite porous lime-based mortars with moderate consistency due to decay from environmental factors, such as sea aerosol, humidity, intense winds, and water capillarity absorption. All samples show nitrate and chloride attack, as evidenced by FTIR, XRF and conductivity measurements.
- FTIR and XRD analyses indicated the presence of aluminosilicates and magnesium compounds, apart from calcium carbonate, in a smaller percentage, a fact that can be attributed to the use of soil material and ceramics of different gradation during the composition of mortars.
- According to the results of the grain size distribution, only the original Venetian mortars have a proper sigmoidal distribution of aggregates. The aggregates are both of calcareous and siliceous nature with round grains, so they originated from river sand.
- There is a differentiation of mortars by the degree of hydraulicity, which is attributed to the use of moderate hydraulic lime manufactured from the firing of marly limestones found in the area, but also to the use of artificial pozzolans, such as fragments of fine-grained or crushed ceramic.
- The use of hydraulic lime could be assigned to Venetian technology, whereas the high addition of fine-grained ceramic powder belongs to the Ottoman technology, as assumed by the constructions' phases.

- The Venetian mortars, despite the decay they suffered, show better cohesion than the Ottomans which are brittle.

The above conclusions can lead to an effective design of repair mortars, compatible to the original Venetian ones with enhanced durability.

References

1. Lane, F.C.: Venice, A Maritime Republic. Johns Hopkins University Press, ISBN 978-0801814600 (1973)
2. Maravelaki, P.N., Theologitis, A., Budak Unaler, M., Kapridaki, C., Kapetanaki, K., Wright, J.: Characterization of Ancient Mortars from Minoan City of Kommos in Crete. Heritage 4(4), 3908–3918 (2021). https://doi.org/10.3390/heritage4040214
3. Stefanidou, M., Pachta, V., Konopissi, S., Karkadelidou, F., Papayianni, I.: Analysis and characterization of hydraulic mortars from ancient cisterns and baths in Greece. Mater. Struct. 47(4), 571–580 (2013). https://doi.org/10.1617/s11527-013-0080-y
4. Velosa, A.L., Veiga, R., Coroado, J., Ferreira, V.M., Rocha, F.: Characterization of ancient Pozzolanic mortars from Roman times to the 19th century: compatibility issues of new mortars with substrates and ancient mortars. In: Dan M.B., Přikryl R., Török Á. (eds) Materials, Technologies and Practice in Historic Heritage Structures. Springer, Dordrecht (2010). https://doi.org/10.1007/978-90-481-2684-2_13
5. Maravelaki-Kalaitzaki, P., Bakolas, A., Moropoulou, A.: Physico-chemical study of cretan ancient mortars, cement and concrete research, vol. 33, no. 5, pp. 651–661 2003. ISSN 0008-8846, https://doi.org/10.1016/S0008-8846(02)01030-X
6. Kapetanaki, K., Kapridaki, C., Maravelaki, P.-N.: Nano-TiO$_2$ in hydraulic lime-metakaolin mortars for restoration projects: physicochemical and mechanical assessment. Buildings 9, 236 (2019)
7. Maravelaki-Kalaitzaki, P., Karatasios, I., Bakolas, A., Kilikoglou, V.: Hydraulic lime mortars for the restoration of the historic masonry in Crete. Cem. Concr. Res. 35, 1577–1586 (2005)
8. Farmer, V.C.: Infrared Spectra of Minerals. Mineralogical Society, London (1974)
9. Maravelaki-Kalaitzaki, P., Agioutantis, Z., Lionakis, E., Stavroulaki, M., Perdikatsis, V.: Physico-chemical and mechanical characterization of hydraulic mortars containing nano-titania for restoration applications (). Cement Concr. Compos. 36(1), 33–41 (2013)
10. Zhu, W., Chen, X., Struble, L.J., Yang, E.H.: Characterization of calcium-containing phases in alkali-activated municipal solid waste incineration bottom ash binder through chemical extraction and deconvoluted Fourier transform infrared spectra. J. Clean. Prod. 192, 782–789 (2018)
11. Moropoulou, A., Bakolas, A., Michailidis, P., Chronopoulos, M., Spanos, C.: Traditional technologies in Crete providing mortars with effective mechanical properties. Trans. Built Environ. 15 (1995)

Analysis of the Behavior of Original Air Lime Mortars Used in Structural Brick Masonry Walls of Ancient Buildings

Ana Isabel Marques[1]([⊠]) [iD], Maria do Rosário Veiga[1] [iD], António Santos Silva[2] [iD], João Gomes Ferreira[3] [iD], and Paulo Xavier Candeias[4] [iD]

[1] Laboratório Nacional de Engenharia Civil, Building Department, Lisbon, Portugal
aimarques@lnec.pt
[2] Laboratório Nacional de Engenharia Civil, Materials Department, Lisbon, Portugal
[3] CERIS, Instituto Superior Técnico - University of Lisbon, Lisbon, Portugal
[4] Laboratório Nacional de Engenharia Civil, Structures Department, Lisbon, Portugal

Abstract. The growing interest in preserving the built heritage is a driving force towards the search for new rehabilitation solutions compatible with the original construction techniques of ancient buildings. For the design of an adequate reinforcement solution, it is necessary to know in detail the building to be rehabilitated, as well as its original constructive solutions and materials.

This paper presents the results of an experimental campaign on samples of air lime-based laying and coating mortars, extracted from an old masonry building in the historic center of Lisbon, built in 1910, during the rehabilitation works. The different parameters analyzed allow for the composition characterization and evaluation of their mechanical, physical and chemical properties. Based on this characterization, the influence of these mortars on the overall behavior of load-bearing walls of buildings belonging to the typology under study is also evaluated.

Considering the results obtained in the characterization tests for the mortars in study, it was verified that the binder used in both mortars was air lime. The values for compressive strength, modulus of elasticity and the curves obtained in the capillarity and drying tests, are also compatible with this type of mortars. It was determined that the mortars are similar in terms of physical and mechanical characteristics, to Portuguese mortars studied in current buildings of the same historical period. This type of information is crucial in a structural analysis and allows to identify materials compatible with the original ones that can be used in rehabilitation interventions.

Keywords: Conservation and Rehabilitation · Ancient Buildings · Structural Brick Masonry Walls · Air Lime Mortars · Characterization Tests

1 Introduction

In portuguese cities, sturdy masonry buildings were built, approximately, until the 1930s. In the mid-20th century, the use of reinforced concrete became significant and marked the

© The Author(s), under exclusive license to Springer Nature Switzerland AG 2023
V. Bokan Bosiljkov et al. (Eds.): HMC 2022, RILEM Bookseries 42, pp. 66–81, 2023.
https://doi.org/10.1007/978-3-031-31472-8_6

beginning of a short transition period, in which the construction of buildings put aside traditional techniques and materials, increasingly resorting to, at the time innovative, structural solutions using reinforced concrete.

The structural behavior of masonry buildings is very much conditioned by the behavior of the walls, which are the most important structural elements. Masonry walls can be considered as elements with high compressive strength. Besides, they can also guarantee an adequate global behavior of the building, when properly constructed and connected to each other and to the floors, creating a box-like behavior, usually known as *box-behaviour*. As mentioned by Costa [1], the main reasons for the poor structural behavior of masonry buildings in response to seismic actions are mostly related to their heterogeneity, anisotropy, poor tensile strength and limited shear strength.

Earthquakes of great magnitude occur in Portugal, although they are distributed with long time intervals between events. Therefore, the occurrence of intense earthquakes in the future is expected. Recent seismic events around the world have shown that the consequences of earthquakes on the existing building stock can be very devastating. In Italy and New Zealand, the masonry buildings typology is quite similar to the typologies that exist in Portugal. Therefore, considering the damage observed in previous earthquakes, namely the L'Aquila (Italy, 2009), Amatrice (Italy, 2016) and Christchurch (New Zealand, 2011 and 2016) earthquakes, a higher seismic vulnerability is observed in old masonry buildings. In this respect, it is necessary to develop and validate seismic rehabilitation and strengthening techniques for masonry constructions, taking into account the preservation of old buildings and the built historical heritage.

The chosen solutions for the strengthening of old masonry buildings can be somewhat complex, as this type of building is characterized by a considerable uncertainty associated with the behavior of their elements, the properties of the materials in their current state, and the evolution of their performance over time. Although old masonry buildings generally consist of load-bearing masonry walls, timber floors, and wood-beamed roofs, their structural functioning and the definition of the most appropriate intervention solutions need further investigation. Experimental testing in situ or in laboratory is the best way to overcome the lack of knowledge about the behavior of the materials and structural elements as a whole.

In the old masonry buildings built in Portugal until the first half of the 20th century, lime-based mortars were used in the construction of the masonry walls. These mortars are used to join the stones or bricks of masonry or as coating of exterior and interior walls. The old mortars worked both as an element of the structure and as a protection of the masonry. In other cases they had a decorative function, imitating materials such as marble or tiles [2].

The internal regularization layers used in old masonry buildings consist of air lime mortars, possibly with mineral additions and organic additives. The internal layers have a coarser grain size distribution than the external ones, with a progressively higher deformability and porosity from the internal layers to the external ones. The finishing layers consist of fine mortar of lime and sometimes gypsum plaster, applied with decreasing thickness from the innermost to the outermost.

Wall coverings, due to their great exposure to external actions and their role of protection of masonry, are the elements that are most subjected to degradation and are, therefore, the most frequently addressed in the interventions. However, the same reasons that lead to repair or replace them also justify great care in these interventions, and a sound knowledge of the composition and behavior of old coatings is fundamental.

This characterization constitutes a first step in the study of different reinforcement solutions based on reinforced plasters, setting a reference to later compare with different reinforcing solutions. The definition of these solutions and their application methodologies will be integrated into a general strategy for the conservation and rehabilitation of these buildings, taking into account aspects such as material compatibility, feasibility and sustainability of application, degree of intrusiveness and impact on elements with potentially high cultural value.

2   Characterization of the Building

The case study of this research work consists of a set of three buildings located in the center of Lisbon. These residential buildings, very similar to each other, were built in 1904 as a rental property investment. Their architecture has a Parisian influence, with eclectic facades that combine *Neo-Renaissance* and *Art Nouveau* details.

Considered at the time one the best residential buildings of the kind in Lisbon, they have rooms with high ceilings decorated with plaster and *boiseries* over the fireplace in perfect condition, testifying the wealth of a bourgeoisie that lived in Lisbon in the late 19th and early 20th centuries.

Originally, these buildings had a ground floor and five raised floors, were separated by large transverse internal open spaces, and each building had a small vertical shaft for ventilation and lighting of the inner rooms [3]. Each floor had two apartments with thirteen rooms and an area of about 300 m^2 with high ceilings with decorative gypsum plaster. Figure 1 shows the picture of the facade of the buildings in their original state.

Fig. 1. General view of the facades of the studied buildings [4].

Two of the buildings were extremely damaged, so the demolition of their core was necessary. Most of the coating and laying mortar samples were extracted from these buildings. Solid areas were chosen, away from areas subjected to weathering (because part of the buildings' roof was missing), so that the mortars would be as original and intact as possible. Regarding mortar extraction, preference was given to the interior zones of the masonry joints in order to avoid contamination with other materials. Thicker horizontal joints were also chosen to facilitate removal without fracturing the mortar, and to obtain samples of sufficient size for testing, as shown in Fig. 2a. Concerning coating mortars, random areas were chosen from inside the building because, in general, these mortars were in good condition, as shown in Fig. 2b.

3   Materials and Methods

3.1   Samples

The collected samples were identified as either coating or laying mortars and were then packed in a wooden box in bubble wrap to cushion the impacts and prevent them from breaking during transport to the laboratory, as illustrated in Fig. 2c. The mortar collection took place during the relatively dry summer weather.

(a)   (b)

(c)

Fig. 2. Sampling in the building studied: (a) example of a laying mortar sample removal zone; (b) example of a coating mortar sample removal zone (c) mortar conditioning.

The two types of mortar samples collected were divided into smaller specimens, whose surfaces were gently flattened and cleaned with a brush to remove loose particles and less adherent biological colonization. These specimens were divided into two groups: specimens for the composition analysis and specimens for physical and mechanical characterization.

The mortar specimens for the composition analysis were submitted to macroscopic analysis, mineralogical analysis by X-ray diffractometry (XRD), simultaneous thermogravimetric and differential thermal analysis (TG/DTA), wet chemical analysis and grain size distribution of the sand present in the mortars.

Mortar specimens for physical and mechanical characterization were submitted to compressive strength, open porosity, bulk density, ultrasonic velocity, capillary absorption and drying.

3.2   Analysis of the Composition of Laying and Coating Mortars

Both types of mortar samples have a beige color with some dark colored aggregates and small ochre brown stones. Rolled grains up to 3 mm and fine sand apparently of siliceous nature are visible. When handled, the apparently cohesive samples release a fine dust that is presumed to be from the binder paste.

The mortars are clean and free of biological colonization and show plenty of small white lime nodules, as shown in Fig. 3.

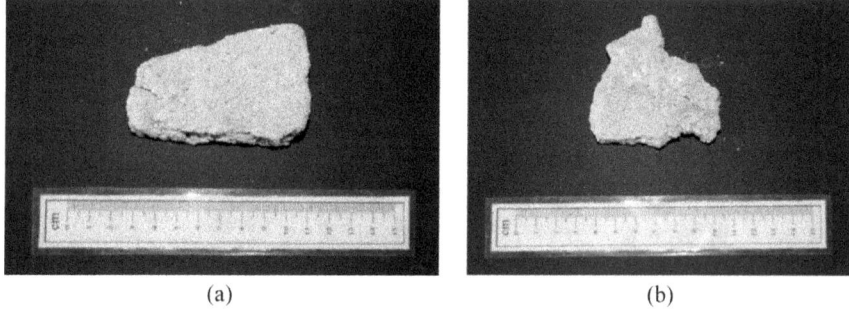

(a)   (b)

Fig. 3.   Selected samples for the composition analysis tests: (a) coating mortar; (b) laying mortar.

The XRD analysis was done in a PHILIPS PW3710 X-ray diffractometer, using 35 kV and 45 mA with a scanning rate of $0.05°\ 2\theta/s$ to determine the mineralogical composition of the mortar constituents (binder, sand, alteration and neoformation products).

Simultaneous thermogravimetric and differential thermal analysis (TG/DTA) were carried out using a thermal analyser SETARAM TGA92 apparatus. The samples were placed in a platinum crucible and heated from room temperature to 1000 °C at a uniform rate of 10 °C/min under argon atmosphere (3 L/h). This analysis enables to obtain the mass variations associated with dehydration, dihydroxylation and decarbonation processes, occurring at specific temperature intervals. Decarbonation generally occurs in the temperature range of 550–900 °C, which enables to obtain the carbonates content, expressed as a percentage of $CaCO_3$, and which can be attributed, in the absence of limestone aggregates, to the carbonated lime content.

The siliceous sand content was obtained after an hydrochloric acid attack, using an acid solution with a ratio of 1:10 (H_2O:HCl), according to a procedure described elsewhere [5]. The insoluble residue (IR) value obtained after acid attack, which roughly translates the actual sand content in the analyzed samples, was calculated and then separated by sieving, using sieves with mesh sizes of 5.00 mm, 2.50 mm, 1.25 mm, 0.630 mm, 0.315 mm, 0.160 mm and 0.075 mm.

3.3 Physical and Mechanical Characterization of Laying and Coating Mortars

To quantify the compressive strength of irregular mortar specimens, the standard method presented in EN 1015–11 [6] was adapted for samples collected in situ, based on the methodology proposed by Valek & Veiga [7] and Magalhães & Veiga [8]. Five specimens of coating mortar and seven specimens of laying mortar were tested.

Three specimens of coating mortar and four specimens of laying mortar were tested for open porosity and apparent density, whose test procedures were based on the European standard EN 1936 [9].

The ultrasonic velocity measurement test was performed to evaluate the mortars' compactness and stiffness. This test was performed using the indirect method (using two alignments on each type of specimen) and the direct method (three points on each type of specimen). In this test, the standard method set forth in EN 12504–4 [10] was adapted for samples collected in situ, based on the methodology proposed by Valek & Veiga [7] and Magalhães & Veiga [8].

The water behavior of the mortars was analyzed through tests to determine the capillarity coefficient and to evaluate the drying rate in two specimens of coating mortars and three specimens of laying mortars. The test methodology was based on the procedures described in the European standards EN 15801 [11] and EN 16322 [12] and the testing technique was adapted for irregular and friable samples [8, 13].

4 Analysis of Results

4.1 Composition of Laying and Coating Mortars

Figure 4 shows the diffractograms obtained for the analyzed mortars and Table 1 the corresponding qualitative mineralogical compositions.

The mineralogical composition of the mortars (Table 1) is rich quartz and feldspars, which associated with mica and traces of kaolinite, is indicative of sands of siliceous nature with a clay component. In terms of the binder paste compounds, calcite is presented in medium proportion which indicates that they are calcium air lime mortars.

From the TG/DTA/DTG charts (Fig. 5) and according to the mineralogical compositions of the mortar specimens four mass losses ranges were considered (Table 2): 25 °C o 150 °C – mass loss due to dehydration of free or absorbed water; 150 °C to 400 °C – mass loss due essentially to zeolitic water and hydrated iron and/or aluminum oxides; 400 °C to 550 °C – mass loss due essentially to dihydroxylation of clay minerals; and 550 °C to 850 °C – mass loss due to decarbonation of calcium carbonate [14, 15].

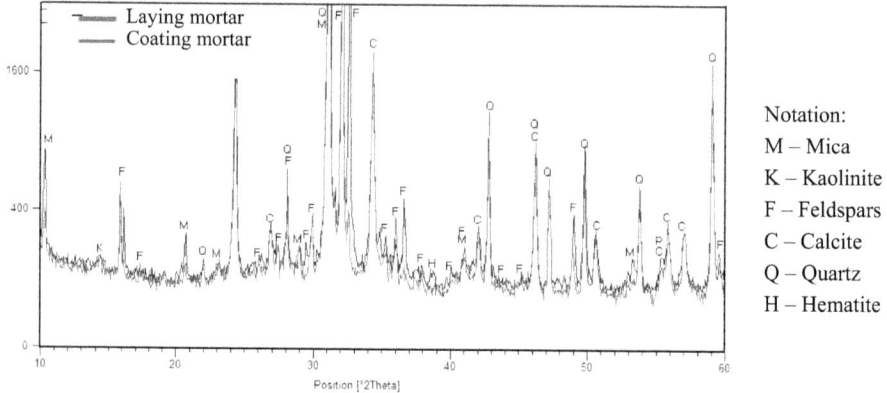

Fig. 4. Diffractograms of laying and coating mortars.

Table 1. XRD composition of the mortars.

Identified crystalline compounds	Coating mortars	Laying mortar
Quartz (SiO_2)	+++	+++
Feldspars ($KAlSi_3O_8 - NaAlSi_3O_8 - CaAl_2Si_2O_8$)	++/+++	++/+++
Mica - moscovite/biotite ($KAl_2(Si_3Al)O_{10}(OH,F)_2$ / $K(Mg,Fe^{2+})_3[AlSi_3O_{10}(OH,F)_2]$)	+	+/++
Calcite ($CaCO_3$)	++	++
Kaolinite ($Al_2Si_2O_5(OH)_4$)	Traces	Traces
Hematite (Fe_2O_3)	Traces	-

Notation: +++ high proportion; ++ medium proportion; + low proportion; - undetected.

The analysis of the TG/DTA/DTG results confirms the XRD results, namely with the main mass loss resulting from the decarbonation of the calcium carbonate [15, 16], proving that the binder used in the mortar was a calcitic air lime.

Figure 6 shows the graph of the sands' grain size distribution curves of the coating (blue) and laying (red) mortars.

In general, the mortars have well graded sands, less than 5.0 mm in size, and fine grain content (less than 0.075 mm in size) of 1.65% for the coating mortar and 7.76% for the laying mortar.

Table 3 shows the mass contents of the different constituents and the binder/aggregate ratios obtained.

The characterization shows that the composition of the coating and laying mortars is very similar. Both mortars are made of calcitic air lime, with a well graded sand and fine grain content (smaller than 0.075 mm) of about 1.65% for the coating mortar and 7.76%

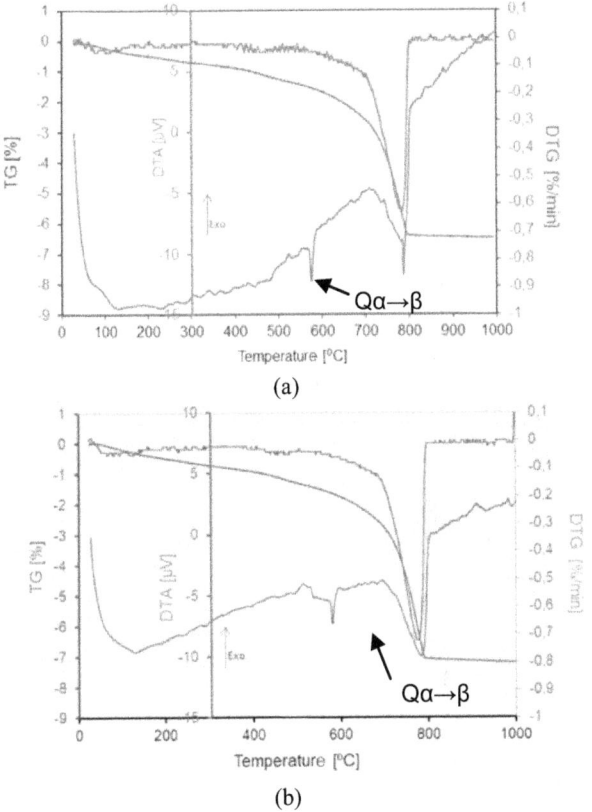

Fig. 5. TG/DTA/DTG charts: (a) coating mortar; (b) laying mortar.

Table 2. Mortar samples mass losses obtained by TGA.

Sample identification	Mass loss [%]				Loss of ignition [%][1]
	25–150 °C	150–400 °C	400–550 °C	550–850 °C	
Coating Mortar	0.49	0.25	0.68	5.03	6.50
Laying Mortar	0.46	0.43	0.61	6.28	7.82

[1] Sample mass loss between 25 and 1000 °C.

for the laying mortar. The mineralogical nature of the sand used is rich in quartz and feldspar which, associated with mica and traces of kaolinite, is consistent with a sand of siliceous nature, probably coming from a sandpit due to the presence of kaolinite. The use of sand with some clay is quite common in this type of mortars, as it brings some "gum", much appreciated by the craftsmen who applied it, as indicated by some old artisans.

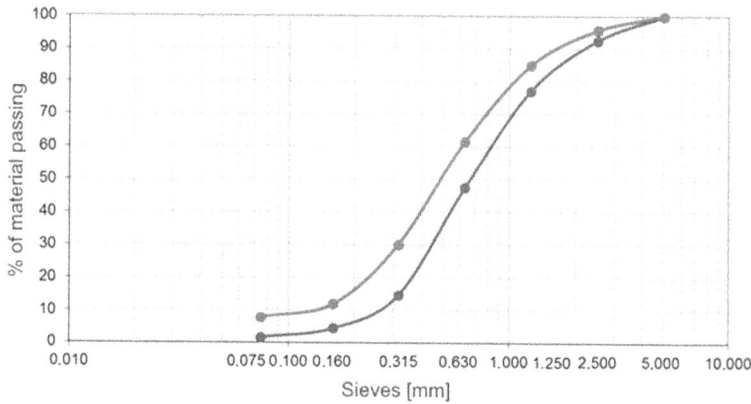

Fig. 6. Grain size distribution curves of sand in mortar samples. Red – laying mortar; blue – coating mortar. (Color figure online)

Table 3. Composition of mortar samples.

Sample identification	Mass content [%]		Lime:Aggregate ratio	
	Sand[1]	Calcite[2]	in mass	in volume
Coating mortar	83	11.4	1: 9.8	1: 3.2
Laying mortar	81	14.3	1: 7.7	1: 2.5

[1]Value corresponding to the insoluble residue content; obtained by acid attack
[2]Value obtained by TGA analysis.

The mortar ratios vary between 1:9.8 and 1:7.7 (hydrated lime: sand) in mass which corresponds approximately to 1:2.7 and 1:3.4 (hydrated lime: sand) in volume, calculated considering that a powder hydrated lime was used as they did in the period of this construction and that the ratio between the bulk density of the aggregate and the air lime is very close to three.

4.2 Physical and Mechanical Characterization of Laying and Coating Mortars

Regarding the compressive strength results of the two types of mortars, the values of maximum force, compressive strength stress, ultimate deformation, and ultimate strain, are summarized in Table 4. It should be noted that in this test it was not necessary to use a "regularization mortar" to adjust the dimensions and shape of the specimens, since the selected specimens had dimensions and flatness adequate for the application of the compressive force.

The coating and laying mortars have moderate strengths in the order of 1.5 MPa and relatively high strains at break in the order of 10%, as it is expected in this type of mortars [8, 17]. In air lime mortars there is not a direct correlation between binder content and mechanical characteristics. In fact, air lime matrix is very porous, so, more lime may not result in higher compaction and mechanical strength. However, it may determine

Table 4. Values obtained in the compressive strength test for the two mortars.

Identification of the type of mortar		Maximum force reached [N]	Compressive strength [MPa]	Deformation at rupture [mm]	Strain at rupture [-]
Average value	Coating	**2747**	**1.72**	**2.09**	**0.079**
Standard deviation		1251	0.78	0.53	0.020
Coefficient of variation		0.46	0.46	0.25	0.256
Average value	Laying	**2019**	**1.27**	**2.26**	**0.111**
Standard deviation		337	0.20	0.31	0.018
Coefficient of variation		0.17	0.16	0.14	0.164

better durability and long-term performance [18]. Additionally, micro cracks may have been produced in the laying mortar due to high compressive stresses resulting from the masonry weight.

The deformations in the load application zone were measured through the internal LVDT of the test machine, so they may include some deformability of the load application mechanisms. The elongation at break was determined taking into account the initial thickness of each specimen. The value obtained (despite its inaccuracy) indicates a high material deformability.

The results obtained for the open porosity and bulk density of the two types of mortars are summarized in Table 5.

The results obtained for open porosity and bulk density are consistent with the similarity of the two types of mortar, and with the type of lime mortars used at the time of construction of the building under study, with very high porosities [18]. The average values obtained for the coating and laying mortars of the bulk density are between 1600 kg/m^3 and 1700 kg/m^3 and the open porosity values are between 32 and 34%.

Regarding the results obtained at the ultrasonic velocity measurement, Fig. 7 shows, as an example, the results obtained for the laying mortar by the indirect method. Table 6 summarizes the results obtained by the two methods and for the two types of mortars.

The average values of the dynamic modulus of elasticity determined by the indirect ultrasonic velocity measurement are 2499 MPa for the coating mortar and 1745 MPa for the laying mortar. Using the direct method, the values obtained for the dynamic modulus of elasticity are 2606 MPa for the coating mortar and 1628 MPa for the laying mortar, thus confirming that these two methods produce similar results, mutually reinforcing their reliability [19]. These values are characteristic of a lime mortar in good condition [17, 20, 21].

Table 5. Results obtained of the open porosity and bulk density for the coating and laying mortar.

Identification of the type of mortar		Open pore volume [mm³]	Apparent volume [mm³]	Real density [kg/m³]	Bulk density [kg/m³]	Open porosity [%]
Average value	Coating	**15.15**	**47.94**	**2494**	**1702**	**32**
Standard deviation		7.7	24.76	20	26	2
Coefficient of variation		0.51	0.52	0.01	0.02	0.05
Average value	Laying	**7.39**	**21.99**	**2426**	**1616**	**34**
Standard deviation		2.22	5.82	19	43	1
Coefficient of variation		0.30	0.26	0.01	0.03	0.04

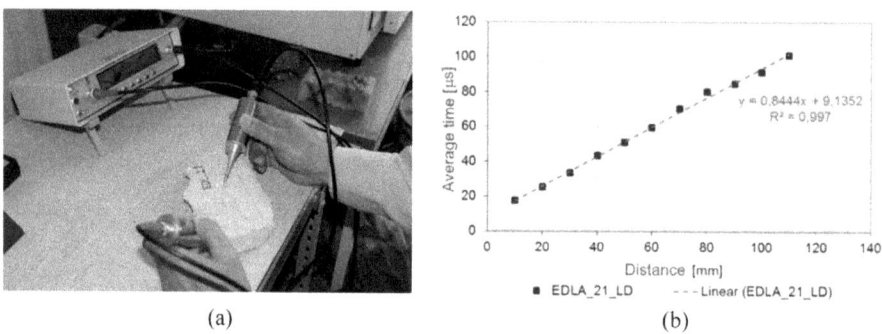

(a)   (b)

Fig. 7. Ultrasonic velocity measurement test: (a) carrying out the test by the indirect method; (b) test result.

Regarding the results obtained in the test capillary absorption test and in the drying test, Fig. 8 shows, as an example, the results obtained for the laying mortar. The results obtained in the test for the two types of mortars are summarized in Table 7.

The capillarity coefficient values for the laying mortar are lower relative to the coating mortar's. This may be due to: i) the laying mortar is subjected to a much higher stress level than the coating mortar due to the weight of the overlying masonry, making the mortar pores smaller and thus hindering the circulation of water in liquid form inside these mortars and ii) the laying mortar is richer in binder, which may also contribute to this difference in results.

Regarding the drying behavior, the values of the first drying phase (D1), which are related to the transport of liquid water towards the surface, occurs more slowly for the laying mortar (evidenced by the lower value of D1), probably due to the lower volume

Table 6. Results obtained in the ultrasonic velocity measurement tests for the coating and laying mortar.

Identification of the type of mortar		Indirect method		Direct method	
		Average velocity [m/s]	Modulus of elasticity [MPa]	Average velocity [m/s]	Modulus of elasticity [MPa]
Average value	Coating	**1275**	**2499**	**1304**	**2606**
Standard deviation		110	433	13	52
Coefficient of variation		0.09	0.17	0.01	0.02
Average value	Laying	**1091**	**1745**	**1058**	**1628**
Standard deviation		132	418	22	66
Coefficient of variation		0.12	0.24	0.02	0.04

of capillary porosity as indicated by the capillarity coefficients. The values of the second drying phase (D2), which are related to water in vapor form, follow the same trend, with the lowest values belonging to the laying mortars. This reduction of water transport may be related to the fact that the laying mortar is subjected to a much higher level of stress than the coating mortar.

Considering all the tests performed, Table 8 shows a summary of the average results obtained in the experimental tests for the characterization of coating and laying mortars. The standard deviation values are shown in brackets.

Considering the results obtained, it can be concluded that these mortars have air lime as the only binder. These mortars are similar, both in terms of composition and physical and mechanical characteristics, to Portuguese mortars studied in current buildings of the same historical time [8, 22].

The lowest values of compressive strength and ultrasonic dynamic modulus of elasticity are obtained for the laying mortars. On the one hand, considering the "stronger" mortar mixtures of the laying mortars, the opposite result was expected. On the other hand, this ratio is coherent with the lower bulk density value and the higher open porosity of the laying mortars compared to the coating mortars. As an explanatory hypothesis, a higher water/binder ratio may have been used in the laying mortars to obtain higher workability. Despite being a stronger mix, it led to a reduction in its compactness (lower density mass and higher open porosity) and related mechanical characteristics (compressive strength and modulus of elasticity). Another hypothesis for these results is related with the fact that laying mortars have less contact with CO_2 and, consequently, a slower carbonation over time.

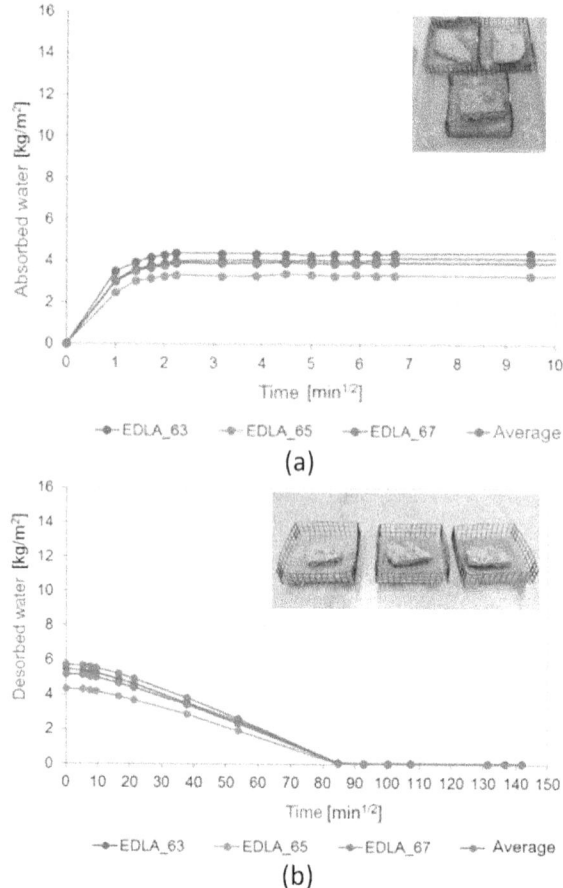

Fig. 8. Curves obtained in the test for the laying mortar: (a) capillarity; (b) evaluation of the drying rate.

The values obtained in the tests performed on the original mortars taken from the building under study are very consistent with the values obtained in the work of Damas *et al.* [23], which characterizes a set of samples composed of air lime and siliceous aggregates, usually with added clay. In addition to the values obtained, the capillary water absorption and drying curves are typical of an air lime mortar [24].

5 Final Considerations

The laboratory testing of mortar samples extracted from ancient buildings allowed for the measurement of physical and mechanical properties, as well as to obtain their composition. This type of information is crucial in a structural analysis and allows to identify materials compatible with the original ones that can be used in rehabilitation interventions.

Table 7. Results obtained in capillarity tests and evaluation of the drying rate of tested mortars.

Identification of the type of mortar		Capillarity coefficient (Cc) [kg/m^2.min$^{0.5}$]	D1 [kg/m^2.h]	D2 [kg/m^2.h$^{0.5}$]
Average value	Coating	**2.70**	**0.14**	**1.65**
Standard deviation	t$_{avg.}$ = 2.9 mm	0.21	0.03	0.21
Coefficient of variation		0.08	0.20	0.13
Average value	Laying	**1.73**	**0.11**	**0.53**
Standard deviation	t$_{avg.}$ = 1.7 mm	0.22	0.01	0.08
Coefficient of variation		0.13	0.09	0.14

t$_{avg.}$ Represents the average thickness of the test specimens for each type of mortar.

Table 8. Synthesis of the average results of the characterization tests of the original coating and laying mortars and their respective (standard deviation).

Property		Coating mortar	Laying mortar
Lime:Aggregate ratio in volume		1:3.2	1:2.5
Type of mortar		Calcium air lime mortar with well graded sand	Calcium air lime mortar with well graded sand
Compressive strength [MPa]		1.72 (0.78)	1.27 (0.20)
Open porosity [%]		32 (2)	34 (1)
Bulk density [kg/m^3]		1702 (26)	1616 (43)
Dynamic modulus of elasticity by ultrasonic (direct method) [MPa]		2606 (52)	1628 (66)
Capillarity coefficient [kg/m^2.min$^{0.5}$];		2.70 (0.21)	1.73 (0.22)
Drying	D1 [kg/m^2.h]	0.14 (0.03)	0.11 (0.01)
	D2 [kg/m^2.h$^{0.5}$]	1.65 (0.21)	0.53 (0.08)

To evaluate the effectiveness of a reinforcement solution for a structural element, an adequate characterization of the element in its original state is required. Only under these conditions it is possible to evaluate this effectiveness, numerically quantifying the increase in strength (and deformation capacity, energy dissipation, etc.) in comparison with the non-reinforced element.

As a general principle, the material properties assumed in the design models, and corresponding constitutive laws, should replicate as faithfully as possible the actual characteristics of the materials. In interventions on existing buildings, this aspect becomes even more relevant as the parameters used in the numerical models must reliably represent the real structural system and the properties of the existing materials [25]. When this is not the case, unrealistic assumptions can be made that result in a lack of safety if the capacity of the structure is overestimated, or unnecessary, uneconomic, or excessively intrusive interventions, when the capacity is underestimated.

Acknowledgment. This work was financed by: LNEC - Research and Innovation Project 2013-2020 "REuSE - Revestimentos para Reabilitação: Segurança e Sustentabilidade"; FCT – Fundação para a Ciência e a Tecnologia Project PTDC/ECIEGC/30567/2017 "RESIST-2020 – Seismic Rehabilitation of Old Masonry-Concrete Buildings"; project PO-CI-01-0145-FEDER-031612 CEMRESTORE: Mortars for early 20th century buildings' conservation: compatibility and sustainability. The authors would like to thank the study and project company A2P Consult, Lda., and the companies Coporgest and HCI Construções, S.A. for their collaboration.

References

1. Costa, A.A: seismic assessment of the out-of-plane performance of traditional stone masonry walls. Doctoral Thesis, Faculty of Engineering of the University of Porto, Porto, Portugal (2012)
2. Veiga, M.R.: Mortars for historic masonry. Functions and characteristics (In Portuguese). In: CIRea2012 – International Conference on Rehabilitation of Ancient Masonry Structures, Nova School of Science and Technology, pp. 17–27. Lisbon, Portugal, ISBN: 978–989–20–3080–7 (2012)
3. Appleton, V., Sousa Dias, J., Almeida, I.A., Appleton, J.: Rehabilitation of the set of buildings on Avenida Duque de Loulé, n. 86, n.º 90 and n.º 94 in Lisbon (In Portuguese). In: BE2016 - Encontro Nacional de Betão Estrutural, Faculty of Sciences and Technology of the University of Coimbra, Coimbra, Portugal (2016)
4. Google Maps. https://goo.gl/maps/1Z46ib1M796RyNsD7. (last accessed on May 2015)
5. Gomes, R.I., Santos, S.A., Gomes, L., Faria, P.: Fernandina wall of Lisbon: Mineralogical and chemical characterization of rammed earth and masonry mortars. Minerals **12**, 241 (2022)
6. CEN. Methods of test for mortar for masonry - Part 11: Determination of flexural and compressive strength of hardened mortar. EN 1015-11:1999, European Committee for Standardization, Brussels, Belgium (1999)
7. Valek, J., Veiga, M.R.: Characterisation of mechanical properties of historic mortars-testing of irregular samples. Adv. Archit. **20**, 365–374 (2005)
8. Magalhães, A., Veiga, R.: Caracterización física y mecánica de los morteros antiguos. Aplicación a la evaluación del estado de conservación. Materiales de Construcción **59**(295), 61–77 (2009). https://doi.org/10.3989/mc.2009.41907
9. CEN. Natural stone test methods-determination of real density and apparent density, and of total and open porosity. EN 1936:2007, European Committee for Standardization, Brussels, Belgium (2007)
10. CEN. Testing concrete. Part 4: determination of ultrasonic pulse velocity. EN 12504-4:2004, European Committee for Standardization, Brussels, Belgium (2004)

11. CEN. Conservation of cultural property. Test methods. Determination of water absorption by capillarity. EN 15801:2009, European Committee for Standardization, Brussels, Belgium (2009)
12. CEN. Conservation of cultural property. Test methods. Determination of drying properties. EN 16322:2013, European Committee for Standardization, Brussels, Belgium (2013)
13. Veiga, M.R., Magalhães, A., Bosilikov, V.: Capillarity tests on historic mortar samples extracted from site. Methodology and compared results. In: 13th International Brick and Block Masonry Conference, Amsterdam, Netherlands (2004)
14. Alessandrini, G., Bugini, R., Negrotti, R., Toniolo, L.: Characterisation of plasters from the church of San Niccolò di Comelico. Eur. J. Mineral. **3**(3), 619–628 (1991). https://doi.org/10.1127/ejm/3/3/0619
15. Santos, S.A., et al.: Characterization of historical mortars from Alentejo's religious buildings. Int. J. Architectural Heritage **4**(2), 138–154 (2010)
16. Santos Silva, A., Santos, A.R., Veiga, M.R., Llera F.: Characterization of mortars from the Fort of Nossa Senhora da Graça, Elvas (Portugal) to support the conservation of the monument. In: Proceedings of the HMC2016 - 4th Historic Mortars Conference, pp. 42–49. Santorini, Greece (2016)
17. Veiga, M.R., Fragata, A., Velosa, A.L., Magalhães, A.C., Margalha, M.G.: Lime-based mortars: viability for use as substitution renders in historical buildings. Int. J. Architectural Heritage **4**(2), 177–195 (2010). https://doi.org/10.1080/15583050902914678
18. Veiga, M.R.: Air lime mortars: what else do we need to know to apply them in conservation and rehabilitation interventions? (A review). Constr. Build. Mater. **157**, 132–140 (2017)
19. Marques, A.I., Morais, J., Santos, C., Morais, P., Veiga, M.R.: Mortar dynamic elasticity module (In Portuguese). CONSTRUINDO Magazine, Belo Horizonte, Brazil, vol. 11, Special Edition: TEST&E 2019: Conference on Testing and Experimentation in Civil Engineering, pp. 63–78, ISSN: 2318–6127 (online), ISSN: 2175-7143 (print) (2019)
20. Ghiassi, B., Lourenço, P.B.: Long-term Performance and Durability of Masonry Structures - Degradation Mechanisms, Health Monitoring and Service Life Design. Woodhead Publishing Series in Civil and Structural Engineering. ISBN: 978-0-08-102110-1 (2019)
21. Veiga, M.R. Conservation of historic renders and plasters: from laboratory to site. In: Válek, J., Hughes, J., Groot, C. (eds.) Historic Mortars, RILEMBookseries, vol. 7, pp. 207–225. Springer, Dordrecht (2012) DOI: https://doi.org/10.1007/978-94-007-4635-0_16
22. Velosa, A.L., Veiga, M.R. Historical heritage mortars: knowing how to conserve and rehabilitate (In Portuguese). In: CINPAR - XII International Conference on Structural Repair and Rehabilitation, Porto, Portugal (2016)
23. Damas, A.L., Veiga, M.R., Faria, P., Santos, S.A.: Characterisation of old azulejos setting mortars: a contribution to the conservation of this type of coatings. Constr. Build. Mater. **171**, 128–139 (2018). https://doi.org/10.1016/j.conbuildmat.2018.03.103
24. Veiga, M.R., Faria, P.: The role of mortars in the durability of old masonry (In Portuguese). In: Pinho, F., Lúcio, V., Lourenço, P., Machado, L. (eds.) In CIRea2018 – International Conference on Rehabilitation of Ancient Masonry Structures, pp. 1-15. Nova School of Science and Technology, Lisbon, Portugal (2018)
25. ICOMOS. Recommendations for the Analysis, Conservation and Structural Restoration of Architectural Heritage. International Council on Monuments and Sites, Paris, France (2018)

Mineral, Chemical and Petrographic Characterization of Hydraulic Mortars & Chronological Building Correlation of the Baths of Porta Marina in Ostia Antica (Italy)

Sarah Boularand[1]([⊠]), Marcello Turci[2], and Philippe Bromblet[1]

[1] Centre Interdisciplinaire de Conservation Restauration du Patrimoine, Marseille, France
{sarah.boularand,philippe.bromblet}@cicrp.fr
[2] Institut für Antike, Universität Graz, Graz, Austria
marcello.turci@gmail.com

Abstract. The Baths of Porta Marina were an imperial public complex located in the maritime suburban district of Ostia, the city harbor of Rome. During its large period of use from the 2nd century AD until the 6th century AD, several modifications of capacity and thermal path took place, resulting in architectural modifications. Previous studies based on archaeological evidence, written sources, architectural and stratigraphic analyses have provided thorough chronological data for construction phases. Characterization of hydraulic coatings and bedding mortars has been carried out focusing on waterproofing processes of antique hydraulic facilities and to propose a timeline of raw materials uses for this particular building and for others hydraulic facilities preceding the thermal complex (1st century AD). Mineralogical, chemical, and petrographic analyses were performed on several mortar samples from pools and cisterns from different construction phases of the Baths of Porta Marina. They confirm that mortars of Roman waterproof coverings, made with pozzolanic aggregates, show a persistent composition over centuries, hardly distinguishable macroscopically. Nonetheless, petrographic examination has been an interesting tool to highlight differences among the mortars. Some distinctive features of volcanic aggregates allow us to identify different depositional units from Alban Hills volcanic district as sources.

Keywords: Ancient Roman mortars · pozzolanic aggregates · volcanic clasts · Alban Hills deposits · Roman thermal baths · petrographic analysis

1 Introduction

The *Water Traces* project (*Water Traces between Mediterranean and Caspian Seas before 1000 AD: From Resource to Storage*) is an international and interdisciplinary research project focused on water management in past societies. One of its research lines is centered on studying waterproofing processes of all kinds of antique hydraulic facilities and aims to collect data on materials and technological strategies for this purpose. In this

V. Bokan Bosiljkov et al. (Eds.): HMC 2022, RILEM Bookseries 42, pp. 82–96, 2023.
https://doi.org/10.1007/978-3-031-31472-8_7

context, characterization of hydraulic mortar coatings from several pools and cisterns of Ostia Antica archaeological site has been carried out. Part of this study, outlined herein, concerned the Baths of Porta Marina, which were located in the maritime suburban district of Ostia, the city harbor of Rome. These baths were an imperial public complex, built during the city expansion in its economic apogee in the 2nd century AD and remained in use, at least, until the beginning of the 6th century AD. This large period of use over centuries implied several architectural modifications due to capacity changes and thermal path transformations. Thereby, this specific building displays a range of hydraulic structures from various construction phases which allows to investigate the evolution of mortar composition and to establish a timeline of raw materials uses. Moreover, an access to previous thermal facilities from 1st century AD, in a level below, enlarges the overview of the collected data. Bedding mortars of pools masonry has been also included in the study in order to relate chronological data more accurately between coating mortars composition and architectural observations.

2 Archaeological Data: New Studies and Chronological Overview

Knowledge of the Baths of Porta Marina dates back to the antiquarian excavations of the 18th century. The excavations carried out by the Scottish painter Gavin Hamilton between 1774 and 1776 identified indeed a thermal complex, located near the ancient shoreline, in relation to the brick pillars. It had been called 'Porta Marina' since then. The building, known as *Thermae maritimae* on the basis of epigraphic evidence (*CIL*, XIV 137) [1] from the late 4th century AD, was excavated again by Pietro Campana during the first half of the 1830s and partially investigated in the 1920s and 1930s by Guido Calza[1]. The excavation was completed in the 1970s under the direction of Maria Floriani Squarciapino [3, 4].

The resumption of investigations by the writer, within the framework of a PhD research on the thermal development of the coastal sector of Ostia [5, 6], started from an analytical study of the monument and from an archive research based on the documentation of the largely unpublished Squarciapino's excavations[2]. This work allowed to reconsider the chronological sequence of the baths' architectural phases, as well as the urban and political-social framework in which the different building phases are placed. This also enabled to follow their development from the final decades of the 1st century AD to the threshold of the 6th century AD.

In terms of methodology, a new survey of the building and a technical analysis of the structures were carried out and gave the ability to make up the architectural organism. In addition, stratigraphic analyses at key points of the complex and an internal sequence of the masonry were achieved. This phase of the work, combined with a punctual verification of all the observations on the documentation and the excavation reports, allowed to define a relative chronology. Through the study of the epigraphic material, the brick stamps found during the Squarciapino's excavations and style mosaics; it has been made possible to date the building phases precisely (Fig. 1).

---

[1] A series of imperial portraits were found in the 1920s [2] (pp. 313–318).

[2] On the mosaics of the *frigidarium* see [7–9]. For a study of the architectural material see [10]; for the sculptures see [2, 11, 12].

As documented by the dedicatory epigraph *CIL*, XIV 98 [9], the Baths of Porta Marina were begun by Hadrian and completed by Antoninus Pius in 139 AD. This building, which was part of a wide-ranging program of urban renewal in the coastal district of Ostia, replaced an earlier bath complex dating from the Domitian period (phase 0/3). A portion of the *frigidarium* of this building, whose occupation levels are one meter below the levels of the Hadrianic period, has been discovered during the Squarciapino's excavations. In particular, along the eastern side, the building is provided with an entrance staircase coated with marble slabs and a 5 m wide pool (pool 5). In addition, the excavations brought to light a southern portion of a large basin (*natatio* 6) with columned projections and corridors situated at the north of the building. This installation insists on structures from the Julio-Claudian period in *opus reticulatum* (phase 0/1) and *opus testaceum* in which it is possible to identify a cistern paved in *opus spicatum* (cistern 1) and a small mosaic room (phase 0/2).

The Antonin building, equipped with a large gymnasium surrounded by a porch, is organized with a large *frigidarium* with two opposing pools (I and F). This also includes lateral colonnaded halls: one in the east with an elliptical pool L, and the other in the west paved with an athletic mosaic. The changing rooms (*apodyteria*), in relation to the entrances from the gymnasium, are located along the wings of the *frigidarium* (C and O). The heated sector, which is accessible from rooms J and S (*tepidaria*), is composed by three large rooms located along the southern side of the complex, following a paratactic path from east to west. A corridor along the eastern side of the heated sector connects the thermal spaces with two entrances on the eastern and southern side. Finally, the service rooms, the *praefurnia*, a cistern (Y) for supplying the three *caldarium* (K) basins and the pool F are located along the south and west sides. A second cistern (RR) was added later, probably during phase 2. Phase 1B includes the decorative and sculptural elements, the achievement of the gymnasium's porch, annexes along the eastern side of the gymnasium, as well as some of the doorframes inside the bath.

The life of the complex is marked by two phases of architectural renovation: the first in the Severan period (phase 2) and the second one in the 4th century AD (phase 4). This last phase can be related to an inscription of the emperors Constantius and Constans (*CIL*, XIV 135). Restorations accompany these major phases in the final decades of the 3rd century (phase 3), the final decades of the 4th century (phase 4) and in the Theoderic period (phase 6).

At the beginning of the 3rd century (phase 2), the building was profoundly renovated. The *frigidarium* was covered by an imposing cross vault on pillars, the gymnasium recovered with mosaic and the heated sector modified internally to allow bathers to reach the *tepidarium* J without having to retrace their steps. The north wall of room Z (*laconicum*) was also re-lined and the vault of the caldarium consolidated.

The restoration of phase 3 affected mainly the eastern wing of the *frigidarium*: elimination of the elliptical pool L replaced by a marine mosaic, eastward extension of pool I and addition of room T, along the eastern side of room S. These works, dated to the final decades of the 3rd century AD, can be related with two inscriptions of the emperor *Probus*: a dedication reused during the later phase in the pool M [13] and *CIL*, XIV 134.

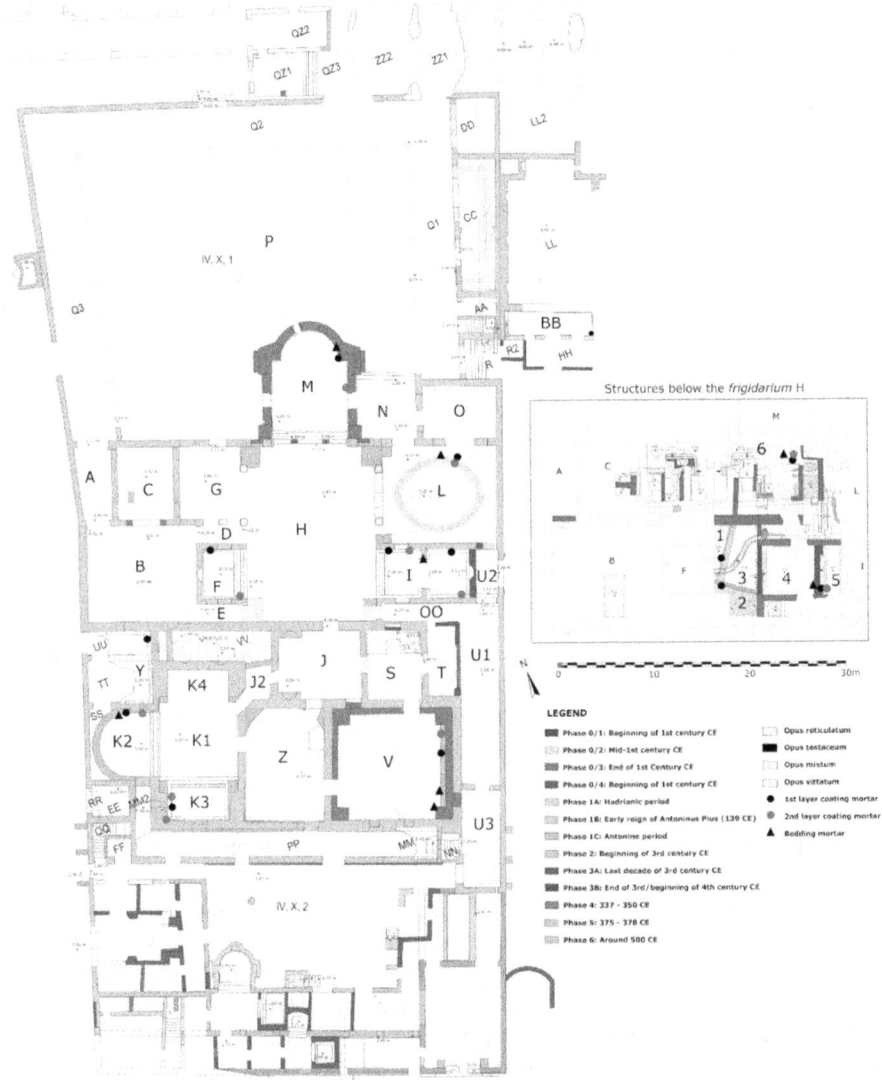

Fig. 1. Planimetry of Porta Marina Baths showing the successive construction phases and the samples location.

The renovation by Constantius and Constans, which can be dated on the basis of the inscription *CIL*, XIV 135 to the years 337–350 AD (phase 4), involved a radical renovation of the *frigidarium* with the addition of a monumental apsidal pool M along the north side and the repaving of the central hall. At the same time, along the eastern side of the gymnasium, several changes were implemented: addition (or perhaps expansion) of a latrine, consolidation of the southern wall in the heated sector, and reconstruction of the cross-ribbed roof of room V, as indicated by the corner pillars.

The following phase 5 probably is connected with the inscription *CIL*, XIV 137, dated to the years 375–378 AD. It was limited to some interventions within the heated sector: enlargement of pool K2 with the addition of an apse, abolition of pool K1 and consolidation of the east wall of room V. The repaving of the gymnasium with reused material has also been initiated during this phase.

Finally, the last phase, documented by the discovery of numerous brick stamps and the dedicatory inscriptions of King Theodoric (Ostia inv. no. 7312), involved the restoration and consolidation of the roofs (pillar in room C). The terms *tubuli* and *ferrum*, which appear in the epigraphic text, could be respectively related to terracotta vaulting tubules found during the last excavations and consolidation of roofs with metal chains [14].

3   Geological Settings

As far as the present study is concerned, the geological environment surrounding Ostia is mainly the result of ancient volcanic activity of Albans Hills and Monti Sabatini volcanoes during mid-Pleistocene age, interbedded with paleo-Tiber River sedimentary deposits. During the last major explosive phase Tuscolano-Artemisio, five main eruption cycles in the Albans Hills volcanic district led to the formation of several pyroclastic deposits: *Tufo Pisolitico di Trigoria* (561 ± 2 ka), *Tufo del Palatino* (530 ± 2 ka), *Pozzolane Rosse* (456 ± 3 ka), *Pozzolane Nere* (407 ± 2 ka) and *Villa Senni* (365 ± 4 ka), which includes the upper loose unit *Pozzolanelle* and the lower lithified unit *Tufo Lionato* [15, 16]. All these deposits provided an important source of building materials in ancient Roman period such as masonry tuff stones from lithified units [17–19] while uncoherent loose ash-flow deposits were excavated to produce pozzolanic aggregates for mortars and concretes [20–23]. Monti Sabatini eruptive geological history and resulting volcanic products won't be discussed herein but accurate information can be found in previous works on the subject [17, 24, 25].

Albans Hills products display a very distinctive composition with mafic, high-potassic, silica-depleted and K-foiditic components [15, 16] which allows to discriminate the provenance from others volcanic districts in the peninsula. Moreover, in several works, Marra and Jackson describe morphological and petrographic characteristics of granular volcanic ash deposits of Alban Hills and propose distinctive features for precise pyroclastic-flow deposit identification [20, 21, 26]. The present study relies on these criteria for the identification of pozzolanic aggregates provenance in the coating and bedding mortars of the Baths of Porta Marina.

4   Materials and Methods

Sampling has been carried out under supervision and with the authorization of *Parco Archeologico di Ostia Antica*. All samples were extracted mechanically with hammers and chisels after a thorough examination of the complete structure. The selection and extraction were achieved following conservative criteria that imply restricted but representative sizes of samples and an adequate and still representative state of conservation of the studied material. For coating mortars, samples were extracted separately for each layer with the complete thickness of the layer.

The whole study deals with 26 samples of coating mortars from the Baths of Porta Marina (including 7 samples from previous structures below elevations) and were completed further by 8 bedding mortar samples (2 from below structures). A detailed map of sampling location is reported on Fig. 1. Not all samples have been prepared and analyzed the same way depending on the degree of precision and nature of the information required. All coating mortars samples were analyzed by X-ray diffraction (XRD) and prepared into cross-sections and a selection of them were prepared into thin sections for petrographic examination. Additional Raman analyses and scanning electron microscope SEM-EDX analyses could have been achieved on thin sections or cross-sections to complete or confirm petrographic results. For bedding mortars, a routine protocol including XRD analyses, thin sections preparation and SEM-EDX imaging and analyses was conducted for all samples.

Petrographic studies were performed on an Olympus BX60 optical polarized microscope on polished thin sections. X-Ray Diffraction analyses were done with a D8 Advance (powder method) by Bruker in a θ-2θ configuration with a Co tube at 35 kV x 40 mA. Raman microspectrometry analyses were conducted using an In Via Renishaw instrument equipped with a 785 nm laser, a ½ inch CCD detector and a Leica microscope. SEM-EDX imaging and analyses were performed at 20 kV under low vacuum conditions (0.2 Torr) with the Philips XL30 or the EVO 15 Zeiss Environmental Scanning Electron Microscope (ESEM) of the Plateforme de Recherche Analytique Technologique et Imagerie (PRATIM, FSCM, Aix-Marseille University, France), both equipped with a LaB_6 filament and Energy Dispersive X-Ray analysis (EDAX Genesis equipped with an Apollo 10 SDD detector).

5 Results

5.1 Macroscopic Observations of Mortar Samples

All the pools' internal walls were covered with a marble slab finishing supported by two layers of pozzolanic mortar (and exceptionally 3 layers) while the cisterns were only waterproofed by a one-layer pozzolanic mortar. These pozzolanic mortars are always composed by a mixture of crushed ceramics and volcanic clasts in a lime-based matrix and we don't observe any relevant macroscopic changes among the samples in a chronological perspective. Bibliographic study confirms this observation: the composition of Roman hydraulic mortar remained unchanged over centuries [27]. The most noteworthy observation drawn from macroscopic examination of pools mortar cross-sections is the selective use of crushed brick limited to the first layer directly applied on the brick masonry. Indeed, the second layers of coating mortar in the pools never exhibit large-size crushed brick fragments nor a large amount of this material. The pozzolanic aggregates of these layers are mainly composed by volcanic aggregates with only rare crushed ceramic.

5.2 Petrographic Observations

Volcanic Deposits Identification. Volcanic components were examined considering specific characteristics such as vesicle shapes, color of glassy matrix, nature, occurrence and development of minerals (phenocrysts), degree of alteration and secondary minerals. In all cases, only Alban hills distinctive volcanic products have been found in the pozzolanic aggregates with one exception discussed further. *Pozzolane Rosse* (PR) specimens have been discriminated by the presence of starry-like leucite in red or black scoriae with amoeba vesicles (Fig. 2a). *Pozzolanelle* (PL) exhibits a high porphyric index due to the presence of leucite, biotite and clinopyroxene crystals in the scoriae with amoeba vesicles (Fig. 2b). *Tufo Lionato* (TL) contains aphyric vitreous scoriae with elongated or rounded vesicles (Fig. 2c). Slightly differences have been observed in some glassy scoriae that can probably be related to different facies of *Tufo Lionato* such as *Aniene, Monteverde* or *Portuense* facies [18]. Starry-like leucites observed in some glassy tuff-like scoriae suggest an intermediate feature between *Tufo Lionato* and *Pozzolanelle* [22] or might be attributed to *Tufo del Palatino* [18] or to another local tuff [19, 28]. Due to ambiguous features to characterize these scoriae it has been described as unidentified tuff (T) until further research. Lava clasts of skeletal leucite, clinopyroxene and magnetite are also commonly observed (Fig. 2d). They have been described as italite or leucitite by some authors (respectively [20, 29, 30] and [31]) and are usually associated with *Pozzolanelle* deposit. All depositional units identified for each sample are summarize in Table 1.

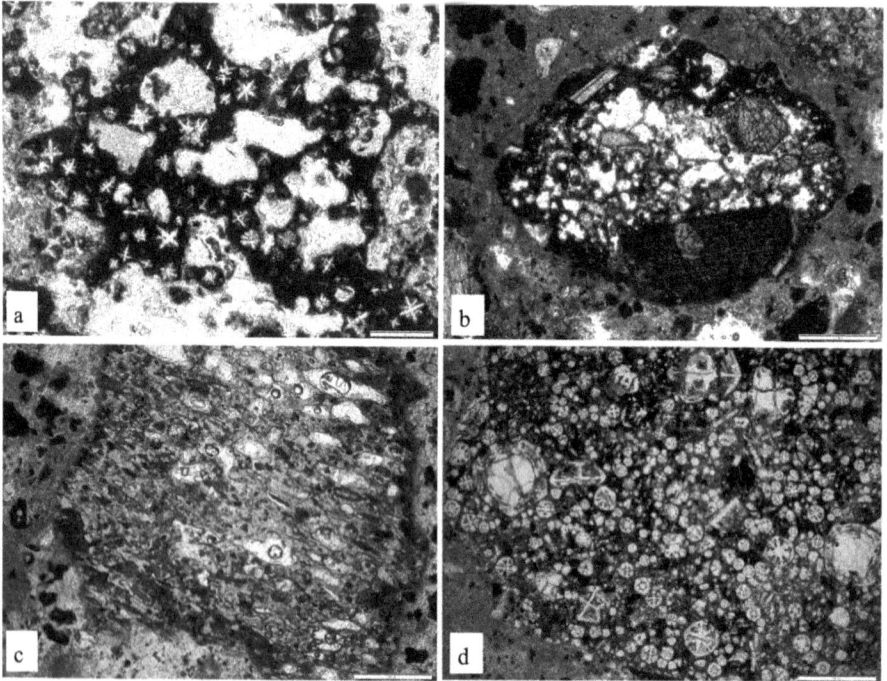

Fig. 2. Microphotographs of mortar samples. (a) Typical *Pozzolane Rosse* scoria showing the characteristic starry-like leucite crystals; (b) Clinopyroxene and biotite crystals within a *Pozzolanelle* scoria with amoeba vesicles; (c) Glassy aphyric scoria of *Tufo Lionato* with elongated vesicles; (d) *Italite* clast showing skeletal leucite and clinopyroxene crystals.

Table 1. Mortar aggregates composition identified for each sample with detailed pyroclastic depositional units from Alban Hills districts and other distinctive features.

Structure and age	Type of mortar	Pyroclastic AH deposits for volcanic aggregates	Other aggregates and distinctive features
Cistern 1 1st century (around 50)	Internal CL	PR, PL, TL	CB, Si, rlc, fLe
	External CL	PR, TL	CB, red ochre painting, fLe
Pool 6 End of 1st century	Bedding mortar	PR, T	fLe
	C 1st layer	PR, PL, TL/T	WP, LC, F, CB, rlc, fLe
	C 2nd layer	PL, TL/T, rare PR	Rare CB, fLe, Si
Pool 5 End of 1st century	Bedding mortar	PR	fLe, L2
	C1st layer	PR, TL	CB, fLe
	C 2nd - 3rd layer	TL, PL, rare PR	CB, fLe
Pool L 139	Bedding mortar	TL, PL, T	CB, rlc, A
	C 1st layer	TL, PL, T	CB, A
	C2nd - 3rd layer	TL, PL, T	F, Si, rare CB, L2, rlc, A
Pools I and F 139	Bedding mortar	PR, PL, TL	Si, rlc, fLe
	C 1st layer	PR, TL	CB, rlc, Si, fLe
	C 2nd layer	PR, PL, TL	Rare CB, A
Cistern BB 139	Coating layer	PL, PR, T	CB, rlc, fLe
Cistern Y 139	Coating layer	T, TL, PR, PL	CB, rlc, F, Si, L2, fLe
Pool M 337–350	Bedding mortar	PR	rlc (D), A
	C 1st layer	PR	CB, A
	C2nd layer	PR	Rare CB, rlc, Si, A
Heated room V (pillar) 337–350	Bedding mortar	PR	L1, L2, rlc (D), Si, CB, A
Heated room V (wall) 375–378	Bedding mortar	PL, TL, T, PR	L2, rlc, Si, A
Pool K2 375–378	Bedding mortar	PR, PL, T	Rare CB, L1, Si
	C 1st layer	PR, PL, T	CB, rlc, A, L2
	C 2nd layer	PR, PL, T	CB, L2, rlc (D), A

Abbreviations in Table 1: CL: coating layer, C coating, AH: Alban Hills, PR: *Pozzolane Rosse*, PL: *Pozzolanelle*, TL: *Tufo Lionato*, T: unidentified volcanic tuff, CB: crushed bricks, WP: white pumices, LC: lava clasts, F: feldspar, Si: silica, fLe: fresh Leucite, A: analcime, L1: limestone type 1 (micritic limestone with bioclasts and sparitic filling/grainstone), L2: limestone type 2 (micro-sparitic limestone), rlc: relict lime clasts, D: dolomitic rhombohedral shape.

In pool 6, a *natatio* built during Domitian age (end of 1st century AD) and belonging to the below level structures, some volcanic specimens not related to Alban Hills materials have been found additionally to *Pozzolane Rosse* scoriae and *Villa Senni* specimens (*Pozzolanelle* and *Tufo Lionato*). Petrographic studies allow to identify white pumices (Fig. 3a) and lava clasts with sanidine, clinopyroxene and biotite phenocrists (Fig. 3b). A similar mixture of *Pozzolane Rosse* with feldspar-bearing pumices has been described for the forum and markets of Trajan [32] and for the forum of Caesar [24], where a mix of Campanian and Monti Sabatini pumices occurs [33]. Moreover, white pumice from Vesuvian provenance have been proved to be employed in Diocletian Baths, in the Temple of Minerva, as well as in concretes of Colosseum [34]. Without further analyses such as geochemical analyses on traces elements, accurate provenance identification cannot be drawn for white pumices from Domitian *natatio*. Nevertheless, a Monti Sabatini provenance or Vesuvian origin can be hypothesized while Vico and Vulsini volcanic districts, Pontine Islands or Aeolian Islands are others possible but more unlikely sources for pumice [33].

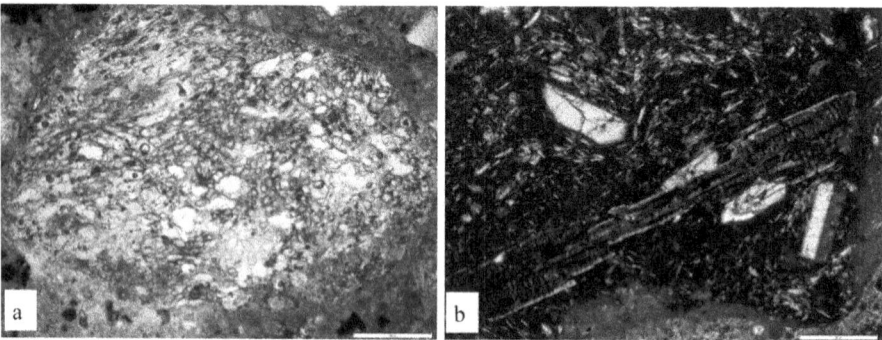

Fig. 3. Microphotographs of mortar samples from the Domitian *natatio* (pool 6), showing volcanic clasts which are not products from Alban Hills eruption. (a) White pumice glassy scoria; (b) Lava clast showing sanidine, clinopyroxene and biotite crystals at crossed Nichols.

Limestone, Lime lumps, Silica and Feldspar. All materials present within the aggregates such as limestone and relict lime clasts have been also considered and reported in Table 1. Two types of limestone have been observed: the first type displays a micritic texture with bioclasts and sparitic veins (grainstone type) while the second type is a micro-sparitic limestone. Both types have been found in the same samples. Pyroclastic deposits such as *Pozzolanelle* or *Pozzolane Rosse* may contain sedimentary rock or limestone clasts which were swept into the eruptive mixture and baked by its heat [20], but limestone fragments seem to have been also intentionally added to the mortar mixture, in particular for the pool K2 and room V wall where limestone fragments are present in high concentration and can be related to a distinctive feature of phase 5 (375–378 AD).

Relict lime clasts are often present in the samples (Fig. 4a), and they reveal that the lime has been produced with a traditional technology. However, a surplus of lumps can indicate a poor production technology [35]. The cracking in the lumps is due to the shrinkage phenomena [35] and allow to identify them easily and to distinguish them from limestone fragments.

Some relict lime clasts exhibit rhombohedral shapes (Fig. 4b) which seem to be related to an under burnt fragment of dolomite [35]. The presence of under burnt fragments indicates strong inhomogeneity of temperature in the kiln, lack of adequate sieving of the time after slaking and a difficulty in the calcination of the stone because of its composition [35].

Silica (SiO_2) and felspars have been also identified in several samples. They might be intentionally added to the mortar mixture, especially for bedding mortars, as they are

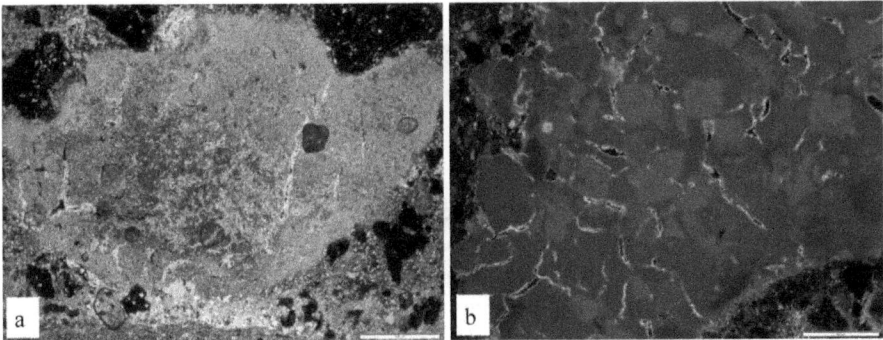

Fig. 4. Microphotograph of a two types of relict lime clasts (cross-polarized light): (a) cracked lump (b) lime lump showing rhombohedral shapes identified as dolomitic ghosts.

not commonly formed in Alban Hills deposits. Silica can be found as polycrystalline quartz or amorphous silex.

Leucite Phenocrists Alterations. Leucite is a feldspathoid and one of the major components formed in the silica undersaturated conditions of Alban Hills eruption. Therefore, it is present in all pyroclastic depositional units within scoria groundmass and as loose crystal fragments. It has been observed showing dendritic crystal development so-called starry-like shape in *Pozzolone Rosse* scoriae but also in its euhedral 8-sided form exhibiting polysynthetic twins (Fig. 5a) in several samples. Nevertheless, in some cases, leucite crystals show the presence of other crystals in a venal growth pattern inside the leucite euhedral crystals (Fig. 5b) associated with smoother edges and a complete loss of the twinned aspect. This altered appearance results in a zeolitisation of leucite ($KAlSi_2O_6$) into analcime ($NaAlSi_2O_6 \cdot H_2O$) by a process of dissolution and replacement [20, 21]. This transformation can be highlighted by petrographic observations but also with Raman microspectroscopy as leucite and analcime display distinctive Raman spectra. SEM-EDX is also a useful tool to discriminate leucite from analcime and to

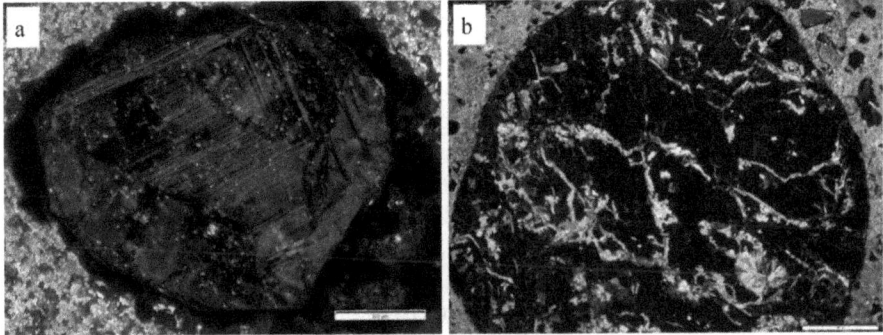

Fig. 5. Microphotographs of mortar samples (cross-polarized light). (a) Fresh leucite crystals showing polysynthetic twins; (b) Altered leucite in a zeolitisation process with an analcime composition and smooth edges.

detect in-progress zeolitisation by the evidence of both sodium and potassium and by a specific morphology. Presence of fresh leucite or analcime is reported in Table 1.

6 Discussion

Bedding mortars have been included in the study because any changes in the coating mortars (like removing or adding a mortar layer during a renovation) might not be observed as clearly as any transformation in the building masonry. Comparison of the aggregates' components between bedding mortars and first layer coating mortars can bring valuable information in the eventual covers' rehabilitations. Except for pool 6 (the Domitian *natatio*), results always show similarities of volcanic aggregates sources used for the bedding mortar and the first layer of the coating mortar which suggests that no restoration process in the pools covers caused the removing of the first layer of coating mortar.

Pozzolane Rosse scoriae are observed in all coating and bedding mortars except in the elliptical pool L which is the only one that contains only *Pozzolanelle, Tufo Lionato*, and another local tuff without any *Pozzolane Rosse* components. This pool was part of the Antonin building achieved in 139 AD and remained in use until its removal by recovering it with a mosaic at the end of 3rd century (phase 3). The lithified source of volcanic stones such as *Tufo Lionato* or any other local tuff implied a grounding process to produce it specifically as fine aggregates [36] or could be the remnant of tuff stone production as cutting waste. All other structures from 139 AD (pools I and F and cisterns BB and Y) contain a mix of *Pozzolane Rosse* scoriae and tuffs fragments and sometimes *Pozzolanelle* scoriae. This difference pointed out herein between pool L mortars and others Antonin structures mortars can't be explained by scientific data but some hypotheses involving social arguments or aleatory circumstances might be formulated. For instance, this difference might be due to a construction of these structures by different builders' teams or at a different moment or one can speculate a momentary run out of *Pozzolane Rosse* granular ashes.

At the contrary we observe a selective and exclusive use of *Pozzolane Rosse* scoriae in the 337–350 AD construction phase (phase 4) which concerns pool M construction and confirm the chronological relation with corner pillars of room V built during the same phase for the cross-ribbed roof reconstruction.

Furthermore, from this construction phase, dolomitic ghosts in relict lime clasts are identified. As they are an indicator of an under burnt fragment of dolomite [35], therefore it suggests the use of dolomitic limestone as a source for lime production from this period and after. Before 337–350 AD, as we don't observe these dolomitic ghosts in lime lumps in any samples, calcitic limestones seem to be used as unique carbonate rock source for lime production.

The study of fresh leucite presence seems to indicate a preferential occurrence in the oldest structures. Some authors [20, 21] pointed out three alteration facies of *Pozzolane Rosse* ignimbrite with some differences in chemical characteristics and scoriae colors along vertical sections. Greatest alteration facies occur at the uppermost horizon while least altered facies take place in the lower horizon. Among chemical differences between the three grades of alteration facies, leucite zeolitisation is highlighted and fresh leucite appears to be mostly well conserved in the least altered layer while in the greatest alteration facies leucite crystals are wholly dissolved or substituted by analcime and in the intermediate alteration facies, coarse fresh leucite crystals remain well-preserved while small crystals have been dissolved and replaced. A selective mining exploitation of *Pozzolane Rosse* quarries or outcrops in the least altered facies seems to have taken place in early construction phases (phases 1 and previous) but might not be still available sometimes between the 2nd century (after phase 1) and the end of the 3rd century (phase 3). Pools I and F show fresh leucite crystals in the bedding and first layer coating mortars but analcime has been detected in the second layer coating mortar even if the mix of scoriae used in the three mortars comes from similar depositional units (*Pozzolane Rosse, Pozzolanelle, Tufo Lionato*). As the extension of pool I occurred during phase 3 (end of the 3rd century AD) we can assume a rehabilitation of the marble slab finishing and only a change of the second layer coating occurred at this time in both pools I and F. Note that the elliptical pool L, dated from 139 AD, displays analcime contents but does not contain any *Pozzolane Rosse* specimens.

7 Conclusion

Petrological observations have been a useful tool to carry out the thorough study of pyroclastic depositional units as an indicator of source provenance for fine aggregates in the coating and bedding pozzolanic mortars. This information has been valuable to confirm or specify chronological relations between architectural structures and to point out some particularities.

In all cases, local volcanic aggregates were used, and we can highlight some chronological distinctive features such as a specific mining exploitation of *Pozzolane Rosse* least altered facies in early construction phases, a selective use of *Pozzolane Rosse* scoriae during phase 4, the use of dolomitic limestone as carbonate rock source for lime production since phase 4 and the presence of limestone fragments added in the aggregates in phase 5.

The Baths of Porta Marina have constituted a perfect case of study to establish a timeline of raw materials used in Roman pozzolanic mortars and to study waterproofing technological strategies with a chronological perspective. These inferences, though, might not be extended to other hydraulic structures from Ostia without a meticulous scientific approach and remain a non-exhaustive overview of materials and technologies in ancient Roman mortars.

Acknowledgement. This work was part of the *Water Traces* research project supported by A*Midex Foundation. It was funded by an Aix-Marseille University – CNRS contract and beneficiated from the collaboration of Centre Camille Jullian and CICRP. Sophie Bouffier, chief and coordinator of the *Water Traces* research project is gratefully acknowledged for her constant support in this study and her kind interest.

We also thank Alessandro D'Alessio, director of the Archeological Park of Ostia Antica, and Claudia Tempesta, archaeologist officer, manager of the Archeological Park of Ostia Antica, for their valuable assistance and authorization in collecting the mortar samples.

References

1. Turci, M.: Le iscrizioni delle Terme di Porta Marina rinvenute da Gavin Hamilton. In: Caldelli, M.L., Laubry, N., Zevi, F. (eds.) Ostia, l'Italia e il Mediterraneo: Intorno all'opera di Mireille Cébeillac-Gervasoni. Atti del Quinto seminario ostiense, Roma-Ostia, 21-22 febbraio 2018, pp. 133–146. Publications de l'École française de Rome (2021). https://doi.org/10.4000/books.efr.14319

2. Valeri, C.: Brevi note sulle Terme a Porta Marina ad Ostia. ArchCl **52**, 306–322 (2001)

3. Squarciapino, M.F.: Terme di Porta Marina o della Marciana. FA, nr. 8343 (1969–70)

4. Mannucci, V.: Restauro di un complesso archeologico: le terme di Porta Marina ad Ostia. Archeologia Laziale **3**, 129–132 (1980)

5. Turci, M.: Lo sviluppo termale del settore costiero della città di Ostia. Riesame della documentazione e nuove indagini alle *Thermae Maritimae* (IV, X, 1) e alle c.d. Terme Marittime (III, VIII, 2). Ph.D. Thesis, Aix-Marseille/Sapienza Universities, BABESCH Supplements, Maastricht (2023)

6. Turci, M.: Lo sviluppo del settore costiero della città di Ostia tra III e VI secolo d.C. In *Trade and Commerce in the Harbour Town of Ostia,* In: Landskron, A., Tempesta, C. (eds.) the XIXth International Congress of Classical Archaeology and Economy in the Ancient World, Cologne/Bonn, 22–26 May 2018; Keryx 7, Graz, pp. 137–171 (2020)

7. Squarciapino, M.F.: Un altro mosaico ostiense con atleti. RendPontAc **59**, 161–179 (1988)

8. Olevano, F., Rosso, M.: Il mosaico a grandi tessere marmoree delle terme "della Marciana" a Ostia. In: Guidobaldi, F., Paribeni, A. (eds.) Atti del VIII Colloquio AISCOM, Firenze 21–23 febbraio 2001; Ravenna, pp. 561–572 (2002)

9. Turci, M.: Nuove proposte di contestualizzazione dei mosaici pavimentali delle Terme di Porta Marina di Ostia. Revue archéologique **n°** **67**(1), 49–90 (2019). https://doi.org/10.3917/arch.191.0049

10. Pensabene, P.: Ostiensium marmorum decus et decor: studi architettonici, decorativi e archeometrici, Roma (2007)

11. Valeri, C.: Le portrait de Plotine et la décoration statuaire des Thermes de la Porta Marina. In: Descœudres, J.-P. (ed.) Ostia, port et porte de la Rome antique, Genève, pp. 303–307 (2001)

12. Valeri, C.: Arredi scultorei dagli edifici termali di Ostia. In: Bruun, C., Gallina Zevi, A. (eds.) *Ostia e Portus nelle loro relazioni con Roma*, Atti del Convegno all'Institutum Romanum Finlandiae, Roma 3–4 dicembre 1999; ActaInstRomFin 27, Roma, pp. 213–228 (2002)

13. Laubry, N., Poccardi, G.: Une dédicace inédite à l'empereur Probus provenant des Thermes de la Porta Marina à Ostie. ArchCl **60**, 275–305 (2009)

14. Caldelli, M.L., Turci, M.: Ostia: un'iscrizione inedita e i restauri di età teodericiana alle Terme di Porta Marina. ArchCl **72**, 267–296 (2021)

15. Marra, F., Karner, D.B., Freda, C., Gaeta, M., Renne, P.: Large mafic eruptions at alban hills volcanic district (central Italy): chronostratigraphy, petrography and eruptive behavior. J. Volcanol. Geothermal Res. **179**(3–4), 217–232 (2009). https://doi.org/10.1016/j.jvolgeores.2008.11.009

16. Giordano, G., the CARG TEAM: Stratigraphy, volcano-tectonics and evolution of the Colli Albani volcanic field. In: Funiciello, R., Giordano, G. (eds.) The Colli Albani Volcano, vol. 3, pp. 43–98. Geological Society, London, UK, Special Publications of IAVCEI (2010)

17. Jackson, M., Marra, F.: Roman stone masonry: volcanic foundations of the ancient city. Am. J. Archaeol. **110**, 403–436 (2006)

18. Brocato, P., Diffendale, D.P., Di Giuliomaria, D., Gaeta, M., Marra, F., Terrenato, N.: Previously Unidentified tuff in the archaic temple podium at Sant'omobono, Rome and its broader implications. J. Mediterr. Archaeol. **32**, 114–136 (2019)

19. Jackson, M., Marra, F., Hay, R., Cawood, C., Winkler, E.: The judicious selection and preservation of tuff and travertine building stone in ancient Rome. Archaeometry **47**, 485–510 (2005)
20. Jackson, M., Marra, F., Deocampo, D., Vella, A., Kosso, C., Hay, R.: Geological observations of excavated sand (*harenae fossiciae*) used as fine aggregate in Roman pozzolanic mortars. J. Roman Archaeol. **20**, 25–53 (2007)
21. Jackson, M., Deocampo, D., Marra, F., Scheetz, B.: Mid-Pleistocene pozzolanic volcanic ash in ancient Roman concretes. Geoarchaeology **25**, 36–74 (2010)
22. Marra, F., D'Ambrosio, E., Gaeta, M., Mattei, M.: Petrochemical identification and insights on chronological employment of the volcanic aggregates used in ancient roman mortars. Archaeometry **58**, 177–200 (2016)
23. D'ambrosio, E., Marra, F., Cavallo, A., Gaeta, M., Ventura, G.: Provenance materials for Vitruvius' *harenae fossiciae* and *pulvis puteolanis*: geochemical signature and historical–archaeological implications. J. Archaeol. Sci. Rep. **2**, 186–203 (2015)
24. Marra, F., Deocampo, D., Jackson, M., Ventura, G.: The Alban Hills and Monti Sabatini volcanic products used in ancient Roman masonry (Italy): an integrated stratigraphic, archaeological, environmental and geochemical approach. Earth Sci. Rev. **108**, 115–136 (2011)
25. Marra, F., Sottili, G., Gaeta, M., Giaccio, B., Jicha, B.: Major explosive activity in the Monti Sabatini Volcanic District (central Italy) over the 800–390 ka interval: geochronological–geochemical overview and tephrostratigraphic implications. Quatern. Sci. Rev. **94**, 74–101 (2014)
26. Marra, F., Danti, A., Gaeta, M.: The volcanic aggregate of ancient Roman mortars from the Capitoline Hill: petrographic criteria for identification of Rome's "pozzolans" and historical implications. J. Volcanol. Geoth. Res. **308**, 113–126 (2015)
27. Schmölder-Veit, A., Thiemann, L., Henke, F., Schlütter, F.: Hydraulic mortars in the imperial residence. In: Fumadó Ortega, I., Bouffier, S. (eds.) Mortiers et hydraulique en Méditerranée antique, pp. 99–110. Presses universitaires de Provence (2019). https://doi.org/10.4000/books. pup.39905
28. Diffendale, D., Marra, F., Gaeta, M., Terrenato, N.: Combining geochemistry and petrography to provenance Lionato and Lapis Albanus tuffs used in Roman temples at Sant'Omobono, Rome, Italy. Geoarchaeology **34**, 187–199 (2019)
29. Mirigliano, G.: Italite tipica fra i prodotti dell' eruzione del Vesuvio nel 79 d. C. Bull Volcanol **5**(1), 65–70 (1939)
30. Gaeta, M., Fabrizio, G., Cavarretta, G.: F-phlogopites in the Alban Hills Volcanic District (Central Italy): indications regarding the role of volatiles in magmatic crystallization. J. Volcanol. Geoth. Res. **99**, 179–193 (2000)
31. Website of Dr Alessandro Da Mommio: Alex Strekeisen I vetrini della mia fantasia, https://www.alexstrekeisen.it/english/vulc/leucitite.php. Last accessed 1 Apr 2022
32. Jackson, M., et al.: Assessment of material characteristics of ancient concretes, Grande Aula, Markets of Trajan, Rome. J. Archaeol. Sci. **36**, 2481–2492 (2009)
33. Marra, F., D'Ambrosio, E., Sottili, G., Ventura, G.: Geochemical fingerprints of volcanic materials: Identification of a pumice trade route from Pompeii to Rome. GSA Bull. **125**, 556–577 (2013)
34. Lancaster, L., Sottili, G., Marra, F., Ventura, G.: Provenancing of lightweight volcanic stones used in ancient Roman concrete vaulting: evidence from Rome. Archaeometry **53**, 707–727 (2011)

35. Pecchioni, E., Fratini, F., Cantisani, E.: Atlante delle malte antiche in sezione sottile al microscopio ottico. Atlas of ancient mortars in thin section under optical microscope. Nardini Editore, Firenze, Italy (2014)

36. Lancaster, L.C.: Pozzolans in mortar in the roman empire: an overview and thoughts on future work. In: Fumadó Ortega, I., Bouffier, S. (eds.) Mortiers et hydraulique en Méditerranée antique, pp. 31–39. Presses universitaires de Provence (2019). https://doi.org/10.4000/books. pup.39835

Characterization of Old Mortars for the Formulation of Replacement Mortars

Isabel Torres[1,2(✉)], Gina Matias[2,3], and Nilce Pinho[4]

[1] Department of Civil Engineering, University of Coimbra, CERIS, Coimbra, Portugal
itorres@dec.uc.pt
[2] Itecons, Coimbra, Portugal
[3] University of Coimbra, CERIS, Coimbra, Portugal
[4] University of Coimbra, Coimbra, Portugal

Abstract. The exterior coatings of old buildings were mostly made with lime-based mortars. As coverings for the external facades of buildings, they are elements that are highly exposed to several actions such as climate actions, mechanical actions, etc., and therefore, they are the first ones to need conservation, rehabilitation or replacement interventions.

The conservation, rehabilitation or replacement of lime-based coatings are actions that involve the complicated task of choosing a new mortar that should not only be compatible with the existing mortar, but also compatible with the substrate itself.

There are two possible ways to properly choose the mortar to use, whether you choose to reproduce the old mortar, thus seeking to ensure its compatibility and its proper functioning, or you formulate a compatible mortar with an adequate behavior for the building in question, with an appearance that preserves the building's image.

In practice, the first way is impossible on its own as it is not yet possible to accurately determine the "evolution" of the mortar (dynamic processes in constant evolution – crystallization, dissolution and recrystallization, etc.) and it is also difficult to identify the technologies used in its execution and application and we are also unaware of the climatic conditions that may have influenced the curing process. The ideal will be a process involving the two mentioned ways.

For this, the first task should be to know the characteristics of the existing mortars and then select a similar mortar with an adequate behavior for the building.

What is presented in this work is a campaign carried out in order to choose replacement mortars for several old buildings located in the center of Portugal. For this, the existing mortars were characterized, suitable replacement mortars were selected and applied to the existing buildings. After they had been cured, an in situ experimental campaign was carried out in order to assess the suitability of the chosen mortars for the walls of the buildings.

Keywords: Mortars · old mortars · replacement mortars

V. Bokan Bosiljkov et al. (Eds.): HMC 2022, RILEM Bookseries 42, pp. 97–107, 2023.
https://doi.org/10.1007/978-3-031-31472-8_8

1 Introduction

Lime-based mortars have been used for thousands of years, and their use is present in the history of construction to the present day. However, with the advent and popularization of Portland cement, traditional knowledge of lime technology was gradually lost.

Many scientific studies have identified that one of the main problems linked to the degradation of old buildings is the physical-chemical incompatibility between support and mortar.

In this sense, this research intends to study, in situ, the old mortars of houses from centenary villages, located in Vila Nova, Miranda do Corvo, central Portugal, where currently the process of eviction is around 70% of the properties. This number can be confirmed by the INE – Instituto Nacional de Estatística, through the demographic census, which shows us that in 1911 Vila Nova was inhabited by 2,402 people, and 110 years later, in 2021, it had only 793 inhabitants, and only 15% of this total is composed of young people up to 25 years old or children.

The work in progress seeks to relate the compatibility between the old mortars of buildings, aged between 100 and 300 years, through physical-chemical and mechanical analyses, showing that it is possible to define a compatible replacement mortar with the houses of the villages, in order to draw attention to a less costly form of mass rehabilitation.

Despite the age of the properties, it is observed that their coating mortars have performed well in the protection role of the respective substrates. In these villages, according to the reports of the residents, the traditional mortars were made with sand from local streams or came from the floods of the Serra da Estrela (Mondego river) thaw. Lime was taken from the ravines and burned in homemade ovens in the region or it could also be purchased from vendors who brought it in bags from the factory (Fig. 1).

Fig. 1. Housing studied.

One of the intentions of this work is, in addition to identifying the characteristics and behavior of mortars, to attract attention to this type of material heritage, whose embedded immaterial value is little studied.

By showing this possibility of a collective process of real estate recovery, an unusual and unexplored resource is created in order to bring new residents to the place (which is less than 30 km from Coimbra), reducing the process of desertification of these villages, even being able to culminate in more frequent interventions in other Portuguese villages.

2 Methodology

Mortar samples were taken from the properties and characterized in the laboratory. A natural hydraulic lime mortar was formulated in the laboratory and a pre-dosed replacement mortar was selected, both compatible with the original mortars. In possession of these two mortars, an experimental campaign was carried out to analyze the compatibility between mortars and substrates, both in the laboratory and in situ.

In laboratory, specimens with dimensions in accordance with the applicable standards were executed, on which the characterization tests were performed.

The existing mortar was partially removed from the original supports of each property, creating uncoated panels, where both the mortar formulated in the laboratory and the pre-dosed replacement mortar were applied.

Several tests were performed on these panels, in order to know the behavior of these mortars in service. This behavior was compared with the behavior of the original mortars and with the behavior of the new mortars obtained in the laboratory.

2.1 Characterization of Existing Mortars

The first phase of characterization of the existing mortars consisted of the determination of the respective traces by acid dissolution, according to a procedure based on the procedure of RILEM TC 167-COM (Characterisation of old mortars with respect to their repair). The collected samples were disaggregated, fractions of 35 g were separated and oven dried for 3 days. Subsequently, they were immersed in hydrochloric acid solution, remaining at rest for 24 h. After this time, the insoluble portion was washed with deionized water. The washing liquid containing the fines was filtered and dried (Fig. 2). Based on the values of the initial and final weighing of the sands and fines, the proportion between the binder and the aggregate was determined through the expression:

$$l/a = (mins - ma)/ma \tag{1}$$

where:

$mins$ – dry mass of water-insoluble material, in g;
ma – dry mass of acid-insoluble material, in g.

In a second phase, the granulometry of the aggregate resulting from the process described above was determined and the obtained granulometric curve is shown in Fig. 3, which also presents the granulometric curve obtained for the aggregate used in the formulation of the new hydraulic lime mortar.

The porosimetry test by mercury intrusion of the existing mortar was also carried out. For this purpose, a Micromeritics mercury porosimeter, model Autopore IV 9500, was used and the tests were performed in accordance with the ISO 15901-1:2005 standard. Figure 4 presents the results obtained.

2.2 Formulation of Replacement Mortars

For the study of replacement mortars, two different solutions were analysed: a solution composed of a mortar formulated in the laboratory with NHL 3.5 hydraulic lime and

Fig. 2. Determination of traces by acid dissolution.

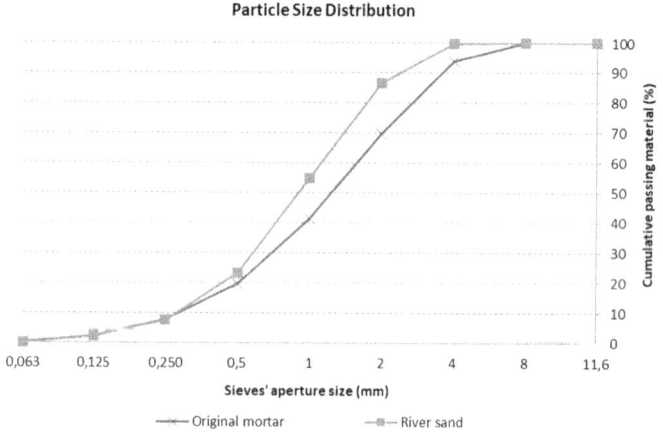

Fig. 3. Particle size distribution of the aggregates.

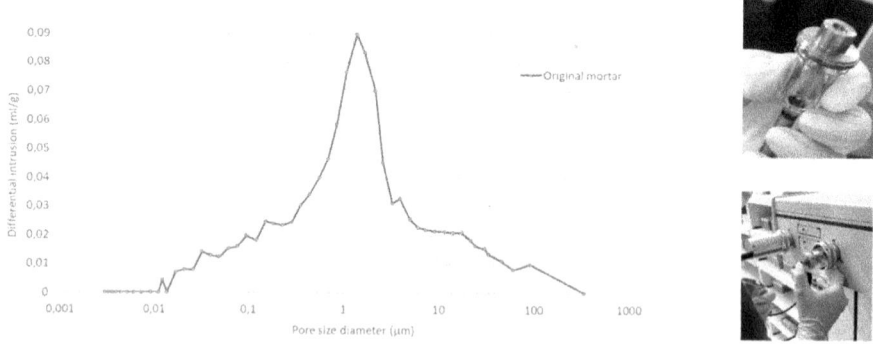

Fig. 4. Porosimetry of the existing mortar.

river sand, with a volumetric proportion of 1:4 (MNHL), and a solution consisting of pre-mixed mortars, specific for rehabilitation, with a levelling layer (MRB) and a finishing layer (MCS).

The selection of river sand to be incorporated into the MNHL mortar was made considering the closest possible proximity to the existing sands. As can be seen from the granulometric curves presented in Fig. 3, the curves of the existing sand are very similar to the selected river sand.

Prismatic specimens measuring $4 \times 4 \times 16$ cm and circular specimens measuring 10 cm in diameter and 1.5 cm in thickness were prepared, according to the indication of the respective standards (Fig. 5).

Fig. 5. Executed specimens.

The tests carried out on the replacement mortars and the standards followed are shown in Table 1.

Table 1. Tests carried out on the replacement mortars.

Test	Standard
Consistency (flow table value)	EN 1015-3:1999
Capillary absorption	EN 1015-18:2002 (mortars of general)
Water vapor permeability	EN 1015-19:2008
Flexural and compressive strengths	EN 1015-11:2019

On the real supports, after removing the existing mortar, cleaning with a wire brush to remove loose grains and wetting with water, the two mortar solutions were applied, which were subjected to curing conditions in an outdoor environment (Figs. 6 and 7).

After the curing period of the applied mortars, the following tests were carried out in situ, on these and on the old mortars:

Water Absorption with Karsten Tubes

This test, which measures the amount of water absorbed at low pressure, was performed in accordance with ASTM E514. Three test tubes are used simultaneously for each experimental panel (Fig. 8). The place for fixing the tubes was previously cleaned with a brush, then the tube was fixed to the wall with the help of suitable fixing material. The tube was then filled with water to the 4 ml level. For each tube, the amount of water absorbed at 5, 10, 15, 20, 30 and 60 min was recorded. In this test, water will be absorbed more slowly the less porous the tested mortar is.

Superficial Resistance. This test was carried out with a Model OS-120 pendulum hammer, following the RILEM – 1997 standard. The hammer allows the measurement

Fig. 6. Cleaning of the support (before and after).

Fig. 7. Preparation and application of new mortars.

Fig. 8. Preparation and application of new mortars.

of superficial hardness from the recoil of an incident mass after impact with the surface to be tested. It is based on the principle that the reflection of an elastic mass thrown against a surface depends on the superficial hardness of the material to be analysed. The softer the material, the greater the amount of energy absorbed and the smaller the height of the shoulder.

Martinet-Baronnie Apparatus. The Martinet-Baronnie apparatus makes it possible to evaluate the plasters in terms of their resistance to the impact of hard bodies.

- *Ball-shock:* gives information on the deformability of the coating. Consists of the application of a 3 J hard body shock, through a steel ball of about 50 mm in diameter, which is dropped from the horizontal position of the apparatus. At the end of each test, the diameter of the dent produced by the impact of the sphere is recorded, with a proper ruler with a precision of 0.1 mm (Fig. 9).
- *Controlled penetration:* allows evaluating the resistance of the innermost layers of the renders. An accessory with a small hole is attached to the opening in the "foot of the apparatus", which allows the placement of a steel nail in contact with the surface. The sphere is replaced by a body of known mass, which is dropped from the horizontal position of the apparatus (Fig. 10). The penetration depth of the nail into the coating is recorded.

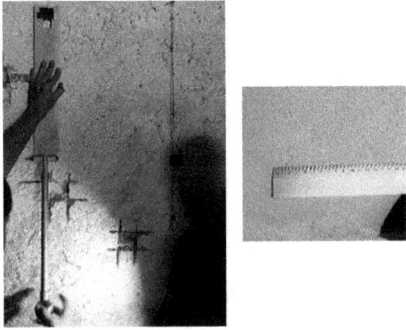

Fig. 9. Martinet-Baronnie: ball-shock

Fig. 10. Martinet-Baronnie: controlled penetration

3   Results

At this point, all the results obtained in the laboratory and in situ campaigns are presented and analyzed.

3.1   Laboratory Campaign

Table 2 presents the results obtained for the three replacement mortars characterized in the laboratory on specimens and under curing conditions in accordance with the aforementioned standards.

Table 2. Results of laboratory tests carried out on replacement mortars.

Test	Average Values		
	MNHL	MRB	MCS
Consistency (flow table value)	121,6	163,0	136,6
Capillary absorption	1,15	1,28	0,09
Water vapor permeability	$1,75E^{-11}$	$1,17E^{-11}$	$1,77E^{-11}$
	0,4	0,8	1,0
Flexural and compressive strengths	1,1	3,9	2,1

3.2   *In Situ* Campaign

After the curing period of the mortars applied in situ, the previously mentioned tests were carried out, on these and on the existing mortars. The results obtained are shown in the following tables.

Water Absorption with Karsten Tubes. The following graphs show the results obtained for the volume of water absorbed as a function of time, for the existing mortar (Fig. 11), for the MNHL replacement mortar (Fig. 12) and the pre-dosed mortar MRB + MCS (Fig. 13). The values shown are the average of three tests performed.

Superficial Resistance. Table 3 presents the results obtained for the average superficial resistance for the existing mortar (ME), replacement mortar (MNHL) and pre-dosed mortar (MRB + MCS). The values presented are the average of three tests performed.

Martinet-Baronnie Test. As mentioned above, two types of tests were carried out with the Martinet-Baronnie apparatus: the ball-shock test and the controlled penetration test, for the existing mortar and for the two replacement mortars. For the first, the average diameter of the dent produced by the impact of the sphere is presented and for the second, the average penetration depth of the nail in the coating. The values presented are the average of three tests performed in each situation.

The results obtained are shown in Table 4.

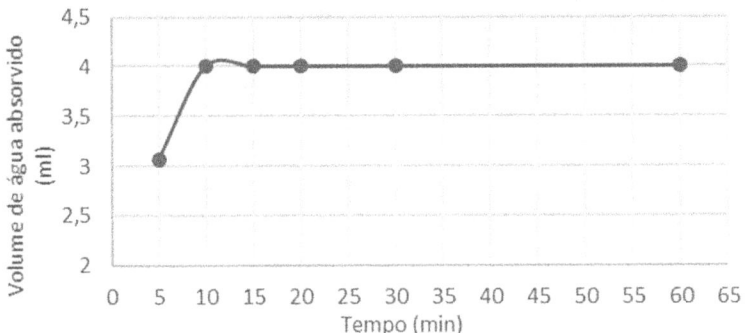

Fig. 11. Volume of water absorbed – existing mortar

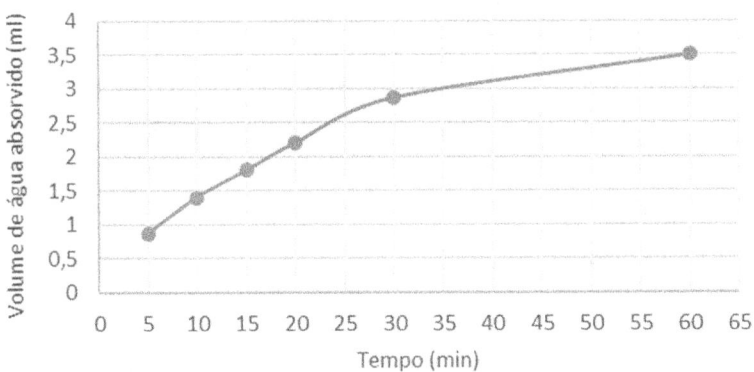

Fig. 12. Volume of water absorbed – NHL mortar

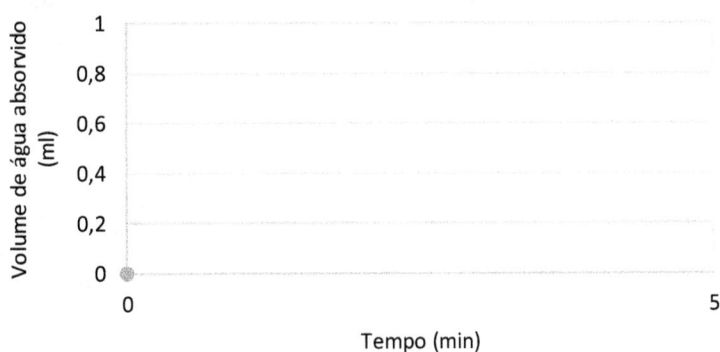

Fig. 13. Volume of water absorbed – pre-dosed mortar

3.3 Analysis of Results

In view of the results obtained for the analyzed case, it can be concluded that replacement mortars have adequate characteristics for the property in question.

Table 3. Superficial resistance for mortars ME, MNHL e MRB + MCS

Mortar	Superficial resistance – Wm (N/mm^2)
ME	0,90
MNHL	1,50
MRB + MCS	1,60

Table 4. Results of the tests Martinet-Baronnie apparatus.

Mortar	Diameter of the dent (cm)	Penetration depth (cm)		
		1st impact	2nd impact	3rd impact
ME	1,37	0,50	0,70	0,80
MNHL	1,43	0,50	0,70	0,80
MRB + MCS	1,37	0,20	0,40	0,50

In fact, regarding the behavior of the mortars in the presence of water, we can mention that the water absorption of both replacement mortars is lower than that of the existing mortar, and that the MNHL mortar has a slightly better behavior (less water absorption observed in situ with Karsten tubes).

For the water vapor permeability, which is intended to be as high as possible, the result obtained for the MNHL mortar was also slightly better than that of the final layer of the solution with the MRB + MCS mortars. The differences between these solutions may be associated with the granulometric distribution of the mortar aggregates, which, in the case of pre-measured mortars, appeared to have a lower maximum dimension than the hydraulic lime mortar.

As for the surface resistance, which can give some indication of the mechanical strength of the mortar, values lower than those obtained for the replacement mortars were obtained for the existing mortar. In these, it was found that the MNHL obtained lower values (confirmed by the values obtained in the laboratory) and more adequate to the situation.

With the results recorded for the Martinet-Baronnie apparatus, it was possible to confirm the aforementioned: the impact and penetration resistance are similar for the existing and replacement mortar and the values are lower for the MNHL mortar.

4   Conclusions

The rehabilitation of the built heritage has undergone enormous development with numerous old properties being rehabilitated instead of proceeding with their demolition.

In this path towards greater sustainability in construction, it is very important to rehabilitate the building in a correct and lasting way.

One of the building elements that suffers early degradation and needs rehabilitation/replacement are the exterior coverings, as they are exposed to harmful actions of the environment, pollution, vandalism, etc.

The old buildings were mostly coated on the outside with lime-based plaster. The adequate rehabilitation/replacement of these elements is a complex task that implies the knowledge of the characteristics of the existing coating and of the respective support. Only with this knowledge can we proceed to the correct choice of repair/replacement mortar.

With the research in development, of which a small part is presented, it is intended to make another contribution so that the rehabilitation of old buildings is carried out in an adequate and lasting way.

The research, which is being carried out in Vila Nova, Miranda do Corvo, Coimbra District, Central Zone of Portugal, aims to deepen the knowledge of the characteristics of the coating mortars of the numerous existing properties in need of rehabilitation, in order to find the most suitable rehabilitation/replacement mortars compatible with the respective substrates, so that they can be selected for application in an easy and quick way.

This article presents one of the first properties studied. The existing mortar was characterized, two possible repair/replacement mortars were selected, one pre-dosed and the other formulated in the laboratory and were also characterized in the laboratory.

After this characterization, the replacement mortars were applied on the property.

After 28 days of curing, several in situ tests were carried out.

After analysing the results obtained, it was possible to conclude that any of the replacement mortars is suitable and compatible with the support, and the laboratory formulated mortars (MNHL) obtained slightly more adequate results.

The investigation is in progress with the study of several properties in the region.

References

1. Lino, G.A.L.: Caracterização In Situ de rebocos com base em cal e metacaulino aplicados em muretes experimentais "Dissertação para obtenção do grau de mestre em engenharia civil – perfil da construção" Universidade Nova de Lisboa, pp. 50–51 (2013)
2. Loureiro, S., Paz, P.A., Angélica, S.: How to estimate the binder: agregate ratio from lime-based historic for Restoration. Front. Mater. 7, 1–13 (2020)
3. Mata, V.L.G.: Caracterização de Meios Porosos – Porosimetria, p. 52. Universidade do Porto p (1998)
4. Silva, T.R.: "Comportamento mecânico de argamassas de reboco com regranulado de cortiça" Dissertação para obtenção do grau de mestre em Engenharia Civil. Instituto Técnico de Lisboa, p. 45 (2014)

Characterization of "Terranova" Render Samples as a Contribution to XX Century Heritage Conservation

Cesare Pizzigatti[ID] and Elisa Franzoni[✉][ID]

Department of Civil, Chemical, Environmental and Materials Engineering (DICAM), University of Bologna, Via Terracini 28, 40131 Bologna, Italy
elisa.franzoni@unibo.it

Abstract. The "Terranova" render was a ready-mix dry colored mortar which diffused in Europe in the first half of the XX Century. During the 1920s and 1930s, it was largely adopted in Italy as a finishing solution for the façades of rationalist architecture. It quickly became very popular thanks to its excellent aspect and high durability. Despite its wide diffusion, its formulation and technical properties remain basically unknown, also due to some patents protecting it. For this reason, Terranova renders are often not recognized in restoration interventions. In this study, different samples of supposedly Terranova render were collected from three rationalist buildings in Ferrara and Forlì (Italy) dating back to the Thirties and a thorough characterization was carried out to investigate their formulations and features. A comparison was also made with a Terranova sample collected from a rationalist building in Bologna (Italy), analysed in a previous study. The aim was to characterize this material and disclose, if existing, a recurrent formulation, to explain its outstanding properties and advance the design of compatible repair mortars. Despite the recurrence of some features, defining a common formulation seemed to be hard, suggesting that alternative recipes were used in manufacturing this industrial render.

Keywords: Terranova · Special render · Dry-mix mortar · Rationalist architecture

1 Introduction

Renders played a crucial role in modern architecture for both aesthetic expressiveness and protection of façades. Among the most innovative materials, the so-called "special renders", i.e. dry-mix colored mortars that spread across Europe mainly in the first half of the XX Century, had particular importance. Their aesthetic features, stated high performances, such as durability and mechanical strength, low cost (also due to the limited thickness) and fast preparation and application made them attractive for modern architecture, where they were extensively applied [1].

In Italy, special renders were widely used in the façades of Rationalist architecture (during the 20s and 30s), and some of them, being manufactured at a national

V. Bokan Bosiljkov et al. (Eds.): HMC 2022, RILEM Bookseries 42, pp. 108–119, 2023.
https://doi.org/10.1007/978-3-031-31472-8_9

scale, also became paradigms of the autarchic policy promoted by the Fascist regime. "Terranova", "Jurasite", "Silitinto", "Terrasit" and "Pietranova" are some particularly worth of note examples [1, 2]. The Terranova render was probably the most popular and important, being extensively advertised as durable, resistant, lightweight, low-cost, easy to apply, and characterized by aesthetic features that perfectly fit modern architectural language [1–3]. The wide range of colors available (over ninety), different grain sizes (fine, medium and coarse) and different application techniques (trowelling and spraying) allowed to obtain a significant range of solutions in terms of texture, roughness and thickness [1, 3, 4].

The Terranova render was patented by Carl August Kapferer in Germany in 1896, as a dry-mix colored mortar for stone-imitating renders [5]. In 1916, Kapferer and Weber patented a method for increasing the permeability of the mortar to air, by adding a dilute emulsion of fat or oil [6]. This render, which quickly became very popular in Europe, diffused in Italy thanks to Aristide Sironi. Through his collaboration with Griesser, owner of the "S. A. Italiana Manifatture Griesser", he met Kapferer, who filed the patent of Terranova in Italy (n° 247015) in 1928 [5]. This document confirmed the main features of this material, also claimed in the previous patents. According to some authors, the admixture was characterized by a binder manufactured by slaking lime with the addition of wax, oil and glycerine, and the formation of calcium silicates (insoluble in water) occurred by mixing silicic acid and sodium fluoride or silicon fluoride to the aggregates [1, 3]. In 1932, Sironi, Griesser and Kepferer established the "S. A. Italiana Intonaci Terranova", starting the manufacturing in Italy [1, 5]. This render, launched with an extensive advertisement, had enormous success during the 1930s [5]. In 1945 Sironi acquired the shares of Griesser and Kepferer and replaced the "S. A. Italiana Intonaci Terranova" with the "Italiana Intonaci Terranova S.p.A.", which was sold to the Austrian "Montana Aktiengesellschaft" company in 1987, already owner of the Austrian and German Terranova industries. In 1993 the entire Terranova company was acquired by the French company "Weber&Broutin". Nowadays, the Terranova brand is owned by Saint-Gobain Weber S.p.A., which commercializes it in different versions, among which is the alleged original one [3, 5].

Despite its wide use in modern architecture, the historic Terranova render has been scarcely investigated so far. Its technical features, properties and formulation are basically unknown, also due to the presence of patents aimed at protecting rather than disclosing the actual mortar formulation. Besides, in contrast to what was declared there, hydraulic binders and/or cement seem to be present sometimes in this render, according to some studies [1, 4], which suggested the possible presence of different recipes or the alteration of the formulation directly on-site, before the application. Even concerning the aggregates, information is still lacking. Quartz and silicate aggregates were found [1], and the presence of mica seemed typical for this mortar, providing a sparkling effect [3, 4].

A deeper knowledge of this material is fundamental for designing suitable conservation and restoration interventions, and could help unveil the reasons for some particularly good performances of this render. This study aims at characterizing samples of Terranova render, to disclose, if existing, a recurrent formulation, to favour the design of compatible repair mortars and to provide useful insights into the design of new high-performance renders.

For these purposes, some original samples of supposedly Terranova renders were investigated and the results were compared to those obtained in a previous study on the original Terranova render of another Rationalist building, the Engineering Faculty in Bologna [6], where it showed an excellent conservation state (Figs. 1a–b). To explain this durability, in the aforementioned study a thorough characterization of the render was conducted, and the following features were highlighted:

- it consisted of a single, colored layer about 12 mm thick, with a very rough surface;
- the aggregates were quite coarse (up to 5 mm) and mainly composed of quartz, feldspar and mica;
- the binder was constituted by dolomitic lime with traces of white cement;
- large voids were present (larger than those detectable by mercury intrusion porosimetry), suggesting the use of air-entraining agents.

This sample, characterized in a previous study, is labelled as ING in this paper and it is used for comparison.

(a)   (b)

Fig. 1. (a) Engineering Faculty building, Bologna (1935), at present; (b) The Terranova render sample.

2 Materials and Methods

The samples investigated were collected from three Rationalist buildings in Emilia-Romagna, dating back to the Thirties:

- The former "Mercato Ortofrutticolo (MOF), located in Ferrara and built in 1937 (Fig. 2). The building, which hosted offices, a café and a trading hall, maintained its original use until 1989 when the Municipality began to use it as an archive, then it was abandoned in 1995. It was restored in 2016, as a part of a recovery project of the whole surrounding area [7], and the samples were collected during the restoration works. Now it hosts the Urban Center and the headquarters of the association of architects in Ferrara.

- The former "Gioventù Italiana del Littorio" (GIL), built in Forlì in 1935 and destined to activities for young people promoted by the fascist regime (Fig. 3) [8]. The building was restored in the 2010s and presently hosts sport and cultural activities as well as the headquarters of the ATRIUM Association (Architecture of Totalitarian Regimes of the XX Century in Europe's Urban Memory). The latter is a European cultural route promoted by the Council of Europe, focusing on fostering the knowledge of architecture built by European totalitarian regimes in the XX Century, to define a safeguard approach towards "dissonant heritage", removing its ideological meanings and highlighting those cultural and historical aspects that are worthy of preservation [9].
- The former "Asilo Santarelli", located in Forlì and built in 1936 (Fig. 4). The nursery school, which underwent various retrofit interventions, kept its original function until 2012, when it was closed [10]. Presently the building is subjected to a restoration project.

All the samples collected were made of two well-adherent layers: the external one, colored and with a rough surface, was applied on a grey substrate having a quite smooth surface. In particular:

- in the samples from the former MOF (labelled as "MOF-1"), the upper layer was green and very thin (about 3 mm thick) (Fig. 5);
- in the sample from the former GIL (labelled as "GIL-1"), the upper layer was red and quite thick (about 8 mm) (Fig. 6);
- in the samples from "Asilo Santarelli" (labelled as "ASL-1"), the upper layer was dark orange and very thin (about 3 mm thick) (Fig. 7).

The cross-section of the samples was observed by an Olympus SZX10 stereo-optical microscope (SOM) to investigate the layers, their interface and the aggregates. Each sample was previously embedded in a resin (Technovit 4004, Kulzer), and then, after some minutes of hardening, the surface under inspection was lapped and polished.

Concerning all the other tests, only the colored layers were investigated, since they were considered the most interesting ones. The two parts were separated manually by a saw, along the interface.

The elemental chemical composition and the morphology were analysed by a Philips XL20 SEM scanning electron microscope (SEM) equipped with EDS microanalysis, previously sputtering the samples with aluminium to make them conductive.

Water absorption and bulk density were calculated by determining the dry mass (after oven drying at 70°C until constant mass) and the saturated mass of the samples (by immersion at room pressure), and measuring the volume through hydrostatic weighing [11].

Fig. 2. Former MOF, Ferrara (1937). State in 2016, before restoration (courtesy of Leonardo srl, Casalecchio di Reno, Italy).

Fig. 3. Former GIL, Forlì (1935). Current state, after restoration.

The capillary absorption coefficient was measured by sorptivity test according to EN 15801:2009 [12], placing the external surface of the render on a layer of cellulose pulp and gauze (allowing a better contact than filter paper suggested by the standard, due to the roughness of the renders) saturated with deionized water. The mass increase of the samples with time per unit area was plotted in a graph versus the square root of time, calculating the capillary absorption coefficient from the slope of the first linear part of the curve.

The microstructure of the renders was also investigated by mercury intrusion porosimetry (Thermo Scientific Pascal Units 140 Series and 240 Series, overall range of investigated pore diameters: 116-0.0074 μm), determining their open porosity, median

Fig. 4. Former Asilo Santarelli, Forlì (1936). Current state.

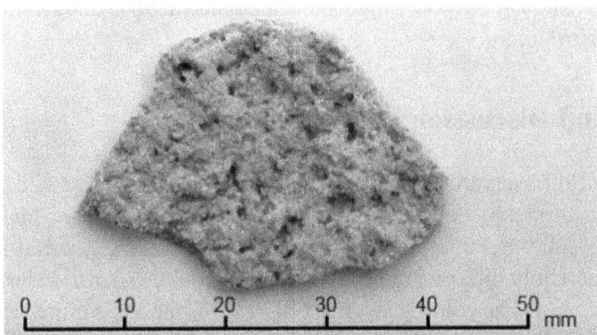

Fig. 5. Top view of MOF-1.

Fig. 6. Top view of GIL-1.

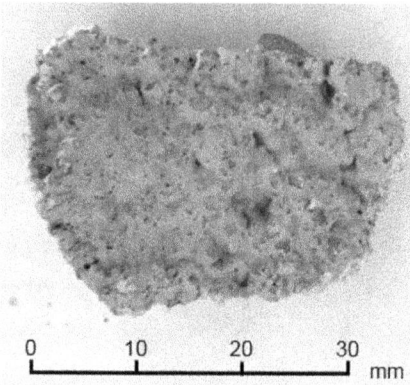

Fig. 7. Top view of ASL-1.

pore diameter (defined as the pore dimension calculated at 50% of the total pore volume) and pore size distribution.

3   Results and Discussion

According to SOM observations (Fig. 8), the base layers of MOF-1 and GIL-1 samples appear to be constituted by an ordinary mortar, grey in color, with ordinary sand aggregates and without any notable features. At the same time, the colored layers of these samples are remarkably different not only in terms of color but also in terms of type and size of the aggregates. The colored layer of GIL-1 (Figs. 8c–d) is quite thick (8 mm) and is characterized by large aggregates (up to 5 mm), while the colored layer of MOF-1 (Figs. 8a–b) is thinner (3 mm), with aggregates characterized by small size (<1 mm) and white or light grey color. Mica flakes appear in both colored layers, similar to the previously characterized Terranova sample (ING). According to these observations, the colored Terranova renders could have been applied over all-purpose base mortars. Concerning the ASL-1 sample (Figs. 8e–f), no particular difference was observed in the aggregates between the colored and the base layers, being the sand similar (size < 1 mm and no particular aesthetic features) and the mica absent. Furthermore, a paint was applied over the colored layer in ASL-1, probably during past maintenance works. This is suggested by the presence of a thin grey layer (likely some deposits) between the mortar and the paint. Homogeneously distributed large voids were detected by SOM in the colored layers of MOF-1, GIL-1 and ASL-1 samples (Fig. 8), approximately in the range 0.01–0.2 mm, 0.02–0.6 mm and 0.02–0.6 mm, respectively. These voids frequently exhibit a circular shape. Consistently, high open porosity and relatively low bulk density were found in these renders (Table 1), which could be ascribed to the use of air-entraining agents in the mix, which was also postulated for the previously characterized ING render [6]. Notably, the porosity of these samples is much higher than the values reported in the literature for cement-based renders (15–18%) [13, 14], indicating that the Terranova render has completely different features compared to ordinary cement-based ones, and more similar to lime-based render.

The pore size distribution (Fig. 9) did not allow to detect the presence of macropores in the colored layers, because MIP is not suitable for large voids (diameter > 100 μm); hence different kinds of analysis would be necessary. However, the renders' pores are mostly between 0.1 and 10 μm. Hence they can be considered quite coarse, especially compared to ordinary cement-based renders, whose size is reportedly in the range 0.002–0.1 μm [15].

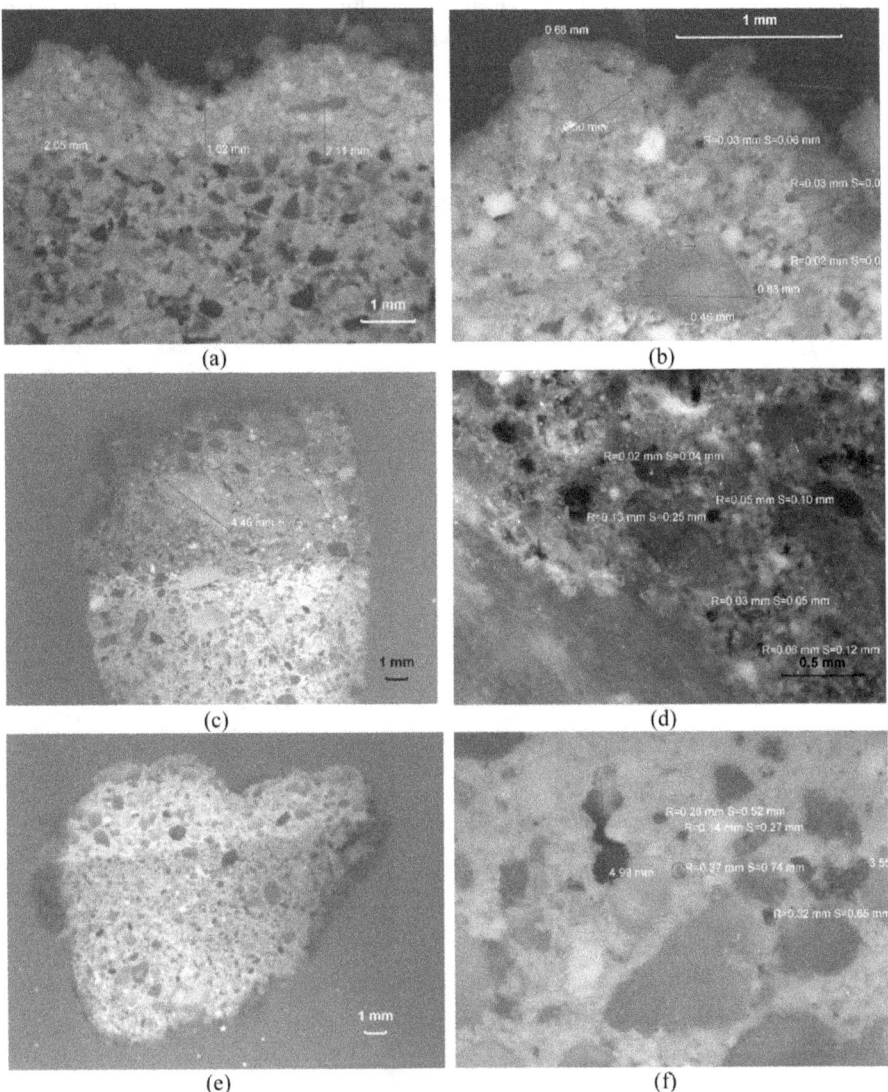

Fig. 8. Stereo-optical microscope images of the three renders under investigation: (a), (b) MOF-1; (c), (d) GIL-1; (e), (f) ASL-1.

In terms of capillary absorption rate, GIL-1, MOF-1 and ING samples are basically identical, while ASL-1 sample exhibit a bit higher value (Table 1). Interestingly, all the values found are quite low (e.g., values in the range 0.07–0.11 $kg/m^2s^{1/2}$ are reported for ordinary cement-based renders [16]), which could be ascribed to the presence of large voids that slow down the capillary absorption flow.

Concerning the chemical composition, no differences between the three renders resulted from the EDS analysis, apart from the elements ascribable to the pigments (Cr and Fe were related to the green and red colors, respectively). All the mortars investigated exhibit the presence of C, O, Mg, Si, K and Ca (Table 2). The significant presence of magnesium could be consistent with magnesium compounds in the binder and/or in the mica aggregates. Magnesian compounds, such as hydromagnesite, brucite and magnesite, were found in ING render, suggesting the presence of dolomitic lime in the binder. Further analyses are running for the characterization of the binder.

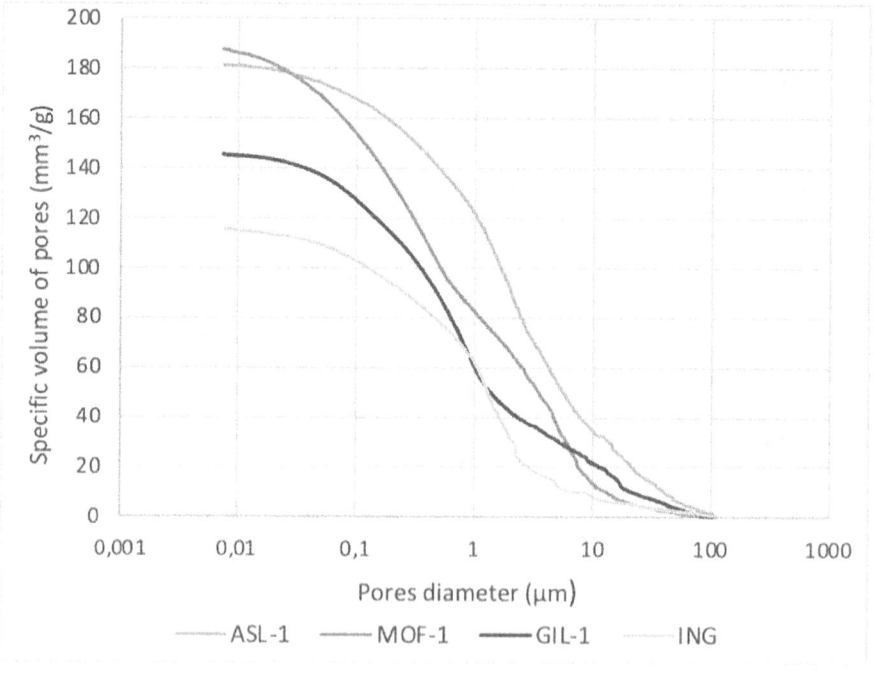

Fig. 9. Pore size distribution curves of the colored layers of the four renders, determined by MIP.

Table 1. Features of the colored layers of the samples.

	Bulk density [a] (g/cm^3)	Water absorption [a] (%)	Open porosity [b] (%)	Median pore diameter [b] (μm)	Capillary absor. coefficient (Kg/m^{2}s$^{1/2}$)
MOF-1	1.70	20.0	32.9	0.621	0.068
GIL-1	1.87	13.6	28.3	0.749	0.056
ASL-1	1.68	18.8	31.3	1.997	0.128
ING	1.82	12.6	24.3	1.288	0.068

(a) determined by water saturation at room pressure and hydrostatic weighing; (b) determined by MIP

Table 2. EDS analyses of the samples (colored layers).

	MOF-1	GIL-1	ASL-1
C	++	++	++
O	+++	+++	+++
Mg	++	++	++
Si	++	++	++
K	+	+	
Ca	+++	+++	+++
Cr	+		
Fe		+	

+++ dominantly present; ++ present; + present in a small amount

4   Conclusions

Given the results exposed above, the following observations can be made about the colored Terranova renders:

- Some common features seem present in the investigated samples, such as the low bulk density, the high porosity, the presence of big pores and large voids, and the presence of magnesium compounds whose attribution to magnesian lime in the binder is still under investigation. However, the different color, thickness and possibly also application technique (not known) of the samples reflect into differences in the pigments and the aggregate used, both in terms of nature and grain size distribution. In two out of three samples, mica aggregates were found.
- The identification of a unique and constant formulation for Terranova render appears quite challenging, and probably a range of different recipes was used. However, the render of the former "Gioventù Italiana del Littorio" (GIL) seems very similar to the Terranova render of the Engineering Faculty in Bologna (ING).

- From the point of view of performance, the high porosity and low density of the renders make them similar to lime-based mortars, in contrast with the low capillary absorption rate, which was ascribed to the presence of large voids owing to the use of air-entraining agents

Quantitative data about the formulation and behaviour of historic Terranova render are still very few. A better knowledge of this material could contribute to its proper preservation and conservation in the restoration of Rationalist architecture.

References

1. Di Resta, S., Favaretto, G., Pretelli, M.: Materiali autarchici. Conservare l'innovazione. Il Poligrafo, Padova, Italy (2021)
2. Fontana, C., Di Battista, V.: L'intonaco Terranova nella Milano del Moderno. In: Biscontin G, Driussi G (eds.) Atti del Convegno di studi, Bressanone 13–16 luglio 2004. Architettura e materiali del Novecento: conservazione, restauro, manutenzione, pp. 1171–1180. Arcadia Ricerche, Marghera (Italy) (2004)
3. Di Battista, V., Gasparoli, P.: Qualità e affidabilità dell'intonaco 'Terranova'. In: Biscontin, G., Driussi, G. (eds.) Atti del Convegno di studi, Bressanone 13–16 luglio 2004. Architettura e materiali del Novecento: conservazione, restauro, manutenzione, pp. 785–794. Arcadia Ricerche, Marghera (Italy) (2004)
4. Danzl, T.: I materiali costitutivi degli edifici del Bauhaus a Dessau tra tradizione e innovazione. Sviluppo di un metodo di restauro conservativo (1998-2004). In: Biscontin, G., Driussi, G. (eds.) Atti del Convegno di studi, Bressanone 13–16 luglio 2004. Architettura e materiali del Novecento: conservazione, restauro, manutenzione, pp. 105–118. Arcadia Ricerche, Marghera (Italy) (2004)
5. Di Battista, V., Cattanei, A.: Intonaco Terranova: storia e attualità di un materiale. La Litografica, Carpi (Italy) (2005)
6. Franzoni, E., Leemann, A., Griffa, M., Lura, P.: The 'Terranova' render of the Engineering Faculty in Bologna (1931–1935): reasons for an outstanding durability. Mater. Struct. **50**(5), 221 (2017)
7. Capozzi, L.: Relazione generale dell'intervento. Reperimento prescrizioni progetto preliminare _ Relazione storica. https://servizi.comune.fe.it/8065/attach/benimonumentali/docs/relazione_generale-exmof.pdf (2014). Last accessed 14 Jan 2022
8. Pagano, G.: Tre anni di Architettura in Italia. Casabella **110**, 2–5 (1937)
9. ATRIUM: Guidelines for the Restoration of Modern Architecture (2015)
10. Progetto ATRIUM: http://atrium.comune.forli.fc.it/asilo-santarelli/. Last accessed 14 Jan 2022
11. EN 1936:2006, Natural stone test methods - Determination of real density and apparent density, and of total and open porosity
12. EN 15801:2009, Conservation of cultural property - Test methods - Determination of water absorption by capillarity
13. Silva, B.A., Ferreira Pinto, A.P., Gomes, A.: Natural hydraulic lime versus cement for blended lime mortars for restoration works. Construct. Build. Mater. **94**, 6805 (2015)
14. Franzoni, E., Bandini, S., Graziani, G.: Rising moisture, salts and electrokinetic effects in ancient masonries: from laboratory testing to on-site monitoring. J. Cult. Heritage **15**(2), 112–120 (2014)

15. Mosquera, M.J., Benítez, D., Perry, S.H.: Pore structure in mortars applied on restoration: effect on properties relevant to decay of granite buildings. Cement Concr. Res. **32**(12), 1883–1888 (2002)
16. Bartos, P., Groot, C., Hughes, J.J. (eds.) PRO 12: International RILEM Workshop on Historic Mortars: Characteristics and Tests. RILEM Publications (2000)

A Discussion on Service Life Prediction Methodologies for External Mortar Cladding in Historic Buildings

Eudes Rocha[✉] and Arnaldo Carneiro

University Federal of Pernambuco, Pernambuco, Brazil
{eudes.rocha,arnaldo.carneiro}@ufpe.br

Abstract. Study models of buildings' degradation allow the estimation of their service life, assisting in planning their maintenance. However, determining an accurate method of applying these study models may be challenging, since there are unknown variables involved, such as the materials and construction techniques, besides interventions and environmental factors that impacted the building along the years. In historic buildings, it is difficult to know the mechanisms of deterioration, and to obtain reliable information about each building. The facade of historic buildings ends up being a great source of information though, since it is the building's element most exposed to degradation. Therefore, studying the deterioration of external mortar coatings of historic buildings may improve the quality of the information about them, and ensure heritage preservation. This paper discusses different methods used to study the service life of building elements and presents a review of mathematical models developed that not yet been directly applied to study the degradation of facades of historic buildings, identifying the main variables used, and indicating which model would be the best suited for applied in external mortar cladding in historical buildings.

Keywords: Historic buildings · Facade degradation · Mathematical models · Service life

1 Introduction

The building and its components are subject to a set of actions and aggressive agents along its lifetime, aging individually or in combination, causing anomalies, damage, or pathological problems that promote the loss of performance and the reduction of durability and service life of the building [1, 2].

The action of the various agents and mechanisms of deterioration, such as atmospheric pollution, water dissolution, frost action or thermal actions, among others, is aggravated when considered facades of buildings, which are generally areas subject to more intense natural deterioration and are exposed to a combination of these agents that often contributing to the evolution of the problem and consequently deterioration of the system [3–5].

V. Bokan Bosiljkov et al. (Eds.): HMC 2022, RILEM Bookseries 42, pp. 120–130, 2023.
https://doi.org/10.1007/978-3-031-31472-8_10

Furthermore, for heritage buildings the situation is grave, due to the heterogeneity of materials and the diversity of construction techniques which difficult the process of maintenance and conservation.

In this regard, relevant research is increasingly emerging in academia, each contributing in its own way to the evolution of knowledge in the area of degradation of coatings [3, 5–8]. These studies seek alternatives that aim to predict the behavior of buildings when exposed to deterioration processes and are based on mathematical models that determine an approximate value of the useful life of the building or any of its various components.

Initially, the application of mathematical models to predict the useful life may seem a bit pleonastic, however, when the models are fed with real and regular data obtained from inspections, it is possible to understand the functioning of the degradation agents assuming these same behaviors for buildings with similar characteristics and that have not been inspected, which would bring some speed to the preventive maintenance strategies of the assets, contributing to their conservation.

Another advantage of using these models is that the cost of their application is much cheaper when compared to an on-site inspection. However, for these models to predict well the lifetime of buildings and their components, it is necessary that they are fed with a database of previously performed inspections.

Only when we have enough data to supply reliable information to the models it is possibly to discard regular inspections, and this would be the main disadvantage of applying mathematical models to predict the service life of facades coated with historic mortars, since this database of information is still insufficient and needs further research.

This study provides a review of the main mathematical models used to predict the service life of various buildings, not necessarily focused on historic buildings. The intention is to identify what are the main difficulties and similarities to implement these models in historic buildings covered with mortar. And also indicate which model would be, in our analysis, the best suited for the study in progress.

2 Mathematical Models for Studying Construction Degradation

The numerous studies in the field of service life prediction, calculation methods are quite advanced nowadays, although, due to the many factors involved, service life still cannot be predicted with complete accuracy. The publication of ISO 15.686 was a major step forward in obtaining a prediction method for building elements and materials, but for many researchers, there are still important issues to be discussed, including:

- The creation of a reference database for the service life of component materials and building elements.
- Degradation studies that present deterioration mechanisms and factors to be characterized.
- Guidelines for applying the service life estimation method that include detailed examples of its application.

There are many discussions about the reliability of mathematical models, since the interpretations about the predictability of the service life of the constructions and its components depend on the understanding of how the elements behave under the action of deterioration mechanisms. Most models work with statistical tools that assign values or criteria of relevance for each data. For this reason, it is important to have a multidisciplinary approach that judges well these criteria so that the interpretation of the phenomena is as close as possible to reality [9].

The following is a summary of the methods that are most accepted and studied by the scientific community:

- Deterministic Methods:

    They are based on studies of the factors that influence the degradation of construction components, understanding the action of degradation agents, and transforming their effects into functions of degradation levels. The factors are converted into formulas that express their action over time. The best known model among the deterministic methods is the factor method included in ISO 15686 – part 1 (ISO, 2000), which will be discussed further.

- Probabilistic Methods:

    Probabilistic methods understand the degradation process as a stochastic process, that depends or results from one or more random variables, which makes it possible to define the probability of degradation of each component when submitted to a certain period [10]. However, the application and veracity of this method requires the use of realistic information collected in the field over a considerable period [9]. The most popular of these methods is the Markov model [9].

- Engineering Methods:

    Engineering methods usually combine the probabilistic and deterministic methods. Like the factor method, these are easy to apply and also describe the degradation processes in the same way as in probabilistic methods [10]. Engineering methods can be used to identify degradation processes analytically, which means that they can be controlled through design, planning and maintenance [11]. Some of the most well-known engineering methods include failure mode effects analysis (FMEA) and performance limit methods (PLM).

    Deterministic methods can also be used to obtain approximate factors until probabilistic and engineering methods reach a sufficient level of development. However, it is important to consider that the life estimated by these methods is not adequate to assess the risk of not reaching the project (service) life [9].

2.1 Deterministic Method (Factor Method)

One of the most recognized methods in academia is the factor method, which consists of a methodology with a systematic approach that is easy to apply in real projects.

It was initially proposed in the document "Principal Guide for Service Life Planning of Buildings" proposed by the Japan Institute of Architecture in 1993 and later adopted by ISO 15 686-8 (building and construction assets – service life planning – part 8: reference service life and service-life estimation) [12]. This method can also be applied in different ways, from a simple checklist to complex calculations, and further emphasizes that the level of application of this model depends on the purpose of the estimate (what's the main purpose of the inspection), the type and quality of the data collected (the way the information was obtained), the skills and expertise of the professional responsible for the estimates and also depends on the time available to perform the calculations [12]. The general formula of the factor method is expressed by the following equation:

$$ESL = RSL \times A \times B \times C \times D \times E \times F \times G \tag{1}$$

where: ESL – Estimated useful life; RSL – Reference useful life; A – material characteristics (quality, treatment, finishing); B – project characteristics or quality level of project; C – execution characteristics or execution level quality; D – internal environment characteristics; E – external environment characteristics; F – use conditions; G – maintenance levels.

The ISO 15686-8 (2008) [12] recommends that the durability factors adopted must present realistic values, close to the unit, proposing the adoption of values in the range from 0.8 to 1.2, and more preferably the durability factors should be between 0.9 and 1.1.

Among the disadvantages of this method is the empirical way in which the model is presented, which depends substantially on data collection [13]; moreover, it is verified:

- The lack of priorities and absence of hierarchies in the adopted variables, because all the adopted factors present the same weight (influence) in the sizing;
- The considerable dependence on modifying the adopted factors and the difficulty in quantifying and organizing them following a hierarchy;
- The high sensitivity and adoption of a constant degradation rate, which often prevents the understanding and characterization of degradation components over time [14–17];

However, despite being somewhat subjective, the factor method is one of the most accepted methods for studying degradation of construction elements among the academic community, being one of the most accepted and studied models with several variations [16].

Mainly because this method can reduce the costs of an on-site inspection, since from a pre-existing database, the model can estimate the useful life of the building and its components assisting in preventive maintenance. Once the useful life of the elements is defined, maintenance strategies can be better programmed and more direct, contributing to patrimonial conservation.

A study applied the factor method testing variables in predicting the service life of 274 facades of buildings in Portugal and concluded that the method is interesting; however, the study stresses that although the quantification of durability factors is subjective, due to the attribution of standard values that do not represent the real condition to which the building is subjected, results with a certain degree of reliability can be obtained [18].

However, the authors still point out that one of the limitations of this methodology is the lack of sensitivity regarding the multiplicity of results of the phenomena involved in the degradation of coatings. Therefore, it would be interesting to integrate into the study more sensitive analyses to the degrees of risk and levels of uncertainty associated with the durability parameters and levels of degradation studied [11].

A way to reduce the subjectivity of the factor method in the study of mortars in historic buildings is when we use the Delphi method, which brings together experts on the subject who propose weights (0.8 to 1.2) for each of the items analyzed. This way we have a pondered classification for each phenomenon that can serve as a guide for other classifications assigned.

Another way to reduce uncertainty in this method is through the creation of sub-factors that will bring the model closer to reality, since the more specific conditions are created, the wider the scope of the model and the more conditions that can be studied.

And finally, we can also apply probability distributions (Montecarlo simulation, Weibull Model, among others) because we can treat the classification of the factors statistically, increasing confidence.

2.2 Probabilistic Method (Markov Method)

The Markov method is a stochastic approach used to simulate the transition from one degradation level to another as a function of time. This model assumes that deterioration is a stochastic process governed by random variables, which in turn defines probabilistic parameters that affect a degradation curve. The random variables considered are the performance criteria defined and their probabilistic parameters correspond to levels scaled from 1 to n (considering failure levels to excellence) according to the degradation aspects considered. The result is a matrix formed from the combination of the number of parameters and the number of levels per parameter that represents the number of states of a given element or part of the building. From the elaborated matrix, the probability of passing from one state to another is determined, per unit of time, for each proposed level [17, 18].

The Markov model considers progressively degrading systems, where for each property or level established, in each period, a degradation probability is defined [13, 15]. Thus, it becomes a complex method and depends, therefore, on extensive field work and the accuracy of the input data, because any equivocation may present an erroneous prediction of the studied element. In order for the data feeding the Markov model to be validated, it needs to be taken from regular inspections conducted every two or trhee years, otherwise the model does not become appropriate for the study.

Table 1 below summarizes some major applications of the Markov Method:

Table 1. Some examples of Markov Chain Applications.

Author	Overview of the study ǀ Material Studied
Helland (1999)	Concrete structures resisting chloride penetration. The intention of this study was to study the penetration of chloride ions into concrete structures considering the reduction of the diffusion of this ion over time
Wisemann (1999)	He studied parking lot structures in Canada subjected to various climatic variations and excessive use of deicing salts. The author devoted himself to the analysis of degradation models for service life prediction, especially of structures using innovative materials
Tepply (1999)	Study of the degradation phenomena of reinforced concrete beams, in this case specifically the phenomena of carbonation and chloride attack. The authors evaluated the process of reinforcement depassivation and the consequent corrosion process over time by comparing the proposed numerical models within situ monitoring
Hong (2000)	It analyzed the deterioration process of reinforced concrete structures subjected to aggressive environments and loads in service for an evaluation and prediction of the service life of these structures
Shekarchi et al. (2009)	They studied the degradation mechanisms and service life of reinforced concrete structures located in the Persian Gulf by proposing a probabilistic analysis model that provides realistic results for studying the durability and service life of these components
Vargas et al. (2015)	The Markov chain can be used to simulate the total time to complete an excavation by comparing it with a real tunneling case. By incorporating variability into planning, it is possible to determine with greater certainty the critical activities and the tasks with slack. In addition, the financial risks associated with planning errors can be reduced and the exploitation of resources maximized and optimized
Lantukh – Lyashchenko et al. 2020	Applied Markov chain for studies of service life and durability of highway bridges in Ukraine and presented a universal model guided by one parameter (Hazard rate) that increases the reliability of the model and extends its applicability to highway bridges built in reinforced concrete. According to the authors, a mathematical model with a single control parameter (the crack opening) allows predicting the service life of reinforced concrete bridge elements at all stages of their service life cycle and can simplify the process of assessing their technical condition

[1] Adapted from [19].

For the application of this method to study degradation of external mortar coatings on historic buildings it will be needed data collected from site inspections conducted with certain frequency and also update laboratory analysis. Therefore, the various applications of the Markov Chain Method in the table above cannot be extrapolated for use in the study of historic facade coatings, but serve as indications of the possibilities of using this method.

In the case of the Markov method, we realize that its application to study the performance of facade coverings and consequently their service life depends on several unpredictable or difficult to determine variables. Factors such as the wetting cycle caused by rain, insolation and atmospheric pollution, for example, are difficult to predict and highly variable criteria, requiring in-depth laboratory studies and regular field inspections, making it difficult to apply this method for this purpose without these data or analysis.

2.3 Engineering Design Methods

These are methods that combine the advantages of the methods previously described and use simplified methodologies, present in deterministic methods, but considering probabilistic variability, without becoming excessively complex [19]. They can be used to identify building degradation mechanisms and performance reduction, feeding information for corrective actions and maintenance plans [19, 20]. Among the main models studied are:

- Delphi Method;
- Failure Modes and Effects Analysis (FMEA);
- Performance Limit Method (PLM);

The Delphi method was initially conceived in the 1950s by mathematicians Norman Dalkey and Olaf Hermes to study the impact that technology would have on the world. This technique is based on the qualitative discussion of a certain subject, bringing relatively precise assumptions or predictions about the studied subject, helping in the decision-making process.

It also presents a more objective view of the investigation, since it is based on the experience of a group of experts and not on the view of a single individual. However, its use for life expectancy prediction and performance study of building materials will depend on the experience of the experts involved, since the Delphi methodology groups the answers of the questionnaires presented in a statistical form and, therefore, both the representativeness of the sample and the opinion of the selected experts will bring a higher or lower degree of agreement or reliability to the study.

Failure Modes and Effects (FMEA) or Failure Modes Effects and Criticality Analysis (FMECA) was developed in the 1960s in Aeronautics studies. When applied to engineering it consists of a risk analysis method, whose principle is to define the potential degradation modes, their origins, causes and consequences under the component or building [21].

This methodology is also indicated in the building inspection standard of the Brazilian Institute of Engineering Appraisals and Expertise (IBAPE) and consists in a risk management tool through the effect and failure analysis method, or even, through the criticality list resulting from the building inspection.

FMEA can determine scenarios of degradation of the building or component, in an iterative way, where possible evolutions of the problems are defined systematically, from the beginning of the operation, and in each step, the possible consequences and causes of degradations are determined [19]. After making explicit for each component the evolution of degradation over time and unifying the information, the estimated useful life will correspond to the degradation scenario with the shortest duration.

Thus, in the FMEA method, there is a quantitative analysis of the performance of building components based on the knowledge of the relationship between the states of degradation of phenomena and the performance levels of the required functions [21]. Table 2 summarizes some of the main applications of this method for civil construction.

Despite its diverse applicability, the FMEA is a method with high application cost and is quite exhaustive due to a routine of repetitive steps. Another limitation is that the method is not adapted to consider dependent failures or those resulting from a succession of events, requiring evaluations with probabilistic methods, cited previously [17, 19].

On the other hand, FMEA is easy to apply, versatile and can be adapted to different construction systems. Its degree of accuracy and reliability is being improved, and it was recently complemented with a criticality indicator that allows the evaluation of all flaws identified in the method.

However, the application of this method to historic constructions requires further studies since the climatic variables and the combination of deteriorating agents successively may represent a low reliability in the interpretations of this model.

The Performance Limit Method (PLM) is a method developed by the Polytechnic of Milan (Durability of Building Components Group) that consists of predicting service life based on the correlation between users' needs and the measured deterioration of the performance characteristics of building elements [19].

Table 2. FMEA Construction Applications.

Autor	Síntese do Estudo	Material Estudado
Frangopol (1999)	Applied the FMEA to create a model to study highway bridges. The objective of the research was to analyze the reliability of the method and thus optimize preventive and corrective maintenance plans
Carlsson et al. (2002)	Used FMEA in conjunction with accelerated aging tests to study the service life of selective solar absorber surfaces for domestic hot water production
Hans; Chavalier (2005)	They improved the methodology by initially indicating the structural and functional analysis of the element to be studied and then applying the FMEA. It is believed that this way all possible failure scenarios of the product in use are obtained, considering the action of the environment to which it is inserted. They suggest the elaboration of the "Failure event graph" which consists of a graph with the various phases of degradation of a material under a single request
Talon et al. (2008)	They support the idea that FMEA is a methodology that should be extended to other case studies in order to develop a knowledge base that allows characterization of all building components in all types of environments. To this end, they studied brick walls covered by insulation
Wehbe; Hamzeh (2013)	The study contributes to construction management by creating a framework and planning to identify and mitigate potential risks and allocating contingencies. This proposed framework was analyzed by FMEA combined with Last Planner System (LPS)
Lee; Kim (2017)	The study attempted to identify why the costs for modular construction in Korea are so much higher than constructions designed in reinforced concrete, so it employed the FMEA method to find critical factors responsible for causing the increased costs in modular construction
Ma; Wu (2020)	The paper used the FMEA method to evaluate the construction quality of 311 apartments in Shanghai and proposed a risk correlation model that used the FMEA method in conjunction with a large collection of construction quality data. The proposed model can predict the relationship between the planned schedule and the project quality risk using multiple variables such as performance index and the budget costs

[1] Adapted from [19–24].

For this it is initially based on the relationship: agents → actions → effects → degradation to evaluate the phenomenon of degradation acting on the structure analyzed and once this relationship is defined, the designer will have enough data to adopt corrections and prevention measures for the occurrence of damage.

Then, with the performance and functionality data properly identified, the simulation of this evolution chain is proceeded. Laboratory aging tests are generally used to provide subsidies to the information adopted. And, finally, the evaluation is done to determine the useful life of the construction system or subsystem analyzed. Figure 1 schematically presents the methodology adopted in this model.

In comparison with FMEA, this method better describes the evolution process of the deterioration of elements, as well as deterioration mechanisms, since it discards probabilistic methods, requiring less information and simplifying the process [25]. However, it is important to emphasize that the method depends on the correct monitoring and definitions about the collection of data on the mechanisms of deterioration and the evolution of damage, in addition to the need for wider applications focused on buildings and building subsystems.

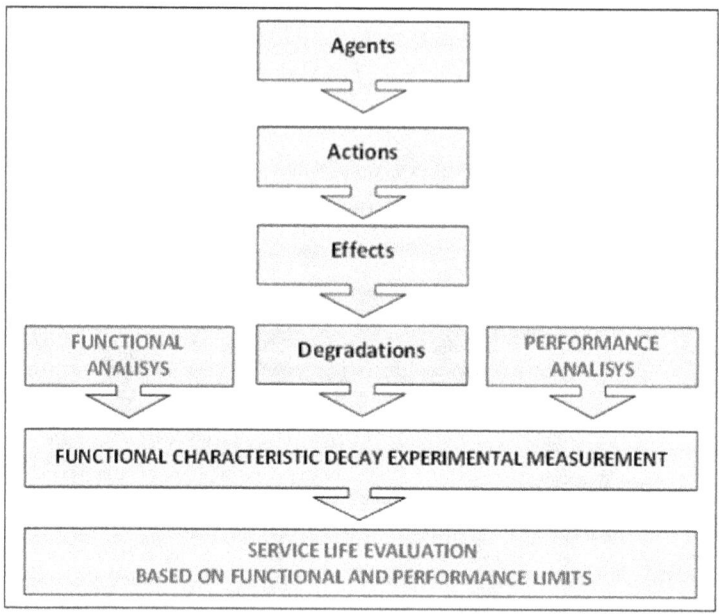

Fig. 1. Analyses scheme for building components' durability evaluation. Source: [11].

3 Concluding Remarks

Most of the models presented obtain the service life prediction through methods of measuring degradation in constructions and defining the maximum acceptable limit of degradation and need to be fed by updated field data that consider aspects of degradation of the different building components and of the environment where they are placed.

Initially for studies of degradation of facades coated with historic mortars we have the difficulty of lack of updated data that can feed the mathematical models. In this case it is suggested to invest in regular inspections in old buildings so that we can have relevant data about the performance of materials under the action of degradation mechanisms. Then, with this information, a database can be created to feed the models in a reliable way and, thus, we can extrapolate the predictions of service life for buildings that present similar characteristics, such as the same construction period, similar construction methodologies or even exposure to the same deterioration agents.

The most relevant variables to be considered in the study of mortar facade coatings on historical buildings are those related to the age of the buildings, the degree of protection to which they are submitted, and the geographic orientation of the facade, which significantly influences the performance, durability and life time of the materials. In this case, it is understood that the most adequate model would be the factor method, because it allows to explain, through the adopted factors, the degradations of the facade coverings and to incorporate the notion of uncertainty, providing a probabilistic view of its durability and useful life.

Finally, further studies are suggested, including the testing of these models, because we believe that the application of mathematical models for predicting the service life of building materials, when based on data obtained from a database of inspections and laboratory analysis, can help in the regular maintenance program of historic buildings with similar characteristics, contributing to faster and less expensive maintenance actions when compared to an on-site inspection.

References

1. Cavalagli, N., Kita, A., Castaldo, V.L., Pisello, A.L.: Hierarchical environmental risk mapping of material degradation in historic mansonry buildings: an integrated approach considering climate change and structural damage. Constr. Build. Mater. **215**, 998–1014 (2019)
2. Ramirez, R.A.L., Maljaee, H., Ghiassi, B., Lourenço, P.B., Oliveira, D.V.: Bond behavior degradation between FRP and mansonary under aggressive environmental conditions. Mech. Adv. Mater. Struct. **26**, 6–14 (2019)
3. Silva, M.N.B.: Avaliação quantitativa da degradação e vida útil de revestimentos de fachada – Aplicação ao caso de Brasília/DF, p. 217. Universidade de Brasília. Faculdade de Tecnologia, Tese de Doutorado em Estruturas e Construção Civil (2014)
4. Bauer, E., Souza, J.S., Piazzarollo, C.B.: Application of the degradation measurement method in the study of facade service life. Build. Pathol., Durability Serv. Life **12**, 105–119 (2020)
5. Bauer, E., Souza, J.S., Mota, L.M.G.: Degradação de fachadas revestidas em argamassas nos edifícios de Brasília, Brasil. Ambiente Construído **21**(4), 23–43 (2021)
6. Zurbrügg, P.: Simulation des phénomènes de dégradation d'éléments de construction. Application d'une méthode basée sur la modification des performances de matériaux et la propagation des contraintes, p. 251, Programme Doctoral em Architecture et Sciences de la Ville. École Polytechnique Fédérale de Lausanne (2010)
7. Munduruca, E.A.M.B.: Reabilitação em edifícios e monumentos Art-Déco – Métodos de avaliação dos revestimentos de fachadas. Dissertação de Mestrado em Geotecnia, Estruturas e Construção Civil, Universidade Federal de Goiás, p. 78 (2013)
8. Bauer, E., Zanoni, V., Silva, M., Castro, E.: Perfis de degradação de fachadas de edifícios em Brasília-Brasil. In: XIII Congresso Latino-Americano da Patologia da Construção, Lisboa. https://www.researchgate.net/publication/281845021 (2015). Acesso: Jan 2020
9. Ortega, L., Serano, B., Fran, J.: Proposed method of estimating the service life of building envelopes. J. Constr. **14**(1), 60–68 (2015)

10. Cecconi, F.: Perfomance lead the way to service life prediction. In: 9th International Conference on Durability Building Materials and Components (DBMC). Brisbane (2002)
11. Daniotti, B., Spagnolo, S.L.: Service life estimation reference service life databases and enhanced factor method. In: 11th International Conference on Durability of Buildings Materials and Components (DBMC), Istanbul, Turkey. 11–14 may, T41 (2008)
12. ISO 15686–8: Buildings and constructed assets, Service Life Planning. Part 8: Reference Service Life and Service Life Planning, International Organization for Standardization, Switzerland (2008)
13. Hovde, P.J.: The factor method – a simple tool to service life estimation. In: Proceedings of the 10th International Conference on Durability of Building Materials and Components pp. 522–529, Lyon, France, 17–20 Apr (2005)
14. Rudbeck, C.: Assessing the service life of building envelope construction. In: Lacasse, Vanier (eds.) Proceedings of the 8th International Conference on the Durability of Building Materials and Components, pp. 1051–1061. NRC Research Press, Canada, Ottawa, 30 May–3 June (1999)
15. Moser, K.: Engineering design methods for service life prediction, in Performance Based Methods for Service Life Prediction – State of the Art Report, Report N° 294, Part B, March, CIB (2004)
16. Gaspar, L.P., Brito, J.: Service life estimation of cement-rendered facades. Build. Res. Inform. **36**(1), 44–55 (2008). https://doi.org/10.1080/09613210701434164
17. Silva, A., Brito, J., Gaspar, L.P.: Stochastic approach to the factor method: durability of rendered facades. J. Mater Civ. Eng. **28**(2), 04015130 (2016). https://doi.org/10.1061/(ASC E)MT.1943-5533.0001409
18. Marques, C., de Brito, J., Silva, A.: Application of the factor method to the service life prediction of ETICS. Int. J. Strategic Property Manag. **22**(3), 204–222 (2018). https://doi.org/10.3846/ijspm.2018.1546
19. Santos, M.R.P.: Metodologias de previsão da vida útil de materiais, sistemas ou componentes da construção. Revisão bibliográfica. Dissertação de mestrado em engenharia civil – construções. Faculdade de Engenharia da Universidade do Porto, Portugal (2010)
20. Quintela, M.: Durabilidade de revestimentos exteriores de parede em reboco monocamada. Dissertação de mestrado em engenharia civil – construções. Faculdade de Engenharia da Universidade do Porto, Portugal (2006)
21. Talon, A., Chevalier, J., Hans, J.: State of the art Report on Failure Modes Effects and Criticality Analysis Research for Application to the Building Domain. CIB Report: Publication 310, Roterdam, (2006)
22. Wehbe, F.A.; Hamzeh, F.R.: Failure model and effect analysis as a tool for risk management in construction planning In:. Proceedings IGLC-21, July – Fortaleza, Brazil (2013)
23. Lee, J.-S., Kim, Y.-S.: Analysis of cost-increasing risk factors in modular construction in Korea using FMEA. KSCE J. Civ. Eng. **21**(6), 1999–2010 (2017). https://doi.org/10.1007/s12205-016-0194-1
24. Ma, G., Wu, M.: A big data and FMEA-based construction quality risk evaluation model considering project schedule for Shanghai apartment projects. Int. J. Qual. Reliab. Manag. **37**(1), 18–33 (2020). https://doi.org/10.1108/IJQRM-11-2018-0318
25. Sousa, R.: Previsão da vida útil dos revestimentos cerâmicos aderentes em fachada. Dissertação de mestrado em engenharia civil, Instituto Superior Tecnológico – IST, Lisboa, Portugal (2008)

Historic Production, Processing and Application of Mortars, Renders and Grouts. Lime Technologies

Mortars and Binders During a Time of Emerging Industries: 19th Century Austro-Hungarian Fortifications in Montenegro

Johannes Weber[1]([✉]), Lilli Zabrana[2], Andrea Hackel[3], Susanne Leiner[3], and Farkas Pintér[1]

[1] Institute of Conservation, University of Applied Arts, Vienna, Austria
johannes.weber@uni-ak.ac.at
[2] Department of Historical Archaeology, Austrian Archaeological Institute/Austrian Academy of Sciences, Vienna, Austria
[3] Arbeitsgemeinschaft Steinrestaurierung, Vienna, Austria

Abstract. A number of military fortresses built between 1853 and 1914 by the Austrian-Hungarian Empire in today's Republic of Montenegro have been investigated by the Austrian Archaeological Institute. The ongoing survey first focused on three sites dating between 1858 and 1897. Mortars had been diversely used in all of them for different purposes of construction.

Analyses were performed on mortar samples by thin-section microscopy and SEM. The results reveal use of several lime, natural "Roman" cement and Portland cement materials, depending on their application in the building and on the period of construction. All three fort buildings contain mainly air lime mortars as bedding and filling of the stone masonry, while pointing on exterior walls and tamped concretes are usually based on Portland cements. The microscopic features of the Portland cement (PC) clinkers reflect the typical conditions of early PC production. Roman cement (RC) mortars were only occasionally found.

Our contribution includes a presentation of the fortresses and their construction principles, followed by a discussion of the mortars by their binder constituents and aggregates. In the context of the state of the Austro-Hungarian cement industry in the 19th century, the observations reveal one of the earliest applications of Portland cement in the whole Empire.

Keywords: Historic Army Construction Sites · Montenegro · Austro-Hungarian Empire · Early Portland Cement · Historic Mortars

1 Introduction

The imperial Austro-Hungarian system of fortifications in the today's Republic of Montenegro was largely built between 1853 and 1914; it was the former southern extremity of the Empire. As well as marking out the southern boundary with the Ottoman Empire (until 1878) and Montenegro, its purpose was to defend part of the Adriatic fleet based in the Bay of Cattaro (today Kotor). The fortresses and the pathways linking them were

V. Bokan Bosiljkov et al. (Eds.): HMC 2022, RILEM Bookseries 42, pp. 133–149, 2023.
https://doi.org/10.1007/978-3-031-31472-8_11

largely abandoned after the First World War, so that most of the remaining structures and materials date back to the time of construction in the 2nd half of the 19th century. For the same reason, and due to the impacts during WW-I, most buildings are today in an advanced state of decay.

Inspired by an initiative of the Ministry of Culture of Montenegro in 2017, the Austrian Archaeological Institute of the Austrian Academy of Sciences started its scientific activities within an ongoing project related to the Austro-Hungarian fortification buildings in Montenegro.

This contribution focusses on the material aspects of a larger research project aimed to raise awareness of the existence of these defensive military structures, and in doing so, alert the need to protect and conserve this cultural heritage for future generations.

2   State of Research and Previous Work

Apart from two existing studies on the military architecture around the bay of Kotor which provide a broad overview of the existent building diversity and construction phases [1–3], neither detailed building analysis nor extensive historical background research has been published until present. In the autumns of 2018 and 2019, two short on-site campaigns were performed with a team of experts in the areas of building research, stone restoration and surveying. Initially, the topographic peculiarities of the bay of Kotor, surrounded by steep mountain ridges, were examined regarding the positioning of the Austro-Hungarian fortification buildings in their various phases of development. Out of a total of more than eighty objects, the fortifications of Kosmač (1858), Goražda (1884–1886) and Vermač (1894–1897) were selected for a detailed examination (Fig. 1). These three fortification complexes cover a time range of 40 years of structural engineering, a period of rapid advances in technological and industrial development within the 19th century. Though in remote positions, they all are characterised by their vicinity to the heavily touristed destinations of Kotor and Budva, which is why these forts possess great potential to make history come alive to educators and students, tourists, and the broader public. At the same time these fortifications are mostly endangered by possible future construction measures.

In 2020, project activities focused on the preparation of building survey plans for the Kosmač, Goražda and Vrmač fortresses, which had been investigated since 2018 (Fig. 2), and on the analysis of a set of stone and mortar samples taken during the course of both field campaigns. As sampling for laboratory analyses had to follow certain constraints, careful selection of samples was based on the idea to achieve insight into the major construction elements of each of the sites.

An important piece of information related to one of the forts could be found in the Vienna Kriegsarchiv (war archives) of the Austrian State Archives in Vienna [4]: a hand-written record on building materials and techniques of construction of Vermač fort (1894–1897), drafted by the Kotor branch of the Army Engineering Department ("Geniedirektion"). Amongst others, this document provided valuable insight into the types and approximate quantities of mortar binders used their proveniences and producers, as well as information on the origin and type of sand and aggregate. The observations made by laboratory mortar analyses will be discussed later with regard to this record.

Fig. 1. Location of the former Austrian naval base of Cattaro/Kotor and the three investigated fortresses Kosmač, Goražda and Vermač (© Google Maps, supplemented by the authors).

By comparison with another record [5] on the construction of the nearby Battery of Vermač (also known as Battery of Skaljari, built 1884–1886), which so far has not been investigated by the authors, conclusions can be drawn on the change of building materials and their suppliers within a time span of 10 years.

Material scientific studies were thus performed on samples taken from selected elements of the three buildings. They comprised concretes, masonry bedding and pointing mortars, exterior and interior renders, as well as natural stones. The analyses were aimed at providing information to the building history including construction processes, origin and procurement routes of the materials as well as achieving basic knowledge needed for their conservation and restoration.

Given the strategic importance of the forts and the administrative management by a big public institution like the Austro-Hungarian Army Command, it can be assumed that building materials and construction methods were at the highest available standards of the time. It seemed of particular interest that the period of construction was marked by significant progresses in the manufacture of cementitious building materials.

3 The Forts of Vermač, Goražda and Kosmač - Inventories and Building Description

In the following, the three fort complexes are briefly described in reverse order, i.e., starting with the latest for which written information is available.

3.1 Vermač (Vrmač)

Constructed in the period 1894–1897, Vermač/Vrmač is the youngest of the fortifications examined. Consisting of several buildings, the complex, located on a ridge above Kotor,

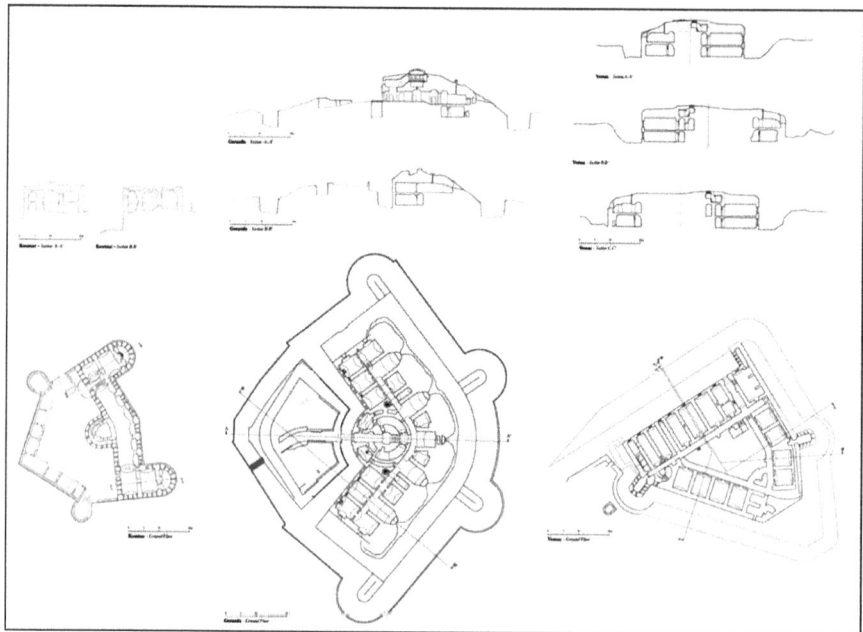

Fig. 2. The fortresses of Kosmač (1858), Goražda (1884–1886) and Vermač (1884–1897), ground plans and sections (© ÖAW-ÖAI, drawing by Nicola Math based on 3-D laser scans by Christian Kurtze).

had to protect and defend the town against possible attacks from across the Montenegrin border situated on the opposite slopes of the Bay of Kotor. It was largely built below the present ground level which had been partly raised to protect particularly the southern fronts of the fortification. The top deck was later covered with a massive "concrete bonnet". A several meters wide moat supported by stone walls runs around the central fortress building.

As mentioned above, written records describe in detail the materials and techniques of construction of Vrmač [4]. Similar reports would most probably exist for all the other Austro-Hungarian military buildings in the area, though so far they could not be traced by the authors, with the exception of the nearby Battery of Vermač [5] which was built about 10 years earlier than the fort there (Fig. 3).

Like the older fort buildings, most walls of the Vermač fortress were constructed as core-and-veneer with shells of regular stone ashlars and a core of rubble stone and mortar. Stone was also employed for most of the door and window framings. The barrel vaults of the ground floor were made of brick, while tamped concrete vaults cover the rooms of the first floor. Presumably in a later stage, the second upper floor was spanned by means of a construction of steel girders and tamped concrete.

As we know from [4], while the rubble quarry stone described as a hard and brittle limestone was extracted in the immediate surroundings, much better workability was assigned to limestone varieties from the island of Curzola/Korčula which were used as

Fig. 3. Vermač, view of the fortification in the terrain with its "block-like" appearance (© ÖAW-ÖAI, C. Kurtze)

dressing stone. Several stone samples investigated by thin section microscopy confirmed the report, though this is beyond the scope of the present contribution.

Bricks forming e.g., the vaults of less exposed rooms of the basement and ground floor were set in lime mortar. The source of all bricks used in the Vrmač constructions was the factory of Count Caboga in Kupari near Ragusa/Dubrovnik. The report mentions their low quality due to high magnesium contents, a statement which was not verified during our survey.

Regarding the mortars used in Vermač, it is evidenced both by the historical source of [4] and by the present survey that Portland cement mortars strongly prevail over Roman cement ones. In this respect, the building of Vrmač differs from the older fortresses where more Roman cement was used.

The Portland cement of the Vermač fortress is reported to originate from Lengenfeld in Ober-Krain/Dovje-Mojstrana, Kranjska Gora. The cement plant owned by Ammann & Co. had started its production in 1893 [6]. Though no more detailed information on their products could be found in the literature, it is likely that there "natural" Portland cement was produced from locally extracted marlstones.

Significantly more information is available on the producer of natural "Roman" cement: the company Gilardi- & Betizza in Spalato/Split [7]. Their Roman cement production started in 1871 and was complemented by – most probably again natural – Portland cement after 1880. Both products were certified and recommended for the construction of fortresses in the Kotor area. While Gilardi & Betizza had delivered Roman cement for both Vermač buildings, i.e., the Battery (1884–1886) and the Fort (1894–1897), their Portland cement was supplied only to the earlier Battery [5]; 10 years later, when the Fort of Vermač was built, the army administration had switched to the producer of Kranjska Gora [4] who had by then started their production of Portland cement.

Lime was purchased as lump quicklime from the locations Krašić and Mali Pristan on the peninsula of Luštica situated at the entrance to the bay of Kotor. Its slaking was done at the building site and is not further described, and no pit is mentioned in the report. Accordingly, the binder microstructure of the lime mortar samples points to a short-term slaking (see 4.2.2).

The costs of 100 kg Portland cement delivered to the building site is given in [4] with 3.4 Fl. (Austrian gulden) compared to 2.8 Fl. For Roman cement. According to the present-day currency [9], this amounts to roughly 72 Euro to 60 Euro, respectively. In comparison, quicklime delivered to Kotor by boat costed between 1.6 and 2 Fl. Per 100 kg, corresponding to approx. 23 to 30 Euro.

As to their use in the constructions, Portland cement-based mortars were used in the following elements of the fortress complex: cisterns, sinkholes and sewer tunnels, imposts of those vaults in the ground floor where the shielded batteries were placed, abutments of gun carriages, and the masonry adjacent to chimneys and window or door openings. For a number of elements designed to be artillery-proof or have heavy load bearing functions, concrete elements were called for, even naming the mixing ratios but not specifying the type of cement used; for the Battery of Vermač built 10 years before the Fort, the use of Roman cement in concrete construction is stated [5].

Concerning the aggregate for mortars and concrete of Fort Vermač [4], gravel was obtained by crushing the quarry stone from the nearby outcrops, whilst coarse sand was supplied from torrents of Bianca/ Bijela near Castelnuovo/Herceg Novi, and fine sand from the beaches next to Budva. The latter was stored outdoors over the winter period in order to have the salts washed out.

Samples taken from mortar elements of the Vrmač fortress in course of the present survey included pointing mortars of stone masonry outdoors and indoors, outdoor renders as well as plasterwork on interior walls. No concrete element was sampled so far. The results confirm the use of binders as stated by the report [4]: lime mortars were found just in masonry beddings and interior plaster, while pointing as well as renders revealed based on Portland cement. More details are reported in 4.2.2.

3.2 Goražda

The turret fortress of Goražda is situated a few km SSE of Vermač on a plateau-like extension of the same ridge. By its position it was meant to control a wide area which included Kotor, the roads from Budva to Kotor and Cetinje, the slopes of the Lovćen and the bay of Tivat down to the southern end of historical Dalmatia. Constructed in 1884–86, Goražda is about ten years older and significantly larger than the neighbouring fortification of Vermač. Similar to the latter, Goražda was also partly built below the ground level by backfilling the terrain around the structure. A several metres wide moat, supported by massive stone walls and accomplished by four projecting carponniers, surrounds the building with its casemates and the central armoured turret originally equipped with two siege guns, which is still preserved up to date.

Other than Vermač which suffered several heavy damages during WW I and was then abandoned, the fortification of Goražda was in use beyond the fall of the Empire in 1918, even during WW II and further in the post-war period until the 1980s. This is confirmed by a variety of repair works as well as by conversions and additions to the original structure as well as by a high amount of paints, whitewashes and graffiti, including more elaborate wall paintings present in different area of the complex.

In contrary to the Vermač fortress, no contemporaneous report specifying details of construction and building materials could be found so far for Goražda. The survey revealed that natural stone – in all places compact limestone of varying lithological features – is the dominant building material of the wall constructions. Analogies with the ashlar stone of the Vermač fortress supplied from Korčula are unlikely under the premise that so far no in-depth analyses were performed to characterize the variety of the Goražda masonry stone. The prevailing construction scheme of most walls is again the core-and-veneer technique; regular dressed stone courses have their joints precisely trimmed in different ways. In most areas of the building interior walls received particular attention by careful execution of the pointing in a raised ribbon style. The structural elements of the walls are beyond doubt from the time of origin, including the core infills of rubble stone and mortar. Most door and window reveals are also original, they are built from equal dimensioned stone ashlars. The barrel vaults in the interior have lunette caps from natural stone, except for those areas spanned and strengthened by steel girders, where ceilings and floors consist of tamped concrete.

Samples taken from mortar elements of the Goražda fortification during this survey comprised tamped concrete, bedding of the stone masonry, pointing on different types of stone masonry outdoors and indoors, outdoor renders as well as plasterwork on interior walls. The results of analysis reveal the use of dolomitic lime mortars for masonry beddings and interior plasters, while the selection of binders for masonry pointing seems to depend on the type of masonry: Roman cement was found in pointing mortars of interior and free-standing single shell stone walls, while Portland cement-based pointing was mainly detected on exterior masonry, repairs and on later added interior walls. Also the concrete samples proved the presence of Portland cement. More details are reported below (Fig. 4).

Fig. 4. Fortification of Goražda (constructed 1884–1886) (© ÖAW-ÖAI, C. Kurtze)

3.3 Kosmač

The fortification of Kosmač was built in a prominent position at 800 m above sea level above the coastal town of Budva. This complex formed the southernmost fortress of the Austrian-Hungarian Empire at the borders of Montenegro and (till 1878) the Ottoman Empire. The construction started in the 1840s and was completed in 1858. By the end of WW I in 1918, the Austrian troops retreated from Kosmač and blew up the fortification, thus laying the ground for the ruinous condition in which it appears till our days, even if it had been garrisoned again by Italian troops during WW II.

Due to its earlier date of erection, the position and the strategic disposition, the construction plan and appearance of the Kosmač fortress differ completely from the above addressed forts of Vermač and Goražda. The three-storey high building originally covered by a hipped roof is formed by two symmetrical wings, each with a semi-circular extension at its end. The central tower faces seaward while large arched gun ports face out over the hinterland (parts of the above description are taken from [10]).

The building was almost completely executed in natural stone in the form of uniform courses of dressed stone ashlars. Like the other two fortresses described above, the unrendered masonries were constructed in a core-and-veneer mode with two parallel shells filled with quarry stones and mortar. A greyish limestone of local origin was used. The walls in the interior of the building are comparatively less regularly textured, built with quarry stones and chips of stone and platy bricks filling the wide joints. It is thus assumed that the interior walls were plastered and probably whitewashed at least in parts, which is confirmed by a number of mortar residues preserved on the stone surfaces. All interior spaces are covered with barrel vaults made of a lightweight porous karst limestone; just the uppermost ceiling had been reinforced with a thick layer of tamped concrete. Most of this layer had been destroyed by the 1918 explosion, so that fragments of this concrete can be found in the debris which today covers the site, along with stones from the masonry and chips of slate from the former roofing (Fig. 5).

Apart from the stone material originating from the masonry and vaults, a few mortar elements were sampled from the site of Kosmač with its limited accessibility to the ruins. They comprised bedding mortars of the masonry along with interior plaster, pointing mortars from the exterior stone masonry, concrete from the collapsed ceiling reinforcement, and a mortar of unknown origin, the two latter samples collected from the debris on the ground. The laboratory analyses revealed feebly hydraulic lime as binder of bedding mortars, air lime in the interior plasterwork, and Portland cement as binder of the tamped concrete as well as of the exterior pointing. Given the early date of construction (1858), the use of PC was somewhat surprising since it appears just a few years after initiation of the first (natural) Portland cement in the Austro-Hungarian Empire (1856) [12].

More details on the mortars of Kosmač are given in 4.2.2.

Fig. 5. Fortification of Kosmač (constructed 1858) 2018 (© ÖAW-ÖAI, C. Kurtze)

4   Mortar Analysis

4.1   Methods

All materials sampled at Vermač, Goražda and Kosmač were analyzed by polarized light microscopy (PLM) and by scanning electron microscopy (SEM) combined with energy-dispersive X-ray spectroscopy (EDS). This approach was considered appropriate to obtain comprehensive information on the type of binders and the composition of mortars. The analyses were performed on polished thin sections made from vacuum-impregnated samples; a blue dye was added to the resin in order to facilitate the visibility of pores.

SEM observations were made under low vacuum to avoid irreversible coating with carbon. Under these conditions, no quantitative EDS analyses could be achieved, though the spectra were reliable enough to identify the nature of binder components and aggregates by their chemical composition.

4.2   Results

Aggregates in Mortars. In the present context, aggregates were not considered of primary significance for the classification of the mortars though they form important components of any mortar material and, by their petrographic nature, are of interest in respect to their provenance and the mixing regime. This section will therefore start with a general view on the aggregates found in the mortar samples, before dealing with the binders as the major item of their classification.

As a consequence of the topographic conditions of Montenegro with its rugged slopes and the seashore at close distance, local stone material was used to provide the aggregate of all types of mortar and concrete. The geological situation, with a clear predominance of calcareous rock outcrops, is reflected by the petrographic nature of the carbonate aggregates found in all of the samples. In addition, the presence of radiolarian-rich layers in most of the calcareous bedrocks adds important amounts of chert to the aggregate mix. It is evident that the high weathering resistance of this microcrystalline silica variety accounts for its frequent accumulation in sediments such as gravel and sand.

Considering Vermač Fort, the army report on its construction [4] also refers to the type and origin of aggregates supplied to the site, without further specifying for which type of mortar they were used. It distinguishes between gravel, coarse, and fine sand, and each class was obtained from a different source. It is of interest to classify the aggregate found in the samples in view of the written information.

Coarse gravel for the construction of Vermač was obtained from a nearby quarry by crushing the rock to the required size, usually up to several cm. This kind of aggregate is supposed to be unsorted. It is explicitly stated that attention was paid to use just sound and uncontaminated rock material and that therefore no careful washing was required. The angular chips obtained by crushing reflect the immediate geological environment of each fortress – compact limestone of locally differing lithological character. Silica layers of dark radiolarite are always frequently intercalated, though the form a minor constituent and were probably rather avoided by the stone workers. Such limestone chips (>1.5 cm) eventually mixed with river gravel were found in one concrete sample from Kosmač (Fig. 6), while the concrete of Goražda contains much smaller (<5 mm) subangular gravel of chert with just occasional carbonate gravel.

It should be noted that all micrographs in the Figures of this section are scaled to the same size of magnification in order to make them comparable.

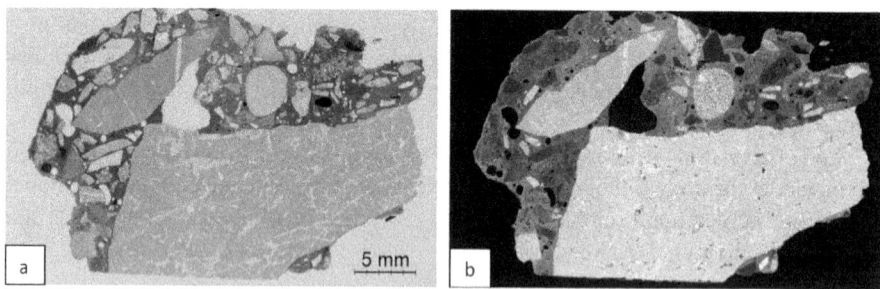

Fig. 6. Limestone chips as coarse aggregates along with a few river or sea gravels (rounded grains) from carbonate and silica (chert) in a PC-based tamped concrete from Kosmač; thin section scan at plane polarized light, PPL (a) and crossed polarized light, XPL (b).

Coarse sand as understood by the authors of [4] includes in fact gravel in the size of a few mm up to about 1 cm in diameter. According to the report, it was taken from some torrents near the coast, at least as far the Vermač fort is concerned. This river sand appears usually well rounded (limestone) to (sub)angular (chert); depending on the source rock of the sediment, it is in many cases chert dominate here thanks to its higher resistance.

River gravel as aggregate was observed in many of the mortar samples from all three sites. They can be attributed either to bedding mortars of the rubble core in cavity walls, or to indoor plaster mortars (Fig. 7), or, as mentioned above, to the tamped concrete of Goražda (Fig. 8, with predominance of chert components).

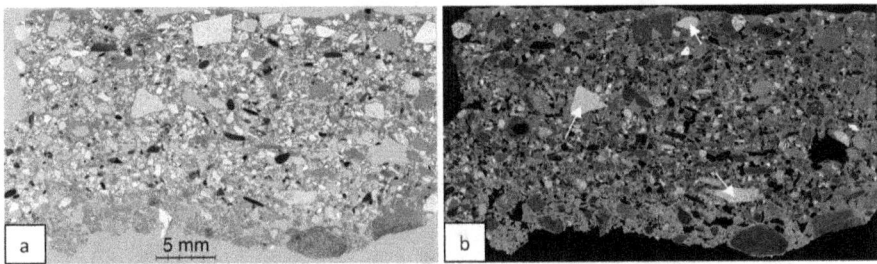

Fig. 7. Unsorted, mostly angular to subangular river sand of limestone (white arrows) and chert (red arrows) in a lime-based indoor plaster from Kosmač; thin section in PPL (a) and XPL (b). (Color figure online)

Fine sand for the Vermač Fort construction was taken from the beaches near Budva as stated by the army report [4]. It can be assumed that the same holds for the other two buildings. The Vermač report mentions that this sand, due to its salt content, was deposited outdoors for a year in order to have the salts washed out by percolating rain water. This beach sand is quite characteristic by the roundness of aggregates and a somewhat broader petrographic composition than the coarser inland counterparts.

Fig. 8. River sand and gravel from subangular chert with just occasional well-rounded limestone grains (centre) in a PC-based tamped concrete of Goražda; it is likely that chert-rich sand was intentionally selected to make this concrete harder; thin section in PPL (a) and XPL (b).

A sediment sample collected from a beach near Budva matches the sand aggregate of the historic mortars to a high extent (Fig. 9): again, limestone and chert form the predominant components of this moderately well sorted sediment, but the former appears spherical with high roundness, while the latter is also isometric though just subangular. Such "fine sand" forms the aggregate in most of the pointing mortars studied, while in some other mortars it was probably mixed with "coarse sand" for a broader grain size distribution (Fig. 10).

Mortars Classified by Binder Types. Air lime mortars were consistently found in samples from interior plasterwork in all of the fort buildings. In the case of Goražda,

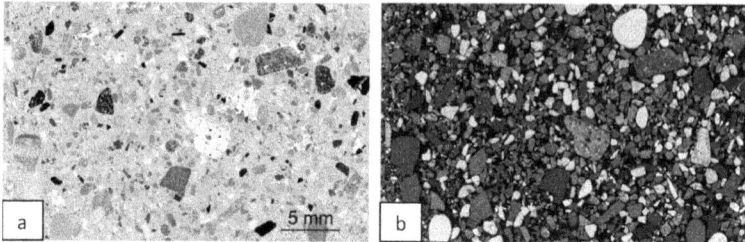

Fig. 9. Sea sand from Budva beach, composed of limestone and chert components; according to the army report on Vermač [4], such sand was used as "fine sand" for construction purposes; thin section in PPL (a) and XPL (b).

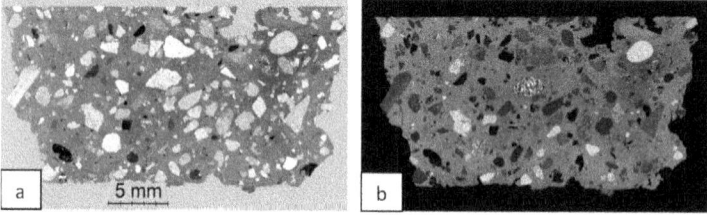

Fig. 10. Fine sand presumably from the seashore of the Budva area, composed of limestone and chert, in a RC-based indoor pointing mortar of Goražda; thin section in PPL (a) and XPL (b).

Mg-rich inclusions exhibiting a spotted appearance clearly indicated the dolomitic nature of the lime binder (compare e.g., the lumps in Fig. 11 to those in Fig. 12), hence the origin of the raw material differs from the other two fortresses despite the close vicinity to Vermač. The same dolomitic lime constitutes the bedding mortar of the stone masonry of Goražda, while the bedding mortars of Vermač are based on pure lime - occasional lime lumps in these mortars point to a quick or "dry" process of slaking [8] in accordance with the fact that no lime pit is mentioned in the otherwise rather detailed report on the construction of this fort. Kosmač, on the other hand, has slightly hydraulic bedding mortars.

As indicated above, a group of samples which might be related to the intentional use of reactive chert as aggregate can be found in some mortar samples from Kosmač. These mortars, particularly rich in chert, appear as bedding, or core filling mortars, respectively, of the stone masonry. It can be assumed that prolonged conditions of moisture in the specific positions of the wall have contributed to the development of hydrated rims around the chert, as illustrated by Fig. 13 and 14. Thus, even if these mortars are extremely leached, they appear as extraordinary hard and strong.

c = mortar fracture face.

Roman cement mortars were not found in any of the samples from Kosmač, despite of the fact that these binders had their heydays in Austria-Hungary about a decade before the fort was built. The apparent lack of Roman cement in our samples there may be due to the ruinous conditions of that fortification, with a limited accessibility to several structures restricting the possibility to sample all types of building elements. Also in Vermač, where the use of minor amounts of Roman cement from Split is known

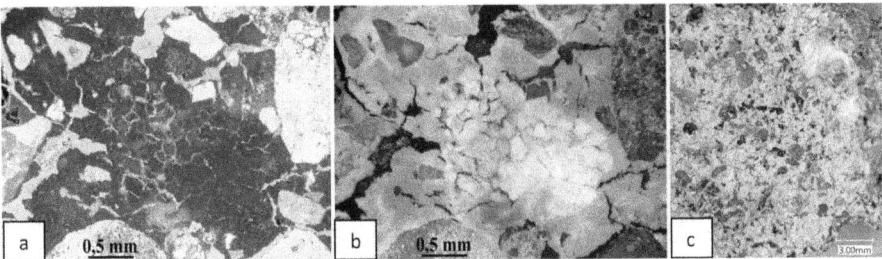

Fig. 11. Mg-rich lime lump in a dolomitic lime bedding mortar of Goražda; thin section in PPL (a) and XPL (b); c = mortar fracture face.

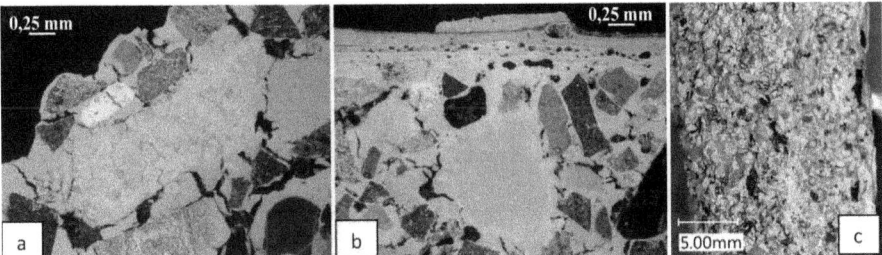

Fig. 12. Binder-related particles in indoor lime plaster samples from Vermač where quicklime was obtained from fossil limestone calcined by local producers in Korčula and short-term slaked on the construction site; residual structure of the raw feed stone (a), and undispersed lime lump (b). Thin section at transmitted light, dark field; in the diffuse light of the dark field mode, microporous areas appear bluish whilst white and compact areas show bright;

although no specific application is mentioned, no such binder was traced in any of the samples. Its use was probably restricted to some minor building elements which have not been studied yet.

However, in Goražda, the largest of the fortresses from which the most samples were taken, Roman cement mortars were found, as pointing mortars from interior and exterior masonries surfaces (see Fig. 15). The typology of binder related nodules found points to a relatively low burned product, where the higher burned portion present in any Roman cement of that period had most probably been removed in the factory.

An exception to Roman cement in the Goražda pointing was found in a sample from a retention wall constructed with stone ashlars, where the otherwise similarly dressed pointing was made with Portland cement.

Portland cement mortars proved to be prevalent in all three of the forts under investigation: in 1890s' Vermač they are in all indoors and outdoors pointing mortars and at least some of the exterior renderwork, while in 1880s' Goražda Portland cement was used in parts of the exterior masonry pointing and renders on concrete, which itself revealed to be based on this binder. Also Kosmač exterior pointing made use of Portland cement, although this may be due to maintenance interventions possibly in times of the reinforcement of the roof in the early 20th century.

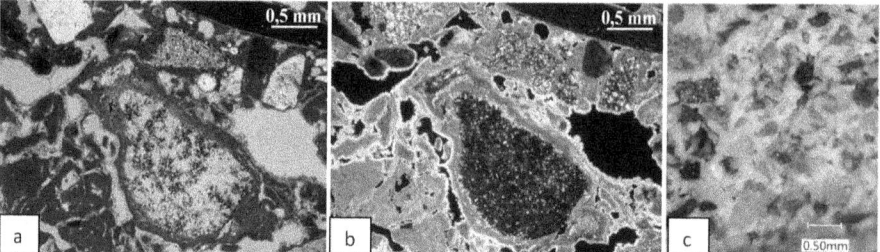

Fig. 13. Reactive chert aggregate in a porous lime-based core filling mortar of the Kosmač stone masonry; thin section in PPL (a) and XPL (b); c = mortar fracture face.

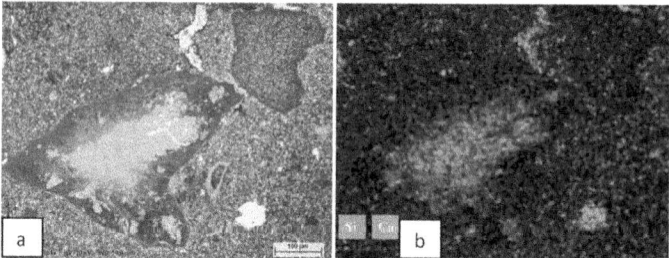

Fig. 14. SEM micrograph of a chert aggregate with leached outer reaction zone in the same mortar as Fig. 13; the binder matrix reveals tiny bright spots of secondary calcium carbonate, a characteristic feature of aged hydraulic binders [13]; SEM-BSE (a) and EDS spot map of Si and Ca (b).

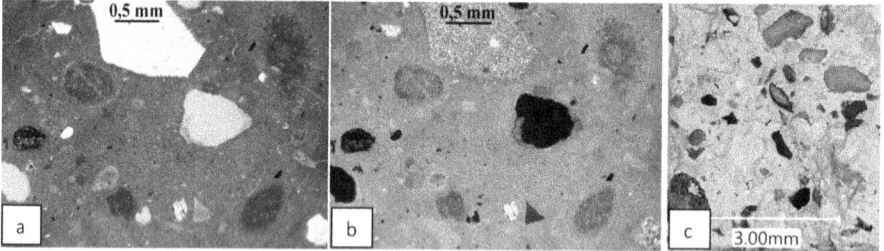

Fig. 15. Binder related nodules characteristic of Roman cement in an exterior pointing mortar of the Goražda stone masonry; thin section in PPL (a) and XPL (b); c = mortar fracture face.

By microscopy, all the PC-mortars studied clearly revealed their early age of production by the coarseness of clinker residues in the binders, and by their internal phase structure, such as coarse aluminoferrite and calcium silicates pointing to the use of vertical shaft kilns with prolonged times of residence and slow cooling (Fig. 16). Moreover, the use of a single, natural raw feed to calcine the cements is suggested by the range of varying clinker microstructures containing insufficiently calcined residues of the raw material [11] in most of the samples (Fig. 17). This means of production was quite common for the majority of cement plants in the territory of the Empire thanks to the frequent occurrence of appropriate marlstone in the Alpine and Dinaric regions. In particular, Alois Kraft in Kirchbichl/Tyrol, the only known Austrian producer in the 1850s [12], as well as Gilardi-Betizza in Split, who had this binder available since 1880 [7] and would have been an ideal supplier through boat transport to Kotor, were offering natural Portland cement.

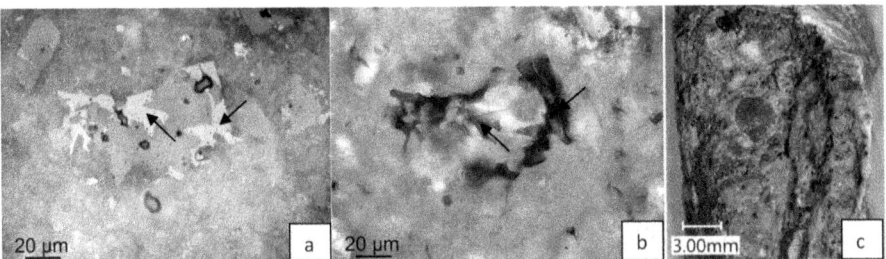

Fig. 16. Coarse PC clinker residues in a late 1890s pointing mortar of a stone masonry in Vermač; due to an improved quality of calcination, the variety of clinker types is less pronounced here than in the earlier PC clinkers of Kosmač; coarse interstitial phases (mostly calcium-aluminoferrite, see arrows), (a, b) point to low cooling rates of a shaft kiln; thin section at incident light (a) and PPL (b); c = mortar fracture face.

Given the generally early age of the fort buildings in respect to the timeline of the production and use of Portland cement in the Austrian-Hungarian Empire, which started in 1856 and developed just slowly till the 1890s in steadily successful competition to Roman cement, it seemed unlikely to find a significant number of Portland cement mortars in all three of the studied sites, in particular in the earliest one, Kosmač from 1858. No use of this then modern binder was reported so far from any Austrian building of the 1850s, even if some amounts of Portland cement were imported from England or Germany, which has likely occurred for a few applications. However, it seems improbable that the army of a big Empire with plenty of production plants would have imported building materials from abroad, while it seems highly probable that they were among the first users of novel materials at a time when civic architecture and engineering was still hesitant to use them.

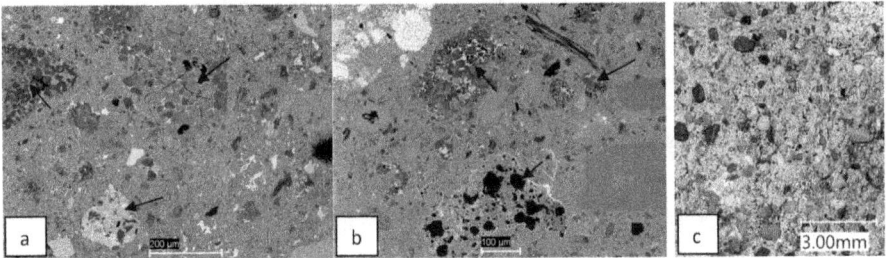

Fig. 17. PC binder matrix of a pointing mortar on an exterior stone wall of Kosmač, both micrographs (a and b) show clinker residues of strongly varying internal texture and composition, a feature characteristic of natural Portland cements; SEM-BSE (a, b), c = mortar fracture face.

5   Conclusions and Outlook

The versatile use of different mineral building materials in the construction works of the 19th century Austro-Hungarian army forts of Montenegro sheds light on the conscious selection of mortar constituents and recipes by the military engineers. Through our approach to identifying binders and aggregates through microscopic laboratory analyses of samples, paralleled by information extracted from the available written sources, it was shown that the most novel cementitious products at their time were employed only for certain elements of construction. This refers in particular to Portland cements, the newest and most expensive of all binders used: their application was, on the one hand, in concrete members of a fortress supposed to carry heavy loads or withstand the enemy's artillery, or to those areas exposed to the heavy autumn and winter rainfalls, i.e. the joints of the otherwise compact exterior stone faces of the masonries. In contrast, the bulk of mortars used to fill the core cavities of these masonries were prepared with much cheaper, locally available air lime. Roman cement, in turn, was only used in limited quantities, quite different from many civic engineering constructions of that time, where such low calcined natural binders were even employed in concretes.

It has to be acknowledged, however, that the above general conclusions are only preliminary since they are based just on a limited amount of analytical data on just three objects surveyed so far by the authors.

In order to ensure the continuation of the successfully started project, an application for further funding was submitted and recently approved. The planned future research, carried out in close cooperation with Montenegrin colleagues of the University of Montenegro in Podgorica, locally coordinated by Prof. Ilja Lalosević, will focus not only on the fortification architecture, but also on main settlements of the Bay of Kotor area, in order to study the Habsburg defence systems in context to the historical setting of the bay, its surrounding region and the whole Austro-Hungarian Empire. This will also include aspects of architectural transition from the Venetian to the Habsburg settings by necessary archival and library research in Montenegro and Austria. The link between history, architecture and material sciences is believed to contribute to a better understanding of the historical/architectural heritage of the Bay of Kotor.

Acknowledgements. The Ministry of Culture of Montenegro supervised planning, organisation, and implementation of the campaign from the outset. Further, colleagues from the Centre for Conservation and Archaeology in Cetinje and from the Directorate for the Protection of Cultural Property in Montenegro supported the fieldwork. We would also like to thank the Austrian Embassy in Montenegro for its financial support in the year 2018.

Thanks are due to Christian Kurtze who performed the 3-D laser scans of the buildings and to Nicola Math for the drawings based on them, as well as to Jovo Miladinović for the discussion of the complex historical background.

References

1. Rolf, R.: Festungsbauten der Monarchie: Die k.k.- und k.u.k.-Befestigungen von Napoleon bis Petit Trianon. Prak, Middelburg (NL), 2011, pp. 69–75 and 242–259 (2011)
2. Pachauer, V.K.: Blaues Meer und dunkle Bauten. eine umfassende baugeschichtliche Bestandsaufnahme und kulturhistorisch-denkmalfachliche Bewertung der Befestigungsanlagen des ehemaligen k. und k. Kriegshafens Cattaro. Unpublished thesis, TU Graz (2008)
3. Pavićević, R.R.: Werk: austro-ugarske tvrđave u Crnoj Gori, Nova pobjeda, Herceg Novi (2019)
4. N.N.: Statistischer Baubericht Nr. 6 über den Bau des Forts Vermač bei Cattaro. Austrian State Archives – Kriegsarchiv, Bibliothek, Fd 15 1/8, Wien (1898)
5. N.N.: Statistischer Baubericht Nr. 6 über den Bau der Batterie Vermač bei Cattaro. Austrian State Archives - Kriegsarchiv, Bibliothek, Fd 15 1/8, Wien (1890)
6. http://museums.si/sl-si/Domov/Prispevki/Prispevek?id=20063. Last visited 4 Feb 2022
7. La prima fabbrica dalmata di cemento Portland Gilardi $ Bettiza-Spalato. https://www.gilard ibettiza.it/la_fabbrica.html. Last visited 24 Nov 2022
8. Hughes, J.J., Leslie, A.B., Callebaut, K.: The petrography of lime inclusions in historic lime based mortars. Annales Geologiques des pays Helleniques, Edition Speciale **XXXIX**, 359–364 (2001)
9. https://www.eurologisch.at/docroot/waehrungsrechner/#/. Last visited 02 2022
10. https://en.wikipedia.org/wiki/Fort_Kosma%C4%8D. Last visited 03 2022
11. Pintér, F., Gosselin, C.: The origin, composition and early age hydration mechanisms of Austrian natural Portland cement. Cem. Concr. Res. **110**, 1–12 (2018)
12. Kölblinger, F.: Darstellung der wirtschaftlichen und technologischen Entwicklung der Zementindustrie unter Berücksichtigung der Gmundner Zementwerke Hans Hatschek Aktiengesellschaft. Diploma Thesis, Wirtschaftsuniversität Wien (cited in: Riepl, F. Die wirtschaftliche und technologische Entwicklung der Zementindustrie unter besonderer Berücksichtigung der Verdienste von Hans Hauenschild. Master Thesis, University of Vienna, 2008) (1983)
13. Weber, J., Baragona, A., Pintér, F., Gosselin, C.: Hydraulicity in ancient mortars: its origin and alteration phenomena under the microscope. In: Copuroglu, O. (ed.) Proceedings of the 15th Euroseminar on Microscopy Applied to Building Materials, Delft, The Netherlands, pp. 147–156, 17–19 Jun 2015

Limewashes with Vegetable Oils: Water Transport Characterisation

Cristiana Nunes[1(✉)], Paulina Faria[2], and Nuno Garcia[2]

[1] Institute of Theoretical and Applied Mechanics of the Czech Academy of Sciences, Prague, Czech Republic
nunes@itam.cas.cz

[2] CERIS, NOVA School of Science and Technology, Nova University of Lisbon, Lisbon, Portugal

Abstract. Limewashes have been used as finishing coats for walls since ancient times. Its protective, aesthetic, antiseptic properties and cost-efficiency are the ground for its worldwide application. The main drawback of lime-based paints is their low durability towards the action of water, particularly wind-driven rain. Additives that grant water-repellent properties have been added to these paints to overcome this issue. Among these additives, vegetable oils have been reported worldwide in ancient documents. In this work, three vegetable oils have been selected based on their composition and promising results in previous studies, global availability, and cost-efficiency: rapeseed, sunflower, and sunflower oil with high oleic acid content. Additionally, a commercial water-repellent lime putty with the addition of olive oil was included to prepare a limewash and compare it with the lab-prepared paints. Two types of stone with very different porous structures were used as substrates to compare the effect of the paints on their water transport properties. The substrate with higher porosity and wider pores showed promising results in terms of water-repellence and drying. In contrast, the stone with lower porosity and fine pores did not show good results. Based on this study, suggestions for further research to improve the performance of the paints in substrates with low porosity and narrow pore size distribution are given.

Keywords: Lime · Water-repellent additives · Hydrophobicity · Damp · Drying · Porous materials

1 Introduction

Limewash or whitewash is a dilution of lime putty with water that has been used as a paint since Classical times. It has been commonly used as a finishing coat worldwide for thousands of years, thanks to the global geological availability of limestone. However, little research has been made about this type of coating [1]. With the advent of the Industrial Revolution, these traditional paints were substituted by synthetic ones. However, synthetic paints are not always compatible with traditional porous materials, mainly because they tend to lower their permeability, thus worsening dampness problems and, consequently, the need for new interventions within a short-term period [2,

V. Bokan Bosiljkov et al. (Eds.): HMC 2022, RILEM Bookseries 42, pp. 150–161, 2023.
https://doi.org/10.1007/978-3-031-31472-8_12

3]. Additionally, damaged synthetic paints need to be entirely removed before applying a new one. In the case of lime-based paints, only the damaged parts need to be removed before reapplying new coats. In Portugal, the traditional annual maintenance with lime-wash overpaints was mostly done during spring (mild temperatures and few rainy days); more than a yearly maintenance intervention, it was also a social practice passed by generations [4].

In Portugal and the Czech Republic, this traditional simple "home-made" paint by local populations was mostly lost during the 60 s [5, 6]. Nowadays, the value of these traditional paints is regaining recognition (Fig. 1) thanks to their low cost, antiseptic properties upon application, compatibility with old and new constructions made of porous building materials, and eco-friendly nature. The final finishing is matte and highly permeable to water. Recent studies have reported that lime paints can enhance the drying of some porous materials [7–10]; this effect has been assigned to the increment of the effective evaporative surface of the substrate.

Fig. 1. Limewashing in S. Miguel island in the Azores during the spring of 2019.

The main drawback of using lime-based paints to protect walls is their low durability towards the action of water, particularly wind-driven rain, which mainly results in the loss of cohesion of the paint, e.g., by chalking and scaling [3]. To overcome this issue, additives that can improve cohesion and grant water-repellent properties have been added to these paints since ancient times. Among these additives, vegetable oils have been reported worldwide in old documents to improve the durability towards the action of water by imparting water-repellence to the paint [1]. The oils were naturally selected by their local availability, e.g., linseed and olive oil in Europe [6, 11], Tung oil in China [12], Areca nut in India [13].

In this work, three types of vegetable oils have been selected as water-repellent additives for improving the durability of limewashes towards the action of water while ensuring that the paint does not suppress the drying of damp porous materials. The vegetable oils were selected based on their composition, global availability, and cost-efficiency: rapeseed oil, sunflower oil, and sunflower oil with high oleic acid content. The paints were prepared by mixing 1.5 wt.% of oil with pure lime putty (by mass). Moreover, a commercial lime putty with the addition of olive oil was also included to prepare a limewash and compare it with the lab-prepared paints. Two types of natural stone with a very different porous structure were used as case study substrates to compare the effect of the paints on the water transport properties.

2 Experimental Campaign

2.1 Materials

Lime Putties and Oils
Two types of lime putties classified as CL90 according to EN459-1 [14] have been used to prepare limewashes: (1) a pure lime putty prepared by storing air lime hydrate in powder (Čertovy schody, Czech Republic) under water for 1 year, and (2) a water-repellent lime putty (Fradical, Portugal) stored under water for over 5 years. According to the producer, the Fradical lime putty is prepared by adding olive oil during the lime slaking process, following a traditional way of preparation. This lime has been selected to assess its performance as a limewash because its water-repellence properties are assigned to the addition of a vegetable oil. To the authors' knowledge, the Fradical lime has never been studied when used to prepare limewash.

Three types of vegetable oils have been selected to add to the pure lime putty: rapeseed oil, sunflower oil, sunflower, and sunflower oil with high oleic acid content (Sovena Oilseeds, Portugal). Rapeseed oil was selected based on the promising results obtained in a previous study focused on the analysis of the microstructure and water-repellence of oils added to lime pastes [15]. Raw sunflower oil and sunflower oil with high oleic acid content were selected based on another study in which these oils were added to lime mortar [16]. Additionally, these oils were selected given their worldwide availability and low cost. The fatty acids composition of the oils provided by the producer (Sovena Oilseeds, Portugal) is given in Table 1.

Table 1. Component fatty acids present in the oils (in wt.%) provided by Sovena Oilseeds Portugal.

Oil	Palmitic C16:0	Stearic C18:0	Oleic C18:1	Linoleic C18:2	Linolenic C18:3
Rapeseed	4.60	1.60	63.90	18.50	8.20
Sunflower	6.80	3.10	38.90	47.90	0.10
Sunflower high oleic	3.82	3.17	84.80	5.84	0.09

Substrates. Two types of natural stone with different porous structure were selected as substrates to study the water transport behaviour of the limewashes. Both stones are from the Czech Republic and have been widely used in architectural heritage.

Mšené sandstone, commonly known as Prague stone, has high porosity and pore size; ca. 30% open porosity and main pore size diameter centred around 27 μm [9]. Opuka is a marlstone with a lower porosity and much smaller pore size than Mšené; ca. 24% open porosity and main pore diameter centred around 0.2 μm [9].

Tablet specimens with 10 mm × 50 mm × 50 mm were used to determine the contact angle with water drops. Cubic specimens with 50 mm edge were used for the water absorption and drying tests. Four sides of the cubic specimens were sealed with epoxy resin; in the case of Opuka stone, which has visible sedimentary bedding planes, the resin was applied on the sides perpendicular to the bedding planes.

2.2 Limewash Preparation and Application to the Substrates

The content of lime in the putties was gravimetrically determined by weighting small portions (three replicates of ca. 50 g) of decanted fresh lime putty and drying at 60 °C to constant mass. The mass of lime in both putties was ca. 50 ± 3 wt.%. This value is close to the one obtained by Faria et al. [17] with lime putty produced with finely crushed quicklime stored under water for 10 months (53 wt.% of lime) and not far from a lime putty made with quicklime after 16 months of storage under water (60%). In the case of pure limewash and Fradical limewash, water was added to the lime putties to obtain lime suspensions with 20 wt.% of lime and 80 wt.% of water. In the case of the limewashes with additives, 1.5 wt.% of each oil (to the lime weight) was added to the lime putty and manually mixed for 5 min with a spoon. Afterwards, water was added and the suspension was mixed for further 5 min.

Four limewash coats were applied with a brush (with soft to medium bristles) criss-crossed on one of the specimens' surfaces. In the case of specimens treated with lime-washes with oils, the first two coats applied were of pure limewash to serve as a primer for the limewashes with additives. Before applying the first limewash coat, the stone surface was wetted with water with a brush to facilitate the penetration and adherence of the limewash and avoid crazing of the paint. The limewash coatings were applied with a minimum time interval of 24 h between coats. All specimens (uncoated and coated) were left to cure for 8 days in a room with 17 ± 3 °C and 70 ± 5% relative humidity (RH).

2.3 Testing Procedures

The specimens were dried at 40 °C to constant mass after curing and the dry mass was registered before testing. The tablet specimens were used for determining the static contact angle according to EN 15802 [18] using a goniometer. Three water drops were measured on each specimen using 4 replicates.

The cubic specimens were used for the water absorption by capillarity test using four replicates of each specimen type; the test was conducted according to EN 15801 [19]. Following the water absorption by capillarity test, the specimens were further immersed in water until constant mass to ensure saturation and begin the drying test. The drying test was performed according to EN 16322 [20]. Due to a technical problem with the acclimatisation system in the room, the drying test was performed at different thermohygrometric conditions than those described in the standard, hence at $17 \pm 3\ °C$ and $70 \pm 5\%$ RH.

3   Results and Discussion

The results of static contact angle with water drops are given in Table 2. The uncoated specimens and those coated with pure limewash absorbed the water drop upon 10 s of contact with the surface. The values obtained show a very high standard deviation, probably due to improper homogenisation of the additives in the mixes. Therefore, new limewash mixtures were prepared by mixing the lime and oil suspensions mechanically using a magnetic mixer at high speed.

The results of the contact angle of the limewashes prepared via mechanical mixing are also given in Table 2, except for the Fradical limewash that was mixed with the additive during slaking. The values show that mechanical mixing reduces the standard deviation and increases remarkably the contact angle of the limewashes applied to MS stone, indicating that the mixes prepared manually were not adequately homogenised. The highest increment was registered in the case of LSO (ca. 138% increment), followed by LS (125%) and LR (92%). The cubic specimens used in the water absorption by capillarity and drying tests were treated with manually mixed suspensions of limewashes. Therefore, the results can be somewhat biased by the improper homogenisation of the additives in the mixtures. This must be taken into account when interpreting the results presented next.

Table 2. Contact angle with water drops measured on the limewashed stone surfaces using manual or mechanical mixing - average of 3 measurements performed in 4 replicates of each limewash type ($\pm$ standard deviation).

Stone/Limewash[1]	Contact angle Manual mixing	Contact angle Mechanical mixing
MS/LR	52.3 ($\pm$23.4)	100.0 ($\pm$14.4)
MS/LS	19.1 ($\pm$16.6)	43.0 ($\pm$1.3)
MS/LSO	28.7 ($\pm$28.8)	68.3 ($\pm$13.6)
MS/LF	138.7 ($\pm$2.7)	*
OP/LF	131.7 ($\pm$4.8)	*

[1] The specimens not listed absorbed the water drop within 10 s after the drop hit the surface
* Not determined (additive added during lime slaking)

All oils imparted water-repellence to the limewash applied to the most porous stone (MS). In the case of the less porous stone (OP), only the Fradical limewash granted water-repellence. In fact, the Fradical limewash produced hydrophobic coats on both stones, i.e., the water drop contact angle was higher than 90°. Rapeseed oil also imparted hydrophobicity, but only to MS stone coated with the mechanically mixed suspension.

Two previous studies [15, 21] comparing the effect of several oils on the degree of water-repellence in lime pastes and Portland cement mortars, respectively, have indicated that oils with a high content of monounsaturated fatty acids are more effective in imparting water-repellence compared to oils with high amount of polyunsaturated fatty acids. In both studies, rapeseed oil, which contains ca. 65% of oleic acid, was considered a good additive for imparting water repellence to the mentioned materials. Therefore, this study comprised oils with a high oleic acid content (Table 1). Though sunflower oil with high oleic acid content has the highest amount of monounsaturated fatty acids, the degree of water-repellence granted to the limewash is much lower than that granted by rapeseed oil. Based on the fatty acid composition of the oils given in Table 1, a small amount of polyunsaturated fatty acids (ca. 20% linoleic and 10% linolenic) can also play a synergistic role in granting water-repellence to lime.

Fradical lime putty contains olive oil, which is known to contain a high content of oleic acid (ca. 78%) and a small amount of linoleic acid (ca. 7%) [22]. However, the amount of oil added to the Fradical lime during slaking is unknown, and this can significantly influence the degree of water-repellence.

The water absorption by capillarity curves of each of the most representative specimen type is given in Fig. 2, and the capillarity coefficients calculated from the curves are given in Tables 3 and 4 for Mšené and Opuka stones, respectively. The curves of the uncoated specimens (R) show an initial linear section that develops into an asymptote at a defined interval, indicating that the stones are homogenous and have a well interconnected pore network. Pure limewash reduces the water absorption rate in the most porous stone (MS) compared to the uncoated specimens. The reduction is slightly enhanced with the addition of oils, except in the case of LSO, which shows a similar absorption rate as pure limewash (Fig. 2.a); as mentioned, this result may be due to the improper mixing of the oil with the lime suspension. The Fradical limewash is the most effective in reducing the water absorption rate, followed by the limewashes with sunflower (LS) and rapeseed (LR) oils. In the case of Opuka stone, only the LF and LS limewash clearly reduced the water absorption rate (Fig. 2.b). The total water absorbed is lower for OP coated specimens compared to MS coated ones, although that value is similar for all coats applied to the same substrate.

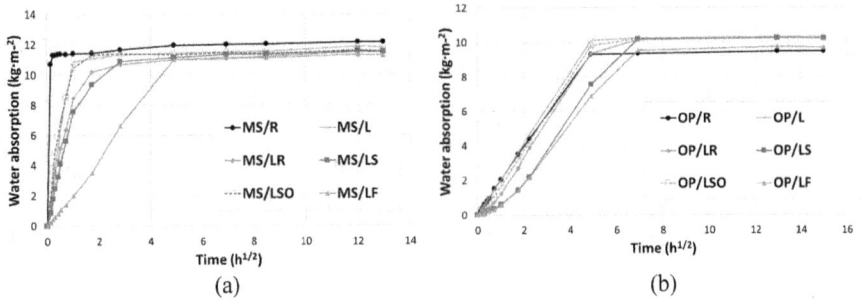

Fig. 2. Water absorption by capillarity curves of: (a) Mšené stone; (b) Opuka stone.

Table 3. Water absorption by capillarity coefficient, drying rate in stage I, and drying index determined in Mšené stone; average of 4 specimens (± standard deviation).

Stone/Limewash	Coefficient of water absorption by capillarity (kg·m^{-2}·h$^{-1/2}$)	Drying rate Stage I (kg·m^{-2})	Drying index (-)
MS/R	51.57 (±1.04)	0.045 (±0.004)	0.32 (±0.01)
MS/L	11.27 (±0.42)	0.053 (±0.007)	0.26 (±0.03)
MS/LR	8.25 (±0.73)	0.050 (±0.003)	0.27 (±0.02)
MS/LS	3.85 (±0.31)	0.057 (±0.005)	0.25 (±0.01)
MS/LSO	13.36 (±2.61)	0.059 (±0.006)	0.23 (±0.03)
MS/LF	1.81 (±0.50)	0.064 (±0.001)	0.22 (±0.01)

Table 4. Water absorption by capillarity coefficient, drying rate in stage I, and drying index determined in Opuka stone; average of 4 specimens (± standard deviation).

Stone/Limewash	Coefficient of water absorption by capillarity (kg·m^{-2}·h$^{-1/2}$)	Drying rate Stage I (kg·m^{-2})	Drying index (-)
OP/R	1.95 (±0.10)	0.025 (±0.001)	0.42 (±0.02)
OP/L	1.99 (±0.12)	0.027 (±0.001)	0.43 (±0.01)
OP/LR	1.88 (±0.16)	0.027 (±0.001)	0.43 (±0.04)
OP/LS	1.54 (±0.11)	0.026 (±0.001)	0.44 (±0.01)
OP/LSO	2.00 (±0.09)	0.025 (±0.001)	0.44 (±0.02)
OP/LF	1.16 (±0.28)	0.026 (±0.001)	0.42 (±0.01)

During the water absorption by capillarity test with OP stone, peeling of the paint layer was observed in one specimen of L and LR (Fig. 3.a). Chalking of the paint layer occurred in all OP specimens and more predominantly in LF and LSO specimens (Fig. 3.b), probably due to abrasion caused by patting the specimens dry with a wet cloth during the capillarity test to remove the excess water along weighing. The observed damage during the capillarity test foresees the low durability of the paints applied to Opuka stone after prolonged contact with water (72 h). These alterations are likely to influence the results of further tests performed on the same specimens; in this study, specimens used in the capillarity test were subsequently used to evaluate their drying behaviour, so this aspect was given attention in the interpretation of the results of the drying test.

The drying rate in stage I and the drying index determined in Mšené and Opuka stones are shown in Tables 3 and 4, respectively. The comparison of the drying rate and drying index obtained in both stones is given in the graphs in Fig. 4. The drying of a porous material saturated with water generally comprises two stages: stage I, in which

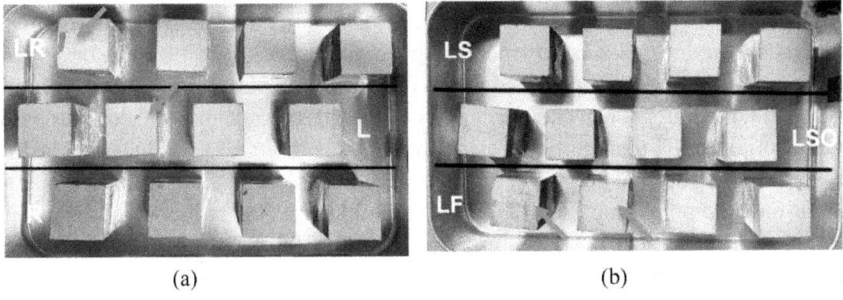

(a)   (b)

Fig. 3. Alterations observed on Opuka limewashed stone surface after the water absorption by capillarity test: (a) aspect of R, L, and LR specimens showing peeling of the paint layer on one specimen coated with L and LR limewashes (indicated with blue arrows); (b) aspect of LF, LS, and LSO specimens showing wearing of the paint, most evident on two LF specimens (indicated with blue arrows).

the water front is at the surface of the material and the supply of water to the surface is steady and governed by capillary water transport; and stage II, when the supply of water to the surface can no longer compensate for the water loss by evaporation, so the water front recedes into the material and water migrates mostly via vapour diffusion. The drying rate was calculated from the Stage I slope of the curve. The drying index was calculated using the entire drying period, i.e., until the specimens reached constant mass (552 h in MS and 792 h in OP). The drying index reflects the global drying kinetics (integral of the drying curve); the higher the drying index, the slower the drying.

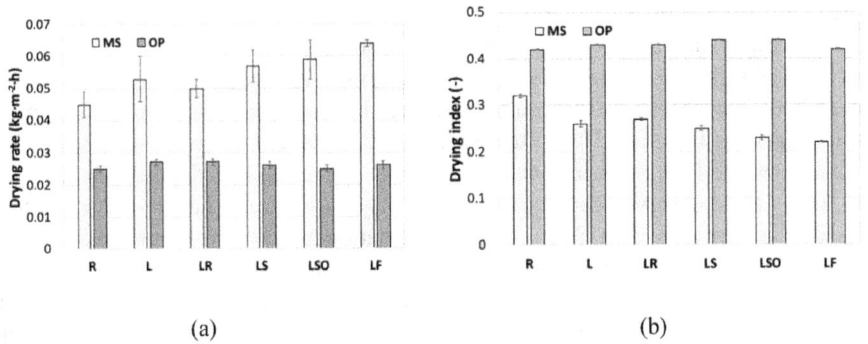

(a)   (b)

Fig. 4. Drying test results of Mšené and Opuka stones: (a) stage I drying rate; (b) drying index.

As expected, the stone with higher porosity and pore size distribution (MS) dries faster than the less porous one (OP). In MS stone, all limewashes promoted a clear increment of the drying rate and reduction of the drying index. As mentioned, the acceleration of the drying rate in porous substrates coated with lime-based paints is related to the increment of the effective evaporative surface and has been reported in other studies using MS stone [9, 10]. The increment was highest in the case of the Fradical limewash (LF), which also showed the highest degree of water-repellence. This is in contrast with previous studies with limewashes with linseed oil [9, 10] that showed that the higher the degree of water-repellence (by increasing the oil dosage), the slower the drying rate. However, LF limewash was prepared with a different lime putty (lime slaked with olive oil stored under water for over 5 years). It is generally known that slaked lime putties have higher plasticity than putties prepared with dry hydrate [23] and that the plasticity and workability improve upon long-term storage under water [24]. Though this effect depends on the type of lime used [25], it is likely related to the reduction of the calcium hydroxide particles' size [24]. Therefore, it is reasonable to assume that the calcium hydroxide particles in the Fradical lime putty have probably smaller sizes than those in the putty prepared from dry hydrate, not only assigned to the preparation process but also to the much longer storage time under water. The smaller size of the lime particles is likely to increase the effective evaporative surface of the porous substrate more effectively. This might be why the Fradical limewashed MS specimens show a higher drying rate than those coated with pure limewash made of dry hydrate, despite LF's hydrophobic properties.

The drying rate of MS specimens coated with limewashes with sunflower oils also registered an increment of the drying rate with respect to the reference (L). However, their water repellence degree is much lower than LF (Table 2), and the high standard deviation is within the reference values. In the case of LR, the drying rate is slightly reduced compared to the pure limewash (L), but it is still above the drying rate of the uncoated stone (R).

In the case of OP stone, the differences registered in both the drying rate and index are virtually negligible compared to those registered in MS stone, which is in line with previous studies using the same stones [9, 10]. The very fine porous structure of Opuka stone (main pore size around 0.2 μm) hinders the penetration of the lime particles into the stone matrix. Scanning electron microscopic observations have shown that limewashes prepared with a similar lime putty do not bond well to the surface of Opuka stone [10], which explains the detachment of the paint during the capillarity test. As the paint coats remain at the surface, they create a thin layer (ca. 150 μm) that is more porous than the underlying stone material [10], so the drying kinetics are not much altered. In the case of MS stone, the lime particles penetrate into the matrix up to a depth of ca. 200 μm [10]. Despite the observed damage to the paint layer observed in a few OP specimens after the capillarity test (Fig. 3), the standard deviation is very low, which reassures that the limewashes have little effect on the drying behaviour of this stone.

During the drying test, one MS specimen coated with LR and LS limewashes developed mildly visible moulds (Fig. 5), which is probably related to the presence of vegetable oils and foresees the biosusceptibility of these coatings, especially if applied in interior walls of low ventilated rooms. It should be taken into account that the drying test was conducted in a room with low intensity of artificial light or sunlight and at very high RH (70 $\pm$ 5%), which can support the development of this type of biological colonisation. Nevertheless, tests promoting biological colonisation and outdoor exposure tests should be considered in the future when evaluating coatings with organic additives. Additionally, according to a recent review on ageing tests [26], tests dealing with biological colonisation are strikingly the less frequent ageing tests performed in lab research studies, so further efforts should be made to overcome this research gap in the field.

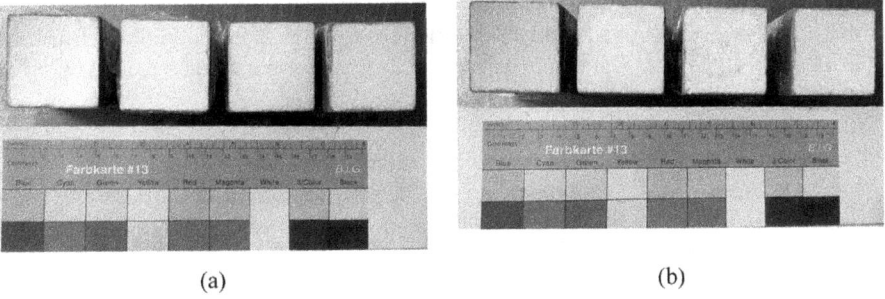

(a)   (b)

Fig. 5. Aspect of moulds (indicated with red arrows) developed during the drying test on Mšené stone coated with: (a) LR limewash; (b) LS limewash.

4   Conclusions

This work studied the performance of five types of limewashes, four of which with the addition of vegetable oils: pure limewash, limewash with rapeseed oil, limewash with sunflower oil, limewash with sunflower oil with a high oleic acid content, and a commercial limewash (Fradical) that is produced with the addition of olive oil. The limewashes were applied to two porous substrates with a different porous structure to study the influence of the limewashes on their water transport properties. The main conclusions of this study are summarised as follows:

- The method of preparation of the limewashes with additives significantly influences the properties of the final coat. Traditionally, oils were mixed with the additives during the slaking process, like the Fradical lime putty with olive oil used in this study. Mixing the additives with a ready-made lime putty manually can lead to a poor homogenisation of the additives in the lime and produce coats with heterogeneous properties. Ideally, mixing oils with lime should be done during the slaking process. When using lime putty, it is advisable to use mechanical mixing.
- The Fradical water-repellent lime putty showed the most promising results in terms of water-repellence and drying behaviour. The limewashes prepared with this type of lime provided hydrophobic coats on both substrates and enhanced the drying rate of the most porous substrate.
- Lime putties prepared with the addition of 1.5 wt.% of rapeseed oil, sunflower oil, and sunflower oil with high oleic acid content granted water-repellence properties to the limewashes without hindering the drying rate of the substrates so that they can be considered as an alternative to more expensive and less available oils like olive oil. Particularly, rapeseed oil should be considered in further studies using traditional preparation methods (mixing the oil during the lime slaking process).
- None of the limewashes (including pure limewash) was found suitable for application in the less porous substrate used in this study (Opuka marlstone). This can be attributable to the low adhesion of the limewashes to this material, as observed by the peeling of the paint in some specimens during the capillarity test. This is probably assigned to the very fine porosity of this stone that hampers the penetration of the lime particles into the porous network.

- Limewashes with rapeseed and sunflower oil showed the development of moulds during the drying test conducted in conditions of low light and ventilation. Therefore, tests promoting biological colonisation and onsite exposure should be done to unveil the biosusceptibility of this type of coat.
- This work showed that the limewashes with vegetable oils have a good performance in regard to water transport when applied on a porous material with a coarse porous network. Further research to improve the suitability of limewashes to substrates with lower porosity and narrow pore size distribution should be undertaken. Using lime putties composed of calcium hydroxide with the lowest particle size as possible and with a lower lime:water ratio should be considered.

Acknowledgements. This research study was funded by the Czech Science Foundation (GAČR) grant reference 18-28142S, and by the project Human Resources Development of the Institute of Theoretical and Applied Mechanics of the Czech Academy of Sciences no. CZ.02.2.69/0.0/0.0/18_053/0016918. The authors are grateful for the Foundation for Science and Technology's support via funding UIDB/04625/2020 from the CERIS research unit.

The authors thank Eng. Marina Reis from Sovena Oilseeds Portugal for kindly providing the vegetable oils and their composition, and Mr. João Melo from Fradical – Fábrica de Transformação de Cal (Portugal) for kindly providing the water-repellent Fradical lime putty used in this study.

References

1. Taliaferro, S.: Documentation and Testing of Nineteenth-century Limewash Recipes in the United States. MSc Thesis, Columbia University, Columbia, USA (2015)
2. Gonçalves, T.D., Pel, L., Rodrigues, J.D.: Influence of paints on drying and salt distribution processes in porous building materials. Constr. Build. Materials **23**(5), 1751–1759 (2009). https://doi.org/10.1016/j.conbuildmat.2008.08.006
3. Silva, J.M., Moura, A.R.: Defects of inadequate paintings over mortars in old facades. In: Proceedings of the 1st Historical Mortars Conference, Lisbon, Portugal, 24–26 Sep 2008
4. Faria, P., Tavares, M., Menezes, M., Veiga, R., Margalha, G.: Traditional Portuguese techniques for application and maintenance of historic renders. In: Proceedings of the 2nd Historic Mortars Conference HMC2010 and RILEM TC 203-RHM Final Workshop, pp. 609–617. Prague, Czech Republic, 22–24 Sep 2010
5. Gil, M., et al.: Colour assays: an inside look into Alentejo traditional limewash paintings and coloured lime mortars. Color. Res. Appl. **36**(1), 61–71 (2010). https://doi.org/10.1002/col.20584
6. Hošek, J., Ludvík, L.: Historical Plasters: Research, Repair, and Typologies. Prague, Czech Republic, Grada Publishing (2007). (In Czech)
7. Brito, V.; Diaz Gonçalves, T.: Artisanal lime paints and their influence on moisture transport during drying. In: Proceedings of the 3rd Historic Mortars Conference HMC2013, Glasgow, UK, 11–14 Sep 2013
8. Ferreira Pinto, A.P., Passinhas, H., Gomes, A., Silva, B.: Suitability of different paint coatings for renders based on natural hydraulic lime. In: Rorig-Dalgaard, I., Ioannou, I. (eds.) Proceedings of the International RILEM Conference on Materials Systems and Structures in Civil Engineering, pp. 175–184. RILEM Publications S.A.R.L., Lyngby, Denmark (2016)

9. Nunes, C., Mlsnová, K., Válek, J.: Limewashes with linseed oil: Effect on the moisture transport of natural stone. In: Siegesmund, S., Middendorf, B. (eds.) Proceedings of the 14th International Congress on the Deterioration and Conservation of Stone, Göttingen, Germany, pp. 591–596. Mitteldeutscher Verlag, Halle, Germany, 7–12 Sep 2020

10. Nunes, C.L., Mlsnová, K., Slížková, Z.: Limewashes with linseed oil and its effect on water and salt transport. Buildings **12**(4), 402 (2020). https://doi.org/10.3390/buildings12040402

11. Sá, A.F.G.: Aerial hydrated lime mortar with fat addition and its usage as plaster. MSc thesis, Technical University of Lisbon, Portugal (2002)

12. Fang, S., et al.: A study of Tung-oil-lime putty: a traditional lime based mortar. Int. J. Adhes. Adhes. **48**, 224–230 (2014)

13. Gour, K.A., Ramadoss, R., Selvaraj, T.: Revamping the traditional air lime mortar using the natural polymer – Areca nut for restoration application. Constr. Build. Mater. **164**, 255–264 (2018). https://doi.org/10.1016/j.conbuildmat.2017.12.056

14. EN 459-1: Building Lime. Part 1: Definitions, Specifications and Conformity Criteria (2015)

15. Nunes, C., Viani, A., Ševčík, R.: Microstructural analysis of lime paste with the addition of linseed oil, stand oil, and rapeseed oil. Constr. Build. Mater. **238**, 117780 (2020). https://doi.org/10.1016/j.conbuildmat.2019.117780

16. Pahlavan, P., Manzi, S., Rodriguez-Estrada, M.T., Bignozzi, M.C.: Valorization of spent cooking oils in hydrophobic waste-based lime mortars for restorative rendering applications. Constr. Build. Mater., **146**, 199–209 (2017)

17. Faria, P., Henriques, F., Rato, V.: Comparative evaluation of lime mortars for architectural conservation. J. Cult. Heritage **9**(3), 338–346 (2008). https://doi.org/10.1016/j.culher.2008.03.003. http://run.unl.pt/handle/10362/11387

18. EN 15802: Conservation of cultural property – Test methods – Determination of static contact angle. Brussels, CEN (2010)

19. EN 15801: Conservation of cultural property. Test methods. Determination of water absorption by capillarity. Brussels, CEN (2009)

20. EN 16322: Conservation of Cultural Heritage – Test methods – Determination of drying properties. Brussels, CEN (2013)

21. Justnes, H., Ostnor, T.A., Villa, N.B.: Vegetable oils as water repellents for mortars. In: Proceedings of the 1st International Conference of the Asian Concrete Federation, pp. 689–698. ACF, Chiang Mai (2004)

22. Gunstone, F.D.: Fatty Acid and Lipid Chemistry. Aspen Publishers, Maryland, USA (1996). https://doi.org/10.1007/978-1-4615-4131-8

23. Hansen, E.F., Rodríguez-Navarro, C., Balen, K.: Lime putties and mortars. Stud. Conserv. **53**(1), 9–23 (2008). https://doi.org/10.1179/sic.2008.53.1.9

24. Rodriguez-Navarro, C., Hansen, E., Ginell, W.S.: Calcium Hydroxide Crystal Evolution upon Aging of Lime Putty. J. Am. Ceram. Soc. **81**(11), 3032–3034 (2005). https://doi.org/10.1111/j.1151-2916.1998.tb02735.x

25. Ruiz-Agudo, E., Rodriguez-Navarro, C.: Microstructure and rheology of lime putty. Langmuir **26**(6), 3868–3877 (2009). https://doi.org/10.1021/la903430z

26. Alves, C., Figueiredo, C., Sanjurjo-Sánchez, J.: Rock features and alteration of stone materials used for the built environment: a review of recent publications on ageing tests. Geosciences **10**(3), 91 (2020). https://doi.org/10.3390/geosciences10030091

Mortars in Archaeological Sites. Construction History. Archaeometry. Dating of Historic Mortars

Chemical and Mineralogical Characterization of Lime Plaster from 6th Century Stone-Chamber Tomb of Baekje, Republic of Korea

Eunkyung Kim and Soyeong Kang$^{(\boxtimes)}$

National Research Institute of Cultural Heritage, Daejeon 34122, Republic of Korea
{eunrima,soyeong.kang}@korea.kr

Abstract. This study investigated the characterization of lime collected from the tomb of Songje-ri in Naju, Republic of Korea. The site is highly significant to understand the burial culture of ancient period in Korean history. The lime remained partially on the wall inside the tomb. The mineralogical and chemical properties of lime samples were characterized by digital microscope, X-ray diffraction analysis (XRD), thermal analysis (TG-DSC), X-ray fluorescence analysis (XRF) and mercury intrusion porosimetry (MIP). As a result of analysis, the sample is a plaster consisting only of lime, which was manufactured by shells. Traces of a small amount of organic matter are confirmed, but further study is required for artificial addition.

Keywords: lime · plaster · shell · tomb of Songje-ri · Republic of Korea

1 Introduction

Lime plaster has been used in various fields for centuries—from ancient Korean murals to modern and contemporary masonry buildings. Among the murals of ancient Goguryeo tombs, Anak Tomb No. 3 (4th century AD), Tonggu Tomb No. 12 (4–5th century AD), and Ssangyeongchong Murals (5th century AD) are representative. Lime plaster is presumed to be a mix of mineral materials such as clay, sand, drying oil, and organic matter such as herbaceous plants [1]. Lime plaster made from oyster shells was used for the base layer of the mural painting in the Gaya tomb (6th century AD) in Goa-ri, Goryeong [2]. Among the tombs of Baekje, traces of lime plaster were found in the tomb in Gamil-dong, Hanam (4th century AD), the tomb in Beopcheon-ri, Wonju (4–5th century AD), the tomb in Pangyo, Seongnam (5th century AD), and the tomb in Songsan-ri, Gongju (6th century AD) [3–6]. As such, lime plaster is a material that was frequently used in Korean ancient burial culture.

However, the research results of lime in the ancient period of Korea are insufficient due to the lack of related records and the difficulty in obtaining samples. To the extent that lime plaster is a material that requires professional knowledge and skill in its production and construction, if scientific analysis were considered along with an archaeological

V. Bokan Bosiljkov et al. (Eds.): HMC 2022, RILEM Bookseries 42, pp. 165–171, 2023.
https://doi.org/10.1007/978-3-031-31472-8_13

analysis, it would be possible to understand the level of construction technology at the time, the burial culture, and the exchange patterns through which the technology was introduced.

In this paper, the characteristics of lime plaster used in the burial culture in the period of Baekje are reviewed through a scientific analysis of the lime plaster used as a finishing material inside the stone burial chambers.

2 Materials and Methods

2.1 Materials

The lime samples collected from a stone chamber tomb in Songje-ri at Naju. As a result of excavation, it is estimated that this site is the tomb of a local leader class and was built in the early to mid-6th century [7]. Most of the lime plaster has been lost and partially remain on the surface or between the stone walls (Fig. 1). Samples were obtained from the wall where the lime remained for each orientation.

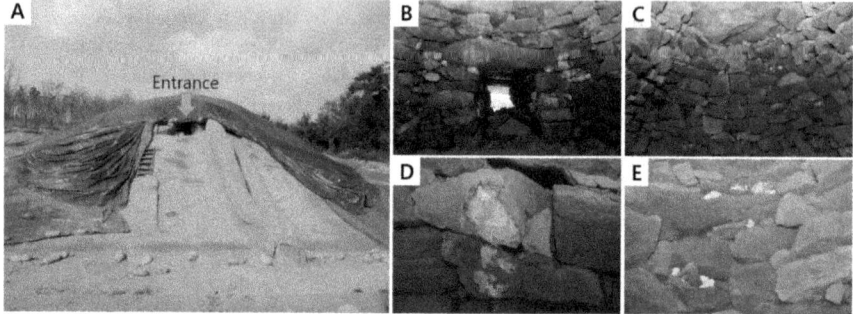

Fig. 1. General and internal view of stone chamber tomb in Songje-ri, Naju. (A) front view (B), (C) inner wall (D), (E) residual state of traditional lime samples.

2.2 Methods

A mineralogical and chemical analysis was conducted to understand the material properties of the lime plaster used as a finishing material inside the Baekje stone burial chambers. First, the collected sample of lime plaster and the tissue morphology of the shells were observed with a digital microscope (DVM6, Leica). The observed sample was powdered, and its constituent minerals were investigated by X-ray diffraction analysis (EMPYREAN, PANalytical, scan range $5°–60°$, step size 0.04, 200 s/step), infrared spectroscopic analysis (Nicolet iS7, Thermo Fisher Scientific, $4,000–600 \ cm^{-1}$, resolution $4 \ cm^{-1}$, 64 scans/s), and thermogravimetric/differential thermal analysis (LABSYS Evo, Setaram, N_2 gas, $30–1,000 \ °C \ 10 \ °C/min$). To determine each sample's chemical characteristics, an X-ray fluorescence analysis (MXF-2400, Shimadzu) was conducted. In addition, pore characteristic of the lime samples was investigated by mercury intrusion porosimetry.

3   Results

3.1   Microscopic Observation

The sample exhibited a white porous matrix and the surface is contaminated with soil and dead plants. Aggregates such as sand were not confirmed. The shells mixed with lime plaster had a gray-blue color. It has a size between 0.5 mm and 10 mm when observed on the surface. In addition, a small amount of organic material indentation was observed (Fig. 2).

Fig. 2. Observations of the shape of the lime and shell samples.

3.2   Mineralogy

The results of the X-ray diffraction analysis identified calcite, which is commonly composed of calcium carbonate ($CaCO_3$), as the main constituent mineral in the samples. This suggests that all samples were plastered with lime as the main material. The confirmed quartz peak was found to be part of the soil remaining on the sample's surface (Table 1).

Table 1. Mineralogical composition of the lime samples by XRD.

	NS-1	NS-2	NS-3	NS-4	NS-5	NS-6
Calcite	++	++	++	++	++	++
Quartz	+	+	+	+	+	+
Feldspars	+	–	–	–	–	–

Peak intensity: ++ Major + present – undetected

As a result of Fourier transform infrared spectroscopy, bands from calcium carbonate and silicate minerals were found in the lime plaster samples (Fig. 3A). The sharp absorption bands around 2500 cm^{-1}, 1404 cm^{-1}, 872 cm^{-1}, and 712 cm^{-1} were due to the overtone, stretching, and bending bands of carbonate ions, respectively. The absorption band appearing in the vicinity of 1000–900 cm^{-1} was due to the Si–O–Si asymmetric stretching band of quartz [8–11].

As a result of measuring the thermogravimetry/differential scanning calorimetry of the lime plaster sample, the main weight reduction and the maximum endothermic reaction were measured in the 700–800 °C range due to the decarbonation of calcium carbonate, which is calcite (Fig. 3B).

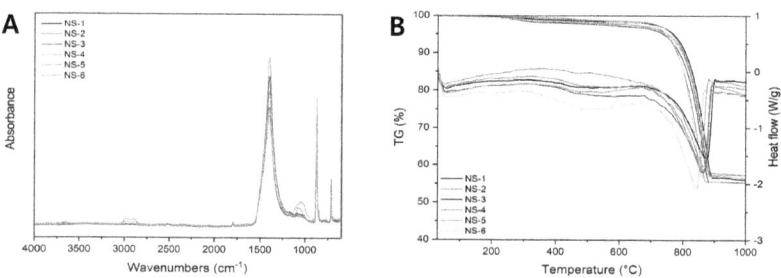

Fig. 3. Comparison of the spectroscopy and thermal analysis of lime samples (A) FTIR spectrum (B) TG-DSC curve.

3.3 Chemical Composition

The X-ray fluorescence analysis of the main chemical components of the samples of lime plaster found CaO 48.72–53.44 wt.%, L.O.I. (loss on ignition) 39.31–43.42 wt.% and SiO_2 1.64–9.08 wt.% in the sample from stone chamber tomb. The results showed that the CaO and L.O.I. content was high under the influence of calcite, which was the main constituent mineral of the sample. In terms of chemical composition, the content of the samples was similar, and some NS-1, NS-4 had higher SiO_2 and Al_2O_3 content than other samples. This is presumed to be the effect of the soil remaining in the lime plaster sample, as in the results of X-ray diffraction analysis (Table 2).

Table 2. Results of analysis on the main chemical composition of the lime samples.

	SiO_2	Al_2O_3	Fe_2O_3	CaO	MgO	K_2O	Na_2O	TiO_2	MnO	P_2O_5	L.O.I
NS-1	9.08	1.3	0.19	48.72	0.26	0.5	0.09	0.01	0.01	0.06	39.31
NS-2	1.64	0.43	0.14	53.44	0.25	0.05	0.04	0.02	0.01	0.09	43.42
NS-3	5.61	1.32	0.34	50.58	0.28	0.23	0.09	0.04	0.02	0.09	41.11
NS-4	2.15	0.92	0.32	52.87	0.29	0.08	0.07	0.03	0.02	0.11	42.76
NS-5	3.38	0.75	0.25	52.03	0.25	0.1	0.05	0.03	0.22	0.2	42.34
NS-6	1.97	0.65	0.20	52.82	0.25	0.08	0.05	0.02	0.02	0.11	43.4

3.4 Pore Characteristics

The porosity and pore size distribution measured on lime plaster samples. The samples show similar porosity of 47.3–56.4% and average pore diameter of 0.45–0.82 μm

(Table 3). The pore size distributions of lime plasters have a similar distribution with maxima at about 1.5–2 μm (Fig. 4).

Table 3. Pore characteristics of lime samples.

	NS-1	NS-2	NS-3	NS-4	NS-5	NS-6
Porosity (%)	52.13	47.28	56.41	55.79	49.19	55.44
Avg. pore size diameter (μm)	0.45	0.63	0.82	0.67	0.53	0.63

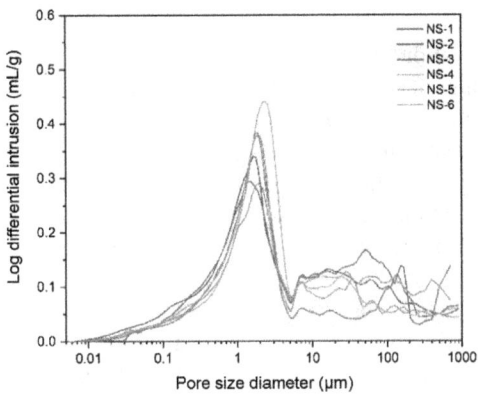

Fig. 4. Pore size diameters of lime samples.

4   Discussion

Regarding the lime plaster used as a finish for the walls of the stone-chamber tomb excavated from the Songje-ri, Naju, small amount traces of herbaceous plants were confirmed in the form mixed with shells.

The lime sample did not contain aggregates such as gravel or sand, and calcite was detected as the main mineral. The amount of $CaCO_3$ in the sample was calculated using the weight loss in 600–900 °C range in TG and CaO content in XRF. As a result of the analysis, the $CaCO_3$ content was confirmed to be 90% or more, and the results were also similar teach other (Table 4). Therefore, the CaO identified by XRF was from lime and also the samples were manufactured by the shell.

The material properties of the lime plaster used in the tombs of Songje-ri were similar to those in Gamil-dong tombs of Baekje estimated to be from the middle of the 4th century to the early 5th century. Lime made from the shells was also used as an internal finishing material for Gamil-dong tombs. The shells found in the lime plaster of Gamil-dong tombs had a layered structure and a thick, curved, radial rib shape. The shells were presumed to be of bivalve oysters. In addition, as a result of observing the

Table 4. Comparative of the %CaCO$_3$ contents obtained by the TG and XRF

	NS-1	NS-2	NS-3	NS-4	NS-5	NS-6
TG	94.5	93.7	94.5	90.6	95.7	93.8
XRF	87	95.4	90.3	94.4	92.9	94.3

microstructure of the shell surface, it was confirmed that the amorphous form deformed by recrystallization during the firing process and pores through which carbon dioxide escaped [12].

Lime plaster was also excavated from stone burial chamber No. 9 in Pangyo, Baekje (the late 4th century to the late 5th century). As a result of the X-ray fluorescence analysis [6], a chemical composition of CaO 53.75 wt.%, SiO$_2$ 0.90 wt.%, and L.O.I. 40 wt.% was observed and thus, only lime plaster seems to have been applied. Further, as a result of treating the lime plaster with hydrochloric acid, shell fragment was confirmed. Accordingly, the raw material of lime plaster was estimated to be shell. Based on this, it is possible to deduce that the Baekje Period possessed the lime plaster production technology and construction method using shells, which was used for the construction of tombs. However, further research is still needed on whether shell was used with lime as an aggregate.

On the surface of the lime plaster of the tomb in Songje-ri, indentations and fluids presumed to be stems of herbaceous plants remain. However, only a very small amount was observed, so it is unclear whether artificial mixing. Regarding the mixing of organic matter, traces remain in the mural paintings of the Gaya tomb in Goari, Goryeong and Gamil-dong tombs [12, 13]. Moreover, it is known that vegetable fibers such as reeds, rice straw, and kiwi stems were added to lime plaster used as a base layer for murals in Goguryeo tombs, Sindap-ri tomb, Yeoncheon and Tonghyeon-ri tomb, Yeoncheon of Baekje [1, 6]. Therefore, if a species analysis of a plant sample with original remains was performed, it could be used as specific data on the construction technology of ancient tomb.

5   Conclusion

Mineralogical and chemical analyses of samples from the stone chamber tomb in Songje-ri, Naju exhibits the characteristics of a lime plaster of 6th century of Baekje. As a result, it was confirmed that lime was made from shells and used as plaster in the Baekje period. Through this analysis of the lime plaster, it is possible to accumulate important data that reveal the lime plaster manufacturing techniques and construction technology. In this respect, the study is of great archaeological significance. If a species identification of the shells and herbaceous species is carried out by securing additional samples in the future, information on lime plaster manufacturing and construction technology in the Baekje dynasty could be identified.

Acknowledgments. The work was supported by the Cultural Heritage Research and Development of the National Research Institute of Cultural Heritage, the Republic of Korea.

References

1. Chae, M.Y.: A study on the base material of Goguryeo Mural Tomb -Focused on lime mortar-. KoguryoBalhae Yonku **9**, 45–107 (2000)
2. Han, K.S., Lee, H.S.: Production technique. In: Han, K.S (ed.) Study on the Conservation for Mural Painting of Goari Tomb in Goryeng, Gaya, pp. 79–94. Goryeong-Gun, KonKuk University, Chungju (2013)
3. Kang, T.H.: The excavation results of Gamil-Dong tumuli of Baekje Hanam city. BaekjeHakbo **27**, 107–126 (2019)
4. Song, E.J., Yoo, H.W.: Beopcheon-ri I. National Museum of Korea, Seoul (2000)
5. Research Division of Artistic Heritage, National Research Institute of Cultural Heritage: The tomb murals in South Korea, pp. 52–67. National Research Institute of Cultural Heritage, Daejeon (2019)
6. Sim, G.Z.: A study of fortress construction method called Jeongto Chukseong used in Hanseong Baekje. HyangtoSeoul **76**, 41–92 (2010)
7. Naju National Research Institute of Cultural Heritage: The summery report on the 1st excavation of the Songje-ri tomb in Naju. Naju National Research Institute of Cultural Heritage. Naju (2019)
8. Laycock, E.A., Pirrie, D., Clegg, F., Bell, A.M.T., Bidwell, P.: An investigation to establish the source of the Roman lime mortars used in Wallsend, UK. Constr. Build. Mater. **196**, 611–625 (2019)
9. Singh, M., Kumar, S.V., Sabale, P.D.: Chemical and mineralogical investigations of lime plasters of medieval structures of Hampi, India. Int. J. Archit. Herit. **13**(5), 725–741 (2019)
10. Diniilia, S., Minopoulou, E., Andrikopoulos, K.S., Tsakalog, A., Bairachtari, K.: From Byzantine to post-Byzantine art: the painting technique of St Stephen's wall paintings at Meteora, Greece. J. Archaeol. Sci. **35**(9), 2474–2485 (2008)
11. Genestar, C., Pons, C., Más, A.: Analytical characterisation of ancient mortars from the archaeological Roman city of Pollentia (Balearic Islands, Spain). Anal. Chim. Acta **557**(1–2), 373–379 (2006)
12. Kim, E.K., Kang, T.H., Kang, S.Y.: Characteristics of lime used in stone chamber tombs of Baekje-Hanseong period -Focused on Gamil-dong tumuli, Hanam-. Baekje Hakbo **30**, 113–140 (2019)
13. Lee, H.S., Han, K.S.: Study on the manufacturing technology of ancient mural tomb in Korea. J. Cult. Herit. **14**, 147–178 (2014)

Characterization of Historic Mortars Related to the Possibility of Their Radiocarbon Dating, Mikulčice and Pohansko Archaeological Sites

Petr Kozlovcev[1]([envelope]) [ORCID], Kristýna Kotková[1] [ORCID], Dita Frankeová[1] [ORCID], Jan Válek[1] [ORCID], Alberto Viani[1] [ORCID], and Jana Maříková-Kubková[2]

[1] Institute of Theoretical and Applied Mechanics of the Czech Academy of Sciences, Prague, Czech Republic
kozlovcev@itam.cas.cz
[2] Institute of Archaeology of the Czech Academy of Sciences, Prague, Czech Republic

Abstract. The archaeological sites of Mikulčice and Pohansko (South Moravia – the Czech Republic) belong to the oldest and the most important localities of Slavic settlement in Central Europe. A number of historic mortars sampled here were collected in order to study their composition, mortar structural characteristics and raw materials provenance. The aim of this study was not only the comparison and characterisation of the historical mortars from these archaeological sites, but we also evaluated the suitability of these mortars to be dated by ^{14}C analysis.

The samples were characterised by several analytical techniques as polarised light and scanning electron microscopy, thermal analyses or quantitative X-ray diffraction. Stable isotope analyses and cathodoluminescence were also performed. The collected mortars contained a considerable amount of lime particles that can adversely affect the possibility of ^{14}C radiocarbon dating. According to the results of the analyses, the samples from both localities had a similar character. Mortars were very rich in the binder and contained unburnt limestone fragments that occurred frequently. Unburnt fragments were classified mostly as a micritic limestone contained bioclastic fragments. This material was determined as Ernstbrunn limestones according to the composition and structure. Stable isotope analysis also suggested that all studied lime samples came from a single source. The presence of geogenic carbonates (not fully burnt lime) affects the resulting ^{14}C age of the analysed samples fundamentally. The character of mortars leads to a discussion on how to adapt the separating procedures of the individual fractions and avoid geogenic carbon contamination.

Keywords: Historic mortars · ^{14}C dating · hydraulic properties · limestone raw material · Mikulčice and Pohansko

1 Introduction

1.1 Possibilities of Mortar Dating

Radiocarbon dating of mortars is an analytical method that allows specifying the age of buildings for which more accurate information and connections from the time of their construction have not been preserved. For buildings from the period of the early

V. Bokan Bosiljkov et al. (Eds.): HMC 2022, RILEM Bookseries 42, pp. 172–189, 2023.
https://doi.org/10.1007/978-3-031-31472-8_14

medieval Moravia, we have only general ideas about their age from detailed archaeological research (9th century). However, a more precise timeline is missing. From this point of view, mortars from the early medieval Moravian buildings represent an interesting study material.

The number of publications proving that the dating of mortar binders is possible under certain conditions and providing relevant results has increased significantly in the last two decades [1–4]. The dating of the carbonate binder of a mortar is based on the technology of lime production and application as a binder. This process starts by burning limestone (mostly $CaCO_3$) in a kiln when it decomposes into quicklime (CaO) and carbon dioxide (CO_2). When the quicklime is slaked (transformed to $Ca(OH)_2$) CO_2 is absorbed back from the air (carbonation process). The result is the original calcium carbonate ($CaCO_3$), the so-called anthropogenic carbonate. It differs from the original geogenic carbonate in its isotopic composition and ^{14}C activity. This corresponds to the activity of atmospheric $^{14}CO_2$ and the isotopic composition of the air at the time of the carbonation reaction. Based on the knowledge of the half-life of ^{14}C, the so-called "conventional radiocarbon age" of the sample can be expressed. This can then be subsequently converted to a calibrated (real) age using a calibration curve [5]. However, this idealised procedure has several pitfalls.

1) Raw materials for lime production have a varied composition, which subsequently affects the carbonisation process in different ways (results in younger age).
2) The burning conditions in a kiln are not uniform and a certain portion of CO_2 of the geogenic origin remains in the lime binder (results in older age).
3) CO_2 must pass through the porous system of the material during carbonation; the carbonation rate slows down with depth (samples from the core of thick structures result in older age).
4) Mortar components may contaminate the sample, e.g. carbonate filler (results in older age
5) The presence of mineral phases, e.g. phyllosilicates binds younger CO_2 (results in younger age).
6) Possible dissolution of carbonate and recrystallises (results in younger age).

Therefore, proper sampling and appropriate procedures of sample pre-treatment are necessary to separate the anthropogenic CO_2 and exclude the contaminants.

The currently used mortar characterisation methods can be used to evaluate the suitability of archaeological mortars for radiocarbon dating. Based on the composition, structure, and other material properties, the presence of compounds or mineral phases that could affect radiocarbon dating can be evaluated. Thorough characterisation thus always represents a fundamental contribution to the successful dating of historical mortar samples.

For carbon separation, the following methods are mainly used: the method based on mechanical separation (modified CryoSonic – Cryobreaking, Sonication, Centrifugation) [6], the method of chemical decomposition by acid hydrolysis is also used [4, 7], and the method of thermal decomposition [8, 9]. The modified CryoSonic method was developed in the CIRCLE Centre (Italy) and it is based on gentle mechanical disintegration, ultrasonic disintegration, centrifugation, sedimentation and filtration [10]. The obtained fine carbonate fraction is dissolved in heated phosphoric acid. This methodology can be well adapted to the character of the studied samples. The previous mortar characterisation is fundamental for the next steps because the mortar composition (matrix particles and fillers) affects the sample preparation significantly. Especially the number of fractions collected for analysis [14]C [11]. The composition of the mortar matrix has a direct effect on dating options [12]. The presence of non-carbonate impurities affects CO_2 binding during carbonisation [13]. The structure and granularity of the binder present reflect the properties of the original raw material, technological processing parameters and carbonisation and recrystallisation conditions [14]. Also the presence of carbonate filler particles and unburned binder pieces of the original carbonate material has a significant impact on dating options [15–17]. They contain geogenic carbonate, which introduces an error in the dating results, which ultimately leads to an incorrect interpretation of the age of the studied sample.

Overall, the main aims of this study were to use knowledge in the field of analysis of historic materials for the comparison and characterisation of the historical mortars from archaeological sites Mikulčice and Pohansko and to evaluate the suitability of these samples to be dated by ^{14}C analysis.

1.2 Historical Background

Radiocarbon dating of mortars is an analytical method that allows specifying the age of buildings for

Mikulčice and Pohansko are the most important sites in the territory of early medieval Moravia. [18], located in the area of the middle and lower reaches of the Morava river [19]. This area, inhabited by Slavs at the time of the migration period, was Christianized in the first half of the 9th century. The area was suitable for settlement and agricultural production, but its undeniable advantage and source of wealth was its location on important trade routes. A state unit soon began to take shape here which played an important role in the early history of Central Europe, and it is considered the ancestor of the following medieval states [20].

Throughout the 9th century, the Moravian territory developed within the framework of political, economic and cultural relations with the Frankish Empire. Along with Christianization, innovations in the field of art and culture also came here, including stone architecture and lime mortar. The history of early medieval Moravia ends with the invasion of nomadic Hungarians at the beginning of the 10th century [21].

1.3 Archaeological Sites of Mikulčice and Pohansko

The early medieval fortification of Mikulčice is situated south-east of the village Mikulčice in the South Moravia Region. This fortification formed the core of an extensive settlement complex referred to as the Mikulčice-Kopčany agglomeration during the 9th century. The castle (fortified core) is situated on the right bank of the Morava River in the Czech Republic. The extra mural settlement and the hinterland of the centre are situated on both banks of the Morava river. On the left bank, in Slovakia, close to Kopčany village, still stands the pre-Romanesque Church of St. Margaret of Antioch dated to the 9th century [22]. The first stage of the archaeological excavations in the Mikulčice-Kopčany area were the large-scale excavations during 1954 and 1992 [23].

They were characterised by the extensive fieldwork that continued uninterrupted for 38 seasons and explored an area of almost 5 ha. Thanks to this huge works were Mikulčice settlements' remains discovered and became more famous [24]. During 1993 and 2003 were fieldworks assessed and became the systematic processing of the previous research results. A new information system was set up, and basic guides were prepared. Moreover, the former fieldwork base was transformed into an academic research centre [25]. In 2004 began the last step to process sources and verify old research. For this purpose, the most important archaeological objects and the well-documented areas were chosen [26]. The previous results were revived and updated with new fieldwork results. At the beginning of 2008, the Slavic fortification in Mikulčice was revitalised.

The second important archaeological area is Pohansko near Břeclav in the South Moravia region, near the Czech-Austrian borders. It is a large Slavonic fortification from 9th century in the vicinity of the confluence of the rivers Morava and Dyje [27]. Archaeological excavations have been conducted in this area since 1958. Excavations uncovered many scattered archaeological findings from various periods of the prehistorical age, but the main and continuous settlement there can be dated between 6th and 10th centuries [21]. The first research was focused on the area around the north-western part of the stronghold, where the highest concentration of mortars and plasters was detected. During 90s many small-scale excavations were conducted to identify the extent of settlement in the whole agglomeration. On the basis of archaeological and geophysical surveys, the maximum extent of Pohansko is currently estimated to be around 60 ha (in the 10th century) [28]. Today approximately a quarter of the total extent of the stronghold is examined, so the research is by far not finished [29].

1.4 Sampling Sites

The studied samples were sampled in the original excavations, mostly from church relics (Mikulčice, Church number II, III and IX, Pohansko, Church number 1 and 2, Mikulčice, Palace number 2, 3, 4, 5, 6, 9 and 10). There were mortars from foundations, fragments of floors, wall pieces with bedding mortars and mortars with surface treatments (Table 1).

Table 1. The list of studied samples

ML 10; ML 11; ML 12; ML 13; ML 14; ML 15; ML 16; ML 17; ML 18; ML 19; ML 20; ML 21; ML 22; ML 23; ML 24; ML 25; ML 26; ML 27; ML 28; ML 29; ML 30; ML 31; ML 32; ML 33	Mikulčice, historic mortars, mortar and rock fragments. Samples from archaeological probes and relicts of historic masonry
POH 1; POH 2; POH 3; POH 4; POH 5; POH 6; POH 7; POH 8; POH 9; POH 10; POH 11; POH 12; POH 13; POH 14; POH 15	Pohansko, historic mortar fragments, rock fabrics, mortar and plaster layers

2 Methodology

The main object of the research was the characterisation of mortar samples with a special focus on the present binder related particles (BRP). All samples were described macroscopically and photographically documented. The mortar samples were then divided according to the planned analytical procedures. One part was used for microscopy, the other part was finely crushed and sieved through a 63 μm sieve. The sub-sieve fraction was used for binder analysis (designation B) by thermal analyses and X-ray diffraction. BRP and limestone fragments occurred in all studied samples. Some of these particles (designation L) were selected and analysed by thermal analysis, X-ray diffraction and mass spectroscopy (used for the determination of carbon (^{13}C) and oxygen (^{18}O) isotope composition).

The samples for **polarised light microscopy (PLM)** were cut with a saw and impregnated under vacuum with epoxy resin. Polished thin-sections approx. 35 μm thick were studied using a polarising light microscope Olympus BX53M equipped with a digital camera Olympus DP27. The samples were observed in a plain polarised light (PPL) and a crossed polarised light (CPL).

Cathodoluminescence (CL) was used for the identification of BRP and for the detection of unburnt limestone particles. Polarising light microscope Olympus BX53M with a digital camera Olympus DP74 and "cold cathode" type Mk 5-2 (CITL Company) was used for observation.

Polished thin-sections were carbon coated and analysed by **scanning electron microscopy with an energy dispersive spectrometer (SEM-EDS)** by Tescan MIRA II LMU coupled with a spectrometer from Bruker AXS. The EDS analysis was used for the micro-chemical analysis of fine-grained calcitic binder (mortar matrix). According to the chemical composition, the cementation index (CI) was calculated, describing the hydraulic properties of the studied mortars [30].

Thermogravimetric analysis (TA) was performed on an SDT instrument Discovery 650 from TA Instruments in the temperature range 25–1000 °C. Approx. 20 mg of sample was heated in a ceramic crucible in a nitrogen atmosphere at a heating rate of 20 °C per minute.

In order to characterise the mineral composition of the studied mortars, samples were analysed by **powder X-ray diffraction (XRD)** using a D8 Bruker Instrument. The measuring conditions were the following: generator settings to 40 mA, 40 kV; tube position - line focus; soller slits - 2.5°; divergence slits - 0.6 mm; angular range (2Θ) - 5–82°; step size (2Θ) - 0.01°; counting time/step was 0.04 s; anode material - Cu; spinning - 15 rqm; sample holder - aluminium. 10 wt. % of NIST676a Al2O3 was added as an internal standard before data collection. The mineral phases of the insoluble residues were determined by quantitative phase analysis (QPA) using the Rietveld method [31].

For **leaching off the samples in acid and sieve analyse**, 1 M solution of acetic acid (CH$_3$COOH) was used. The leaching time was over a minimum period of 12 h (according to the weight). The samples were then washed in distilled water. The remaining solid matter was dried to a constant weight. After that it was sieved through a set of standard mesh size 8; 4; 2; 1; 0.5; 0.25; 0.125; 0.063 mm.

The **isotopic composition** of carbon (δ^{13}C ‰) and oxygen (δ^{18}O ‰) was determined by decomposition in 100% phosphoric acid under a vacuum at 25 °C, the decomposition time was 24 h. The composition of carbon and oxygen isotopes in the released CO_2 was measured on Delta V mass spectrometer with a total measurement error being $\pm 0.1\%_0$. The composition of carbon isotopes is related to the international PDB standard, the values of oxygen isotope composition related to the SMOW standard are determined from the measured values against the PDB standard by recalculation. The isotopic composition was carried out in the Laboratory of the Czech Geological Survey, Prague Barrandov.

3   Results

3.1   Macroscopic Description of the Samples

The studied mortars consisted of a fine-grained carbonate binder and macroscopically visible clusters of quartz grains. Fine-grained white to beige limestone fragments (size up to 50 mm) were also relatively common.

3.2   Microscopic Characterisation of the Samples (PLM, CL and SEM-EDS)

According to the PLM, CL and SEM-EDS methods, quartz grains represented the predominant filler component. The most common were semi-sharp-edged or sharp-edged

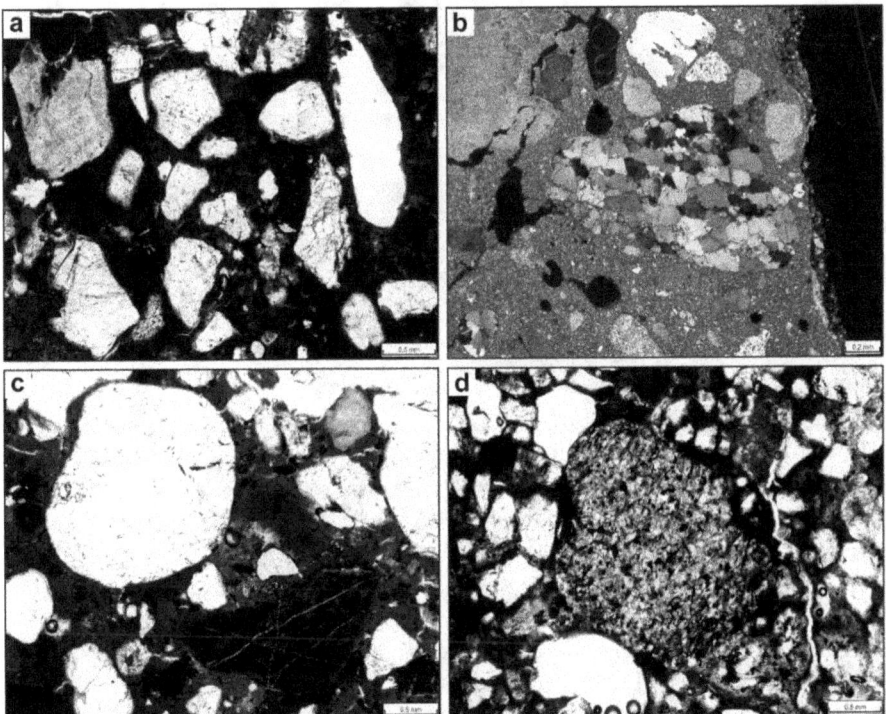

Fig. 1. Microphotographs observed using PLM: (a) Quartz grains typical for mortar samples (ML 29, PPL); (b) Polycrystalline quartz grain (ML 33, CPL); (c) Monocrystalline quartz grain (ML 29, PPL); (d) Fragment of fine-grained calcareous sandstone (ML 27, PPL).

grains (Fig. 1a). Polycrystalline quartz also occurred (Fig. 1b). The size of most grains ranged from 0.3 to 0.7 mm. However, the presence of monocrystalline quartz grains larger than 1 mm was no exception (Fig. 1c). The largest quartz grain was 5 mm in diameter.

Feldspars (sodium and potassium) occurred less frequently, than usual quartz grains. The presence of feldspars was evident, especially in the CL. These components were characterised by bright blue luminescence (Fig. 2a, 2b, 2c). The size of feldspar grains usually ranged from 0.3 to 0.6 mm.

Other minerals such as elongated mica, apatite grains (typical green colour in CL) (Fig. 2d) and small opaque minerals also occurred in studied mortars. The fragments of fine-grained calcareous sandstone (lithoclasts) were formed by a mixture of small quartz and feldspar clusters cemented with calcite. They had a regular oval shape and a size of up to 2 mm (Fig. 1d).

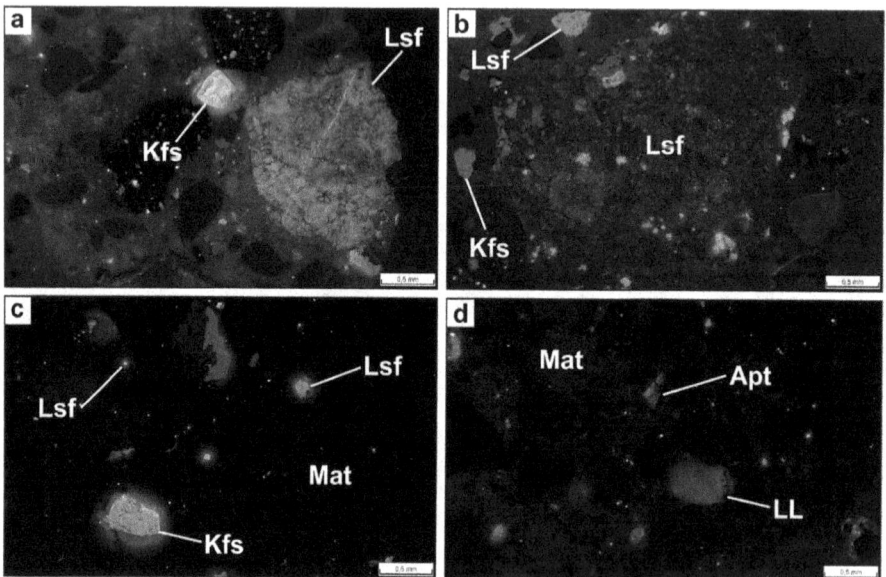

Fig. 2. Microphotographs observed using CL: (a) Limestone fragment (Lsf) and small grain of K-feldspar (Kfs) (ML 29); (b) Large limestone fragment (Lsf) with dark red cathodoluminescence (ML 33); (c) K-feldspars (Kfs) and small bright coloured limestone fragments (Lsf) in matrix (Mat) (ML 22); (d) Lime-lump (LL) and apatite (Apt) grains in matrix (Mat) (ML 25).

Fragments of fine-grained limestone were found in all studied samples (Fig. 3a, b). These particles were irregular and sharp-edged, from 0.3 to 3 mm in size. However, significantly larger fragments were not exceptional (e.g. a 14 mm limestone fragment was found in sample ML 31). Small isolated limestone particles in matrix were the most common (Fig. 2c). The limestone consisted of a predominant fine-grained micritic matrix with coarser sparite zones. Veins and pressure-solution seams were filled with secondarily precipitated sparitic calcite. Relics of microfossils, mostly coral remains, were identified in these limestone fragments (Fig. 3a, b). Classification: biomicritic limestone (according to the Folk classification system) [32], mudstone or wackestone (according to the Dunham) [33]. All limestone fragments had quite a uniform character. Slight differences were observed in CL. Bright red or orange luminescence was typical for most of the limestones (Fig. 2a), but also dark red luminescence occurred (Fig. 2b).

The binder of all studied samples consisted of a fine-grained carbonate matrix. It filled the inter-granular space between the clastic filler grains and limestone fragments (Fig. 3c). The structure of the binder was uniform, without any significant transformation or traces of degradation. Cathodoluminescence of the matrix was very weak; the matrix appeared as black background (Fig. 2c, d).

Binder related particles (BRP) frequently occurred in the studied samples (Fig. 3d). These particles were derived from a binder and were of interest because some of them clearly reflected the original raw material, and others were in the form of lime lumps sensu stricto, as described by [34]. The size of the BRP was variable, small particles of 0.2 mm were more frequent, but there were also large particles of about 6 mm (samples ML 32 and 33). Cathodoluminescence of the BRP was orange or slightly red (Fig. 2d). Partially burnt particles (containing visible unburnt core) had a brighter cathodoluminescence colour (Fig. 4a, b).

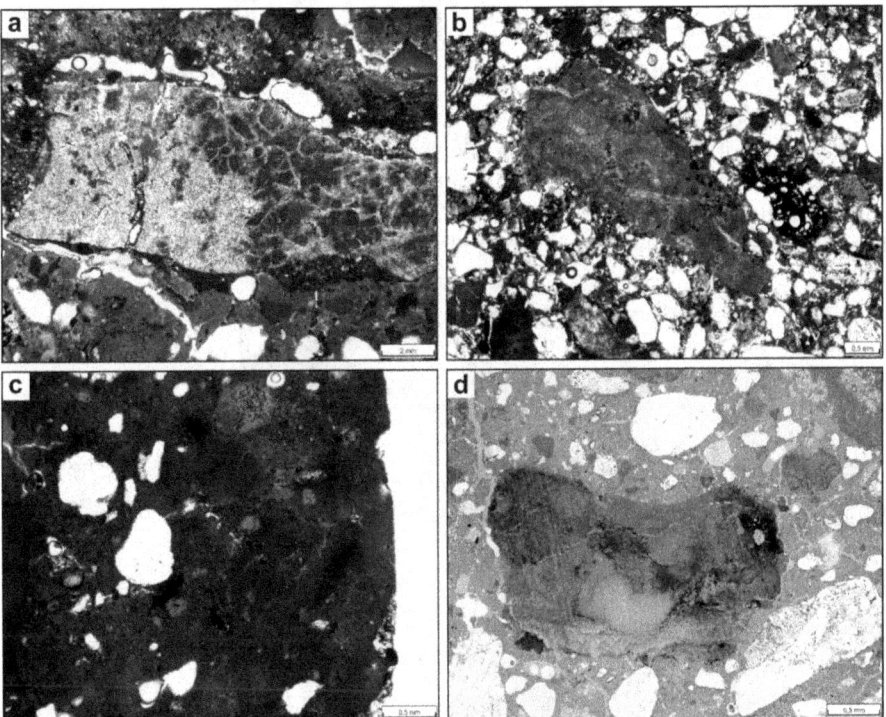

Fig. 3. Microphotographs observed using PLM: (a) Limestone fragment (ML 29); (b) Limestone fragment (POH 3); (c) Fine-grained calcite matrix (ML 22); (d) Binder related particle - lime-lump (ML 33).

The structure of the mortars was mostly compact, but there were also small pores and cracks. The ratios between the filler, the binder and the pore spaces were mostly constant in the samples. The filler represented 35 to 50%, the binder 40 to 65% and the pore spaces 20 to 30% of the sample volume.

The EDS was mainly used to characterise the binder and BRP (Fig. 5). Both the binding matrix and the BRP were rich in CaO content (Table 2). The calculated cementation index (CI) does not exceed 0.16, which corresponds to air lime.

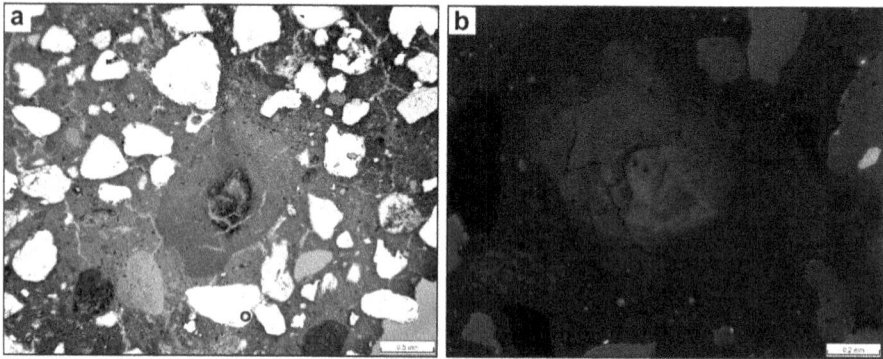

Fig. 4. Binder related particle contained visible core observed using PLM (a) and using CL (b).

Fragments of limestones, PBP, lime lumps and matrix differed in the presence of chlorite. Chlorite was typical for almost all BRP (e.g. lime lumps) and the matrix. It was due to the presence of salts in the binder. In contrast, significantly less chlorite was detected in the unburnt or partially burnt limestone fragments.

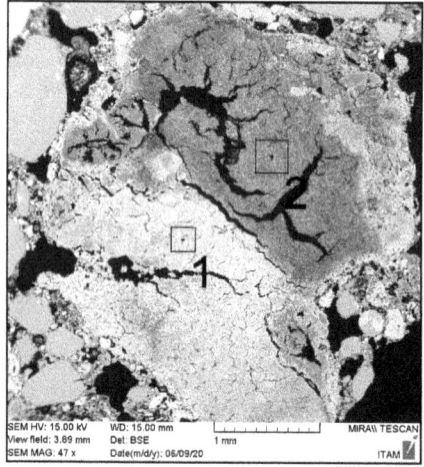

Fig. 5. Positions of SEM-EDS microanalyses in ML 19.

Table 2. Selected SEM-EDS microanalyses of sample ML 19 – difference between two binder related particles (BRP)

Analysis	1	2
CaO	99.13	96.00
SiO_2	0.18	0.86
Al_2O_3	0.17	0.07
MgO	0.37	0.45
Na_2O	0.09	-
SO_3	-	0.09
Cl	0.04	1.17
Fe_2O_3	-	0.26
P_2O_5	-	1.10
Sum	99.98	100
CI	**0.01**	**0.03**

3.3 Microscopic Characterisation of the Samples (PLM, CL and SEM-EDS)

Based on the macroscopic and microscopic evaluation, the studied materials were divided into four groups:

(A) limestone fragments, possibly including raw material and unburnt raw material particles,
(B) Binder related particles - BRP, including lime lumps,
(C) binder, represented by fractions up to 63 μm,
(D) calcareous sandstone and other lithoclasts.

The TA analyses confirmed some differences among these groups (Fig. 6). The limestone samples showed a weight loss in the temperature interval 600–950 °C (decomposition of CaCO3) of about 43 wt. %. Which is significant for high-purity limestones (Table 3). The loss between temperatures 50–600 °C connected to the release of physically and chemically bound water did not exceed 0.5 wt. % (Table 3). On the other hand, group B and C samples mostly had a weight loss in percentage units within this interval (Fig. 6). It corresponded to the loss of moisture contained in the mortar aggregate or moisture contained in the hydrated binder phases. In addition, the technological particles (group B) were less contaminated than the samples of the binders (group C).

Table 3. Weight loss of selected samples, groups A, B, C, and D and estimated CaCO3 content

Sample	Weight loss [wt.%]			
	50–600 °C	600–950 °C	$CaCO_3$ [wt. %]	
ML 12L1	0.27	43.67	99.3	Limestone (A)
ML 14L1	0.22	43.64	99.2	Limestone (A)
ML 17L2	0.25	43.38	98.6	Limestone (A)
ML 19L1	0.19	43.46	98.8	Limestone (A)
ML 20L1	0.44	42.35	96.3	Limestone (A)
ML 13L1	2.85	34.84	79.2	BRP (B)
ML 13L2	2.07	39.17	89.0	BRP (B)
ML 16L1	2.20	33.61	76.4	BRP (B)
ML 17L1	2.13	36.94	84.0	BRP (B)
POH 2L1	0.6	42.33	95.0	BRP (B)
POH 6L1	1.0	42.51	95.0	BRP (B)
POH 7L1	0.3	42.84	96.4	BRP (B)
POH 8L1	2.2	41.80	92.7	BRP (B)
ML 11b	3.46	33.05	75.1	Binder (C)
ML 13b	4.28	33.85	76.9	Binder (C)
ML 16b	3.64	31.71	72.1	Binder (C)
ML 17b	3.85	35.47	80.6	Binder (C)
ML 20b	3.03	34.85	79.2	Binder (C)
POH 6b	1.7	33.79	74.9	Binder (C)
POH 7b	2.0	38.69	86.2	Binder (C)
ML 18	0.56	23.92	54.4	Sandstone (D)

3.4 X-Ray Diffraction (XRD)

The samples analysed by the TA were also analysed by XRD. The major component of all materials was calcite. Dolomitisation was low, and the highest dolomite content was detected in ML 20L1 (Table 4). Quartz, feldspar and muscovite were mostly of aggregate origin. Raw materials (group A) were rare. Contamination of binder samples (group C) with aggregate ranges from 5 to 18% by weight. A comparison of the X-ray results and the thermal analysis results of the raw materials (group A) suggests that a large part of the amorphous fraction is probably less crystalline calcite, which appears amorphous on the diffractogram.

3.5 Leaching off in Acid and Sieve Analyse

The ratio of soluble and insoluble mortar components was influenced by the number of limestone fragments and BRPs. These components were soluble in acid as well as

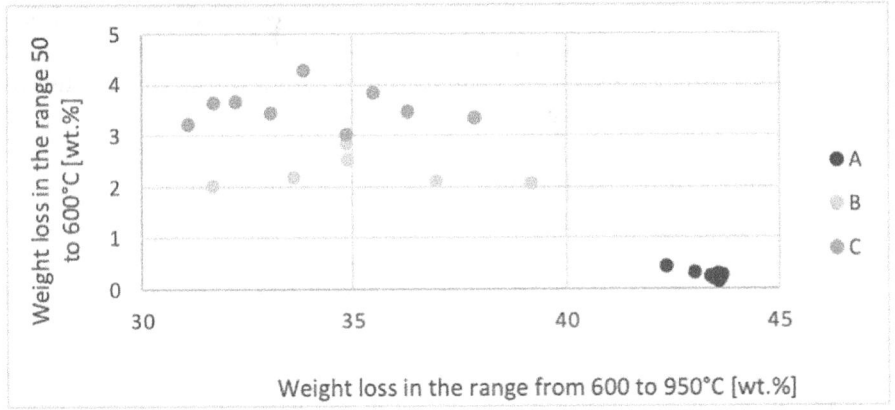

Fig. 6. Comparison of the weight losses in the ranges 50 to 600 °C and 600 to 950 °C for the groups A, B and C.

Table 4. Mineralogical composition of selected samples (XRD), Q - quartz, Calc - calcite, Feld - feldspars, Mus - Muscovite, Chlo - chlorite, Dol - dolomite and Am - amorphous

Sample	Q	Calc	Feld	Mus	Chlo	Dol	Am
ML 14L (A)	<0.5	80.8	-		-	-	19.0
ML 17L2 (A)	<0.5	71.5	-	-	-	0.7	28.0
ML 19L2 (A)	0.6	80.5	-	-	-	-	19.0
ML 20L1 (A)	<0.5	73.2	-	-	-	1.8	25.0
ML 13L1 (B)	8.6	60.9	1.4	0.6	-	-	28.4
ML 15L1 (B)	15.1	50.0	2.6	0.5	-	-	31.9
ML 16L1 (B)	10.3	55.6	1.5	0.4	-	-	32.0
ML 17b (C)	5.6	64.1	2.6	1.1	<0.5	-	26.4
ML 20b (C)	5.7	63.4	1.7	0.9	-	-	28.0
ML 29b (C)	3.3	77.6	1.3	0.6	-	-	17.2
POH 2b (C)	8.7	87.3	3.2	0.8	-	-	-
POH 5b (C)	41.6	27.2	16.0	5.9	4.2	-	1.3
POH 6b (C)	9.6	85.1	3.1	0.8	0.7	<0.5	0.6
POH 7b (C)	5.4	90.9	1.4	1.4	<0.5	<0.5	0.7
POH 8b (C)	12.6	77.3	3.6	1.6	1.5	0.7	-

the binding matrix. Larger limestone fragments and BRP were possible to remove but the majority were left and thus included in the soluble component. The weight ratio of insoluble and soluble components was in the range of 1: 0.5–2.1 (Table 5). The mortars were rich in the binder, but the soluble part was higher due to the presence of particles, which, although added to the mortar as a binder, did not behave as effectively due to their size. The sands obtained by dissolving the mortar samples had similar characteristics. Only sample ML 15 contained significantly coarser fractions. This possibly corresponds to some specific use of this mortar.

Table 5. Selected weight ratios of insoluble and soluble components

Sample	Sample weight before dissolution [g]	Insoluble components A [wt. %]	Soluble components B [wt. %]	A/B ratio
ML 10	50.7	68.2	31.8	2.1
ML 11	95.6	44.4	55.6	0.8
ML 13	75.1	58.1	41.9	1.4
ML 15	81.8	46.1	53.9	0.9
ML 16	125.9	34.7	65.3	0.5
ML 19	61.8	65.7	34.3	1.9
ML 20	89.6	60.7	39.3	1.5

3.6 Leaching off in Acid and Sieve Analyse

The ideal value of carbonate binder $\delta^{13}C$ V-PDB, in which atmospheric CO_2 was absorbed by calcium hydroxide, is between $-20‰$ and $-27‰$ [35]. Moreover, $\delta^{13}C$ fluctuate to both positive and negative values in the presence of contaminants, typically geogenic carbonates or layered silicates, which absorb recent atmospheric CO_2 [36]. The presence of these contaminants or geogenic carbon can cause inaccurate results in radiocarbon dating. Sound and hard-looking white particles were removed from mortars and their isotope content was determined, mainly aiming to trace their origin.

The content of oxygen and carbon stable isotopes was determined for 12 samples (Fig. 7). The variance of $\delta^{13}C$ V-PDB values was between 1.6 ‰ and -17.7 ‰ (Table 6). This content suggests that the selected samples are all limestones, POH 2L could be partially burnt and re-carbonated limestone. They most probably contain geogenic carbonate. For dating purposes, this material must not be included.

The differences between the materials from Mikulčice and Pohansko were also observed. The results suggest that the raw material sources were probably not identic but isotopically related. Only the sample POH 2 showed significantly different values, which indicate carbonation without geogenic impurity. Probably it was a partially burned material.

4 Discussion

All studied samples from Mikulčice and Pohansko were of a similar nature. No secondary recrystallisation was observed, which suggests that the samples may be suitable for ^{14}C dating in terms of sample rejuvenation.

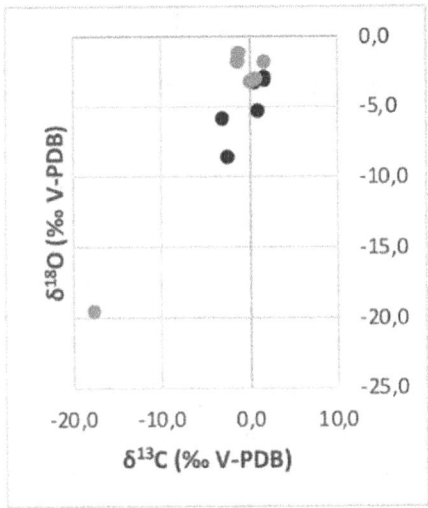

Fig. 7. $\delta^{13}C$ and $\delta^{18}O$ isotopes of selected samples.

Table 6. $\delta^{13}C$ & $\delta^{18}O$ isotopes of selected samples.

Sample	$\delta^{13}C$ (‰ V-PDB)	$\delta^{18}O$ (‰ V-PDB)	$\delta^{18}O$ (‰ V-SMOW)
ML 10L1	0.5	−3.3	27.7
ML 10L2	1.6	−3.2	27.7
ML 12L1	1.0	−5.3	27.6
ML 14L1	1.6	−2.9	27.8
ML 20L1	−3.0	−5.8	26.0
ML 30L	−2.4	−8.6	27.4
POH 1L1	0.6	−3.1	13.7
POH 2L	−17.7	−19.5	9.3
POH 3L	0.1	−3.2	9.4
POH 9L	1.6	−1.8	18.0
POH 14L	−1.2	−1.2	13.2
POH 15L	−1.4	−1.9	12.9

The individual components (groups A, B and C) were distinguishable based on macroscopic and microscopic observation and the results of thermal analysis (TA). A slight difference was found in the amorphous phase content between the samples from Mikulčice and Pohansko. The samples from Mikulčice had a higher amorphous content. Based on the TA analysis, the amorphous phase includes calcite (see group A), which may behave differently during the acid dissolution process to collect CO_2 fractions.

The binder of all samples was a calcitic lime with a minimal admixture of non-carbonate components. According to the SEM-EDS and XRD analyses, the cementation index describing the hydraulicity of the binder and BRP was very low (on average 0.02), which corresponds to air limes. Therefore, it is likely that the raw material for the binder production was very pure limestone without impurities. This is also in agreement with the results of the TA analysis of BRP and limestone fragments. The stable isotope analysis suggests that the limestone could come from similar sources, possibly a single source.

The mortars were characterised by a high binder content and a significant amount of limestone fragments and BRP. Most of the limestone fragments do not show signs of temperature changes. Some particles (especially in the ML 29 sample), which retained their original structure and shape, showed the presence of fine cracks and were composed of fine-grained carbonate mass. These BRP were most probably partially burnt and thus, they represented the composition of the original raw material. At the same time, their isotope $\delta^{13}C$ content was shifted by the carbonation process (POH 2L).

Based on the composition and identical structural features [37], the so-called Ernst-brunn limestones of the Jurassic Period can be identified as a potential source of these limestones, which form significant ridge formations in the Pálava area. Differences in stable isotope composition in limestone material from the Mikulčice and Pohansko localities indicate the same limestone type.

As stated by Leslie a Hughes [15], Ingham [16] or Elsen [17, 34], the presence of carbonate filler particles and unburnt binder fragments of the original limestone material has a significant impact on dating options. The presence of geogenic carbonate introduces an error in the dating results and leads to an incorrect interpretation of the age of the studied mortar. However, such mortar particles are relatively abundant in samples from Mikulčice and Pohansko. They were well identified using by OM and CL. It is, therefore, crucial to choose an adequate pre-processing of these mortars focusing on an appropriate separation method of the limestone particles. The limestone fragments could be physically removed by a mechanical procedure [35]. Alternatively, the fractionation during the acid dissolution can be used to exclude this geogenic carbon [11]. The suitability of the mechanical or chemical separations needs to be verified further.

In addition to the already mentioned limestone fragments, the presence of all carbonate particles is very important. The BRP present in the mortars were of various types, including partially burnt lumps, not disintegrated lime fragments and lime lumps sensu stricto. The TA results also showed that some of these BRP were less pure than the present limestone fragments, which suggests their contamination. Some of them can possibly be slightly hydraulic. These particles may also form non-carbonate clusters that do not reflect representative CO_2 in the atmosphere at the time of the construction process. In order to use the BRP for dating, as suggested by [13], their identification would have to be improved by further studies.

5 Conclusion

Based on the characterisation of the historical mortars, the following points should be considered if their radiocarbon dating is to be carried out: All mortars are made of carbonated calcitic air lime and non-carbonate sand. Studied samples contain carbonate particles: limestone fragments, BRP and calcareous sandstone. These components are of various sizes and cannot be simply separated as individual particles. Therefore, a geogenic carbon present can negatively affect the dating results.

The limestone and calcareous sandstone fragments are clearly distinguishable. They differ from the binding matrix, favouring their separation during the pre-treatment phase carried out prior to the CO_2 extraction for the dating.

The CL demonstrated the presence of carbonates of geogenic origin. The bright red colour was assigned to unburnt limestone; the dark red was typical for particles that probably originated from the fragmentation of partially burnt particles. They varied in size ranging from several to hundred micrometres, this makes them harder to separate from the binding matrix.

The binder related particles in the mortars comprised (i) partially burnt limestone, (ii) fragments of not disintegrated but carbonated lime and (iii) lime lumps sensu stricto.

The lime lumps sensu stricto can in theory be used directly for the ^{14}C dating; however, the study showed that their differentiation from the other types of BRP requires analysis of their structure by microscopy.

Acknowledgement. The research was carried out with the support of the Czech Ministry of Culture, as a contribution to project DG20P02OVV028. The authors would like to thank the colleagues from the Department of Stable Isotopes of the Czech Geological Survey who carried out the MS analysis.

References

1. Ringbom, Å., Gustavsson, K., Lindroos, A., Heinemeier, J., Sveinbjörnsdôttír, Á.: Mortar dating – a method with a potential for the future. In: Proceedings of the 2nd European Congress of Medieval studies of the Fédération Internatiole des Instituts d' Ètudes Médievales, Barcelona (1999). Author 1, A., Author 2, B.: Title of the chapter. In: Editor 1, A., Editor 2, B. (eds.) Book Title, 2nd ed., vol. 3, pp. 154–196. Publisher, Publisher Location, Country (2007)
2. Hale, J., Heinemeier, J., Lancaster, L., Lindroos, A., Ringbom, Å.: Dating ancient mortar. Am. Sci. **91**, 130–137 (2003)
3. Lindroos, A.: Carbonate Phases in Historical Lime Mortars and Pozzolana Concrete: Implications for ^{14}C Dating. Department of Geology and Mineralogy, Åbo Akademi University, Åbo (2005)
4. Heinemeier, J., Ringbom, Å., Lindroos, A., Sveinbjörnsdóttir, Á.E.: Successful AMS ^{14}C dating of non-hydraulic lime mortars from the medieval churches of the Åland Islands, Finland. Radiocarbon **52**(1), 171–204 (2010)
5. Reimer, P.J., et al.: IntCal13 and Marine13 radiocarbon age calibration curves 0–50,000 years cal BP. Radiocarbon **55**(4), 1869–1887 (2013)
6. Marzaioli, F., et al.: Accelerator mass spectrometry ^{14}C dating of lime mortars: Methodological aspects and field study applications at CIRCE (Italy). Nucl. Instrum. Methods Phys. Res. **294**, 246–251 (2013)
7. Lichtenberger, A., Lindroos, A., Raja, R., Heinemeier, J.: Radiocarbon analysis of mortar from Roman and Byzantine water management installations in the Northwest Quarter of Jerash, Jordan. J. Arch. Sci.: Rep. **2**, 114–127 (2015)
8. Ponce-Antón, G.: Hydrotalcite and hydrocalumite in mortar binders from the medieval castle of Portilla (Álava, North Spain): accurate mineralogical control to achieve more reliable chronological ages. Minerals **8**, 326 (2018)

9. Toffolo, M.: Radiocarbon dating of anthropogenic carbonates using the thermal decomposition method. In: MoDIM 2018. Mortar Dating International Meeting. Université Bordeaux Montaigne, Bordeaux (2018)

10. Arnau, A.Ch.: New strategies for radiocarbon dating of mortars: multi-step purification of the lime binder. In: Proceedings of the 4th Historic Mortars Conference, Santorini, pp. 665–672 (2016)

11. Hajdas, I., et al.: Preparation and dating of mortar samples - mortar dating inter-comparison study (MODIS). Radiocarbon **59**(5), 1–14 (2017)

12. Hayen, R., et al.: Analysis and characterisation of historic mortars for absolute dating. In: HMS 2016, 4th Historic Mortars Conference, pp. 656–664 (2016)

13. Bakolas, A., Biscontin, G., Moropoulou, A., Zendri, E.: Characterisation of the lumps in the mortars of historic masonry. Thermochim. Acta **269**(270), 809–816 (1995)

14. Pesce, G.L., Ball, R.J.: Dating of Old Lime Based Mixtures with the "Pure Lime Lumps", p. 40 (2011)

15. Leslie, A.B., Hughes, J.J.: Binder microstructure in lime Mortars: implications for the interpretation of analysis results. Q. J. Eng. Geol. Hydrogeol. **35**, 263–275 (2002)

16. Ingham, J.P.: Investigation of traditional lime mortars – the role of optical microscopy. In: Proceedings of the 10th Euroseminar on Microscopy Applied to Building Materials, pp. 1–18 (2005)

17. Elsen, J., Brutsaert, A., Deckers, M., Brulet, R.: Microscopical study of ancient mortars from Turnay (Belgium). Mat. Character. **53**(2–4), 289–294 (2004)

18. Poláček, L.: Hradiště Mikulčice-Valy a Velká Morava. Svazek: II, Archeologický ústav AV ČR, Brno, v. v. i., Brno 2016, p. 151 (2016)

19. Galuška, L., Poláček, L.: Církevní architektura v centrální oblasti velkomoravského státu. In: Sommer, P. (ed.) České země v raném středověku, Praha, pp. 92–153 (2001)

20. Wihoda, M.: Od Moravanů k Velké Moravě a zpět. In: Koupřil, P. (ed.) Velká Morava a počátky křesťanství, Brno, pp. 46–50 (2014)

21. Měřínský, Z.: České země od příchodu Slovanů po Velkou Moravu, p. 564. Libri, Praha (2009)

22. Baxa, P.: K významu kostola sv. Margity Antiochijskej v dejinách Kopčan, okr. Skalica. In: Pamiatky Trnavy a Trnavského kraja 3, Trnava, pp. 44–47 (2000)

23. Poláček, L.: 50 let výzkumu v Mikulčicích. Přehled výzkumů **46**, 324–327 (2005)

24. Poláček, L.: Terénní výzkum v Mikulčicích. 2. Rozšířené vydání, ARUB, Brno (2006)

25. Baxa, P., Gregorová, J., Poláček, L.: Projekt archeologického parku Mikulčice-Kopčany. Pamiatky a múzeá **2003**(1), 53–57 (2003)

26. Poláček, L., Mazuch, M., Baxa, P.: Mikulčice-Kopčany. Stav a perspektivy výzkumu. Archeologické rozhledy **58**, 623–642 (2006)

27. Dreisler, P.: Opevnění Pohanska u Břeclavi. Masarykova Univerzita Brno, p. 274 (2011)

28. Macháček, J.: Zpráva o archeologickém výzkumu Břeclav-Líbivá 1995–1998. In: Archaeologia mediaevalis Moravica et Silesiana I, pp. 39–62. Masarykova univerzita, Brno (2000)

29. Dresler, P.: Analysis of archaeological field prospection in the hinterland of the early medieval centre Pohansko near Břeclav. In: Abstracts book/13th Annual Meeting of the European Association of Archaeologists. Zadar, pp. 128–129 (2007)

30. Eckel, E.: Cements, Limes and Plasters. A Facsimile of the 3rd ed. Donhead Publishing, Dorset (2005)

31. Rietveld, H.M.: A profile refinement method for nuclear and magnetic structures. J. Appl. Crystallog. **2**, 65–71 (1969)

32. Folk, R.L.: Practical petrographic classification of limestones. AAPG Bull. **43**(1), 1–38 (1959)

33. Dunham, R.J.: Classification of carbonate rocks according to depositional textures. Mem. Am. Assoc. Pet. Geol. **1**, 108–121 (1962)

34. Elsen, J.: Microscopy of historic mortars—A review. Cem. Concr. Res. **36**, 1416–1424 (2006)
35. Addis, A., et al.: Selecting the most reliable ^{14}C dating material inside mortars: the origin of the Padua Cathedral. Radiocarbon **61**(2), 375–393 (2016)
36. Dotsika, E., Psomiadis, D., Poutoukis, D., Gamaletsos, P.: Isotopic analysis for degradation diagnosis of calcite matrix in mortar. Anal. Bioanal. Chem. **395**, 2227–2234 (2009)
37. Kozlovcev, P., Válek, J.: The micro-structural character of limestone and its influence on the formation of phases in calcined products: natural hydraulic limes and cements. Mater. Struct. **54**(217), 1–27 (2021)

Historic Renders and Plasters. Gypsum-Based Plasters and Mortars. Adobe and Mud Mortars. Rammed Earth Constructions. Natural and Roman Cement

Repair Mortar for a Coloured Layer of Sgraffito Render – A Technological Copy

Jan Válek[1]([⊠]) [iD], Olga Skružná[1] [iD], Zuzana Wichterlová[2] [iD], Jana Waisserová[3], Petr Kozlovcev[1] [iD], and Dita Frankeová[1] [iD]

[1] Institute of Theoretical and Applied Mechanics of Czech Academy of Sciences, Prague, Czech Republic
valek@itam.cas.cz
[2] Faculty of Restoration, University of Pardubice, Litomyšl, Czech Republic
[3] Freelance Restorer, Zahořany, Czech Republic

Abstract. Sgraffito technique was used to decorate renders by scratching the top layer of lime wash in the Renaissance time. In order to contribute to the preservation of surviving sgraffiti in the town of Slavonice in the Czech Republic a study was carried out assessing the possibility to replicate the original materials and the application techniques. Historical sgraffito layers were sampled in situ and studied in a laboratory by commonly used analytical methods - OM, TA, XRD, SEM-EDS. The raw materials, lime binder and sand, were characterised and the mixing proportion app. 1 to 0.7 (vol.) of lime putty to sand was determined. Based on the character of the local limestone, a similar raw material was obtained and burnt in an experimental lime kiln. The sand was obtained locally from a disused pit quarry. Obtaining the raw materials from similar sources as the historic ones allowed studying possible production technologies and application techniques. Appropriate and probable techniques were verified by a series of practical experiments. These considered: lime putty v. dry slaked hydrate; thickness of a layer; trowelling and final finishing; timing of drawing, scratching application of lime wash. The performance of the produced mortar mix was assessed by compressive and flexural strengths, capillary absorption, drying index, open porosity and water vapour diffusion coefficient. The mortar, designed as a material replica of the original, was used in a conservation project on a façade of a house, where missing parts of a sgraffito render were reconstructed.

Keywords: repair mortar · technological copy · sgraffito · limestone provenance · lime burning · production methods

1 Introduction

Historic mortars are characterised in order to describe their composition and physical properties in relation to design suitable repair procedures [1–3]. When a new repair mortar is designed, conservation professionals often advocate a "like for like" approach suggesting that the new mortar recipe should follow that of the original [4, 5]. From the analytical point of view, a copy based on composition is theoretically possible, but there is

© The Author(s), under exclusive license to Springer Nature Switzerland AG 2023
V. Bokan Bosiljkov et al. (Eds.): HMC 2022, RILEM Bookseries 42, pp. 193–206, 2023.
https://doi.org/10.1007/978-3-031-31472-8_15

also a technological part that makes it quite a complicated task. Creating a technological copy of a work of art requires the use of the same materials and techniques that were used for the original.

In the historical centre of Slavonice, on the side facade of the house No. 545, there were found fragments of a Renaissance sgraffito with figures of Landsknechts. (A term used for mercenary soldiers in the period of the Renaissance in the territory administered by the Habsburg emperors. The term has a deeper meaning: a servant of the country fighting for Christian ideas.) The sgraffito used to be covered with a plaster, which protected it from the weather, and thanks to this protection, it has survived in an authentic state, which documents the original craftsmanship and art techniques. These sgraffito fragments offered an ideal opportunity to study the originally used materials and applied technologies. The subject of the research were the figural sgraffito fragments on the right side of the façade depicting a Landsknecht parade. Five figures have been preserved in their entirety, of the other two only fragments of their body (hand) and parts of armament (sword) have been preserved (Fig. 1).

Fig. 1. State before reconstruction of the facade with the Landsknechts. On both sides of the secondarily placed window there are fragments of the original Renaissance sgraffito. The sgraffito pattern on the left side of the picture belongs to the later Renaissance decoration.

Sgraffito, or "scratching" in render, or plaster, is usually defined as an engraved figural and ornamental decoration of the exteriors, and exceptionally of the interiors of buildings. The period in which sgraffito originates influences the character of artistic appearance and techniques. Sgraffito as a distinctive decorative technique, as found on the Renaissance facades in the Czech Republic, comes from the Italian Renaissance [6]. In the Renaissance, the authors of contemporary texts describe the sgraffito technique and point out its painting aspects, which consist of shading. Girogio Vasari (1511–1574) and Filippo Baldinucci (1625–1696) describe details of a technique called "scraping on facades" such as hatching (*tratteggiare*), "pushing" (*aggravare*) and adding "light shadow" (*chiari e scuri*) or "semi-coloured" (*mezzo colore, tinta di mezzo*), which is clearly a description of shading [7]. Both authors also suggest a further colouring or supporting of shadows with watercolour as part of the technique [7] (Fig. 2). Most

Renaissance sgraffiti in the Czech Republic consist of only one fine-grained mortar layer, called *intonaco* in Italian terminology, with a lime wash on the surface [8]. When the texture of this render is revealed by scraping, the colour of the base becomes apparent. Italian contemporary terminology refers to black intonaco (*intonaco nero*), coloured with charcoal. However, the base mortar most commonly has a colour of a filler, which is than simply called a coloured layer/intonaco.

Fig. 2. Detail showing three basic surface finishing techniques of the sgraffito. In the upper third, the lime wash is untreated and this finish acts as the brightest one, in the middle third, the lime wash is partially removed and smoothed with a spatula. This surface was subsequently decorated with hatching. The darkest part at the bottom was created by completely scraping away the lime wash and partially also the mortar surface.

2   Original Sgraffito Technique

The intention of the restoration of the sgraffito was that, if the appropriate graphic originals could be found, the figures would be completed and the fragments could be perceived as more complete scene. The original parts were to be conserved based on the principle of minimal intervention. The newly reconstructed parts of the sgraffito were to be made as a material and technological copy of the original, while maintaining the differences between the reconstructions and the original. By comparing the original Landsknecht figures traced on transparent plastic sheets with a large number of contemporary illustrations, it was found that the prints from the digital collection of the Herzog-Anton-Ulrich Museum in Braunschweig corresponded best to them. However, the Erhard Schön's original illustrations for Hans Sachs's poems about the Landsknechts had to be adapted to the scale of the surviving fragments of sgraffito figures.

2.1   Material Characterisation

The survey investigated the material composition and layering of sgraffito. The parameters assessed and the methods used are summarized in Table 1.

Table 1. Analytical procedures used in the study of the Landsknecht sgraffito mortar system and the summary of the analytical results based on samples.

Thickness of layers and their composition Methods used: *in-situ* survey, OM	The studied sgraffito consists of three layers: an underlying background layer, a 5–10 mm thick coloured intonaco and a 0.3–0.6 mm thick lime wash coating. The intonaco is sound and contains only lime and sand with no other colouring additives (Fig. 3). The lime wash does not contain any filler, only isolated binder related particles up to 0.4 mm in size. The lime wash is a single coat, no application sublayers were detected. A good bond between the lime wash and the substrate is probably due to fresh-on-fresh application. The surface area of the lime wash is enriched with the binder. The lime wash shows brush stroke-marks and different variations of thinning and smoothing down to the substrate
Characteristics of the binder Methods used: OM, SEM-EDS, TA, XRD	The same or very similar calcitic binder was used for both the coloured layer and the lime wash (Figs. 4 and 5). The coloured layer contains macroscopically visible light-coloured binder related particles up to a size of 4 mm. They are mainly composed of partially burnt binder related particles formed by imperfect slaking or during mortar preparation. There are also calcium silicate particles (lesser extent) and their composition differs from the binding matrix. These are very probably relics of silicate minerals from the original raw material, crystalline limestone. These particles potentially add hydraulicity to the binder
Characteristics of the agg Methods used: OM, SEM-EDS Sieve analysis	The aggregate of the coloured layer consists of sand with majority of quartz and gneiss, feldspars and light and dark micas. The particles are unworked, sharp-angled, with an uneven surface. This indicates a short transport or formation by in situ weathering. The sand is less than 4 mm in size, the fraction under 0.063 mm is up to 5 wt. %. The sand was probably treated by sieving. There was no evidence of organic admixtures
Binder to aggregate ratio Methods used: AD, OM	The mortar is riche in binder. The mass ratio of the soluble component to the insoluble component in the coloured intonaco is 37.3: 62.7. The gauging proportions calculated for the materials used to make the trial panels and the technological copy are summarised in Tables 2 and 3
Organic additives Methods used: OM, GC-MS, IR, TA	Traces of the milk protein (casein) were detected in the lime wash in sample OSLB 4 using nanoliquid chromatography. This finding was not confirmed in the other samples and its presence was not identified by IR. The same sample also contained calcium oxalate. It was detected by TA and IR only in this sample. Its presence was interpreted as a result of the action to biological growth (Fig. 4)

[*]*OM – optical microscopy; SEM-EDS – scanning electron microscopy; TA - thermal analysis; XRD - x-ray powder diffraction; AD - acid dissolution; GC-MS - liquid chromatography with mass spectrometer; IR - infrared spectroscopy; CI - cementation index.*

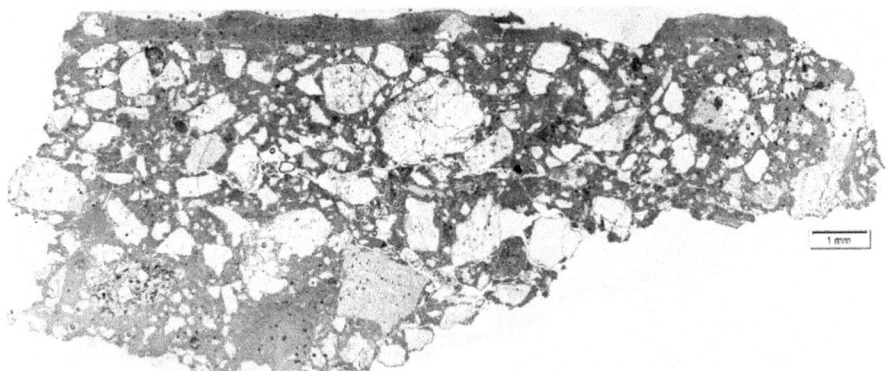

Fig. 3. The coloured intonaco layer is composed of fine-grained binder with occasional binder related particles and an aggregate - predominantly quartz and gneiss. The lime wash does not contain filler and appears as one about 0.2–0.4 mm thick single layer. Sample OSL 4, PPL.

Table 2. Gauging proportions of lime and sand base on acid dissolution test.

	Soluble v. insoluble ratio	Binder - lime putty	Binder – hydrated lime powder	Binder – lump quicklime
	Content [wt.%]	Vol. ratio	Vol. ratio	Vol. ratio
Binder	37.3	1.0	1.0	1.0
Sand	62.7	0.7	0.9	1.4

Table 3. Properties of lime binder (Nedvědice quarry) and aggregate (Slavonice quarry) used to convert mass to volumetric proportions.

	Bulk density [kg/m^3]	Dry matter [hm. %]
Lime putty	1350	40
Powder hydrate	740	
Lump quicklime	850	
Sand	1850	

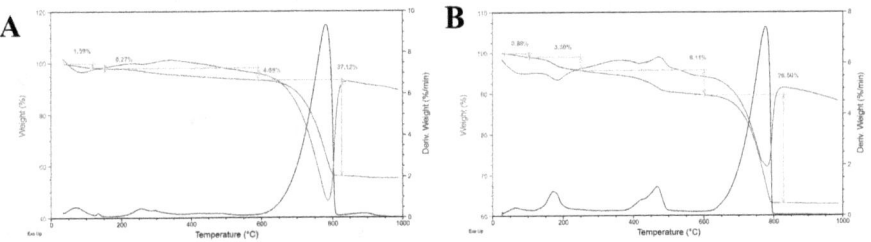

Fig. 4. TA analysis of a binder related particle from the coloured intonaco from the OSL 3 sample (A) and from the OSL 4 sample (B). Beside the carbonate decomposition band, there are two bands corresponding to the contamination of the OSL 3 sample by calcium oxalate.

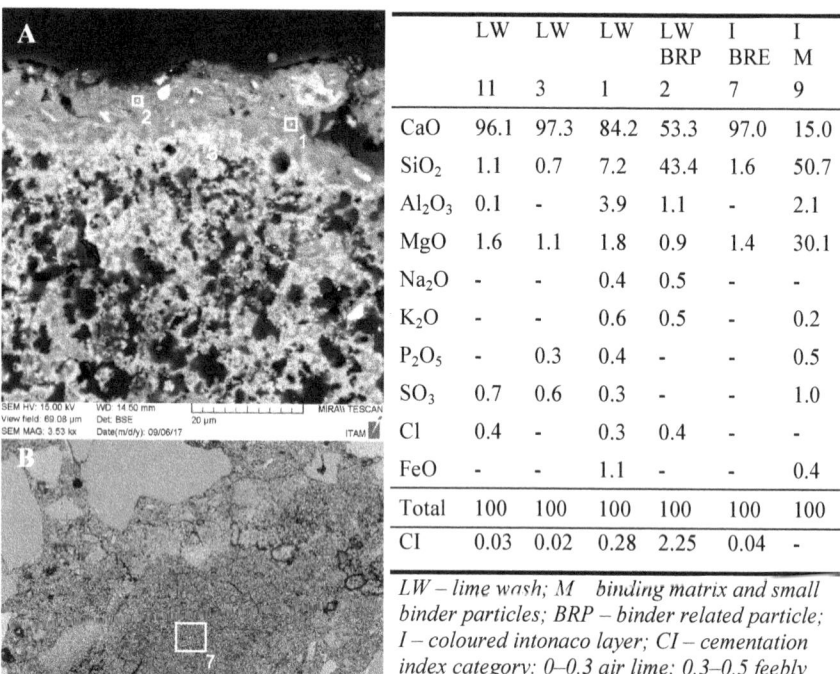

	LW	LW	LW	LW BRP	I BRE	I M
	11	3	1	2	7	9
CaO	96.1	97.3	84.2	53.3	97.0	15.0
SiO$_2$	1.1	0.7	7.2	43.4	1.6	50.7
Al$_2$O$_3$	0.1	-	3.9	1.1	-	2.1
MgO	1.6	1.1	1.8	0.9	1.4	30.1
Na$_2$O	-	-	0.4	0.5	-	-
K$_2$O	-	-	0.6	0.5	-	0.2
P$_2$O$_5$	-	0.3	0.4	-	-	0.5
SO$_3$	0.7	0.6	0.3	-	-	1.0
Cl	0.4	-	0.3	0.4	-	-
FeO	-	-	1.1	-	-	0.4
Total	100	100	100	100	100	100
CI	0.03	0.02	0.28	2.25	0.04	-

LW – lime wash; M binding matrix and small binder particles; BRP – binder related particle; I – coloured intonaco layer; CI – cementation index category: 0–0.3 air lime; 0.3–0.5 feebly hydraulic lime; 0.5–0.7 moderately hydraulic lime; 0.7–1.1 eminently hydraulic lime; > 1.1 natural / roman cement.

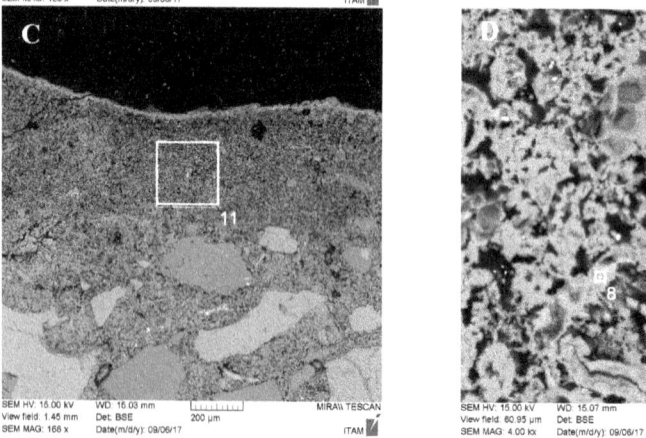

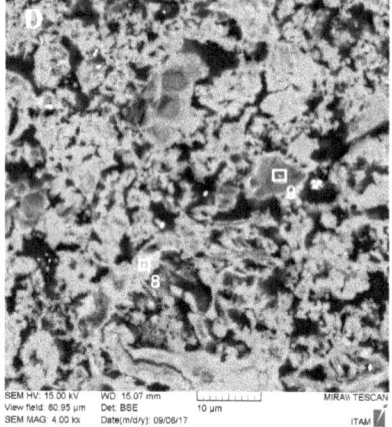

Fig. 5. Surface of lime wash with contamination (A). A binder cluster in the coloured intonaco layer (B). Lime wash and coloured intonaco layer (C). The lime wash is well bonded with the background, no individual sub-layers are visible in the lime wash coating. Matrix of coloured intonaco with small binder particles varying in composition (D). Sample OSL 4, SEM – BSE. The table present the EDS analysis of the binding matrix and the binder related particles.

3 Design of a Repair Mortar for Coloured Intonaco Layer as a Material Copy

3.1 Lime

The town of Slavonice is located in the Moldanubic area with magmatic and meta-morphosed rocks, including crystalline limestone or marbles. Knowledge of the local geology suggests that crystalline limestone was very likely the raw material used for the lime production, which is in line with the findings of the material analysis. In the search for a suitable raw material, several historic limestone quarries in the area were explored [9]. The nearest possible sources could not be used for various reasons (aban-doned quarries were mostly backfilled). The quarry in Municipal Quarry in Nedvědice is about 120 km away but is located in the same geological formation, and sufficient quantities of good quality stone can still be found there. This raw material fulfilled the basic material characteristics determined by the analysis of the binder and was the most readily available of the identified historic quarries. It is a crystalline limestone with calcium-silicate admixtures, clasts of which were present in the mortar binder and in the lime wash coating. The chemical composition of the samples taken on the spot indicated a raw material suitable for the production of air lime (cementation index $< 0,3$; $CaCO_3$ content $> 94\%$), which was also confirmed by the XRD analysis of the raw material and the burnt quicklime samples, see Table 4.

Table 4. XRD quantitative phase analysis of the raw material (white marble) and three quicklime samples of white Nedvědice marble burnt at 1050 °C with residence times 3 h, 5 h and 8 h.

	NB Raw material	NB_3h	NB_5h	NB_8h
Calcite	94.7			
Lime		62.5	57	61.9
Dolomite				0.3
Portlandite		3.5	7.1	4.5
Periclase				0.1
K-feldspar	0.8		1.3	0.3
Mica	1.4			
Clinopyroxene		1.5		0.9
Diopside	3.0			
Amorphous	n.d	32	34	32

3.2 Sand

In the vicinity of Slavonice, there are eluvial deposits of gneiss massif. The sand is dominated by sharp-angled quartz particles, gneiss, light and dark micas. Samples taken from extinct sand pits in several places near Slavonice corresponded in their mineralogical composition to the sand used in the Slavonice sgraffito. Apart from the regional affiliation, the selection of suitable sand took into account grain size distribution and mineralogical composition, as well as the proportion of fine and clay particles. The workability of mortar, its colour and the appearance of the sgraffito lines and surface finishes were also crucial in the evaluation. In terms of colour and plasticity, the mortar made with sand that had the lowest content of the finest fraction among the sands from the local pits was the most suitable. The granulometric curve of this sand also corresponded best to the original, both in the amount of fine fraction and in the majority of the 0.5–1 mm fraction. This sand from the now abandoned sandpit in Slavonice was chosen as a suitable one (Fig. 6). It contained a very small amount of organic particles and therefore did not need to be cleaned by washing, only the max. Grain size was limited by sieving. The content of Cl^-, NO^{2-}, SO_4^{2-} anions was check and was found to be low.

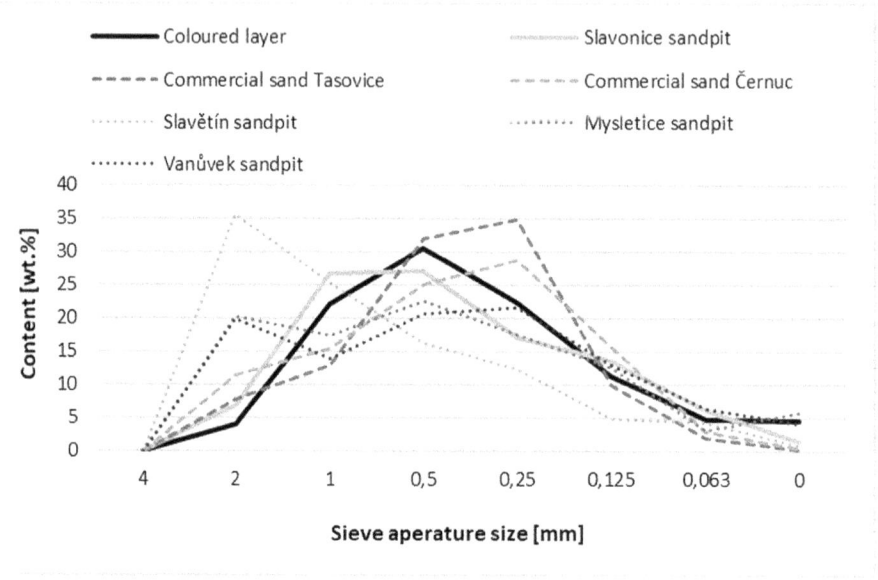

Fig. 6. Comparison of the sand used in the coloured intonaco layer with the samples from Slavonice sandpit, two commercially available sands and other local sandpits.

3.3 Production and Processing Technologies

The aim was to produce a mortar that replicated the original in as many details as possible. The selected raw material was burnt in a wood-fired flare lime kiln according to a procedure that reflected the traditional lime burning in the area [10]. By calcining the Nedvědice marble, we obtained quicklime in lumps, whose further processing had to be decided. The texture of the original coloured intonaco layer contains white "lime particles", which are sometimes clearly visible on the scraped surfaces of the original sgraffito. This led to a suggestion that lime was dry slaked to powder before its use as a binder. This idea was supported by the fact, that the present lime-based particles were limited in size to 4 mm and smaller as if the binder was sieved through the same sieve as the sand. After practical tests and taking into account the above-mentioned aspects, the lime intended for the preparation of the coloured intonaco was dry-slaked to powder. A traditional way was followed, when the quicklime is immersed in a basket for about 15 s to water and then it is left to slake on a heap [11]. After slaking, the grain size was then adjusted by sieving to obtain particles below 4 mm. For the lime wash, the quicklime was slaked in excess water in a wooden slaking vessel. The hydrated lime and the putty were prepared two weeks prior to use, ensuring that all binder particles had enough time to react with water. The intonaco mortar was prepared from lime and sand in a vol. ratio 1:1.

4   Assessment of Physical Properties of the Material Copy

The anticipated advantage of using a technological replica of the original mortar is the guarantee that it will appropriately complement the existing historic materials. An inappropriate material addition could result in a shortened life span of the original or even in accelerated degradation. To assess the copy of the intonaco, its mechanical and physical properties were determined on laboratory specimens according to EN standards. The testing procedure is published elsewhere [12]. Mortar testing was conducted in two sets of different consistencies (initial water contents). The mortar properties obtained were compared with some published values for lime-based mortars, see Table 5. The strength of the intonaco replica is higher than the predicted strength for air lime mortar. The determined strengths correspond to a weakly hydraulic lime, category NHL 2. The capillary absorption, drying rate and water vapour permeability are also consistent with NHL mortars. The open porosity is slightly higher than the values published for naturally hydraulic lime mortars.

SEM analysis of a sample taken from a replicated sgraffito showed an increased level of various mineral relicts with reaction rims compared to the original mortar (Fig. 7). The composition of lime wash was similar to the original. The heavier particles probably sedimented when the putty was left to rest and were not present in the paint.

5   Practical Use and Evaluation by Experiments

The designed mortar was tested on trial panels in order to assess its properties and practical suitability for application in the proposed restoration. The following parameters were considered: workability and plasticity, the speed of setting in relation to the quality of sgraffito drawing, and last but not least, the final appearance and its surface texture. In practical tests, it was necessary to maintain similar layer thicknesses as on the original for both the coloured intonaco and the lime wash. During the tests, the timing of the individual steps was assessed; e.g., sufficient setting of the coloured intonaco for the application of the lime wash, the optimum state of the lime wash paint for the sgraffito drawing and shading; the presence of shrinkage cracks. The first practical tests took place outside the building itself. Once the technological procedure had been refined,

Table 5. Characteristic values of the designed mortar compared with published values. The properties were determined after 120 days of curing on standardized specimens (40 mm × 40 mm × 160 mm). B: A = binder: filler.

	Designed mortar		Reference mortars from literature	
Water content	29 wt. %	26 wt. %	Hydraulic limes (NHL)	Air limes (AL)
Compressive Strength [MPa]	5.9 ± 0.4	6.8 ± 0.3	7.5–10.5 (B:A 1:1 vol.) after 180 days [13]	3.5–4.2 (B:A 1:1 vol.) after 1 year [14]
Flexural strength [MPa]	3.0 ± 0.1	3.3 ± 0.3	1.2–2.5 (B:A 1:1 vol.) after 180 days [13]	0.9–1.1 (B:A 1:1 vol.) after 1 year [14]
Capillary absorption Cc 10′–2′ [kg.m^{-2}.h$^{-1/2}$]	6.61 ± 0.33	5.73 ± 0.36	5.64 (B:A 1:3 vol.) after 3 years [15]	8.34 (B:A 1:3 vol..) after 3 years [15]
Drying index [−]	0.196 ± 0.004	0.179 + 0.016	0.139–0.117 (B:A 1:3 vol.) after 3 years [15]	0.168 (B:A 1:3 vol.) after 3 years [15]
Total open porosity [vol. %]	35.7 ± 0.4	34,5 ± 0.2	28.0–30.0 (B:A 1:1 vol.) after 1 year [13]	24.1–27.1 (B:A 1:1 vol.) after 1 year [13]
Water vapour diffusion coefficient μ [−]	10.1 ± 2.1	11.5 ± 2.5	11.1–17.5 after 3 years [15]	23.3 after 3 years [15]

further tests were carried out on site, which allowed the timing of the individual steps in accordance with the local conditions. The choice of sand was verified on site in relation to the original mortar appearance. Its colour and effect on the drawing pattern were assessed, and the tendency to crack at different thicknesses of coloured intonaco was noted and evaluated.

The following practical conclusions regarding the application technique were drawn:

- The choice of processing lime to a powdered hydrate proved as practical as there were fewer cracks compared to mortars made with putty. The local sand from the Slavonice sandpit was also suitable in terms of the required colour and the fineness of drawing.
- The mortar should be prepared at least four days in advance to ensure the completion of the slaking process.

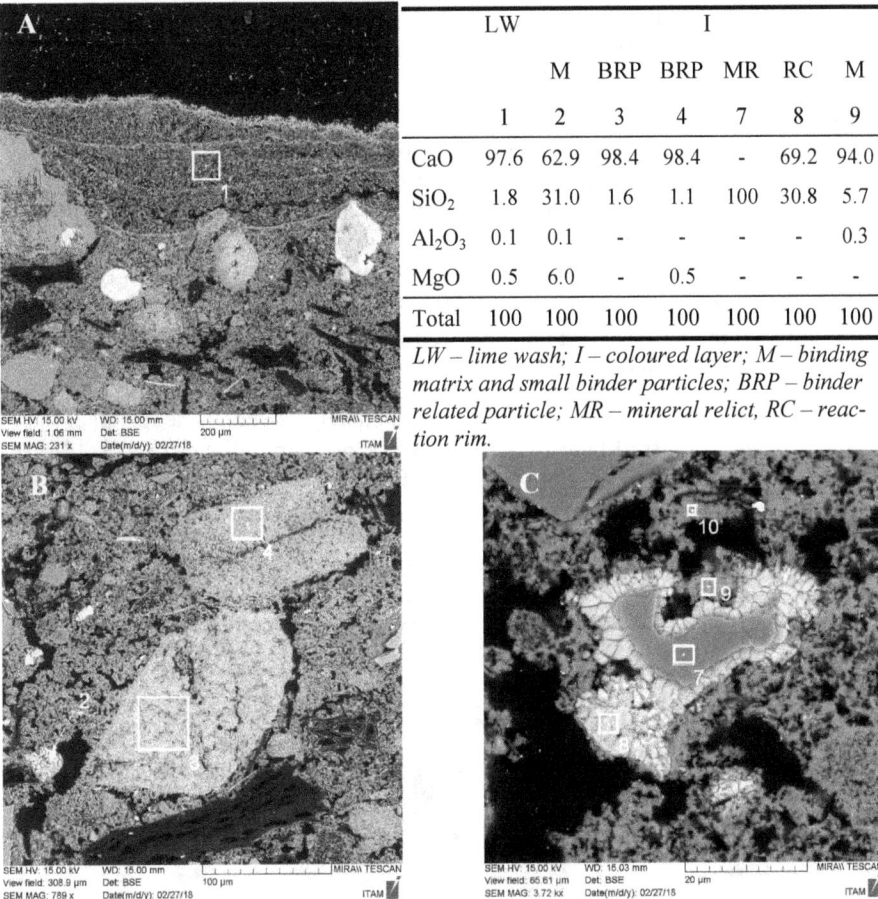

	LW		I				
	M	BRP	BRP	MR	RC	M
	1	2	3	4	7	8	9
CaO	97.6	62.9	98.4	98.4	-	69.2	94.0
SiO$_2$	1.8	31.0	1.6	1.1	100	30.8	5.7
Al$_2$O$_3$	0.1	0.1	-	-	-	-	0.3
MgO	0.5	6.0	-	0.5	-	-	-
Total	100	100	100	100	100	100	100

LW – lime wash; I – coloured layer; M – binding matrix and small binder particles; BRP – binder related particle; MR – mineral relict, RC – reaction rim.

Fig. 7. Replicated sgraffito. Lime wash applied in three coats (A). Binder related particles of various composition in the replicated coloured intonaco (B). A mineral relict with reaction rim and the surrounding binding matrix in the replicated coloured intonaco (C). Sample SGSLU, SEM – BSE. The table present the EDS analysis of the binding matrix and the binder related particles.

- It was confirmed that the analytically determined proportions of powdered hydrate to sand in a volume ratio of 1:1 were suitable for the coloured intonaco. The lime-rich mixture had good water retention and the desired thickness was achieved with a minimum of cracking.
- The absorption capacity of the substrate significantly influenced the timing of the sgraffito application procedure. Adequate pre-wetting of the substrate allowed the speed of the process to be controlled. In parallel, the climatic conditions must be taken into account during the application process.
- Due to the tendency to crack, it is better to apply the mortar in two layers. It seems that the first layer is preferable to be slightly thinner as it then adheres better to absorbent rough substrate.

- Lime wash is prepared from lime putty diluted by water. The thinning of the paint corresponded to the degree of intonaco setting, masonry absorption capacity and weather conditions. It is desirable to apply it only after the primary drying cracks of the coloured layer appear and are sealed. The drying cracks were evident in all tests and, if pushed back at the right time, they did not further appear.
- It is preferable to apply the lime wash in several coats with minimum time intervals, however, the bottom coat must be set before application of the next layer.
- During the reconstruction of the sgraffito, it is desirable to apply the paint with a brush at the moment when the surface does not react to the fingerprint of the hand and therefore the paint does not mix with it. The first coat should be thinner, as with the coloured intonaco, so that it blends well with the substrate. A thicker coat can be applied in the next layer.

The findings related to the sgraffito technique itself, as well as the restoration procedures, are not discussed here in details but can be found elsewhere [16, 17]. The restoration project was successfully completed. The new sgraffito parts were slightly retouched with pigments from the local sand to reduce the contrast between the old and new areas (Fig. 8).

Fig. 8. The reconstructed sgraffito facade after two years. Slavonice, house No. 545.

6 Concluding Remarks

The article presents the procedure used to create a technological copy of a Renaissance sgraffito. This procedure combined several sub-tasks related to the analysis of historic sgraffito and the interpretation of the analytical results for building conservation practice. This holistic approach, which links the quality of the material with the craftsmanship and final appearance of the sgraffito, proved to be an interesting way to deepen the understanding of these topics, which are often studied independently.

In terms of technology, there are some points which should be assessed further in future. Dry slaked lime to powder hydrate was used as there were binder related particles found in the historic mortar. Slaking lime to putty did not provide so many binder related particles. The use of powder hydrate allowed for better workability adjustment of mortar. The water content of lime putty is typically around 50% by weight or more, which makes the lime-rich mortar too thin and runny. Reducing water content in lime putty mortars is possible, for example by drying or mechanical squeezing, but in our case, this seemed less practical.

The dry slaked lime had a higher heterogeneity in particle size than the currently available standard commercial limes would have. This affected the workability of the mortar. The larger particle size of lime, up to 4 mm according to the original, also meant a higher proportion of binder in determining the binder to sand ratio. The gauging ratio, if calculated based on the dissolution of the mortar in acid, should ideally also be related to the size distribution of the binder particles. However, as this is not practically possible, a minimum knowledge of the raw materials used and the differences between lime production technologies in the past and today is necessary. The current EN 459-1 standard requires at least 85% (wt.) particles to pass a 90 μm sieve. Historical technologies very probably produced coarser particles. Replicating the technological processes of the past could provide binders comparable to historical ones.

TA and SEM-EDS analyses of the binder indicated that a relatively pure lime was used to produce the original mortar. This resulted in a search for a calcitic marble with a high calcium carbonate content. A suitable source of raw material was found and the marble was burnt in a wood-fired kiln in a manner that would have been consistent with the 16th century in then Bohemia. However, it was found later, when the mechanical properties were assessed, that the mortar made from this lime had parameters comparable to natural hydraulic limes. The XRD of quicklime did not show presence of any typical hydraulic phases such as C_2S [18], nor any quartz relics, probably due to the fact, that the analysed samples were not fully representative of the whole batch. The SEM-EDS analysis of the replicated mortar revealed a number of calcium-silicate particles and quartz particles with reaction rims. It is possible that the localized calcium-silicate impurities in marbles reacted and were activated at high temperatures during calcination (900–1200 °C) and subsequently caused the formation of hydraulic bonds in the mortar. This could have been the case also of the historical mortar, but the determination of hydrated hydraulic phases in historic mortars is quite a complex task, especially when they are exposed to moisture cycles and their carbonation is promoted over a long time [19, 20].

Knowledge of the technological processes of lime production and mortar preparation cannot be obtained by studying samples alone. Here, an approach was chosen which consisted in combining the findings from historical samples based on laboratory analyses with possible technological processes, which were then verified by practical experiments. Such a procedure led to new insights in the field of the original Renaissance sgraffito techniques. Finally, it was thus possible to apply these materials in the reconstruction carried out to complete the missing parts of the Landsknecht figures. The historical sgraffito was conserved, the additions were realised as a technological copy.

Acknowledgements. The research has been carried out with the support of TAČR TL03000603 and NAKI DG20P02OVV028 projects. The authors would like to thank to Dr. Štěpánka Kučková for analysis of organic additives and proteins and Dr. Alberto Viani for XRD-QPA of the samples.

References

1. Groot, C., Ashall, G., Hughes, J.: Characterisation of Old Mortars with Respect to their Repair. RILEM (2007)
2. Elert, K., Rodrigues-Navarro, C., Pardo, E.S., Hansen, E., Cazalla, O.: Lime mortars for the conservation of historic buildings. Stud. Conserv. **47**(1), 62–75 (2002)
3. Elsen, J.: Microscopy of historic mortars – a review. Cem. Concr. Res. **36**, 1416–1424 (2006)
4. Bell, D.: Technical Advice Note 8: The Historic Scotland Guide to International Charters, Historic Scotland, Edinburgh (1997)
5. Jokilehto, J.: The context of the Venice Charter (1964). Conserv. Manag. Archaeol. Sites **2**(4), 229–233 (2013)
6. Waisser, P.: Renesanční sgrafito: Technika a technologie. Sgrafita zámku v Litomyšli. Univerzita Pardubice, Litomyšl (2011)
7. Waisserová, J.: Figurální sgrafitová výzdoba s landsknechty na fasádě domu čp. 545 v ulici B. Němcové ve Slavonicích. Univerzita Pardubice, Litomyšl (2020)
8. Wichterlová, Z.: Průzkum techniky renesančního sgrafita. Univerzita Pardubice, Litomyšl (2015)
9. Válek, J., et al.: GIS of historic and current raw material sources and technologies for production of lime binders (2015). http://www.calcarius.cz/gis-calcarius/. Accessed 21 Dec 2022
10. Válek, J.: Lime technologies of historic buildings. ÚTAM AV ČR, Praha, Czech Republic (2015)
11. Wacha, R., Pintér, F.: Lime slaking in baskets. In: Proceedings of the 4th HMC, Santorini, 2016, pp. 155–162 (2016)
12. Válek, J., Skružná, O.: Performance assessment of custom-made replications of an original historic render – a study of application influences. Constr. Build. Mater. **229**, 116822 (2019)
13. Lanas, J., Pérez Bernal, J.L., Bello, M.A., Alvarez Galindo, J.I.: Mechanical properties of natural hydraulic lime-based mortars. Cem. Concr. Res. **34**(23), 2191–2201 (2004)
14. Lanas, J., Alvarez Galindo, J.I.: Masonry repair lime-based mortars: factors affecting the mechanical behavior. Cem. Concr. Res. **33**(11), 1867–1876 (2003)
15. Silva, B.A., Ferreira Pinto, A.P., Gomes, A.: Natural hydraulic lime versus cement for blended lime mortars for restoration works. Constr. Build. Mater. **94**, 346–360 (2015)
16. Wichterlová, Z., Waisserová, J., Skružná, O., Válek, J.: New shading technique revealed through reconstructing the sgraffito technology used north of the Alps during the Renaissance. In: Weyer, A., Klein, K. (eds.) Sgraffito in change. Materials, Techniques, Topics, and Preservation, pp. 124–137. Michael Imhof Verlag, Petersberg (2019)
17. Válek, J., Skružná, O., Waisserová, J., Wichterlová, Z., Maříková-Kubková, J., Kozlovcev, P.: Podle starého vzoru: rekonstrukce malt, sgrafit a štuků. ÚTAM AV ČR, Praha (2021)
18. Alvarez, J.I., et al.: RILEM TC 277-LHS report: a review on the mechanisms of setting and hardening of lime-based binding systems. Mater. Struct. **54**(2), 63 (2021)
19. Frankeová, D., Koudelková, V.: Influence of aging conditions on the mineralogical microcharacter of natural hydraulic lime mortars. Constr. Build. Mater. **264**, 120205 (2020)
20. Elsen, J., Balen, K.V., Mertens, G.: Hydraulicity in historic lime mortars: a review. In: Válek, J., Hughes, J., Groot, C. (eds.) Historic Mortars. RILEM Bookseries, vol. 7, pp. 125–139. Springer, Dordrecht (2012). https://doi.org/10.1007/978-94-007-4635-0_10

Evaluation of the Hygroscopic and CO_2 Capture Capacities of Earth and Gypsum-Based Plasters

Tânia Santos[1] ⓘ, António Santos Silva[2](✉) ⓘ, Maria Idália Gomes[1,3] ⓘ, and Paulina Faria[1] ⓘ

[1] CERIS, Department of Civil Engineering, NOVA School of Science and Technology, NOVA University Lisbon, Caparica, Portugal

[2] National Laboratory for Civil Engineering (LNEC), Lisbon, Portugal
ssilva@lnec.pt

[3] Department of Civil Engineering, Instituito Superior de Engenharia de Lisboa (ISEL), Instituto Politécnico de Lisboa (IPL), Lisbon, Portugal

Abstract. Earth mortars and gypsum mortars present ecological advantages compared to mortars made with other common binders. When applied as plasters, they are also referred as having advantages in improving comfort and indoor air quality. For earth plasters, this improvement is associated with the hygroscopic capacity of the clay minerals, which promotes high sorption and desorption capacity of water vapor. So, earth plasters can contribute to the regulation of the indoor relative humidity. Another important advantage of plasters could be their ability to capture carbon dioxide (CO_2). In the present study, the sorption and desorption performance, and the capacity to capture CO_2 by earth and gypsum plasters are evaluated. It is confirmed that the earth plaster has the greatest sorption and desorption capacity, but also higher CO_2 capture capacity than gypsum plaster. This confirmation opens new perspectives for the use of functionalized plasters that guarantee greater control of air quality inside buildings.

Keywords: CO_2 Capture · Clayey Earth · Gypsum · Hygroscopicity · Indoor Air Quality · Mortar

1 Introduction

Currently, and with urban development, people spend a large part of their time indoors. The time spent inside buildings increased compared to what was already usual due to the global pandemic of COVID-19 and, consequently, the lockdown and teleworking. Therefore, indoor air quality inside the buildings is quite important since exposure to poor indoor air quality has direct effects on the health of occupants as it may cause several health problems, depending on the exposure time and the concentration of pollutants [1, 2].

Plasters can significantly contribute to indoor air quality, as they cover a large part of the indoor surface of buildings, on walls and ceilings.

V. Bokan Bosiljkov et al. (Eds.): HMC 2022, RILEM Bookseries 42, pp. 207–215, 2023.
https://doi.org/10.1007/978-3-031-31472-8_16

Earth as a construction material is composed by different types and fractions of clay, silt, sand and gravel, and has several ecological advantages, as it is a natural material, non-toxic, reusable when not chemically stabilized, recyclable, with low carbon dioxide (CO_2) emissions [3]. Earth mortars are produced with earth mixed with water; when the clay content is high, complementary sand is added. Earth plasters have several advantages, namely the fact that they improve comfort and indoor air quality due to the hygroscopic characteristics of clays, which allows these types of plasters to adsorb and release a high amount of water vapor [4–6]. Therefore, these plasters contribute significantly to passively balancing indoor relative humidity (RH) and temperature [7, 8]. Earth plasters are frequently applied without any finishing, to profit from the natural color they present.

Gypsum plasters are composed by gypsum hemi-hydrate (produced with low temperatures of 120–180 °C) and sand but frequently they have additions such as air lime or retarders, as the gypsum produces speed hardening. Gypsum plasters are frequently complemented with a finishing coat. They are sometimes referred as having good hygroscopicity but it is low when compared to earth plasters [9].

In the specialized literature in this area of knowledge, it is often said that earth plasters contribute to clean polluted indoor air [10, 11]. However, to the authors' knowledge, this statement has not yet been scientifically proven [10, 11]. According to Reeves et al. [12], the absorption of toxins may be related to the porosity of the earth. It is even said that the earth plasters can absorb and bind indoor air pollutants which are dissolved in water vapor [13]. A team of scientists has shown that CO_2 can penetrate the inner layers of some clay minerals [14].

The quantification of CO_2 capture can be carried out through thermogravimetric analysis (TGA), where the thermal decomposition of carbonates leads to a tangible weight loss that can be accurately recorded [15].

The present study intends to evaluate the sorption and desorption capacity, as well as the CO_2 capture, through TGA analysis, of earth and gypsum plasters.

2 Mortars and Specimens

Two mortars were tested: a pre-mixed earth-based (E) and a pre-mixed gypsum-based (G) plasters. The E mortar was produced by Embarro company and was composed by a reddish clayey earth (with illite-kaolinite composition) from Algarve region (South of Portugal), additional fine sand with 0–2 mm and plant fibers, although the proportion of each constituent was not known [3]. The G mortar was produced by Sival company, being the proportions of each constituent also not known. The mortars were prepared only by addition of water to the pre-mixed dry mixture. The water content of E and G mortars was 15% and 43%, respectively, and was determined by the percentage loss of mass of the fresh mortar samples, after oven drying [3]. Different types of flat specimens were prepared and applied: in metallic molds measuring 200 mm × 500 mm with 15 mm of thickness [3, 16] and on the back of ceramic tiles with 450 mm × 450 mm with 20 mm of thickness. After drying, the finishing coat was applied on the gypsum specimens with a maximum thickness of 2 mm.

More details on mortar preparation, fresh state and mechanical characterizations, linear shrinkage and dry bulk density were presented elsewhere [3, 16].

After more than one year of mortars production, the plaster specimens applied in the ceramic tiles were placed in sealed rectangular cells with 1.0 m × 0.5 m × 0.5 m, covering the sides and the bottom, simulating the area of walls and ceiling of a room. These specimens were exposed to 28 cycles of CO_2 injection, injecting an approximate value of 38,000 ± 6,600 ppm. After exposure to these CO_2 injection cycles, which took approximately 8 months, the cells were opened, and samples were taken from the plasters for TGA analysis.

3 Methods

3.1 Sorption and Desorption

The sorption capacity of each plaster was determined based on DIN 18947 [17], specific for unstabilised earth plasters: the specimens in the metallic molds were placed in a climatic chamber at 23 °C and 50% RH until constant mass; then were exposed to 80% RH at 23 °C; the water vapor gain (in g/m^2) was determined by weighing after 1, 3, 6, 12 and 24 h. The weighing at 0.5 h was not performed as it was considered too initial and in order to minimized interference in the stabilization time of the climatic chamber during the weighing process [6]. The desorption capacity of plasters (not defined by DIN 18947 [17]) was determined by a reverse process [3, 6, 16]: the RH inside the climatic chamber was decreased from 80% to 50%, evaluating the decrease of water vapor content (in g/m^2) at the same defined periods of time for sorption (1, 3, 6, 12 and 24 h).

3.2 Thermal Analysis

To analyze the impact of exposure to CO_2 on the composition of the E and G plasters and their ability to capture CO_2 present in the indoor air, simultaneous thermogravimetric analysis with differential thermal analysis (TGA/DTA) were carried out on samples taken from each plaster, before and after CO_2 exposure. XRD was also performed, but it was chosen to present only the TGA/DTA results because in this technique all the CO_2 captured by the existing compounds is counted, regardless of whether they are crystalline or not.

Samples for TGA/DTA were crushed and passed in a 106 μm sieve. TGA/DTA was performed in a SETARAM TGA92 apparatus, using argon atmosphere (3 L/h), with an heating rate of 10 °C/min from room temperature to 1000 °C [18].

4 Results

4.1 Sorption and Desorption

The average curves of sorption and desorption of the E and G plasters and the limits of sorption classes defined by DIN 18947 [17] after 12 hours (WSI for sorption ≥ 35 g/m^2; WSII for ≥ 47.5 g/m^2; WSIII for ≥ 60 g/m^2) are shown in Fig. 1. These results were previously presented by Santos et al. [3, 16] but in comparison with other plasters.

Observing the curves, it is possible to conclude that the E plaster presents significantly higher sorption capacity in comparison with the G plaster, which adsorbs approximately

4.7 times more after 24 h. After 12 h the E plaster adsorbed about 78 g/m^2 and can be classified in the WSIII class. On the other hand, after 12 h the G plaster adsorbed 21 g/m^2, not being able to reach the limits defined by DIN 18947 [17] standard (which is not specific for gypsum-based plasters). However, these same mortar after 1 h of testing achieved the limit of class WSII (adsorbing about 11 g/m^2 for a limit of ≥ 10.0 g/m^2) and after 3 h is in the limit of WSI class (adsorbing about 17 g/m^2 for a limit of ≥ 13.5 g/m^2). Therefore, it seems the gypsum plaster as an initial good sorption capacity but rapidly (after 3–6 h) that capacity is saturated. In contrast, the earth plaster shows an increasing sorption behavior over the 24 h of the test, and it is not even saturated after all that period.

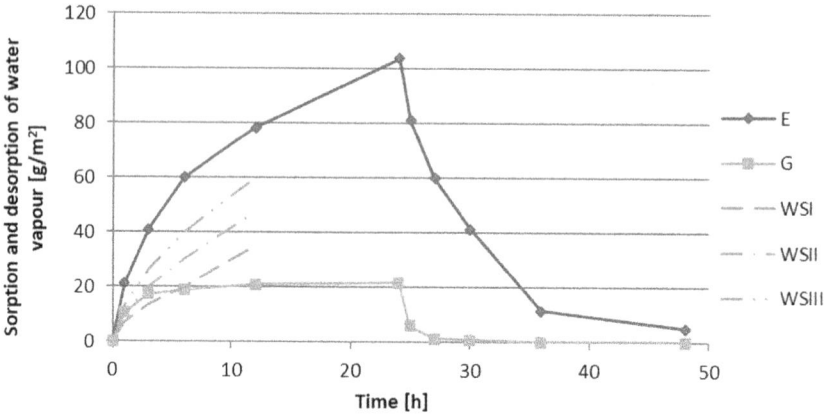

Fig. 1. Sorption and desorption of water vapor of plasters and sorption limits (WSI, WSII and WSIII) defined by DIN 18947 [17].

The high sorption capacity demonstrated by the earth plaster should be related to the high hygroscopic capacity of clays, in particular the illite clay present in this earth [3, 6]. In addition to the type of clay, the presence of plant fibers can also explain the high sorption capacity of this mortar [19].

In desorption, both plasters present a good behavior, as they desorbed almost all water vapor they had adsorbed. Nonetheless, after 24 h the E plaster shows some hysteresis, that is, it desorbed a little less water vapor than it had adsorbed. However, the desorption tendency of this plaster is still decreasing and is not stabilized after 24 h; so, if the test was carried out for a few more hours, likely the water vapor desorption would likely reach the initial value.

Lima et al. [6] analyzed the influence of the type of clay on the sorption and desorption capacity of plasters and concluded that a montmorillonite plaster presents the best behavior, an illite plaster the intermediate behavior and a kaolinite plaster the worst behavior. These authors reveal that after 24 h, the illitic plaster had a sorption of approximately 80 g/m^2. In the present study, after 24 h the E plaster presents a sorption of approximately 100 g/m^2. The high sorption capacity of the E plaster may probably be attributed to the presence of plant fibers, as mentioned above.

In another study, Lima et al. [20] analyzed the influence of the addition of 5, 10 and 20% of gypsum on the sorption capacity of earth plaster and concluded that the higher the percentage of gypsum addition, the lower sorption capacity of the earth-based plasters. After 24 h, the earth plaster without gypsum had a sorption of about 75 g/m^2, while the earth plaster with 20% of gypsum presented a sorption of about 65 g/m^2. Minke [10, 11] analyzed the sorption capacity of a gypsum plaster using specimens with 15 mm of thickness, at 21 °C, raising the RH from 50 to 80% and obtained sorption of 30 g/m^2 and 40 g/m^2, after 24 h and 48 h, respectively. The gypsum plaster analyzed in the present study has a lower sorption capacity, which can be justified by having some unknown additions and so, a different composition, because it is an industrial pre-mixed mortar.

4.2 Thermal Analysis

The TGA/DTA graphs, as well as the derivative thermogravimetric curve (DTG), of the samples before and after CO_2 exposure are presented in Fig. 2. Observing in more detail the DTG graphs of the plasters (Fig. 3) it is possible to conclude that the E plaster after CO_2 exposure shows higher mass losses in the ranges of 25–200 °C, 200–500 °C and 500–900 °C, compared to the E plaster before (without) CO_2 exposure. Regarding the G plaster, no significant mass variations are detected before or after CO_2 exposure.

Table 1 presents the mass loss obtained in the three temperature ranges where mass variations were detected (which correspond to the peaks in the DTG curves). The range 25–200 °C corresponds to mass loss due to the loss of moisture and crystallization water (physical and chemical sorption); 200–500 °C is associated with the loss of water from hydrated compounds, such as clay minerals (chemical sorption); and 500–900 °C is due to loss of CO_2 caused by the decomposition of carbonates (sorption by carbonation).

The clay minerals present in general three main mass loss (Fig. 2 and Table 1): at low temperature, in the interval of 25–200 °C, there is a release of free or sorbed water; at a mid-range temperature, in the interval of 200–500 °C, loss of bound water or dissociation of hydroxyls from the lattice occurs, inducing its amorphization; and at high temperature, in the interval of 500–900 °C, there is a final breakdown of the residual clay lattice and recrystallization of new mineral and/or glass phases [6, 21].

The gypsum plaster presents two mass losses: the first one in the interval of 25–200 °C due to the decomposition of water of crystallization from gypsum, and the second one in the interval of 500–900 °C due to the decarbonation of carbonates (release of CO_2). So, it is likely that the pre-mixed gypsum plaster is composed by a mix of gypsum and calcitic lime (frequently used as a retarder of hardening and to control gypsum expansion). It should be remembered that the plasters were tested for CO_2 capture after more than one year, so the carbonation of that lime had already occurred.

Regarding the mass loss in the interval 500–900 °C due to the CO_2 release, the E plaster presents the higher variation after CO_2 exposure, which means that it presents the best CO_2 capture.

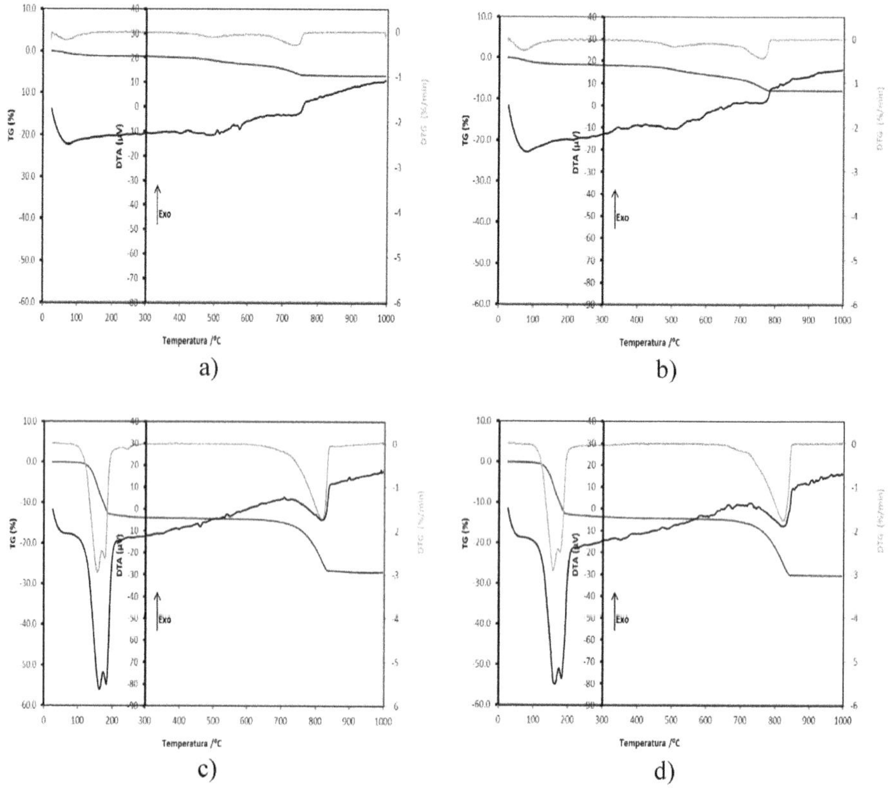

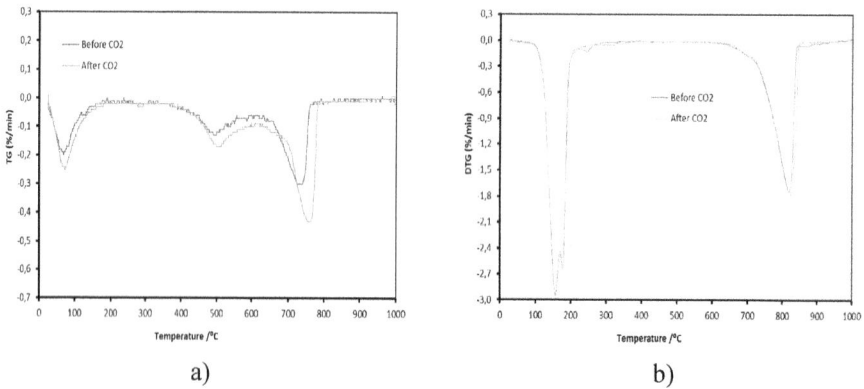

Fig. 2. TGA/DTG/DTA graphs of the plasters: a) E plaster before CO_2 exposure; b) E plaster after CO_2 exposure; c) G plaster before CO_2 exposure; d) G plasters after CO_2 exposure.

Fig. 3. DTG graphs of the E (a) and G (b) plasters before and after CO_2 exposure.

Table 1. Mass loss by TGA and variation of CO_2 content for plasters after CO_2 exposure.

Plaster		Mass loss [%]			Variation of CO_2 after carbonation [%]
		25–200 °C	200–500 °C	500–900 °C	
E	Before CO_2	1.34	1.63	2.98	51.3
	After CO_2	1.64	1.88	4.51	
G	Before CO_2	13.21	0.71	13.05	6.7
	After CO_2	13.13	0.83	13.92	

5 Conclusions

Indoor air quality is essential for the comfort and health of building occupants. The present study analyzed the sorption and desorption capacity, as well as the CO_2 capture capacity, of earth and gypsum pre-mixed plasters. The results obtained demonstrated that:

– The earth plaster (E) shows higher sorption and desorption capacity than the gypsum plaster (G). Furthermore, after 24 h, the E plaster shows an increasing behavior, with no tendency to stabilize, while the G plaster shows a stable behavior after approximately 6 h.
– The results of TGA, before (without) and after plasters exposure to CO_2, shows that there is a higher mass loss variation in the E plaster compared to the G plaster, which may prove the greater CO_2 capture by the E plaster.
– Since the literature mentions that clayey earth captures and binds indoor air pollutants dissolved in water vapor, the high sorption and desorption capacity of the E plaster may justify the ability of earth plasters to capture CO_2 present in the indoor air. In this way, indoor air pollutants may dissolve in water vapor and, subsequently, be adsorbed by the earth plasters (with higher sorption capacity). The CO_2 could become bound in the intercalated clay layers and difficult to be released again into the indoor air. These findings justify further studies to confirm the effect earth plasters may have on indoor air quality, namely contributing to control levels of CO_2.

Acknowledgments. The authors acknowledge the Portuguese Foundation for Science and Technology (FCT) for the support through funding UIDB/04625/2020 from the research unit CERIS. T. Santos also thanks FCT for their PhD fellowships SFRH/BD/147428/2019. The authors also acknowledge Vitor Silva for his participation in the mortar preparation and CO_2 testing.

References

1. Jacobson, T.A., Kler, J.S., Hernke, M.T., Braun, R.K., Meyer, K.C., Funk, W.E.: Direct human health risks of increased atmospheric carbon dioxide. Nat. Sustain. **2**, 691–701 (2019). https://doi.org/10.1038/s41893-019-0323-1

2. Gomes, M.I., Lima, J., Santos, T., Gomes, J., Faria, P.: The benefits of eco-efficient plasters for occupant's health—A case study. In: Malik, J.A., Marathe, S. (eds.) Ecological and Health Effects of Building Materials, pp. 383–404. Springer, Cham (2022). https://doi.org/10.1007/978-3-030-76073-1_20

3. Santos, T., Gomes, M.I., Santos Silva, A., Ferraz, E., Faria, P.: Comparison of mineralogical, mechanical and hygroscopic characteristic of earthen, gypsum and cement-based plasters. Constr. Build. Mater. **254**, 119222 (2020). https://doi.org/10.1016/j.conbuildmat.2020.119222

4. Emiroğlu, M., Yalama, A., Erdoğdu, Y.: Performance of ready-mixed clay plasters produced with different clay/sand ratios. Appl. Clay Sci. **115**, 221–229 (2015). https://doi.org/10.1016/j.clay.2015.08.005

5. Maskell, D., Thomson, A., Walker, P., Lemke, M.: Determination of optimal plaster thickness for moisture buffering of indoor air. Build. Environ. **130**, 143–150 (2018). https://doi.org/10.1016/j.buildenv.2017.11.045

6. Lima, J., Faria, P., Santos Silva, A.: Earth plasters: the influence of clay mineralogy in the plasters' properties. Int. J. Archit. Herit. **14**, 948–963 (2020). https://doi.org/10.1080/15583058.2020.1727064

7. Bruno, A.W., Gallipoli, D., Perlot, C., Mendes, J.: Effect of stabilisation on mechanical properties, moisture buffering and water durability of hypercompacted earth. Constr. Build. Mater. **149**, 733–740 (2017). https://doi.org/10.1016/j.conbuildmat.2017.05.182

8. Randazzo, L., Montana, G., Hein, A., Castiglia, A., Rodonò, G., Donato, D.I.: Moisture absorption, thermal conductivity and noise mitigation of clay based plasters: the influence of mineralogical and textural characteristics. Appl. Clay Sci. **132–133**, 498–507 (2016). https://doi.org/10.1016/j.clay.2016.07.021

9. Lima, J., Faria, P., Veiga, R.: Comparison of an earth mortar and common binder mortars for indoor plastering. In: Horszczaruk, E., Brzozowski, P. (eds.) ICSEFCM2021 – Proceedings of the 2nd International Conference on Sustainable, Environmentally Friendly Construction Materials, pp. 71–76, Szczecin (2021)

10. Minke, G.: Building with earth. Design and Technology of a Sustainable Architecture, p. 15. Publishers for Architecture, Birkhäuser (2006)

11. Minke, G.: Earth Construction Handbook: The Building Material Earth in Modern Architecture, p. 14. WIT Press, Boston (2000)

12. Reeves, G.M., Sims, I., Cripps, J.C.: Clay Materials Used in Construction, vol. 21, p. 399. The Geological Society, London (2006)

13. Morton, T.: Earth Masonry. Design and Construction Guidelines, p. 11. IHS BRE Press, United Kingdom (2008)

14. Carbon Dioxide can penetrate the inner layers of clay minerals in the Earth's deep subsurface. https://foundry.lbl.gov/2018/02/28/a-team-of-foundry-scientists-and-users-have-shown-that-carbon-dioxide-can-penetrate-the-inner-layers-of-some-non-swelling-clay-minerals-which-make-up-the-dominant-clays-in-the-earths-deep-sub/. Accessed 18 May 2022

15. Zhang, D., Ghouleh, Z., Shao, Y.: Review on carbonation curing of cement-based materials. J. CO2 Util. **21**, 119–131 (2017). https://doi.org/10.1016/j.jcou.2017.07.003

16. Santos, T., Faria, P., Gomes, M.I.: Earth, gypsum and cement-based plasters contribution to indoor comfort and health. In: Pereira, E.B., Barros, J.A.O., Figueiredo, F.P. (eds.) RSCC 2020. RB, vol. 32, pp. 105–117. Springer, Cham (2021). https://doi.org/10.1007/978-3-030-76547-7_10

17. DIN 18947: Earth plasters – Terms and definitions, requirements, test methods (in German). DIN, Berlin (2018)

18. Parracha, J.L., Santos Silva, A., Cotrim, M., Faria, P.: Mineralogical and microstructural characterisation of rammed earth and earthen mortars from 12th century Paderne Castle. J. Cult. Herit. **42**, 226–239 (2020). https://doi.org/10.1016/j.culher.2019.07.021

19. Ashour, T., Georg, H., Wu, W.: An experimental investigation on equilibrium moisture content of earth plaster with natural reinforcement fibres for straw bale buildings. Appl. Therm. Eng. **31**, 293–303 (2011). https://doi.org/10.1016/j.applthermaleng.2010.09.009
20. Lima, J., Correia, D., Faria, P.: Rebocos de terra: influência da adição de gesso e da granulometria da areia. In: Argamassas 2016 – II Simpósio Argamassas e Soluções Térmicas Revestimento. ITeCons, Coimbra (2016)
21. Bernal, S.A., et al.: Characterization of supplementary cementitious materials by thermal analysis. Mater. Struct. **50**, 26 (2017). https://doi.org/10.1617/s11527-016-0909-2

Influence of Natural Sand Replacement by Mineral Wastes on Earth and Air Lime Plastering Mortars, and Professionals Training

Tânia Santos⬤ and Paulina Faria[✉]⬤

CERIS, Department of Civil Engineering, NOVA School of Science and Technology, NOVA University Lisbon, Caparica, Portugal
paulina.faria@fct.unl.pt

Abstract. It is increasingly necessary to rethink constructive solutions to minimize the construction industry's environmental impact without compromising technical characteristics, including conservation. The replacement of natural sand by ceramic waste and clayish earth was analyzed in the fresh and hardened state of earth and air lime mortars. Seven different mortars with clayish earth and air lime as binders were produced and characterized: two earth mortars with added sand and with brick waste instead of sand; five air lime mortars with sand, with clayish earth, or with brick waste instead of sand. The replacement of sand by brick waste promotes a decrease in the mortars' dry bulk density and mechanical strengths. For the air lime mortar, the replacement of sand by brick waste promotes an increase in dynamic modulus of elasticity, compared with the reference mortar with low flow. The air lime mortar with clayish earth presents lower flexural strength and dynamic modulus of elasticity. Many of the mortars with alternative added aggregate fulfil the requirements for plastering old and new buildings and, based on old practices, can be a more sustainable alternative with natural pigmentation. However, training is needed for professionals using this type of mortars.

Keywords: Air Lime Mortars · Earth Mortars · Excavation Earth · Natural Sand · Red Ceramic Waste

1 Introduction

Two of the main problems faced by the construction industry are the over-exploitation of available resources and negative environmental effects, with 10% of the global emission of carbon dioxide (CO_2) due to the supply of construction materials, of which 85% concerns cement [1].

In Portugal, the construction industry is responsible for the production of 25–40% of the waste produced and 24% of the natural resources extracted [2]. In the European Union (EU), construction and demolition wastes (CDW) represent more than a third of the waste generated [3]. CDW contain a wide variety of materials such as concrete, mortars, bricks, tiles, excavation earth, wood, glass, metals, and plastic, among others [2, 3]. Excavation earth represents a high percentage of all the generated CDW.

The use of environmentally friendly materials with low CO_2 emissions and embodied energy are increasingly important to help reducing the environmental impact of the construction industry. On the other hand, the reuse and/or recycling of the materials, such as CDW, other wastes, and by-products of other industries, reduce the use of natural resources and the volume of waste sent to landfills. Consequently, that reduces the environmental impact associated with the construction industry. According to Farinha et al. [4], red or white ceramic, powder glass, CDW, several types of polymers and stone powder, among other wastes, can be incorporated into mortars to achieve a good technical performance.

Clay, gypsum and air lime are the oldest binders used in mortars [5–7]. Earth is a natural building material, generally non-toxic and reusable (provided it is not chemically stabilized). This material requires low energy for its extraction, transport and preparation, compared to cement and air lime, and consequently, presents low embodied energy and CO_2 emissions [8]. High volumes are generated when excavation works occur (namely for metro lines construction) which, if not used, are transported to the landfill.

Earth is composed by clay, silt, sand and coarser aggregates. For mortar's production, the coarser aggregates are removed by sieving. Many excavation earths have high clay contents and, for this reason, it is common to add additional sand or other types of aggregates to reduce the clay content, control drying shrinkage and improve mortars' strength.

When air lime became available and particularly when resistance towards the water was required, most probably a volume of air lime was added to a higher volume of earth to produce lime-earth mortars [9]. With time, the earth was then replaced by natural sand for what is known now as air lime mortars.

From the end of the 19th century, there was a gradual replacement of the use of these traditional binders by cement in the production of mortars [5]. However, air lime requires less energy for its production compared with cement since air lime is produced at a lower temperature (about 900 °C compared to about 1500 °C for cement production).

The knowledge for the preparation and application of earth and air lime mortars, which guarantees good durability and performance, was passed on from generation to generation [10, 11]. With the disuse of the earth and air lime as binders, this artisanal knowledge was discontinued and that creates difficulties in its practical use. For this reason, professional training is very important for the choice and good application of these mortars, namely in terms of water content, good workability, and adequate mixing of the constituents, among others.

In a mortar, sand is usually the component in higher proportion. Natural sand is a non-renewable resource, and its extraction can have negative consequences for the environment. Therefore, it would be beneficial to replace the natural added sand by by-products or wastes in mortars. Different waste materials can be reused as artificial sand to increase mortars' circularity [4].

A significant amount of waste for disposal is produced by the ceramic industry, which has negative impacts on the environment. The incorporation of thermally treated clays in mortars can be found in archaeological and historic buildings [12]. Ceramic wastes (fragments together with dust) were used in Roman air lime mortars to partially replace aggregates, and the slightly hydraulic properties achieved by those mortars were recognized [12]. The use of ceramic waste may thus provide environmental, economic and technical benefits to mortars, especially when pozzolanic reactions are considered [12].

Mortars have many different applications, namely in screeds, as masonry bedding mortars, for pointing joints, renders and plasters. For plastering, high strength cementitious mortars are not required nor compatible, and earth and air lime mortars are more eco-efficient.

However, the sustainability of those types of mortars can be further improved using ceramic waste and clayish earth as aggregates in natural sand replacement. These alternative mortars can be applied in the conservation and rehabilitation of buildings, thus avoiding severe anomalies due to differences in the rigidity and permeability to water vapor namely between the plasters and the substrate. Furthermore, these alternative mortars may also be applied in new construction [13].

The present study aimed to develop and characterize highly ecological mortars, assessing the influence of the replacement of raw sand by brick waste or by clayish earth in earth and air lime mortars, to be applied as plasters or renders in the rehabilitation and conservation of old buildings, as well as in new constructions.

2 Materials, Mortars and Methods

2.1 Materials and Mortars

Seven different mortars with clayish earth and air lime, as binders, and natural sand, clayish earth and red brick waste, as aggregate, were produced and characterized (Table 1).

Table 1. Seven different mortars produced and characterized.

Mortars	Description
1E_0.5S	Earth (E) mortar with sand (S)
1E_0.5BW	Earth (E) mortar with brick waste (BW) instead sand
1CL_2S	Air lime (CL) mortar with sand (S)
1CL_2E	Air lime (CL) mortar with clayish earth (E) instead of sand
1CL_3S_hf	Air lime (CL) mortar with sand (S), with high flow (hf)
1CL_3S_lf	Air lime (CL) mortar with sand (S), with low flow (lf)
1CL_3BW	Air lime (CL) mortar with brick waste (BW) instead of sand

The natural earth (E) was composed by fractions of clay, silt and sand. The sand (S) was a river sand. The air lime (CL) was a CL90-S (EN 459-1 [14]) produced by Lusical – Lhoist Group. The red brick waste (BW) was obtained by milling broken hollow bricks and was mainly composed by fragments and some dust resulting from the milling process. The loose bulk density (LBD) of the materials is 1.33 kg/dm^3 for

Table 2. Mortars' compositions (in volume and weight) and fresh state characterization.

Mortars	Volumetric proportions				Weight proportions				Mixing water[1] [%]	Flow [mm]
	E	CL	S	BW	E	CL	S	BW		
1E_0.5S	1	-	0.5	-	1	-	0.6	-	15	159
1E_0.5BW	1	-	-	0.5	1	-	-	0.4	22	167
1CL_2S	-	1	2	-	-	1	9.3	-	17	163
1CL_2E	2	1	-	-	7	1	-	-	23	154
1CL_3S_hf	-	1	3	-	-	1	12.4	-	19	185
1CL_3S_lf	-	1	3	-	-	1	14.0	-	15	136
1CL_3BW	-	1	-	3	-	1	-	8.7	36	160

Note: [1]percentage added, considering the total mass of mortars' dry constituents

E, 1.54 kg/dm^3 for S, 0.37 kg/dm^3 for CL and 1.13 kg/dm^3 for BW. The mortars' formulation is presented in Table 2.

The volumetric and weight proportions of binder: aggregate of 1E_0.5S and 1E_0.5BW mortars cannot be exactly determined because the earth (E) is composed by different fractions of clay, silt and sand. Thus, it is considered that clay acts as binder and silt, sand (added or not) and brick waste, as aggregate.

The mixing of mortars was carried out manually by students of a Conservation and Restoration Program during laboratory classes, trying to reproduce what some people without much experience can produce *in situ*. The mixing procedure was the following: all dry constituents of mortars (binder and aggregates) were manually homogenized after weighting for exact proportions; the mixing water was added gradually to achieve good workability; manual mixing stopped when each mortar was considered workable for rendering/plastering.

The mixing water of mortars is presented in Table 2. Two mortars (1CL_3S) were produced with the same composition of dry materials but different water contents, as they were produced by not highly skilled professionals.

For each mortar three prismatic specimens with 40 mm × 40 mm × 160 mm were produced in metallic molds. They were filled in two layers mechanically compacted with 20 stokes each and manually levelled. The specimens were kept in the molds in laboratory conditions (20 ± 5 °C and 65 ± 15%) and demolded after approximately 8 days. The hardened specimens were tested after 28 to 106 days.

2.2 Methods

In the fresh state, the mortars were characterized by flow table consistence, based on EN 1015-3/A1/A2 [15] but using a flow table in accordance with the previous version of this standard. The wet bulk density was assessed according to EN 1015-6/A1 [16].

In the hardened state, an evaluation of color and drying shrinkage was performed through photography and visual observation. Furthermore, the mortars were characterized by dry bulk density according to DIN 18947 [17] and based on EN 1015-10/A1

[18], by the ratio between the dry mass and the volume of each prismatic specimen with a digital caliper and a 0.001 g precision digital scale. Dynamic modulus of elasticity (Ed) was determined based on EN 14146 [19], using a Zeus Resonance Meter ZMR 001 equipment with its specific software. Flexural (FStr) and compressive (CStr) strengths were determined based on DIN 18947 [17] and EN 1015-11 [20] with load cells of 2 kN and velocity of 10 N/s or 0.2 mm/min for flexural strength, and with load cells of 2 kN or 50 kN and velocity of 50 N/s or 0.7 mm/min for compressive strength, as two different equipment were used: a ETI HM-S and a Zwick Rowell Z050, respectively.

Three prismatic specimens of each mortar were characterized for each property in the hardened state. However, for the 1CL_2S mortar, only two specimens were used, since one specimen broke during demolding. For this reason, for this mortar, only the average results of the two specimens will be presented, without standard deviation.

3   Results

3.1   Fresh State Characterization, Color and Drying Shrinkage

The fresh state characterization of mortars is presented in Table 2 and Fig. 1.

The use of brick waste or clayish earth instead of raw sand promotes the increase of the mixing water necessary for the good workability of the earth and air lime mortars. That is expected by the replacement of a more absorbent aggregate.

The earth mortars present flow table consistence of 163 ± 4 mm. The air lime mortars present flow table consistence of 160.5 ± 24.5 mm for all mortars and 158.5 ± 4.5 mm if the mortars with high and low flows are not included. The increase of mixing water necessary due to the replacement of sand by brick waste in the mortars (1E_0.5BW and 1CL_3BW) promotes a slight increase in the flow table consistence of these mortars comparing with 1E_0.5S and 1CL_3S_lf mortars. In contrast, the increase of mixing water promotes by the replacement of sand by clayish earth promoted a slight decrease in the flow of the 1CL_2E mortar. Clay has a significant water retention capacity [21]. The decrease in the flow table consistence of the mortar with added earth as aggregate, despite the increase in mixing water, can be explained by the retention of water by the clayish earth.

It can be observed that the higher and lower flow table consistence shown by the 1CL_3S_hf and 1CL_3S_lf mortars is directly associated with the higher and lower mixing water added by the unskilled professionals.

Analyzing the wet bulk density of mortars (Fig. 1), the use of brick waste in the earth mortar (1E_0.5BW) promotes a decrease in wet bulk density compared with the reference mortar with raw sand (1E_0.5S). The same occurres with the use of clayish earth in air lime mortar (1CL_2E). This can be justified by the lower loose bulk density of the clayish earth (1.33 kg/dm^3) and brick waste (1.13 kg/dm^3), compared with the sand (1.54 kg/dm^3). In the air lime mortar with a volumetric ratio of 1:3 (1CL_3S_hf, 1CL_3S_lf and 1CL_3BW), this change in wet bulk density is not so significant.

Figure 2 presents the different specimens of mortars, showing the different colors presented: the 1CL_2S and 1CL_3S mortars show a white tone, while the 1E_0.5S mortar shows a reddish color. Furthermore, the replacement of sand by clayish earth

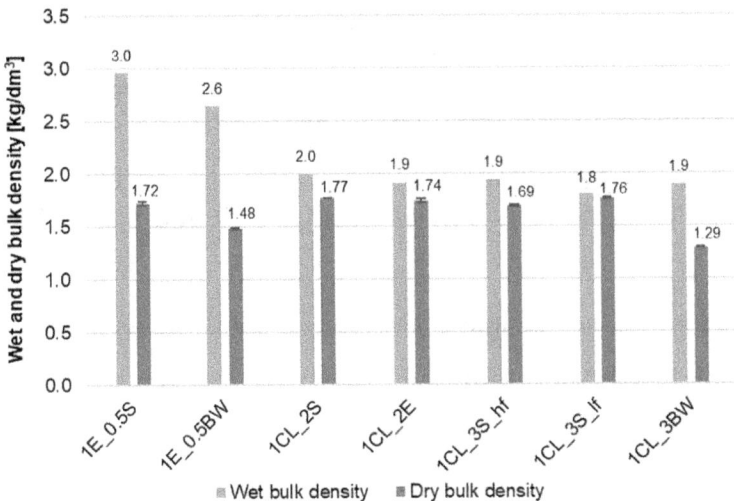

Fig. 1. Wet and dry bulk density of mortars.

and brick waste also causes this change of color in the air lime mortars, making it salmon. The replacement of sand by brick waste on the earth mortar does not cause a color change. Natural pigmentation can be an advantage when the plastering mortars are applied unpainted. By visual observation of the dried mortars in the molds, no significant linear shrinkage is observed.

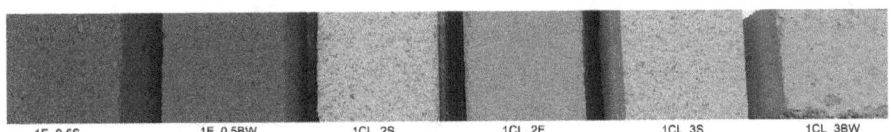

Fig. 2. Different colors of the mortar specimens.

3.2 Dry Bulk Density

The dry bulk density of mortars is presented in Fig. 1.

The replacement of sand by the brick waste in earth and air lime mortars promotes a decrease in the dry bulk density of the 1E_0.5BW and 1CL_3BW mortars compared to their reference mortars (1E_0.5S, 1CL_3S_hf and 1CL_3S_lf). This can be justified by the lower loose bulk density of the brick waste (1.13 kg/dm^3) compared to the sand (1.54 kg/dm^3). On the other hand, the replacement of sand by clayish earth does not significantly influence the dry bulk density of the mortars. This can be due to the more similar loose bulk density of sand and clayish earth (1.33 kg/dm^3). The dry bulk density of reference earth and air lime mortars is similar. The 1CL_3S_hf mortar presents a slightly lower dry bulk density compared with 1E_0.5S, 1CL_2S and 1CL_3S_lf mortars and this can be justified by the higher mixing water, which may have promoted a higher porosity of this mortar, and, consequently, lower dry bulk density.

By comparing wet and dry bulk density of mortars (Fig. 1), it is possible to conclude that they present a similar tendency for the earth mortars: the decrease in wet bulk density verified between the 1E_0.5S and 1E_0.5BW mortars is also verified in the dry bulk density. The same occurred when comparing the 1CL_2S and 1CL_2E mortars. This tendency is not verified when comparing the 1CL_3S_hf, 1CL_3S_lf and 1CL_3BW mortars.

3.3 Dynamic Modulus of Elasticity and Flexural and Compressive Strength

The dynamic modulus of elasticity (Ed) and flexural (FStr) and compressive (CStr) strength of mortars are presented in Fig. 3.

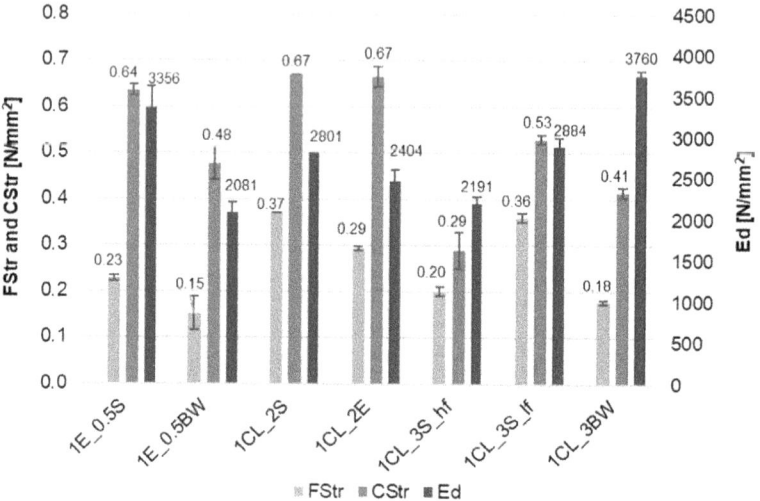

Fig. 3. Dynamic modulus of elasticity (Ed) and flexural (FStr) and compressive (CStr) strengths of mortars.

In general, it can be observed that the earth mortars present mechanical properties in the same range of air lime mortars with 1:2 to 1:3 proportions of lime:aggregate and that the replacement of sand by earth and by brick waste may promote the decrease of mechanical properties. However, the effect may be lower than the one obtained by using a higher volume of mixing water.

In a more detailed analysis, the high mixing water in air lime mortar (1CL_3S_hf) leads to a decrease in the mechanical performance of this mortar compared to 1CL_3S_lf mortar with lower mixing water. This shows the high influence proper mixing can have on air lime mortars, which can only be ensured by trained professionals. Therefore, specific training for professionals should always be considered, as it includes not only the mixing but also the application and the needed re-compaction after initial drying shrinkage [10].

The earth mortar with sand (1E_0.5S) has a higher dynamic modulus of elasticity compared with air lime mortars (1CL_2S, 1CL_3S_hf and 1CL_3S_lf). Air lime mortars with sand (1CL_2S and 1CL_3S_lf) present high flexural strength compared to earth mortar with sand (1E_0.5S), except the air lime mortar with high water content (1CL_3S_hf). On the other hand, the earth mortar with sand presents compressive strength similarly to the mortar with higher air lime content (1CL_2S) but higher compressive strength compared to mortars with lower air lime content (1CL_3S_hf and 1CL_3S_lf).

In the earth mortars, the replacement of the sand by brick waste promotes a decrease of Ed, FStr and CStr. On the other hand, in the air lime mortars, this replacement also promotes a decrease of FStr and CStr (when comparing the 1CL_3S_lf and 1CL_3BW mortars). However, there is an increase of Ed. When comparing the 1CL_3S_hf mortar, with higher water content, with the mortar with lower water content (1CL_3S_lf), it is verified that 1CL_3S_hf mortar presents a lower Ed, FStr and CStr. The same happens when comparing the 1CL_3S_hf mortar with 1CL_3BW mortar, except in FStr, that is slightly higher for 1CL_3S_hf mortar. This can be justified by the high mixing water of the mortar, which may have some influence on the increase in the porosity of the mortar since there is also a decrease in the dry bulk density. Therefore, it will be important to evaluate the mortars' microstructure.

The decrease in the dry bulk density of the 1E_0.5BW, 1CL_2E and 1CL_3BW mortars compared to their reference mortars (1E_0.5S, 1CL_2S and 1CL_3S_lf) may justify the decrease in the FStr and CStr of these mortars. The same for the 1CL_3S_hf mortar compared to the reference one (1CL_3S_lf).

The replacement of sand by clayish earth in the air lime mortar (1CL_2E) promotes a decrease of Ed and FStr and maintains the CStr concerning its reference mortar (1CL_2S). Compared with the air lime mortar with brick waste (1CL_3BW), the 1CL_2S mortar presents higher FStr and CStr. On the other hand, the 1CL_2E mortar presents a lower Ed compared to the 1CL_3BW mortar. However, it is necessary to consider that the volumetric ratios of these mortars are different.

It is important to remember that the mortars were mixed manually by not skilled professionals (Conservation students), so this may influence on the correct mixing of the mortar components and, consequently, on the characteristics of the mortars.

Santos et al. [22] analyzed earth mortars with different compositions, with fine and coarse sand, phase change materials and natural fibers, mixing water content of 17–25% and obtained dry bulk density of 1.18–1.79 kg/dm^3, Ed of 370–4331 N/mm^2, FStr of 0.09–0.24 N/mm^2 and CStr of 0.28–0.58 N/mm^2. In the present study, the earth mortars present results in the same range of these authors, except for the 1E_0.5S mortar which has a slightly higher CStr. Being earths composed by different types and contents of clay, silt and sand, in comparison to more standardized materials, a higher range of results may be expected.

In another study, Santos et al. [23] analyzed a pre-mixed earth mortar and another earth mortar with coarse and fine sand, with water content (determined by the difference of mass of wet and dry samples) of 10–15% and obtained dry bulk density of 1.77–1.82 kg/dm^3, Ed of 3781–4267 N/mm^2, FStr of 0.20–0.25 N/mm^2 and CStr of $0.96-1.01$ N/mm^2. In the present study, all earth mortars present lower hardened characteristics, except the 1E_0.5S mortar that presents FStr in the same range.

Faria [9] characterized air lime mortars with 1:2 and 1:3 volumetric proportions (lime:river sand) and obtained dry bulk density of $1.59-1.60$ kg/dm^3, Ed of $2902-3243$ N/mm^2, FStr of $0.30-0.33$ N/mm^2 and CStr of $0.66-0.70$ N/mm^2. In the present study, air lime mortars with the same 1:2 and 1:3 volumetric ratios with brick waste, earth or sand present higher dry bulk density and lower Ed, except the 1CL_3BW mortar that presents lower dry bulk density and higher Ed. For FStr, no mortar presents results in the same range obtained by Faria [9]. On the other hand, all air lime mortars present lower CStr, except the 1CL_2S and 1CL_2E mortars that present CStr in the same range obtained by Faria [9].

Matias et al. [24] analyzed air lime mortars with 1:3 volumetric proportion, with replacement of 0, 20 and 40% of sand by brick waste, and obtained FStr of $0.2-0.3$ N/mm^2 and CStr of $0.2-1.3$ N/mm^2. Compared to air lime mortar without brick waste: the FStr increased with the replacement of 20% sand by brick waste; however, the replacement of 40% promoted only a very slight increase; the CStr increased with the increasing replacement of sand by brick waste. In the present study, the air lime mortars with volumetric proportions 1:3 binder:aggregate, with and without replacement of sand by brick waste (1CL_3S_hf, 1CL_3S_lf and 1CL_3BW), present FStr and CStr in the same range, except the 1CL_3BW mortar which presents slightly lower FStr.

General requirements, such as the dynamic modulus of elasticity and flexural and compressive strength, of rendering and plastering mortars to be applied in historic buildings were defined by Veiga et al. [25]: Ed of $2000-5000$ N/mm^2; FStr of $0.2-0.7$ N/mm^2; CStr of $0.4-2.5$ N/mm^2. All mortars analyzed in the present study meet the defined limits, except the earth and air lime mortars with brick waste (1E_0.5BW and 1CL_3BW) which have a slightly lower FStr, and the 1CL_3S_hf mortar which has a CStr lower than the defined limit. The EN 998-1 [26] defines that plastering and rendering mortars must be classified by categories according to the range of compressive strengths after 28 days. All mortars analyzed in the present study can be classified as CSI class, the lowest class with a CStr between 0.4 and 2.5 N/mm^2, except the 1CL_3S_hf mortar with a CStr of 0.29 N/mm^2. Nevertheless, the mortars present Ed, FStr and CStr results very close to the limits defined by Veiga et al. [25] and EN 998-1 [26] for the application of these mortars as plasters (the recommended possibility for the unstabilized earth mortars) or renders.

4   Conclusions

The influence of the replacement of mortars' natural sand, a non-renewable material, by clayish earth and brick waste was assessed, following past technologies and giving these residues a use instead of sending them to landfill. The characteristics of earth and air lime mortars in the fresh and hardened state were evaluated to assess the feasibility of these eco-efficient replacements for application as plasters and renders.

Through this study, it is possible to obtain different conclusions:

– Air lime mortars are workable even with low flow; a too high water content impairs their characteristics in the fresh and hardened state, presenting lower mechanical

strengths. The result of an inaccurate mixing procedure is an example of what may occur *in situ* when mixing is carried out by inexperienced professionals.
- The change in color given to the air lime mortars by the replacement of natural sand by brick waste or clayish earth can be considered an advantage, providing natural pigmentation.
- No significant linear drying shrinkage was observed in the earth mortar and the mortars with the replacement of natural sand by clayish earth or brick waste. However, further studies are recommended.
- The total replacement of natural sand by brick waste or clayish earth in earth and air lime mortars generally promotes a decrease in the mechanical strength of the mortars. Nevertheless, most meet the limits defined for application as plaster in historic buildings as well as in new construction, thus allowing their applicability in different substrates.

Due to their vulnerability in contact with water, non-stabilized earth mortars can only be applied indoors. But most probably adequate protection of the most exposed areas (such as corners) could be enough to guarantee its applicability with adequate durability, for example, as archaeological sacrificial plasters in sheltered sites.

It is known that, for the conservation of historic built heritage, mortars must not exceed the mechanical properties of the substrate on which they will be applied, ensuring long-term compatibility between the mortar and the substrate. That is easily achieved with earth and air lime mortars. However, without jeopardizing the previous, mortars' durability is also needed. In future works, it is intended to evaluate the durability and relate it with the microstructure of the analyzed mortars.

Acknowledgments. The authors acknowledge the Portuguese Foundation for Science and Technology (FCT) for the support through funding UIDB/04625/2020 from the research unit CERIS. T. Santos also thanks FCT for their PhD fellowships SFRH/BD/147428/2019. The authors also acknowledge Vitor Silva and the Conservation and Restoration Program students who participated in the mortar mixing laboratory classes.

References

1. Kappel, A., Ottosen, L.M., Kirkelund, G.M.: Colour, compressive strength and workability of mortars with an iron rich sewage sludge ash. Constr. Build. Mater. **157**, 1199–1205 (2017). https://doi.org/10.1016/j.conbuildmat.2017.09.157
2. Cunha, C.: Resíduos de construção e demolição (RCD's). https://www.semural.pt/residuos-de-construcao-e-demolicao/. Accessed 11 Mar 2022
3. European Commission: Construction and demolition waste. https://ec.europa.eu/environment/topics/waste-and-recycling/construction-and-demolition-waste_pt. Accessed 11 Mar 2022
4. Farinha, C.B., Silvestre, J.D., Brito, J., Veiga, M.R.: Life cycle assessment of mortars with incorporation of industrial wastes. Fibers **7**(7), 59 (2019). https://doi.org/10.3390/fib7070059
5. Barbero-Barrera, M.M., Maldonado-Ramos, L., Van Balen, K., García-Santos, A., Neila-González, F.J.: Lime render layers: an overview of their properties. J. Cult. Herit. **15**, 326–330 (2014). https://doi.org/10.1016/j.culher.2013.07.004

6. Bruno, P., Faria, P., Candeias, A., Mirão, J.: Earth mortars use on pre-historic habitat structures in south Portugal. Case studies. J. Iber. Archaeol. **13**, 51–67 (2010)

7. Freire, M.T., Veiga, M.R., Santos Silva, A., Brito, J.: Studies in ancient gypsum based plasters towards their repair: physical and mechanical properties. Constr. Build. Mater. **202**, 319–331 (2019). https://doi.org/10.1016/j.conbuildmat.2018.12.214

8. Melià, P., Ruggieri, G., Sabbadini, S., Dotelli, G.: Environmental impacts of natural and conventional building materials: a case study on earth plasters. J. Clean. Prod. **80**, 179–186 (2014). https://doi.org/10.1016/j.jclepro.2014.05.073

9. Faria, P.: Argamassas de cal e terra: características e possibilidades de aplicação. Ambiente Construído **18**, 49–62 (2018). https://doi.org/10.1590/s1678-86212018000400292

10. Veiga, R.: Air lime mortars: what else do we need to know to apply them in conservation and rehabilitation interventions? A review. Constr. Build. Mater. **157**, 132–140 (2017). https://doi.org/10.1016/j.conbuildmat.2017.09.080

11. Schroeder, H.: The new DIN standards in earth building - the current situation in Germany. J. Civ. Eng. Archit. **12**, 113–120 (2018). https://doi.org/10.17265/1934-7359/2018.02.005

12. Matias, G., Faria, P., Torres, I.: Lime mortars with heat treated clays and ceramic waste: a review. Constr. Build. Mater. **73**, 125–136 (2014). https://doi.org/10.1016/j.conbuildmat.2014.09.028

13. Santos, T., Faria, P., Silva, V.: Can an earth plaster be efficient when applied on different masonries? J. Build. Eng. **23**, 314–323 (2019). https://doi.org/10.1016/j.jobe.2019.02.011

14. EN 459-1: Building lime – Part 1: Definitions, specifications and conformity criteria. CEN, Brussels (2015)

15. EN 1015-3/A1/A2: Methods of test for mortars for masonry – Part 3: Determination of consistence of fresh mortars (by flow table). CEN, Brussels (1999/2004/2006)

16. EN 1015-6/A1: Methods of test for mortars for masonry – Part 6: Determination of bulk density of fresh mortars. CEN, Brussels (1998/2006)

17. DIN 18947: Earth plasters – Terms and definitions, requirements, test methods (in German). DIN, Berlin (2018)

18. EN 1015-10/A1: Methods of test for mortars for masonry – Part 10: Determination of dry bulk density of hardened mortar. CEN, Brussels (1999/2006)

19. EN 14146: Natural stone test methods. Determination of the dynamic modulus of elasticity (by measuring the fundamental resonance frequency). CEN, Brussels (2004)

20. EN 1015-11: Methods of test for mortar for masonry – Part 11: Determination of flexural and compressive strength of hardened mortar. CEN, Brussels (2019)

21. Faria, P., Lima, J.: Earth Plasters. Earth Construction Notebooks 3, p. 48. Argumentum, Lisbon (2018). (in Portuguese)

22. Santos, T., Faria, P., Santos Silva, A.: Eco-efficient earth plasters: the effect of sand grading and additions on fresh and mechanical properties. J. Build. Eng. **33**, 101591 (2021). https://doi.org/10.1016/j.jobe.2020.101591

23. Santos, T., Gomes, M.I., Santos Silva, A., Ferraz, E., Faria, P.: Comparison of mineralogical, mechanical and hygroscopic characteristic of earthen, gypsum and cement-based plasters. Constr. Build. Mater. **254**, 119222 (2020). https://doi.org/10.1016/j.conbuildmat.2020.119222

24. Matias, G., Faria, P., Torres, I.: Lime mortars with ceramic wastes: characterization of components and their influence on the mechanical behaviour. Constr. Build. Mater. **73**, 523–534 (2014). https://doi.org/10.1016/j.conbuildmat.2014.09.108

25. Veiga, M.R., Fragata, A., Velosa, A.L., Magalhães, A.C., Margalha, G.: Lime-based mortars: viability for use as substitution renders in historical buildings. Int. J. Archit. Herit. **4**, 177–195 (2010). https://doi.org/10.1080/15583050902914678

26. EN 998-1: Specification for mortar for masonry – Part 1: Rendering and plastering mortar. CEN, Brussels (2016)

Evaluation of Physical and Mechanical Parameters in Commercial NHL-Based Green Plaster for the Preservation of Historical Buildings

Cristina Tedeschi[(✉)] and Maria Cecilia Carangi

Politecnico di Milano, Milan, Italy
{cristina.tedeschi,mariacecilia.carangi}@polimi.it

Abstract. The need to develop products for the conservation of the architectural heritage, in particular binders for plasters, is increased nowadays and those products are required to be in line with green production system and materials in full compliance with current environmental sustainability criteria. It has already been established that Natural Hydraulic Lime (NHL) is a good material for restoration; it is compatible with ancient materials and respects the environmental requirements. The use of green technologies often involves the partial use of environmental low impact materials that do not always find practical use in restoration. The starting point of the research is the selection and study of NHL-based commercial plasters used for the conservation of historic masonries and their optimization through additives cement-free in order to create green products for applications in the conservative restoration of historical heritage. The five different types of formulations chosen for this work, has been evaluated and analyzed from a physical and mechanical point of view, through dynamic elastic module, compression and flexion tests, porosity and density analysis. During the execution of the research, the focus was on the effect of microstructural features of porous material, investigated in laboratory with porosimetry test as they are fundamental parameters for studying compatibility with pre-existing materials and durability over time.

Keywords: Commercial NHL plaster · conservation mortars · porosity · physical-mechanical parameters · compatibility

1 Introduction

The ONU 2030 Agenda proposes a global approach to the production system that could pay attention not only to the economic aspect of the manufacturing process but also to the social and environmental ones. The development of new products, obtained from sustainable raw material worked with green manufacturing process and with an accurate life cycle management is, nowadays, the only way to address the problems deriving from greenhouse gas emissions which, if not stopped, could make Earth unlivable in the future [1]. A lot of factors like air, water, pollutants, have a strong influence on

© The Author(s), under exclusive license to Springer Nature Switzerland AG 2023
V. Bokan Bosiljkov et al. (Eds.): HMC 2022, RILEM Bookseries 42, pp. 227–239, 2023.
https://doi.org/10.1007/978-3-031-31472-8_18

the durability of construction materials in different ways [2, 3]. To repair and protect the structural integrity and functionality of the masonry system, or the continuity of the finishing surfaces, a lot of different operations are performed to accomplish these tasks, like a partial re-building, substitution of construction elements, integration of deteriorated mortars joint and re-pointing [4, 5].

A critical point is the interaction between original and new material, the compatibility of the intervention must always be guaranteed [6]. Restoration mortars should be as durable as possible without causing a damage in the original material, in a direct or indirect way. It is customary to say that compatibility between the pre-existing materials and those applied during restoration must be chemical, physical and mechanical. Therefore, it is necessary to evaluate characteristics such as surface features, molecular composition, mechanical strength, elastic modulus, porosity and density, of both the ancient mortars and those formulated for the intervention must be as similar as possible [7–9]. Compatibility is not only a purely physical factor but also an aesthetic one; in fact, the final appearance must be as similar as possible to the pre-existing one but, at the same time, the characteristics of the historic mortars must be sufficiently distinguishable from the new ones to ensure that authenticity is respected [10]. The extensive use of cement-based mortars inferred a lot of conservation works, causing detrimental effects and irreversible damage in the original building materials due to their different physicochemical characteristics and potential in comparison to the pre-existing mortars [11]. Cement has also a high environmental impact. It has been estimated that its production and raw material extraction accounts around 7 to 8% of global CO_2 emission and 12 to 15% of global industry energy consumption [12, 13]. On the other hand, Natural Hydraulic Lime (NHL) seem to be an ideal binder for the production of conservation mortars and it can be optimized with a wide range of additives. NHL can set underwater and gain strength by both hydratation and carbonatation reactions, it is faster and stronger than air-lime but have a greater permeability and reduced stiffness in comparison to cement based mortars. The current European standard accept the use of NHL 2, NHL 3.5 and NHL 5 as binders, differentiated by the compressive strength developed after 28 days curing and to the $Ca(OH)_2$ content [14, 15].

The ideal way to obtain an adequate repair-mortar is the study of the ancient materials of the building case by case and formulate the optimal mixture to applicate [10]. There are a lot of historic buildings that need to be preserved, but the case-by-case approach results to be complex to implement because it requires a multidisciplinary analysis method. Various types of ready-mixed mortars have therefore been formulated and marketed for preservation purposes; their preparation then involves the addition of an amount of water, written in the product data sheets, to the mixture of binder, aggregate and any additives to be then applied.

The microstructure of the mortars, via indirect techniques, using the Mercury Intrusion Porosimeter (MIP), will be explored in depth, because pore size distribution plays a key role in mechanical properties and compatibility with integration mortars [16].

In the present work will be characterized a selection of hardened commercial ready-mixed mortars used as a conservation plaster; the dynamic elastic module, compression and flexural test and the porosity and density analysis will be evaluated. All of them are totally cement-free, with a minimum content of soluble salts and based on NHL binders. Five mortars are NHL 5 based and one has NHL 3, 5 as binder.

2   Materials and Methods

2.1   Products

Five ready-mixed commercial NHL-based mortars samples, Table 1, specifically formulated for restoration were selected. All of the mortars are said to be based on Natural Hydraulic Lime 5. According to European Standards [17] the samples sets refer to 4 × 4 × 16 prismatic specimens.

Table 1. List of the samples according to the type of substrate

Sample	Purpose	Specimens' surface aspect
M1	General indoor and outdoor	
M2	General indoor and outdoor	
M3	Indoor and outdoor renovation and thermal insulation	
M4	General indoor and outdoor	
M5	Indoor and outdoor thermal insulation	

2.2   Analytical Techniques

Dynamic and Static Elastic Module. The dynamic elastic module has been determined with non-destructive methods, using the Resonance Frequency methodology (DME-RF), according to the standard EN 14146 (2004) [18]. With this method, it is possible to calculate the longitudinal dynamic modulus of elasticity in MPa (Ed_L), determined from the longitudinal fundamental resonance frequency and the flexural dynamic modulus of elasticity in MPa (Ed_F), determined from the flexural fundamental resonance frequency. The Ultrasonic test is based on instantaneous excitation and the Direct Ultrasonic dynamic methodology, according to standard EN 12504-4 (2005) [19], performed with CONTROLS ultrasonic equipment using 50 kHz probes. The static elastic module tests has been performed with a destructive methodology, according to ISO 6784:1982 [20].

Mechanical Test. Flexural and compressive strength of mortars were studied by means of an INCOTECNIC-Matest hydraulic press, following the EN 1015-11 standard [21]. Compressive strength resistance has been studied on the two fragments of each specimen resulting of the previous flexural test.

Mercury Intrusion Porosimetry (MIP). MIP is based on the Washburn equation, which relates the contact angle (θ) formed by a non-wetting liquid (like Hg) and a surface of the material under study, with the surface tension (γ) of the liquid itself and the pressure (P) applied. In this way is possible to measure the equivalent pore radii (r) using Eq. (1) [22, 23]:

$$r = -2\gamma \cos\theta/p \tag{1}$$

By the way MIP shows some limitations, like the alteration of the samples through the formation of cracks and ruptures caused by high pressure the impossibility to use again the sample after the analysis, because it is fully filled with mercury and the so-called "ink-bottle effect". When large internal pores are only accessible by narrower pore throats and so the intruded volume that should determinate the pore size distribution, misrepresents the size of these large internal pores [24].

Mortar porosity and pore size distribution were studied by means of Mercury Intrusion Porosimetry (MIP), using a Micromeritics AutoPore IV 9500 porosimeter (ability to measure pore diameters from 0.003 to 360 μm). Prismatic sample pieces of approximately 1–1,5 cm were oven-dried at 70 $\pm$ 5 °C for 8 h prior to the analysis.

Density Analysis. Density measurement of materials has been carried out using the bulk density data obtained by Mercury Intrusion Porosimeter (MIP) described in 2.2.3. Those results has been compared with the values of Volumic mass calculated following the formula described in UNI EN 12390-7:2002 [25] standard Eq. (2):

$$D = \frac{m}{V} \tag{2}$$

where D is the volumic mass expressed in Kg/m^3, m is the speciemen's mass in Kg and V is the material volume in m^3. To allow comparisation with the density values determinated by Mercury Intrusion Porosimeter, data were transformed into g/mL.

3 Results

This section contains the results obtained by the experimental tests described before.

3.1 Mechanical Tests

Flexural and Compression. By observing the results obtained from the compression and flexural tests (shown in Table 2), it is possible to note a lower strength in the mechanical tests for samples M3 and M5. The sample with the highest resistance to mechanical testing turns out to be M2, while M5, consistent with what might be expected by observing the mass of the sample recorded the lowest resistance to compression and, together with sample M3 are the least resistant to the flexural test. All the results has been compared to those found in literature [26, 27].

Table 2. Mechanical tests: flexion and compression.

Material	Flexural strength (N/mm^2)	Compressive Strenght (N/mm^2)
M1	0.27	1.55
M2	0.22	1.91
M3	0.32	1.24
M4	0.91	1.83
M5	0.32	0.65

Elastic Modulus. The values obtained from the static and dynamic elastic modulus analysis, shown in Table 3, are consistent with those obtained from the compression and flexural tests. M4 recorded the highest values in these tests and the mortars M5 and M3 recorded the two lowest static and dynamic elastic modulus. Apparently, another inconsistency can be observed by comparing M1 and M2: the former has a higher value of the dynamic elastic modulus while the latter recorded a maximum in the static elastic modulus curve. The results of static elastic modules are shown in Fig. 1.

Table 3. Mechanical tests: elastic modulus.

Material	Longitudinal dynamic modulus of elasticity - Ed_L (MPa)	Flexural dynamic modulus of elasticity - Ed_F (MPa)	Static elastic moduls (MPa)
M1	2939	3188	1463.1
M2	2498	3066	1713.1
M3	603	902	1097.5
M4	3403	3877	1841.3
M5	917	995	673.7

Ultrasonic Tests. The ultrasonic tests, shown in Table 4, once again, point out a higher propagation speed of the emitted wave in the case of sample M5. The lowest velocity was recorded with sample M3.

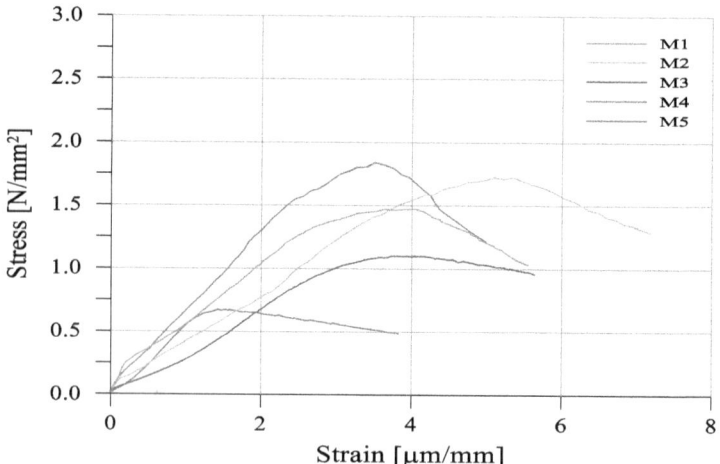

Fig. 1. Stress-strain diagrams

Table 4. Ultrasonic test

Material	Transmitted velocity (m/s)
M1	1509.57
M2	1579.65
M3	1185.85
M4	1602.15
M5	1620.26

Porosity. Using the Mercury Intrusion Porosimeter, it was possible to estimate the total porosity and porous structure of the samples. Results about pores area and the percentage of material porosity are shown in Table 5. By studying the graphs comparing pore diameter with mercury intrusion and the logarithm of mercury intrusion, collected in Fig. 2a, 2b, it is possible to divide the space represented by the axes into three zones:

1. Between 0.01 mm and 0.1 mm (microporosity);
2. Between 0.1 mm and 10 mm (mesoporosity);
3. Between 10 mm and 100 mm (macroporosity).

From the analysis made using the mercury intrusion porosimeter, Table 5, the porosity of the samples is between 34 and 42%; the average diameter is between 0.1437 mm and 4.8360 mm. The lowest value of porosity was recorded in sample M5, although the average pore diameter is 4.8360 mm. The highest value was detected in sample M3. The samples with the lowest pore area are M5 followed by M3, both of which have the lowest percentage of micropores.

Specifically, as can be seen in Fig. 2a, 2b, all samples except M3 and M5, had a maximum in the graph in the region corresponding to the mesopores and peaks in the region of the micropores. In contrast to these, samples M3 and M5 show more or less intense signals in the macropores region. The low porosity of sample M5, could in fact be caused by the majority of macroporosity characteristic of the sample.

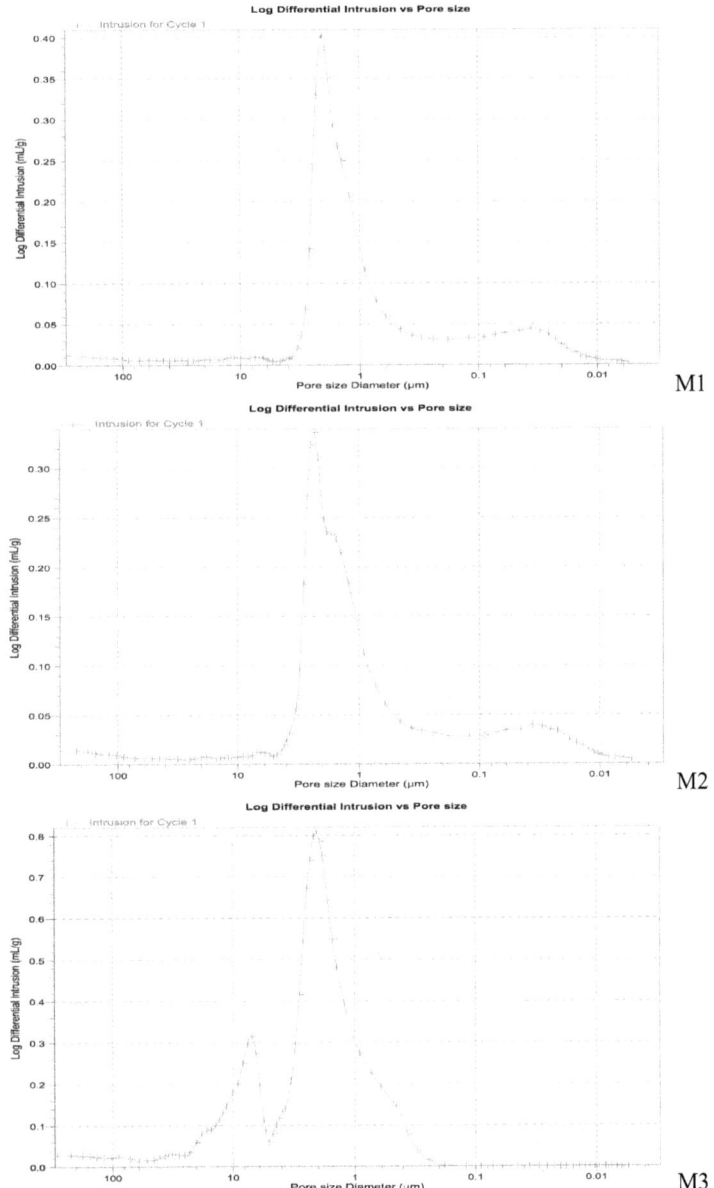

Fig. 2. a) Distribution of porosity as a function of pore diameter in samples M1, M2 and M3. b) Distribution of porosity as a function of pore diameter in samples M4 and M5.

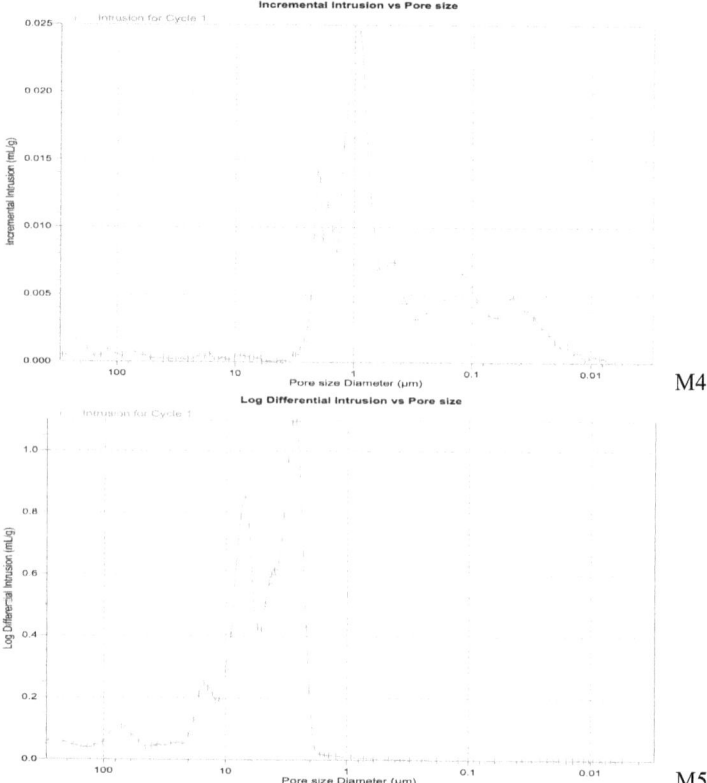

Fig. 2. (*continued*)

Table 5. MIP analysis

Material	Pores area (m²/g)	Average pore diameter (μm)	Porosity %
M1	5.196	0.1692	35.6711
M2	5.581	0.1559	36.6223
M3	1.332	1.5926	42.3130
M4	5.161	0.1798	37.3286
M5	0.531	4.8360	34.7788

Density. By comparing the density values obtained from the two methods, in Table 6, results were almost comparable for all samples. For both methods, the lowest density value correspond to sample M5, followed by M3. From the calculation of density, the highest result was obtained from M1, followed by M2. In contrast, MIP recorded a higher density value for M2 followed by M1. This result could be due to a possible higher presence of closed pores in sample M1 that cannot be determined by MIP.

Table 6. Volumic mass of the speciemens and the bulk density obtained by Mercury Intrusion Porosimeter (MIP).

Material	Volumic mass (g/mL)	Bulk density at 0.25 psia (g/mL)
M1	1.65166	1.6227
M2	1.61133	1.6382
M3	0.81469	0.7976
M4	1.60874	1.6087
M5	0.49830	0.8299

4 Discussion

The analyses carried out allowed an in-depth study of the plaster samples and clarified the main characteristics. The results obtained were then compared with what is stated in the technical data sheets of the individual products. The mechanical compression, flexural, elastic modulus and ultrasonic tests gave consistent results, with the exception of some apparently non-compliant results listed above. In all cases, M3 and M5 recorded the lowest test resistance values, as is possible to observe in Fig. 3 and Fig. 4. When comparing the results obtained from the laboratory tests and the data sheets provided by the company, the results relating to the flexural tests, apart from in the case of M4, were not compliant but very different, while the range of values relating to the compression test was so wide that it is possible to state that the results are compliant according to standard labelling.

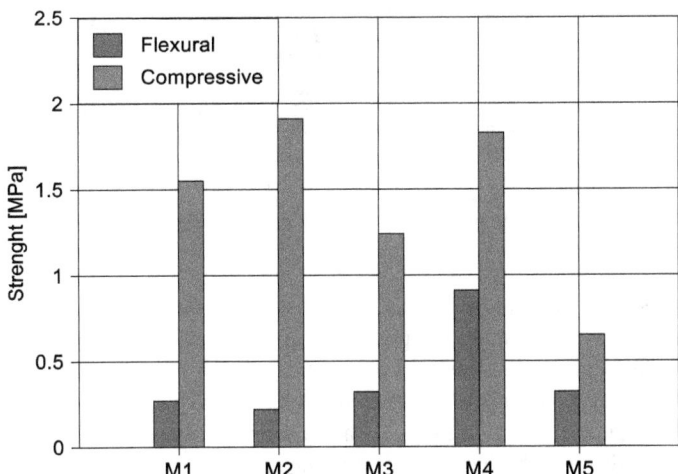

Fig. 3. Comparison of the flexural (blue) and compressive (red) strength measurement after 28 days.

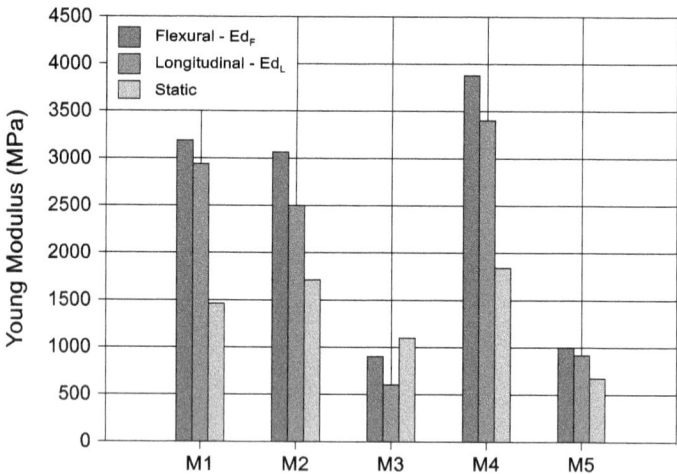

Fig. 4. Comparison of the three elastic modulus measurements: dynamic (Ed$_F$ in blue, and Ed$_L$ in red) and static (green) after 28 days.

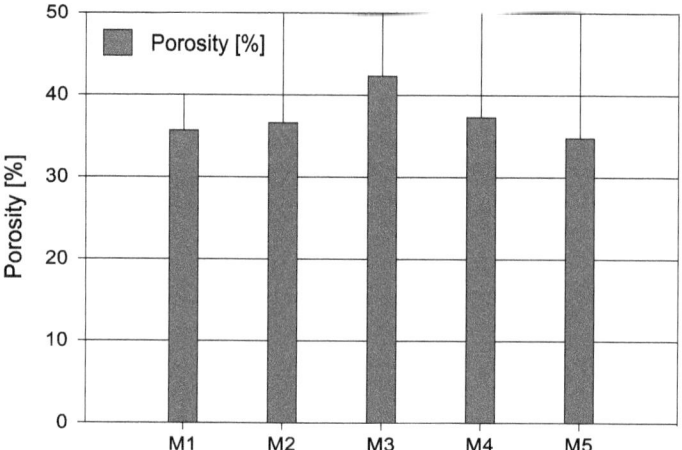

Fig. 5. Comparison of the porosity values obtained with MIP analysis after 28 days.

Data sheets allow companies to indicate a class of mortar with a very wide range and therefore do not give the restorer the possibility of being able to choose the compatible mortar. As in fact our results show, the data are compliant, but often the correct mortar cannot be chosen because apparently even though they have highly different character-istics, they are standardized by the classification and the very wide range. Some mortars are significantly different but are classified in the same class. The microstructure of the samples was investigated using the mercury intrusion porosimeter (Fig. 5), which pro-vided data on the percentage of the porosity of the samples and the average diameter of the mercury-accessible pores and density of the sample.

Comparing the porosity values with the densities and mass of the individual samples (Fig. 6), one would have expected a higher porosity in the case of samples M3 and M5, as they were found to be the least dense samples with the highest presence of macropores.

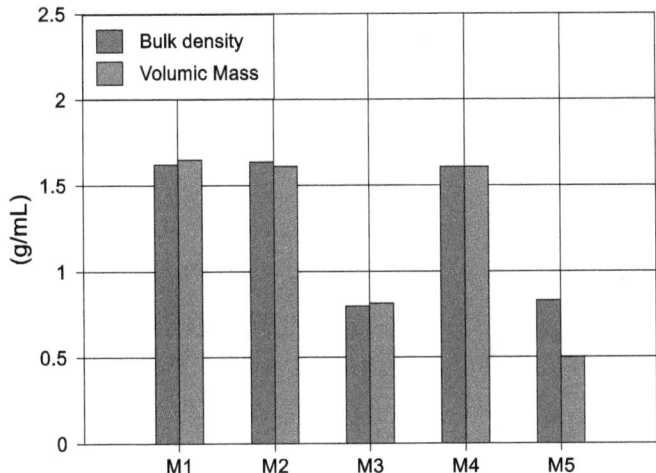

Fig. 6. Comparison of the volumic mass calculated, in red, and the bulk density of the speciemens in blue.

M5 did not confirm the initial assumptions as it seems to be the sample with the lowest porosity percentage, this could be caused by the excessive presence of inaccessible voids. In the case of all other samples, a higher percentage of mesopores followed by micropores was observed from the graphs. What was observed from the analyses, despite the instrumental disadvantage related to the impossibility of considering non-accessible, or closed pores, was confirmed in all cases by density and/or bulk density analyses.

5 Conclusions

Analyses has been carried out on a selection of ready-mixed NHL-based commercial plasters for conservation of historic masonries. This work focused on the evaluation of the physical and mechanical characteristics of the materials with testing their properties and investigate their microstructural characteristics in function of their porosity.

The following conclusions were drawn from the discussion of the results obtained from the analyses:

- Some analyses needs to be replied (such as bending, compression, static elastic modulus) and so, the three replicates provided by the standards are not enough.
- The results obtained could be further investigated by carrying out microscopic analyses on cross-section in order to observe the porosity of the samples and assess the possible presence of closed pores and confirm what the porosimeter showed.
- Chemical analyses could also be carried out to study the mineralogical composition and the presence of any organic additives probably not included in the data sheet of the companies

- Physical and mechanical analyses allowed the identification of different microstructural characteristics that may affect durability. It is therefore necessary to continue the research with tests on the durability of these materials and the variation of these characteristics over time, and to assess their compatibility as a function of physical and mechanical parameters.
- The drop test for studying the water-repellent of the surface and Karsten's tube test to verify the presence of water-repellent on the surface should be evaluated in future.
- The current classification method of compressive strength (CS I and CS II) covers too wide a range of values, preventing the restorer from being able to choose the most suitable product.
- Although all the samples analyzed are plaster mortars made with a hydraulic binder (declared by the Company), the mechanical and physical characteristics of the samples are very different, so the chemical characteristics of the binder seem to be the least influential as regards the evaluation of compatibility.

References

1. La Scalia, G., Saeli, M., Adelfio, L., Micale, R.: From lab to industry: scaling up green geopolymeric mortars manufacturing towards circular economy. J. Clean. Prod. **316**, 128–164 (2021)
2. Baronio, G., Binda, L.: Studio sull'alterazione delle murature in laterizio. L'edilizia el industrializzazione, Milan, n. 6, pp. 313–320 (1989). ISSN 1593-3962
3. Siegesmund, S., Snethlage, R.: Stone in Architecture. Properties, Durability. Springer, Heidelberg (2011). https://doi.org/10.1007/978-3-642-14475-2
4. Gulotta, D., Goidanich, S., Tedeschi, C., Nijland, T.G., Toniolo, L.: Commercial NHL-containing mortars for the preservation of historical architecture. Part 1: compositional and mechanical characterisation. Constr. Build. Mater. **38**, 31–42 (2013)
5. Gulotta, D., Goidanich, S., Tedeschi, C., Toniolo, L.: Commercial NHL-containing mortars for the preservation of historical architecture. Part 2: durability to salt decay. Constr. Build. Mater. **96**, 198–208 (2015)
6. Henriques, F.M.A.: Challenges and perspectives of replacement mortars in architectural conservation. In: International RILEM workshop on repairs mortars for historic masonry, vol. 67, pp. 143–161. RILEM Proceeding Pro, Delft (The Netherlands) (2005)
7. Van Hees, R.: Damage diagnosis and compatible repair mortars. In: Bartos, P.J.M., Groot, C., Hughes, J.J. (eds.) International RILEM Workshop on Historic Mortars: Characteristics and Tests, pp. 27–35. RILEM Publications, Paisley (Scotland) (1999)
8. Groot, C., Ashall, G., Hughes, J.J.: Characterization of old mortars with respect to their repair. Final Report of RILEM TC 167-COM, pp. 1–178 (2004)
9. Groot, C.: RILEM TC 203-RHM: repair mortars for historic masonry. Mater. Struct. **45**(9), 1277–1285 (2012)
10. Schueremans, L., Cizer, Ö., Janssens, E., Serré, G., Balen, K.V.: Characterization of repair mortars for the assessment of their compatibility in restoration projects: research and practice. Constr. Build. Mater. **25**(12), 4338–4350 (2011)
11. Apostolopoulou, M., Aggelakopoulou, E., Bakolas, A., Moropoulou, A.: Compatible mortars for the sustainable conservation of stone in Masonries. In: Hosseini, M., Karapanagiotis, I. (eds.) Advanced Materials for the Conservation of Stone, pp. 97–123. Springer, Cham (2018). https://doi.org/10.1007/978-3-319-72260-3_5

12. Faridmehr, I., Huseien, G.F., Baghban, M.H.: Evaluation of mechanical and environmental properties of engineered alkali-activated green mortar. Materials **13**(18), 4098 (2020)
13. Ali, M.B., Saidur, R., Hossain, M.S.: A review on emission analysis in cement industries. Renew. Sustain. Energy Rev. **15**(5), 2252–2261 (2011)
14. Barr, S., McCarter, W.J., Suryanto, B.: Bond-strength performance of hydraulic lime and natural cement mortared sandstone masonry. Constr. Build. Mater. **84**, 128–135 (2015)
15. Lanas, J., Pérez Bernal, J.L., Bello, M.A., Alvarez Galindo, J.I.: Mechanical properties of natural hydraulic lime-based mortars. Cem. Concr. Res. **34**(12), 2191–2201 (2004)
16. Brunello, V., et al.: Understanding the microstructure of mortars for cultural heritage using X-ray CT and MIP. Materials **14**(20), 5939 (2021)
17. UNI EN 196-1:2005. Methods of testing cement - Part 1: Determination of strength (2005)
18. EN 14146:2004: Natural stone test methods. Determination of the dynamic modulus of elasticity (by measuring the fundamental resonance frequency), European Committee for Standardization
19. EN 12504-4:2004: Testing concrete. Part 4: Determination of ultrasonic pulse velocity, European Committee for Standardization
20. ISO 6784:1982: Concrete - Determination of static modulus of elasticity in compression, International Organization for Standardization
21. EN 1015-11:1999/A1: Methods of test for mortar for masonry, Part 11: Determination of flexural and compressive strength of hardened mortar, European Committee for Standardization (2019)
22. Klobes, P., Riesemeier, H., Meyer, K., Goebbels, J., Hellmuth, K.-H.: Rock porosity determination by combination of X-ray computerized tomography with mercury porosimetry. Fresenius J. Anal. Chem. **357**, 543–547 (1997)
23. Moro, F., Böhni, H.: Ink-bottle effect in mercury intrusion porosimetry of cement-based materials. J. Colloid Interface Sci. **246**, 135–149 (2002)
24. Münch, B., Holzer, L.: Contradicting geometrical concepts in pore size analysis attained with electron microscopy and mercury intrusion. J. Am. Ceram. Soc. **91**(12), 4059–4067 (2008)
25. EN 12390-7:2002: Testing hardened concrete – Density of hardened concrete
26. Cnudde, V., Cwirzen, A., Masschaele, B., Jacobs, P.J.S.: Porosity and microstructure characterization of building stones and concretes. Eng. Geol. **103**, 76–83 (2009)
27. Cizer, O., Van Balen, K., Van Gemert, D.: Competition between hydration and carbonation in hydraulic lime and lime-pozzolana mortars. In: 7th International conference on structural analysis of historic constructions, SAHC, Shanghai, pp. 241–246 (2010)

Historic Portland Cement-Air Lime Mortars. Historic Portland Cement Mortars

Characterization of Mortars and Concretes from the Mirante da Quinta da Azeda, Setúbal (Portugal). A Case Study from the Beginning of the 20th Century

Luís Almeida[1]([✉]) [iD], Ana Rita Santos[2] [iD], António Santos Silva[1] [iD],
Maria do Rosário Veiga[2] [iD], and Ana Velosa[3] [iD]

[1] Materials Department, National Laboratory for Civil Engineering, Av. do Brasil, nº 101, 1700-066 Lisbon, Portugal
lalmeida.geo@gmail.com

[2] Buildings Department, National Laboratory for Civil Engineering, Av. do Brasil, nº 101, 1700-066 Lisbon, Portugal

[3] Department of Civil Engineering, University of Aveiro, Campus Universitário de Santiago, 3810-193 Aveiro, Portugal

Abstract. The Mirante da Quinta da Azeda, in Setúbal (Portugal), is a peculiar observation tower built in the early 20th century, and one of the first examples in which reinforced concrete was applied in Portugal. It has an unusual architectural configuration, displaying elements of great slenderness. In the scope of the CemRestore research project - Mortars for the conservation of early 20th century buildings: compatibility and sustainability, several mortar and concrete samples were collected from this structure and were characterized using a combination of mineralogical, microstructural, physical, and mechanical techniques, including XRD, petrography, SEM-EDS, open porosity, capillarity coefficient, compressive strength, and ultrasonic pulse testing. In this paper, the main characterization results are presented and discussed. The results show that all structural and decorative samples are made with Portland cement, while one rendering mortar is lime-based. The sand is mostly siliceous whereas pebbles and crushed limestone can be found as coarse aggregates in concrete samples. This characterization allows for broadening the scientific knowledge about the materials of that period used in Portugal, also enabling the establishment of the requirements to be met by mortars and concrete to be used in the repair of this distinct structure.

Keywords: 20th Century · mortar · concrete · characterization · Portugal

1 Introduction

The built heritage of the early 20th century is characterized by a diversity of architectural styles ranging from Art Nouveau and Art Deco to Modernism. This was a time of transformation in construction practice, mainly fostered by changes in binders, from air lime to the wide use of Portland cement.

© The Author(s), under exclusive license to Springer Nature Switzerland AG 2023
V. Bokan Bosiljkov et al. (Eds.): HMC 2022, RILEM Bookseries 42, pp. 243–257, 2023.
https://doi.org/10.1007/978-3-031-31472-8_19

Some scientific knowledge about historically emblematic concrete structures and buildings from the 19th century onwards has been carried out and reveals that a diversity of materials was employed in their construction [1–8]. One of these new materials is natural cement, which was used on a massive scale in Europe for the economic and easy manufacture of decorative elements and renders; this hydraulic binder, produced from the calcination of marlstones, was of great help in buildings' construction, particularly because of its properties (rapid setting and waterproof properties), until the development of artificial Portland cements [9]. In Portugal, few buildings from this period have been studied and the lack of specific details of mortars and concretes used makes it essential to know the characteristics of these materials, including the type of binder used.

The Mirante da Quinta da Azeda (Fig. 1), located in the city of Setúbal (Portugal), is one of the first building structures from the early use of reinforced concrete in Portugal [10]. It presents an unusual architectural configuration, displaying elements of great slenderness that testify the strong influence of the dominant iron architecture of its construction period, and it was erected above a water cistern. Its construction dates back to the early 20th century. However, Soares and Silva [10] refer to the existence of a late nineteenth-century residence, which presented a balcony, limited by concrete railings, from which the owners had access to the observation tower, at the first-floor level through a walkway (Fig. 1b), which nowadays is laid on the ground, next to the tower, since the residence was demolished. The construction process and the construction phases of the observation tower are unknown up to now. Currently, the structure has safety issues that have led to its shoring, being the access restricted. Most of the anomalies observed are related to the corrosion of the rebars (Fig. 2).

(a)   (b)

Fig. 1. Images of Mirante da Quinta da Azeda in 2014. Credits by Bruna Pereira de Oliveira. (a) General view from the outside; (b) View of the first-floor level, walkway to the tower and the cistern wall below the observation tower at the ground floor.

(a)   (b)   (c)

Fig. 2. Images of Mirante da Quinta da Azeda in 2021, during the sampling campaign. (a) General view of the observation tower after the shoring; (b) Vertical cracking in a structural column; (c) Corrosion of reinforcement and delamination of concrete cover.

Within the scope of the CemRestore project, funded by the Portuguese Foundation for Science and Technology, to study the mortars from the early 20th century in Portugal, and evaluate the changes that occurred in terms of the use of hydraulic materials and their production process, the present publication aims to provide information about the mortars and concrete materials used on this peculiar structure of the early 20th century. This study allowed a better understanding of the composition and the physical-mechanical properties of the materials applied in this period, in order to select the most suitable restoration products to be used in future repair works.

2   Characterization of Samples

Several samples from the Mirante were collected to allow the compositional characterization. The selection of samples was made according to LNEC methodology [11–13] and considering the importance of the different functions (structural or decorative use) and materials.

The samples were collected from the tower structure (main building – 4 specimens), the cistern (located at the bottom of the tower – 2 specimens), and the walkway (4 specimens). Most of the samples were taken by hand, however, the samples from the slab of the walkway were taken using a core drilling machine. Table 1 shows the list of the collected samples, their locations, and their description.

Table 1. List of the samples, general description, and their location in the main structure

Samples	Specimen	Location	General description	Images (samples and collection site)
M1	M1-A	Walkway railing	Handrailing - grey mortar with roughened surface	
	M1-B		Balustrade - precast reinforced decorative element	
M3	M3-A	Walkway slab	Slab coating- grey mortar covering the concrete slab	
	M3-B		Structural slab - reinforced concrete with crushed coarse limestone aggregates	
M4	M4	Ceiling of the first level structure	Whitish thin render with white lumps	
M5	M5	Decorative element from a column	Chapiter precast reinforced decorative element	
M6	M6	Decorative element from a column	Chapiter precast reinforced decorative element (similar to M5)	
M8	M8	Column base (structural element)	Reinforced concrete with coarse pebble aggregates	
M12	M12-A	Cistern's structure	Render - grey mortar covering the concrete support	
	M12-B		Reinforced concrete with crushed coarse limestone aggregates	

3 Experimental Program

With the main aim to characterize the mortars and concrete materials and to identify the differences among them, mineralogical, microstructural analysis, physical, and mechanical tests were performed. The characterization methodology comprises a wide range of techniques that complement each order, including XRD, optical microscopy by petrography, SEM-EDS, water absorption capillarity coefficient, open porosity, ultrasonic pulse velocity, and compressive strength [11–13].

3.1 Mineralogical and Microstructural Analysis

To determine the binder and sand mineralogical composition, XRD analysis was performed with a Malvern Panalytical Aeris X-ray diffractometer with 40 kV and 15 mA, using Copper Kα radiation (λ = 1.5406 Å). Diffractograms were recorded in the range 5–85 °2θ, at a step size of 0.20°/s.

Two types of fractions were analysed: the fraction corresponding to the samples as collected, designated as the overall fraction, was obtained by grinding to pass through a 106 μm sieve, and a fraction designated as fine fraction, which has a higher binder concentration, that was obtained by extracting the fines passing a 106 μm sieve directly from the bulk mortar. Semiquantitative XRD analysis was done by Rietveld phase analyses on the overall fraction.

Moreover, microscopic studies were also carried out as a powerful tool for identifying the type of binder and complementing data on the chemical composition of the studied materials. Thin and polished sections of the samples were prepared with vacuum impregnation with an epoxy resin. These were observed on an Olympus BX60 polarizing microscope and under Scanning Electron Microscopy (SEM) Tescan Mira3 coupled with a Bruker energy dispersive X-ray spectrometer (EDS) XFlash 6|30, using an electron beam voltage of 20 kV.

3.2 Physical and Mechanical Characterization

In order to assess the physical-mechanical characterization of the samples and their functional capacity to support requirements for new compatible repair materials, some mechanical and water behaviour tests were performed. Since the samples collected *in situ* are of non-standardized dimensions and are irregular, the laboratory characterization requires some special test methods, developed and validated in previous scientific works [14–16].

The capillary absorption by contact test consists of periodic weighing of samples until the total saturation. The absorbed water is determined through the difference between the weights measured periodically and the initial weight and the capillarity coefficient is determined as the slope of the initial straight part of the absorption curve in the function of the square root of time [14, 15]. In general, to perform this test, the samples are put in a basket with wet geotextile gauze and weighed together, however since the samples, in general, present high cohesion, this procedure was only used on the M4 sample (Fig. 3a).

For the compressive strength test a direct compression test was carried out, giving compressive strength values by the division of the compressive force that produces rupture of the sample by the 40 mm × 40 mm area of force application [15, 16], since the samples have the regular shape and dimensions (approx. 40 mm × 40 mm or a little more) necessary to use the standardized equipment, non "shaping mortar" is used in this case study (Fig. 3b). An electromechanical testing machine ETI HM-S was used, with a load cell of 200 kN.

Additionally, ultrasonic pulse testing was used, based on EN 12504-4 [17], to evaluate the compactness and stiffness of the samples [18]. This technique was performed with an Ultrasonic Tester Steinkamp BP-7 model and is based on measuring the speed of propagation of longitudinal ultrasonic waves, in microseconds, through specimens, using the transducers positioned, in this study case, on the same surface (indirect transmission), using a minimum distance of 10 mm between them and successively increasing the distance by displacing one of the transducers (Fig. 3c). The ultrasonic impulse velocity is determined as the slope of the distance/timeline obtained with the measurements.

The pore structure was evaluated through the total open porosity by immersion and hydrostatic weighing, based on EN 1936 [19].

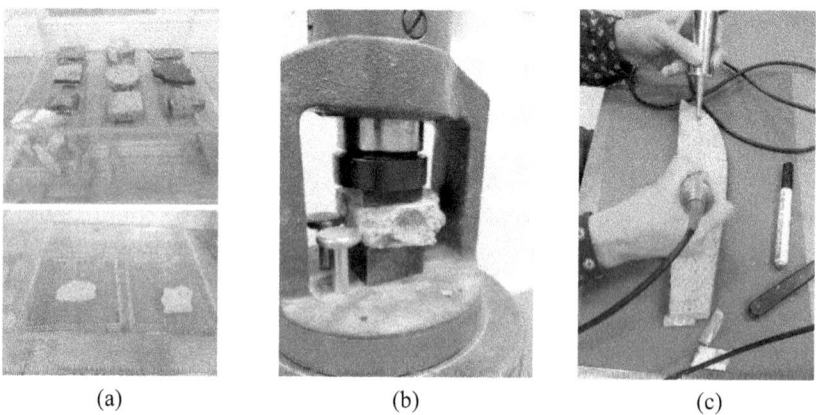

(a)   (b)   (c)

Fig. 3. Physical and mechanical tests performed at the samples. (a) water absorption by contact; (b) compressive strength; (c) ultrasonic test.

4   Results and Discussion

The results are presented and discussed considering the typology of the samples. The samples were grouped into the three main sets of construction materials, i.e., rendering mortars (M1-A, M3-A, M4 and M12-A), precast decorative concrete elements (M1-B, M5 and M6), and structural concrete (M3-B, M8 and M12-B).

4.1   Mineralogical Analysis and Semi Quantification of Crystalline Compounds

Table 2 shows the semiquantitative results obtained by Rietveld analysis for the overall fraction. The main compounds are related to the aggregates, which are mainly quartz and k-felspar (microcline), comprising most of the siliceous sand in all groups of samples analysed. Decorative and structural concrete show a relevant presence of calcite, (between 18.7% and 37.8%) presumably due to the contribution of carbonate aggregates, whilst the rendering mortars, with the exception of M12-A, show a lower presence of this carbonate phase (between 7.9% and 10.7%). The lowest value is shown in M4 mortar, simultaneously showing the highest presence of sand by means of siliceous phases identified.

Table 2. Crystalline compounds identified by XRD and Rietveld semiquantitative analyses of the overall sample fractions (values in percentage)

Crystalline compounds	Rendering mortars				Precast reinforced decorative element			Structural concrete		
	M1-A	M3-A	M4	M12-A	M1-B	M5	M6	M3-B	M8	M12-B
Quartz (SiO_2)	58.9	66.5	80.5	56.1	38.2	62.0	55.8	49.7	58.4	69.3
Microcline ($KAlSi_3O_8$)	14.9	10.9	6.6	12.3	14.6	9.9	7.1	12.2	8.3	9.8
Calcite ($CaCO_3$)	8.8	10.7	7.9	28.2	37.8	23.7	30.2	31.9	31.5	18.7
Ettringite ($Ca_6Al_2(SO_4)_3(OH)_{12}.26(H_2O)$)	2.1	2.2	-	-	1.7	-	-	-	-	0.6
Muscovite ($KAl_2(Si_3Al)_{10}(OH,F)_2$)	1.2	0.9	1.0	0.9	1.3	0.5	0.7	0.1	1.1	0.2
Gypsum ($CaSO_4.2H_2O$)	1.4	1.4	2.1	-	-	1.8	1.7	1.1	-	0.9
Portlandite ($Ca(OH)_2$)	4.2	3.3	-	-	1.3	-	-	0.4	-	-
Brucite ($Mg(OH)_2$)	0.8	0.6	<1.0	-	0.7	-	-	-	-	-
C_3S - Alite (Ca_3SiO_5)	2.4	1.0	-	0.9	0.9	0.1	0.1	1.1	0.3	0.2
C_4AF – Brownmillerite ($Ca_2(Al,Fe^{3+})_2O_5$)	1.6	1.1	-	0.5	1.3	0.6	1.0	<0.1	0.4	0.1
C_2S – Larnite (belite - Ca_2SiO_4)	1.2	0.3	-	0.8	0.3	0.1	0.5	1.4	<0.3	<0.1
Hydrocalumite ($Ca_2Al(OH)_7\cdot3H_2O$)	0.3	0.1	-	<0.3	0.9	0.4	0.4	<0.1	<0.3	0.2
Vaterite ($CaCO_3$)	-	-	-	-	<0.3	0.9	2.5	-	<0.3	-
Dolomite ($CaMg(CO_3)_2$)	-	-	-	0.3	-	-	-	-	-	-
Aragonite ($CaCO_3$)	-	-	-	-	-	-	-	2.0	-	-
Syngenite ($K_2Ca(SO_4)2\cdot H_2O$)	-	-	1.9	-	-	-	-	-	-	-

(-) compound not detected

Regarding the binder (fine fraction), Fig. 4 shows the occurrence of C_3S, C_2S, C_4AF, AFt, and CS phases, which indicates the use of Portland cement as the binder in all the mortars and concretes studied, except for M4 rendering mortar, which exhibits brucite in the fine fraction suggesting the use of dolomitic air lime as binder.

In respect to other crystalline compounds, it should be pointed out the presence of salts (gypsum and syngenite) in M4 mortar. This occurrence may be due to some contamination coming from the concrete support materials. Calcium-carbonate polymorph vaterite is found in all decorative concrete elements and in structural concrete sample M8. Besides this secondary carbonate can be found in shells as a biomineral or upon carbonation of CSH [20], since these elements are environmentally exposed it can be related to a dissolution/chemical precipitation cycle of the pre-existing carbonates (or the carbonated paste) due to water action.

4.2 Microstructural and Chemical Composition

The petrographic assessment by optical microscopy showed a variety of aggregates which confirms the mineralogical analysis by XRD. Rendering mortars have, in general,

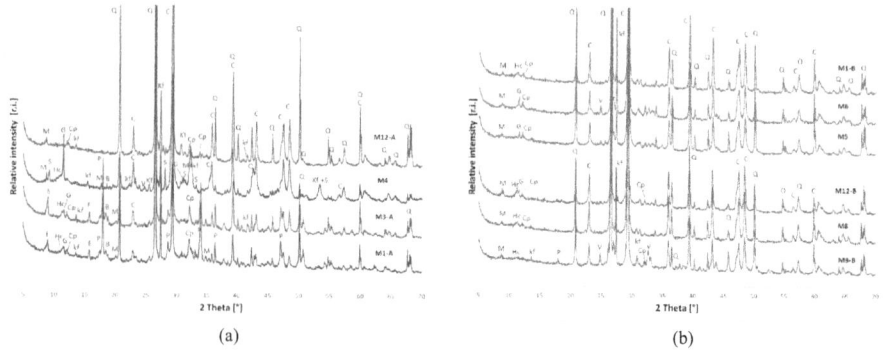

Fig. 4. Fine fraction diffractograms of: (a) rendering mortar samples, (b) decorative (red plots) and structural concrete (green plots) samples. Notation: Q – Quartz; C – Calcite; kf – Microcline; M – Muscovite; G – Gypsum; P – Portlandite; E – Ettringite; B – Brucite; Cp – Portland Clinker minerals (Alite+Larnite+Brownmillerite); Hc – Hydrocalumite; V – Vaterite; S – Syngenite (Color figure online)

a limited range of aggregate types. Siliceous sand mainly comprises quartz in mono or polycrystalline varieties (Fig. 5a, b). Decorative precast elements feature coarser aggregates of limestone (M1-B) and carbonate shells (M5 and M6) (Fig. 5c, d). Rounded pebbles observed in hand specimens from structural concrete (sample M8) are of quartzite and limestone nature (Fig. 5e), while coarse crushed aggregates in the other analysed concrete samples from the walkway slab and cistern structure (samples M3-B and M12-B) share the same lithological nature, both are limestones (Fig. 5f). The absence of quartzite pebbles in these samples is a distinguishing feature and could be associated with a different construction period.

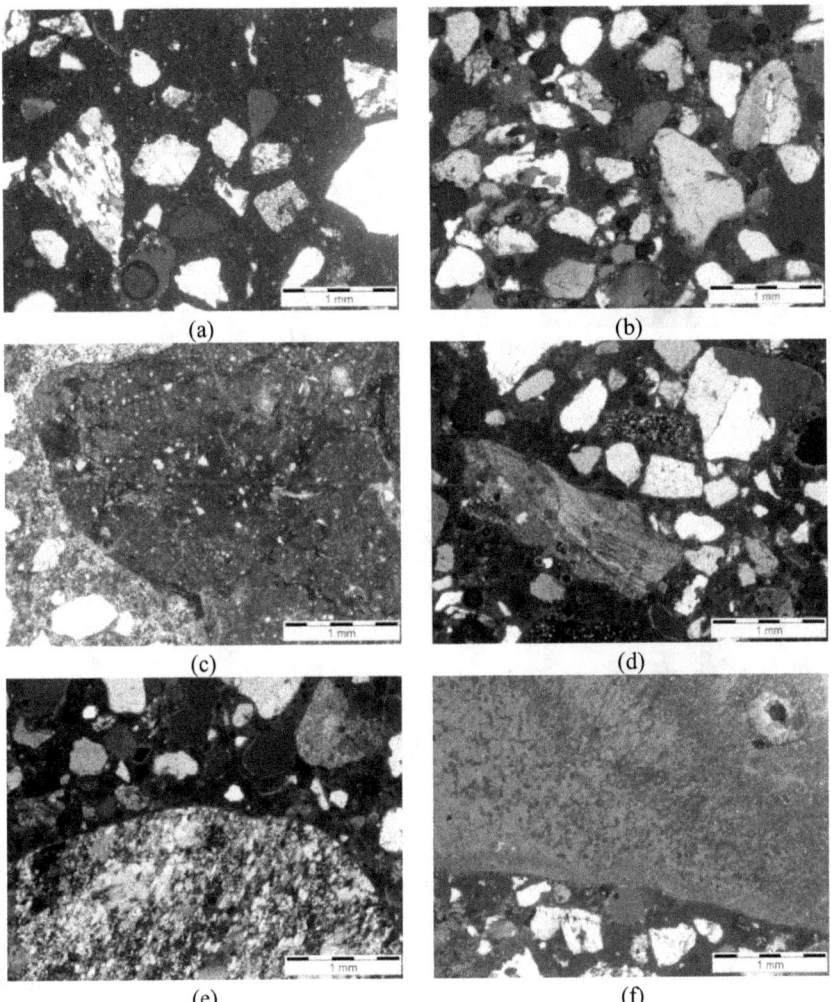

Fig. 5. Main aggregates' features from the three sets of analysed materials. Micrographs in crossed polarized light. Mono and polycrystalline quartz are the main sand aggregates in (a) sample M1-A and in (b) sample M4. Sample M1-B (c) shows a wide limestone grain (micritic texture with a ferrous cement) and in sample M5 (d) it can be seen an elongated bioclast carbonate shell in the bottom left. A quartzite round aggregate grain is shown in sample M8 (e - bottom) and crushed micritic limestone in sample M3-B (f - top).

Regarding the binders, the EDS analysis carried out in the sample M4 binder, provides sufficient data to clarify its nature. The presence of Mg in the EDS spectra (Fig. 6) proves that the binder is dolomitic air lime. Moreover, it was previously suggested by the presence of the brucite phase in XRD of the fine fraction and by the absence of clinker Portland compounds (Fig. 4a). No traces of hydraulicity were found in this sample through microscopic analysis.

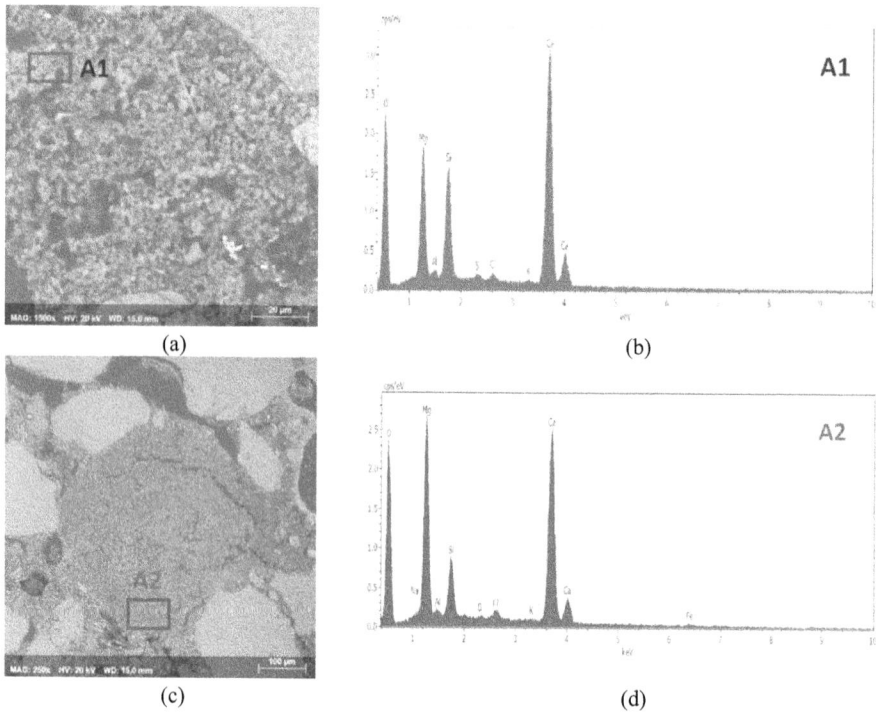

Fig. 6. Backscattered images of M4 sample. (a) Aspect of binder texture and chemical composition of area A1 by EDS (b), and unhydrated dolomitic lime lump (c) with correspondent EDS elemental composition of the area A2 (d).

Apart from sample M4, clinker remnants or residues are often found in all sets of samples (Fig. 7), regardless of the carbonation grade of mortars and concretes' matrix. Under the optical microscope, it can be seen the carbonation phenomenon affecting some of the studied materials, which is materialized by the chemical alteration of belite and alite caused by the calcium depletion in unhydrated clinker residues [21, 22] (Fig. 7c, d).

Further microscopy studies (SEM-EDS) were performed to complement and support the petrographic ones. Figure 8 shows some clinker remnants images combined with the chemical data obtained on several unhydrated clinker residues for each set of materials thus validating the occurrence of the principal crystalline compounds of the clinker Portland cement. Remarkable is the frequent presence of the C_3S phase (alite) in every sample (Fig. 8d, e, f), which is consistent with the use of Portland cement as this phase clearly distinguishes Portland cement from other historical binders [23].

4.3 Physical and Mechanical Characterization

The results of physical and mechanical properties are summarized in Table 3.

As can be seen from Table 3, two distinct rendering mortar compositions were used: the mortar M4, which presents the highest capillarity coefficient, the highest open

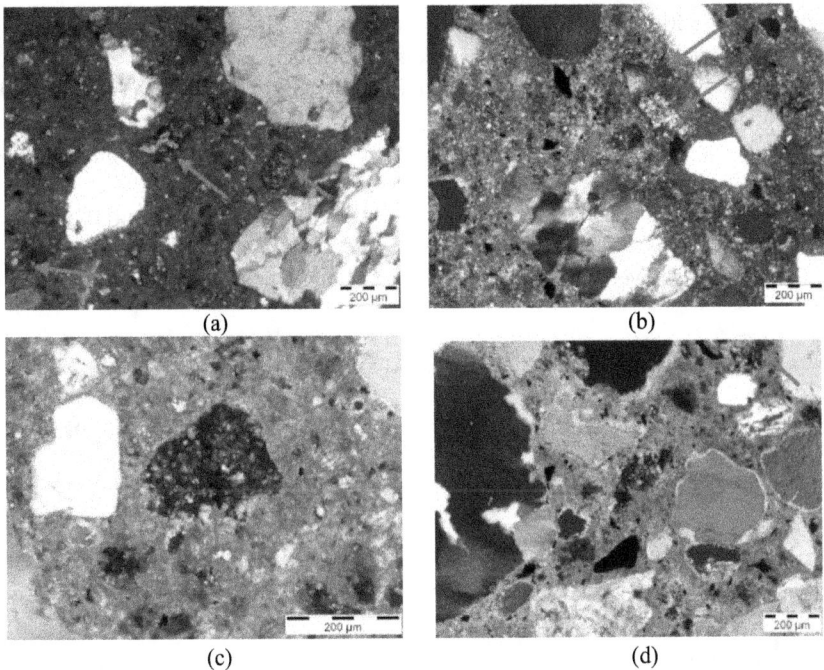

Fig. 7. Micrographs in crossed polarized light showing clinker remnants (arrows) inside a hydrated cement matrix in (a) sample M1-A and in (b) sample M12-B. Precast concrete element M5 (c) and structural concrete M8 (d) show alteration in belite and alite inside clinker residues (arrows), which corresponds to silica enrichment due to the carbonation phenomenon.

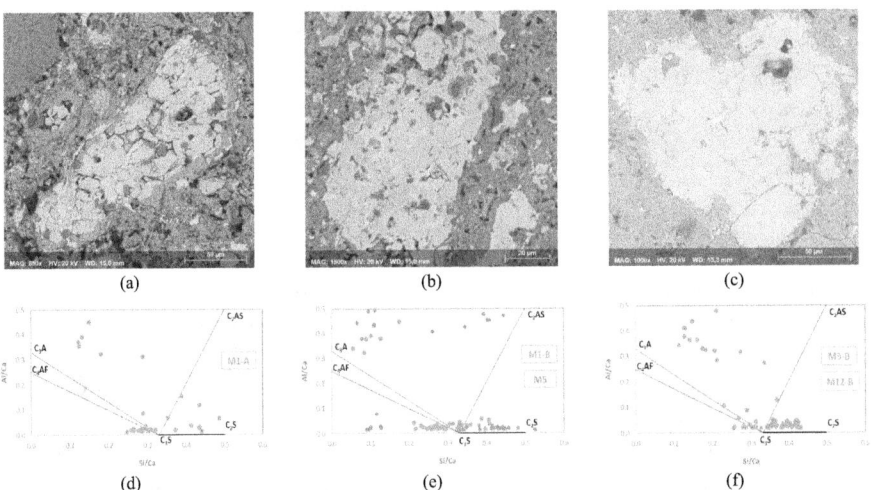

Fig. 8. Examples of unhydrated clinker residues from the different sets of mortars and concretes. (a) rendering mortar, (b) decorative precast and (c) structural concrete. (d, e, f) The respective set phase plots.

porosity, and the lowest bulk density and dynamic modulus of elasticity. These values are typical of air lime mortars and consistent with old lime mortars in a good state of conservation or with a low degree of degradation [15], despite of being contaminated by soluble salts.

Unlike the mortar M12-A which shows a low capillarity coefficient and the lowest open porosity values, with the highest apparent density and dynamic modulus of elasticity values. Despite the lower rate of absorption observed in the M3-A sample, which can be related to the finishing of this render that could delay the water absorption (Fig. 9), M1-A and M3-A showed similar behaviour and composition, as previously stated.

Table 3. Results of physical and mechanical characterization

Properties	Rendering mortars				Precast reinforced decorative element		Structural concrete		
	M1-A	M3-A	M4	M12-A	M1-B	M6	M3-B	M8	M12-B
Bulk density (kg/m^3)	2 096	2 000	1 833	2 224	2 043	2 184	2 212	2 177	2 186
Open porosity (%)	12	15	22	6	15	10	12	10	11
Dynamic modulus of elasticity by ultrasonic pulse test (MPa)	27 714	20 483	5 806	28 998	22 210	10 381	18 033	-	-
Compressive strength (MPa)	15	31	-	44	50	59	17	28	-
Water capillarity coefficient by contact (kg/m^2.min$^{1/2}$)	0.15	0.01	1.51	0.03	0.25	0.09	0.12	0.08	0.09

(-) Tests not carried out

Regarding the other samples (decorative precast and structural concrete), the bulk density is homogeneous between them (2043–2212 kg/m^3), and all samples present low values for water absorption coefficient (<0.25 kg/m^2.min$^{1/2}$), low open porosity values (<15%), high compressive strength values (>17 MPa) and, in general, high values for the dynamic modulus of elasticity.

Moreover, the structural ones showed the maximum values of total absorption (Fig. 9) and relatively lower compressive strength values (by comparison), probably due to the presence of coarse aggregates, that could increase the water absorption (by the aggregates) and decrease the mechanical strengths, due to ITZ zones with a larger volume.

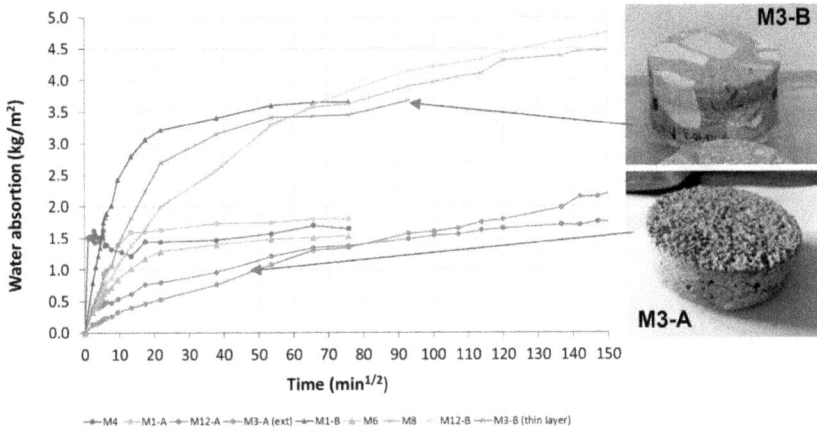

Fig. 9. Capillary water absorption curves.

5 Conclusions

Three sets of construction materials, i.e., rendering mortars, precast reinforced decorative concrete, and structural reinforced concrete from a singular architectonical observation tower in Setúbal, Portugal, built in the early 20th century, were characterized. The results obtained for the analytical characterization led to the main following conclusions.

Rendering mortars:

- Mainly composed of siliceous aggregates (quartz and K-feldspar). Only the sample from the cistern shows a significant increase in calcite, probably associated with a contribution of biogenic and carbonate aggregates, which may suggest that this rendering mortar is not contemporary with the other ones.
- Ordinary Portland cement was the binder used, except for sample M4, which was formulated with dolomitic aerial lime. The results of the physical and mechanical tests corroborate this characteristic.
- The sample M4 revealed a higher presence of soluble salts (gypsum and syngenite), most likely due to contamination from the upper concrete support materials.

Precast reinforced decorative concrete:

- Mainly composed of siliceous aggregates (quartz and K-feldspar). Calcite is the second crystalline compound identified which is associated with limestone aggregates in the sample M1-B and with the presence of carbonate shells in the M5 and M6 samples which may be an indicator of a local source.
- Ordinary Portland cement is the identified binder in all samples.

- The physical and mechanical results are in accordance with cement composite materials.

  Structural reinforced concrete:

- Mainly composed of siliceous aggregates (quartz and K-feldspar). Calcite is the second crystalline compound identified which is associated with coarse limestone aggregates in M3-B and M12-B samples. The calcite content is generally close to that of the precast reinforced concrete decorative samples, however with a higher maximum dimension.
- Aggregates from the observation tower structure is slightly different from the other structures' concretes. The coarse aggregates are rounded pebbles mainly of quartzite and limestone nature.
- Ordinary Portland cement is also the identified binder in all the concretes analysed.
- The physical and mechanical results are in accordance with cement composite materials with low values of open porosity and water absorption and very high compressive strength values.

Furthermore, despite the different uses and functionalities, the samples M5, M6, and M8 show similar compositions and physical performances - distinct from the samples M1, M3, and M12, so we can conclude that the Mirante probably had at least two different construction times or simply two sets of materials were applied by different people or contractors, in which Portland cement was used.

Acknowledgments. The authors wish to thank Bruna Pereira de Oliveira for all the collaboration, namely for being the interlocutor to the Setúbal City Hall and for providing the old Mirante's photos. The authors also wish to thank the Setúbal City Hall, the current owner of the Mirante da Quinta da Azeda, for authorizing this study and finally the Portuguese Foundation for Science and Technology (FCT) for the financial support within the scope of the Research Project CemRestore - Mortars for early 20th-century buildings conservation - Compatibility and Sustainability - POCI-01-0145-FEDER-031612.

References

1. Gosselin, C., Verges-Belmin, V., Martinet, G.: Natural cement and stone restoration of Bourges Cathedral (France). Conservar Património **7**, 5–19 (2008)
2. Haspel, J., Petzet, M., Zalivako, A., Ziesemer, J.: The Soviet Heritage and European Modernism. Heritage at Risk. Hendrik Bässler Verlag, Berlin (2007)
3. Cannata, M.: The repair and alterations of the De La Warr Pavilion. J. Archit. Conserv. **12**, 81–94 (2006)
4. Buck, S.: A material evaluation of the Gropius house: planning to preserve a modern masterpiece. APT Bull. **28**, 29–35 (1987)
5. Boothby, T., Parfitt, M., Roise, C.: Case studies in diagnosis and repair of historic thin-shell concrete structures. APT Bull. **36**, 3–11 (2005)
6. Allanbrook, T., Normandin, K.: The restoration of the fifth avenue facades of the metropolitan museum of art. APT Bull. **38**, 45–53 (2007)

7. Dell'Acqua, C., Franzoni, E., Sandrolini, F., Varum, H.: Materials and techniques of Art Nouveau architecture in Italy and Portugal: a first insight for a European route to consistent restoration. Restor. Build. Monum. **15**, 129–144 (2009)

8. Sandrolini, F., Franzoni, E., Varum, H., Nakonieczny, R.: Materials and technologies in Art Nouveau architecture: façade decoration cases in Italy, Portugal and Poland for a consistent restoration. Inf. Constr. **63**, 5–11 (2011)

9. Kozłowski, R., Hughes, D., Weber, J.: Roman cements: key materials of the built heritage of the 19th century. In: Dan, M.B., Přikryl, R., Török, Á. (eds.) Materials, Technologies and Practice in Historic Heritage Structures. Springer, Dordrecht (2010). https://doi.org/10.1007/978-90-481-2684-2_14

10. Soares, J., Silva, T.: Quintas de Setúbal: Valores Culturais. DEPA, Setúbal (1985)

11. Santos Silva, A.: Caracterização de Argamassas Antigas – Casos Paradigmáticos. Caderno de Edifícios Nº2 – Revestimentos de paredes em edifícios antigos. LNEC, pp. 87–101 (2002)

12. Santos Silva, A., Paiva, M.: Aplicação da Caracterização Mineralógica no estudo da Degradação de Argamassas Antigas. LNEC, Relatório, vol. 196 (2004)

13. Veiga, R., Aguiar, J., Santos Silva, A., Carvalho, F.: Methodologies for characterization and repair of mortars of ancient buildings. In: 3rd International Seminar Historical Constructions, Guimarães, pp. 353–362 (2001)

14. Veiga, R., Magalhães, A., Bosilikov, V.: Capillarity tests on historic mortar samples extracted from site. Methodology and compared results. In: 13th International Brick and Block Masonry Conference, Amsterdam (2004)

15. Magalhães, A., Veiga, R.: Physical and mechanical characterisation of ancient mortars. Application to the evaluation of the state of conservation. Mater. Constr. **59**(295), 61–77 (2008)

16. Válek, J., Veiga, R.: Characterization of mechanical properties of historic mortars - testing of irregular samples. Adv. Archit. Ser. **20**, 365–374 (2005)

17. European Committee for Standardization: EN 12504-4, Testing concrete in structures; Part 4: Determination of ultrasonic pulse velocity, Brussels (2004)

18. Galvão, J., Flores-Colen, I., Brito, J., Veiga, R.: Variability of in-situ testing on rendered walls in natural aging conditions - rebound hammer and ultrasound techniques. Constr. Build. Mater. **170**, 167–181 (2018)

19. European Committee for Standardization: EN 1936, Natural stone test methods; Determination of real density and apparent density, and of total and open porosity, Brussels (2006)

20. Elsen, J., Van Balen, K., Mertens, G.: Hydraulicity in historic lime mortars: a review. In: Válek, J., Hughes, J., Groot, C. (eds.) Historic Mortars. RILEM Bookseries, vol. 7. Springer, Dordrecht (2012). https://doi.org/10.1007/978-94-007-4635-0_10

21. Shtepenko, O., Hills, C., Brough, A., Thomas, M.: The effect of carbon dioxide on dicalcium silicate and Portland cement. Chem. Eng. J. **118**(1), 107–118 (2006)

22. Weber, J., Baragon, A., Pintér, F., Gosselin, C.: Hydraulicity in ancient mortars: its origin and alteration phenomena under the microscope. In: 15th Euroseminar on Microscopy Applied to Building Materials, Delft, The Netherlands (2015)

23. Walsh, J.: Petrography: Distinguishing natural cement from other binders in historical masonry construction using forensic microscopy techniques. J. ASTM Int. **4**(1), 1–12 (2006)

Early Age Properties of Hydraulic Lime Mortar Prepared Using Heavy Metal Contaminated Aggregate

Tilen Turk, Violeta Bokan Bosiljkov, Maks Alič, and Petra Štukovnik[✉]

Faculty of Civil and Geodetic Engineering, University of Ljubljana, 1000 Ljubljana, Slovenia
petra.stukovnik@fgg.uni-lj.si

Abstract. Traditional stone aggregates used to prepare mortars are generally considered inert fillers that do not interact with a binder. The European Green Deal, on the other hand, encourages the use of secondary material as aggregate. This secondary material is rarely an inert part of lime-based composites. This study focuses on heavy metal contaminated aggregate, for which has already been confirmed to alter microstructure development and lower the early strength of cement composites. Its influence on the early age properties of hydraulic lime mortar was studied through microstructure development recorded using transmission ultrasound technique and tensile and compressive strength after 1, 3 and 7 days. Results are then compared to those of cement composites. The obtained results show that heavy metals delay microstructure development of hydraulic lime mixture during the first 60 h, but do not decrease the tensile and compressive strength of the mortar after 3 and 7 days, as is the case with cement mortar.

Keywords: hydraulic lime · ultrasound · early age · heavy metals

1 Introduction

Traditional stone aggregate is considered inert filler that does not interact with the early processes of forming the microstructure. However, when designing a mixture, aggregate needs to be carefully selected to withstand environmental conditions where abrasion and erosion can occur, especially in the case of renders. A designer can choose between naturally (gravel, sand) or mechanically formed stone aggregates. The latter is produced in quarries by mining and then crushing large rocks to chosen fractions. Therefore, its impact on the availability of natural resources and the environment is not neglectable. Metal mining also includes crushing before chemical decomposition and galvanising metal-rich rocks to extract metal [1]. As a result, a large amount of tailing (non-metal part of the rock) is left over. As a consequence of the European Green Deal, tailing is often used as a source of cement composite aggregates.

Different studies have shown that in a highly alkaline medium metal-contaminated aggregate forms various hydroxide metal complexes. These hydroxide complexes can be formed by free metals leaching out during mixing or by galvanic processes between different cement constituents. The process known as chelation prevents calcium saturation and causes a delay in microstructure formation in cement composites [2]. Mines also use the galvanic processes where, for example, Fe^{3+} (originating from cement constituent Brownmillerite) ions oxidise Zn from sphalerite to form Zn2+ ions [3]. Metals can originate from the rock crystallisation process known as metamorphosis, where isomorphic substitution occurs. Isomorphic substitution is a substitution of atoms inside the crystal unit cell. It is only possible when the substitute has the same or smaller size and same charge as an atom that it substitutes. With advanced analytical methods and their coupling, it is possible to detect metals even in concentrations of order ppm or ppb. Most frequently used analytical methods for their detection are ICP -MS (inductively coupled plasma mass spectroscopy) as well as Raman spectroscopy, XRF (X-Ray Fluorescence), XRD (X-Ray Diffraction) and SEM-EDS (Scanning Electron Microscope coupled with Energy Dispersive Spectroscopy) and TEM (Transmission Electron Microscope) [4].

Isomorphic substitutions are present in all rocks, but in such low quantities that the development of cement or lime binder microstructure is not influenced. Ordinary dolomite aggregates may, for example, contain eight metal isomorphs listed in Table 1.

Table 1. Heavy metals found in traces in Dolomite aggregate [1]

Element	Quantity	Element	Quantity
As	1 ppm	Ni	20 ppm
Cr	11 ppm	Pb	9 ppm
Cu	4 ppm	V	20 ppm
Mo	3 ppm	Zn	20 ppm

To our knowledge, there are no relevant studies about the influence of heavy metals on the early age properties of hydraulic lime mortars used for restoration and repair of historic buildings. Therefore, this research aimed to study microstructure development using transmission ultrasound waves and to characterise the early mechanical properties of hydraulic lime mortar prepared using aggregates contaminated with heavy metals.

2 Materials and Methods

Two different aggregates were used in this study. Aggregate A is a reference aggregate that generally does not alter the early age microstructure development of mineral binder. Aggregate B is contaminated with heavy metals and, according to previous studies, alters the microstructure development of cement composites. With these two aggregates 1:2:9 (volume ratio cement: lime: aggregate) hydraulic lime mortars were prepared and tested in this study (HL_A and HL_B). The obtained results were compared to those of the cement mortars tested using the same testing protocols. The water-to-cement ratio of the cement mortars was 0.45, and aggregate occupied 60% of the volume of mortars.

2.1 Binder Properties

Hydraulic lime mixtures were prepared using CL70-S hydrated lime and cement CEM I. The lime: cement volume ratio was 2:1. The compositions of binders are presented in Table 2.

Table 2. Binder compositions according to XRD analysis

Binder	SiO_2	Al_2O_3	Fe_2O_3	CaO	SO_3	MgO	Na_2O	K_2O	Cl^-	SO_2	LOI
CEM I [5]	19.33	5.62	2.70	62.06	3.23	2.07	0.35	0.75	0.009	/	/
Lime [6]	/	0.6	0.19	71.25	0.06	2.09	/	/	/	0.79	25.69

2.2 Aggregate Properties

Contaminated aggregate B was characterized according to standard EN 1097, where organic matter, sieve analysis and density were determined. The XRD analysis of aggregates A and B was performed using Cu Kα1 radiation. Diffractograms were refined using the Rietveld method.

2.3 Ultrasound Measurements

The microstructure formation was studied by measuring transmission times of ultrasound P-waves using the Proceq Pundit PL-200 instrument and 54 kHz transducers. Measurements were performed on mortar cast into the 10 cm × 10 cm × 10 cm mould. The interval between subsequent measurements was set to 5 min. The setting time was estimated at a pulse velocity of around 500 m/s, according to [7].

2.4 Tensile and Compressive Strength

Tensile Strength. Tensile test was performed on three samples after 1, 3 and 7 days, using the MTS Exceed Series 43 universal testing machine with force capacity between 5 N and 10 kN. The load was applied at a constant rate of the crosshead displacement equal to 0.06 mm/s.

Compressive Strength. Compressive test was performed on six samples after 1, 3 and 7 days. The testing machine and the loading rate were the same as for the tensile test.

3 Results and Discussion

3.1 Aggregate Properties

Aggregate A has a density of 2850 kg/m^3 and, according to XRD, consists of the mineral dolomite (99.5%) and the mineral calcite (0.5%) (Fig. 1). Aggregate B is without organic matter (Fig. 2), and its density is 2730 kg/m^3. Aggregate B consists predominantly of the mineral dolomite (61.2%) and the mineral calcite (35.6%). It also contains lead (1.3%), sphalerite (1%) and flint (0.8%). Aggregate A is produced in the quarry, while aggregate B is produced with tailing crushing. The density of aggregate A is higher than that of aggregate B, because of the mineral dolomite which occurs in higher concentrations (Fig. 2). Sieve analysis gave similar grain size distributions of aggregates A and B. However, aggregate A appears to contain slightly more particles below 0.063 mm (Fig. 3).

Fig. 1. Organic matter test with 3% NaOH

Properties of fresh hydraulic lime mortars HL_A and HL_B are presented in Table 3. From the properties, it can be seen that heavy metals increase the workability of mortars. The density of HL_A is higher than that of HL_B due to the higher density of aggregate A.

3.2 Ultrasound Measurement

Ultrasound measurements show that the initial setting time of HL_B, when the load-bearing skeleton of the mortar starts to form, is delayed about 10 h compared to that of mortar HL_A. For comparison, in the case of cement mortars (C_A and C_B), aggregate B delayed the initial setting time by about 15 h (Fig. 4). From the ultrasound

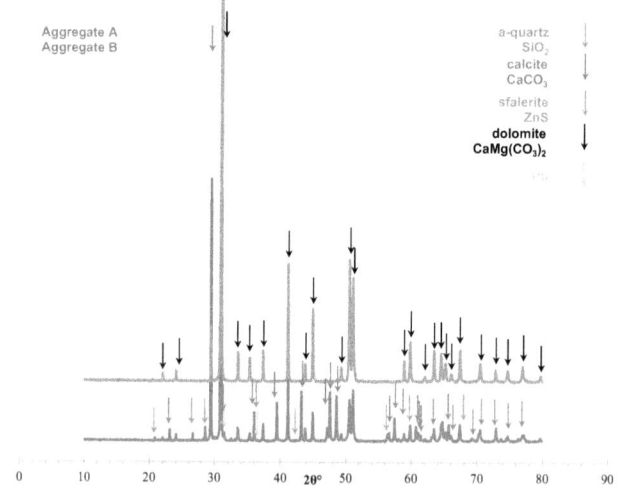

Fig. 2. X-Ray Diffraction analysis of aggregate A and aggregate B

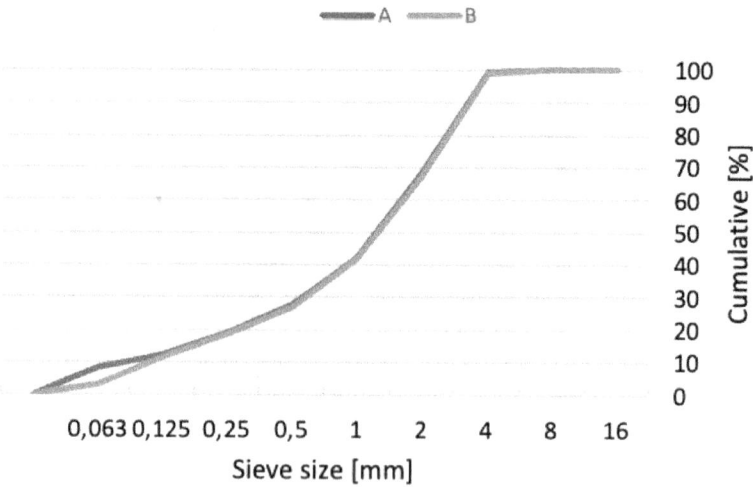

Fig. 3. Grain size distribution of aggregate A and aggregate B

Table 3. Properties of fresh hydraulic lime mortars

Mixture	Flow value [mm]	Density [kg/m³]	Superplasticiser [g]	Added Water [g]
HL_A	113	2210	3	980
HL_B	117	1830	3	980

measurements one can see that the impact of heavy metals on the microstructure development is less pronounced in hydraulic lime mortar compared to cement mortar. This can be due to higher calcium availability originating from the hydrated lime. However, a larger scatter was observed in the ultrasound velocities for mortars HL, which can be attributed to the slower microstructure formation in the composite lime-cement binder, where only the cement hydration products influenced the obtained results during the 3 days measurements.

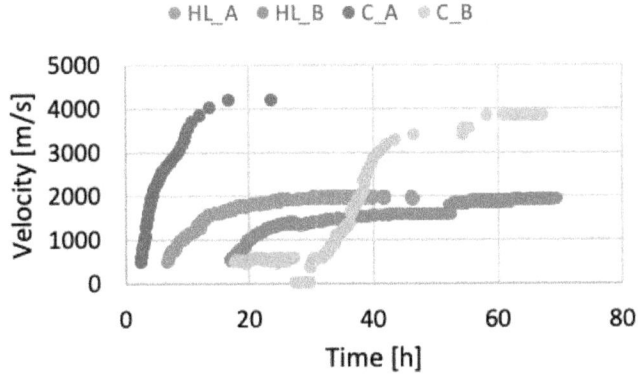

Fig. 4. Development of ultrasound velocities for hydraulic lime and cement mortars.

3.3 Tensile and Compressive Strength

Tensile Strength. After one day, the average tensile strength of the mortar HL_B (0.16 MPa) is significantly lower than that of mortar HL_A (0.41 MPa). The difference is caused by delayed microstructure formation in the mortar HL_B, which was confirmed by ultrasound measurements. After three days, the average tensile strength of mortar HL_B (1.04 MPa) exceeded the same strength of mortar HL_A (0.88 MPa) and remained higher also after 7 days (HL_B = 1.59 MPa and HL_A = 1.32 MPa) (Fig. 5 left). Compared to cement mortar, heavy metals seem to have less influence on tensile strength when calcium-rich binder such as lime-cement binder is used (Fig. 5 left, Fig. 5 right).

Compressive Strength. After one day, the average compressive strength of mortar HL_B (0.71 MPa) was lower than that of mortar HL_A (1.69 MPa). After three days, there is no significant difference in average compressive strength between the two mixtures (HL_A = 3.59 MPa, HL_B = 3.41 MPa). After seven days, average compressive strength of mortar HL_B (5.38 kPa) exceeds that of mortar HL_A (4.97 MPa) (Fig. 6 left). Cement mortar C_B does not reach compressive strength of the reference (C_A) even after 7 days (Fig. 6 right). We can conclude that using the studied aggregate contaminated with heavy metals in the hydraulic lime mortar results in higher compressive and tensile strengths compared to the reference mortar. Therefore, its use in lime-cement mortars would not be as problematic as it is in the case of cement binder.

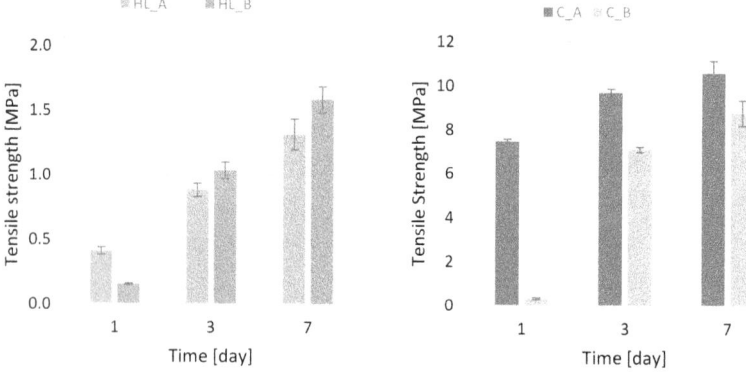

Fig. 5. Tensile strength development during seven days for hydraulic lime mortars (left) and cement mortars (right).

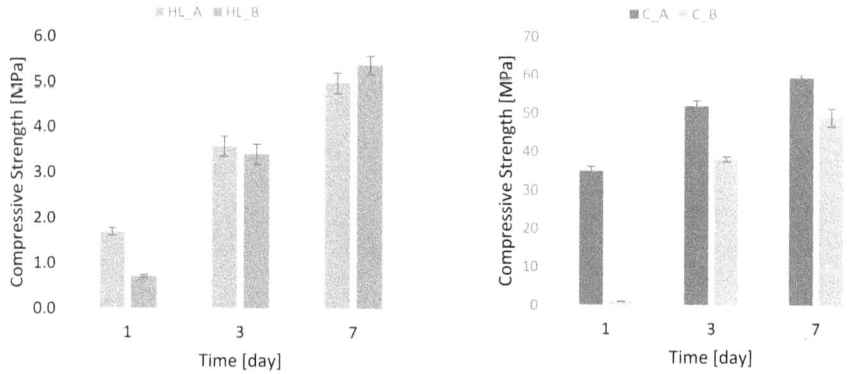

Fig. 6. Compressive strength development during seven days for hydraulic lime (left) and cement (right) mortars.

Acknowledgment. Authors acknowledge the financial support from the Slovenian Research Agency through Tilen Turk's PhD project and research programme P2-01285.

References

1. Bradl, H.B.: Sources and origins of heavy metals. Interface Sci. Technol. **6**, 1–27 (2005)
2. Gutsalenko, T., et al.: Impact of trace metals Zn, Cu, Cd and Ni on the reactivity of OPC and GGBS-based hydraulic binders at early age for sediment stabilization. Constr. Build. Mater. **346**, 128406 (2022). https://doi.org/10.1016/j.conbuildmat.2022.128406
3. Ahmed, I.A.M., Hudson-Edwards, K.A., European Mineralogical Union, Mineralogical Society (Great Britain): Redox-reactive minerals : properties, reactions and applications in clean technologies, vol. 447 (n.d.)

4. Rabenau, A.: R. C. Evans: Einführung in die Kristallchemie. Übersetzt und bearbeitet von J. Pickardt und E. Riedel, Walter de Gruyter, Berlin 1976, XVI, 329 Seiten, Preis: DM 36. Berichte Der Bunsengesellschaft Für Physikalische Chemie **80**(10), 1034–1034 (1976). https://doi.org/10.1002/BBPC.19760801025

5. Štukovnik, P., Prinčič, T., Pejovnik, R.S., Bokan Bosiljkov, V.: Alkali-carbonate reaction in concrete and its implications for a high rate of long-term compressive strength increase. Constr. Build. Mater. **50**, 699–709 (2014). https://doi.org/10.1016/j.conbuildmat.2013.10.007

6. Padovnik, A., Bokan-Bosiljkov, V.: The influence of dry hydrated limes on the fresh and hardened properties of architectural injection grout. Materials **14**(19), 5585 (2021). https://doi.org/10.3390/ma14195585

7. Serdar, M., Gabrijel, I., Schlicke, D., Staquet, S., Azenha, M. (eds.): Advanced Techniques for Testing of Cement-Based Materials. STCE, Springer, Cham (2020). https://doi.org/10.1007/978-3-030-39738-8

Conservation Issues Concerning Mortars, Plasters, Renders and Grouts. Diagnosis. Decay and Damage Mechanisms. Case Studies

Gaji, a Gypsum-Earth Plaster in the Wall Painting Technology of the Church of St. Demetrios of Thessaloniki, David Gareji, Kakheti, Georgia

Mariam Sagaradze[1], Joshua A. Hill[2,3]($\boxtimes$), Sophia Mikaberidze[1], Nana Khuskivadze[1], Manana Kavsadze[4], Stephen Rickerby[5], and Lisa Shekede[5]

[1] Apollon Kutateladze Tbilisi State Academy of Arts, 22 Alexander Griboedov St., 0108 Tbilisi, Georgia

[2] School of Science and Technology, Nottingham Trent University, Nottingham NG11 8NS, UK
joshua.hill@ntu.ac.uk

[3] Department of History of Art, University College London, Gower Street, London WC1E 6BT, UK

[4] Alexander Janelidze Institute of Geology, Ivane Javakhishvili Tbilisi State University, 5 Ana Politkovskaia St., 0186 Tbilisi, Georgia

[5] Rickerby and Shekede Wall Painting Conservation, Ross-On-Wye, UK

Abstract. Gaji is the Georgian name both for a gypsum-and-clay-containing soft rock found in eastern Georgia, and for the plaster traditionally derived from it. Gypsum-earth plasters have a long history of use across Caucasia, central Asia, and the Middle East. Gaji has been used in the preparation of mortars, plasters and decorative finishes, and as a support for Georgian religious and secular wall paintings together spanning the 9th – 20th centuries. Little research has been conducted into its properties, however, and within the context of wall painting conservation neither its influence on condition nor its implications for treatment have been adequately explored. Using the Church of St. Demetrios of Thessaloniki, rediscovered in 2015, at the monastic complex of David Gareji as an example, this paper explores the use of gaji plaster technology and the conservation challenges it presents.

Keywords: wall painting · plaster · gypsum · earth · gaji

1 Introduction

Global wall painting technologies exhibit enormous diversity in their materials and technology of execution [1]. One widely used classification can be made based upon the apparent binder of the plaster layer; most commonly lime, gypsum, earth, or cement [2]. However, the label of a "gypsum plaster" may belie crucial differences between superficially similar plasters, whether this be in terms of composition, material characteristics, or production technology. To complicate matters further, some plasters may have multiple binders or exhibit ambiguity over whether particular components are an aggregate or a binder [3, 4].

© The Author(s), under exclusive license to Springer Nature Switzerland AG 2023
V. Bokan Bosiljkov et al. (Eds.): HMC 2022, RILEM Bookseries 42, pp. 269–287, 2023.
https://doi.org/10.1007/978-3-031-31472-8_21

Such complex plasters are not uncommon. Many will contain calcium carbonate, calcium sulfate (hydrates), as well as clay minerals within the same material, for example in the Neolithic plasters of Çatalhöyük [3] and the plasters used in the New Kingdom of Ancient Egypt. The latter were commonly derived from calcitic earth deposits at the foot of the Theban Mountains, containing clay and gypsum components [4]. With evidence that both clay and gypsum may contribute binding properties to these plasters, it is also apparent that the function and composition of these plaster components was controlled in a sophisticated manner to modify properties [4, 5]. Alongside important implications for understanding the production technology of the plasters, and of the paintings as a whole, these different scenarios have direct implications for investigating their deterioration and planning their effective conservation [6].

Gypsum-earth plasters have a long history of use across Caucasia, central Asia, and the Middle East; compositions may vary depending on the geology of the source quarry and the function of the plaster. For example, in Transcaucasia the earth is typically clay-rich, while in central Asia it is loess, leading to different material characteristics associated with local soils [7]. Examples of the gypsum-earth material called 'ganch' in Uzbekistan attest to the use of different grades of material for plasters serving different functions within sophisticated decorative schemes [8, 9]. Other local names for gypsum and gypsum-earth plasters include 'ganch' in Russian, Kazakh, Kyrgyz, and Uzbek, and 'gaaj' in Tajik [10]. According to Soviet literature, in the Middle East and Caucasia this material is also variously known as 'gaja', 'gachi', 'arzak', 'gaji', 'white earth', 'gypsum earth' and 'clay-gypsum' [7].[1]

Gaji is the Georgian name both for a gypsum-and-clay-containing soft rock found widely in Georgia, and for the plaster traditionally derived from it. In Georgia, gaji has been used in the preparation of mortars, plasters and decorative finishes, and as a support for religious and secular wall paintings spanning the 9th – 20th centuries. Little research has been conducted into its properties, however, and within the context of wall painting conservation neither its influence on condition nor its implications for treatment have been adequately explored. These properties should form the basis of the overall conservation approach; whether in preventive, passive, or remedial interventions it is essential to have an adequate understanding of the technology, characteristics, and condition of the plaster. Without this understanding it is often impossible to accurately identify the causes and activation mechanisms of deterioration [11].

The need to address these omissions came into focus in 2015, when the Church of St. Demetrios of Thessaloniki was discovered at Dodorka in the monastic complex of David Gareji. Condition and technical examination of the paintings identified a gaji plaster support in a critical state of delamination and decohesion. In response, a research project was launched to investigate the composition and properties of both the original plasters and currently produced gaji as a basis for creating compatible grouts and repairs. Qualitative and semi-quantitative data were obtained on porosity and geological components using X-ray diffraction, petrographic techniques, and chemical analyses. Using

---

[1] It should be noted here that "gaza", as mentioned in documentation of the St. Petersburg Cement Plant as a component of Portland cement [7], does not refer to a gypsum-clay material, but rather to a loose clay limestone.

the Church of St. Demetrios of Thessaloniki as an example [12], this paper explores the use of gaji plaster technology and the conservation challenges it presents.

2   Use of Gypsum-Earth Plasters in Georgia

In Georgia the term gaji is used both for a naturally occurring gypsum-and-clay-containing soft rock, as well as the material for plasters, mortars and/or other architectural-decorative finishes. Modern gaji quarries are spread in various regions of Georgia, including Samegrelo-Zemo Svaneti, Kvemo Kartli and Kakheti (Gardabani, Kumisi, Badiauri, Magaro-melaani, Marneuli), as well as Tbilisi (Dampalo, Orkhevi, Grmagele, Soganlugi), although the use of this material is mainly confined to eastern Georgia, particularly Tbilisi and Kakheti. Historic quarry sites are often found near these modern quarries and, in the absence of archival evidence, these provide the best available evidence for historic sources [7, 13]. The raw gaji material varies between quarries with some key distinguishing differences including its colour. Depending on the quarries and locations within the stratigraphy, gaji is found in various shades of white, grey, brown, and yellow. The colour of gaji is mainly determined by the composition and ratio of clay types, calcium sulfate hydrates, and the nature of impurities.

In eastern Georgia, gaji continues to be used as a plaster material both for religious/ecclesiastic structures and in secular residential buildings. The use of gaji as a wall painting support dates from at least the medieval period. The earliest known wall paintings with gaji-based plaster date from the 9th century and can be found in the medieval David Gareji Monastery complex, Kakheti, Georgia [14, 15]. The David Gareji Monastery complex is located on the semi-arid slopes of Mount Gareja, on the edge of the Lori plateau and partly extends across the border into Azerbaijan. It is only 70 km away from Tbilisi, the capital city of Georgia. The Gareji semi-desert territory covers around 1000 km^2 and includes twenty-one known extant monastery complexes and at least 5000 individual rock-cut cells and sanctuaries. In the first half of the 6th century, the Monastery (later called Lavra) was founded by David Garejeli, one of the 13 Assyrian Fathers [16, 17]. In Gareji additional monasteries named Dodorka (the horn of Dodo) and Natlismtsemeli (the Baptist) were founded by David Garejeli's disciples Dodo and Luciane, respectively. In a period of activity in the 10th – 12th centuries, the Monastery complex was further extended and Bertubani, Chichkhituri, and Udabno monasteries, among others, were founded. The Gareji complex has astonishing wall paintings not only here, but in Qolagiri, Tetriudabno, Sabereebi, and Tsamebuli Monasteries [15–17].[2]

Technical investigations of wall painting technology in the David Gareji monastery complex are scant. However, some research does identify gaji as a plaster support for wall paintings. According to the investigations and observations of Georgian restorers at the end of the 20th century, the 9th century and later wall paintings of the Gareji Monastery complex are executed on gaji plaster (described by the authors as comprising a loam soil with a high proportion of gypsum) [14]. According to laboratory analyses undertaken in the 1990s, the 10th – 11th and 12th – 13th – century wall paintings in the

---

[2] Three monasteries are located in present day Azerbaijan – Udabno and Chichkhituri in part and Bertubani in its entirety.

churches of Udabno, Bertubani, and Natlismtsemeli monasteries are executed on clay-gypsum plaster, which the author refers to as 'gazha' [18]. A recently discovered 12th – 13th – century painted church at Dodorka monastery, discussed below, is also based on gaji plaster.

The large-scale use of gaji as a building material in the urban development of Tbilisi is evident in buildings of the 18th – 19th centuries. From this period onwards, gaji was the main plastering material for the interiors of both civil and industrial buildings. Gaji has been found in historical buildings of Kldisubani, Bethlehem district, Ortachala and other parts of Tbilisi [19]. Gaji as a decorative wall-painting support and as an interior plaster material was used in 19th – 20th centuries buildings located on Lermontovi Str N1 and N3, Abo Tbileli Str N6, and Rustaveli Str N37 among others [20, 21]. The interiors of the historic building of Tbilisi State Academy of Arts at Griboedovi Str N22 have gaji-based plasters, mouldings, and decorative wall paintings (Fig. 1) [22]. The gaji-based wall paintings were mostly executed in common areas of the historic residential buildings, such as entrances and stairwells. The paintings were also found in the rooms of the flats, mostly on the ceilings and on some areas of the walls. In Tbilisi, the Church of Transfiguration at the Palace Complex of Queen Darejani is another example of gaji-based wall paintings from the early 20th century (Fig. 2) [23].

Fig. 1. Decorative stucco and wall paintings undertaken on gaji support in a building, today housing part of the Tbilisi State Academy of Arts, that was originally built as a residential palace by an Armenian trader [22].

Gaji is still widely produced and used in Georgia. However, current production methods and materials may not closely match historic production. Today, gaji is typically produced using gas-fired furnaces, while historic firing processes likely used kilns fuelled with wood [24].[3] Soviet era material specifications stipulated that unburnt, natural gaji must contain at least 38% gypsum [7]. Unfortunately, in present-day gaji production plants, laboratory analyses of composition are not routinely performed and other technical data are not commonly reported. These differences may also result in significant divergence in the properties of modern and traditionally produced gaji.

The continuing use of gaji in Georgia can be explained by its excellent working properties. It provides a smooth, even finish, and due to its plasticity and hardening rate can readily be moulded to produce relief decorative elements. Its fine-grained texture also makes it an appropriate painting support. However, gaji plasters are susceptible to environmental deterioration which may manifest in delamination, deformation, decohesion, cracking, and loss. In order to preserve this important aspect of Georgian technology and culture, understanding of the nature and performance of gaji materials is urgently required.

Fig. 2. The Church of Transfiguration at the Palace Complex of the Queen Darejani. Gaji-based wall painting on the west wall.

3 The Church of St. Demetrios of Thessaloniki

The Church of St. Demetrios of Thessaloniki, belonging to the Dodorka monastery complex at David Gareji, became one of the first case studies to focus on investigating gaji plaster technology and its conservation challenges. A collaborative research project

---

[3] Personal communication with master craftsmen who, before gas-fired factory production became the norm, used to process gaji using traditional methods.

was established between Tbilisi State Academy of Arts, Faculty of Restoration, Art History and Theory; The Technical University of Georgia; the Conservation Centre of Fine Arts Ltd. (co-funder); Geoengineering Ltd.; the Materials Research Laboratory, History of Art Department, University College London; and UK-based wall painting conservators Rickerby & Shekede (conservation supervision of the project). The project was funded by the International Education Center Alumni Association in November 2019. Unfortunately, the project was curtailed due to the Covid-19 pandemic, therefore only the initial stages of the project have so far been undertaken.

The Dodorka Monastery, which was founded by Dodo – one of the disciples of David Garejeli – preserves several unique painted churches containing wall paintings dating to as early as the 9th – 10th century (Fig. 3) [25].

Fig. 3. 9 – 10th century wall paintings in the conch of the small church in Dodorka Monastery Complex [25].

The cultural, artistic, and historical significance of the monastery was enhanced by the chance discovery in 2015 of the rock-cut church of St. Demetrios of Thessaloniki while monks were removing debris. Palaeographical evidence provided by graffiti on the interior suggests that the church had become inaccessible, possibly due to landslide, at some point in the 18th century. The construction date of the church is unknown, but the astonishing wall paintings can be dated stylistically to the 12th or 13th century. These depict scenes of St Demetrios of Thessaloniki, a unique iconographical scheme for Georgian medieval wall paintings [26]. The church's eponymous patron saint is depicted among other warriors on the north wall, distinguished by his larger size and golden halo. The scenes from the "Martyrdom" of St Demetrios are depicted in chronological order starting from the west end of the south wall and continuing clockwise (Fig. 4) [26].

On discovery, the interior was found to be filled with earth to a height of 1.5 m above the floor. Additionally, the paintings were extensively covered with plant roots, and

Fig. 4. Church of St. Demetrios Thessaloniki north wall: scenes from the Life of St Demetrios.

the plaster was delaminating and powdering. In 2015 E. Privalova's Technical Research Centre for Paintings 'Betania' undertook emergency interventions including the removal of plant roots and temporary edge stabilisation of detached plaster using gypsum plaster and bamboo sticks [27].

Four years later the Conservation Center of Fine Arts Ltd. together with international consultants (Lisa Shekede and Stephen Rickerby) undertook a further conservation programme including environmental monitoring, detailed condition assessment, graphic documentation, research into the physical history, and remedial interventions (Fig. 5) [28].

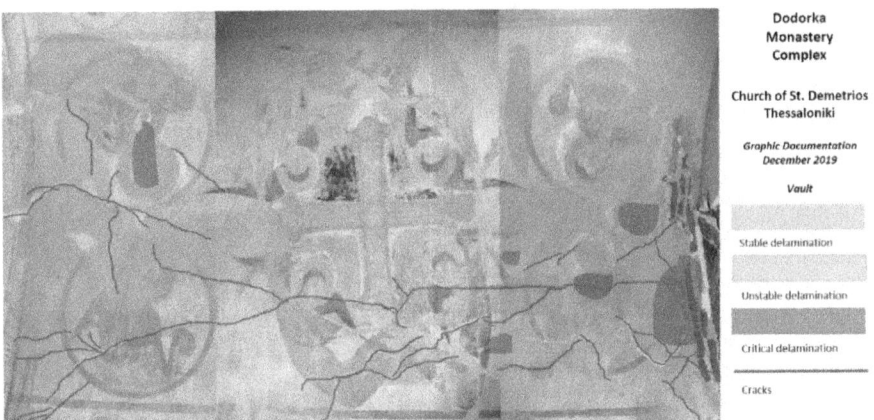

Fig. 5. Graphic documentation of the condition phenomena of the vault.

The church is a single-nave, hybrid structure of approximately 3 m × 5.5 m × 4 m. The northern part of the structure is hewn from the tuff-sandstone cliff-face, the vault is constructed from brick, and the south side of the church is built with irregularly shaped cut stone masonry. The only entrance is the south door, which is framed with brick. All three materials serve as primary supports for the wall paintings. The topography of the interior walls is uneven and fully covered with plaster, which forms the secondary support for the paintings.

At least two gaji plaster layers are present, distinguishable by subtle differences in colour, texture, and particle characteristics. The upper layer is smooth in texture, pale yellow-white in colour, up to 0.5 cm thick, and contains fine inorganic aggregates along with smaller charcoal inclusions. The lower plaster layer is up to 8 cm thick and darker in colour. Aggregates are more plentiful, and more irregular in shape and colour than in the upper layer; charcoal inclusions range from 1 to 7 mm (Fig. 6).

Adhesion failure between the plaster and primary supports was identified as one of the most prominent deterioration phenomena, in association with stress cracking, through the entire fabric. These conditions had facilitated, and been further exacerbated by, networks of plant roots spreading both behind areas of delaminating plaster and across the surface of the paintings, additionally causing mechanical damage to the paint layer. Microbiological staining and insect infestation were also visually evident. Failure of cohesion both of the plaster and the rock support was identified as a further major

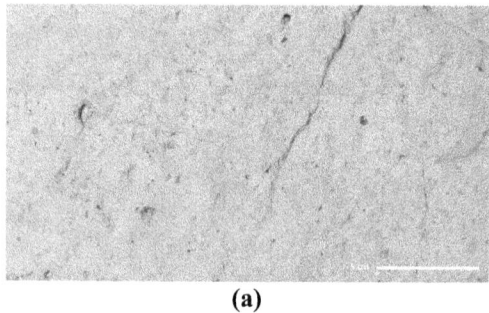

(a)

(b)

Fig. 6. (a) Black charcoal inclusion of upper plaster, North wall of the Church of St Demetrios Thessaloniki and (b) from the same church, a fallen fragment of plaster with various sizes of charcoal inclusions.

problem. It was considered possible that environmental conditions may be a contributory factor in this form of deterioration, so environmental monitoring was carried out as part of the programme.

The climate of Gareji varies, depending on altitude, from sub-tropical to temperate [29], the David Gareji monastery complex being located in an area classified as *Cfa* (humid subtropical) in the Köppen-Geiger climate classification. Environmental monitoring, undertaken between September 2019 and September 2020 recorded external temperatures as low as −5 °C and as high as 36 °C, with RH varying between 17% and almost 100% (Fig. 7).[4] The church is thermally well buffered against the exterior (range: 9 °C to 24 °C), with interior environmental change manifesting mainly as a gentle, incremental increase over the course of the summer months. However internal RH fluctuations are greater and more rapid, and in the summer months RH often oscillates at the activation boundaries of many common salts and hygroscopic clays within the painting support (annual range 44% to 83%) [30].[5] Internal conditions are also generally favourable for microbiological activity [31]. The interior generally displays high relative humidity levels; elevation of interior absolute humidity values above those of the exterior suggest that there may also be additional source of moisture.

4   Analytical Programme

For the current study, analysis of the historic gaji plasters at St Demetrios was undertaken for the purposes of understanding their original technology, composition, and deterioration problems with the ultimate aim of identifying or developing suitable materials for their conservation. Findings were to be assessed within the context of previous historic plaster studies, specifically those of the medieval David Gareji churches of Bertubani, Udabno, and Natlis-Mtsemeli [18].

10 plaster samples were collected from the Church of St Demetrios of Thessaloniki. These comprise: 4 samples (S02, S03, S04, S05) from the lower register/dado level, where the paint layer was lost; 5 plaster samples (S06, S1/2015, S2/2015, S5/2015, S6/2015) of unpainted fallen fragments from the dado level which were stored in the niches;[6] and painted fragment S07 with full stratigraphy (at least two plaster layers along with paint layers) fallen from the upper register of the church, from the St Demetrios figure on the North wall. In addition, two samples of modern processed gaji powder (i.e.

---

[4] Environmental monitoring (RH and AT) was undertaken using U12 Hobo internal sensors and UX23 Pro V exterior sensors.

[5] There were no visual signs of salt efflorescence or crystallisation within the painting stratigraphy. Preliminary salt (anion) analysis was performed using MQuant ion test strips on stone, stone masonry, brick, and plaster samples. Chloride was below the level of detection in all cases; sulfate was found in plasters and mortar consistent with containing calcium sulfate; nitrate was found at moderate concentrations in some plaster samples and in one stone masonry sample.

[6] S/2015 samples derive from the clearing of the church in 2015. The interior of the church of St Demetrios Thessaloniki was found filled with earth to a height of 1.5 m above the floor. Unfortunately, while removing the soil, some original plaster fragments fell from the dado level, which were then stored by church authorities in the niches of the church.

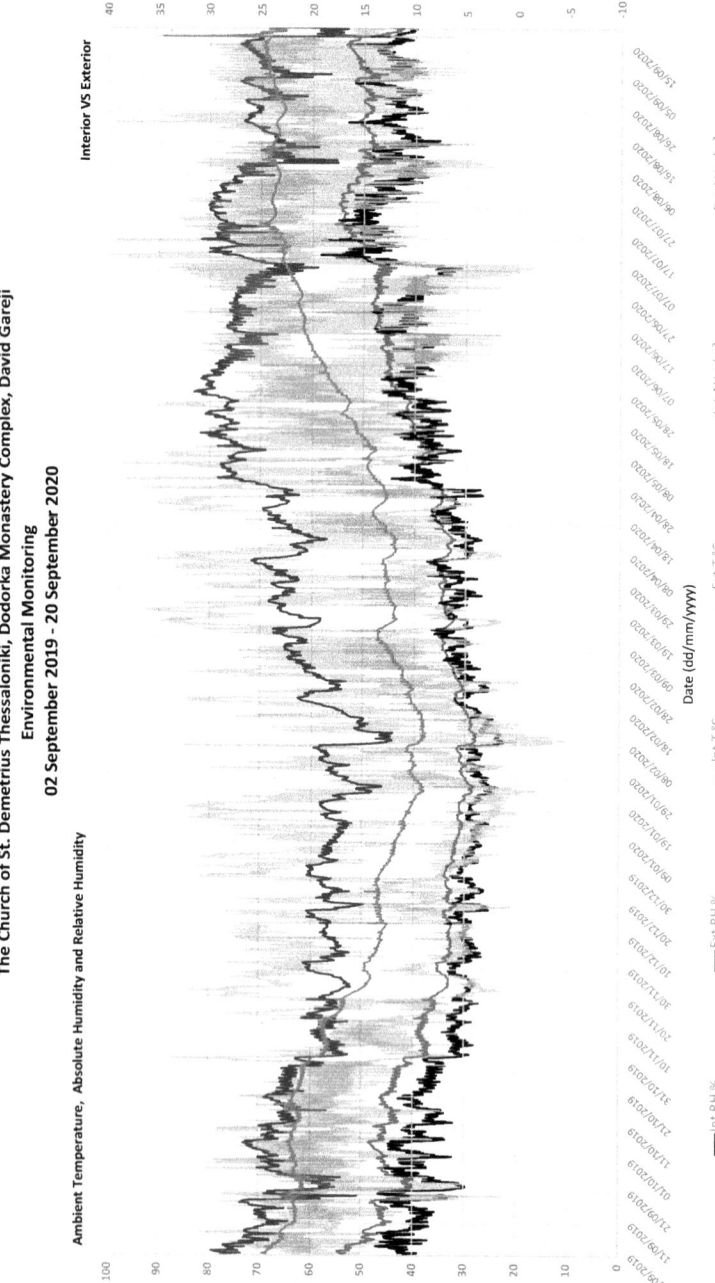

Fig. 7. Environmental data from the Church of St Demetrios. The graph presents internal and external temperature, relative humidity, and absolute humidity values recorded (AH calculated from concurrent T and RH values) between September 2019 and September 2020.

unhydrated, not yet made into a plaster) were obtained for analysis to assess their compatibility with historic gaji plasters, and to establish its potential for use as a conservation material (Sample C1 – Didostati Ltd and C2 – Ginu Ltd)[7].

Granulometric analysis was conducted for historic gaji plasters (except S05 and S07[8]) to determine particle size distribution. Petrographic thin-sections[9] were prepared to establish qualitative and semi-quantitative geological content of all historic samples except S02 and S04;[10] To allow comparison in thin-section with historic plaster, modern processed gaji (samples C1 and C2) in powder form was mixed with water to make solid plaster samples. Chemical analyses (moisture content, water of crystallisation, acid insoluble residue (AIR), loss on ignition (LOI), CaO, SO_3)[11] and X-ray diffraction (XRD)[12] were undertaken on historic plaster of S07 (for XRD - two readings: upper and lower layers) and commercial processed gaji samples C1 and C2 in powdered, unhydrated form.

Petrographic analysis established that the historic gaji plaster is polymictic (containing different types of minerals), clastic sedimentary rock composed of gypsum, bassanite, clay minerals, calcite, and other silica components such as plagioclase, quartz, and pyroxene. All samples consist predominantly of gypsum (calcium sulfate dihydrate)

---

[7] Raw gaji samples C1 – Didostati Ltd and C2 – Ginu Ltd were collected from production plants. The plants and quarries are located very close to each other (ca. 10 km) in Garadabni municipality near villages Axal Samgori and Gamarjveba, Georgia.

[8] Granulometric analysis was not undertaken on painted sample S07 to protect the painting surviving on the fragment nor on sample S05, due to potential interference of primary support (natural rock) within the sample.

[9] Historic gaji plaster samples, as well as commercial gaji samples (C1 and C2) prepared as plasters by mixing with water, were hot-mounted (150–200 °C) in adhesive (colophony resin and/or epoxy resins) and then made into thin-sections using a grinding and polishing machine. Thin-sections were examined under a polarizing microscope 'Optika B-383POL' (Italy). This method was adopted due to resource constrains, though we note that the high temperatures involved in preparation of thin sections may influence the hydration state of calcium sulfates within the samples; alternatives methods will be pursued in future.

[10] The decohesive nature of these sample prevented their preparation as thin-sections with the methods available.

[11] Chemical analysis was conducted according to USSR standard: 1. Moisture content GOST 22688–77, gravimetric analysis; 2. Crystalisation water GOST 23789–79, gravimetric analysis; 3. AIR GOST 23789–79, gravimetric analysis. 4. LOI GOST 5382–91 P.7, gravimetric analysis, 5. CaO GOST 5382–91 P.10, volumetric analysis, 6. SO_3 GOST 5382–91 P.9.3, gravimetric analysis. The historic sample S07 was ground and analysed, while the commercial processed (unhydrated) gaji samples were already in powder form. The data of the given chemical analysis was used to calculate the percentage composition of gypsum, bassanite, clay content, calcite, and insoluble fraction of the samples.

[12] For XRD analysis, the historical plaster sample was ground to a powder and the modern gaji sample was analysed in raw powder form. Analysis was performed using diffractometer Дрон-4.0, НПП "Бу-ревестник" with copper anode and nickel filter, U-35kV, I-20mA, scan rate 2 degree/min. λ = 1.54178 Å.

and clay minerals. Additionally, the samples were found to contain large quantities of charcoal inclusions of varying size (mainly >2 mm) (Figs. 8 and 9).[13]

Granulometric analysis (S03, S04, S06, S1/2015, S2/2015, S5/2015, S6/2015) established that historic gaji samples have an aleuropelitic (silty-clayey) structure.[14] The ratio between very fine and coarse sizes varies from 1:1 to 1:2. Particle size distributions differ between samples of upper and lower plaster layers. The matrix of the upper layer of gaji plaster (S5/2015) consists of silt-sized particles (up to 0.1 mm) while that of the lower layer is coarser-grained (up to 0.5 mm) (S6/2015).

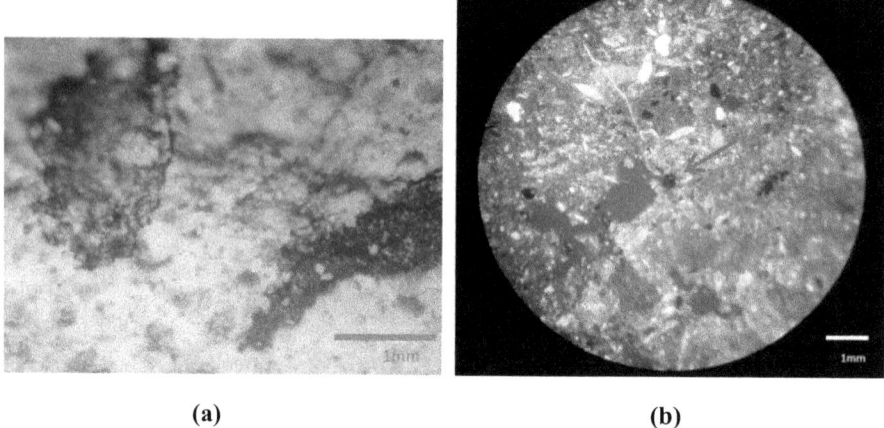

(a)   **(b)**

Fig. 8. Charcoal inclusions from the plaster sample S6/2015. (a) under binocular microscope; (b) thin-section under polarized light microscope (40× magnification), charcoal indicated by red arrow.

---

[13] It is currently unclear whether the charcoal was a deliberate addition to the plaster or present as a consequence of the firing process, due to the accidental inclusion of natural vegetation during quarrying of the raw materials, or derived from fuel used for firing the gaji oven. In February 2020, members of the research project interviewed the director of 'Ostatis Saxli' Ltd. (gaji producer). According to the source, the historic firing process of gaji obtained from the quarried rock involved the construction of 'ovens' from the raw material. The structure contained holes into which wood fuel was placed. After the gaji was fully burnt it was crushed together with any remaining fuel or charcoal and then sieved.

[14] For samples S05 and S07, the aleuro-pelitic structure was evident through petrographic analysis as granulometric analysis was not undertaken (see footnote 8).

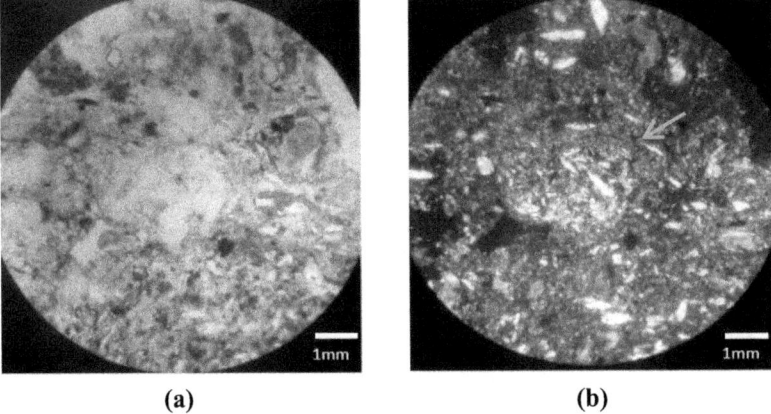

(a)   **(b)**

Fig. 9. Thin-section of lower layer of historic gaji plaster S07 under a) plane polarizers and (b) crossed polars (40× magnification) showing the aleuropelitic structure. There are gypsum crystals concentrated in the area indicated by red arrow. The cementitious material is clay-carbonate pelitic mass.

Like the historic plaster samples, the commercially produced modern gaji was found to be polymictic, with an aleuropelitic structure (Fig. 10).[15] Furthermore, comparison between thin sections taken from the historic plasters and plasters prepared from commercial gaji established that both contain:

- Gypsum, formed as lens-shaped and rhombic mono-crystals, as well as mosaic distributions;
- Feldspars (mostly plagioclases) characterized by secondary alterations;
- Small quantities of unaltered quartz;
- Few inclusions of pyroxenes and amphiboles;
- Lithoclastic materials (highly altered volcanic rock shards);
- Carbonate materials, of which the majority are pleomorphic and organic limestone fragments, rarely calcite shards.

Results of petrographic analysis are consistent with XRD undertaken on upper and lower layers of historic plaster from S07. Both layers contain gypsum ($CaSO_4 \cdot 2H_2O$) with traces of quartz and feldspar. Chemical analysis of S07 has also revealed a composition of 55% gypsum, 34% insoluble fraction (clays and silicates) and up to 1% calcite (8% mass unaccounted for in analysis). The two samples of commercial processed gaji in unhydrated form have almost identical composition. The XRD of samples C1 and C2 showed a high proportion of bassanite, trace of quartz and feldspar, and C1 additionally

---

[15] The plasters made from commercial gaji powder contained unexpectedly high proportions of bassanite. The reason for this remains unclear and will be investigated in future work. This plaster was prepared one month before the thin-section preparation and so we expect the bassanite to have largely transformed into gypsum. However, it is possible that some of this gypsum (only in the modern plaster and not the historic plaster) was transformed due the high temperatures used in the thin section preparation (see footnote 9). We note that other possible sources of bassanite include "over-burned" bassanite that hydrates very slowly.

had a trace quantity of gypsum. This is consistent with chemical analysis which showed 65–67% bassanite, 0–2% gypsum, 14–19% insoluble fraction (clays and silicates), and 14–17% calcite. If prepared as plasters this would theoretically become 70–71% gypsum and 13–17% clay content on the assumptions of complete conversion (hydration) of bassanite (Table 1).

Table 1. Chemical analysis results from an historic gaji plaster and two samples of contemporary commercially produced gaji in powder form. For commercial gaji samples these analyses were carried out on the processed gaji powder (i.e. not hydrated). Presumed gypsum content in commercial gaji is reported here (noted as calculated) assuming complete conversion of bassanite to gypsum on addition of water together with original data. *8% unaccounted for.

Mineral content (%)	Gypsum $(CaSO_4 \cdot 2H_2O)$	Bassanite $(CaSO_4 \cdot 0.5H_2O)$	Insoluble fraction	Calcite $(CaCO_3)$
S07 – Church of St. Demetrios Thessaloniki gaji *	55	–	34	<1
C1 – commercial gaji, Didostati Ltd (unhydrated powder)	2	65	14	17
C1 plaster (calculated)	72	–	13	15
C2 – commercial gaji, Ginu Ltd (unhydrated powder)	–	67	19	14
C2 plaster (calculated)	70	–	17	12

These findings show that all gaji materials share some broad characteristics, but there were significant differences in the ratios of these materials:

Both layers of historic gaji plaster are composed primarily of gypsum with some bassanite and anhydrite. The two commercially produced samples of gaji, when prepared as plasters, can be expected to contain more calcium sulfate and calcium carbonate and significantly less clay minerals as compared with the historic samples.

5   Discussion and Suggestions for Further Study

Comparison of the current findings with previous analytical studies carried out on medieval plasters from David Gareji churches (Bertubani, Udabno, and Natlis-Mtsemeli) note all examined sites as having two gaji plaster layers, the upper layer containing finer, fewer aggregates than the lower [18]. The plasters are described as being "gypsum-clay plasters" with a pelitic structure, containing 10–30% aggregate (mainly quartz) and 70–90% binder (principally gypsum and a few percent of natural calcite, though we note

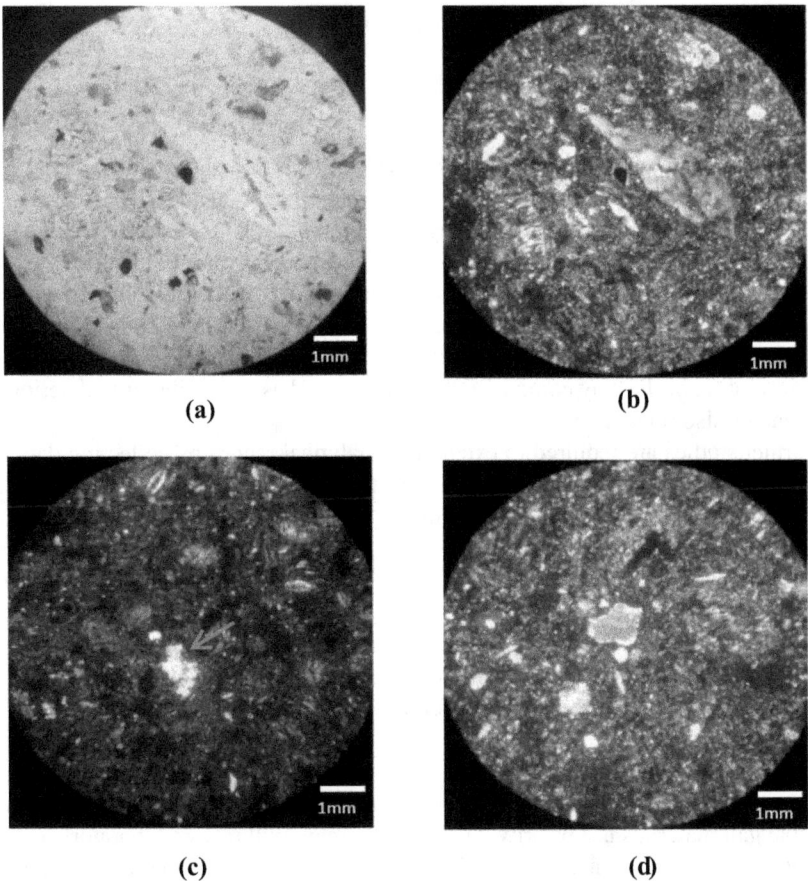

Fig. 10. Thin-section of plaster made from commercial gaji under (a) plane polarizers and (b) crossed polars (40× magnification), in which rhombic and fibrous dehydrated gypsum crystals (red arrow) can be observed within a cementitious clay-carbonate pelitic matrix; (c) elsewhere in the sample, possibly showing gypsum transforming into hemihydrate; (d) detail of a high interfacial anhydrite aggregate.

that the latter is likely better considered an aggregate). The author also notes that microscopic observations showed carbon (presumably charcoal) and "grog" with quartz and plagioclase grains.[16]

Based on the Bertubani, Natlis-mtsemeli, Udabno, and St Demetrios plaster analysis the following characteristis are common:

- pelitic structure;
- mainly a mixture of gypsum and clay-minerals;

---

[16] Ref. 18 identifies an organic binder in the plasters; this possibility has not been investigated at St Demetrios church. Organic additives to plasters are historically common and are worthy of future investigation.

- gypsum is the principal component by mass;
- contain small amounts of calcite;
- contain small amounts of fine (0.1–0.3 mm) quartz and plagioclase grains;
- contain charcoal particles.

Although there are clear similarities between these gaji plasters, firm conclusions cannot yet be drawn regarding sourcing, content, and processing due both to differences in approach and analytical methodology, and the small sample set. Other identified problems are those concerning comparison of historic gaji plasters with modern raw materials, particularly regarding significant variations in composition ratios and calcium sulfate hydration states. It remains unclear to what extent these are due to differences in source materials, production practices, and natural aging processes. Research into the contribution of the clay component as a binder, and as a contributor to deterioration mechanisms also remains unexplored.

Further studies are required to extend the analytical record by collecting data from other historic sites using an approach which facilitates more direct data comparison. The current research highlighted problems in distinguishing original source materials from the effects of processing and the use of additives in historic gaji plaster production. This can be partly resolved through the identification of specific historic gaji sources, for example through fieldwork, geological and historical research, and partly through further research into historic gaji production methods including firing temperatures, and perhaps the collection of evidence of recent non-industrial practices. This may also help in the characterisation of, and differentiation between, intentional additives and accidental inclusions.

As the main purpose behind this research is the development of compatible conservation materials for gaji plasters, further testing, analysis, and adaptation of currently available gaji materials is a priority. Such development will rely on laboratory and field tests with broad scope, including of the physical-mechanical properties of repair and original plasters.

6 Conclusion

The enduring and important role of gypsum-earth plasters in the development and adornment of Georgia's magnificent built heritage is indisputable. Nevertheless, the complexity of this technology remains largely unacknowledged. Despite its use as a support for some of Georgia's most important and most ancient paintings, scant attention has been paid to the particular vulnerabilities of aged historic gaji plasters. Furthermore, insufficient specialist attention has been paid to understanding of the plaster technology to facilitate effective conservation. The consequences are significant, since in the absence of knowledge in this area of technology, not just in Georgia but wherever gypsum-clay plaster technologies are found, conservators tend to address problems specific to gaji using methods and materials intended for quite different technologies, which may prove to be actively harmful. The current study has gone some way to addressing these issues, not least by illustrating, through the case study, some of the conservation issues associated with these plasters and the pressing need for both short-term stabilisation and long-term strategies for their conservation. It has also, crucially, highlighted some of the significant difficulties which stand in the way of identifying historic gaji sources and the extent to which raw gaji material was adapted for use as a plaster and why. The complex behaviours of gypsum and clay minerals are also acknowledged as an important future area of research, and the need for further testing of modern gaji materials for potential use as a conservation material is also demonstrated.

The small advances in understanding that have been achieved during the course of this study are significant, but the scope of research yet to be done is both huge and vital. It is to be hoped that in focusing on the integral role of gaji plasters in the creation of Georgia's ancient wall painting heritage and on their susceptibilities, this paper will stimulate renewed interest in the study and preservation of this fascinating technology.

Acknowledgements. The authors would like to thank the International Education Center Alumni Association, Georgia for its support. This paper presents the early stages of a project that was curtailed due to the pandemic.

References

1. Aoki, S., et al. (eds.): Conservation and Painting Techniques of Wall Paintings on the Ancient Silk Road. CHS, Springer, Singapore (2021). https://doi.org/10.1007/978-981-33-4161-6
2. Torraca, G.: Lectures on Materials Science for Architectural Conservation, pp. 38–54. The Getty Conservation Institute, Los Angeles (2009)
3. Çamurcuoglu, D., Siddall, R.: Plastering the prehistory: marl as a unique material to cover, maintain and decorate the neolithic walls of Çatalhöyük. In: Proceedings of the 4th Historic Mortars Conference HMC2016, 10th–12th October 2016, Santorini, Thessaloniki, Greece, pp. 482–489 (2016)
4. Wong, L., Bicer-Simsir, B., Porter, J., Rickerby, S., El-Din, A., Sharkawi, M.: Analytical challenges in the study of New Kingdom plasters from tombs in the Valley of the Queens, Luxor, Egypt. In: Proceedings of the 4th Historic Mortars Conference HMC2016, 10th–12th October 2016, Santorini, Thessaloniki, Greece, pp. 67–76 (2006)
5. Rickerby, S., Wong, L.: The technology of royal tomb decoration. In: Wilkinson, R.H., Weeks, K. (eds.) The Oxford Handbook of the Valley of the Kings (Oxford Handbooks in Archaeology), pp. 137–152. OUP, Oxford (2016)
6. Wong, L., Rickerby, S., Rava, A., Sharkawi, A., Alaa El-Din, A.: Developing approaches for conserving painted plasters in the royal tombs of the Valley of the Queens'. In: Proceedings of Terra 2012, the 11th International Conference on the Study and Conservation of Earthen Architectural Heritage, Lima, Peru, 22–27 April 2012 (2012)
7. Природные ресурсы Грузинской ССР: Т. 2: Неметаллические полезные ископаемые, 68–75. Издательство академии наук СССР, Москва. (Natural Resources of the Georgian SSR (1959), vol. 2: Non-metallic useful excavations, pp. 68–75. Publishing Academy of Sciences of the USSR, Moscow) (1959). in Russian
8. Laue, S.: Detailed studies of gypsum plasters from the Ihrat Khana mausoleum in Samarkand, Uzbekistan. In: Proceedings of the 5th Historic Mortars Conference, Pamplona, Spain, 19–21 June 2019, pp. 248–258 (2019)
9. Laue, S., Kleinmann, P.: Ganch as historical building material and the Kundal wall painting technique in the Mausoleum Ishrat Khana, Samarkand, Uzbekistan. In: Zhang, L., Krist, G. (eds.) Archaeology and Conservation Along the Silk Road, vol. 16, pp. 49–62. Vandenhoeck & Ruprecht, Göttingen (2019)

10. Fodde, E.: Traditonal earthen building techniques in Central Asia. Int. J. Archit. Herit. **3**(2), 145–168 (2009)
11. Cather, S.: Assessing causes and mechanisms of detrimental change to wall paintings. In: Gowing, R., Heritage, A. (eds.) Conserving the Painted Past: Developing Approaches to Wall Painting Conservation, pp. 64–74. James & James Science Publishers Ltd., London (2003)
12. Mikaberidze, S., Sagaradze, M., Khuskivadze, N., Hill, J.A.: Understanding gaji, a gypsum-earth plaster in Georgian Wall Paintings. In: Proceedings of International Conference Davit Gareja Multidisciplinary Study and Development Strategy, Tbilisi, pp. 71–81. Tbilisi State Academy of Arts, Tbilisi (2020)
13. ქალაქ თბილისის მიწათსარგებლობის გენერალური გეგმის საპროექტო მომსახურება, ტომი 1. ზოგადი დებულებები: (Project service of the general plan of land use of the city of Tbilisi, Volume 1, General provisions (2017)) (2017). in Georgian
14. Noll-Minor, M.: Wandmalereien Des Höhlenklosters David Gareja in Georgien: Entwicklung Eines Konzeptes Zur Langfristigen Erhaltung. Beiträge Zur Erhaltung Von Kunst- Und Kulturgut, 1, 14–24. (Noll-Minor, M. (2003) Murals of the David Gareja Cave Monastery in Georgia: Development of a Concept for Long-Term Conservation. Contributions to the preservation of art and cultural property, 1, 14–24.) (2003). In German
15. Georgian Mural Paintings. http://arthistory.tsu.ge/murals/locations/georgia/. Accessed 22 Feb 2022
16. Eastmond, A.: The cult of St Davit Garejeli: patronage and iconographic change in the Gareja Desert'. In: Skhirtladze, Z. (ed.) Desert Monasticism: Gareja and the Christian East 2001, pp. 220–239 (2001)
17. David Gareji Monasteries and Hermitage. https://whc.unesco.org/en/tentativelists/5224/. Accessed 22 Feb 2022
18. Dneprovskaya, M.B.: Medieval pigment and plaster technology in the XII-XIII Century mural paintings at David-Garedji, Georgia. In: Proceedings of Material Research Society Symposium, vol. 352, pp. 727–732 (1996)
19. Kavsadze, M.: Mineralogical-Petrological analysis of the plasters and mortars in the historic buildings of Tbilisi. Unpublished project reports (2010–2021). in Georgian
20. Sagaradze, M.: Investigation of the Historic buildings in Tbilisi. Unpublished report (2019)
21. Sagaradze, M., Mikaberidze, S., Papiashvili, N., Khachidze, O.: Conservation research of Melik Azarianci Building. Unpublished report (2021). in Georgian
22. Kuprashvili, N., Papiashvili, N., Sagaradze, M., Koberidze, N.: Diagnostic Study of the interior decoration at the building of Tbilisi State Academy of Arts. Unpublished report (2014). in Georgian
23. Ninoshvili, L.: The Wall Paintings Conservation of the Church of Transfiguration at the Palace Complex of Queen Darejani, ICOMOS Georgia. Unpublished report (2019). in Georgian
24. Ashkaveti Ltd., Gaji Workshop/Factory, Environmental Audit Report (2018). https://mepa. gov.ge/ge/Files/ViewFile/2251. Accessed 22 Feb 2022
25. Tumanishvili, D., Mikeladze, K., Didebulidze, M. (eds.): Georgian Christian Art, Cezanne, Tbilisi (2008)
26. Bulia, M.: Vita cycle of St Demetrius of Thessaloniki and his holy relics at Dodorka Monastery. In: Soltes, O.Z. (ed.) Proceedings of International Conference Davit Gareja Multidisciplinary Study and Development Strategy, Tbilisi, Georgia, pp. 95–107 (2020)
27. Kuprashvili, N., Rubashvili, A., Akhalashvili, S., Liluashvili, T.: Diagnostic study and emergency stabilizaton of wall paintings in church of St Demetrius of Thessaloniki at Dodorka Monastery. In: Soltes, O.Z. (ed.) Proceedings of International Conference Davit Gareja Multidisciplinary Study and Development Strategy, Tbilisi, Georgia, pp. 192–196 (2020)
28. Conservation Centre of Fine Arts Ltd and Rickerby and Shekede, The National Agency For Cultural Heritage Preservation of Georgia: Wall Painting Conservation at Church of St Demetrius of Thessaloniki at Dodorka Monastery. Unpublished report (2019). in Georgian

29. Rehabilitation of Cultural Heritage Site Infrastructure at David Gareji Monastery Complex (ER), Review of the Impact on the Environment, Regional Development project funded by the World Bank, Tbilisi, Georgia (2015). http://mdf.org.ge/?site-lang=ka&site-path=docume nts/&id=229. Accessed 22 Feb 2022. in Georgian

30. Wong, L., Agnew, N. (eds.): The Conservation of Cave 85 at the Mogao Grottoes, Dunhuang: Development and Implementation of a Systematic Methodology to Conserve the Cave Wall Paintings and Sculpture, pp. 232–234. Getty Conservation Institute, Los Angeles, CA; Dunhuang Academy: Mogao Grottoes, People's Republic of China (2011)

31. He, D., et al.: Insights into the bacterial and fungal communities and microbiome that causes a microbe outbreak on ancient wall paintings in the Maijishan Grottoes. Int. Biodeterior. Biodegrad. **163**, 105250 (2021)

Performance Evaluation of Patch Repairs on Historic Concrete Structures (PEPS): An Overview of the Project Methodology

Simeon Wilkie[1]([✉]) [iD], Ana Paula Arato Goncalves[1], Susan Macdonald[1],
Elisabeth Marie-Victoire[2,3], Myriam Bouichou[2,3], Jean Ducasse-Lapeyrusse[2,3,4],
Nicki Lauder[5], David Farrell[6], Paul Gaudette[7], and Ann Harrer[8]

[1] Getty Conservation Institute, Los Angeles, California, USA
swilkie@getty.edu
[2] Laboratoire de Recherche des Monuments Historiques, Paris, France
[3] Centre de Recherche sur la Conservation, Sorbonne Universités, Paris, France
[4] Comue Paris Est Sup, Marne-La-Vallée, France
[5] Historic England, Swindon, UK
[6] Rowan Technology Ltd., Manchester, UK
[7] Wiss, Janney, Elstner Associates, Chicago, IL, USA
[8] Wiss, Janney, Elstner Associates, Los Angeles, CA, USA

Abstract. The development of reinforced concrete through the 20th century has resulted in a wealth of culturally significant concrete structures around the world. However, as a relatively modern material durability issues were not fully understood at the time of construction, and many of these structures require ongoing interventions as a result. While there is now a general acceptance of the importance of concrete heritage from this era, there few widely accepted guidelines on the approach to its preservation and conservation. In particular, despite many studies and published guidance on concrete repair and the performance of concrete repairs, there are few on the long-term performance of patch repairs designed to match the aesthetic of the original while simultaneously keeping loss of the original fabric to a minimum. As a response to this challenge, three institutions, the Getty Conservation Institute (GCI), Historic England (HE) and Laboratoire de Recherche des Monuments Historiques (LRMH) are collaborating on The Performance Evaluation of Patch Repairs on Historic Concrete Structures (PEPS) to produce practical guidance that will help those repairing historic concrete. This paper provides an overview of the assessment methodology that has been developed as part of this international collaboration.

Keywords: Historic concrete · concrete repair · conservation · patch repair

1 Introduction

Within the field of conservation, concrete heritage is becoming an area of growing concern as the number of culturally significant concrete structures which require ongoing repair and maintenance continues to increase. The development of reinforced concrete

© 2023 J. Paul Getty Trust
V. Bokan Bosiljkov et al. (Eds.): HMC 2022, RILEM Bookseries 42, pp. 288–299, 2023.
https://doi.org/10.1007/978-3-031-31472-8_22

through the late-19th and 20th centuries provided a new material that revolutionized the construction industry and provided a means of achieving new architectural and structural feats. The arrival of this material came at a time of unprecedented technological, social, and economic change, with huge-scale rebuilding following two world wars, rapid population growth, mass migration and the expanse and development of urban societies [1]. This new material was quickly and widely adopted in numerous different architectural styles and functions and provided not only a new tool for artistic expression but also for social reform through the creation of new spaces to live and work. As a result, there is now a diverse wealth of culturally significant 20th-century concrete structures around the world that present new and unique conservation challenges.

In the past, concrete durability issues were not as well understood [2]. In many early publications and promotional materials, it was widely believed that this new material would be permanent and maintenance-free. Unfortunately, this is not the case and a combination of factors, such as corrosion of reinforcement, high water/cement ratios (w/c), the use of reactive aggregates and unsuitable mix designs, have resulted in concrete deterioration at many of these structures requiring ongoing interventions. While there is now a general acceptance of the importance of concrete heritage from the 20th century [3], there are few widely accepted guidelines on the approach to its preservation and conservation [4–7]. While it is understood that the fundamentals of best practices and performance criteria in concrete repair also apply to heritage concrete, the many studies and published guidance [8–14] on the performance criteria of concrete repairs

Fig. 1. Example of a repair which has failed to weather and has a negative impact on the historic aesthetic of a culturally significant concrete structure.

are not always familiar to the conservation field and, in some cases, promote irreversible changes to the structure that are in conflict with traditional conservation principles such as minimal intervention and retreatability.

There are few studies on the long-term performance of patch repairs designed to preserve the aesthetic significance of the original fabric. As a result, there has been a history of repairs carried out without application of good concrete repair fundamentals and had an unacceptable impact on the aesthetic significance of the concrete structure. As a result, these poor-quality repairs have short life cycles and, ultimately, cause physical damage to the heritage structure (Fig. 1).

2 Project Background

An international experts' meeting focused on conserving concrete heritage took place at the GCI, Los Angeles, in 2014 [15], during which the materials and techniques used to carry out patch repairs on culturally significant concrete structures were determined to be key issues that require further investigation. More specifically, the experts agreed that, within the field of concrete conservation, there was a need for both an improved understanding of the patch repair process, and for better information on repair materials and their long-term performance.

While conservation work often aims for 'like-for-like' repairs [4], which replicate the original material in both composition and aesthetics, this is not always feasible [15]. For example, when compared to polymer-modified and proprietary repair materials, some like-for-like repairs may not have mechanical properties which are as desirable, may have more issues bonding to the original substrate, and may not provide the same level of durability or protection to the steel reinforcement. Additionally, while conservation work aims to integrate the principles of minimal intervention and retreatability, this can be at odds with contemporary concrete repair guidance [9] which may recommend sacrificing more of the original fabric in exchange for increased performance of the repair.

There needs to be additional guidance for the conservation community on selection of repair materials and the approach to patch repair work on culturally significant concrete heritage. As a response to this specific challenge, three institutions, the GCI, Historic England (HE) and the Laboratoire de Recherche des Monuments Historiques (LRMH) commenced work on an international collaborative research project, 'Performance Evaluation of Patch Repairs on Historic Concrete Structures' (PEPS). The project team is also joined by expert consultants from Rowan Technologies (England) and Wiss, Janney, Elstner Associates, Inc. (USA). Begun in 2018, the PEPS project aims to produce practical guidance to help those repairing historic concrete through the process of the selection of appropriate repair approaches, procedures, and materials. The scope of work includes the assessment of case studies in the USA, England, and France within a variety of climatic and environmental conditions, typologies, and repair approaches.

3 Phase I: Project Development

The first phase of the research began in 2018, and had 3 key outcomes:

3.1 The Development of the Research Proposal

After reviewing the existing needs of the field, it was determined that the project should specifically aim to produce better understanding on:

- Current repair methods and practices;
- Performance of patch repairs typically undertaken on historic concrete;
- Efficacy and durability of patch repair materials currently used in historic concrete;
- Efficacy of current repair techniques.

Additionally, it was agreed that, in the long-term, the project could service the concrete conservation community through the development and dissemination of:

- Guidance on selection of appropriate repair materials;
- Practical guidance on repair methods;
- A methodology for recording, monitoring, and evaluating repairs to historic reinforced concrete.

3.2 The Development of the Assessment Methodology

While there is clear guidance on carrying out assessments of concrete structures [16–18], there were no pre-existing protocols that detailed the scope of work required of the PEPS research. As such, a new assessment methodology had to be designed specifically for the project. This methodology, developed by an interdisciplinary team of professionals working in the field of concrete conservation, includes a variety of traditional and non-traditional, non-destructive, mechanical, and chemical and electrochemical characterization and diagnostic techniques. In addition to the specification of individual evaluation protocols, a shared way of recording results and reporting observations also had to be developed to ensure data was collected consistently and could be unified for further analysis.

The methodology was developed with aim of answering the following questions:

- What materials and protection systems were used in the repair and adjacent substrate?
- How was the repair area prepared and how has this impacted the repair bond?
- How is the repair performing technically?
- How is the repair performing aesthetically?

3.3 The Selection of Case Studies

Up to ten case studies were identified in each country (USA, England, and France) where patch repairs had been applied to culturally significant concrete structures and both the aesthetic and technical performance had been a consideration in the repair strategy. It was concluded that the case studies selected should include repairs in a wide range of conditions and environments and using different materials and techniques – reflecting the diversity of approaches and contexts for repair that currently exist. Specific criteria for selection included:

- Concrete buildings or structures, where the surface is exposed or untreated;
- Repairs performed with the goal of conserving historic and aesthetic values;

- Access to documentation, and professionals who worked on the repair (architects, engineers, tradesmen, and craftsmen);
- Easy access to the site and patches to be evaluated;
- Consent must be given from site owners/managers for access to site and data;
- Case studies should have diverse ages of original concrete and ages of repair;
- Include repairs made with various levels of craftsmanship;
- Include repairs that used different materials (concrete, mortar, like-for-like, proprietary, etc.) and placement techniques (hand troweled, poured);
- Case studies can present combined use of patch repairs and other protection or treatment techniques;
- Complementary techniques applied should not obscure patches, such as coatings;
- Case studies should represent a good range of environmental conditions (coastal, urban, etc.) and aspects.

4   Phase II: Preliminary Assessment

4.1   Establish Case History

During the preliminary assessment, all of the selected case studies were evaluated. Before conducting site visits, desk studies were undertaken on each site to establish each site's case history and provide data on how the repair strategies were developed and executed, and support the interpretation of current conditions observed on site. The desk study included a review of existing building specifications and repair documentation, complemented by interviews with owners, asset managers and other relevant individuals who could validate the history of the repairs.

4.2   Initial Site Visits

Preliminary Investigations. During the first site visits, approximately 10 patches were assessed per case study using only non-destructive tests (NDT) and documentation. The selection of patch repairs to include in the study was done following an initial walk through the site to identify, whenever possible, patches representing different degrees of deterioration, exposure, and vintages. The field assessment was then completed using the following techniques:

- Visual and tactile observations;
- Sounding to determine delamination, poor consolidation, or voids;
- Photographic documentation using scale, color checker, and crack gauge;
- Water spray to test repellence and possibility of hydrophobic treatments;
- Rebound hammer and scratch test to determine differences in surface hardness between repairs and adjacent substrate;
- Covermeter survey of patch and adjacent substrate to identify reinforcement location and depth of cover.

   Determination of Patch Performance. Upon completion of the Phase II site visits, each patch was assigned a designation (good, fair, poor, very poor) based on its aesthetic and technical performance. However, there are no widely accepted criteria for judging both the aesthetic and technical performance, so an agreed criteria had to be developed. The criteria are described in detail in Table 1, and examples of patches with different performances are shown in Fig. 2.

Table 1. Criteria for judging the aesthetic and technical performance of patch repairs

Aesthetic		Technical	
Good		**Good**	
Great care and effort paid off. Expert has trouble finding it, layperson cannot find it.		No deterioration immediately observed, minor shrinkage cracks observed on close inspection.	
Colour	Match	Bonding	Sound
Texture	Match	Cracks	Minor shrinkage cracks
Profile	Match	Efflorescence	No sign
Weathering	Match	Corrosion signs	No sign
Fair		**Fair**	
Good attempt but could be improved. Expert takes time to find it, layperson cannot find it		Showing minor signs of deterioration, little to no risk to safety or to the structure.	
Colour	Match	Bonding	Minor area of detachment
Texture	Match, but could be better	Cracks	Minor hairline cracks
Profile	Match, but could be better	Efflorescence	Minor deposits
Weathering	No match	Corrosion signs	No sign
Poor		**Poor**	
There was an effort but ultimately failed. Expert can immediately point it out, layperson notices it.		Showing clear signs of deterioration that will evolve to failure.	
Colour	No match, but not terrible	Bonding	Detachment detected
Texture	No match, but not terrible	Cracks	Large cracks
Profile	No match	Efflorescence	Significant deposits
Weathering	No match	Corrosion signs	Minor signs of corrosion
Very Poor		**Very Poor**	
Little to no effort. Layperson can immediately point it out.		Advanced deterioration is observed. Spalling has occurred or is imminent.	
Colour	No match	Bonding	Spalled or significantly detached
Texture	No match	Cracks	Many large cracks
Profile	No match	Efflorescence	Significant deposits
Weathering	No match	Corrosion signs	Reinforcement corroding

5 Phase III: Detailed Diagnostic

Following a review of the results of Phase II, 5 English, 3 American, and 4 French case studies were selected for further assessment in Phase III. The aim of this phase is to carry out a more detailed investigation to detect underlying deterioration which may not have

Fig. 2. Examples of a technically and aesthetically good patch (top left), an aesthetically good but technically poor patch (top right), technically good but aesthetically very poor patches (bottom left), and a patch that is aesthetically poor and technically very poor (bottom right).

been identified during Phase II and to identify material characteristics that could help explain the performance of the repairs. Phase III testing is being performed on 5 patches chosen from the group assessed previously in Phase II and their adjacent areas to detect any significant differences in behavior between the repair and concrete substrate. During these site visits, a detailed procedure including both NDT and invasive testing is being conducted, and samples are being removed for laboratory analyses. Laboratory analyses were carried out internally at the labs of LRMH and the GCI, with some additional testing carried out by external laboratories in the USA and France.

5.1 Detailed Site Investigation

Detailed site investigation focused on adding more measurements to preliminary investigations and introducing additional evaluation techniques on key aspects of aesthetic and technical performance. Evaluation of historic concrete structures can be challenging due to a lack of reliable information about the materials and techniques used. Furthermore, removing material from culturally significant structures for testing in the laboratory can be a challenge due to legal, ethical and financial restrictions on the removal of material, and as the amount of material that can be removed is often minimal, this can result in laboratory samples which are not representative of the material as a whole [19]. As such, it was determined that the testing methodology should utilize non-destructive means where possible to maximize the amount of data obtained without causing unnecessary damage to the structure. However, it was also recognized that it was critical to include some invasive techniques to properly assess the technical performance of patch repairs, and so sites were selected where this would be possible (Table 2).

Table 2. Overview of field tests carried out.

Non-Destructive Testing	Standard	Invasive Testing	Standard
Surface observations & documentation		Opening inspection	
Sounding		Bond strength of repair (pull-off)	[20]
Colorimetry	[21]	Half-cell potential	[22, 23]
Water absorption (sponge test)	[24]	Linear polarization resistance	[25]
Surface hardness (rebound hammer)	[26]	Sample removal	
Surface hardness (Mohs' hardness)			
Cover depth & reinforcement position	[27]		
Electrical resistivity	[28]		

While the site testing included NDT techniques and tools which are standard for concrete investigations, such as sounding, rebound hammer, covermeter and electrical resistivity (Wenner probe), it also included additional tests which are commonly applied to cultural heritage. Colorimetric measurements were taken on-site using Konica Minolta CR-410 colorimeters with D65 CIE standard illuminant, 2° CIE standard colorimetric observer angle, and measurement area of ⌀ 50 mm, with the results recorded in CIELAB. The number of measurements taken per site was dependent on the size of the patch and the variability of the material. However, a minimum of 6 representative areas were measured per patch, with a further minimum of 6 measurements taken on the original concrete adjacent to each patch. This allowed a more objective determination of differences in color between the patch repairs and the original concrete. Surface observations were also recorded using Dino-Lite field microscopes (20–220x), which aided in determining the causes of color variation.

Initial water absorption was determined by the contact sponge method. Contact sponge measurements were carried out on 5 patches per site, with 3 measurements per patch, and a further 3 measurements taken on the original concrete adjacent to each of the 5 patches. While the amount of water and contact time of the sponge must be determined on a case-by-case basis depending on the material, a time of 90 s with 4 ml of water was typically found to be suitable in this study. This provides a much quicker alternative to traditional water absorption techniques, such as the Karsten/RILEM tube.

Necessary invasive procedures included traditional electrochemical techniques, half-cell potential and linear polarization resistance, which were carried out on-site to assess the probability of reinforcement corrosion – one of the key causes of repair failure due to the expansion of corroding steel, which induces tensile stress in concrete, in turn causing cracking and spalling of the concrete cover. In addition to these, pull-off tests were conducted to assess the bond strength of the repair materials, and inspection openings allowed visual observations of the steel to determine its condition and whether any protective coatings had been applied (Table 3).

Table 3. Overview of samples to be removed.

Area	Objective	Sample	No.
Good Patch 1	Interface characterization (thin section)	Core/Fragment	1
	Bond strength (*in situ* pull-off test)	Core	3
	External laboratory testing	Core	1
	Internal laboratory testing	Core	2
Good Patch 2	Bond strength (*in situ* pull-off test)	Cores	3
	Internal laboratory testing	Cores	2
Fair Patch	Interface characterization (thin section)	Core/Fragment	1
	Bond strength (*in situ* pull-off test)	Core	3
	External laboratory testing	Core	1
	Internal laboratory testing	Core	2
Poor Patch 1	Interface characterization (thin section)	Core/Fragment	1
	External laboratory testing	Core	1
	Internal laboratory testing	Core	2
Poor Patch 2	Internal laboratory testing	Core	2
Original concrete	Internal laboratory testing	Core	2
	External laboratory testing	Core	1

5.2 Laboratory Testing

Analyses of the samples removed from the sites were carried out by the authors in the laboratories of both LRMH and the GCI, with petrographic examinations of thin sections performed by specialists at WJE, and some additional specialist analyses subcontracted to external laboratories in France.

Analyses of historic concrete in the laboratory are complicated due to the limited amount of material that is often available and both the chemical and physical alterations which occur over time [19]. As a result, several of the test procedures had to be slightly modified to account for this, and an optimized system had to be developed to maximize the amount of data obtained from a minimal number and volume of samples. In most cases, the modification of the test procedure was limited to using samples that were below the size requirements or older than the age limit specified. In all tests where the relevant standard required the samples to be dried, this was done at no higher than 45 °C to prevent any further chemical or physical degradation or the test would be considered destructive. Where temperatures above this were absolutely necessary, these tests were performed last, and the sample was not used in any further testing.

Additional complications arose from the fact that two fundamentally different types of patch repair were encountered on the sites studied – 'form-and-pour concrete' and 'hand-applied mortars' – which resulted in significant differences in the depth of repairs and size of aggregates used. As such, two slightly different processes for analyzing samples had to be adopted to accommodate these differences.

An overview of the specific laboratory tests carried out and the relevant standards is given in Table 4.

Table 4. Overview of laboratory tests being carried out.

Internal Laboratory Testing	Standard	External Laboratory Testing	Standard
Carbonation depth	[29]	Elemental analysis	
Static contact angle	[30]	Chloride content	[31]
Ultrasonic pulse velocity (UPV)	[32]	Alkali content	
Capillary absorption	[33]	Sulfate content	
Density and porosity accessible to water	[34]	Determination of insoluble residue	
X-ray diffractometry (XRD)		Thermogravimetric analysis (TGA)	
Microscopy		Differential thermal analysis (DTA)	
		Determination of mix design	

6 Conclusion

The PEPS project goal is to improve repair of heritage concrete by producing practical guidance to help those repairing historic concrete in the process of selecting appropriate repair approaches, procedures, and materials. It is assumed this will be built on the basis of good concrete repair practice and craftsmanship. This will be informed by field and laboratory assessments of previous repairs that have been carried out on culturally significant concrete structures in England, France, and the USA. This paper provides an overview of the project methodology and outlines the project background, development, and assessment phases currently underway, while final results and conclusions will be presented in future publications.

References

1. Macdonald, S., Arato Goncalves, A.P.: Concrete conservation: outstanding challenges and potential ways forward. Int. J. Build. Pathol. **38**(4), 607–618 (2020)
2. Neville, A.: Consideration of durability of concrete structures: past present, and future. Mater. Struct./Matériaux et Constructions **34**, 114–118 (2001)
3. ICOMOS International Committee on Twentieth Century Heritage: Approaches to the conservation of twentieth-century cultural heritage – Madrid-New Delhi Document (2017). http://www.icomos-isc20c.org/pdf/madrid-new-delhi-document-2017.pdf. Accessed 26 Jan 2022
4. Odgers, D. (ed.): English Heritage: Practical Building Conservation: Concrete, 1st edn. Ashgate, London (2012)
5. Urquhart, D.: Historic Concrete in Scotland Part 3: Maintenance and Repair of Historic Concrete Structures. Historic Scotland, Edinburgh (2014)
6. Gaudette, P., Slaton, D.: Preservation Brief 15: Preservation of Historic Concrete. National Park Service, U.S. Department of the Interior, Washington D.C. (2007)
7. Macdonald, S., Arato Gonçalves, A.P.: Conservation Principles for Concrete of Cultural Significance. Getty Conservation Institute, Los Angeles (2020)
8. ACI: ACI 546.3R-14. Guide to Materials Selection for Concrete Repair. American Concrete Institute, Farmington Hills (2014)
9. ACI: ACI 546R-14. Guide to Concrete Repair. American Concrete Institute, Farmington Hills (2014)
10. ACI: ACI 563-18. Specifications for Repair of Concrete in Buildings. American Concrete Institute, Farmington Hills (2014)
11. Tilly, G.P., Jacobs, J.: Concrete Repairs: Performance in Service and Current Practice. IHS BRE Press, Bracknell (2007)
12. The Concrete Society: Technical Report No. 69: Repair of concrete structures with reference to BS EN 1504. The Concrete Society, Camberley (2009)
13. CEN: EN 1504-10:2017. Products and systems for the protection and repair of concrete structures - Definitions, requirements, quality control and evaluation of conformity – Part 10: Site application of products and systems and quality control of the works. European Committee for Standardization, Brussels (2017)
14. CEN: EN 1504-9:2008. Products and systems for the protection and repair of concrete structures - Definitions, requirements, quality control and evaluation of conformity – Part 9: General principles for the use of products and systems. European Committee for Standardization. Brussels, Belgium (2008)
15. Custance-Baker, A., Macdonald, S.: Conserving Concrete Heritage: Experts Meeting. Getty Conservation Institute, Los Angeles (2015)
16. ACI: ACI 201.1R-08. Guide for Conducting a Visual Inspection of Concrete in Service. American Concrete Institute, Farmington Hills (2008)
17. Currie, R.J., Robery, P.C.: Repair and Maintenance of Reinforced Concrete. Building Research Establishment, Watford (1994)
18. The Concrete Society: Technical Report No. 54: Diagnosis of deterioration in concrete structures - identification of defects, evaluation, and development of remedial action. The Concrete Society, Crowthorne (2000)
19. Wilkie, S., Dyer, T.: Challenges in the analysis of historic concrete: understanding the limitations of techniques, the variability of the material and the importance of representative samples. Int. J. Archit. Herit. **16**(1), 33–48 (2020)
20. CEN: EN 1542:1999. Products and systems for the protection and repair of concrete structures - Test methods - Measurement of bond strength by pull-off. European Committee for Standardization, Brussels (1999)

21. CEN: EN 15886:2010. Conservation of Cultural Property - Test methods - Colour measurement of surfaces. European Committee for Standardization, Brussels (2010)
22. ASTM: ASTM C876-15. Standard Test Method for Corrosion Potentials of Uncoated Reinforcing Steel in Concrete. ASTM International, West Conshohocken (2015)
23. RILEM TC 154-EMC: 'Electrochemical Techniques for Measuring Metallic Corrosion'. Half-cell potential measurements – Potential mapping on reinforced concrete structures. Mater. Struct. **36**, 461–471 (2003)
24. UNI: UNI 11432:2011. Cultural Heritage. Natural and artificial stone. Determination of the water absorption by contact sponge. Ente Nazionale Italiano, Milan (2011)
25. RILEM TC 154-EMC: 'Electrochemical Techniques for Measuring Metallic Corrosion'. Test methods for on-site corrosion rate measurement of steel reinforcement in concrete by means of the polarization resistance method. Mater. Struct. **37**, 623–643 (2004)
26. CEN: EN 12504-2:2012. Testing Concrete in Structures - Part 2: Non-destructive testing Determination of rebound number. European Committee for Standardization, Brussels (2012)
27. BSI: BS 1881-204:1988. Testing Concrete. Recommendations on the use of electromagnetic covermeters. British Standards Institution, London (1988)
28. RILEM TC 154-EMC: 'Electrochemical Techniques for Measuring Metallic Corrosion'. Test methods for on-site measurement of resistivity of concrete. Mater. Struct. **33**, 603–611 (2004)
29. CEN: EN 14630:2006. Products and systems for the protection and repair of concrete structures - Test methods - Determination of carbonation depth in hardened concrete by the phenolphthalein method. European Committee for Standardization, Brussels (2006)
30. CEN: EN 15802:2009. Conservation of cultural property - Test methods - Determination of static contact angle. European Standardization Committee, Brussels (2009)
31. CEN: EN 14629:2007. Products and systems for the protection and repair of concrete structures - Test methods - Determination of chloride content in hardened concrete. European Committee for Standardization, Brussels (2007)
32. CEN: EN 12504-4:2021. Testing concrete - Part 4: Determination of ultrasonic pulse velocity. European Standardization Committee, Brussels (2021)
33. CEN: EN 15801:2009. Conservation of cultural property - Test methods. Determination of water absorption by capillarity. European Committee for Standardization, Brussels (2009)
34. AFNOR: NFP 18 459:2010. Concrete - Testing hardened concrete - Testing porosity and density. Association Francaise de Normalization, Paris (2010)

Traditional Techniques on Post-Civil War in Spanish Modern Architecture: The Case of the Ceramic Wall on OSH Pavilion in the Casa del Campo (Madrid)

María del Mar Barbero-Barrera[1]([✉]) [ID] and José de Coca Leicher[2] [ID]

[1] Department of Construction and Technology in Architecture,
Universidad Politécnica de Madrid, Madrid, Spain
mar.barbero@upm.es
[2] Department of Architectural Graphical Ideation, Universidad Politécnica de Madrid, Madrid,
Spain
jose.decoca@upm.es

Abstract. The rise of cement at the beginning of the XX century provoked the lime mortars decay. In the case of Spain, they were revival after the Spanish Civil War (1936–1939) due to the insolation of the country (autarky period). It is especially interesting since traditional materials defined new styles. An example of this duality is the *Obra Sindical del Hogar* (OSH) pavilion, at *Casa de Campo* exhibition area in Madrid. The ceramic wall placed on its entrance is a perfect example of modernity in the graphical representation of the scenes, using traditional techniques and materials: porous ceramic and lime mortars. As part of the restoration of this mural, a complete analysis of the original mortars and ceramic had been performed. The results showed a careful selection of the materials and the knowledge of the workforces. Porous ceramic calcined at low temperature as support of the glassy treatment and their compatibility with the aerial lime mortar defined its state of conservation, in spite of the abandon suffered for more than 40 years. Air lime mortars showed a high compactness and a reduced porosity mainly because a careful selection of the aggregates and care in their execution and installation.

Keywords: Lime mortars · Ceramic wall · Casa de Campo · Modernity · Recovery · Traditional

1 Introduction

The change of living models along the twenty century implied the loss of traditional craft and artisanship by the imposition of new materials and techniques. Against this trend, in the case of Spain, the Spanish Civil War (1936–1939) and the following isolation period imposed by the authoritarian regime provoked a considerable delay in about 30–40 years in relation to the neighbouring countries, in such a way that there are numerous examples of using traditional techniques and materials along 40's and 50's.

© The Author(s), under exclusive license to Springer Nature Switzerland AG 2023
V. Bokan Bosiljkov et al. (Eds.): HMC 2022, RILEM Bookseries 42, pp. 300–312, 2023.
https://doi.org/10.1007/978-3-031-31472-8_23

The Architects who worked in this time had been trained under the 1914's study plan which impelled the use of new language while reinforcing the construction subjects, as answer to the social demands [1]. Another significant fact is the emergence of young architects graduated between 1941 and 1946 known as "the first post-war generation" [2]. They were pioneers of modern ideas inspired by Italian and Scandinavian references, like Miguel Fisac, Francisco de Asís Cabrero, Alejandro de la Sota or Rafael Aburto. All of them, in Madrid and in the Feria del Campo, introduced modern language in a fairground today considered as an ensemble or "laboratory of modern architecture" [3] (Fig. 1).

In this period, the importance of the use and knowledge of materials is clear with the implementation of the Building Materials Laboratory at the Technical School of Architecture, but also for the scientific character that many of the architects trained in this period achieved. As Zaballa mentioned "Architecture is the Belle Art of the Construction [...] and the construction is the scientific, spatial and complete technique, the art of the construction is, in general, the Engineer, and the Belle Art of the construction is the Architecture" [4]. The latter showed the importance given to the combination in the language which imprinted the character of the novel architects, which would be boosted in the following 1932's study plan disrupted by the Civil War and recovered after this.

In addition, the scarcity of the resources after the Civil War because of the autarchy economy of the authoritarian regime provoked the look towards the traditional materials and techniques that began to be abandoned in the previous stage. It must be highlighted the recovery of traditional techniques as part of the experimental character of the dwellings developed in the 40's and 50's as reflected in the paper published by Moya (1943): "the project tries to systematize the development with popular character to achieve an economic solution" [5]. Indeed, the use of traditional materials is collected by different documents in this time. Concerning this, the Basque-Navarre Official School of Architects (1949) published a paper in which collected the opinion of technicians and experts about the most suitable materials and constructive systems to improve the building of economic dwellings [6]. Most of the authors agreed the use of traditional materials because of their availability and their economy to solve the problem of dwellings [7]. They used materials taken from the tradition and the popular architecture such as limewashed white walls, exposed brick walls and ceramic tile roofs, to which were added baseboards and masonry walls made of natural stone: granite and "colmenar" stone, most of the time dry-laid or with air lime and sand mortar. What is more, during this time, literature focused on the fact that the use of novel materials such as cement or iron, without an extended local production and dependent of transport, must be used only in the applications in which would be essential such as the slabs. Among the traditional materials, lime mortars are an essential part of the constructive systems since "there is plenty of stones apt to obtain lime and gypsum, and so on" [7]. However, it must highlight the statement: "the use of lime in renders and mortars is traditional in many Spanish regions, and in them, it is worked with such perfection that it worth it its study and dissemination to widen their use in the whole national scale in this type of economic buildings" [8]. Together with the air lime, ceramic pieces were a perfect support to combine tradition and innovation following the words of Rafael Domenech: "the tradition is not a jump back but a continuum: I am a link in the traditional chain of my family, but I can not be my grandfather" [9]. Indeed, the authors recover a traditional technique devaluated by the excessive use pre-civil war [10] and put it to the service of the new style. In words of Alejandro de la Sota, it is a process of "assimilation of the known or creation of the unknown [...] the past, the constant, is translated into the future, the creative" [11]. This dichotomy was clearly reflected in the exhibition areas such as the *Casa de Campo*.

This area born with the idea of exhibiting the advances in the livestock and agriculture along Spain but also the uses and customs of the different regions in such a way that the area converted into a rural, folklore and gastronomy tour. An example of this was the pavilion of the *Obra Sindical del Hogar* to show the advantage in the design and promotion of dwellings in Spain (Fig. 2).

The building was designed and built in 1956 by the architects Francisco Asís Cabrero and Felipe Pérez Enciso, with the idea of showing some of the elements of the popular architecture from the courtyard, ponds and lattices for solar protections integrated with a novel language with pure lines, in which the decorative elements were introduced as part of the architecture. The incorporation of the decorative elements is part of the architecture reading. In this case, a ceramic wall placed at the entrance will show the visitor the different alternatives in the dwellings (Fig. 3). In the design of this ceramic wall, the architects counted with two young artists: Amadeo Gabino and Manuel Suárez Molezún which introduce a new language [12].

After its preliminary used, it was abandoned for 40 years up to 2018 when a project to recover the area as Associative and Cultural Campus was approved. This one searches the recovery of different pavilions as headquarters of neighbourhood associations by enhancing the XX century Architecture. In the context of the recovery project [13], a complete documental and experimental analysis was performed in order to evaluate the most suitable interventions [14, 15]. On the documental study, of the most representative elements in the pavilion was the ceramic wall, designed by Amadeo Gabino and Manuel Suárez Molezún, both young artists and close to the aforementioned architects. It is important to highlight that traces of both artists can be observed in the ceramic wall, on one side, the iconography of the dwelling mural at the OSH pavilion, based on the combination of coloured areas and forms drawn with a black line, is recognisable in Amadeo Gabino travel sketches made in the early 1950s. On the contrary, the painting of Suarez Molezún, co-author of the mural, is dynamic and close to geometric abstraction, dominating the superimpositions of colours, a technique that can also be seen in the in the dwelling mural [14].

2   Ceramic Wall

The recovery of the ceramic wall at the entrance of the pavilion was part of the restoration of the building (Fig. 4). In spite of the abandon of the building, most of the ceramic wall was in perfect state of conservation.

According to the photogrammetric flights, the wall was installed in 1956 attending to the pavilion opening. However, between 1974 and 1985, a canopy was installed over it which, although modified the original aesthetics, considerably allowing to preserve the wall until the renovation works. The rupture and detachment of some pieces was

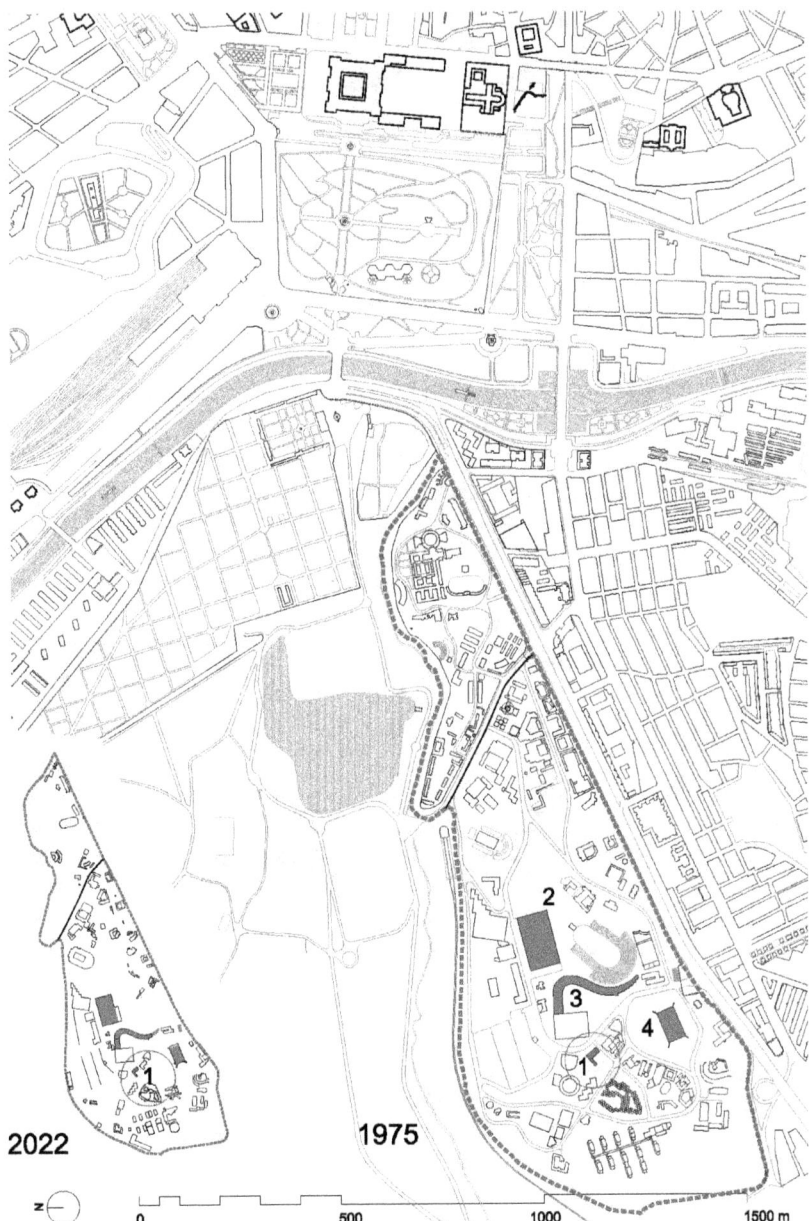

Fig. 1. The fairground and its representative pavilions, the Manzanares river and Madrid.1 OSH Pavilion 2 Crystal Pavilion 3 International Pavilion 4 Palace of Agriculture [12]

mainly observed at the bottom part of the wall due to the rising damp because of the modification of the original height. And, at the top, weld spatter, minimum painting and spots of cement can be observed due to the installation of the structure of the roof.

Fig. 2. The OSH pavilion as published in the review Hogar y Arquitectura, 4, 1956. In the background the ceramic wall.

Fig. 3. The ceramic wall and the OSH pavilion before intervention [14]

The ceramic wall composed by 69 pieces in horizontal and 15 in vertical, so, in total, 1035 pieces of 20 × 20 cm of dimension. The latter is of great interest since attending to the commercial size of this type of glazing ceramic and showed the adequation of the artists to the available manufacture. In addition, visible traces of the firing procedure were observed in such a way that the wall was structured in vertical strips that showed differences in the white colour of the basement. Following the documental analysis, the pieces were manufactured in «Industrias cerámicas Julián Vilar Esteve» in Manises (Valencia), that there is no longer exists, it was placed in Valencia street, number 28.

2.1 Characterization of the Ceramic Pieces

The pieces are traditional ceramic tiles boiled at low temperature over which is applied the glazing with a traditional technique named «a la talaverana».

Fig. 4. The ceramic wall after the restoration process

The traditional technique consisted on painting the surface over a covering. The procedure of manufacture was the former immersion of the porous ceramic support into a white enamel. After drying at ambient temperature, it became to harden enough to be scratched with the nail or other material. This surface is the support of the decoration, which was performed by pigments dissolved into water and applied with brushes. The pigments could be natural oxides or silicates prepared in the laboratory. This technique was continued by the two artists, among others, due to their proximity in time and similarity in graphic technique, the murals on the terraces of the flats on the Paseo de la Castellana in Madrid, partially preserved in the CA2M Centre [16] are particularly noteworthy.

On the opposite to the use of a traditional support, the novelty of the ceramic wall lied in the decorative technique. The use of traditional materials integrated in the creative process which was stated and defended by other contemporary architects such as De la Sota [17]. It was similar to the watercolour in the use of free movements and semi-transparent colours, in such a way that the light tones were firstly applied following by the dark and black colours which were used to outline the motifs. In addition, the artist used the «*esgrafiado*» technique to remove the colours of the surface or even achieve the ceramic support to enhance some areas or create reliefs. In addition, the artists applied pointings or coloured squares (blue or green) over the black surfaces to light the background up. In the pieces, a shrinkage due to an excess of pigment was detected in some of the surfaces (Fig. 5).

Based on the literature [18], among the common pigments used in the ceramic tiles were cobalt (intense blue or violet), cobalt carbonate (light blue or violet blue), copper (green or turquoise), copper carbonate (light green or light turquoise), manganese (violet

Fig. 5. Details of ceramic fragments showing materials and techniques

or brown), iron (reddish brown or beige), vanadium (yellow), chromium (reddish orange or opaque green), bichromate potassium (orange or green), antimonium (opaque yellow or opaque white) and nickel (brown) [14].

Finally, the decorated piece was boiled. Given the data of the ceramic wall, the combustible used would be wood or fuel which implied an abundance of carbon during the combustion. In addition, it was expected that the temperature of glazing would be low (lower than 1000°C). The different shades of the same colour that had been identified could correspond to different batches or even the position inside the kiln.

2.2 Characterization of the Lime Mortars

Lime mortars were used in the ceramic wall with two proposals, as joint mortar in the ceramic bricks used as support of the ceramic tile, as well as to guarantee the adherence of the latter.

Before handling the restoration, a complete analysis and diagnosis was performed in order to analyze the type of damages, their placement and the state of the conservation of the ceramic wall together with the materials and techniques used, among others. The methodology proposed by RILEM Historic mortars group was applied [19]. Both the support and the materials used in the ceramic walls were analyzed and characterized to select the most suitable materials and techniques to use in the restoration. Since they are the aim of the research, they are analyzed in a separated section.

3 Lime Mortars

In all the cases, air lime mortars were used. Following the common practice in Spain due to the absence of hydraulic lime as well as the recovery during this period. In addition, it would be probable that the slaking took place in the construction site, while reducing the cost of the construction. This hypothesis could be corroborated by the high amount of lime lumps found in the composition and most of them visible at naked eyes. They suggested the procedure in the elaboration of the mortar that, in this case, would probably correspond to the dry slaking method [20].

During the post-civil war a common practice in the construction site was to establish "little" manufactures in which the products were assembled in-situ. Following this line, it could be estimated that calcined rocks, with a lower weight, arrived to the construction site where the process of production of the mortar would be finished. It allowed to reduce costs and enhance of efficiencies in times of shortage. About the production of the lime, there had not been due further inquiries, but it is probable that the area that supply the air lime to the center of Madrid would be the Tajuña's river. In this area in the southern of Madrid had a great amount of lime kilns that were used up to 80's–90's [21] to provide lime to the manufactures.

Together with the lime, aggregates of sub-angular siliceous nature were used in the production of the mortars. Their composition as well as the predominant idea of using local products and local builders would justify their extraction from the proximity. According to the IGME, in the municipality of Madrid there were quarries for extraction of sand and gravel that although nowadays are abandoned they could be working in that period [22].

Both components were used in the common lime to aggregate ratio [23], the mortar showed a 1:3, in volume. Furthermore, attending to the X-ray diffraction analysis (Fig. 6), belite and clay were found which justified the addition of the latter to the aerial lime mortar to confer it hydraulic properties. The incorporation of ceramic dust into the lime mortars had been a common practice in the history as had been stated in different international and national researches [24–27]. Once again, a demonstration of the recovery and preservation of traditional techniques applied to the new architecture was shown, with an experimental character to extract the maximum potential of the technique with a modern language [4, 17].

Its use is translated in the development of a compact matrix observed in the samples due not only to the addition of brick dust but also to the particle size distribution of the mortars with a higher amount of fine particles in the aggregates that guarantee a good penetration of the mortar into the ceramic porous structure. In addition, reaction rims were found in the interface between the ceramic and the lime mortar [25].

Regarding this, the use of low temperature ceramic in both the support and the basement of the ceramic tile would be essential to guarantee a suitable adherence and compatibility among the materials.

Then, microscopical observations (Fig. 7) demonstrated a high quality and skills of the builders. The selection of the materials is so care that the limit between the lime mortar area and ceramic tile is difficult to notice. In this regard, the know-how of the workers is clear in the selection of the components as well as the application and placement of the tiles.

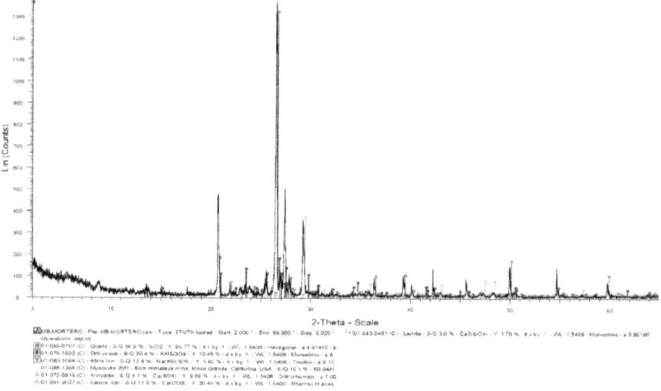

Fig. 6. X-ray diffraction of the lime mortars

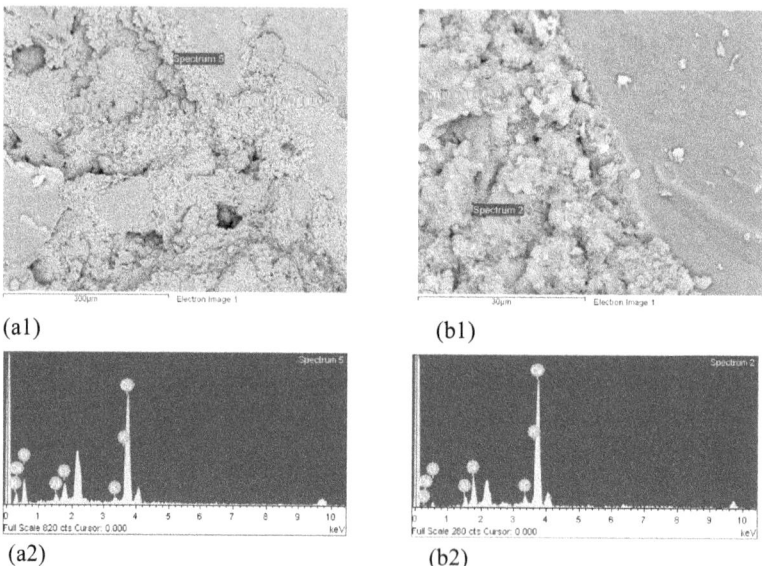

(a1)   (b1)

(a2)   (b2)

Fig. 7. Microscopical observations and EDM analysis of the mortars

Together with the microscopical analysis of the mortars and ceramic tiles and bricks, a macroscopical analysis was developed to evaluate their performance. Regarding this, bulk density of the mortars was of 1746 ± 77 kg/m3, an slightly over the literature [26], in contrast to the bulk density of the ceramic tiles of 1553 ± 11 kg/m3 which is slightly smaller [26]. In line with this, open porosity showed similar values in the lime mortars of 26.4 ± 1.7% which was considerably smaller than the ceramic samples (38 ± 0.84%) which is over the 22–30% of the literature [26–28] or the 14.4 ± 0.9% found in the ceramic bricks used as support.

Dense structure of the mortar was also observed on the absorption coefficient of 13.8% which is smaller than the 19–35% in the literature [29]. It must highlight that, on the contrary, ceramic tiles showed an absorption coefficient of 25% which is over the literature for similar pieces and uses [26–28]. Concerning the sorption coefficient by capillarity (Fig. 8), lime mortars showed 2.23 kg/m^2.min, 0.98 kg/m^2.min and 0.68 kg/m^2.min at 2, 5 and 10 min, respectively, which were considerably lower than the values achieved with the ceramic tiles of 10.30 kg/m^2.min, 6.23 kg/m^2.min y 4.31 kg/m^2.min, respectively. In both cases, 10–90 sorption coefficients were null following the trend of the historical materials [30] and the traditional practices of the construction works. In any case, both of the results were according to the literature for both ceramic tiles [27, 28] and lime mortars [30]. Desorption analysis which followed the procedure 0803/112/19460 LNEC [31], showed a similar performance with a former fast evaporation followed by a decrease from 2–3 h in advanced [31, 32]. Mechanical properties were not performed due to limited amount of pieces used in the characterization as well as the interest to recover the ceramic tiles.

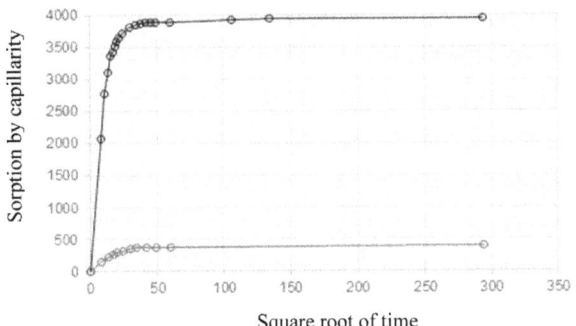

Fig. 8. Sorption by capillarity of both ceramic tiles (in black) and lime mortars (in red) (Color figure online)

Then, microscopy and macroscopy analysis corroborated the delicate care shown along the whole process, from the selection of raw materials, their execution and installation of the compounds. It also enhances the full validity of the traditional techniques and craftsmen in an autarky period in which traditional artisans must search solutions of low cost with high ingenuity and the interest of perpetuity which had been predominant up to this moment.

This understanding of the materials and constructive techniques and systems and the collaborative work between different trades is also found in other notable examples along the postwar period in which the innovation promotes by Architects, Engineers and Artists found a clear support on an skilled craftsmanship. This wall is only an example of this interesting period of the Spanish Architecture.

4   Conclusions

The Spanish post-Civil war period, from 40's till the end of the 50's, can be considered a golden age of the Architecture in which tradition and innovation matched.

The lack of materials and the autarky during this period forced the search of cheap solutions in which, however, a new language was incorporated. During this time the use of traditional materials, among them air lime, arisen.

In this research, a complete characterization of the ceramic wall installed in the OSH pavilion placed in *Casa de Campo* was analyzed as part of the renovation of the building. According to the analysis of both ceramic tiles and lime mortars, an exquisite care on the selection of the raw materials together with the installation showed the know-how of the craftmenship which contrast with the innovative language introduced by architects, engineers and artists. A clear understanding of the materials and constructive techniques and systems and the collaborative work between different trades is also found in other notable examples along the postwar period.

This multidisciplinary work opens new fields of study in other contemporary ceramics made by the artist, such as those of the balconies of the houses in the *Paseo de la Castellana* that are partially preserved in the CA2M [16] center. These have been recently visited and a first observation it gives the impression that the traditional production techniques were perfected with the experience of the artist at OSH mural. We hope that future analyzes and characterization methodologies of the materials as used in this research will confirm our hypotheses. Finally, we would like to emphasize that the field of artistic achievements and their integration into architecture, such as the OSH mural, suggests the collaboration of researchers from different disciplines.

References

1. Navascués Palacio, P.: Introducción. Teodoro Anasagasti y Algán. Apunte biográfico. En: Enseñanza de la Arquitectura: cultura moderna técnico artística. Textos sobre la enseñanza de la arquitectura (2). Instituto Juan de Herrera, Madrid (1995)
2. Flores, C.: Arquitectura Española Contemporánea, I 1880–1950, pp. 241–265. Aguilar Maior, Madrid (1989)
3. de Coca Leicher, J.: El recinto Ferial de la Casa de Campo de Madrid (1950–75). PhD thesis, UPM, Madrid (2013). https://oa.upm.es/19952/
4. Zabala y Gallardo, M.: Informe de la reforma de plan de estudios. Escuela Técnica Superior de Arquitectura, Madrid. ETSAM. Texto original en Biblioteca, Madrid (1921)
5. Moya, L.: Casas abovedadas en el barrio de Usera, construidas por la Dirección General de Arquitectura. Revista Nacional de Arquitectura **14**, 52–57 (1943)
6. Basque-Navarre Official School of Architects: Tema II. Construcción. Revista Nacional de Arquitectura **90**, 256–257, 263 (1949)
7. Vallejo, A.: Materiales más adecuados y métodos constructivos más convenientes para incrementar y mejorar la vivienda popular. Revista Nacional de Arquitectura **90**, 258–261 (1949)
8. Juliá, S., García Delgado, J.L., Jiménez, J.C., Fusi, J.P.: La España del siglo XX, 2nd edn., pp. 536–542. Marcial Pons Historia, Madrid (2003)
9. Domenech, R.: Exposición Nacional de Arte Decorativo, pp. 16–60. Pequeñas monografías de Arte, Madrid (1911)
10. Perla, A.: La asimilación del pasado en el resurgimiento de la azulejería sevillana. In: La azulejería española de los siglos XIX-XX; Proceedings of the VIII Congress of the Ceramic Society, Castellón, Spain, pp. 91–103 (2003)

11. Tomás Roldán, A.: Tradición de futuro. Alejandro de la Sota: lo popular como referente de una nueva arquitectura. In: Couceiro Núñez, T. (ed.) Pioneros de la arquitectura moderna española: vigencia de su pensamiento y obra., coord.: General de ediciones de arquitectura S.L., Spain, pp. 32–51 (2015)

12. de Coca Leicher, J.: El pabellón de la Obra Sindical del Hogar: una fusión entre tradición, modernidad y arte plástico: Francisco de Asís Cabrero y Felipe Pérez Enciso. En: Congreso NACIONAL DE ARQUITECTURA PIONEROS DE LA ARQUITEctura moderna Española, 3, Aprender de una obra: actas. Fundación Alejandro de la Sota, Madrid, Spain, pp. 206–216 (2016)

13. de Coca Leicher, J.: Expanded architectures. The Exhibition Pavilion at the Casa de Campo in Madrid. Proyecto, Progreso, Arquitectura, no. 24, pp. 108–128 (2021)

14. de Coca Leicher, J., del Barbero-Barrera, M.M.: Estudio Mural Pabellón Icona I Recinto Ferial de La Casa de Campo de Madrid. Universidad Politécnica de Madrid, Madrid (2020)

15. de la Colina, L., Plaza, M.: Informe de tratamientos de conservación-restauración de mural cerámico de Amadeo Gabino y Manuel Suárez Molezún, situado en el Pabellón ICONA I de la Casa de Campo. Universidad Complutense de Madrid, Madrid (2020)

16. Esquivias, P.: A veces decorado. Catálogo exposición. CA2M Centro de Arte Dos de Mayo: Comunidad de Madrid, Spain, pp. 82–83 (2013)

17. Alagón Laste, J.M.: Alejandro de la Sota y su aportación a los pueblos de colonización de la cuenca del Ebro (1941–1946). NORBA, Revista de Arte **37**, 279–309 (2017)

18. VV.AA. Taller de esmaltes sin plomo y técnicas de decoración. San Agustín Etla. Innovando la tradición (2009). http://innovandolatradicion.org/

19. Van Balen, K., et al.: RILEM TC 203-RHM: repair mortars for historic masonry (2010)

20. Hughes, J.J., Leslie, A.B., Callebaut, K.: The petrography of lime inclusions in historic lime based mortars. Annales Geologiques des pays Helleniques, Edition Speciale **39**, 359–364 (2001)

21. Barbero-Barrera, M.M., Cárdenas Chávarri, J.: Los hornos de cal periódicos en la comunidad de Madrid: estudio tipológico y nuevas ubicaciones. In: Huerta Fernández, S. (ed.) coord, pp. 103–112. Santiago de Compostela, Actas del Séptimo Congreso Nacional de Historia de la Construcción (2011)

22. IGME. Inventario nacional de balsas y escombreras. Ministry of Industry and Energy, Madrid (1988)

23. Botas, S.M.; Veiga, M.R.; Velosa. A. Adhesion of air-lime based mortars to old tiles: moisture and open porosity influence in tile/mortar interfaces. Am. Soc. Civil Eng. (2015). https://doi.org/10.1061/(ASCE)MT.1943-5533.0001108

24. Moropoulou, A., Bakolas, A., Bisbikou, K.: Physico-chemical adhesion and cohesion bonds in joint mortars imparting durability to the historic structures. Constr. Build. Mater. **14**, 35–46 (2000)

25. Moropoulou, A., Bakolas, A., Bisbikou, K.: Investigation of the technology of historic mortars. J. Cult. Herit. **1**, 45–58 (2000)

26. Cultrone, G., Madkour, F.: Evaluation of the effectiveness of treatment products in improving the quality of ceramics used in new and historical buildings. J. Cult. Herit. **14**, 304–310 (2013)

27. Vaz, M.F., Pires, J., Carvalho, A.P.: Effect of the impregnation treatment with Paraloid B-72 on the properties of old Portuguese ceramic tiles. J. Cult. Herit. **9**, 269–276 (2008)

28. Botas, S., Velosa, A., Veiga, M.R.: Adherence evaluation in tile-mortar interface. Mate. Sci. Forum **730**, 403–408 (2012). https://doi.org/10.4028/www.scientific.net/MSF.730-732.403

29. Santos Silva, A., Santos, A.R., Veiga, R.: O Forte de Nossa Senhora da Graça, Elvas. Pedra e Cal **58**, 22–25 (2015)

30. Magalhães, A., Veiga, R.: Physical and mechanical characterisation of historic mortars: application to the evaluation of the state of conservation. Materiales de Construcción **59**(295), 61–77 (2009)

31. Veiga, R., Santos, D.: Métodos de ensaio revestimentos de paredes existentes. Ensaios in situ e ensaios em laboratório sobre amostras recolhidas em obra. Procedimiento 0803/112/19460. LNEC, Lisboa (2016)
32. Veiga, R.: Air lime mortars: what else do we need to know to apply them in conservation and rehabilitation interventions? a review. Constr. Build. Mater. **157**, 132–140 (2017)

Measuring Water Absorption in Replicas of Medieval Plaster. Assessing their Reliability as Models for Conservation Trials

Mette Midtgaard[1,2][✉]

[1] Environmental Archaeology and Materials Science, National Museum of Denmark, Kongens Lyngby, Denmark
mmm@natmus.dk

[2] Royal Danish Academy - Architecture, Design, Conservation, Copenhagen, Denmark

Abstract. Using laboratory produced mortar samples in wall painting conservation research requires sufficient chemical and physical resemblance to the original object to make the tests on the samples relevant. This study applies the contact sponge method comparing the water absorption rate of limewashed medieval wall painting plaster with mortar samples of varying binder-to-aggregate ratios produced by the hot-mix technique and with lime putty mortar. Different types of limewashes are also examined. Based on this comparison, the study evaluates the samples' suitability as models for wall painting conservation research. Moreover, the study explores the validity of the contact sponge method for measuring highly porous mortars.

The experiment demonstrates that the contact sponge method can be used for measuring the water absorption rate of highly porous mortar samples, and that the fastest water absorption rates are found in the mortar samples with the highest lime content. Moreover, the closer the sample's binder-to-aggregate ratio is to the historical plaster, the more similar the values are. The poorest correlations are found for lime putty mortar samples with a 1:3 ratio heretofore used in conservation trials. Furthermore, the absorption rate is significantly influenced by the type of limewash on the samples.

Keywords: Contact Sponge · Lime-rich mortar · Medieval wall paintings · Water absorption rate · Replicas

1 Introduction

Testing new materials and methods on like-for-like mortar samples/replicas prior to on-site implementation is considered a vital part of wall painting conservation practice. For laboratory tests to be relevant, it is important that replicas, often comprising smaller mortar samples, have a close resemblance to the replicated historical plaster, not only with regard to material composition, but also to physical characteristics, such as capillarity.

Capillarity influences the water absorption rate of a plaster/mortar, a property that plays a significant role in conservation treatments such as cleaning or desalination on

© The Author(s), under exclusive license to Springer Nature Switzerland AG 2023
V. Bokan Bosiljkov et al. (Eds.): HMC 2022, RILEM Bookseries 42, pp. 313–326, 2023.
https://doi.org/10.1007/978-3-031-31472-8_24

lime-based wall paintings. This is especially the case when conducting such operations by poulticing. In such cases, the water absorption rate not only has a determinant effect on the amount of liquid that penetrates the object but also on the performance of the poultice in terms of advection and drying rate. Thus, when performing laboratory cleaning or desalination tests, a close match of the water absorption rate of the laboratory models to the replicated historical mortar/plaster should be strived for. Such a cleaning experiment, where tests on mortar samples are conducted prior to on-site implementation, is illustrated in Fig. 1.

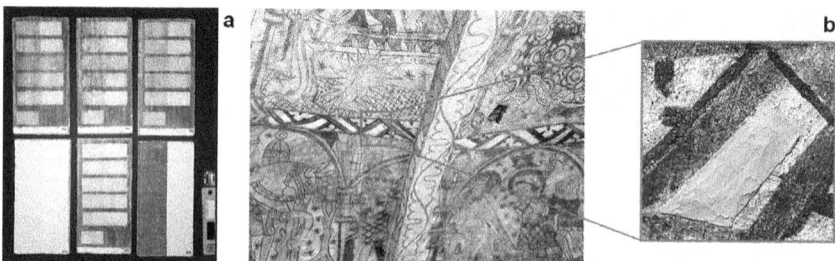

Fig. 1. (a) Cleaning tests conducted on artificially soiled mortar samples; (b) Cleaning tests conducted with poultice in situ on a medieval wall painting.

However, the replication of historical mortar/plaster, such as the very lime-rich plaster constituting the support for numerous medieval wall paintings in Europe, is a challenging and complex procedure. Not only do the mixing proportions and materials used in the initial mortar mix play an important role in the final characteristics of the plaster, but the properties of the replicated plaster are also significantly affected by such factors as processing procedures, application methods and conditions during carbonation. An additional complication is encountered in the replication of Danish medieval wall paintings, as most of these paintings were painted on a limewashed plaster. Thus, a replication of the limewashed ground should also be included in sample production.

Standard mortar samples are frequently made using the contemporarily recommended mortar mixture of one part slaked or hydrated lime to three parts aggregate (volumetric) [1, 2]. Yet, historical mortar and plaster often has a much higher lime content, with a typical binder-to-aggregate (b:a) ratio in the range of 1:1 to 1:2 [3–5]. In the case of medieval wall paintings in Denmark, the lime content often exceeds the quantum of aggregates, with b:a ratios such as 2:1 [6, 7]. The discrepancy of the high lime content in historical mortar versus standard mortar has been previously shown [6]. Similar micrographs, Fig. 2, show how a mortar composed with a 1:3 ratio (Fig. 2a) has a completely different structure to that found in a medieval plaster (Fig. 2b) or in a replicated lime-rich plaster (Fig. 2c).

When the replicated historical mortar or plaster is limewashed, a similar issue arises. Since Gothic limewash usually has a significant thickness, as illustrated in Fig. 3, it is inevitable that the limewash will influence water absorption. Yet, again, in Denmark and probably also in other countries, limewash with very different characteristics to the original is commonly used in conservation experiments.

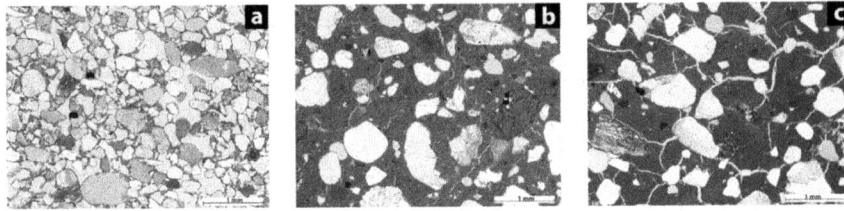

Fig. 2. Micrographs of impregnated thin sections: a) 1:3 lime putty plaster; b) Medieval plaster with an estimated b:a ratio of 2:1; c) Hot-mixed lime plaster with an b:a ratio of 1:1. Brown areas present the lime matrix, grey/white/green/beige areas are aggregates and yellow areas present air-voids/cracks.

At present, the limewash used in historical buildings usually consists of a thin mix of approximately one part lime putty to five parts of water/limewater (volumetric), applied in three to seven layers until the desired coverage is achieved [8–10]. This is done because the same thickness as found in historical limewash can only be achieved by multiple thin layers if extensive shrinkage and crazing is to be avoided. In Denmark, finely graded sand or powdered limestone is often added to one or two layers of the limewash to attain a better coverage. However, these thin layers of limewash, with or without added aggregate, create an entirely different structured limewash than the thick layered limewash found in medieval wall paintings.

The thin-section micrographs shown in Fig. 3 b-c depict the differences between a Gothic limewash dating from ca. 1450 and a new limewash produced for mortar samples used in conservation trials.

Fig. 3. Limewash indicated by red marker: (a) Stereomicrograph showing a sample of an original medieval plaster with several thick layers of limewash. Kongsted Church (ca. 1450); (b) Micrograph of impregnated thin section from Kongsted Church, showing at least three layers of limewash in plane polarised light; (c) Micrograph of impregnated thin section of limewashed 1:3 mortar sample in plane polarised light. Powdered limestone has been added to this limewash to attain greater thickness.

But what are the consequences of using laboratory mortar samples with divergent composition to that found in historical mortar/plaster? Are the results attained on standard laboratory samples still close enough to the results attained in situ, so that the laboratory tests give a valid indication of what to expect when new methods and materials are implemented on the historical mortar/plaster?

This study aims to examine these questions by measuring the water absorption rate in mortar samples with three different compositions and comparing these results to measurements conducted in situ on a Danish Gothic wall painting.

For the investigation, a standard 1:3 lime-putty mortar was compared to two mortars specifically designed to reproduce the plaster typically found in Danish Gothic wall paintings. These two mortars, with b:a ratios (volumetric) of 1:2 and 1:1, were mixed from quicklime and aggregate using the hot-mix method. This was done partly because it is almost impossible to achieve sound mortars with such high binder content when using lime putty mortar [2, 3, 11, 12], but mostly because more recent studies indicate that the vast majority of traditional lime mortars were hot-mixed [6, 13–15]. A detailed description of the production of these samples and the data on the medieval plaster targeted for reproduction are published elsewhere [7].

To study the influence of the type of limewash on the water absorption rate, three limewashes mixed from different lime putties were included in the study. Two of the chosen lime putties are often used in conservation work and in historical buildings in Denmark, such as for limewashing the interior and exterior of medieval churches. A closer match to a Gothic limewash was strived for in the third composition. This was done by using a more than 67-year-old lime putty collected from a lime pit at Roskilde Cathedral. This lime putty has a dense and grainy structure with visible sand particles from the pit and sharp crystals, presumably calcite crystals. Due to its age and the presence of aggregates originating from the pit, the Roskilde lime putty creates a limewash that can be applied thickly, creating an opaque limewash with just a few strokes.

The contact sponge method was used to examine the water absorption rate. This method was developed by *Istituto per la Conservazione e la Valorizzazione dei Beni Culturali* in Florence in 2004, originally for in situ measuring of the effectiveness of water-repellent treatment on stone and plaster but is also used for measuring water absorption of untreated material and for consolidation testing [16–19]. It is a simple and rapid method where the gravimetric changes of a wet sponge pressed against the sample record absorbed water. Compared to the capillary rise method, the contact sponge has the advantage that it is non-destructive and can be used in situ for multiple measurements. Moreover, comparative studies have proven this method more suited than Karsten tube for measuring the initial water absorption [16, 20].

The contact sponge method does, however, have limitations, as the restricted amount of water in the sponge can make the method unsuitable for longer measuring periods and might not work on highly porous material [20, 21]. Since lime-rich mortars, as the ones tested in the present experiment, are highly porous [7, 22], the study also examines the validity of using the contact sponge method for measuring binder-rich historical mortar. In addition, it aims at establishing the most suited measuring time for examining lime-rich mortars. The European Standard for contact sponge measurements (status: in draft for comments) EN 17655 [23] dictates that preliminary tests are undertaken at intervals of 30 s to 3 min in order to determine contact time. However, the present study also included a series of measurements at longer contact times (up to 10 min), to achieve a longer time course of water absorption for the mortar samples. A contact time of 30 s was chosen as the shortest contact time, as it was assessed that shorter contact times than this could result in a considerable time error when conducting in-situ measurements.

2 Materials

2.1 The Choice of Limewashed Mortars

The focus of this laboratory study was to evaluate water absorption rates of mortar samples, whose properties closely resembled those of the mortar encountered in situ. Virtually all of the Danish wall paintings were created on limewashed plaster. Consequently, only limewashed mortar samples were studied in the present work. An overview of the different samples can be found in Table 1.

Table 1. Sample preparation. Slaking technique, ratio and limewash

Sample	Slaking technique, lime form	b:a (v/v)	Lime used for limewash
1:1 Roskilde	Hot-mix, CaO (powder)	1:1	Lime putty from Roskilde Cathedral, >67 years old
1:2 Roskilde	Hot-mix, CaO (powder)	1:2	Lime putty from Roskilde Cathedral, >67 years old
1:2 Rødvig	Hot-mix, CaO (powder)	1:2	Rødvig Lime putty, 23 years
1:3 Horsens	Lime putty mix, $Ca(OH)_2$ (putty)	1:3	Horsens Lime putty, 23 years

Medieval Wall Painting. In situ measurements were conducted on a Gothic wall painting partially uncovered on the vault in Ørslev Church (Skælskør, Denmark) in 2014. The absence of any conservation treatments eliminated any possible errors in measurements caused by non-original materials or due to dirt on the surface. Also, no salt damage is found on this uncovered area.

The ornamental painting, the uncovered area of which measures approximately 1 m², is dated to ca. 1300–1325. Due to the uniqueness of this untouched early Gothic wall painting, non-destructive measures were imperative, excluding sampling. However, it could be observed with the naked eye that a single, quite thin layer of limewash comprised the ground for the painting. The wall painting is illustrated in Fig. 5.

Hot-Mixed Mortars. A micronised Ca-rich quicklime (CL90-Q) from Lhoist was used for mortar 1:1 and 1:2. The aggregates, aiming for similarity to those found in Danish medieval wall paintings, comprised local pit sand from Roskilde (0–3 mm), consisting mainly of quartz and feldspar, and crushed limestone from Lhoist (0.1–2 mm). The mortar was hot-mixed by initially mixing quicklime and sand and then adding water in a vertical axis mortar mixer. The water-to-binder ratio was approximately 2:1 (by weight). After mixing, the mortar was matured in closed plastic buckets for approximately one month.

The mortar samples (7.5 × 15 × 1.5 cm) were prepared manually, applying a 0.7 cm layer of mortar to unglazed ceramic terracotta tiles (Sima Ceramiche) in a similar way as when plastering a wall. To ensure the same height in all samples, wooden frames were used in the production. All mortar samples were cured for more than two years before the testing commenced. Phenolphthalein staining was used to continuously examine the degree of carbonation. The extent of carbonation was verified by X-ray diffraction, showing full carbonation after nine months of curing as no portlandite crystals could be detected at this time.

Lime Putty Mortar. For this mortar, a previously produced batch of limewashed mortar test samples (12 × 12 × 1.5 cm) were used. These mortar samples were produced in 2012 for a wall painting conservation experiment, studying microemulsions ability to remove aged acrylic coatings from wall paintings [24], which makes them suitable as examples of standard mortar samples used in conservation trials.

The mortar samples were produced using 1 part slaked lime (Horsens), 2.5 parts silica sand (grain size 0–0.3 mm) and 0.5 part fine gravel (0–2.2 mm) (Grejsdalens Filterværk ApS). The same mortar application procedure as described in the hot-mixed mortar was used in the production of these samples: an approximately 0.7 cm layer of lime mortar was applied to unglazed terracotta tiles, using wooden frames to secure same overall height.

2.2 Limewash

Three different limewashes were used to limewash replicated specimens.

Roskilde Limewash. For the replication of medieval limewash, a limewash was made from a more than 67-years-old lime putty collected at a lime pit at Roskilde Cathedral. Before mixing with water, the putty was sieved through a 5 mm mesh to remove lime lumps, larger sand particles and calcite crystals. The limewash was mixed with 1 part lime putty to 1 part water. This limewash was applied to five samples of the 1:1 mortar and five samples of the 1:2 mortar, with three layers applied on each sample.

Rødvig Limewash. A limewash produced by a lime putty from Rødvig matured since 1995, was chosen as a standard conservation limewash. The lime putty is produced from lumps of Faxe quicklime that have been slaked with water and stored in underground pits since 1995, and is still commercially available [25]. The limewash was mixed using 1 part lime putty to 1 part water. This limewash was applied to five samples of the 1:2 mortar, with four layers applied on each sample.

The Roskilde limewash and the Rødvig limewash were applied four months prior to the contact sponge measurements. The extent of carbonation was examined by phenolphthalein stain. After four months no staining occurred from the phenolphthalein spray indicating full carbonation.

Horsens Limewash. On the 1:3 samples produced in 2014, a limewash made from Horsens lime putty (matured since 1995) had been applied. Powdered limestone had been added to the limewash to achieve a better coverage. The limewash comprised 1 part lime putty, 1 part powdered limestone and 1 part water. This lime putty is no longer available, thus, this limewash could not be tested on the more lime-rich mortar samples. Five 1:3 mortar samples were used, each containing two layers of limewash.

3   Method

3.1   Contact Sponge

The contact sponge measurements were performed following the European Standard EN 17655 (status: in draft for comments) and the instructions in the CTS Contact sponge kit [23, 26]. The CTS kit used for the measurements contains a sponge (Calypso natural make-up sponge from Spontex®) and a polycarbonate contact-plate, wherein the sponge is placed. The sponge has a surface area of 23.76 cm², which correspond to the inner diameter of the base of the contact-plate. The thickness of the sponge exceeds the height of the contact-plate, thus, when pressing the sponge against an object, the sponge is in full contact with the object. For each measurement, the sponge was immersed in distilled water until full dilatation. The sponge was then wringed and patted dry with a blotting paper. After placing the sponge onto the contact-plate base, 6 ml of distilled water was added to the sponge with a syringe, and it was covered with the contact-plate lid to limit loss of water due to evaporation. The enclosed sponge was weighed on a balance with a precision of 0,0001 g and the initial weight (m_i) was noted. The plate was then placed on a table and the contact plate lid removed. The surface of the sample was pressed down onto the moist sponge until the sample surface touched the borders of the contact-plate base, as it is illustrated in Fig. 4. Placing the contact sponge beneath the sample instead of pressing it unto the sample was done to ensure that the water was only absorbed by capillary suction. After a predefined contact time, the mortar sample was lifted, and the lid was placed back unto the contact plate. The enclosed sponge was then weighed and the final weight (m_f) for this contact time was noted. To continue at longer contact times, the sponge was replaced on the table and the mortar sample was pressed down in the same position as before, and a new measurement was conducted. Measurements were conducted for the following contact times: 30, 60, 90, 120, 150, 300, 450 and 600 s.

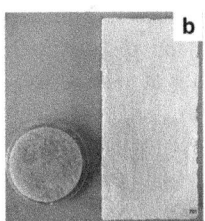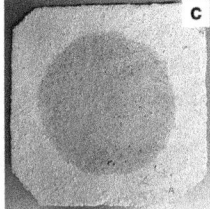

Fig. 4.   Illustration of contact sponge test procedure: (a) Pressing a sample onto the contact sponge; (b-c) 1:1 Roskilde sample and 1:3 Horsens sample after contact sponge measurements.

The measurements were performed in a laboratory environment of 20–22 °C and 65 ± 5% RH. The samples were not dried before testing but were instead stored at a relative humidity of 65 ± 5% RH to mimic the in-situ conditions in Danish churches in order to obtain comparative measurements between mortar samples and medieval plaster.

At Ørslev Church, the measurements were conducted by pressing the sponge onto blank limewashed areas in the painting, taking care that the perimeter of the contact plate

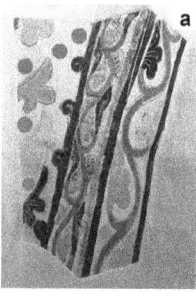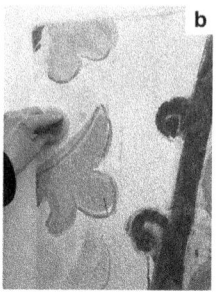

Fig. 5. Contact sponge measurements in situ at Ørslev Church: (a) Uncovered wall painting. Magenta circles indicate tested areas; (b) Contact sponge pressed onto vault.

was in full contact with the test area. The test areas and contact sponge measurements in-situ are illustrated in Fig. 5.

The amount of water absorbed at different time intervals could be calculated by subtracting the mass at the specific time from the initial water content. The water absorption rate (Wa) at the different contact times was calculated by following Eq. (1):

$$\mathrm{W}a\,(g/m^2s) = \frac{(m_i - m_f)}{A \cdot t} \tag{1}$$

where A is the surface area of the contact sponge (0.002376 m^2), t equals the contact time, m_i is the initial mass of the enclosed contact sponge and m_f is the final mass of the enclosed sponge.

4   Results and Discussion

4.1   Overview of Results

Each of the five samples of the 1:1 and 1:2 mortars was measured 5 times at the contact times t = 30 s, 60 s, 90 s, 120 s and 150 s. This resulted in a total of 25 measurements for the individual types of mortar at each of the shorter contact times. For the longer contact times (t = 300 s, 450 s and 600 s), each mortar sample was measured 3–5 times resulting in an average of 18 measurements for each mortar at a specific contact time. In the case of the five 1:3 samples, 3–4 measurements were conducted for the first 30–150 s, while 2–3 measurements were conducted at 300–600 s. At Ørslev Church contact times >150 s were excluded to avoid any unnecessary impact on the original wall painting. Due to the limitations of areas with blank limewash in the wall painting fragment, only six measurements were conducted in situ.

The water absorption rate (Wa) at the different contact times is presented in Table 2. In this table, the mean values and coefficient of variation for the measurements is noted, excluding one outlier in the 1:3 mortar measurements. Table 2 shows the speed at which water is absorbed into the mortar samples. Thus, for example, the 1:2 Roskilde mortar at 60 s contact time shows an average of 14.1 g of water absorbed into 1 m^2 per second, while the speed is 3.7 g/m^{2}s for the 1:3 Horsens samples.

In Fig. 6, the time course is shown of water absorption (in g/m^2) for the various mortars studied.

Table 2. Mean values and coefficients of variation (in parentheses, %) of the water absorption rate (Wa) (g/m^{2}s).

Mortar Samples	Contact time (s) / Coefficients of variation (%)							
	30	60	90	120	150	300	450	600
1:1 Roskilde	29.6 (14.4)	17.7 (19.0)	13.3 (22.6)	10.7 (25.4)	9.0 (27.6)	5.0 (28.1)	3.4 (30.7)	2.5 (30.7)
1:2 Roskilde	22.9 (17.3)	14.1 (21.3)	10.5 (25.1)	8.5 (27.8)	7.1 (32.4)	3.5 (32.0)	2.6 (39.4)	1.9 (36.2)
1:2 Rødvig	8.4 (19.1)	5.9 (26.3)	4.7 (30.9)	4.0 (32.9)	3.6 (33.5)	2.1 (36.4)	1.3 (37.9)	1.0 (37.9)
1:3 Horsens	5.2 (26.8)	3.7 (21.5)	3.0 (20.1)	2.7 (27.3)	2.3 (16.9)	1.5 (18.3)	1.2 (19.1)	1.0 (20.0)
Medieval	21.5 (20.8)	14.8 (16.0)	11.9 (14.2)	10.1 (14.7)	9.0 (16.1)	-	-	-

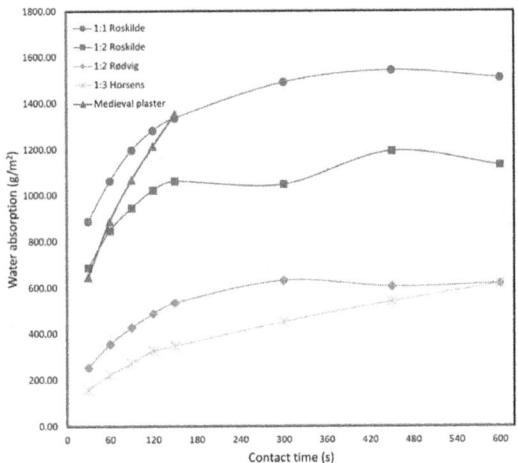

Fig. 6. Time course of water absorption for the mortar samples and the medieval mortar. The curves show mass of water absorbed per surface area at different contact times. Each point was obtained by multiplying the corresponding value from Table 2 by its time of measurement. Since this multiplication leaves the values of the variation coefficients (shown in Table 2) unchanged, these values were not represented in Fig. 6 for the sake of clarity.

4.2 Comparing the Mortar Samples and the Medieval Mortar

The time course of water absorption in four types of artificial mortars tested in the laboratory, as well as in one type of medieval mortar studied in situ (Ørslev Church, Fig. 5), is shown in Fig. 6. As prolonging the measurement up to 600 s was only practical in the laboratory, the longest in situ measurement was done at 150 s. For all four mortar types, the curves obtained in the laboratory display a levelling off with time. In the case of the 1:1 Roskilde and 1:2 Rødvig mortars, a plateau is reached at 300 s, the same being likely true for 1:2 Roskilde mortar, which at this particular time point appears to show a measurement error. In the case of the 1:3 Horsens mortar, while no plateau is reached within 600 s, the absorption rate slows appreciably past the 120 s point, where 58% of the final 600 s value is reached.

In Fig. 6, no levelling off of the water absorption time course was seen in the case of the in situ measurements. However, a tendency to slowing of the water absorption rate appears to be present over the time course up to 150 s. This would be consistent with a plateau occurring at longer times, had longer time points been available.

In the contact sponge method, it is important to ensure that enough water is available for the transfer from the sponge at all times [20]. However, it can be excluded that water amount was the limiting factor responsible for the curves levelling off in Fig. 6. Had this been the case, one would expect the plateau to occur at earlier times for those mortars showing the highest rates of absorption. In contrast, despite a 5.6-fold difference between the highest (1:1 Roskilde) and lowest (1:3 Horsens) absorption rates (Table 2, 30 s), it may be seen that in all cases, levelling off first becomes pronounced between 120 and 150 s. Therefore, the plateau phase of these time courses must reflect the physical properties of water movement into the mortar surfaces.

In order to further verify this argument, the amounts of water (Λm) transferred from sponge to mortar were calculated for contact times 30 s, 150 s and 600 s. These data can be found in Table 3, together with the corresponding percentages of water used. For this calculation, the 6 g of water added to each sponge at the start of the experiment was taken as 100%, although the total initial water content exceeded that amount (see Method section).

It is clear from Table 3 that enough water was available for transfer from the sponges at all times. For the mortar with highest absorbency (1:1 Roskilde), more than 40% of the initial water content was remaining at 600 s.

The plateau-attaining nature of the water absorption time courses, as seen in Fig. 6, implies that the clearest discrimination between absorption rates of various mortars will be obtained at short measurement times. Indeed, the data in Table 2 show this to be the case. For instance, absorption rates of 1:1 Roskilde and 1:3 Horsens differed by factors 5.6, 4.4 and 3.9, as measured at 30, 90 and 150 s, respectively. In conclusion, these data argue for using short measurement times when characterising water absorption properties of mortars.

Table 2 and Fig. 6 show the same relation between water uptake and porosity as seen in other studies [20, 21, 27]. The absorption rate increases as the porosity of the mortar increases. Thus, the highest values are obtained by the mortar samples with the highest lime content. The type of limewash also exerts a significant effect on the absorption rate. The mortar samples with Roskilde limewash have the highest absorption rate, while the samples with Rødvig and Horsens limewash have significant lower values. Since pure limewash has a higher lime content and thus higher porosity than mortar, the high absorption rate found in the samples with Roskilde limewash agree well with the fact that the Roskilde limewash is thicker than the other two limewashes. This is observed by the full opacity of the Roskilde limewash, while the greyish mortar can be detected through the limewash in case of the samples with Rødvig and Horsens limewash, confirmed also by measurements of the thickness of limewash. While the Roskilde limewash has an average total thickness of approximately 0.3 mm, the four layers of Rødvig limewash has resulted in a limewash with an average total thickness of approximately 0.15 mm. The Horsens limewash on the 1:3 samples is difficult to measure because it has penetrated into the mortar as illustrated in Fig. 3c. Hence, in some areas, it is almost non existing, while it has a thickness of up to 0.5 mm in areas where it has penetrated the mortar.

Table 3. Average values of absorbed water Δm (g) $\pm$ standard deviation at contact times 30 s, 150 s and 600 s and percentages of water absorbed from sponge to mortar.

Mortar Samples	Contact time (s)					
	30		150		600	
	Δm (g)	%	Δm (g)	%	Δm (g)	%
1:1 Roskilde	2.11 $\pm$ 0.30	35.2	3.17 $\pm$ 0.88	52.8	3.59 $\pm$ 1.10	59.0
1:2 Roskilde	1.64 $\pm$ 0.28	27.3	2.55 $\pm$ 0.82	42.5	2.69 $\pm$ 0.97	44.8
1:2 Rødvig	0.60 $\pm$ 0.11	10.0	1.27 $\pm$ 0.42	21.2	1.47 $\pm$ 0.56	24.5
1:3 Horsens	0.37 $\pm$ 0.10	6.2	0.82 $\pm$ 0.14	13.7	1.33 $\pm$ 0.52	22.2
Medieval	1.55 $\pm$ 0.35	25.8	3.03 $\pm$ 0.52	50.5	-	-

Nevertheless, it is still surprising to see the degree to which limewash influences water absorption rate. When comparing the two 1:2 mortars, the difference induced by the limewash is significant, with the Roskilde limewashed mortar absorbing water at double the speed as the mortars with Rødvig limewash. Moreover, the differences between the mortars with different limewashes are greater than the mortar composition, as the 1:1 Roskilde and 1:2 Roskilde have quite close values, while the 1:2 Rødvig samples, as mentioned above, have much lower values.

Most important, however, is how the samples compare to the measurements conducted on the medieval wall painting. Figure 6 illustrates how both lime-rich samples with Roskilde limewash have comparable values to the medieval plaster. The 1:2 samples with Rødvig limewash and the 1:3 standard mortar samples have, on the other hand, absorption rates that differ significantly from those of medieval plaster. As expected, the 1:3 samples have the lowest absorption rate values. Compared to the medieval plaster, these samples absorb water at a 4 times slower rate. This emphasises the huge differences between the capillarity and porosity of the standard lime putty mortar samples and the historical plaster or mortar, which they are supposed to imitate.

5 Conclusion

When conducting wall painting conservation research, the creation of mortar samples/replicas with the same characteristics as those found in historical wall painting plaster is essential.

In the present study, the physical resemblance of replicated mortar samples of varying binder-to-aggregate ratios to that of medieval wall painting plaster was examined by measuring the water absorption rate by the contact sponge method. Moreover, the study also aimed at examining the contact sponge method's suitability for measuring the water absorption rate in highly porous plaster/mortar and to establish the most suited measuring time for this method.

The study concludes that the contact sponge method is a valid, non-destructive and easy-to-use method to compare different mortar samples with each other and to evaluate the mortar samples' suitability as test specimens for wall painting conservation trials. The study shows that despite the high porosity of the medieval plaster and the hot-mixed mortar, the contact sponge contains enough water for transfer, even at long contact times. However, the plateau-attaining nature of the water absorption time courses shows that the clearest discrimination between absorption rates of various mortars is obtained at short measurement times, thus, short contact times should be used when characterising water absorption properties of mortars.

The contact sponge method has the advantage that it mainly measures the water absorption of the superficial layer of a sample, while a method as the frequently used Karsten tube measures the water absorption of the full thickness of a sample. The contact sponge method, thus, can be advantageous in a case like present study, where the aim is to measure the water absorption of a thin layer of plaster and limewash and not to measure the absorption of the full construction, including the brickwork beneath the plaster. This advantage is especially pertinent now that this study has shown that the thin layer of limewash has a significant influence on the water absorption behaviour of the test specimens. This influence is so great that it has a comparative impact as considerable as that of binder-to-aggregate ratio in the mortar samples.

The study clearly shows the significant influence of mortar and limewash composition on the water absorption rate in mortar samples. This conclusion is particularly emphasised when comparing the 1:3 samples, a standard mortar that has been used as a laboratory model for conservation trials, with the 1:1 mortar samples with Roskilde limewash, specifically designed to replicate the plaster found in Danish medieval wall paintings. Here it can be seen that in the first 30–150 s of contact time, the 1:1 plaster absorbs four to six times more water than the 1:3 plaster. Comparing the samples with the medieval plaster, a close match of the water absorption rate is found between the medieval plaster and the samples specifically designed to replicate this plaster (1:1; 1:2 Roskilde), while the water absorption rate values found in the 1:3 samples differ significantly from the medieval plaster. Thus, using 1:3 mortar samples for model-based conservation trials, such as cleaning or desalination tests, will most likely lead to results considerably divergent from those attained in situ on historical mortar or plaster.

Hence, this study contributes to the ongoing discussion of the importance of using conservation models with same composition and properties as the objects they are replicating instead of using standard models.

Acknowledgements. Thanks to Torben Seir (Seir-Materialeanalyse A/S) for sparking the author's interest in present research subject. The author kindly thanks Isabelle Brajer (National Museum of Denmark) and Jane Richter (The Royal Academy) for supervising and reviewing the article and Marek Treiman for discussions and review. Also, thanks to master mason Ole Frederik Jensen for producing the hot-mixed mortars and to Kathrine Segel and Anna Katrine Hansen (National Museum of Denmark) for helping with the production of laboratory mortar models. The author is grateful to Ørslev Parish Council for giving access to their paintings and to Roskilde Cathedral for allowing sampling of lime putty from their lime pit. The study was funded by the Augustinus Foundation, whose support is sincerely acknowledged.

References

1. Válek, J., Matas, T.: Experimental study of hot mixed mortars in comparison with lime putty and hydrate mortars. In: Válek, J., Hughes, J.J., Groot, C.J.W.P. (eds.) Historic Mortars, RILEM Bookseries, vol. 7, pp. 269–281. Springer, Dordrecht (2012)
2. Henry, A.: Hot-mixed mortars: the new lime revival. Context **154**, 30–33 (2018). Institute of Historic Building Conservation
3. Wiggins, D.: Historic Environment Scotland Technical Paper 27. Hot-Mixed Lime Mortars: Microstructure and Functional Performance, p. 30. Historic Environment Scotland, Edinburgh (2018)
4. Copsey, N.: Hot Mixed Lime and Traditional Mortars: A Practical Guide to Their Use in Conservation and Repair. The Crowood Press Ltd, Wiltshire (2019)
5. Caroselli, M., et al.: Insights into carolingian construction techniques – results from archae-ological and mineralogical studies at Müstair Monastery, Grisons, Switzerland. In: Álvarez, J.I., Fernández, J.M., Navarro, Í., Durán, A., Sirera, R. (eds.) 5th Historic Mortars Conference, pp. 743–757. RILEM Publications SARL, Paris (2019)
6. Midtgaard, M., Brajer, I., Taube, M.: Hot-mixed lime mortar: historical and analytical evidence of its use in medieval wall painting plaster. J. Archit. Conserv. **26**, 235–246 (2020)
7. Midtgaard, M., Seir, T., Brajer, I.: Replicating medieval wall painting plaster using the hot-mix technique. Stud. Conserv. **66**, 230–243 (2021)
8. Historic Environment Scotland. TAN 15-External Lime Coatings on Traditional Buildings, Technical Advice Note 15, p. 39. Technical Conservation, Research and Education Division, Edinburgh (2001)
9. Copsey, N.: Hot limewashes and sheltercoats. In: The Building Conservation Directory, pp. 136–139 (2017)
10. Brøgger, P., Græber, H., von Jessen, C., Larsen, M.: Kirkens mørtel og kalk. Nordisk arbejdsgruppe for bevaring af kirker, Kirkeministeriet, Copenhagen (1990)
11. Eriksson, J., Lindqvist, J.E.: Lime render, shrinkage cracks and craftsmanship in building restoration. J. Cult. Herit. **37**, 73–81 (2019)
12. Margalha, G., Veiga, R., Silva, A.S., de Brito, J.: Traditional methods of mortar preparation: the hot lime mix method. Cement Concr. Compos. **33**, 796–804 (2011)
13. Oliveira, M., Guimarães, E., Meneghini, A., Carvalho, H.: Simulation of humidity fields in aerial lime mortar. J. Cult. Herit. **53**, 35–46 (2022)
14. Pesce, C., Godina, M.C., Henry, A., Pesce, G.: Towards a better understanding of hot-mixed mortars for the conservation of historic buildings: the role of water temperature and steam during lime slaking. Heritage Sci. **9**(72), 1–18 (2021)
15. Copsey, N.: Historic environment Scotland technical paper 30: historic literature review of traditional lime mortars, p. 291. Historic Environment Scotland, Edinburgh (2019)
16. Vandevoorde, D., Pamplona, M., Schalm, O., Vanhellemont, Y., Cnudde, V., Verhaeven, E.: Contact sponge method: performance of a promising tool for measuring the initial water absorption. J. Cult. Herit. **10**, 41–47 (2009)
17. Gherardi, F., Otero, J., Blakeley, R., Colston, B.: Application of nanolimes for the consolida-tion of limestone from the medieval Bishop's palace, Lincoln, UK. Stud. Conserv. **65**(sup.1), 90–97 (2020)

18. Ribeiro, T., Oliveira, D.V., Bracci, S.: The use of contact sponge method to measure water absorption in earthen heritage treated with water repellents. Int. J. Archit. Heritage **16**, 85–96 (2022)

19. Normand, L., Duchêne, S., Vergès-Belmin, V., Dandrel, C., Giovannacci, D., Nowik, W.: Comparative in situ study of nanolime, ethyl silicate and acrylic resin for consolidation of wall paintings with high water and salt contents at the chapter hall of Chartres cathedral. Int. J. Archit. Heritage **14**, 1120–1133 (2020)

20. Vandevoorde, D., et al.: Validation of in situ applicable measuring techniques for analysis of the water adsorption by stone. Procedia Chem. **8**, 317–327 (2013)

21. Nogueira, R., Ferreira Pinto, A.P., Gomes, A.: Assessing water absorption of mortars in renders by the contact sponge method. In: Lourenco, P.B., Haseltine, B., Vasconcelos, G. (eds.), Proceedings of the 9th International Masonry Conference, Guimarães, Portugal (2014)

22. Arandigoyen, M., Bernal, J.L.P., López, M.A.B., Alvarez, J.I.: Lime-pastes with different kneading water: pore structure and capillary porosity. Appl. Surf. Sci. **252**, 1449–1459 (2005)

23. Standards, E. 21/30431529 DC BS EN 17655. Conservation of cultural heritage. Determination of water absorption by contact sponge method. https://www.en-standard.eu/21-30431529-dc-bs-en-17655-conservation-of-cultural-heritage-determination-of-water-absorption-by-contact-sponge-method/. Accessed 17 Mar 2022

24. Brajer, I., et al.: The removal of aged acrylic coatings from wall paintings using microemulsions. In: ICOM-CC Preprints, 17th Triennial Conference, Melbourne (2014)

25. Rødvig Kulekalk 1995   KALK.dk. https://kalk.dk/product/roedvig-kulekalk/2-95. Accessed 27 Feb 2022

26. CONTACT-SPONGE-KIT, https://webcache.googleusercontent.com/search?q=cache:b8C sHjvF_A0J:https://www.ctseurope.com/img/cms/documentazione/11-4/INGLESE/Con tact%2520Sponge%2520-%2520Kit_tds.pdf+&cd=14&hl=da&ct=clnk&gl=dk. Accessed 8 Mar 2022

27. Anna, A., Giuseppe, C.: The water transfer properties and drying shrinkage of aerial lime-based mortars: an assessment of their quality as repair rendering materials. Environ. Earth. Sci. **71**, 1699–1710 (2014)

The *Sgraffito* in Križanke – Interdisciplinary Approach to the Conservation-Restoration of Coloured Historic Plaster

Maja Gutman Levstik[✉] and Anka Batič

Institute for the Protection of Cultural Heritage of Slovenia, Restoration Centre, Poljanska cesta 40, 1000 Ljubljana, Slovenia
maja.gutman@rescen.si

Abstract. Križanke Outdoor Theatre is a theatre in Ljubljana, Slovenia, used for summer festivals. Between 1952 and 1956, architect Jože Plečnik converted this former monastery into one of the city's most important cultural venues. To the left of the entrance to the courtyard is a small triumphal arch decorated with *sgraffito*. The conservation and restoration of the triumphal arch was carried out in the summer of 2021. During the investigation of the historical documentation and the first steps of surface cleaning, it was found that there had been an intervention in the past. Microscopic analysis revealed that a lime-cement mortar with aggregates consisting mainly of round carbonate grains had been used for the triumphal arch. The original and the reconstructed *sgraffito* plaster consisted of pigmented lime binder. The black pigment in the grey plaster was identified as carbon black. Pigments such as iron oxide (hematite) and iron hydroxide (goethite) with some carbon black are present in the red coloured plaster of the original *sgraffito*. Only hematite was used for the reconstructed red coloured *sgraffito* plaster. The mineralogical-petrographic composition of the aggregates in the two (original and reconstructed) *sgraffito* plasters is similar and consists mainly of carbonate grains. On the surface of the original *sgraffito* plaster there was a gypsum layer as a weathering product. The goal of this research was to study and characterise the plasters used in *sgraffito* decoration. The information was crucial for processes such as the development of repair plasters for the damaged parts.

Keywords: sgraffito · plaster · Jože Plečnik · Križanke · pigments

1 Introduction

The Križanke Outdoor Theatre is a theatre in Ljubljana, Slovenia, used for summer festivals, built in the courtyard of the former Monastery of the Holy Cross. Between 1952 and 1956, architect Jože Plečnik converted this former monastery into one of the city's main cultural venues. To the left of the entrance to the courtyard is a small triumphal arch flanked by two niches with statues by the Austrian sculptor Leopold Kastner, called Learning and Progress (Fig. 1). The arch and the columns supporting it are decorated with *sgraffito* representing the current political symbols, the sickle and the hammer [1].

Sgraffito is an old decoration technique, involves of scratching a pattern into plaster before it sets, using a fine point metal tool, in order to reveal the colour and texture of the underlying plaster (Fig. 2). Once the top layer, usually white with a fine texture, is scraped off, the plaster layer underneath is revealed, usually a different colour and texture. The bottom layer consists of plaster coloured in contrasting colours to the ones on the surface. This is done by adding charcoal (for grey and black colours) or brick dust to make it reddish, or by using different types of sand to achieve a specific coloration [2]. The sgraffited motifs generally express nature, geometry or personal emblems [3]. *Sgraffito* decorations on building walls were popular in many European countries, especially during the Renaissance [4–6].

Although the triumphal arch is partially protected by a concrete roof ceiling, it was still exposed to decay and damage. The plaster was poorly preserved in some parts of the triumphal arch. The damage manifested itself mainly in the form of spalling, layering, falling off of the plaster and in some places in the form of dark coatings and growth of algae and other microorganisms on the plaster.

(a)   (b)

Fig. 1. Triumphal arch with marked sampling locations (a) from the front; (b) from behind.

The Križanke complex was proclaimed a cultural monument of local importance in 1986 and a cultural monument of national importance in 2016. In 2019, a detailed program for the conservation-restoration of the triumphal arch in Križanke was prepared, including all decorative techniques with *sgraffito*, decorative plaster, terrazzo and concrete elements, as well as two statues. This paper deals only with conservation-restoration of *sgraffito* decorative plasters.

The project began with a review of literature, archival documents, and photographic documentation of the *sgraffito* to better understand its history and technique. The state of conservation was assessed and a proposal for intervention was developed. For the proposed methodology, the conservators had to study the material structures of all the

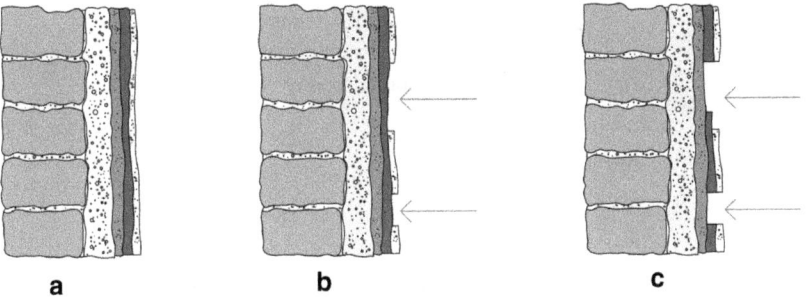

Fig. 2. Making of *sgraffito*: (a) applying the mortars; (b) scratching of the first layer; (c) scratching of the second layer.

plasters used. The methodology of this project will later be used for other facades in Križanke with the same decoration techniques.

In Slovenia, there are quite a few examples of *sgraffito* techniques from the modern period on exterior or interior buildings [7]. Rare examples of *sgraffito* from the Renaissance have been preserved [7]. Until now, there has been no published study of the technique and materials used for *sgraffito* in Slovenia. The study of the *sgraffito* in Križanke provides valuable information about the material used for the *sgraffito* from the 20th century and also for the implementation of appropriate conservation-restoration interventions.

2   Investigation of Historic Samples

2.1   Materials

A total of thirteen samples were collected from the triumphal arch (Table 1). The sampling location was initially chosen at the back of the triumphal arch (samples SLA 1–3) to minimise visual damage caused by sampling (Fig. 1b). Later, during conservation and restoration work, additional samples were then taken from the front of the triumphal arch (samples SLA 4–7) (Fig. 1a). The samples were taken with a chisel and a hammer.

2.2   Methods

In order to characterise the texture of the plasters and determine the content of certain constituents of the aggregate, the grain sizes, and type of the binder, we first examined the thin sections of the plasters using optical microscopy. The polished thin sections of the plaster and the polished cross sections of the coloured plaster samples were examined by optical microscopy using an Olympus BX-60 equipped with a digital camera SC50 (Olympus).

Raman microspectroscopy has been used to identify pigments in coloured plasters. Raman spectra were obtained from the polished cross-sections of the plaster samples using the confocal Raman spectrometer, Sentera II (Bruker). Measurements were performed with a 785 nm laser excitation line, and the Olympus 100 × objective. The spectral resolution was about 4 cm^{-1}.

Table 1. Table of analysed samples.

Sample		Originality	Colour of mortar/plaster	Location
SLA 1	a	reconstructed	white	upper transverse part of the triumphal arch, left side
	b	reconstructed	light grey	
	c	original	dark grey	
	d	original	red	
SLA 2	a	original	dirty white	lower shorter transverse element, left side
	b	original	dark grey	
	c	original	red	
	d	original	white	
SLA 3		reconstructed	bright red	upper transverse part of the triumphal arch, left corner
SLA 4		original	dirty white	ornament on the upper part of the right column
SLA 5		reconstructed	white	ornament on the upper transverse part, right side
SLA 6		reconstructed	white	profile above the arched part, under the roofing
SLA 7		original	dirty white	lower decorative part, between the ornaments

3 Conservation-Restoration

At the beginning of the conservation-restoration project, it was discovered the *sgraffito* had already been restored in the past. After reviewing archival photographs and documentation, it was concluded that the intervention took place in the late 1980s, when a large portion of the original *sgraffito* was removed and reconstructed (Fig. 3) [8]. This led the conservators to adjust the methodology. Moreover, additional samples had to be taken because the execution technique differed from the original one.

Fig. 3. The original *sgraffito* and decorative plaster of the arch (green areas), reconstructed parts in 1980s (blue areas).

The condition of the *sgraffito* was poor. The worst situation was at the top of the arch on the crossbeam, where several parts of the plaster were missing due to water leakage. On the surface of the *sgraffito* were dark stains, biological growth, detachment of various layers of plaster, cracks - all due to inadequate protection of the concrete slabs on the top of the arch. The arch is located under a massive old linden tree, and at times there are a lot of leaves, branches and other residues on the tiles, which lead to excessive moisture and dirt.

The conservation-restoration work included cleaning the surface of dirt and biological growth, local and complete consolidation of the plaster, injection of grout in the empty spaces between each separated plasters, infilling and reconstruction of the missing *sgraffito* areas (Fig. 4). The reconstruction was carried out using the same technique as the original: three different coloured plasters were applied and then incised in the same motif. The problem was to adapt the same surface appearance as a *sgraffito* on the arch, as it had been exposed to environmental influences for several years. When applying the plaster, the restorers sprayed the surface with water to get the roughened appearance.

Fig. 4. Reconstruction of *sgraffito*: (a) before conservation-restoration process; (b, c) application of the mortar; (d) applying the motif; (e) scratching of the mortar; (f) final result of the reconstruction.

4   Results and Discussion

4.1   Historic Plasters and Mortars

Original Plasters. The aggregate of the base mortar is poorly sorted. The size of the aggregate ranges from 0.12 mm to 30 mm, corresponding to the fraction of sand to gravel. Consists mainly of rounded lithic carbonate grains, followed by lithic grains of quartz sandstones, slate siltstones, cherts, claystones, and mudstones. Some angular or sub-angular quartz grains are present. The binder is lime-cement, compact, with visible relicts of cement clinker (Fig. 5a) and lime lumps. Round pores are present.

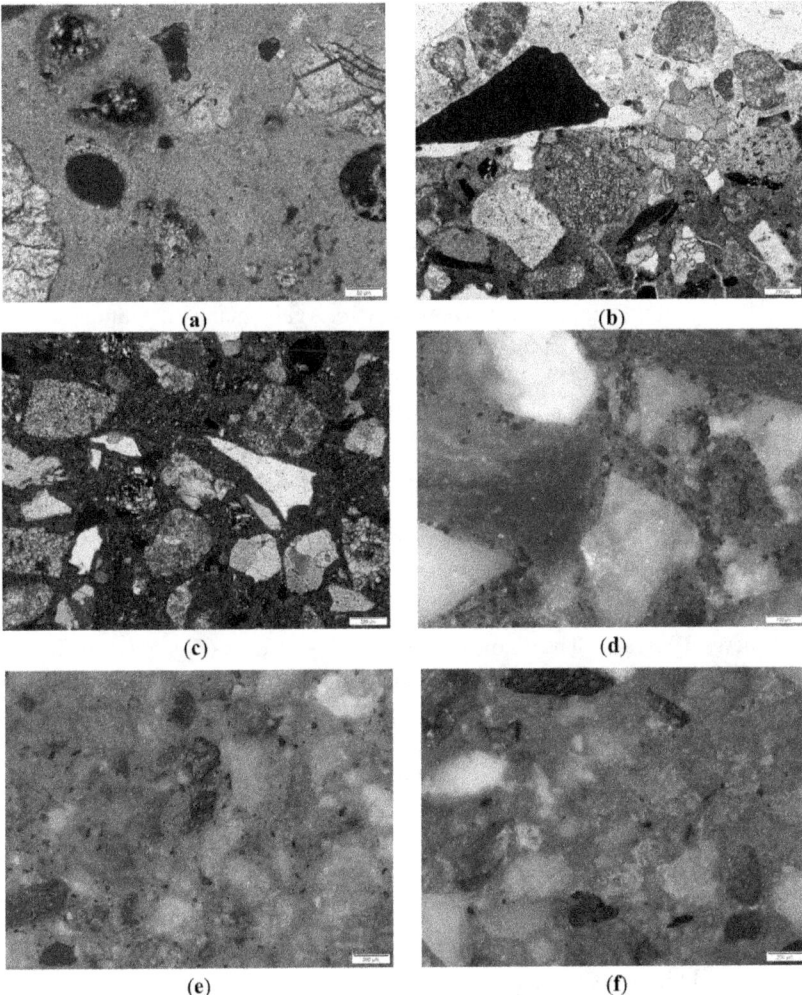

Fig. 5. (a) Relict of cement clinker in base mortar marked by red arrow, sample SLA 2d. Optical microscope, transmitted light, crossed polars; (b) Gypsum on the top of the finish plaster, sample SLA 4. Optical microscope, transmitted light, parallel polars; (c) Predominant carbonate aggregate in reconstructed plaster, sample SLA 5. Optical microscope, transmitted light, crossed polars; (d) Dark grey coloured plaster, sample SLA 1c. Optical microscope, reflected light, parallel polars. (e) Red coloured plaster, sample SLA 1d. Optical microscope, reflected light, parallel polars. (f) Bright red coloured reconstructed plaster, sample SLA 3. Optical microscope, reflected light, parallel polars.

Two coloured layers of plaster and a final white layer were applied on cement-lime mortar. Both coloured layers and the final white layer have similar aggregate composition, but differ in the colour of the binder. The aggregate in the red and grey coloured plaster is medium sized sand with individual larger grains. The grains are mostly rounded and subrounded, some are sharp-edged and measure between 0.13 and 1.2 mm. The aggregate consists of well-rounded lithic grains of limestones, quartz sandstones, slate siltstones, cherts, claystones, mudstones, and siltstones. Some angular or subangular quartz grains are present. The shape of the grains indicates a fluvial origin, most probably the alluvial deposits of the Sava River, composed of light and dark grey limestone and dolomite, sandstone grains, quartz, schists, shales and magmatic pebbles [9]. The binder in red coloured plaster is lime with added pigments such as iron oxide (hematite) and iron hydroxide (goethite) and some carbon black (Fig. 6). Many lime lumps are visible. In dark grey plaster, the lime binder was mixed with black, carbon-based pigment (Fig. 5d). Rare goethite grains are also present here. Again, many lime lumps are visible.

The white finishing layer also contain medium sized aggregates with individual larger grains. The grains are mostly rounded and subrounded, some are sharp-edged and slightly larger than those in coloured plasters. The size of the grains ranges from 0.25 to 1.9 mm. The mineral composition of the aggregates is similar to that of coloured plasters. The binder is lime. As can be seen in Fig. 5b, gypsum has replaced the calcite binder on the top of the finished plaster. Gypsum is known to be the major weathering product of carbonate-based materials, formed by a chemical reaction between calcite and atmospheric sulphur dioxide. Its presence is probably related to the weathering of the outer layer of the *sgraffito*, which was exposed to environmental pollution.

Reconstructed Plasters. The reconstructed red plaster layer (Fig. 5f) is visually more vivid than the original red (Fig. 5e). While the binder of the original red layer is lime with pigments such as iron oxide (hematite), iron hydroxide (goethite) and some carbon black (Fig. 6), only lime with iron oxide (hematite) was used for the reconstructed red plaster. The size of the grains (ranges from 0.12 to 1.2 mm) and also the mineralogical-petrographic composition of the aggregates is similar to that of the original plaster and consists of lithic grains of limestones, quartz sandstones, slate siltstones, claystones, cherts, mudstones and siltstones.

The reconstructed grey plaster layer is very similar to the red plaster in mineralogical-petrographic composition and size of aggregates. The binder is lime with the addition of carbon-based pigments.

The reconstructed white surface layer also contains medium sized aggregates with individual larger grains. The grains are mostly rounded and subrounded, and some are sharp-edged (Fig. 5c). The size of the grains ranges from 0.10 to 1.5 mm and are smaller compared to the grains of the original plaster. The mineralogical-petrographic composition of aggregates is similar to that of the original finish plaster and consists of lithic grains of limestones, quartz sandstones, slate siltstones, cherts, claystones, mudstones, and siltstones. The binder is lime, with visible lime lumps. Some black carbon-based particles were also present. They were probably added to the plaster to achieve a similar colour tone to the original plaster.

Considering the similar mineralogical-petrographic composition of the aggregate in the original and the reconstructed plaster, we assume that the aggregate for both comes from the Sava River. It should be noted that the plaster for the reconstructed parts of the triumphal arch was prepared very carefully during the last restoration-conservation works and is very similar to the original plaster.

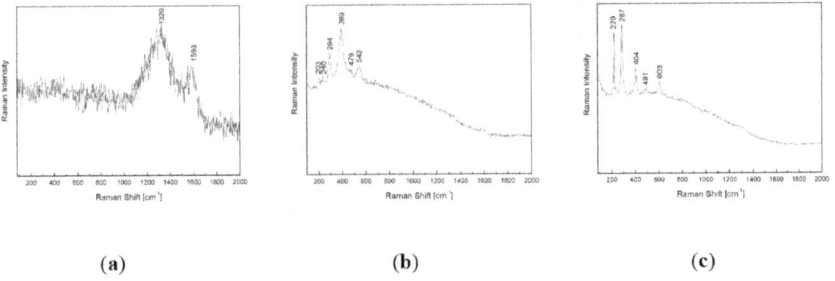

(a)   (b)   (c)

Fig. 6. Raman spectrum of: (a) carbon black; (b) goethite; c) hematite.

4.2   Plasters and Mortars for Reconstruction

Additions to the Reconstructed Plaster in the 1980s. The investigation confirmed that carbon-based particles were present in the newer plaster (Fig. 7a), which mimicked the original *sgraffito* with its black stones (black claystones, slate siltsonestones) from the riverbank. This affected the process of cleaning and consolidating the plaster, as the restorers planned to use ammonium carbonate for desulfination process [10] and later ammonium oxalate for consolidation [11]. Ammonia reacts with graphite and forms some kind a chemical reaction. It was not possible to use it for restoration because the graphite obtained brown, rusty appearance after treatment (Fig. 7b).

(a)   (b)

Fig. 7. (a) Newer plaster of *sgraffito* with graphite; (b) newer plaster of *sgraffito* with graphite after treatment with ammonium carbonate.

Plaster for Reconstruction of *Sraffito*. Prior to on-site work, restorers made 60 samples of different coloured plasters to match visual and structural characteristics with original and recent portions of the *sgraffito*. Depending on availability, the materials found during microscopic examination were matched. The pigments used were mostly the same as in the original. The sand was obtained from the river bank of the Sava (Gramoznica Stari grad, Krško) and carbonaceous sand from Calcit d.o.o. or from quarry in Verd, Vrhnika was used. As a binder, lime putty was an obvious choice, since it was the binder used in the original. The restorers discussed using a smaller amount of cement as well. It was found in the lower layer of the *sgraffito* and usually in the mortars of Jože Plečnik's works [12]. It was assumed that it would give some stability to the plaster in a very unstable environment.

Four experimental mortar samples were prepared for the reconstruction, differing in aggregates and use of cement (Table 2). Various tests were carried out: determination of the flexural and compressive strength of the hardened mortar after 28 and 90 days (Fig. 8), determination of the consistency of the fresh mortar using a flow table (Fig. 9), determination of the capillary porosity and the water absorption coefficient after 24 h [13]. Due to the better results, the lime-cement plaster with aggregates from Stari Grad and Verd was discussed as the best plaster composition for the arch (V2).

The restorers then prepared detailed instructions for the preparation of all layers, coloured plasters and decorative plasters on all surfaces.

Table 2. Table of composition of lime-cement plasters.

Sample	Lime (g)	Cement (g)	Aggregate (g) 1	Aggregate (g) 2	Aggregate (g) 3	Water (g)	Ratio binder/water
V1	420	86	957	236		417	0,82
V2	420	86	957	236		373	0,74
V3	420	86	957		236	373	0,74
V4	420				236	373	0,89

[1]Aggregate from Stari Grad, Krško. [2]Aggregate from Verd, Vrhnika. [3]Aggregate from Calcit d.o.o

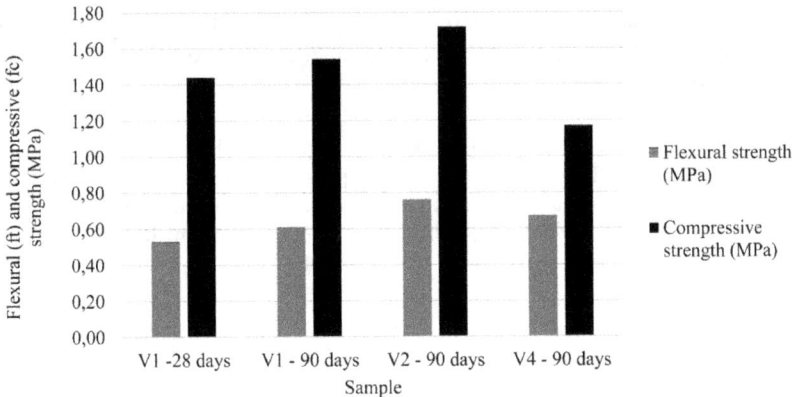

Fig. 8. A table showing the flexural (red) and compressive strength (black) of the hardened mortar [13].

Fig. 9. (a) Determination of the consistency of the fresh mortar using a flow table; (b) determination of flexural strength of the hardened mortar [13].

5 Conclusion

The process of conservation-restoration of the *sgraffito* was carried out with great care, taking into account the results of various studies. The restorers used almost the same materials as the original and respected the ethical requirements for the restoration of an object. At the same time, they were aware of the problems of changing atmospheric conditions, the location of the arch under a tree and its exposure to excessive humidity, and the inadequate protection of the cement tile roof.

The first intervention on the *sgraffito* was made 30 years after its creation and it was probably in a very poor condition to reconstruct almost all areas. In determining the methodology for the restoration in 2021, the restorers consulted conservators and other professionals because of the obvious sensitivity of the materials to the environment. Consolidation was paramount because of the protection of the plaster, as there is no layer of paint. The use of cement was also discussed as the best option for the stability of the mortar.

At the end of the conservation-restoration project, annual monitoring of the arch was proposed and also cleaning of dirt, leaves and other residues on the concrete slabs at least twice a year.

Acknowledgements. Special thanks to Dr. Andreja Padovnik from Faculty of Civil and Geodetic Engineering, University of Ljubljana for testing the experimental mortar samples and preparing a report on them, as well as to leading conservator Irena Vesel head of the IPCHS Kranj Regional Office and conservator-restorer mag. Tjaša Pristov from IPCHS Ljubljana Regional Office.

References

1. Hrausky, A.: Simboli v Plečnikovi arhitekturi, Lili in Roza, Ljubljana, p. 59 (2016)
2. Salema, S., Aguiar, J.: Sgraffito in Portugal: a contribution to its study and preservation. In: 2nd Historic Mortars Conference HMC2010 and RILEM TC 203-RHM Final Workshop, 22–24 September 2010, Prague, Czech Republic (2010)
3. Faria, P., Tavares, M., Menezes, M., Veiga, R., Goreti, M.: Traditional Portuguese techniques for application and maintenance of historic renders. In: 2nd Historic Mortars Conference HMC2010 and RILEM TC 203-RHM Final Workshop, 22–24 September 2010, Prague, Czech Republic (2010)
4. Bartz, W., Rogóż, J., Rogal, R., Cupa, A., Szroeder, P.: Characterization of historical lime plasters by combined non-destructive and destructive tests: the case of the sgraffito in Bożnów (SW Poland). Constr. Build. Mater. **30**, 439–446 (2012)
5. Jagiełło, M.: Sgraffito as a method of wall decoration in the renaissance and mannerist silesia. Arts **11**(1), 25 (2022)
6. Wichterlová, Z., Waisserová, J., Skružná, O., Válek, J.: New shading technique revealed through reconstructing the sgraffito technology used North of the Alps during the renaissance. Sgraffito in Change. Technique, Topics, and Preservation, 2.-4. November 2017, Hildesheim, Deutschland (2017)
7. Šeme, B.: Preservation and conservation-restoration of modern and contemporary wall paintings and mosaics. In: Summaries of the International Meeting of Conservators-Restorers, National Gallery, pp. 54–56, Ljubljana (2019)
8. Archive documentation IPCHS: D. G. Predračun-ponudba, Ljubljana, 26 Sept 1986
9. Premru, U.: Basic geological map of SFRJ, sheet Ljubljana, 1: 100 000, Federal Geological Institute, Belgrade (1980)
10. Coladonato, M.: Cleaning wall paintings: methodological approach with low chemical risk theory and practice, Workshop, Zagreb (2018)
11. Conti, C., Colombo, C., Dellasega, D., Matteini, M., Realini, M., Zerbi, G.: Ammonium oxalate treatment: Evaluation by μ-Raman mapping of the penetration depth in different plasters. J. Cult. Herit. **12**(4), 372–379 (2011)
12. Potočnik, I., et al.: Plečnikova hiša, konservatorsko restavratorski posegi 2012–2015: konservacija in obnova. Hiša Plečnik house: ob 100-letnici nakupa hiše na Karunovi 4 in v letu celovite prenove Plečnikove hiše (1915–2015), Ljubljana: Muzej in galerije mesta Ljubljane, 57–71 (2015)
13. Padovnik, A.: Končno poročilo: Testiranje apneno-cementne malte za obnovo Plečnikovega slavoloka v Križankah, št.: 401–1/2022–5, Fakulteta za gradbeništvo in geodezijo, Ljubljana (2021)

Preservation. Consolidation Materials and Techniques. Development of New Products. Preventive Conservation

Experimental Study on Properties of Hydraulic Mortars with Mixed in Crystallisation Inhibitors

Ameya Kamat$^{(\boxtimes)}$ ⓘ, Barbara Lubelli ⓘ, and Erik Schlangen ⓘ

Delft University of Technology, Delft, The Netherlands
a.a.kamat@tudelft.nl

Abstract. Sodium chloride (NaCl) is one of the most commonly occurring weathering agents, responsible for a progressive damage in mortar. Current solutions to mitigate salt damage in mortar, such as the use of mixed-in water repellent additives, have often exhibited low compatibility with the existing building fabric. In the last years, research has shown promising results in mitigating salt decay by making use of crystallisation inhibitors. Sodium ferrocyanide is one of the inhibitors that has proven to be particularly effective to reduce damage due to sodium chloride crystallisation.

In this research the possibility of developing hydraulic mortars with mixed-in inhibitor (sodium ferrocyanide) for an improved resistance to sodium chloride crystallisation damage is investigated. As a first step, the interaction between the inhibitor and the hydraulic binder: natural hydraulic lime (NHL), was studied; the results are presented in this paper. Various concentrations of sodium ferrocyanide were tested (0%, 0.1% and 1% by binder weight). The effect of the inhibitor on several physical (hydration, water absorption, pore size distribution) and mechanical (compressive and flexural strength) properties was experimentally assessed, using several complementary methods and techniques. The results show that the addition of the sodium ferrocyanide does not affect the fresh and hardened properties of mortar. These results are promising and open new possibilities for the application of inhibitors to improve the durability of hydraulic mortars.

Keywords: Repair mortars · Natural hydraulic lime · crystallisation inhibitor · salt damage · sodium chloride · sodium ferrocyanide · conservation

1 Introduction

Salt crystallisation is a common cause of decay in historic buildings. Among damaging salts found in historic buildings, sodium chloride is one of the frequently occuring, due to its multiple sources [1]. Plasters and renders, as well as pointing mortars, particularly suffer from salt decay. Their high susceptibility is due to several reasons. First of all, they lie at the surface of construction, where most of the salts accumulate, and they are exposed to the environment (wind, sea salt spray, RH and T changes). Moreover, these materials have a bimodal pore size distribution with coarse and fine pores, which can lead to a development of high crystallisation pressure. Traditional air lime-based mortars have a limited mechanical strength, which can be easily overcome by the pressure developed

© The Author(s), under exclusive license to Springer Nature Switzerland AG 2023
V. Bokan Bosiljkov et al. (Eds.): HMC 2022, RILEM Bookseries 42, pp. 341–350, 2023.
https://doi.org/10.1007/978-3-031-31472-8_26

due to salt crystallisation. Salt damage in mortars manifests in the form of loss of surface cohesion (powdering, scaling, spalling etc.) or loss of adhesion between the mortar and the substrate or between the mortar and the paint layer on top.

Current solutions to improve the durability of mortars to salt crystallisation damage involve the use of cement as binder, to increase the mechanical strength, or the addition of water repellent additives in the mass, to reduce the ingress of salt solution in the mortar. However, these solutions have often a low compatibility with the existing, historic fabric, worsening the problem they were supposed to solve [2].

In order to further improve the durability of mortars to salt decay the use of crystallisation inhibitors has been considered in the past years, as a novel approach to mitigate salt crystallisation damage. Crystallisation inhibitors are chemicals that tweak the process of crystallisation by delaying the nucleation and/or by altering the crystal habit [3]. Such crystallisation inhibitors are often salt specific, i.e. they are effective only for a specific salts.

Alkali ferrocyanides (FeCN), which are common anti-caking agents of NaCl, have shown to be particularly effective in inhibiting the growth of sodium chloride (NaCl) crystals [4]. Application of FeCN solution in porous building materials has also shown to modify NaCl crystallisation in a way which can limit salt decay. A study with FeCN solution in NaCl contaminated bricks reported a delay in nucleation of the salt, enabling higher advection of salt ions to the evaporating surface [5]. Other researchers reported an increase in the amount of harmless efflorescence, confirming the above mentioned enhancing of salt transport to the surface [6–8]. FeCN modifies the crystal habit of the NaCl crystal, from cubic to dendritic; this habit change also contributes to enhanced salt transport via creeping (i.e. the self-amplifying transport mechanism by which salt solution is transported along growing crystallite tips) [9]. The application of FeCN in lime mortar was observed to lead to a higher nucleation density and smaller crystals, possibly leading to a lower pore clogging. Thus, leading to a positive effect on salt transport and lowering of crystallisation pressures [10]. In early studies, FeCN was introduced along with NaCl to building materials via capillary suction. However, it was shown that FeCN was effective only during the nucleation stage and had negligible effect on dissolution of already present salt crystals [6]. Thus, addition of FeCN before NaCl ingress can be a solution to prevent salt damage. This has been first tested in hydrated lime-based mortars by adding FeCN directly during the mixing of the mortar [11, 12]. This application of inhibitor offers an advantage that the FeCN ions are already distributed in the mortar and can react to salt ingress at an early stage, preventing or effectively reducing salt crystallisation damage. Granneman et al. report that the mortar with the FeCN inhibitor showed an increased resistance with respect to the reference mortar, after an accelerated crystallisation test, while not showing any change in other relevant physical and mechanical properties [12, 13]. On the other hand, Natural hydraulic lime-based mortars can offer a good compromise, having both sufficient durability and compatibility with the historic, valuable materials [14].

Mortars based on hydrated lime, although ideal from the point of compatibility with historic building fabric, have still a quite limited durability and require thus frequent maintenance, resulting in high costs. Owing to these issues, in the field of conservation of historic buildings, hydraulic binders such as natural hydraulic lime (NHL) are more often used, as they generally provide a better durability while keeping a good compatibility with the existing historic materials. This research investigates the possibility of mixing in FeCN inhibitor in NHL based mortars, and assess whether any interaction between the inhibitor and the hydraulic components of the binder occurs, possibly altering the propertied of the fresh and hardened mortar.

Unlike air lime which gains its strength via carbonation, hydraulic mortars also involve the process of hydration i.e. the exothermic reaction between binder and water. Thus, microstructure development and material behaviour of NHL and air lime are different and the possible effects of the inhibitor is still unknown. This needs to be carefully investigated as first step towards the development of mortars with mixed-in crystallisation inhibitors.

In this paper, the interaction between alkali ferrocyanides and natural hydraulic lime mortars is investigated, by the use of a several complementary techniques and methods.

2 Materials

2.1 Binder, Sand, Inhibitor

Natural hydraulic lime (NHL) with a strength class of 3.5 from Saint Astier was used as the binder. CEN standard sand, with a grain size distribution between 0.08–2 mm, was used as per EN 196–1 [15]. Analytical grade sodium ferrocyanide decahydrate $(Na_4Fe(CN)_6.10H_2O)$ was used as the crystallisation inhibitor. The inhibitor was purchased from Sigma Aldrich.

2.2 Specimen Type, Preparation and Storage

Two different types of specimens were prepared: binder paste specimens and mortar specimens.

Binder Paste Specimens Paste specimens were prepared by mixing NHL with distilled water. The water-binder (w/b) ratio was maintained at 1 by weight. Specimens with mixed-in inhibitor were prepared by first dissolving sodium ferrocyanide in distilled water. Specimens with different inhibitor concentrations were prepared: 0.01%, 0.1% and 1% (weight of inhibitor with respect to the weight of the binder). Additionally, specimens without inhibitor were prepared as reference.

Mortar Specimens

Mortar specimens were prepared according to EN 459–2 [16]. A binder to aggregate ratio (b/a) 1:3 by weight was used. FeCN was added in the amount of 0.1% and 1% with respect to the binder weight. FeCN was first dissolved in water and added to the specimens during the mixing process to obtain a homogenous distribution of the inhibitor in the mortar. The w/b ratio was adjusted to 0.6 as per EN 459–2. Reference specimens were prepared without inhibitor.

Mortar specimens were cast as prisms or as slabs. The prismatic beams were cast in polystyrene mould of 160x40x40 mm and compacted using mechanical vibrating table as per EN 196–1 [15]. Each mortar slab was cast on a red clay brick to obtain properties comparable to that of the mortar when applied in the field [17, 18]. A paper towel was placed on top of the brick before casting the slab, in order to facilitate demoulding. The slabs had a dimension of 200 × 100 × 20 mm and were hand compacted.

All specimens were covered in plastic for the first 24 h. They were demoulded or detached from the brick substrate after 5 days. The specimens were then stored in a curing chamber with a relative humidity greater than 95% and a temperature of 20 °C until tested, in order to minimise carbonation.

3 Test Methods

3.1 Characterisation of the Early Age Properties

Several measurements were carried out to assess the effect of the inhibitor on the properties of the fresh and hardened mortar. An overview of test methods and type of specimens is provided in Table 1.

Table 1. An overview of test methods and specimens

Test method	Measured property	Specimen type	Size/weight	Inhibitor concentration[a]	Replicates
Isothermal calorimetry	Heat of hydration	Binder paste	5.5 g	Ref, 0.01%, 0.1%, 1%	2
Vicat penetration test	Setting time	Binder paste	–	Ref, 1%	2
Mechanical testing	Compressive and flexural strength	Mortar prisms	160 × 40 × 40 mm	Ref, 0.1%, 1%	3
MIP/ N2	Pore size distribution	Mortar slabs	~ 1 cm^3	Ref, 0.1%, 1%	1
Capillary absorption and drying	Water absorption coefficient	Mortar slabs	50 × 20 × 20 mm	Ref, 0.1%, 1%	4

[a]percentage of binder weight

Measurement of Heat of Hydration. The development of microstructure and, in turn, the properties of hardened mortar depend on the hydration reaction (i.e. the exothermic reaction between water and the binder). Therefore, measuring the heat released during the hydration indirectly provides information on the early age hydration products and microstructure development.

Heat evolution and rate of hydration was measured on binder paste specimen by the use of an 8 channel thermometric isothermal calorimeter (TAM-Air). The test procedure is based on EN 196–11 [19]. Each sample consisted 5.5 g of binder mixed with equal amount of water (w/b = 1). The heat evolution was monitored continuously for 168 h and the chamber was maintained at a constant temperature of 20 ± 2 °C. Quartz specimens with a known specific heat capacity were used as a reference to eliminate the background noise.

Setting Time. The setting time, i.e. the time required for the binder to completely lose its plasticity and attain a certain resistance to pressure, was measured using an automated Vicat penetration test. The test was performed in accordance to EN 459–2 on paste specimens [16]. A comparison was made between specimens without inhibitor and specimens with 1% sodium ferrocyanide (as percentage of dry weight).

Workability. Workability is the ease at which mortar can be placed/ compacted during construction; this property is important for an easy application of the mortar in practice. A flow table test, according to EN-1015–3 [20], was performed on the fresh mortar to measure its workability. The freshly mixed mortar was placed on a standard flow table and jolted 15 times at a rate of 1 jolt/ second. The diameter of the resulting mortar was measured at right angles using a calliper.

3.2   Characterisation of the Properties of the Hardened Mortar

Measurement of Mechanical Properties. Mechanical properties provide insights into the load bearing capacities of mortars and thus give a (partial) indication of their expected durability. Compression and flexural strength was measured as per EN 1015–11 on mortar prisms of $160 \times 40 \times 40$ mm [21]. The test was performed at 28 days after casting. The loading rate was 0.1 kN/s and 5 N/s for compressive and flexural strength respectively.

Determination of Porosity and Pore Size Distribution. Open porosity and pore size distribution were measured using Mercury intrusion porosimetry (MIP). The test was performed on samples of about 1 cm^3 collected from the mortar slabs after 28 days of curing. The specimens were dried to a constant weight in a freeze drier, before performing the test. The specimens were subjected to a maximum intrusion pressure of 210 MPa; pore throats between 100 μm and 0.01 μm could be measured my MIP. The contact angle between the Mercury and the mortar samples was assumed to be 141°.

Additionally, N_2 adsorption measurements were performed to obtain information pertaining to pores smaller than 0.01 μm, that could not be assessed using MIP. The sample were prepared as for the MIP test but in this case smaller binder samples (2–4 mm particle size, separated by the use of a sieve) were used. The air in the testing chamber was first evacuated at a rate of 500 mm Hg/min before performing N_2 adsorption. No separate degassing procedure was performed.

Capillary Absorption and Drying Test. The effect of FeCN on moisture transport properties of the mortar was assessed by performing capillary water absorption followed by a drying test.

The water absorption by capillarity of mortar was measured using EN 1925 [22] as a guideline. The test was performed on $50 \times 50 \times 20$ mm mortar specimens, obtained from larger mortar slabs. After the 28 day curing period, the specimens were dried to a constant weight in an oven at 40 °C. The sides of the samples were sealed using a paraffin film to have uni-directional absorption and drying. The water absorption was measured using a weighing scale with a precision 0.01 g at prescribed time intervals as per EN 1925. The water absorption coefficient (WAC) was obtained and used for comparison across different inhibitor concentrations. Following capillary saturation, specimens were dried at 40 °C and a relative humidity of 15 ± 5%. The weight of the specimens was measured at different time intervals to obtain the drying curves.

4   Results and Discussions

4.1   Effect of the Inhibitor on Early Stage Properties

The results obtained from the Vicat penetration test, presented in Fig. 1, show that the penetration curve is unaffected by the addition of the inhibitor. The initial (~420 min) and final set (~1200 min) for both specimens with and without inhibitor are in a similar range.

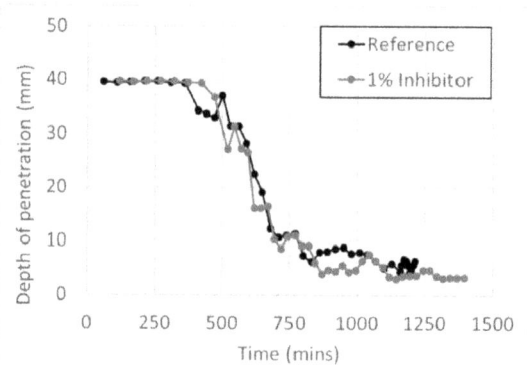

Fig. 1. Comparison of Vicat penetration test results on specimens with and without inhibitor.

The heat of evolution obtained from the isothermal calorimeter is presented in Fig. 2. The results are normalised to the weight of the specimens. It can be seen that there is a negligible difference between reference specimens and specimens with the inhibitor. The cumulative heat of hydration obtained after 168 h is 26 ± 2 J/g, irrespective of the inhibitor concentration. From these results, it seems that the hydration kinetics are not significantly affected by the addition of the inhibitor.

The workability of the fresh specimens was measured using the flow table test. The results, presented in Table 2, show that the flow decreases slightly with the addition of the inhibitor, but not necessarily in proportion to the amount of inhibitor. It should be noticed that there is a large scatter in the data. This is probably due to factors like mixing speed and mixing time, which are not easily controlled and may affect the flow. Considering the variation observed between replicates, the impact of the inhibitor on the flow is minor.

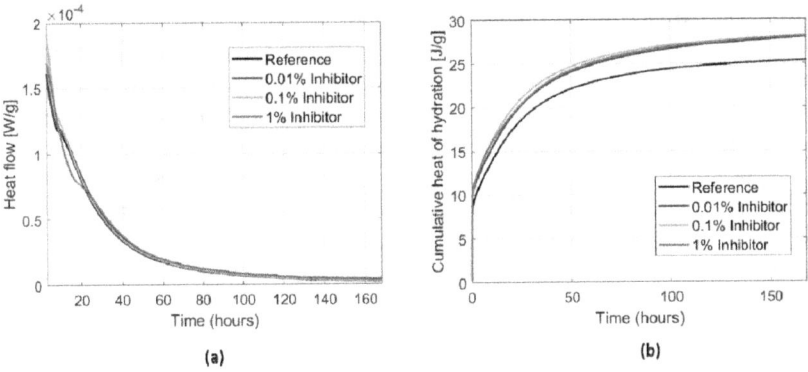

Fig. 2. (a) Heat flow rate (b) cumulative heat of hydration for different concentrations of inhibitor.

4.2 Effect of the Inhibitor on Properties of the Hardened Mortar

The mechanical properties of the mortar containing different amounts of inhibitor are reported in Table 2. The strength of specimens with inhibitor show only minor deviation from reference samples.

The pore size distribution and open porosity of the mortar was measured using MIP, for pores throats between 100 μm and 0.01 μm and N_2 for pore throats smaller than 0.01 μm. The results are presented in Fig. 3. It is possible to conclude that the addition of the sodium ferrocyanide did not significantly affect the pore size distribution, in the range of both larger and smaller pores by the addition of FeCN. The mean diameter as obtained from MIP varies between 0.1 and 0.2 μm. The open porosity of mortar with and without inhibitor as obtained from MIP is also very similar (Table 2).

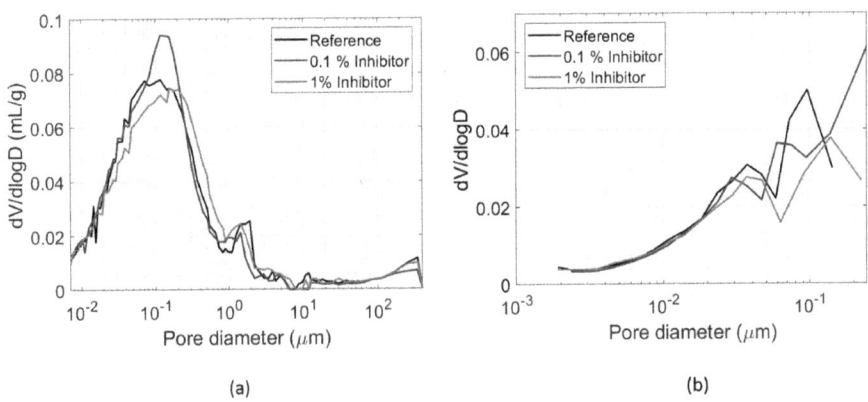

Fig. 3. (a) Pore size distribution as measured by MIP (b) Pore size distribution as measured by N_2 adsorption for pores smaller than 0.01 μm

The water absorption curves of mortar specimens with different concentrations of inhibitor were measured and the results normalized to the dry weight of the specimens.

This was done to take into account for the slight variation in the sample thickness of the specimen, due to hand compaction of the slabs. The intersection point of the two absorption stages, necessary for the determination of the water absorption coefficient (WAC), was obtained by performing a linear regression on the two stages. Figure 4a shows that the water absorption curves and the WAC (Table 2) of mortar specimen with and without inhibitor are similar. The drying curves of specimens with and without the inhibitor, (Fig. 4b) are similar. These results confirm that the moisture transport of the mortar is not significantly affected by the addition of the inhibitor in the studied concentrations.

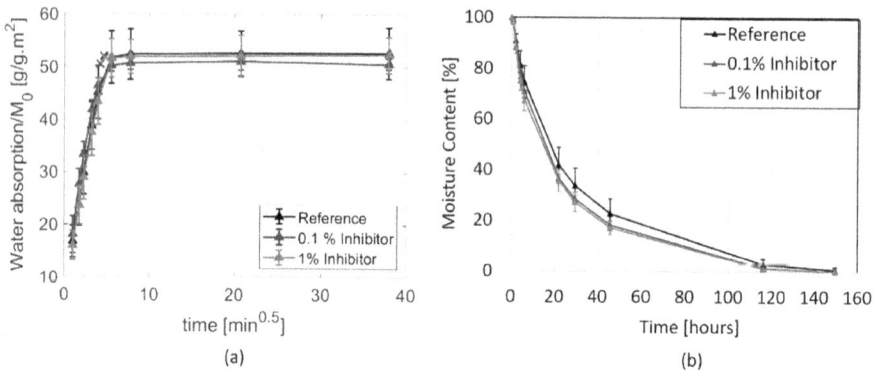

Fig. 4. (a) Capillary absorption **(b)** Drying behaviour of specimens with different inhibitor concentration.

Table 2. Summary of mortar properties for different inhibitor concentrations

Property	Reference	0.1% Inhibitor[a]	1% Inhibitor[a]
Flow table [mm]	143.2 ± 1.8	139 ± 3.7	136.5 ± 2.1
Compressive strength [MPa]	3.22 ± 0.24	3.52 ± 0.06	3.75 ± 0.09
Flexural strength [MPa]	1.30 ± 0.02	1.44 ± 0.13	1.44 ± 0.04
WAC [$g/m^2s^{0.5}$]	137.11	123.46	131.97
Open Porosity [%]	23.44	23.51	23.54

[a]Percentage of binder weight

5   Conclusions and Outlook

In this study, the effect of sodium ferrocyanide (an inhibitor for sodium chloride crystallisation) on various relevant properties of NHL-based mortar was studied. The results clearly show that the addition of sodium ferrocyanide, in concentrations up to 1% of binder weight, do not significantly alter the properties of the mortar, both in its fresh and hardened state. Based on these results, it can be concluded that sodium ferrocyanide remains inert and does not participate in the microstructure development of these mortars. This is a positive result and a significant first step in development of NHL-mortars additivated with crystallisation inhibitors, for application in conservation interventions as well as in new buildings located in areas at risk of high salt loads, such as e.g. coastal areas. In the future, further research is planned to rule out possible chemical binding between the sodium ferrocyanide which could limit its effectiveness. Finally, the durability of the mortar with respect to salt damage will be investigated by means of accelerated salt crystallisation tests in laboratory and application in the field on case studies.

Acknowledgements. This research is carried out within the framework of the project MORtars with mixed-in Inhibitors for mitigation of SALt damage- MORISAL - (project n. 17636), financed by NWO. The authors are grateful to Arjan Thijssen for his help with MIP measurements.

References

1. Charola, A.E., Bläuer, C.: Salts in masonry: an overview of the problem. Restor. Build. Monum. **21**(4–6), 119–135 (2015)
2. Lubelli, B., van Hees, R.P.J., Groot, C.J.W.P.: Sodium chloride crystallization in a "salt transporting" restoration plaster. Cem. Concr. Res. **36**(8), 1467–1474 (2006)
3. Rodriguez-Navarro, C., Benning, L.G.: Control of crystal nucleation and growth by additives. Elements **9**(3), 203–209 (2013)
4. Bode, A.A.C., et al.: Anticaking activity of ferrocyanide on sodium chloride explained by charge mismatch. Cryst. Growth Des. **12**(4), 1919–1924 (2012)
5. Gupta, S., Terheiden, K., Pel, L., Sawdy, A.: Influence of ferrocyanide inhibitors on the transport and crystallization processes of sodium chloride in porous building materials. Cryst. Growth Des. **12**(8), 3888–3898 (2012)
6. Rodriguez-Navarro, C., Linares-Fernandez, L., Doehne, E., Sebastian, E.: Effects of ferrocyanide ions on NaCl crystallization in porous stone. J. Cryst. Growth **243**(3–4), 503–516 (2002)
7. Lubelli, B., van Hees, R.P.J.: Effectiveness of crystallization inhibitors in preventing salt damage in building materials. J. Cult. Herit. **8**(3), 223–234 (2007)
8. Rivas, T., Alvarez, E., Mosquera, M.J., Alejano, L., Taboada, J.: Crystallization modifiers applied in granite desalination: the role of the stone pore structure. Constr. Build. Mater. **24**(5), 766–776 (2010)
9. Van Enckevort, W.J.P., Los, J.H.: On the creeping of saturated salt solutions. Cryst. Growth Des. **13**(5), 1838–1848 (2013)
10. Granneman, S.: Mitigating Salt Damage in Lime-Based Mortars by Built-in Crystallization Modifiers, Delft University Press (2019)
11. Lubelli, B., Nijland, T.G., Van Hees, R.P.J., Hacquebord, A.: Effect of mixed in crystallization inhibitor on resistance of lime-cement mortar against NaCl crystallization. Constr. Build. Mater. **24**(12), 2466–2472 (2010)
12. Granneman, S.J.C., Lubelli, B., van Hees, R.P.J.: Effect of mixed in crystallization modifiers on the resistance of lime mortar against NaCL and Na2SO4 crystallization. Constr. Build. Mater. **194**, 62–70 (2019)
13. Granneman, S.J.C., Lubelli, B., Van Hees, R.P.J.: Characterization of lime mortar additivated with crystallization modifiers. Int. J. Archit. Herit. **12**(5), 849–858 (2018)

14. Maravelaki-Kalaitzaki, P., Bakolas, A., Karatasios, I., Kilikoglou, V.: Hydraulic lime mortars for the restoration of historic masonry in crete. Cem. Concr. Res. **35**(8), 1577–1586 (2005)
15. NEN-EN 196–1: Methods of Testing Cement - Part 1: Determination of Strength, vol. 1 (2015)
16. NEN-EN 459–2: Building Lime-Part 2: Test Methods (2008)
17. Groot, C.J.W.P.: Effect of Water on Mortar-Brick Bond, Delft University of Technology (1993)
18. Wijffels, T.; van Hees, R.: The influence of the loss of water of the fresh mortar to the substrate on the hygric characteristics of so-called restoration plaster. In: International Workshop on Urban Heritage and Building Maintenance VII; pp 49–54 (2000)
19. NEN-EN 196–11: Methods of Testing Cement-Part 11: Heat of Hydration-Isothermal Conduction Calorimetry Method (2019)
20. NEN-EN 1015–3: Methods of Test for Mortar for Masonry - Part 3: Determination of Consistence of Fresh Mortar (by Flow Table) (1999)
21. NEN-EN 1015–11: Methods of Test for Mortar for Masnory-Part 11: Determination of Flexural and Conpressive Strength of Hardened Mortar (2019)
22. NEN-EN-1925: Natural Stone Test Methods- Determination of Water Absorption Coefficient by Capillarity (1999)

Utilization of Lavender Waste in Traditional Mortars

Maria Stefanidou[1]([✉]), Vasiliki Kamperidou[2], Chrysoula Kouroutzidou[1],
and Petrini Kampragkou[1]

[1] Laboratory of Building Materials, School of Civil Engineering, Aristotle University of
Thessaloniki, Thessaloniki, Greece
stefan@civil.auth.gr
[2] Laboratory of Forest Products Technology, School of Forestry and Natural Environment,
Aristotle University of Thessaloniki, Thessaloniki, Greece

Abstract. It is well known that the most effective method to improve the internal tensile strength of lime-based mortar and minimize plastic shrinkage is the incorporation of fibres as reinforcement. Nowadays, commonly used fibres in lime-based mortars are polymeric or synthetic ones, while only recently the building industry's current trend to develop sustainable materials, brought natural fibres to the forefront. Natural fibres are abundant, low-cost, renewable, with a CO_2-neutral life cycle and high filling levels possible, sustainable, energy efficient, biodegradable, non-toxic, of nonabrasive nature, of low weight and density, yielding lightweight composites of low environmental footprint. In this sense, the current research aims to examine the possibility of utilization the solid lignocellulosic waste material generated from the lavender oil extraction plants in lime-based mortars. The lavender fibres were introduced in two ways: as additives in 1.5% b.v. of the mortar and as a layer (net) between two mortar layers. Both techniques were often met in historic structures. Specimens were produced to examine the mechanical, thermal, physical and microstructure characteristics of the composites. The results indicate the utilization potential of aromatic plants' wastes in the building sector contributing to the development of sustainable and energy efficient materials suitable for repair works.

Keywords: lavender fibres · lime mortars · thermal conductivity · mechanical properties

1 Introduction

To minimize waste and meet the EU targets, the construction industry is making efforts to embrace sustainable development and utilize bio-based materials instead of using less environmentally friendly, conventional ones. The bio-based and biodegradable materials seem to be the solution as renewable resources have attracted increasing attention over the last decades due to their low environmental impact and independence from fossil resources [1, 2]. The growing interest in sustainable construction materials has brought back to the spotlight the use of natural fibres as reinforcement of concrete and

© The Author(s), under exclusive license to Springer Nature Switzerland AG 2023
V. Bokan Bosiljkov et al. (Eds.): HMC 2022, RILEM Bookseries 42, pp. 351–358, 2023.
https://doi.org/10.1007/978-3-031-31472-8_27

other building materials [3]. The reinforcement materials should have favorable intrinsic properties (geometry, morphology, length, shape, modulus of elasticity, strength, surface roughness, hygroscopicity, etc.) and demonstrate good compatibility with the matrix while being cost-effective [3].

Based on this approach, research of environmentally friendly materials of bio-fibres is proposed [2, 4]. Unfortunately, the hygroscopic nature of fibres, related to the presence of hydroxyl groups in the fibre's structure, can induce a relatively high-water absorption when added in a mixture, which in fact leads to a reduction of the mixture's consistency, dimensional changes and, consequently, in poor adhesion between the matrix and the fibres. Furthermore, the bio-degradable character of these fibres, as well as the alkaline character of mortar matrix, decrease the natural fibre durability causing deterioration of the fibre's microstructure which results to the deterioration of the composite's performance [5]. Furthermore, the aging behaviour of the bio-fibres has also been identified. In this frame, the research focuses on the utilization of alternative options of waste bio-fibres in mortars [5, 6].

Lavender is considered a shrubby, multi-branched, aromatic plant that grows from 20 to 60 cm, in mountainous and semi-mountainous areas and originates mainly from the mountainous areas of Mediterranean Europe. In 2015, the total number of lavender farmers in Greece was 580, cultivating land of 628 hectares. After the lavender plant is harvested, it is usually dried and the oil is distilled. From the total amount of dry biomass, only the 2–10% corresponds to the oil [7]. It is therefore comprehensive that the amounts of waste that remain after the distillation of lavender plants are very high. Once the extracts of the lavender plant become isolated through the process of distillation, the solid residual material of distillation process remain unutilized, commonly allowed to be decomposed in fields neighboring to the distillation units or is being burnt to contribute to the distillation energy requirements fulfillment. It has been estimated that 4 tons of residual lignocellulosic biomass is being produced per hectare of cultivated area. 30% of the amount of the produced biomass, after processing, is the solid residue [2]. Specifically, it has been estimated that for every 10 hectares of lavender cultivated area, about 4 tons of dry biomass are produced and therefore, about 3.6–3.9 tons per 10 hectares are the quantities of waste after distillation [8]. Since the boiling point of lavender oil is lower than that of water, the distillation process is conducted at low temperatures (<100 °C), and therefore, this solid lignocellulosic biomass material does not undergo any intensive modification and it has not been severely degraded. In Greece, it is estimated that every year 5.787 tons of dry biomass are being produced and therefore, more than 5.000 tons of waste are generated only from the distillation of lavender plant. The highest quantities of lavender waste come from the regions of Western Macedonia, while at the same time, the producers of lavender do not apply any specific treatment or utilization of them. Therefore, an environmental problem is created in the wider area, for which solutions should be found with the goal of sustainable development and environmental protection.

Lime, on the other hand, is a critical binder, used alone or in mixture with other binders in several applications [9]. After its long use in history until the 19th century, strong hydraulic binders, such as Portland cement, gradually replaced lime. Traditional materials, such as lime, have been "re-discovered" due to their compatibility and are used mainly in fields where special attention has to be paid to the physical, chemical and mechanical compatibility of materials, as architectural monuments conservation and rehabilitation field [10].

Lime-based mortars are "soft" materials of relatively low mechanical strength compared to modern binders, having additional limitations concerning the long setting and hardening time, high water absorption capacity through capillarity and major volumetric change. One of the most effective methods to improve the internal tensile strength of lime-based mortar, and minimize plastic shrinkage, is the incorporation of fibres as reinforcement [11]. This method is as old, as human civilization. Traces of hair, eggs, blood, shells and natural fibres such as flax, cotton, straw, silk, wool and wood have been found in constructions of ancient civilizations all over the world [6, 12]. Commonly used fibres in lime-based mortars are polymeric or synthetic ones, while only lately the building industry's current trend to develop sustainable materials, brought natural fibres to the forefront [11]. In this paper, lavender plants wastes were cut and tested as bio-fibres in lime mortars following the principles of EN1015–3. The aim is to explore the possibility of using the highly available residual biomass of aromatic plants distillation in building technology.

2 Materials and Methods

Fresh and harden properties of the lime-based mixtures were tested in order to investigate if they can be incorporated in building materials' technology. The lavender waste material was obtained from a distillation plant in Chalkidiki region (North Greece) from an exposed to climatic factors open area where they were stored for about one week. The fibres are shown in Fig. 1 and two ways of utilization were employed in this work.

(a)   (b)

Fig. 1. (a) Lavender plants after oil extraction (b) after processing-cutting

The lavender fibres were introduced in the lime mortars structure in two ways: the first concerns their use as additives in 1.5% b.v. of the mortar and concerning the second way, they were used as a layer (a net) between two mortar layers (Fig. 2). Both techniques have been often met in historic structures (Fig. 3). The mean length of the fibres was

18 mm. The fibres were kept in water for 24 h to attain the highest level of absorption (44% water absorption capacity. They were surface dried and then they were carefully added in the mortar mixture either as additive or as a net. The moisture content of fibres could act positively towards a closer affinity to the lime matrix.

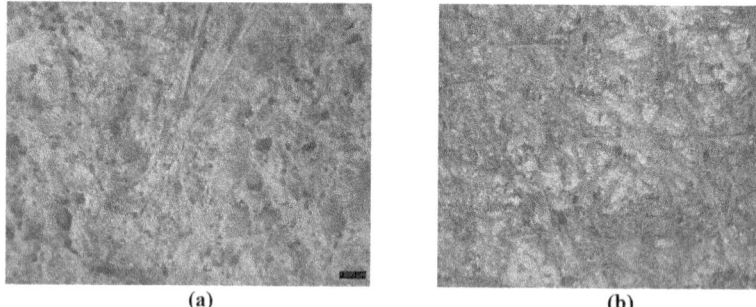

(a)   (b)

Fig. 2. (a) Bio-fibres distributed in a mortar (b) A net of bio-fibres on a lime mortar

Fig. 3. Use of natural fibres as a net in a historic structure

Different samples were produced in order to test mechanical, thermal, physical and microstructural characteristics of the composites. All specimens were kept wet (under wet burlaps) at room temperature until 2 days before tested. Prisms of $(40 \times 40 \times 160)$ mm were used for linear shrinkage tests as well as for the flexure and compression at 28 days. Plate specimens of $(200 \times 200 \times 25)$ mm were produced in order the thermal conductivity achieved at the same testing age to be recorded. Open porosity, water absorption and specific gravity of the samples were also recorded. The compositions produced are reported in Table 1.

Table 1. Constituents of the mortars in parts per weight

Composition	Lime	Pozzolan	Sand (0–4 mm)	w/b	Lavender fibres	Workability (cm)
LA	1	1	4	0.580		14.8
LI	1	1	4	0.605	1.5%b.v	15.5
LP	1	1	4	0.580	Net of fibres (1.5%)	14.7

In compositions were the fibres were used as net, no water excess was required for the mixture and the w/b ratio was not altered or the LA and LP compositions. The composition LP was identical to LA as the fibres were added above the first layer of mortar put in the mold. The thickness of the first layer was 3 cm. The fibres were carefully added and then another layer of 1 cm thickness was placed. The whole system was put to the vibration table for 60 s, in order to establish cohesion. In specimens of fibres used as additives, there was a need for water/binder ratio 0.605. All specimens were cured at 25 °C and 65%RH chamber until tested. After 28 days, the porosity and the mechanical properties were investigated. Three point bending test was performed at 3 prisms and the mean values are exhibited in Table 2. Six specimens, derived from flexure test, were tested in compression forces based on EN1015–3. Average values are also provided. Porosity was recoded based on RILEM CPC11.3.

Table 2. Physico-mechanical properties of the lime mortars

Composition	Flexure (MPa)	Compression (MPa)	Porosity (%)	Spec. Gravity
LA	0.795	4.969	26.94	1.93
LI	0.808	4.331	27.23	1.83
LP	1.863	5.593	29.90	1.84

3   Results

According to the results, it seems that flexure was improved by 2% with the lavender fibres used as additives, while a considerable increase of 134% was recorded when the fibres were added as net. Net fibres formed a stable surface resisting deformation. Compression strength was also improved in this case by 12%. On the contrary, a reduction of 13% was recorded in samples with fibres as additives. Porosity was slightly increased probably due to the lower density of the fibres and the compositions with fibres became lighter, as one of the main advantages of bio-fibres is their lightness [12].

The volume variation measured for each sample is presented in Fig. 4. The reference sample presents a sharp and high-volume change while the samples with the fibres presented a smooth behavior. Especially the distributed in the mass fibres has a lower volume variation.

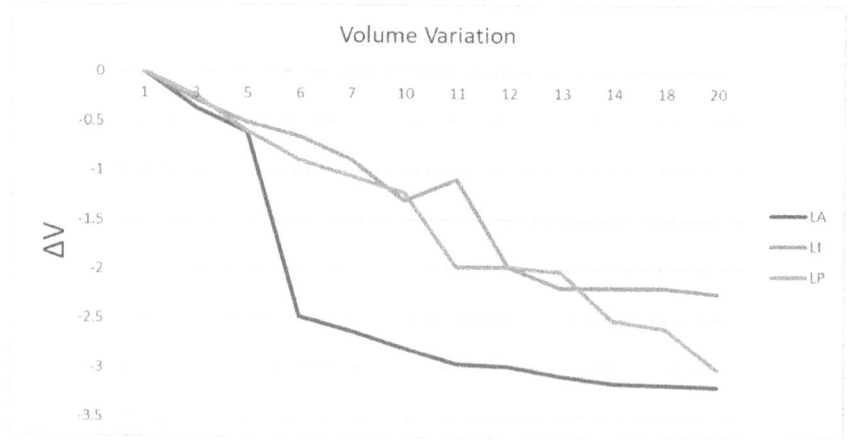

Fig. 4. Volume variation of lime mortars by time

As regards the thermal properties recorded during the measurement of the prepared plates, the results are shown in Table 3. It seems that the presence of fibres has improved the thermal behavior of the composites. The thermal conductivity was improved further when the fibres were applied as net (25% increase in relation to the reference sample at 10 °C).

Table 3. Thermal conductivity of samples at the age of 3 months

Sample	Dimentions [mm]			λ [W/(m*K)]	
	Length	Width	Thickness	10 °C	20 °C
L	200.0	199.5	25.4	0.7246	0.7943
LP	200.0	200.0	24.6	0.5430	0.5647
LI	199.5	197.0	23.0	0.6260	0.6226

4 Conclusions

In this article, the results of a research utilizing lavender wastes as fibres in lime mortars were presented. It constitutes preliminary research in order to investigate the possibility to utilize agricultural waste material of aromatic plants that keep causing extensive environmental problems so far, in building materials. The results indicate their positive role in specific properties considering their organic and hydrophylic nature. The most important outcome is that the mild hydrothermal treatment that the lignocellulosic biomass of lavender plants undergo during the distillation process of oil, seems to favor the fibres properties and therefore, improve the affinity between these bio-fibres and lime matrix.This process seems to make them capable to act in/withstand the highly alkaline environment of lime-based materials and the chemical changes induced should be further investigated.

Two different ways of introducing the fibres into the mortar mixture were tested. The first utilized the fibres as additives in 1.5% b.v. and improved flexure and thermal behavior (lower thermal conductivity) were recorded. The second method is referring to the formation of a fibre-net between two mortar layers. In this case, the benefits of the fibres were much higher, concerning the mechanical performance, but also the physical and thermal properties.

This research field is wide and different sustainable approaches can be followed in order to elaborate the surface of the plants and increase further their adhesion capacity, hydrophobicity or hardness. Utilization of bio-wastes to make them efficient to function as reinforcement in mortars and use them for repair works, fulfilling criteria of compatibility as well as sustainability is of great interest.

References

1. ACI Committee 544, Report on Fiber Reinforced Concrete, ACI 544.1R-96, Reapproved 2002
2. Banthia, N.: Current innovations in fiber reinforced concrete. In: The 3rd ACF International Conference- ACF/VCA (2008)
3. Kesikidou, F., Stefanidou, M.: Natural fiber-reinforced mortars. J. Buil. Engi. **25**, 100786 (2019)
4. Barth, M., Carus, M.: Carbon Footprint and Sustainability, Nova Institute (2015)
5. Akers, S.A.S., Studinka, J.B.: Ageing behaviour of cellulose fibre cement composites in natural weathering and accelerated tests. Int. J. Cement Compos. Lightweight Concr. **11**(2), 93–97 (1989). https://doi.org/10.1016/0262-5075(89)90119-X
6. Thiroumalini, P., Sekar, S.K.: Review on herbs used as admixture in lime mortar used in ancient structures. Indian J. Appl. Res. **3**(8), 298 (2013). https://doi.org/10.15373/2249555X/AUG2013/93
7. Chakyrova, D., Doseva, N.: Analysis of the energy recovery possibilities of energy from lavender straws after a steam distillation process. IOP Conf. Series: Mater. Sci. Eng. **1032**(1), 012023 (2021)
8. Filipović, V.; Ugrenović, V.: The Composting of Plant Residues Originating from The Production of Medicinal Plants. International Scientific Meeting "Sustainable Agriculture and Rural Development in Terms of the Republic of Serbia Strategic Goals Realization within the Danube Region - Achieving Regional Competitiveness" In: Cvijanović, D., Subić, J., Vasile, A.J. (Ed.) The Institute of Agricultural Economics Belgrade, Topola, Serbia, 5–7 December (2013). ISBN 978–86–6269–026–5
9. Moropoulou, A., Bakolas, A., Anagnostopoulou, S.: Composite materials in ancient structures. Cem. Concr. Compos. **27**(2), 295–300 (2005)
10. Faria, P., Silva, V.: Natural hydraulic lime mortars: influence of the aggregates. In: Hughes, J.J., Válek, J., Groot, C.J.W.P. (eds.) Historic Mortars, pp. 185–199. Springer, Cham (2019). https://doi.org/10.1007/978-3-319-91606-4_14

11. Stefanidou, M., Papachristoforou, M., Kesikidou, F.: Fiber reinforced lime mortars. In: Proceedings of the of 4th Historic Mortars Conference, 10–12 October (2016), Santorini, Greece
12. Stefanidou, M., Kamperidou, V., Konstantinidis, A., Koltsou, P., Papadopoulos, S.: Use of Posidonia oceanica fibres in lime mortars. Constr. Build. Mater. **298**(6), 123881 (2021). https://doi.org/10.1016/j.conbuildmat.2021.123881

Restoring Historical Buildings Amid Climate Crisis: Hydraulic, Waste-Based Lime

Jelena Šantek Bajto[✉], Nina Štirmer, and Ana Baričević

Faculty of Civil Engineering, Department of Materials, University of Zagreb, Zagreb, Croatia
jelena.santek.bajto@grad.unizg.hr

Abstract. Reducing greenhouse gas emissions and dependence on fossil fuels is the cornerstone of all European climate and energy strategies. At the same time, waste reduction and recycling are at the top of the European waste hierarchy. Therefore, the European Union (EU) is turning to renewable energy and sustainable options that address complex local conditions and growing energy and material needs. Bioenergy is the most important renewable energy source in the EU, accounting for more than 2/3 of the renewable energy mix. As wood is a carbon neutral energy source, it is one of the predominant sources of biomass for energy production in the EU. However, the continuous expansion of wood biomass power plants leads to the formation of wood biomass ash (WBA), a solid by-product of wood combustion on a large scale. The disposal of WBA is a problem that requires long-term strategic and sustainable solutions. Therefore, WBA is the focus of this paper as an alternative material with pozzolanic and/or hydraulic properties needed for the production of waste-based artificial hydraulic lime (AHL). This type of a novel hybrid WBA-lime binder would serve as an end result for the production of a repair mortar for historical buildings. Two different sources of locally available WBA were considered for the preparation of binders as mixtures of WBA and natural hydraulic lime (NHL) in different ratios. Based on the chemico-mineralogical composition and physical properties, the potential of WBA as a component of an AHL is evaluated. The contribution of the individual binder mixtures to mortar properties was analyzed on the basis of the demonstrated water demand, the flow diameter, and the compressive and flexural strength according to EN 459-1.

Keywords: Wood Biomass Ash · Hydraulic Lime · Alternative Binders · Industrial Waste · Sustainability · Cementless Binder

1 Introduction

After two devastating earthquakes in central Croatia in 2020, national plans and strategies focused on the comprehensive renovation of buildings, but also on promoting the renovation of buildings with cultural and historical value. The renovation of damaged and energy inefficient historical buildings should lead to a reduction in CO_2 emissions and energy consumption, but also contribute to the development of a circular economy and the use of sustainable solutions. This is a small niche where further measures can be included to improve the already mandatory basic requirements for buildings, such as

V. Bokan Bosiljkov et al. (Eds.): HMC 2022, RILEM Bookseries 42, pp. 359–373, 2023.
https://doi.org/10.1007/978-3-031-31472-8_28

alternative waste-based repair materials that contribute to the sustainable use of natural resources. Just as the reuse of waste is at the top of the European waste hierarchy and prevention itself is given priority [1], the rehabilitation of existing buildings must be given priority over new construction [2]. Rehabilitation in the sense of measures that allow us to continue or use a cultural asset in a contemporary way through repairs and modifications should not have a negative impact on its cultural and historical values [3]. 13% of all buildings in Croatia are immovable cultural heritage or are located in protected areas that have not been rehabilitated [4], as modern interventions are considered invasive and a potential threat to cultural heritage. In order to mitigate the damage to the built heritage and the effects of climate change, the right choice of repair materials is crucial. As for the "right choice", traditional and modern materials are often incompatible, making the selection of an appropriate, equivalent material a particular challenge in the rehabilitation of buildings with historical and cultural significance. For this reason, the use of state-of-the-art materials in compliance with applicable codes and the preservation of cultural and historical values often fall by the wayside. Since early building materials such as stone and clay brick were bonded with lime mortar, lime-based materials are the key to material compatibility while preserving historical structures. These are, of course, the materials that were available at the time of construction. Therefore, even today, when cement and cementitious materials are prevalent, there is still a need for lime-based materials, i.e., repair materials that are compatible with the materials originally used. The selection of such compatible materials is extremely important to avoid negative consequences for the historical heritage. But it is also important that these solutions are sustainable, environmentally friendly and socially acceptable, adding a new dimension to the rehabilitation of historic buildings in times of climate and energy crisis. In its efforts to put the European Green Plan [5] into action, Europe is therefore compelled to turn to an innovative model of a resource-efficient economy and decarbonise highly energy-intensive industries such as cement and lime. While turning towards renewable energy and sustainable options, the European Union (EU) is focused on addressing complex local conditions and growing energy and material needs. As atmospheric warming spirals out of control, even greater climate variability is expected in the coming decades. In addition to the financial burden, all industrial sectors face significant challenges in reducing net greenhouse gas emissions, and the lime industry is no exception. Emissions from the downstream processing and hydration of lime account for about 1.5% of total CO_2 emissions [6]. Their competitiveness and sustainability are directly related to the availability of advanced, multifunctional modern lime-based materials that can be used in conventional ways. Europe's key to this goal is the gradual phase-out of coal combustion, relying largely on renewable energy capacity to fill the gap left by the decline in coal combustion [7]. The recast of the Renewable Energy Directive suggests that bioenergy and biomass could make a positive contribution to creating a low-carbon energy system with an 18% share of total energy supply by 2050 [8], provided biomass is managed sustainably [9]. Industrial use of bioenergy is expected to triple to 24 EJ by 2060, accounting for nearly 14% of industrial energy demand [10]. Therefore, biomass-derived materials will play a key role in the transition to a circular economy, which must be based on sustainable consumption and production, while promoting the recovery of key raw materials from waste [11]. Bioenergy is the main renewable energy source in the EU, accounting for more than 2/3 of the renewable energy mix [12, 13]. Since wood is a carbon neutral energy source, it is one of the predominant sources of biomass for energy production in the EU [14]. However, the continuous expansion of wood biomass power plants results in large amounts of wood biomass ash (WBA), a solid by-product of wood combustion [15]. Although WBA contains numerous nutrients such as calcium, potassium, and phosphorus that could be useful in agriculture, the ash also contains heavy metals that pose a particular threat to the environment [16]. While these facts clearly point out the need for strategic foresight in the management of WBA, the common practise in dealing with WBA is to dispose of it directly in landfills or to use it as a soil amendment, often without any monitoring, resulting in clogged and illegal landfills [12, 17, 18]. The disposal of WBA should not be neglected, especially in view of the continuous increase in WBA quantities and disposal costs. With the adoption of the new "Circular Economy Package" as well as the extended Directive 2018/851 on waste and Directive 2018/850 on landfills [1, 19], the EU is determined to minimise the uncontrolled proliferation of waste in the environment and to promote its recovery. In the interest of finding sustainable, cost-effective, and practical functions for WBA in the long term, the vast majority of studies addressed the beneficial use of WBA in cementitious composites and identified it as a raw material substitute in cementitious composites. In contrast, the recognition of WBA as a pozzolanic and/or hydraulic component in the formulation of hydraulic lime, which is the focus of the proposed research, has been little studied. The use of WBA in the formulation of hydraulic lime is consistent with the standard HRN EN 459-1 [20], which distinguishes between natural and artificial lime and defines artificial hydraulic limes (referred to as hydraulic limes (HL) and formulated limes (FL)) as binders containing aerial lime and/or natural hydraulic lime (NHL) in combination with hydraulic and/or pozzolanic materials. Hydraulic limes vary in their degree of hydraulicity, ranging from weak to very hydraulic, but in all of them two setting mechanisms occur simultaneously-a hydraulic reaction, i.e., setting with water, and a carbonation reaction, i.e., absorption of CO_2, which contributes to some of the hardening of the material. The exact use of FL has been known since the Romans used volcanic ash to make lime when they built the Colosseum. Mixing pure lime with clay and/or other silicate/aluminate minerals produces a high performance artificial hydraulic lime that gives strength to the mortar but somewhat reduces breathability and flexibility [21, 22]. According to the chemical composition, calcium oxides (CaO) and silicates (SiO_2) are necessary to achieve hydraulic properties, i.e., the production of hydraulic lime. Accordingly, the same analogy is outlined in the planned research – mixing NHL with WBA as a hybrid AHL. The use of WBA from local production, in which either calcium or silicon oxides usually predominate, would avoid industrial waste while conserving natural resources. At the same time, it would create a waste-based, cement-free binder suitable for mortars for historic buildings.

2   Experimental Work

The experimental part of the research aimed to determine the prospects for the use of WBA in artificial hydraulic lime, taking into account its chemico-mineralogical composition and physical properties. This was to serve as a starting point for modifying and improving the design of the artificial hydraulic lime mortar with WBA to determine the acceptable proportion of WBA and NHL that would ensure the hardening of the AHL mortars. The methodological framework includes locally available WBAs, differing mainly in calcium and silica content, as an alternative material with hydraulic and/or pozzolanic properties needed for the production of hybrid lime-WBA binders and consequently lime-based mortars for repair applications.

2.1 Materials and Methods

Materials. Two WBAs used in this study were collected from power plants in Croatia that use untreated wood chips as fuel and employ grate combustion technology. A single source of commercially available natural hydraulic lime (Baumit NHL 3.5) was selected because it was found that NHL 3.5 is currently the most commonly used in practice (Fig. 1). In the power plants themselves, the bottom ash and the coarse fly ash fraction are collected together and stored in the same closed containers, where the WBA is not in direct contact with weathering, but which are not sealed, so that the carbonation process and the influence of moisture are possible.

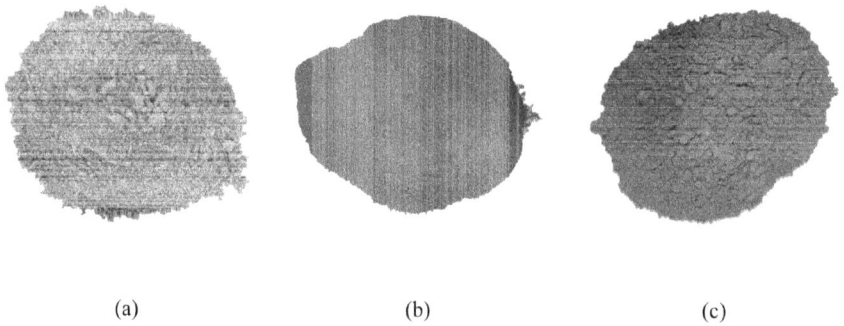

(a)   (b)   (c)

Fig. 1. AHL binder components: (a) NHL 3.5, (b) WBA1, (c) WBA2.

Two types of aggregates were used to make the preparation of mortar mixtures: Quartz sand and dolomite sand, since these sands are commonly used as aggregates in mortars for the preservation of older masonry [23]. Particle size distribution of two types of sand used is presented in Fig. 2.

Mortar Mixtures: Composition, Mixing and Curing Conditions. Two groups of mortars were prepared and tested within the presented research, each of them containing one reference mortar, differing in aggregate type and NHL:WBA ratio. The reference lime mortar mixtures (designated as LM0-1, LM0-2, LM0-3) were prepared using only NHL as binder, while 4 other mixtures contained a hybrid lime binder (designated as LM1, LM2, LM1-1, LM2-1). Within the NHL-WBA system, two NHL:WBA ratios were investigated, 1:1 and 1:0.5 (by volume), with the water and superplasticizer content modified to achieve a flow diameter of 160 to 170 mm, as these flow values are considered to be workable on site. Before preparing the hybrid lime binders, the WBAs were first sieved through a 0.125-mm sieve to remove obvious impurities, such as unburned wood

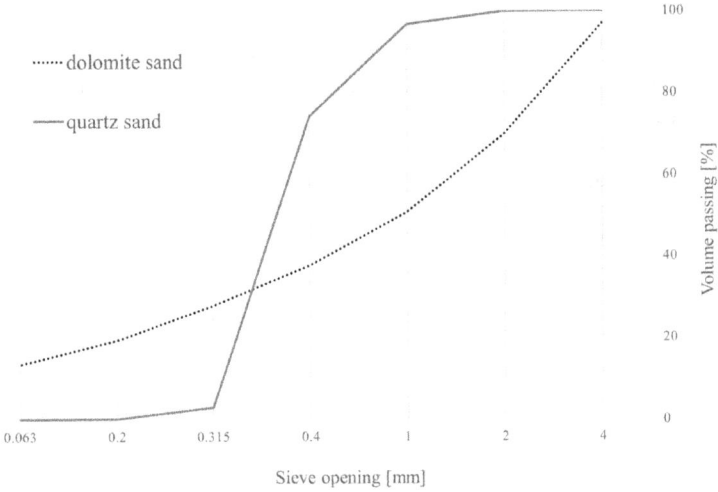

Fig. 2. Particle size distribution of two types of sand used.

pieces or charcoal fragments. This type of selective removal was intended to reduce negative effects on binder properties, such as excessive water demand or reduced performance [22]. The volume ratio of binder to aggregate (b/a) was kept constant at 1:2 in all mixtures. To maintain the same 1:2 volume ratio of binder to aggregates, the weight of sand, WBA, and NHL for each mortar mix was adjusted based on the bulk density. The water to binder (w/b) ratio was kept at 0.60 for all AHL mixtures, except for the LM0-2 mixture, where the water-binder ratio was slightly lower, 0.55. The detailed composition of the AHL mortar mixtures can be found in Table 1.

Table 1. AHL mortar mixtures composition.

Mix ID	Aggregate	Binder		NHL: WBA: Aggregate	b/a ratio	w/b ratio	Additives
LM0-1	Quartz sand	NHL 3.5		1: 0: 2	1/2	0.60	water reducing super - plasticizer
LM0-2				1: 0: 2		0.55	
LM1		NHL 3.5	WBA1	1: 1: 4		0.60	
LM2		NHL 3.5	WBA2	1: 1: 4			
LM0-3	Dolo -mite sand	NHL 3.5		1: 0: 2	1/2	0.60	
LM1-1		NHL 3.5	WBA1	1: 0.5: 3			
LM2-1		NHL 3.5	WBA2	1: 0.5: 3			

The mixing procedure was carried out according to EN 459-2 [20]. Previously homogenised binder combinations were added to water, i.e., the mixture of water and superplasticizer and mixed at low speed for 30 s. Then the aggregate is added gradually over 30 s. Mixing continues for another 30 s at low speed. After the first 90 s of mixing, there is a pause of 30 s, after which mixing continues at low speed for 15 s. The mixing process took a total of 2 min and 15 s. Standard steel moulds for prismatic specimens (40 × 40 × 160 mm) were used to mould the mortar specimens. The mortars were kept in the moulds under humid conditions (relative humidity RH of 95 ± 5% and temperature of 20 ± 2 °C) for a period of 3 to 5 days after mixing, according to EN 1015-11 [24]. This curing regime also complies with the standard EN 459-1[25], according to which a RH of at least 60% is required for the curing of lime specimens. Once demoulding was possible, extended by the WBA content, the mortar specimen were cured for up to 7 days under humid conditions and then for up to 28 days at a relative humidity of 50 ± 10% and a temperature of 20 ± 2 °C. While the specimens with an NHL:WBA ratio of 1:0.5 were demoulded after 3 days of curing, the specimens with an NHL:WBA ratio of 1:1 were too soft to demould after 5 days of curing. Hydration of hydraulic binders requires 95% to 100% of RH, especially in the first days after casting [26].

Methods. The investigation is carried out at the binder and mortar level, testing both individual binder components and binder combinations. The chemical composition of each WBA and NHL was determined before mixing the binders to verify that they comply with the standard EN 459-1. In addition, the contribution of each binder to the mortar properties was analysed using the test methods listed in Table 2. The NHL-WBA mortars were tested in fresh and hardened states. Consistency using the flow table test and air content as well as bulk density were measured on fresh mortars immediately after mixing, while compressive and flexural strength were measured on prismatic specimens after 7 and 28 days of curing.

The development of carbonation was assessed visually after applying a phenolphthalein solution on the broken surface of the specimens at 28 days of age, which were previously used to test compressive strength. The compressive strength was measured on the halves of the prismatic specimens (of three specimens) in a measurement range of 400 ± 40 N/s.

3   Experimental Results and Discussion

3.1   AHL Constituent's Characterisation

The hydraulic nature of both the NHL and WBA samples was assessed by chemical analysis. It is important to emphasise that the chemical composition of WBA can vary greatly depending on the feedstock and combustion technology used [27, 28], as well as from batch to batch. The chemical analysis of NHL 3.5 and the WBAs (before sieving) is shown in Table 3.

When comparing the results of chemical analysis of WBAs with those of NHL, minor differences in the content of individual chemical compounds were found. Therefore, it can be concluded that WBA2 is more similar to NHL. It is also noted that the studied WBAs have similar chemical composition, with the content of CaO and SiO_2 being quite comparable in both WBA samples, while the CaO is the predominant oxide in the NHL sample. The content of the oxides SiO_2 and Fe_2O_3 is almost the same in WBA samples, while the CaO content is slightly lower in WBA1. The slightly elevated SO_3 content in WBA1 does not meet the criteria of EN 459-1, which states that SO_3 should not exceed

Table 2. Test methods for assessment on binder and mortar level.

Level	State	Property	Test period	Standard
Binder	Individual & binder combinations	Chemical composition	Prior to mixing	ASTM D7348-21 ASTM D6722-19 ISO/TS 16996:2015
		Particle size distribution by air-jet sieving		EN 459-2:2021
		Free water content		
Mortar	Fresh state	Bulk density	Immediately upon mixing	EN 1015-6:2000
		Air content		EN 459-2:2021
		Flow diameter		
	Hardened state	Compressive strength	After 7 and 28 days of curing	
		Flexural strength		
		Carbonation	After 28 days of curing	EN 14630:2007

Table 3. Comparison of chemico-mineralogical composition of NHL and the WBAs [wt.%].

NHL/WBA sample	LOI (950 °C)	SiO_2	CaO	SO_3	Al_2O_3	Fe_2O_3	MgO	HI (Eq. 1)	CI (Eq. 2)
NHL 3.5	16.1	35.06	52.81	0.73	5.39	2.66	1.97	0.79	1.91
WBA1	6.6	30.1	28.02	2.19	9.85	3.56	4.43	1.34	2.85
WBA2	6.2	30.2	36.71	1.26	6.17	3.58	4.45	0.97	2.19

2%, but even with an NHL:WBA ratio of 1:1, this is annulled by the weighted average value.

The hydraulic index (HI) and cementation index (CI) are calculated as per Eqs. (1) and (2) respectively [29].

$$CI = (1.1 \times Al_2O_3 + 0.7 \times Fe_2O_3 + 2.8 \times SiO_2)/(CaO + 1.4 \times MgO) \quad (1)$$

$$HI = (Al_2O_3 + Fe_2O_3 + SiO_2)/(CaO + MgO) \quad (2)$$

The evaluation of lime by hydraulic index is formed on the fact that the higher the index, the more hydraulic properties are expected [29]. The same analogy applies to the classification of lime according to the cementation index. The hydraulic index and cementation index, which are above the value of 0.7 as the lower limit for highly hydraulic

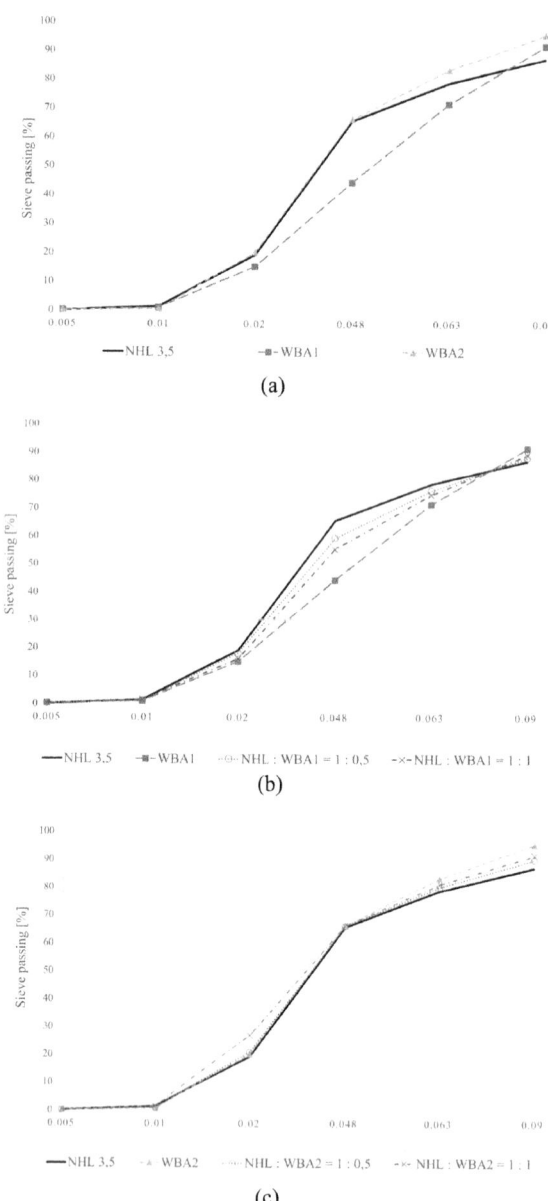

Fig. 3. Particle size distribution of a) 100% NHL/WBA, b) binders with WBA1, c) binders with WBA2.

properties in both WBA samples investigated, suggest that the two WBAs are highly hydraulic materials. Of course, this is not the whole picture, because the mineralogy of hydraulic binders is complex, and the properties of hydraulic binders depend not only on their composition ('CI' and 'HI'), but also on the conditions of their production, i.e. firing temperature and time, as these affect the mineralogy of the final product [30].

The air-jet sieving was performed for individual WBAs and NHL as well as for each binder mixture. The particle size distribution of the previously sieved WBAs and NHL shows that WBA2 is more comparable to NHL (Fig. 3). Nevertheless, all binders with both WBA1 and WBA 2 meet the criteria of EN 459-1, which requires that the residue on the sieve of size 0.09 mm be less than or equal to 15% of the binder mass (Table 4).

Table 4. Residue retained on the sieve of size 0.09 mm after air-jet sieving

Binder mixture	Residue retained [%]	EN 459-1 criteria [25]
NHL 3.5	12.9	≤ 15
WBA1	8.2	
WBA2	4.4	
NHL: WBA1 = 1: 1	10.6	
NHL: WBA1 = 1: 0.5	11.5	
NHL: WBA2 = 1: 1	8.3	
NHL: WBA2 = 1: 0.5	9.9	

3.2 Influence of WBA on Fresh Mortar Properties

Immediately after completion of the mixing process, the properties of the mortar in fresh state were determined. A comparison between the reference mixtures without WBA (LM0-1, LM0-2, LM0-3) and the mixtures with the hybrid lime binder is presented in Table 5.

The flow diameter of the standard mortar was measured on the flow table (Fig. 4) to relate the flow diameter to the amount of water required for the mortar. According to the standard EN 459-2, the flow diameter of the mortar with NHL 3.5 is prescribed to be (165 ± 3) mm, which was exceeded in the reference mortar mixtures with quartz sand, indicating that the w/b ratio needs more adjustment. At the same time, using the same amount of water in the reference mixture LM0-3 mixture with dolomite sand, an adequate flow diameter of 164 mm was obtained. All mixtures with WBA have shown comparable flow values, ranging from 162 mm to 171 mm. Accordingly, the LM1 and LM2 mixtures with WBA and quartz sand as aggregate with flow values of 171 and 165 mm, respectively, may suggest a behavioural tendency where the water demand increases with the amount of WBA, but is also related to the different water absorption of the specific aggregates. Reduced workability, i.e., false setting of the mortar with WBA, was observed in mixtures with WBA, particularly emphasized with WBA1. In

Table 5. Fresh state results.

Mix ID	NHL: WBA: Aggregate ratio	w/b ratio	Flow diameter [mm]	Air content [%]	Density [kg/m^3]
LM0-1	1: 0: 2	0.60	210	7	1950
LM0-2	1: 0: 2	0.55	180	8.5	1960
LM1	1: 1: 4	0.60	171	10.5	1830
LM2	1: 1: 4		165	8.5	1930
LM0-3	1: 0: 2		164	8.0	1990
LM1-1	1: 0.5: 3		165	13.0	1910
LM2-1	1: 0.5: 3		162	8.2	2010

addition to the loss of workability, bleeding of water was observed in mixtures with WBA1, as shown in Fig. 4 (b), which clearly depends on the WBA type. The density of fresh mortar mixtures containing WBA ranged from 1.830 kg/m^3 to 2.010 kg/m^3, while the reference density values in all 3 mixtures containing only NHL were very similar, ranging from 1.950 to 1.990 kg/m^3. Higher values of air content in fresh mortar, correlating with the reduction in density, were found in mixes containing WBA1 and quartz sand (10.5% and 13%), which was influenced by the poor grading of the sand.

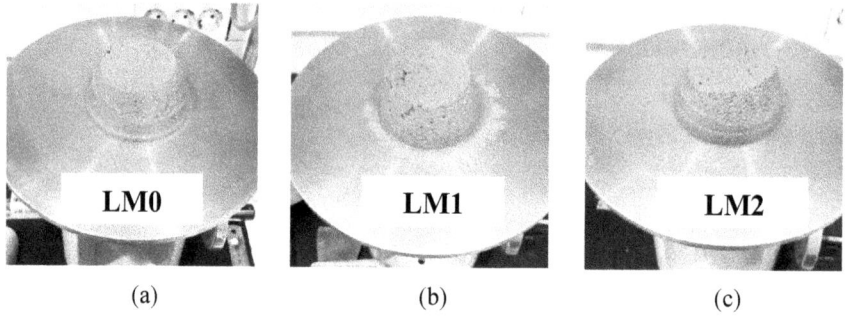

| (a) | (b) | (c) |

Fig. 4. Standard mortar flow diameter measured on the flow table: (a) LM0-1; (b) LM1; (c) LM2.

3.3 Influence of WBA on Hardened Mortar Properties

The effects of the individual WBAs on the mechanical properties of the AHL mortar were analysed based on the demonstrated compressive and flexural strengths. The differences between the individual WBAs have been reflected in the properties of the mortar mixtures. The absolute values of compressive and flexural strength at 7 and 28 days of age are shown in Table 6, categorized by the content of NHL in the binder blends. The relative values of compressive and flexural strength of the mortar mixtures after 7 and 28 days of curing are shown in Fig. 5 and Fig. 6. The values are normalised to the NHL content in the mortar, i.e. to the mass of NHL in the binder blends, in order to better assess the contribution of the WBA to the mortar strength.

Table 6. Compressive and flexural strength of mortar and compressive to flexural strength ratio.

Mix ID	NHL content in binder blends [%]	Compressive strength after 7 days, $f_{C,7}$ [MPa]	Compressive strength after 28 days, $f_{C,28}$ [MPa]	Flexural strength after 7 days, $f_{F,7}$ [MPa]	Flexural strength after 28 days, $f_{F,28}$ [MPa]	$f_{C,28}/f_{F,28}$
LM0-1	100	0.64	2.11	0.39	0.76	2.8
LM0-2		0.80	2.56	0.32	0.83	3.1
LM0-3		0.92	2.58	0.54	1.09	2.4
LM1	50	0.27	0.99	0.09	0.66	1.5
LM2		0.41	1.23	0.14	0.52	2.3
LM1-1	70	0.75	2.22	0.53	1.21	1.8
LM2-1		1.12	3.11	0.58	1.23	2.5

Although the additional distinction between the two mortar groups is the type of sand used, the compressive strength values obtained on the reference specimens are comparable, with indications of a lower water requirement when quartz sand was used. The increase of ca. 18% in the compressive strength of the reference LM0-3 specimen in comparison to LM0-1 can be attributed to a finer grading curve exhibited by the dolomite sand (Fig. 2). In addition, the compressive strength decreases by up to 60% (LM2 with respect to LM2-1) when the NHL:WBA ratio is increased. A similar trend is observed for the mixtures with WBA1 (LM1 with respect to LM1-1) at both ages. When comparing the two types of WBA, the highest values of compressive strength were observed for the AHL mortar specimens prepared with WBA2 (LM2-1), which were even higher than the compressive strength of the reference mixture with dolomite sand. This can be related to favourable chemical composition and particle size distribution of WBA2 but also with good grading of dolomite sand. The results of the flexural strength tests are consistent with the compressive strength values, except for the mixtures with WBA1. The flexural strength after 28 days of curing these specimens is higher (LM1 with respect to LM2) or fairly close (LM1-1 with respect to LM2-1) to the values of the specimens with WBA2, in contrast to the higher compressive strengths observed for the specimens with WBA2.

Silva et al. [31] and Moropoulou et al. [32] and found that the relationship between compressive and flexural strength is proportional to modulus of elasticity in tests lasting from 1 to 6 months, proposing that a low ratio of compressive to flexural strength (f_C/f_F) correlates with a low modulus of elasticity. Therefore, mortars that have a low f_C/f_F ratio are potentially compatible with that of the original material due to their elastic behaviour, ensuring better durability [31]. Accordingly, the f_C/f_F is highlighted as a simplified method for evaluating the potential ductility of mortar. The values of the f_C/f_F

ratio for the mortars with unmixed NHL as binder and the binder blends with WBA are presented in Table 6. It can be seen that the mortar specimens with WBA1 have the lowest compressive to flexural strength ratio and that, as expected, this parameter increases with decreasing WBA content. Although the f_C/f_F values increase with NHL content, the f_C/f_F values for mortars with WBA are within the range determined by Moropoulou et al. [32], where the tested mortars exhibit elastic behaviour compatible with that of historical mortars.

From the relative results of compressive strength per NHL mass, Fig. 5, it can be presumed that the addition of both WBA1 and WBA2 to the NHL-WBA binder system contributes to the increase of compressive strength, where the ratio NHL:WBA is 1:0.5, i.e. the proportion of NHL in the binder mass is 70%. The relative values of mixtures with 50% NHL in the binder blend, are comparable with the reference values.

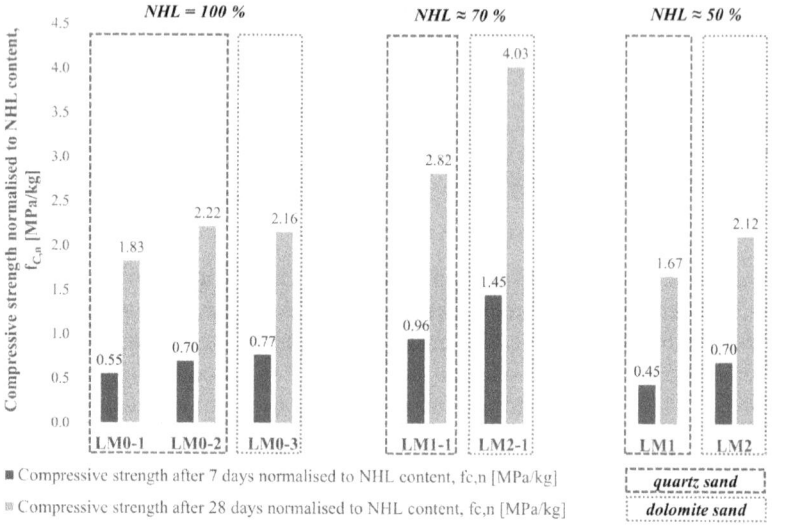

Fig. 5. Compressive strength of mortar mixtures after 7 and 28 days of curing, normalised to NHL content.

The phenolphthalein test showed a greater reduction in alkalinity in mortars with WBA compared to reference mixtures with only NHL binder. Accordingly, the mixtures with the ratio NHL:WBA = 1:0.5 (Fig. 7, (e) and (f)) have demonstrated a smaller reduction in alkalinity than the mortars with the ratio NHL:WBA = 1:1 (Fig. 7, (b) and (c)). Considering that the curing conditions were the same for all specimens, carbonation, i.e., portlandite consumption, is accelerated by the addition of WBA. Since both carbonation and pozzolanic reactions are portlandite consuming reactions, the greater reduction in alkalinity in mortars with WBA could indicate a possible pozzolanic reaction between NHL and WBA.

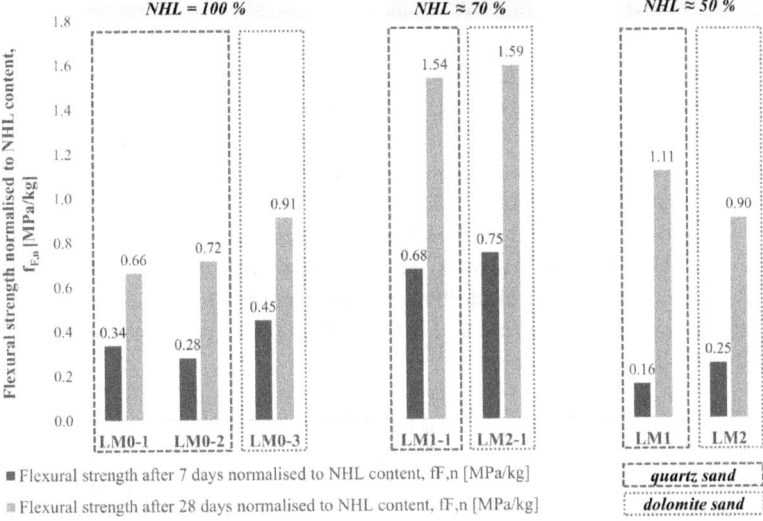

Fig. 6. Flexural strength of mortar mixtures after 7 and 28 days of curing, normalised to NHL share.

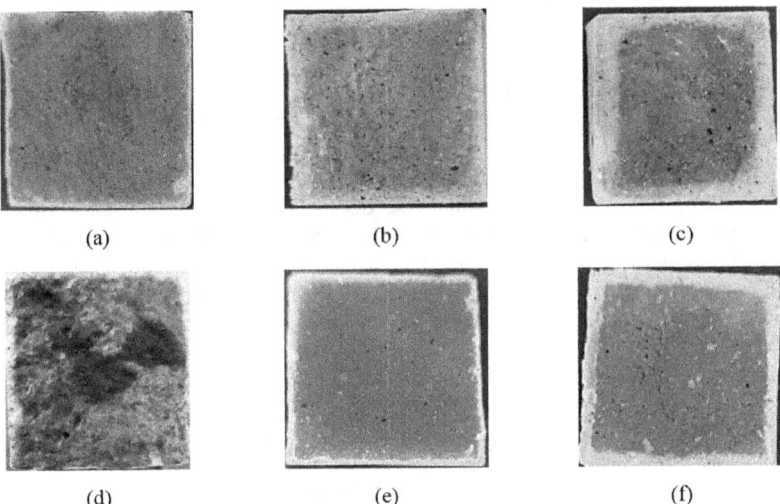

Fig. 7. Carbonation of mortar after 28 days: (a) LM0-1; (b) LM1; (c) LM2; (d) LM0-3; (e) LM1-1; (f) LM2-1.

4 Conclusions

The research presented here is still at a preliminary stage, although certain conclusions can be drawn that may serve as a reference for future experimental work. First, the chemico-mineralogical composition of WBA is highly variable and dependent on the combustion technology. Together with the particle size distribution of the binder, this should be the first step to evaluate WBA as a component of an artificial hydraulic lime and to identify the key factors that determine the engineering behaviour and durability of a new composite. In addition, the results for WBA-containing binders show that a moderate WBA content in the NHL-WBA binder ensures a) lower water demand, needed to achieve workable flow values and b) acceptable mechanical properties. All the properties studied are also strongly dependent on the WBA used. Certain problems arising from the physical and chemical properties of WBA, as well as the selection of a suitable, well-graded aggregate, such as extensive bleeding and false setting, should be the focus of further experimental work. The problems identified could be addressed by appropriate selection of chemical additives and/or by pre-treatments of the ash itself. The focus of the next phase of this research is to confirm a synergistic relationship between WBA and NHL as AHL components.

Acknowledgments. This research work was funded by the Ministry of Economy and Sustainable Development of the Republic of Croatia under the project "Development of Innovative Construction Products with Biomass Ash" based on the grant agreement for projects financed by the European Structural and investment funds for the Period 2014–2020 (NO. KK.01.2.1.01.0049). The work of Jelena Šantek Bajto was supported by the Croatian Science Foundation within the "Young Researchers' Career Development Project - Training of Doctoral Students" (DOK-01-2018).

References

1. European Council Directive 2008/98/CE of the European Parliament and of the Council of 19 November 2008 on waste and repealing certain Directives. Off. J. Eur. Union **L312**, 1–59 (2008)
2. European Commission Communication from the Commission to the European Parliament, The Council, The European Economic And Social Committee and the Committee of the Regions - New European Bauhaus Beautiful, Sustainable, Together (2021)
3. Roca, P., Lourenço, P.B., Gaetani, A.: Historic Construction and Conservation - Materials, Systems and Damage, 1st Edn. Routledge (2021). ISBN 9781032090238
4. Government of the Republic of Croatia National Plan for Recovery and Resilience 2021. - 2026 (2021)
5. European Commission Secretariat-General The European Green Deal 2019, 58, 7250–7 (2019)
6. EuLA - The European Lime Association A Competitive and Efficient Lime Industry - Cornerstone for a Sustainable Europe (2014)
7. Carević, I., Baričević, A., Štirmer, N., Šantek Bajto, J.: Correlation between physical and chemical properties of wood biomass ash and cement composites performances. Constr. Build. Mater. **256**, 14 (2020). https://doi.org/10.1016/j.conbuildmat.2020.119450
8. European Commission European Green Deal: Commission proposes transformation of EU economy and society to meet climate ambitions. EU Comm. - Press release 2021
9. Scarlat, N., Dallemand J.-F., Taylor, N., Banja, M.: Brief on biomass for energy in the European Union. In: Javier, S.L., Marios, A. (eds.) Publications Office of the European Union (2016)
10. IEA Bioenergy Technology Roadmap Delivering Sustainable Bioenergy, p. 89. IEA Publ. (2017)
11. Independent Group of Scientists appointed by the Secretary-General Global Sustainable Development Report 2019: The Future is Now – Science for Achieving Sustainable Development; United Nations, New York (2019)

12. Agrela, F., Cabrera, M., Morales, M.M., Zamorano, M., Alshaaer, M.: Biomass fly ash and biomass bottom ash. In: New Trends in Eco-efficient and Recycled Concrete, pp. 23–58 (2019). ISBN 9780081024805

13. ETIP Bioenergy Bioenergy in Europe (2020)

14. Bioenegy Europe Bioenergy: A renewable energy champion, pp. 1–7 (2021)

15. Milovanović, B., Štirmer, N., Carević, I., Baričević, A.: Wood biomass ash as a raw material in concrete industry. Građevinar **71**, 505–514 (2019). https://doi.org/10.14256/JCE.2546.2018

16. Udoeyo, F.F., Inyang, H., Young, D.T., Oparadu, E.E.: Potential of wood waste ash as an additive in concrete. J. Mater. Civ. Eng. **18**, 605–611 (2006). https://doi.org/10.1061/(ASC E)0899-1561(2006)18:4(605)

17. Freire, M., Lopes, H., Tarelho, L.A.C.: Critical aspects of biomass ashes utilization in soils: Composition, leachability PAH and PCDD/F. Waste Manag. **46**, 304–315 (2015). https://doi. org/10.1016/j.wasman.2015.08.036

18. Pesonen, J., Kuokkanen, T., Rautio, P., Lassi, U.: Bioavailability of nutrients and harmful elements in ash fertilizers: effect of granulation. Biomass Bioenerg. **100**, 92–97 (2017). https:// doi.org/10.1016/j.biombioe.2017.03.019

19. Directive of the European Parliament and of the Council of 30 May 2018 amending Directive 1999/31/EC on the landfill of waste. Off. J. Eur. Union **2018**, 100–108 (2018)

20. The European Committee for Standardization Building lime – Part 2: Test methods (EN 459-2) (2010)

21. Brocklebank, I.: The building limes forum. In: Brocklebank, I. (ed.) Building Limes in Conservation, 1st edn. Routledge (2012). ISBN 978-1-873394-95-3

22. Forster, A.M.: How hydraulic lime binders work - hydraulicity for beginners and the hydraulic lime family (2018)

23. RILEM Technical Committee TC 203-RHM Repair Mortars for Historic Masonry - State of the Art Report of RILEM Technical Committee TC 203-RHM 2016, 178

24. EN 1015-11 Methods of test for mortar for masonry – Part 11: Determination of flexural and compressive strength of hardened mortar

25. The European Committee for Standardization Building lime – Part 1: Definitions; specifications and conformity criteria (EN 459-1:2015), p. 52 (2015)

26. Alvarez, J.I., et al.: RILEM TC 277-LHS report: a review on the mechanisms of setting and hardening of lime-based binding systems. Mater. Struct. **54**(2), 1–30 (2021). https://doi.org/ 10.1617/s11527-021-01648-3

27. Adeleke, A.A., et al.: Sustainability of multifaceted usage of biomass: A review. Heliyon **7**, e08025 (2021). https://doi.org/10.1016/j.heliyon.2021.e08025

28. Baričević, A., Carević, I., Bajto, J.Š., Štirmer, N., Bezinović, M., Kristović, K.: Potential of using wood biomass ash in low-strength composites. Mater. (Basel). **14**, 1–24 (2021). https:// doi.org/10.3390/ma14051250

29. Thirumalini, S., Ravi, R., Rajesh, M.: Experimental investigation on physical and mechanical properties of lime mortar: Effect of organic addition. J. Cult. Herit. **31**, 97–104 (2018). https:// doi.org/10.1016/j.culher.2017.10.009

30. Elsen, J., Van Balen, K., Mertens, G.: Hydraulicity in historic lime mortars: a review. In: RILEM Bookseries (2012)

31. Silva, B.A., Ferreira Pinto, A.P., Gomes, A.: Influence of natural hydraulic lime content on the properties of aerial lime-based mortars. Constr. Build. Mater. **72**, 208–218 (2014). https:// doi.org/10.1016/j.conbuildmat.2014.09.010

32. Moropoulou, A., Bakolas, A., Moundoulas, P., Aggelakopoulou, E., Anagnostopoulou, S.: Strength development and lime reaction in mortars for repairing historic masonries. Cem. Concr. Compos. **27**, 289–294 (2005). https://doi.org/10.1016/j.cemconcomp.2004.02.017

Criteria for the Utilization of Perlite By-products in Traditional Mortars

Maria Stefanidou[1], Fotini Kesikidou[1](✉), Stavroula Konopisi[1],
Eirini-Chrysanthi Tsardaka[1], Vasiliki Pachta[1], Evangelia Tsampali[1],
and George Konstantinidis[2]

[1] Laboratory of Building Materials, School of Civil Engineering, Aristotle University of Thessaloniki, Thessaloniki, Greece
fotinikesi@gmail.com
[2] Perlite Hellas, Volos, Greece

Abstract. Climate change has become the main problems of humanity, affecting all the everyday actions and habits. The decrease of CO_2 emissions is mandatory for the protection of the environment and the construction sector has a 36% share of the global problem. On the other hand, conservation of cultural heritage remains one of the main targets in European level. In the case of conservation works, the materials and the techniques used follow specific regulations included in relative norms. The aim of this work is to combine the needs mentioned above for sustainable and qualitative materials that can be used for restoration works. In this frame, the properties (chemical, physical, mineralogical, mechanical) of four perlite by-products were tested and the results were reviewed to meet the criteria set by the regulations which they will determine the possibility to use them in air lime mortar production. Fineness, color, salt content, chemical composition and reactivity were determined. The alumisilicate content of the binders was measured to comply with regulations for natural pozzolans and the pozzolanicity index of the binders was also tested. The research results to the possibility to follow specific criteria to evaluate industrial by-products, saving natural resources and increasing the environmental profile of air lime-based mortars.

Keywords: perlite · by-products · traditional mortars · pozzolanicity · air lime · criteria

1 Introduction

Nowadays, climate change has become one of the biggest issues of humanity. The need to reduce CO_2 emissions is acute and has led to changes even in everyday life worldwide. Industry and construction sectors adopt new policies and rules to comply with the European regulations and strategies on the protection of the environment. Thus, over the last years, many scientists have focused on the production of "green" concrete, incorporating high volumes of by-products and recycled materials (fly ashes, slags, recycled concrete) [1–3].

© The Author(s), under exclusive license to Springer Nature Switzerland AG 2023
V. Bokan Bosiljkov et al. (Eds.): HMC 2022, RILEM Bookseries 42, pp. 374–385, 2023.
https://doi.org/10.1007/978-3-031-31472-8_29

On the other hand, research on traditional mortars is always on the foreground due to the importance of the preservation of cultural heritage. Materials and techniques used in repairing historic structures should meet certain criteria regarding their compatibility [4, 5] and durability [6, 7] and should follow specific regulations included in relative EN norms. At the same time, RILEM technical committees are publishing guidelines indicating the quality criteria that should be followed for a successful intervention [8–10].

Therefore, under these objectives of conservation and protection of the environment, several supplementary pozzolanic materials have been tested in cooperation with hydrated lime for the repairing of old mortars such as metakaolin [11], brick dust [12], perlite [13], waste building materials [14]. Perlite is a volcanic mineral (alumino-silicate in origin) used mainly in agriculture and in special applications in building materials (gypsum plasters, cement boards, light-weight concrete). Greece is among the first countries in perlite production with 25000 to 30000 tons per year. During its industrial process, an increasing number of by-products (around 10% of the material is a waste) mainly remain unexploited causing severe environmental problems due to their fineness and the difficulty in storing. Previous work of the authors has proven that some of these by-products can be potentially used as replacement of pozzolan in traditional grouts showing good mechanical and physical properties [15].

The aim of this work is to combine the needs mentioned above for sustainable and qualitative materials that can be used for restoration works. In this frame, four perlite by-products (D1S, D1C, D1CS and A3) were investigated to determine their suitability for use in traditional air lime mortars with low environmental footprint. For a material to be characterized as pozzolanic, there are some requirements given in standards, such as the content of aluminosilicate components and the pozzolanicity index of the binder. To this scope, chemical analysis of the binders was conducted and other properties were analyzed, such as color, pH, moisture content, fineness. Mortars with hydrated lime were produced, according to ASTM 593-95 [16], to measure their compressive strength for the pozzolanicity index. Flexural strength, dynamic modulus of elasticity, porosity, absorption and specific gravity of these mortars were also tested.

The research results to the evaluation of the measured properties according to regulative frames and the possibility to use perlite by-products to produce traditional mortars, saving natural resources and increasing the environmental profile of air lime-based mortars used in restoration.

2   Materials and Methods

Perlite production line follows four main steps (mining, drying, heating and packaging). After mining the collected material is crushed and dried. During this stage two types of unexpanded by-products derive (D1S and D1C). Afterwards, the dried perlite is heated and expanded and the remain powder in air filters is A3. The final step is the packing of the material. In this paper, four perlite by-products were tested: D1S, D1C, D1CS (mix of D1S at 10% and D1C at 90% as deriving from the production line) and A3 (expanded) to investigate their use in traditional mortars for repairing historic structures.

The physical (colour, specific gravity, pH, moisture content) and chemical properties of the perlite by-products were studied and pozzolanicity index of the binders was measured. Mechanical characteristics were tested such as dynamic modulus of elasticity – based on BS 1881-203: 1986 [17] – and flexural and compressive strength according to EN 1015-11 [18]. Physical characteristics (open porosity, absorption and apparent specific gravity) of the produced mortars were also determined based on RILEM CPC 11.3 method [19].

The colour of the raw materials was classified according to Munsell colour chart and particle size distribution was determined by Malvern P.S.A Mastersizer 2000 (laser scattering technique). The chemical composition of the raw materials was analysed using Atomic Absorption Spectroscopy (AAS) with AAnalyst 4400 and X-Ray Fluorescence Spectroscopy (XRF) with S8 Tiger, Brucker. The mineralogy of the materials was characterized by X-Ray Diffraction Spectroscopy (XRD – D2 Phaser, Bruker), and the amorphous content was determined with EVA V0.5 Software. Ionic chromatography (IC thermos Scientific DIONEX ICS-1100) was used to detect the soluble salts content of the binders.

To test the reactivity of the by-products, pozzolanicity index was measured with the production of lime-based mortars according to ASTM C593-95 (Standard Specification for Fly Ash and Other Pozzolans for Use with Lime) [16]. Based on the American Standard, mortar mixtures containing 180 g of hydrated lime, 360 g of the tested by-product and 1450 g of standard sand were produced and moulded in prismatic (40 × 40 × 160) mm and cubic (5 × 5 × 5) cm moulds. The produced mixtures (Table 1) were placed in an oven for 7 days at 54 °C, saturated air. After 7 days, the specimens were demoulded and kept in a chamber with 21 °C and RH95% for 28 days. Table 1 indicates the high need for water in the case of the sample containing A3 which is the expanded perlite by-product which is very light and fine material.

Table 1. Water content (w/w) and workability (cm) of fresh mortar mixture according to ASTM C593-95 protocol [16]

Mixture	Water/Binder ratio	Workability (cm)
D1S	0.65	15.5
D1C	0.69	15.0
D1CS	0.61	15.5
A3	1.57	16.0

3 Results

3.1 Physical Properties of Raw Materials

The tested perlite by-products are shown in Fig. 1 (a–d). Colour classification according to Munsell colour chart (Table 2) defines D1S having a yellow hue (2.5Y 8/1) and D1C having a yellow – red hue (7.5YR 8/1). D1CS is a mix of D1S and D1C with a yellow-red hue (10YR 8/1) and A3 is classified white.

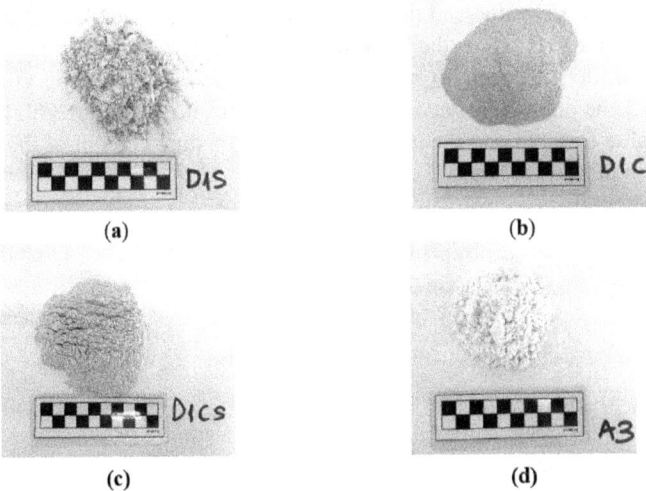

Fig. 1. Macroscopic images of perlite by-products (a) D1S, (b) D1C, (c) D1CS, (d) A3.

Table 2. Colour classification according to Munsell colour chart

By-product	Munsell colour
D1S	2.5Y 8/1
D1C	7.5YR 8/1
D1CS	10YR 8/1
A3	white

Particle size distribution of the by-products presented in Table 3, indicates different finenesses between the binders. For D1S, the distribution of the particles varies among 2–28 μm, whereas the coarser D1C varies between 17–189 μm. Particles of D1CS have a granulometry between 6–171 μm and A3 between 3.8–32 μm. Based on these results, D1S can be characterized as the finer among the four by-products and D1C the coarser one.

Table 3. Particle size distribution of the raw materials

Raw material	D(0.1) (μm)	D(0.5) (μm)	D(0.9) (μm)
D1S	2.202	8.913	28.634
D1C	17.269	78.123	188.891
D1SC	6.082	51.388	171.347
A3	3.810	12.417	32.197

Specific gravity, pH value and moisture content of the perlite by-products are given in Table 4. D1S and D1C show the highest specific gravity values (2.459 to 2.415 g/ml respectively), while A3 seems to be the lightest of the binders (1.462 g/ml) which is expected as it is he only expanded material. The pH results run from 8.80 to 9.82, with the highest value for D1S and the lowest for D1C. Moisture content of the materials is less than 1% in all the binders.

Table 4. Specific gravity, pH value and moisture content of tested by-products

By-product	Specific gravity (g/ml)	pH	Moisture content (%)
D1S	2.459	9.82	<1%
D1C	2.415	8.53	<1%
D1CS	2.343	9.02	<1%
A3	1.462	8.80	<1%

3.2 Chemical Properties of Raw Materials

The results of the chemical analysis by Atomic Absorption Spectroscopy and XRF are displayed in Table 5 and Table 6 respectively. Water soluble salts (%wt) (Table 5) were quantified by using Ion Chromatography. Loss of Ignition (LI%) was determined through the calcification from RT (room temperature up to 1000 °C. For the XRF examination, the deviation of the results of the raw materials led to repetition of the test with the binders grinded under 45 μm. However, the low specific gravity of A3 hampered the necessary process preparation of the sample for the XRF test, thus the test was unable to be performed. Table 7 compares the results of calcium oxide (CaO) and aluminosilicate components ($SiO_2 + Al_2O_3 + Fe_2O_3$) derived from the two methods.

Table 5. Chemical properties of the raw materials (wet chemical analysis assisted by Atomic Absorption Spectrometer)

Sample	SOLUBLE IN ACIDS % w.t.								SOLUBLE SALTS % w.t.		
	Na_2O	K_2O	CaO	MgO	Fe_2O_3	Al_2O_3	SiO_2	L.I.%	Cl^-	NO_3^-	SO_4^{2-}
D1S	2.67	2.21	1.61	0.40	0.99	16.26	71.95	3.91	0.01	–	0.01
D1C	2.35	2.11	1.26	0.34	0.88	16.04	73.95	3.07	0.03	<0.01	<0.01
D1CS	2.24	2.01	0.93	0.24	0.76	12.51	78.39	2.92	0.02	<0.01	–
A3	2.70	2.41	0.92	0.17	0.69	15.55	76.68	0.88	0.02	–	<0.01

Table 6. Results of XRF analysis

Oxide formula	D1S		D1C		D1CS
	raw	<45 μm	raw	<45 μm	raw
CaO (%)	2.31	2.29	1.57	1.57	1.34
Al_2O_3 (%)	10.1	14.20	6.82	14.82	8.67
SiO_2 (%)	57.7	70.85	43.2	71.20	52.7
Fe_2O_3 (%)	1.46	1.48	1.33	1.33	1.22
K_2O (%)	3.01	3.05	2.70	2.70	2.73
Na_2O (%)	2.88	2.81	1.50	1.50	2.13
MgO (%)	0.398	0.398	0.255	0.255	0.210
TiO_2 (%)	0.125	0.125	0.157	0.157	0.176
BaO (%)	0.168	0.168	–	–	–
MnO (ppm)	739	739	601	601	692
P_2O_5 (ppm)	479	479	–	–	–
SrO (ppm)	97.0	97.0	286	286	281
CuO (ppm)	91.5	91.5	–	–	–
Cl (%)	0.121	0.121		932 ppm	
ZrO_2 (ppm)	70.9	70.9	303	303	110

Table 7. Results of AAS and XRF analysis

Sample	CaO		Aluminosilicate components $(SiO_2 + Al_2O_3 + Fe_2O_3)$	
	AA	XRF	AA	XRF
D1S	1.61	2.29	89.2	86.53
D1C	1.26	1.57	90.87	87.35
DICS	0.93	1.34	91.66	62.59*
A3	0.92	–	92.92	–

* raw material without processing

Calcium oxide (CaO) determined by AAS varies from 0.92 to 1.61% for all perlite by-products, with D1S having the highest content. The aluminosilicate content $(SiO_2 + Al_2O_3 + Fe_2O_3)$ of the raw materials ranges from 89.2 (D1S) to 92.92% (A3). XRF analysis indicated similar results for the grinded samples (<45 μm). Calcium oxide (CaO) for D1C and D1S was detected 1.57% and 2.29% respectively and the aluminosilicate content 86.53% for D1S and 87.35% for D1C.

According to ASTM 618-03 (Standard Specification for Coal Fly Ash and Raw or Calcined Natural Pozzolan for Use in Concrete) [20], a material can be classified as natural pozzolan based on its content in aluminosilicate components (SiO_2 + Al_2O_3 + Fe_2O_3), which should be above 70%. Based on the results, it is observed that the aluminosilicate content of the perlite by-products is around 90% and exceeds the 70% limit set by the ASTM Standard. Silicon dioxide content (SiO_2) varies between 72–78%, which also complies with ASTM 618-5 [20]. The soluble salts content of the samples (Table 4) is generally low, with D1C having the highest value around 0.03% - percentage under the 0.15% limit given by ACI Specifications 301-10, 318-14, 329R-14, 332-14 and 349-13 [21–24].

X-Ray diffraction patterns of the samples are given in Fig. 2. Sample A3 can be characterized as amorphous without crystalline phases, whereas D1S, D1C and D1CS binders are amorphous with a small percentage in crystalline content, mainly aluminosilicate based. The amorphous content of D1S and D1C was found 63.6% and 60% respectively.

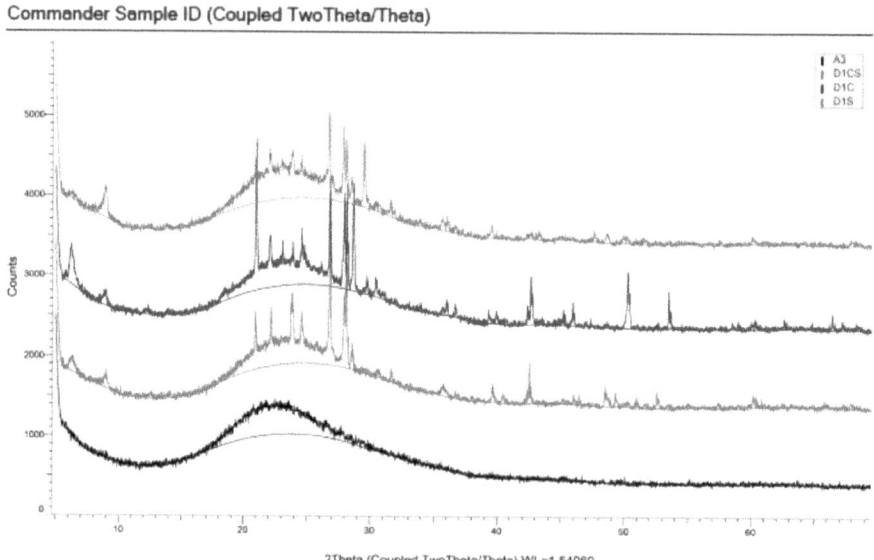

Fig. 2. XRD analysis of the perlite by-products.

3.3 Pozzolanicity Index

Pozzolanicity index was determined by measuring the mechanical properties (dynamic modulus of elasticity, flexural and compressive strength) of the produced mortars based on ASTM C593-95 [16]. The results are given in Figs. 3, 4 and 5 and the physical properties of the mortars (porosity, absorption, specific gravity) are presented in Table 8.

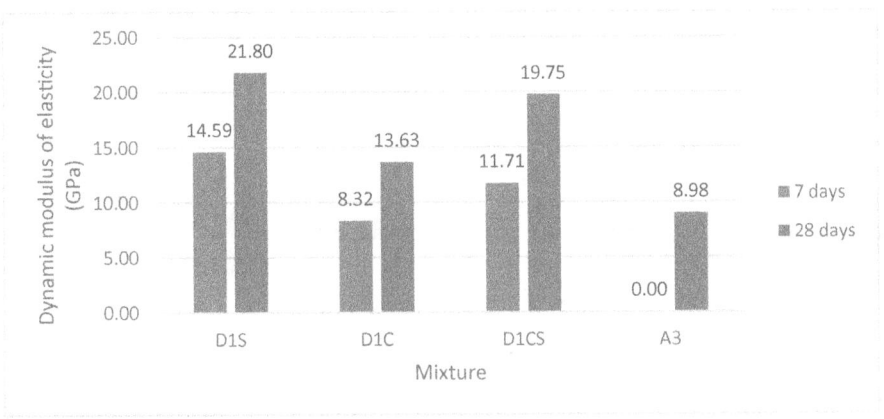

Fig. 3. Dynamic modulus of elasticity of mortars.

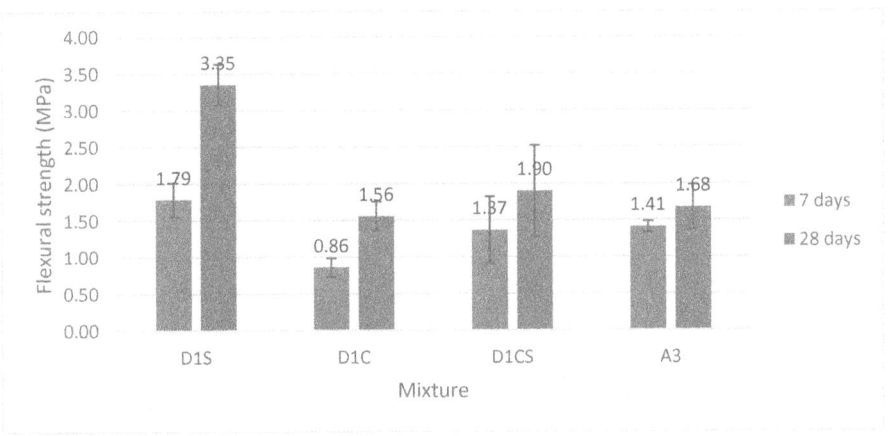

Fig. 4. Flexural strength of mortars.

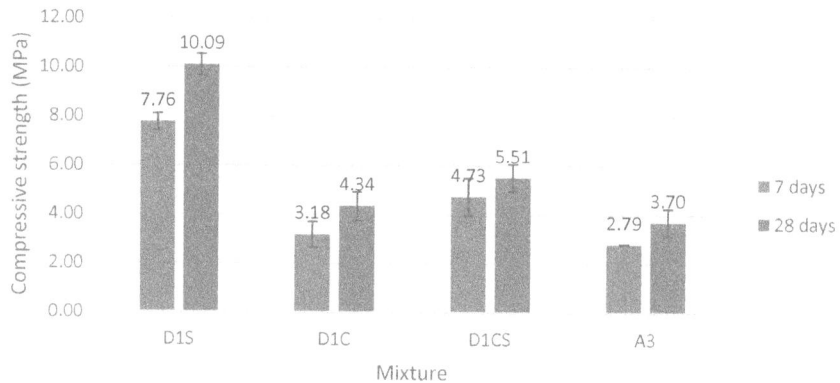

Fig. 5. Compressive strength of mortars.

Table 8. Physical properties of mortars

Mixture	Porosity (%)		Absorption (%)		Ap. Specific gravity	
	7 days	28 days	7 days	28 days	7 days	28 days
D1S	17.45	10.91	9.23	5.63	1.890	1.939
D1C	20.62	18.66	11.42	10.07	1.806	1.854
D1CS	23.77	14.08	18.08	7.32	1.118	1.924
A3	26.69	30.87	18.47	21.62	1.445	1.428

Based on the above, D1S seems to outweigh the other by-products in terms of strength with flexural strength of 28 days at 3.35 MPa and compressive strength of the same age at 10 MPa. D1CS follows with 1.9 MPa under flexure and 5.5 MPa under compression. D1C reaches 1.6 MPa and 4.3 MPa for flexure and compression respectively, whereas A3 1.68 MPa and 3.7 MPa. Therefore, it can be said that D1S, D1C and D1CS mortars comply with the limit of 4.1 MPa of compressive strength according to ASTM C593-95. Therefore, D1S has the highest pozzolanicity index followed by D1CS and D1C.

Porosity values of the mortars range from 10.91% to 30.87%, absorption from 9.23 to 18.08% and apparent specific gravity from 1.428 to 1.939. D1S has the lowest porosity and the highest specific gravity, whereas D1C, D1CS and A3 indicate higher porosity values, with the highest pertained to A3 due to the nature of the material (expanded and under thermal process). Overtime, porosity and absorption of all mortars is reduced.

4 Discussion

Regarding the results for D1S, based on the fineness of the material, the content in aluminosilicate oxides and the high compressive strength (>4.1 MPa), it can be concluded that D1S could be characterized as a binder with high pozzolanic behavior. According to ASTM C618According to ASTM C618-5 [20], it can be classified as a natural poz-zolan (type N) and can be used without any processing in the production of lime-based mortars.

D1C and D1CS indicate a lower pozzolanic behavior, although complying with the limit of compressive strength set by ASTM C593-95 (4.1 MPa) [16]. The binders have a particle distribution around 6-189 µm, which probably affects their reactivity, but their content in aluminosilicate oxides is high. Thus, they could be characterized as binders with a medium pozzolanic behaviour and could be used - without processing - in the production of mortars of low and medium strength. However, after grinding under 45 µm, they could potentially have improved properties.

A3 is an amorphous material, light with no crystalline phases and low reactivity with lime. It is very fine with low specific gravity; hence it could be used mainly as a filler to produce lightweight mortars of low strength. Based on its properties, it is believed that mortars containing A3 by-product could present good thermal properties and are worth to be investigated.

According to the chemical analysis of the materials, it was demonstrated by both AAS and XRF that they are binders with high aluminosilicate content, a high amount of which is amorphous. Moreover, the particle size of the material influences the chem-ical composition concerning only the silicon oxide and aluminium oxide content (XRF results, Table 6).

According to EN 998-2 (Specification for mortar for masonry—Part 2: Masonry mortar)[25], masonry mortars are classified based on their compressive strength in dif-ferent categories (M1, M2.5, M5, M10, M15, M20, Md) with values under compression of 1 MPa, 2.5 MPa, 5 MPa, 10 MPa, 20 MPa and over 20 MPa respectively. In EN 998-1 (Specification for mortar for masonry - Part 1: Rendering and plastering mortar) [26] renders and plasters are classified depending on their compressive strength in four categories CSI (0.4–2.5 MPa), CS II (1.5–5 MPa), CS III (3.5–7.5 MPa), CS IV (≥6 MPa). Consequently, it can be concluded that D1S can be used as a pozzolanic mate-rial in mortars and grouts, producing mortars of classes M5-M10 and renders/plasters type CS III-CS IV. D1C and D1CS by-products can be used in mortars in full or partial substitution of natural pozzolan resulting in mortars M2.5–M5 and renders/plasters of type CS II. Additionally, their use in grouts should be also investigated. A3 could be used in restoration mortars partially substituted natural pozzolan to produce low strength mortars (M1) and renders/plasters (type CS I).

5 Conclusions

In this paper the criteria set by regulations to characterize and use a pozzolanic material in air-lime mortars that could be used for restoration works have been applied in four by-products of a perlite industry. The materials were tested in relation to their physical and chemical properties. So, their fineness, specific gravity, pH and moisture content as well as their chemical and mineralogical composition was determined through a thorough instrumental analysis. Beyond that, the pozzolanicity index of these materials was measured by producing standard air lime mortars based on ASTM C593-95. The mechanical and physical properties of the standard mortars were recorded at 7 and 28 days.

The results show that D1S meets the requirements of all the standards as it is a fine and reactive material that has the potential to be used as a binder with high pozzolanic properties. D1C and D1CS by-products can be characterized as binders with low - medium pozzolanic behavior. They are medium coarse in size. A3 is a very fine and light powder with low pozzolanic reaction. It can be used partially in air lime-pozzolan mortars to produce lightweight mortars probably of low strength but further tests are necessary.

The whole study indicates that the by-products of perlite industry can be utilized in air lime mortar production after thorough examination of the properties these materials possess. The criteria set by regulations can be applied to by-products in order to produce high-quality repair mortars which also have environmentally friendly properties. Further investigation is necessary to investigate the durability and long-term behavior of the mortars when by products are incorporated.

Acknowledgements. Co-financed by the European Regional Development Fund of the European Union and Greek national funds through the Operational Program Competitiveness, Entrepreneurship and Innovation, under the call RESEARCH – CREATE - INNOVATE (project code: T2EDK-01105).

References

1. Papachristoforou, M., Mitsopoulos, V., Stefanidou, M.: Use of by-products for partial replacement of 3D printed concrete constituents; rheology, strength and shrinkage performance. Frat. ed Integrita Strutt. **13**(50), 526–536 (2019). https://doi.org/10.3221/IGF-ESIS.50.44
2. Braga, M., de Brito, J., Veiga, R.: Reduction of the cement content in mortars made with fine concrete aggregates. Mater. Struct. **47**(1–2), 171–182 (2013). https://doi.org/10.1617/s11527-013-0053-1
3. Papayianni, I., Anastasiou, E.: Concrete incorporating high volumes of industrial by-products. In: Role Cem. Sci. Sustain. Dev. - Proc. Int. Symp. - Celebr. Concr. People Pract., pp. 595–604 (2003). https://doi.org/10.1680/rocisd.32477.0058
4. Papayianni, I., Pachta, V., Stefanidou, M.: Analysis of ancient mortars and design of compatible repair mortars: the case study of Odeion of the archaeological site of Dion. Constr. Build. Mater. **40**, 84–92 (2013). https://doi.org/10.1016/j.conbuildmat.2012.09.086
5. Stefanidou, M., Pavlidou, E.: Scanning mortars to understand the past and plan the future for the maintenance of monuments. Scanning **2018** (2018). https://doi.org/10.1155/2018/7838502
6. Strauss, A.: Proceedings PRO 128
7. Arizzi, A., Viles, H., Cultrone, G.: Experimental testing of the durability of lime-based mortars used for rendering historic buildings. Constr. Build. Mater. **28**(1), 807–818 (2012). https://doi.org/10.1016/j.conbuildmat.2011.10.059
8. Hughes, J.J., Lindqvist, J.E.: RILEM TC 203-RHM: repair mortars for historic masonry: the role of mortar in masonry: an introduction to requirements for the design of repair mortars. Mater. Struct. **45**(9), 1287–1294 (2012). https://doi.org/10.1617/s11527-012-9847-9
9. Alvarez, J.I., et al.: RILEM TC 277-LHS report: a review on the mechanisms of setting and hardening of lime-based binding systems. Mater. Struct. **54**(2), 1–30 (2021). https://doi.org/10.1617/s11527-021-01648-3

10. Válek, J., Hughes, J.J., Pique, F., Gulotta, D., van Hees, R., Papayiani, I.: Recommendation of RILEM TC 243-SGM: functional requirements for surface repair mortars for historic buildings. Mater. Struct. **52**(1), 1–18 (2019). https://doi.org/10.1617/s11527-018-1284-y

11. Arizzi, A., Cultrone, G.: Aerial lime-based mortars blended with a pozzolanic additive and different admixtures: a mineralogical, textural and physical-mechanical study. Constr. Build. Mater. **31**, 135–143 (2012). https://doi.org/10.1016/j.conbuildmat.2011.12.069

12. Navrátilová, E., Rovnaníková, P.: Pozzolanic properties of brick powders and their effect on the properties of modified lime mortars. Constr. Build. Mater. **120**, 530–539 (2016). https://doi.org/10.1016/j.conbuildmat.2016.05.062

13. Bulut, Ü.: Use of perlite as a pozzolanic addition in lime mortars. Gazi Univ. J. Sci. **23**(3), 305–313 (2010)

14. Pašalić, S., Vučetić, S., Zorić, D.B., Ducman, V., Ranogajec, J.: Pozzolanic mortars based on waste building materials for the restoration of historical buildings. Chem. Ind. Chem. Eng. Q. **18**(2), 147–154 (2012). https://doi.org/10.2298/CICEQ110829056P

15. Pachta, V., Papadopoulos, F., Stefanidou, M.: Development and testing of grouts based on perlite by-products and lime. Constr. Build. Mater. **207**, 338–344 (2019). https://doi.org/10.1016/j.conbuildmat.2019.02.157

16. ASTM C 593-95: Standard specification for fly ash and other pozzolans for use with lime for soil stabilization

17. BS 1881:203: Testing concrete — part 203: recommendations for measurement of velocity of ultrasonic pulses in concrete (1986)

18. EN 1015-11:2006: Methods of test for mortar for masonry - part 11: determination of flexural and compressive strength of hardened mortar (2006)

19. RILEM TC: CPC 11.3 absorption of water by concrete by immersion under vacuum. RILEM Recomm. Test. Use Constr. Mater., pp. 36–37 (1984)

20. ASTM C618-03: Standard specification for coal fly ash and raw or calcined natural pozzolan for use (2003)

21. ACI Committee 301 and American Concrete Institute: Specifications for structural concrete : an ACI standard. American Concrete Institute (2010)

22. ACI 318-14: Building code requirements for structural concrete. Commentary on building code requirements for structural concrete (2014)

23. ACI 329: Report on performance-based requirements for concrete (2014)

24. ACI 349-13: Code requirements for nuclear safety-related concrete structures and commentary (2014)

25. EN 998-2: Specification for mortar for masonry - part 2: masonry mortar

26. EN 998-1: Specification for mortar for masonry - part 1: rendering and plastering mortar

Development and Testing of Lime Based Mortars Using Perlite By - Products

Maria Stefanidou[1], Vasiliki Pachta[1(✉)], and George Konstantinidis[2]

[1] Laboratory of Building Materials, School of Civil Engineering, Aristotle University of Thessaloniki, Thessaloniki, Greece
vpachta@civil.auth.gr
[2] Perlite HELLAS, Volos, Greece

Abstract. Perlite is a volcanic rock containing high amount of amorphous material as well as alumino-silicate minerals. The outcoming product (expanded perlite) is often used in construction, due to its light weight and insulating properties. During its industrial process, an increasing number of by-products results that mainly remains unexploited as the fineness of these materials renders them difficult to store. In this study, an effort has been made to experimentally study the influence of two by-products (D1S, D1C), in lime based mortars. These materials have been used as binders. To this direction, 9 mortar mixtures where manufactured and tested where natural pozzolan was gradually replaced by the perlite by-products. The Binder/Aggregate ratio in all mixtures was maintained at 1/2, whereas aggregates were natural of siliceous origin and gradation 0–4 mm. The physico-mechanical properties of the specimens were tested at the age of 28 and 90 days. From the correlation of the results, it was asserted that the partial or even total substitution of natural pozzolan by perlite by-products, enhanced the mortars' physical and mechanical properties. It maybe therefore concluded that the exploitation of waste perlite in the construction sector is feasible, leading to the development of effective, low-cost and environmentally friendly products for specific applications.

Keywords: Perlite by-products · Lime mortars · Physical properties · Mechanical properties

1 Introduction

Perlite is an amorphous mineral, coming from volcanic deposits [1–3]. Perlite is a very promising industrial material as during the last 60 years less than the 1% of the existing deposits have been mined. It mainly consists of SiO_2 and Al_2O_3, with low amounts of other compounds (sodium, potassium, iron, calcium, magnesium) [1, 2, 4]. In its crystallized form it has a relatively high water content (2–5% w/w) that is evaporated when it is rapidly heated (900–1200 °C), expanding its volume up to 20 times [1, 2, 4]. Expanded perlite (EP), resulting from the heating process of perlite, has been used in construction during the last decades, due to its light weight and insulating properties [1–3, 5–7].

© The Author(s), under exclusive license to Springer Nature Switzerland AG 2023
V. Bokan Bosiljkov et al. (Eds.): HMC 2022, RILEM Bookseries 42, pp. 386–395, 2023.
https://doi.org/10.1007/978-3-031-31472-8_30

The largest perlite mines are located in China, Greece, US and Turkey resulting in the 95% of the worldwide perlite production (3,4 million tons for the year 2020), whereas Greece is the largest exporting country [8]. In the US, the perlite consumption in 2020 was 610,000 tons, while the applications of EP were for construction products (53%), fillers (16%), horticultural aggregates (16%), filter aids (12%), insulation and other uses (3%) [8]. Perlite mines are generally located in remote areas, characterized by a limited environmental impact and overburden to manage [8, 9]. It is significant to point out that during the last 60 years less than the 1% of the worldwide perlite reserves have been mined and used, rendering it a well promising material for various future uses [9].

Recent research on the influence of perlite in concrete and mortars, proved its beneficial role, due to its low bulk density (32–150 kg/m^3), thermal conductivity (0.04–0.06W·m-1K-1), as well as its high heating resistance (melting point: 1260–1343 °C) [1, 3–5]. It has been either used for substituting cement and/or aggregates in concrete [1, 5, 6, 10, 11] or cement-based mortars [7, 12, 13] and developing elaborated materials for specific applications (i.e. light-weight concrete, fire resistant plasters).

During the industrial process of perlite, a short, however countable amount of by-products results [11, 12]. In the studied plant (Perlite Hellas, Volos, Greece), the production line of perlite includes three stages (crushing, heating and expanding), during which a total amount of 10% of by-products is created. These by-products, obtained by air separation, are in a powder form and are usually deposited in the environment (in large areas near the plant). Taking into account, the increasing production and applications of EP, it is obvious that the secondary products deriving from its process will be also increased, with negative environmental effects. To this direction, their exploitation to various applications is of paramount interest and should be further investigated.

Recent studies [10, 11] have documented that waste perlite, may present induced pozzolanic reactivity and could be used as an alternative pozzolanic material in cement-based mortars and concretes, enhancing their properties. Additionally, the substitution of natural pozzolan with waste perlite in lime-based grouts [2], has proved to be beneficial for their overall performance.

Of natural pozzolan by perlite by-products, led to positive results. To this end it may be stated that the exploitation of waste perlite in the construction sector is feasible, leading to multiple environmental In this paper, the influence of two by-products deriving in the Greek plant (Perlite Hellas), in lime-based mortars was experimentally studied. To this direction, 9 mortar mixtures where manufactured and tested, where natural pozzolan was gradually replaced by perlite by-products. The physico-mechanical properties of the specimens were tested, showing that the partial or even total substitution benefits.

2   Materials and Methods

During this study, two perlite by-products were envisaged, named after D1S and D1C, coming from the second stage of the perlite industrial process in the Greek plant (Perlite Hellas). They are separated during perlite heating, with air, without being further milled or processed. According to their properties (Table 1), D1S concerns a fine material with induced pozzolanic properties (pozzolanicity index: 7.8 MPa), while D1C presents a slightly lower apparent specific density, increased grain diameter, and lower pozzolanicity (3.2 MPa). Regarding their chemical characteristics (Table 2), both waste perlites mainly consist of alumino-siliceous compounds, while their chemical composition slightly differs.

Table 1. Characteristics of the raw materials.

Binder	App. Specific density g/cm^3	Pozzolanic activity index (ASTM C311:77) (MPa)	Particle size analysis (laser granulometry) Grain diameter (μm) of volume fractions (%)	
			D(0.5) (μm)	D(0.9) (μm)
Hydrated lime powder	2.471	-	3.09	10.80
Milos pozzolan	2.403	10.5	4.30	11.60
D1S	2.459	7.76	8.91	28.63
D1C	2.415	3.18	78.12	188.89

Table 2. Chemical characteristics of the binding agents used (wet chemical analysis assisted by Atomic Absorption).

Binder	Total Oxides (% w/w)							
	Na$_2$O	K$_2$O	CaO	MgO	Fe$_2$O$_3$	Al$_2$O$_3$	SiO$_2$	L.I.
Hydrated lime powder	0.22	0.02	73.08	2.85	0.01	0.08	0.07	23.57
Milos pozzolan	1.66	2.67	2.91	0.44	1.41	10.90	71.10	8.89
D1S	2.67	2.21	1.61	0.40	0.99	16.26	71.95	3.91
D1C	2.35	2.11	1.26	0.34	0.88	16.04	73.95	3.07

Taking into account the individual properties of D1S and D1C and in an effort to produce effective modified mortars, nine mixtures were manufactured, according to Table 3, including a reference mortar containing lime and natural pozzolan (LP). In compositions, natural pozzolan was gradually replaced by the perlite by-products in a proportion of 25%, 50%, 75% and 100%. Their code number respectively depicted this replacement, according to the by-product used (i.e. the synthesis D1S-25 refers to the replacement of natural pozzolan by D1S, in a proportion of 25%).

The Binder/Aggregate (B/A) ratio was maintained in all mixtures at 1/2, whereas aggregates were natural of siliceous origin and gradation 0–4 mm. In order to reduce the water demand, a small proportion of polycarboxylate superplasticizer (1% w/w of binders) was added [13–15]. The Water/Binder (W/B) ratio was adjusted for achieving workability 15 ± 1 cm, in accordance to EN1015-3 [16]. EN1015–11 [17] was followed for manufacturing and curing mortars, preparing 10 prismatic specimens (dimensions 4 × 4 × 16 cm) for each composition.

Table 3. Constituents and proportions of the mortar mixtures.

Raw materials	Mortar compositions (parts of weight)								
	LP	D1S-25	D1S-50	D1S-75	D1S	D1C-25	D1C-50	D1C-75	D1C
Hydrated lime powder	1	1	1	1	1	1	1	1	1
Milos pozzolan	1	0.75	0.5	0.25	-	0.75	0.5	0.25	-
D1S	-	0.25	0.5	0.75	1	-	-	-	-
D1C	-	-	-	-	-	0.25	0.5	0.75	1
Siliceous sand (0-4mm)	4	4	4	4	4	4	4	4	4
W/B ratio	0.71	0.62	0.60	0.62	0.57	0.59	0.52	0.53	0.52
Workability (cm) (EN1015–3:1999)	15.4	14.6	14.9	14.7	15.0	14.9	14.5	14.8	15.0

Twenty-eight and ninety days after their manufacture, the physico-mechanical properties of the compositions were determined, regarding porosity, absorption and apparent specific gravity (RILEM CPC 11.3) [18], capillary absorption index (EN 1015–18: 2002) [19], dynamic modulus of elasticity (BS 1881-203:1986) [20], flexural and compressive strength (EN1015–11) [17]. All results were comparatively evaluated in order to proceed to conclusions regarding the overall performance of the modified compositions, as well as the impact of perlite by-products.

3   Results and Discussion

3.1   Water Content

Regarding the water demand of the mixtures (Table 3), it was asserted that the reference mortar (LP) showed the highest W/B ratio (0.71). The gradual substitution of natural pozzolan by perlite by-products reduced the water content around 15–30%. The lowest W/B ratio was attained in the D1C addition, whilst in D1S a significant reduction was also observed. Generally, the higher proportion of by-products led to a lower W/B ratio, with the D1S and D1C mixtures to show the lower water demand.

3.2   Physical Properties

In Table 4, the physical properties of the mortar compositions are presented. According to the results, the 90d values for porosity and absorption were increased, whereas apparent specific gravity was decreased. Generally, porosity ranged from 23.7 to 28.4% at 28 and 27.3 to 33.2% at 90 days, with the lowest values to be attributed to D1S-50 (28d) and D1C-50 (90d). Respectively, the highest values were given by LP and D1S at both ages. Absorption followed porosity trend at both ages, ranging from 13.3 to 17.1% at 28 and 15.6 to 20.2% at 90d. Apparent specific gravity varied from 1.68 to 1.78 at 28 and 1.66

to 1.76 at 90 days. The lowest values were given by D1S at both ages and the highest by D1C.

From the evaluation of the results it was concluded that the replacement of natural pozzolan by D1S up to the proportion of 75%, led to a decrease of porosity and absorption and a respective increase of apparent specific gravity. This may be linked with the lower W/B ratio of the compositions, as well as the larger grain diameter of D1S (compared to natural pozzolan) that could block the smaller voids of the structure.

Table 4. Physical properties of the mortar compositions.

Compositions	Porosity (%)		Absorption (%)		Ap. Spec. Gravity	
	28d	90d	28d	90d	28d	90d
LP	28.44	32.85	16.75	19.79	1.70	1.66
D1S-25	23.74	29.84	13.33	17.38	1.78	1.72
D1S-50	24.54	30.22	13.80	17.51	1.78	1.73
D1S-75	26.32	29.41	15.09	17.10	1.74	1.72
D1S	28.73	33.19	17.11	20.22	1.68	1.64
D1C-25	28.54	28.79	16.43	16.66	1.74	1.73
D1C-50	27.94	27.39	15.89	15.65	1.76	1.75
D1C-75	26.71	29.09	15.36	16.62	1.74	1.75
D1C	27.01	28.50	15.33	16.17	1.76	1.76

The 28d capillary absorption presented in Fig. 1, was also influenced by the by-products addition, resulting to a final index ranging from 1.3 to 3.9 g/cm^2. The highest values were recorded for series D1C and especially in the highest replacement proportions (D1C-75, D1C), whilst the lowest for compositions D1S-25 and D1S-50. It was asserted that the lower proportion of D1S in the mixtures, seemed to block the interconnectivity of the capillary pores, probably due to its grain size, reducing capillary absorption. D1C on the other hand, induced capillary absorption, showing the higher values.

3.3 Physical Properties

Mechanical properties were influenced by the presence of perlite by-products, while their 28d to 90d development was remarkable in most cases (Fig. 2–4). Dynamic modulus of elasticity (Fig. 2) ranged from 2 to 12 GPa at 28 days and 3.5 to 14.5 GPa at 90 days. The highest 28d values were recorded for D1S-25 and LP and the lowest in series D1C (especially with the largest replacement of pozzolan). Generally, when the proportion of by-products was increased the modulus of elasticity decreased. At 90 days, there was a significant increase, except for the reference mortar. This increase, ranging from 20 to 100% was mainly recorded in series D1C. D1S-25 and D1S-75 showed the highest 90d values, while D1S-50 and D1C-25 had similar to LP values.

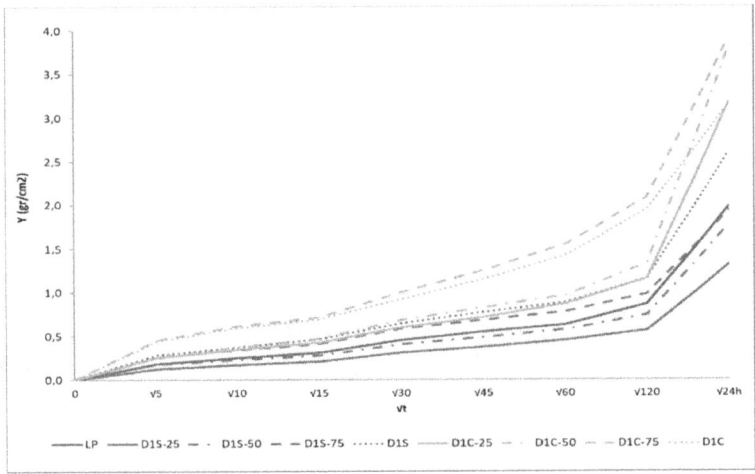

Fig. 1. Capillary absorption of the mortar mixtures.

Flexural strength (Fig. 3) varied from 0.1 to 1.4MPa at 28 days and 0.4 to 3.4 MPa at 90, showing also a great development rate. At 28 days the highest strength was given for compositions D1S-50 and LP, whereas at 90 for mixtures D1S-75, D1S-50 and LP. D1S-25 and D1C-25 presented a comparable to LP strength level. In the D1C series, the amount of D1C seemed to have a proportional negative effect in flexural strength, both in 28 and 90 days, however in D1S series results varied.

Compressive strength (Fig. 4) ranged from 0.7 to 5.1 MPa at 28 days and 1 to 7.8 MPa at 90. The highest 28d values were attributed to LP and D1S-25, whereas the 90d ones to series D1S. D1C-25 and LP values were again similar. In series D1C, the D1C amount also seemed to have a proportional negative effect in compressive strength, in both testing ages. However, the 90d values for series D1S were similar (6.9–7.8 MPa) and higher than LP (6.5 MPa).

Generally, it may be asserted that the strength development rate was significantly lower in the modified compositions, compared to the reference one, influenced by the type and proportion of the perlite by-product. D1S seemed to result in performable mortars, especially regarding the 90d compressive strength level, which in all cases exceeded LP values. All pozzolan substitutions had a beneficial impact, even the total one, showing that D1S can be used as an alternative pozzolanic material in lime-based mortars. D1C on the other hand presented a lower strength level, especially in the cases of a high amount. In the lowest proportion (D1C-25), it presented comparable with the reference mortar results, showing that its use could be feasible in low proportions.

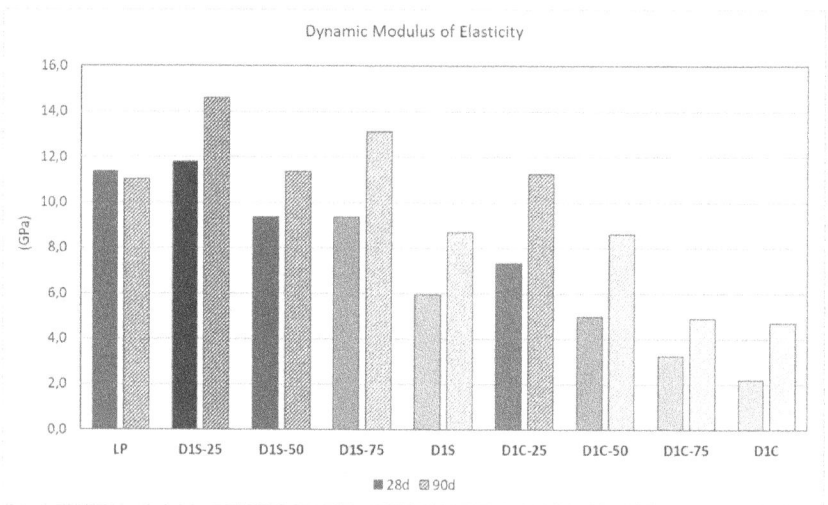

Fig. 2. Dynamic Modulus of Elasticity of the mortar mixtures.

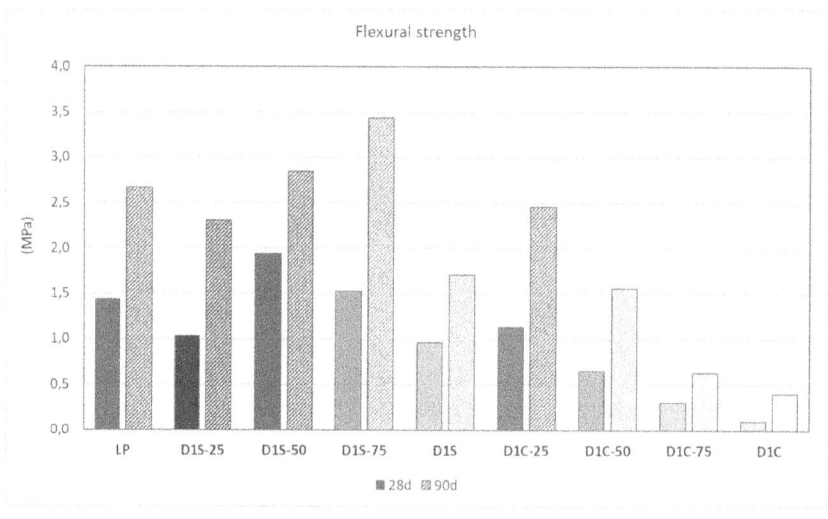

Fig. 3. Flexural strength of the mortar mixtures.

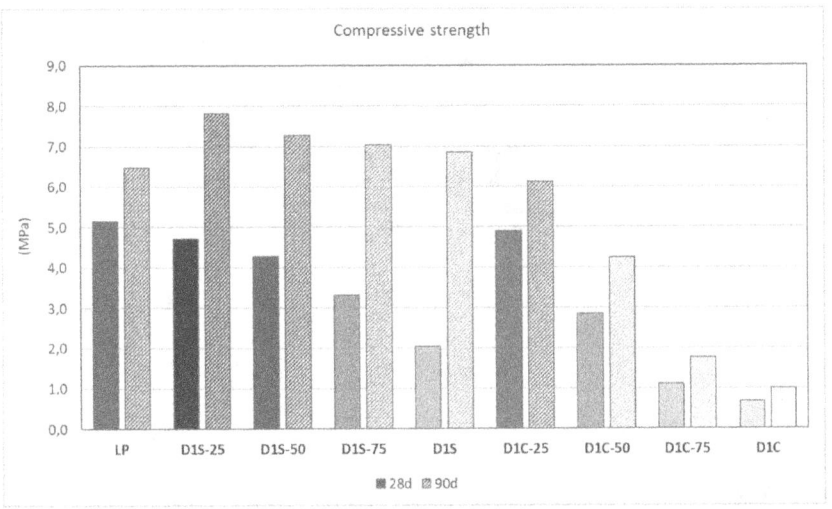

Fig. 4. Compressive strength of the mortar mixtures.

4   Conclusions

Generally, the partial and even total replacement of natural pozzolan by perlite by-products, for manufacturing restoration mortars for historic building, seems to be a feasible and environmental beneficial approach. Regarding the latter, the produced materials could be characterized by a low environmental footprint due to the limitation of the mining of natural pozzolan needed for their manufacture, as well as the exploitation of perlite by-products that would be otherwise deposited outdoor.

The positive influence of perlite by-products on lime-based mortars, could be synopsized as following:

– Reduction of the W/B ratio around 15–30%. The higher proportion of by-products the lower was the water demand in all cases.
– Reduction of porosity, absorption and capillary absorption, especially for the pozzolan partial substitution by D1S (up to 75%).
– The low strength development rate of the modified mortars, led to enhanced 90d mechanical characteristics that should be further envisaged.
– 90d Dynamic Modulus of elasticity was increased by the pozzolan substitution (up to 75%) by D1S, as well as flexural and compressive strength.

From the above- mentioned remarks, it may be asserted that the type and proportion of perlite by-products are of paramount importance, influencing directly the physical and mechanical properties of the modified mortars. The partial substitution of natural pozzolan with D1S (up to 75%), resulted in mortars with a better performance compared to the conventional (LP) mixture. The total substitution had also positive results, showing that D1S could be used as an alternative pozzolanic material in lime-based mortars. D1C on the other hand, presented a lower strength level, especially in the cases of its higher proportion. In the lower one (25%), its behaviour was in a good level, showing that its use could be feasible in a low proportion and for specific applications.

Acknowledgements. The study was implemented during the Project 'PerliMat - Exploitation of industrial products from the perlite process for the development of increased added value, ready mixed restoration materials', Co-financed by the European Regional Development Fund of the European Union and Greek national funds through the Operational Program Competitiveness, Entrepreneurship and Innovation, under the call RESEARCH – CREATE - INNOVATE (project code: T2EDK-01105).

References

1. Rashad, A.M.: A synopsis about perlite as building material - a best practice guide for civil engineer. Constr. Build. Mater. **121**, 338–353 (2016)
2. Pachta, V., Papadopoulos, F., Stefanidou, M.: Development and testing of grouts based on perlite by-products and lime. Constr. Build. Mater. **207**, 338–344 (2019)
3. Papa, E., et al.: Characterization of alkali bonded expanded perlite. Constr. Build. Mater. **191**, 1139–1147 (2018)
4. Maxim, L.D., Niebo, R., McConnell, E.E.: Perlite toxicology and epidemiology - a review. Inhalation Toxicol. **26**(5), 259–270 (2014)
5. Turkmen, I., Kantarcı, A.: Effects of expanded perlite aggregate and different curing conditions on the physical and mechanical properties of self-compacting concrete. Build. Environ. **42**, 2378–2383 (2007)
6. Wang, L., et al.: Strength properties and thermal conductivity of concrete with the addition of expanded perlite filled with aerogel. Constr. Build. Mater. **188**, 747–757 (2018)
7. Palomar, I., Barluenga, G.: A multiscale model for pervious lime-cement mortar with perlite and cellulose fibers. Constr. Build. Mater. **160**, 136–144 (2018)
8. U.S. Geological Survey, Mineral Commodity Summaries 2021, USGS, Virginia (2021)
9. Perlite Institute, Sustainability fact sheet (2010). https://www.perlite.org/wp-content/uploads/2018/03/sustainability-factsheet-perlite.pdf
10. Fodil, D., Mohamed, M.: Compressive strength and corrosion evaluation of concretes containing pozzolana and perlite immersed in aggressive environments. Constr. Build. Mater. **179**, 25–34 (2018)
11. Rózycka, A., Pichór, W.: Effect of perlite waste addition on the properties of autoclaved aerated concrete. Constr. Build. Mater. **120**, 65–71 (2016)
12. Kotwica, L., Pichór, W., Kapeluszna, E., Rózycka, A.: Utilization of waste expanded perlite as new effective supplementary cementitious material. J. Clean. Prod. **140**, 1344–1352 (2017)
13. Stefanidou, M., Pachta, V.: Influence of perlite and aerogel addition on the performance of cement-based mortars at elevated temperatures, SBE19, Sustainability in the built environment for climate change mitigation, Thessaloniki, 23–25 October 2019 (2019)
14. Pachta, V., Konopisi, S., Stefanidou, M.: The influence of brick dust and crushed brick on the properties of lime-based mortars exposed at elevated temperatures. Constr. Build. Mater. **296**, 123743 (2021)
15. Papayianni, I., Pachta, V., Stefanidou, M.: Analysis of ancient mortars and design of compatible repair mortars: the case study of Odeion of the Archaeological site of Dion. Constr. Build. Mater. **40**, 84–92 (2013)
16. BS EN1015–3, Methods of test for mortar for masonry, Part 3: Determination of consistence of fresh mortar (by flow table) (1999)

17. BS EN1015–11, Methods of test for mortar for masonry, Part 11: Determination of flexural and compressive strength of hardened mortar (1999)
18. RILEM CPC 11.3, Absorption of water with immersion under vacuum, Materiaux et Constructions 17, 101 391–394 (1984)
19. BS EN1015–18, Methods of test for mortar for masonry, Part 18: Determination of water absorption coefficient due to capillary action of hardened mortar (2002)
20. BS 1881–203, 1986, Testing concrete. Part 203: Recommendations for measurement of velocity of ultrasonic pulses in concrete

Durability of Lime Mortars Treated with Ammonium Phosphate

Greta Ugolotti, Giulia Masi, and Enrico Sassoni[✉]

University of Bologna, Bologna, Italy
enrico.sassoni2@unibo.it

Abstract. The present paper aims at comparing diammonium hydrogen phosphate (DAP) and nanolimes (NL) as consolidants for the preservation of lime mortars. The durability of the two consolidants was evaluated in terms of resistance to freezing thawing and salt crystallization cycles. The results of the study point out that, compared to the untreated reference and NL-treated samples, DAP-treated mortars showed much improved resistance to freezing-thawing cycles. A lower benefit was found in the case of salt crystallization cycles, as DAP-treated samples exhibited a behavior substantially similar to untreated and NL-treated ones. The different improvement in durability observed for the two weathering processes could be ascribed, on the one hand, to the slight modification of the pore size distribution after treatment (which may lead to increased crystallization pressure) and, on the other hand, to the severity of the salt weathering test (involving the use of a saturated $Na_2SO_4 \cdot 10H_2O$ solution). All things considered, the potential of the DAP treatment for the conservation of lime-based mortars is confirmed.

Keywords: Hydroxyapatite · Ammonium phosphate · Consolidation · Freeze-thaw · Salt weathering · Lime-mortars

1 Introduction

Considering the limitations of traditional consolidants for the conservation of carbonate materials, the use of a new phosphate-based product was proposed [1]. Diammonium hydrogen phosphate (DAP) is used as a precursor to form hydroxyapatite (HAP) on calcitic substrates, which is able to improve the cohesion between grains and thus the mechanical properties of the material.

Effectiveness and compatibility, that have been investigated in previous studies [2], are only two of the main requirements that any consolidant must ideally fulfil [3, 4]. In fact, consolidants must also guarantee a proper durability, maintaining their efficacy and compatibility after exposure to environmental weathering, without leading to release of harmful products because of ageing [3]. After consolidation, actually, the consolidated part should not experience accelerated deterioration compared to the untreated substrate, for instance if a surface crust is formed. To characterize the behavior of consolidated

V. Bokan Bosiljkov et al. (Eds.): HMC 2022, RILEM Bookseries 42, pp. 396–402, 2023.
https://doi.org/10.1007/978-3-031-31472-8_31

materials, accelerated ageing tests are often applied, especially when the performance of a new consolidating products needs to be evaluated [4].

In the case of porous materials, such as lime mortars, the most common and dangerous weathering processes are freezing-thawing cycles and salt crystallization cycles [5]. Therefore, in the present study HAP-treated lime mortars were subjected to accelerated ageing to assess their durability against freezing-thawing cycles and salt crystallization cycles, with the aim of assessing the potential dissolution of newly formed soluble calcium phosphates, as well as evaluating the effects of changes in pore size distribution caused by the consolidant. For comparison's sake, all the tests were performed also on mortar samples treated with a commercial product based on nanolimes (Nanorestore®), one of the most common products for the conservation of lime plasters and mortars.

Considering that national and international recommendations about experimental weathering procedures (e.g. European EN 12371 [6], Italian UNI 1186 [7] for freezing-thawing test, European EN 12370 [8], RILEM MS-A.1 [9], RILEM MS-A.2 [10]) often do not realistically reproduce the transport, crystallization and damage processes encountered in the field [11, 12], the methods adopted in this study partly modified the standard procedures, as described in the following.

2 Materials and Methods

2.1 Lime Mortar Specimens

Mortar samples were prepared using slaked lime and calcareous aggregates according to historic recipes. A binder-to-aggregate ratio of 1:2 v/v (0.41 w/w) and a water-to-binder ratio of 1:1 v/v (0.45 w/w) were used. After curing for 7 months to ensure carbonation, the specimens were treated by brushing until apparent refusal with: (i) a 3 M DAP solution, followed by a limewater poultice [13] (treatment labelled as "DAP"); (ii) a commercial nanolime dispersion in ethanol with 5 g/L concentration (treatment labelled as "NL"). One month after the consolidant application, untreated (UT) and treated samples were subjected to accelerated ageing, as shown in Fig. 1.

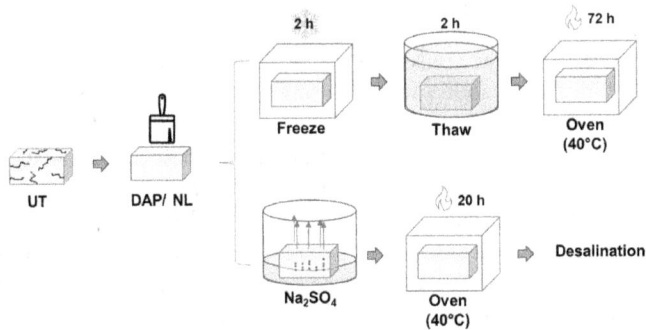

Fig. 1. Schematic illustration of the methodology adopted for accelerated ageing after consolidation treatment.

2.2 Freezing-Thawing Cycles

Freezing thawing (FT) cycles were performed by partially modifying the European [6] and Italian [7] standards, as done in previous studies [4]. Before starting the cycles, specimens were saturated by partial immersion in de-ionized water for 72 h. Then, each cycle consisted in two steps: i) freezing at $-20 \pm 2\,°C$ for 2 h; ii) thawing in water at $+ 20 \pm 2\,°C$ for 2 h. Three cycles were performed each day, leaving the samples submerged in water the whole night. Every 5 cycles, the sample were dried in the oven at 40 °C for 72 h, then visually observed, weighed and characterized, in terms of dynamic elastic modulus (see Sect. 2.4). After that, the cycles were started again. In total 10 cycles were carried, as UT and NL samples exhibited dramatic material loss.

2.3 Salt Weathering Cycles

For the salt weathering (SW) cycles, sodium sulphate was used, because it is known as the most dangerous salt and is widely used in accelerated durability tests [14]. A saturated (14% w/w) solution of $Na_2SO_4 \cdot 10H_2O$ was prepared using de-ionized water at $25 \pm 2\,°C$. To reproduce field conditions where salt accumulates coming from the ground, salt cycles were performed by capillary absorption of the solution via the bottom face, by immersing the sample by half the depth. In this way, the specimens are partially immersed [12] and the solution can evaporate from the upper surface. Each cycle consisted of immersion in the solution for 2 h and drying at 40 °C for 20 h, followed by cooling to room temperature for 2 h. In total, 10 cycles were carried out, because visible damage started to occur, and mechanical testing would have been impossible on samples too heavily damaged.

At the end of the 10 crystallization cycles, the samples were desalinated by poulticing, with the aim of evaluating the changes in pore size distribution and mechanical properties without interference from the salts. The whole surface of each specimen was wrapped with a poultice prepared using cellulose pulp and de-ionized water (1:4 w/w), then the samples were wrapped with a plastic film for 24 h and finally unwrapped and dried at room temperature with the cellulose pulp still in contact with the specimens.

2.4 Sample Characterization

First, the effectiveness of the consolidants was assessed 1 month after the consolidant application (before the durability tests). Curing for 1 month was selected as it is recommended by the technical data sheet of the nanolimes. The effects of the consolidants were evaluated in terms of variations in: (i) dynamic elastic modulus (E_d) determined by ultrasonic testing, (ii) compressive strength by double punch test (DPT) and (iii) pore size distribution by mercury intrusion porosimetry (MIP). The measurements were performed on untreated and consolidated samples, before and after accelerated ageing.

Ultrasonic measurements were performed using a Pundit instrument with 55 kHz transducers, determining the ultrasonic pulse velocity (UPV) and then calculating the E_d, as described in detail in [15, 16].

Compressive strength by DPT was performed by loading the specimens using 20 mm diameter circular steel plates. The resulting compressive strength was calculated as the ratio of the failure load to the loaded area. The maximum load was set to 500 daN, with a speed of 3 daN/sec.

The alterations in open porosity and pore size distribution were evaluated by MIP using a Pascal 140 and 240 instrument (minimum pressure 0.0125 MPa, maximum pressure 200 MPa). The MIP samples (~1 cm^3) were collected by a pincer after mechanical testing.

3   Results and Discussion

Compared to the untreated reference and also to the NL-treated samples, DAP-treated mortars showed much improved resistance to freezing-thawing cycles, as reported in Fig. 2. Both UT and NL samples showed a dramatic weight loss (almost 50% for UT, slightly better for NL) after the freezing-thawing cycles, while the DAP samples almost remained unaltered. This trend was confirmed also by ultrasonic measurements (Fig. 2) and DPT (Fig. 3). Indeed, after the cycles the residual compressive strength of DAP-treated samples was more than three times that of NL.

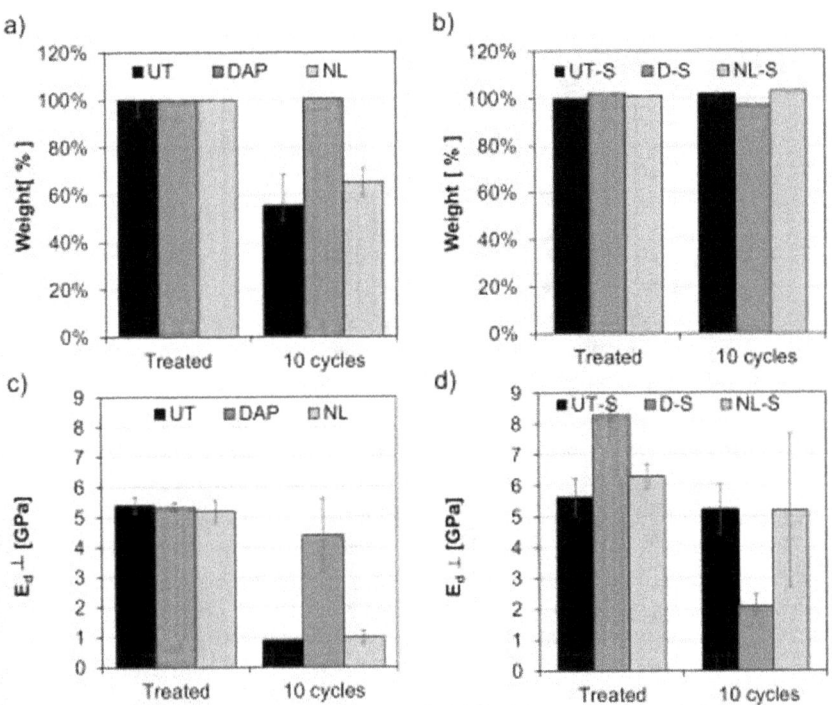

Fig. 2. Variation of weight after treatment and after 10 cycles for a) freezing–thawing cycles; b) salt weathering cycles; Variation of dynamic elastic modulus after c) freezing–thawing cycles; d) salt weathering cycles.

A lower benefit was found in the case of salt crystallization cycles, as DAP-treated samples exhibited a decrease in E_d (Fig. 2), even though the efficacy right after treatment was higher than NL. Nonetheless, after desalination at the end of the salt crystallization cycles, the DAP-treated samples exhibited higher residual compressive strength than the NL-samples (Fig. 3). This discrepancy between E_d and DPT results may be ascribed to surface material loss, which worsens the contact between the samples and the transducers used for the ultrasonic measurements, thus leading to an apparent decrease in E_d.

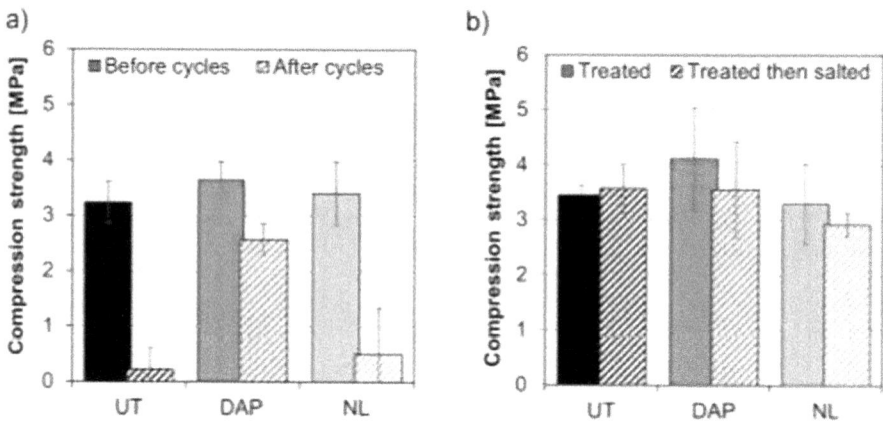

Fig. 3. Variation of the compressive strength a) before and after freezing-thawing cycles; b) before and after salt weathering cycles.

The pore size distribution of untreated and treated samples, before and after accelerated ageing, is shown in Fig. 4. After the freezing-thawing cycles, the UT samples exhibited an increase in coarse pores, which is a sign that new microcracks were formed. Similarly, an increase in pore size was registered also for the NL-treated samples. In the case of the DAP-treated samples, it is noteworthy that a change in pore size distribution was experienced right after treatment, as more abundant fine pores (below 0.1 μm) were present. Even though the ice and salt crystallization pressure is known to be higher in smaller pores, in the case of DAP samples the increase in mechanical properties was apparently sufficient for compensating such potentially negative increase in smaller pores. As a result, the DAP-treated samples resisted well to freezing-thawing cycles (Fig. 3) and the final pore size distribution was not too different from that before the cycles (Fig. 4).

In the case of salt crystallization cycles, the two consolidants seem to have impacted the pore size distribution in different ways: DAP was again responsible for an increase in smaller pores, similar to the previous case, while NL apparently occluded smaller pores (Fig. 4b). After the cycles, the pore size distribution of the NL samples was basically the same as the UT reference, while the DAP ones still exhibited a different pore size distribution, characterized by more abundant fine pores.

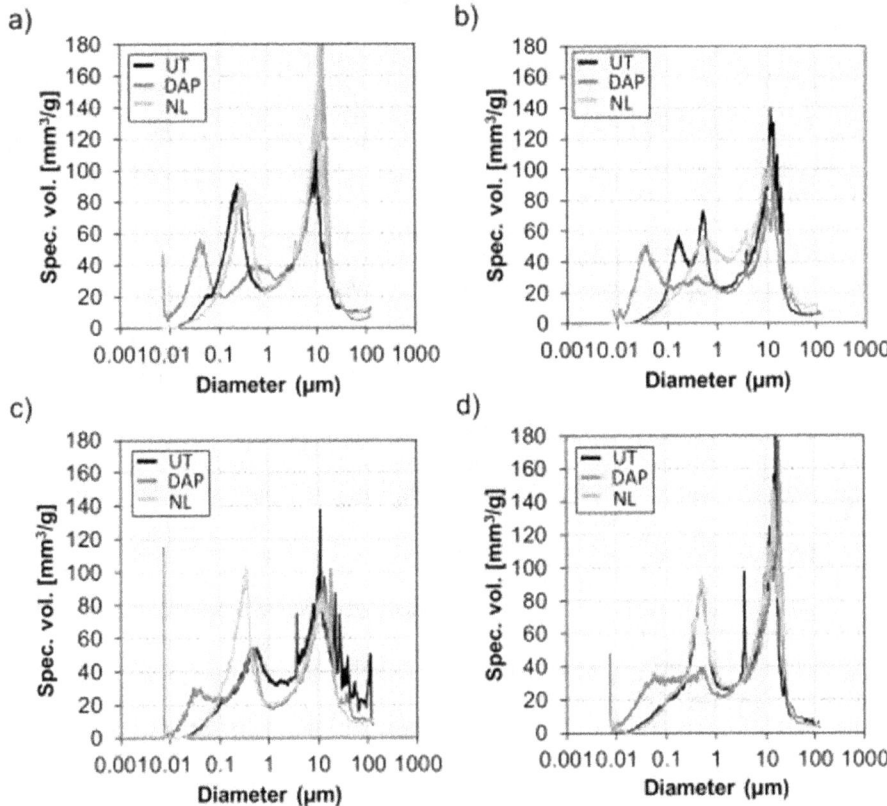

Fig. 4. Pore size distribution of untreated and treated samples before a) freezing–thawing cycles; b) salt weathering cycles, and after c) freezing–thawing cycles; d) salt weathering cycles

4 Conclusions

The present study aimed at evaluating the durability of lime-mortars treated with an innovative phosphate-based consolidant and with traditional nanolimes. In terms of freezing–thawing cycles, DAP-treated mortars showed a higher durability than the untreated reference and also the nanolimes benchmark, in terms of weight loss, dynamic elastic modulus and compressive strength by double punch test. Nonetheless, after salt weathering cycles, the DAP-treated samples exhibited a substantially similar behavior as the other samples.

The different performance of the DAP treatment observed between freezing-thawing cycles and salt crystallization cycles could be ascribed, on the one hand, to the slight modification of the pore size distribution after treatment (which may lead to increased crystallization pressure) and, on the other hand, to the severity of the salt weathering test (involving the use of a saturated solution of $Na_2SO_4 \cdot 10H_2O$ solution). For a more reliable evaluation of the salt resistance of consolidated mortars, additional tests are recommended using a less aggressive and more reliable salt weathering test, like that being developed within the RILEM TC 271-ASC.

References

1. Sassoni, E., Naidu, S., Scherer, G.W.: The use of hydroxyapatite as a new inorganic consolidant for damaged carbonate stones. J. Cult. Herit. **12**(4), 346–355 (2011)
2. Masi, G., Sassoni, E.: Comparison between ammonium phosphate and nanolimes for render consolidation. In: IOP Conference Series: Materials Science and Engineering, vol. 949, no. 1 (2020)
3. Lazzarini, L., Tabasso, M.L.: Il restauro della pietra. Utet Scienze Tecniche, Torino (2010)
4. Sassoni, E., Graziani, G., Franzoni, E.: An innovative phosphate-based consolidant for limestone. Part 2: durability in comparison with ethyl silicate. Constr. Build. Mater. **102**, 931–942 (2016)
5. Scherer, G.W.: Crystallization in pores. Cem. Concr. Res. **29**(8), 1347–1358 (1999)
6. European Standard EN 12371 (2010), Natural Stone test methods - Determination of frost resistance
7. Italian Standard UNI 1186 (2008), Beni culturali - Materiali lapidei naturali ed artificiali - Metodologia per l'esposizione a cicli di gelo e disgelo
8. European Standard EN 12370 (2020) - Natural Stone test methods - Determination of resistance to salt crystallization
9. RILEM Recommendation: MS-A.1, Determination of the resistance of wallettes against sulphates and chlorides. Mater. Struct. **31**, 2–9 (1998)
10. RILEM Recommendation: MS-A.2, Uni-directional salt crystallization test for masonry units. Mater. Struct. **31**, 10–11 (1998)
11. Lubelli, B., et al.: A new accelerated salt weathering test by RILEM TC 271-ASC: preliminary round robin validation. Mater. Struct **55**, 238 (2022)
12. Benavente, D., García Del Cura, M.A., Bernabéu, A., Ordóñez, S.: Quantification of salt weathering in porous stones using an experimental continous partial immersion method. Eng. Geol. **59**(3–4), 313–325 (2001)
13. Franzoni, E., Sassoni, E., Graziani, G.: Brushing, poultice or immersion? The role of the application technique on the performance of a novel hydroxyapatite-based consolidating treatment for limestone. J. Cult. Herit. **16**(2), 173–184 (2015)
14. Dohene, E., Prince, C.A.: Stone Conservation: An Overview of Current Research, 2nd edn. Research in Conservation (2010)
15. Ruedrich, J., Knell, C., Enseleit, J., Rieffel, Y., Siegesmund, S.: Stability assessment of marble statuaries of the Schlossbrücke (Berlin, Germany) based on rock strength measurements and ultrasonic wave velocities. Environ. Earth Sci. **69**(4), 1451–1469 (2013)
16. Sassoni, E., Ugolotti, G., Pagani, M.: Nanolime, nanosilica or ammonium phosphate? Laboratory and field study on consoledation of a byzantine marble sarcophagus. Constr. Build. Mater. **262**, 120784 (2020)

Repair Mortars and Grouts. Requirements and Design. Compatibility Issues. Durability and Effectiveness. Adequacy of Testing Procedures

Long-Term Mechanical Properties and Durability of Lime-Spongilite Mortars

Martin Vyšvařil[✉] ⓘ, Martin Krebs ⓘ, and Patrik Bayer ⓘ

Faculty of Civil Engineering, Brno University of Technology, Brno, Czech Republic
vysvaril.m@fce.vutbr.cz

Abstract. This paper presents a study on partial replacement of lime binder with fine spongilite with the purpose of exploring a new application of this natural material as lime mortar additive. Standard air lime mortars were made by incorporating from 0% to 40% of spongilite powder in replacement to lime and their mechanical performances, microstructure, and durability were determined. The spongilite powder showed similar pozzolanic activity as natural zeolite or waste brick powder predicting an improvement in the mechanical properties and durability of prepared mortars. As the replacement level in lime mortars increased, the amount of mixing water needed for the same mortar consistence decreased, and the performance properties of the mortars improved. The increase in strengths of mortars was manifested mainly in the long term of 180 and 365 days. The incorporation of fine spongilite led to the formation of slightly denser, more water absorptive, however, more frost resistant and salt crystallization resistant structure in air lime mortars. The effective use of spongilite powder as a supplementary material in air lime mortars was assessed to enhance their performance in building practice or to prepare feebly hydraulic mortars used in the past in constructions nowadays considered built heritage.

Keywords: Lime Mortar · Spongilite · Salt Crystallization · Durability · Frost resistance · Desalination

1 Introduction

The protection of historic buildings requires the use of traditional building materials compatible with the historical ones or as close to them as possible. Due to the fact that mortars for restoration purposes must be sufficiently durable and resistant to weather attack, blended mortars are a very good alternative for combining the advantages of air lime (easy water vapor permeability and compatibility with historical mortars) and hydraulic binders and pozzolans (durability in a humid environment and frost resistance). Various additives, admixtures, or fibers are commonly added to improve the mechanical properties and durability of air lime materials. Recent and current research on lime mortars mainly concerns the use of natural or waste materials that act as fillers in mortars or have pozzolanic properties.

Spongilite - the technical geological name of the sedimentary rock of the Upper Cretaceous, a siliceous calcium marlite more commonly but incorrectly called marl,

V. Bokan Bosiljkov et al. (Eds.): HMC 2022, RILEM Bookseries 42, pp. 405–415, 2023.
https://doi.org/10.1007/978-3-031-31472-8_32

marlstone, or clay marl – is a typical representative of organogenic pozzolans. Triaxial needles of dead microscopic marine fungi of the Porifera strain, so-called spongies, play a significant role in the heterogeneous structure of spongilites. The needles of these fungi are formed by opal, but during diagenesis the opal recrystallizes to chalcedony and quartz. The secondary components are usually clay minerals, mica, and calcite [1]. Spongilites are very porous rocks, characterized by a relatively large pore volume with their narrow distribution, usually below 0.1 mm. Spongilite was one of the most widespread building materials of the Romanesque and Gothic period in the Czech Republic. Thanks to its relatively abundant occurrence and easy processing, it has become a popular building and art material. Despite being known for its low resistance to weathering, it was still used in some areas of buildings in the first half of the 20th century [2].

The influence of the very fine fraction of spongilites in building binders has been very little studied so far and there is almost no professional work on this topic. Spongilites were observed as materials that prevent the formation and course of the alkali-silica reaction (ASR) in concrete [3]. It was also found that the replacement of 7% of aggregate by marlstone powder (rock geologically related to spongilite) causes enhancement in the mechanical properties of pervious concrete. The high content of reactive silica in marlstone compared to other stones studied (limestone, mudstone, sandstone, natural sand) contributes to chemical interaction with $Ca(OH)_2$ formed during the hydration of cement to produce additional calcium silicate hydrates (CSH) [4]. The partial replacement of cement by spongilites causes consistency thickening and prolongation of the initial and final setting times of cement pastes with a concomitant increase in viscosity and yield stress, and a decrease in the thixotropy of the pastes [5]. The spongilites in cement causes increase in water retention of mortars, slightly reduction of their bulk density, increase in porosity of mortars due to the growing predominance of capillary pores maintaining sufficient mortars strengths, and slightly increase in the frost-resistance of cement mortars [6].

The use of spongilites in lime mortars has not yet been the subject of any published research, although the partially amorphous nature of spongilites (especially the content of opal and chalcedony) predetermines this material as a pozzolanic additive to lime mortars. In the Czech Republic, there are several localities of spongilite in the overburden of high-quality foundry sands. They are often mined and stored without further use; therefore, this paper includes a study on partial replacement of lime binder with spongilite powder with the purpose of exploring a new application of this unexploited material as lime mortar additive.

2 Materials and Methods

2.1 Materials

Hydrated lime CL90-S (Carmeuse Czech Republic s.r.o. Mokrá, Sivice, Czech Republic) was used as a binder in prepared mortar mixes. Washed quartz sand meeting EN 13139 (fraction 0/2 mm from Filtrační Písky spol. s.r.o., Chlum, Czech Republic) was used as an aggregate, and fine spongilite powder with maximum particle size of 125 µm (Kalcit s.r.o., Brno, Czech Republic) was a pozzolanically active lime substitute in prepared mortars. The chemical composition of all raw materials is given in Table 1. The phase compositions obtained by X-ray diffraction analysis (XRD) using Rietveld method are presented in Table 2. The particle size distribution of initial materials was presented in a previous literature (spongilite DL) [6]. The initial materials were also characterized by their fundamental physical parameters. Loose bulk density, specific gravity, and Blaine specific surface area (S_g) of materials stipulated by EN 196-6:2018 are given in Table 3. The spongilite exhibited high Blaine fineness, which together with high amount of reactive opal, explains its pozzolanic activity that was assessed by the modified Chapelle test method according to NF P 18-513: 2012 (Table 3). The spongilite was not modified in any way before the pozzolanicity test, a fine spongilite powder with a granulometry of up to 125 μm was used for the test. Determination of pozzolanic activity of spongilite was performed after 24 h and also after 5 days of reaction with $Ca(OH)_2$ in an autoclave at 90 °C. The spongilite exhibited similar or slightly higher pozzolanic activity as thermally treated zeolite [7] or waste brick powder [8] but did not reach such high values as fly ashes [9] or waste glass dust [10]. According to Raverdy et al. [11], material is considered as pozzolanically active if its Chapelle reactivity is ≥650 mg of $Ca(OH)_2/g$ of testing material. The spongilite safely fulfilled this condition.

Table 1. Chemical composition of initial materials (wt%).

	SiO₂	Al₂O₃	Fe₂O₃	CaO	MgO	K₂O	Na₂O	MnO	TiO₂	SO₃	L.O.I.
Lime	0.92	0.71	0.39	68.09	1.33	0.48	0.11	0.01	0.10	0.19	27.94
Quartz sand	98.50	0.38	0.15	0.01	0.03	0.09	0.01	0.04	0.09	0.02	0.12
Spongilite	60.37	3.11	1.41	16.12	0.51	1.13	1.09	0.34	0.19	0.06	15.41

Table 2. Mineralogical composition of initial materials (wt%).

Mineral	Lime	Quartz sand	Spongilite
Brucite	0.5	–	–
Calcite	1.8	–	37.5
Cristobalite	–	–	15.5
Glauconite	–	–	3.1
Portlandite	97.1	–	–
Quartz	–	98.3	24.3
Opal	–	–	14.6
Orthoclase	–	1.5	4.8

2.2 Experimental Procedures

Mortar mixtures were prepared using a binder:aggregate volume ratio of 1:1 and different water:binder coefficients for each mixture to achieve similar fresh mortar consistence, 160 ± 5 mm, measured by the flow table test (EN 1015-3). This volume ratio is commonly used in the preparation of lime renders in practice and research and is also supported by the results obtained by Lanas et al. [12]. The used consistency of mortars guarantees

Table 3. Fundamental parameters of initial materials (a – after 1 day treatment, b – after 5 days treatment).

	d_{50} (μm)	Loose bulk density (kg m^{-3})	Particle density (kg m^{-3})	$S_{g\ Blaine}$ (m^2 kg^{-1})	Pozzolanic activity (mg Ca(OH)$_2$/g)
Lime	2.6	465	2420	1320	–
Quartz sand	447	1670	2650	–	–
Spongilite	77.8	770	2570	620	[a]694, [b]910

easy handling, good adhesion to the substrate and at the same time allows their use in machine rendering. Fine spongilite powder was used as a partial replacement of lime in three different dosages (10%, 20%, and 40% of the lime weight). The proportioning of the mortar mixtures is given in Table 4. Bulk density (ρ_{fm}), water retention value (WRV), and entrained air in fresh mortars were set following the standard EN 459-2. The freshly casted samples (40 mm × 40 mm × 160 mm) were freely covered by polyethylene foil in order to avoid their cracking caused by rapid drying. Hardened mortar specimens were demolded after 48 h and then cured in a wet chamber at temperature $T = (22 \pm 3)$ °C and a relative humidity RH $= (95 \pm 5)\%$ for 26 days. The samples were then stored under laboratory conditions at $T = (22 \pm 3)$ °C, RH $= (45 \pm 5)\%$. During the entire ageing period, the samples were placed on plastic grids to make their surface as accessible as possible for carbonation. High relative humidity favoured the pozzolanic reaction, whereas curing under ambient conditions allowed the carbonation of hydrated lime mortars.

The basic physico-mechanical properties of mortars were determined after 28, 90, 180, and 365 days of curing. The bulk density of hardened mortars was estimated following EN 1015-10, and the flexural strength and compressive strength of the samples were determined according to EN 1015-11. For the particular mortar mixture, a set of three prisms was evaluated. The study on the pore structures of the samples including the determination of total open porosity and pore-size distribution was conducted with high-pressure mercury intrusion porosimetry (MIP) using a PoreSizer 9310 (Micromeritics BV, Eindhoven, Netherlands). Frost resistance of the mortars was assessed after 90 curing days according to the modified Czech standard (ČSN 722452). The total test required 15 freeze-thaw cycles. One cycle consisted of 6 h freezing of water-saturated samples at −20 °C and 12 h thawing in a desiccator at a constant relative humidity of 99% and a temperature of 23 °C. The frost resistance coefficient (D_f) was determined as a ratio of flexural strength of specimens after 15 freeze-thaw cycles to the flexural strength of reference specimens. Water absorption of the samples by immersion at atmospheric pressure was calculated according to EN 13755. The transport of liquid water in the studied materials at the age of 90 days was characterized by the capillary water absorption coefficient according to EN 1015-18, sorptivity, and liquid water diffusivity [13]. The salt crystallization resistance of mortars at the age of 90 days was determined using 10% Na$_2$SO$_4$, 3% NaCl, and 3% NH$_4$NO$_3$ solutions. The procedure was performed according to the European standard (EN 12370:1999) and it is described in detail in authors' previous paper [14].

Table 4. Composition and fresh state properties of render mixtures.

	Lime (g)	Quartz sand (g)	Spongilite (g)	H_2O (g)	ρ_{fm} (kg m^{-3})	Air content (%)	WRV (%)
L-ref	100	358	0	100	1920	4.6	97.1
LS10	90	334	10	100	1930	3.7	98.3
LS20	80	315	20	95	1940	2.9	99.3
LS40	60	283	40	88	1960	2.4	99.8

3   Results and Discussion

3.1   Fresh State Properties

The water required to prepare mortars with similar consistency decreased with increasing amount of lime replaced with spongilite. This is due to the larger particle size of spongilite (Table 3) than to lime, and therefore also to the lower specific surface area of the spongilite particles. The decrease in water/binder ratio to achieve the same workability corresponds with the results found in the literature for other pozzolanic additives such as metakaolin, natural zeolite etc. [15–18]. With the gradual replacement of lime with spongilite powder, the water retention (WRV) in mortars increased proportionally from 97.1% to 99.8% (Table 4). According to Faria P. [19], the decrease in water/binder ratio for lime-pozzolan mortars leads to the increase in WRV. A similar increase in WRV was observed in spongilite-blended fine-grained cement mortars [6]. The air content in mortars decreased slightly with increasing spongilite replacement, resulting in a gradually increasing bulk density in the fresh (Table 4) and hardened state and a decrease in the porosity of the mortar (Table 5). The variations in air content are most-likely related to various rheology of binder-paste.

3.2   Mechanical Parameter

Bulk density of hardened mortars slightly increased with an increasing amount of spongilite incorporated (1650 kg m^{-3} for L-ref, 1660 kg m^{-3} for LS10, 1680 kg m^{-3} for LS20, 1720 kg m^{-3} for LS40 at the age of 28 days) due to the reduction of mixing water quantity required for the same fresh mortar consistency. As spongilite content increased, the mortars became denser but still met the recommended bulk density for repair mortars in the range of 1500–1800 kg·m^{-3} [15, 16]. The bulk density of mortars slightly increased over time due to carbonation (1680 kg m^{-3} for L-ref, 1700 kg m^{-3} for LS10, 1720 kg m^{-3} for LS20, 1740 kg m^{-3} for LS40 at the age of 365 days).

Time evolution of mortar strengths is presented in Fig. 1. Initially, mortars with spongilite powder had a lower flexural and compressive strength than the reference mortar, but their strengths increased during ageing, and after 90 d, the mortars with 10% and 20% lime replacement surpassed the reference mortar in their strength. It is evident that the pozzolanic reaction of spongilite powder requires a longer period of time to be fully reflected in the properties of lime mortars. However, it should be noted, that the 28 days strengths can be unfavourably affected by the increased humidity of specimens at the time of testing as they were tested right after the removal from highly humid environment. The flexural strength of LS40 mortar remained at its initial values throughout 365 days and it was below the reference mortar value. In contrast, the compressive strength of this mortar significantly exceeded all other mortars in all monitored ages, and already at 90 days of age reached a value of about 4 MPa. The observed significant increase in compressive strength is comparable to other pozzolanic additives (metakaolin, natural zeolite, silica fume…) [15, 16, 18, 20, 21].

Building practice classifies rendering and plastering mortars based on their 28-days compressive strength (σ_c) according to EN 998-1. In this connection, L-ref, LS10, and LS20 mortars can be classified into category CS I (σ_c in the range 0.4–2.5 MPa) and LS40 mortar belongs to the category CS II (σ_c in the range 1.5–5.0 MPa). Veiga M. [22] recommends compressive strength of repair mortars after 90 days in the same range of 0.4–2.5 MPa, the mortars with 10% and 20% of spongilite powder fulfil this recommendation.

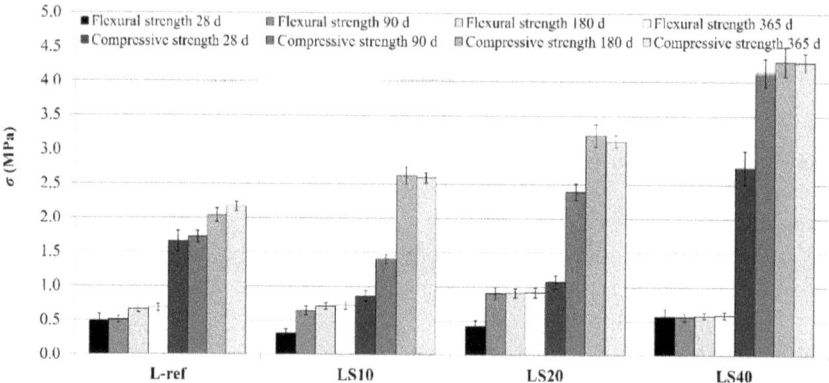

Fig. 1. Time evolution of flexural and compressive strength of mortars.

3.3 Frost Resistance

As the frost resistance test is performed on water-saturated samples, air lime mortars sometimes break up in this test before the completion of 15 cycles. In this study, all tested mixtures withstood 15 freeze-thaw cycles and it was possible to determine their flexural strengths and subsequently evaluate the frost resistance coefficients, D_f (Table 5). The reference lime mortar is not frost-resistant and showed the lowest D_f value. The frost resistance of mortars increased with an increasing amount of spongilite powder in the samples up to 20%. These mortars exceeded the frost resistance coefficient of 0.75 and thus met the frost resistance criterion. This confirmed the effective use of spongilite powder as a supplementary material improving the properties of lime mortars. The improvement in frost resistance by partial replacement of lime with spongilite is more pronounced than in the case of using natural zeolite [21] or foam glass dust [11]. Although the mortar with 40% lime replacement by spongilite narrowly met the criterion of frost resistance, its D_f value was the lowest of the blended mortars. This change in the frost resistance trend is due to the higher total open porosity and water absorption of LS40 mortar. From the point of view of frost resistance, 20% spongilite replacement appears to be the best of the studied lime substitutes.

Table 5. Total open porosity (Ψ), frost resistance coefficient (D_f), water absorption (A_w) and liquid water transport parameters of mortars.

Mixture	Ψ (%)	D_f (–)	A_w (%)	C_m (kg m^{-2} min$^{-1/2}$)	κ (mm^2 min^{-1})	S (mm min$^{-1/2}$)
L-ref	28.64	0.44	15.35	1.534	43.89	1.54
LS10	27.62	0.83	15.54	1.130	19.63	1.13
LS20	26.99	0.97	15.93	0.851	14.28	0.85
LS40	30.32	0.78	17.31	1.306	26.11	1.31
L-ref 15 c	32.24	–	14.07	1.247	34.83	1.25
LS10 15 c	30.94	–	17.43	1.344	25.06	1.35
LS20 15 c	29.70	–	16.62	1.297	22.77	1.30
LS40 15 c	32.57	–	18.58	1.341	22.84	1.34

3.4 Liquid water Transport

Capillary liquid water transport in the mortars was measured by applying sorption experiments before and after the frost resistance test at the age of 90 days (Fig. 2). The results achieved are given in Table 5. Liquid water transport through the mortars representing by capillary water absorption coefficient (C_m), liquid water diffusivity (κ), and sorptivity (S) was slower in the spongilite mortars despite their greater water absorption. The capillary water action slowed down with increasing chalcedonite content up to 20%. There was a significant increase in all monitored parameters for the mortar with 40% lime replacement by spongilite, which is in line with its higher total porosity and water absorption of this mortar (Table 5). The difference was probably caused by the increase in the content of capillary active pores with a diameter below 0.07 μm and their better interconnection (Sect. 3.5). This finding is opposite to the trend which was observed when adding foam glass dust [11] or natural zeolite [21] to lime mortars. All the mortars were characterized by greater water absorption coefficient values than 0.3 kg m^{-2}/24 h, which is considered in EN 998-1 as minimal for renovation mortars.

After the frost resistance test, the spongilite mortars showed higher liquid water transport parameters and water absorption at atmospheric pressure due to the formation of microcracks in the samples during the frost resistance test (Table 5), which was also reflected in the increased total open porosity of the mortars.

3.5 Pore Structure

Porosity of the mortars (obtained by MIP) has been determined before and after the frost resistance test (15 °C) at the age of 90 days (Fig. 3, Table 5). The results show

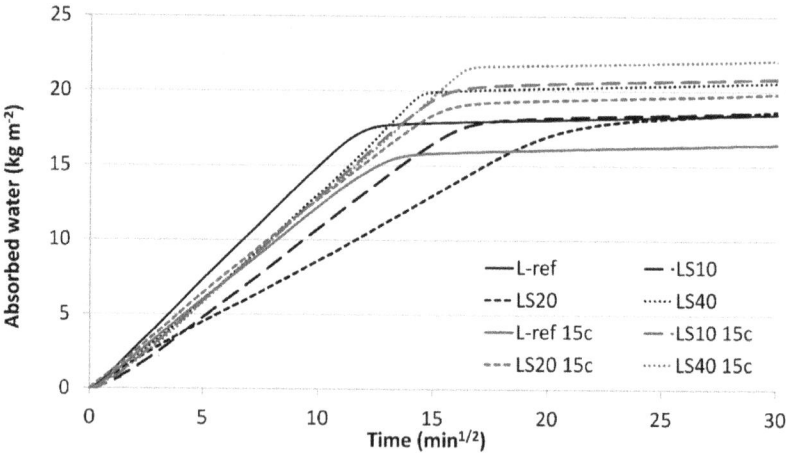

Fig. 2. Capillary water absorption curves for tested mortars; (black) before the frost resistance test (at the age of 90 days), (red) after the frost resistance test.

that as spongilite content increased up to 20%, the total open porosity (Ψ) of mortars decreased. But surprisingly, the mortar with 40% lime replacement by spongilite was the most porous. The change in total open porosity is due to a decrease in the amount of mixing water with increasing amount of spongilite and also to the formation of calcium silicate hydrate (CSH) gel by the pozzolanic reaction of spongilite. In these ways, the added spongilite also affected the pore size distribution in the mortars (Fig. 3). As the spongilite powder content in the mortars increased, the proportion of pores with a diameter of about 0.5 μm gradually decreased and the proportion of pores of 0.25 and 0.05 μm increased. Small pores with a diameter of about 0.01 μm, typical for the CSH gel, appeared to a greater extend in the structure of LS40 mortars which was also reflected in the water absorption of the mortar and increased values of liquid water transport parameters (Table 5). All prepared mortars met the requirement of total open porosity in the range of 20%–40% for possible use as repair mortars of historic buildings [23]. Similar differences in pore size distribution in the lime mortars were observed using natural zeolite [21] or foam glass dust [10] as a partial replacement of lime.

After 15 freeze-thaw cycles (15 °C), the reference mortar and LS10 mortar included a significantly higher proportion of pores with a diameter around 0.5 μm due to the partial destruction of small pores by ice crystallization pressure, while in the LS20 mortar, the pores with a diameter around 0.5 μm were filled by portlandite crystallizing from a pore solution and by a more complete pozzolanic reaction of spongilite powder, and there was an increase in the content of pores with a diameter around 0.01 μm (Fig. 2). Nevertheless, the total open porosity slightly increased in all monitored mortars (Table 5).

3.6 Salt Crystallization Resistance

The content of anions in aqueous leaches of the samples before and after salt crystallization resistance test is presented in Table 6 together with the number of cycles until the sample decomposition. The results show that the concentration of the monitored anions

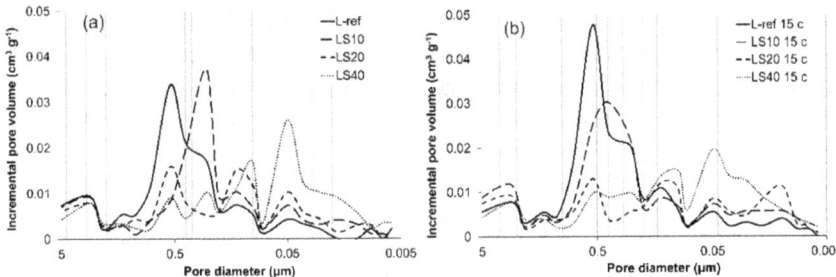

Fig. 3. Pore size distribution in studied mortars in the area of capillary pores; (a) before the frost resistance test (at the age of 90 days), (b) after the frost resistance test.

in the samples after treatment with saline solutions increased more than 100 times; mostly in LS20 sample. The accumulated amount of salts in the samples increased with increasing spongilite content up to 20%. The mortar with 40% lime replacement by spongilite did not show any further improvement in salt accumulation and so it was confirmed that the more porous structure of mortars is not always more advantageous for salt accumulation and that sufficiently strengthened lime mortars by pozzolanic reaction of the binder may have improved salt accumulation abilities, such as when using natural zeolite [21]. The increased amount of nitrates in the structure of the LS10 and LS20 mortars caused earlier decomposition of the samples than in the reference mortar, but from the point of view of salt accumulation (in desalination of masonry) these mortars appear to be more suitable (especially in potential use as sacrificial renders).

Table 6. Concentrations of monitored anions in aqueous leaches of mortar samples before (c_{g0}) and after (c_g) the salt crystallization resistance test, and a number of the test cycles to sample disintegration (10 = intact sample).

Mixture	c_{g0} SO_4^{2-} (g kg^{-1})	c_{g0} Cl$^-$ (g kg^{-1})	c_{g0} NO_3^- (g kg^{-1})	c_g SO_4^{2-} (g kg^{-1})	c_g Cl$^-$ (g kg^{-1})	c_g NO_3^- (g kg^{-1})	Cycle count in Na$_2$SO$_4$	Cycle count in NaCl	Cycle count in NH$_4$NO$_3$
L-ref	0.103	0.055	0.487	21.04	7.73	19.85	3	10	10
LS10	0.090	0.213	0.398	40.18	14.12	32.08	3	10	6
LS20	0.073	0.203	0.285	44.17	15.23	36.77	3	10	6
LS40	0.049	0.196	0.209	38.76	13.95	30.41	3	10	9

4 Conclusions

The mortars with distinct levels of lime replacement with spongilite powder were made and their mechanical performances, microstructure, and durability were determined. Based on the obtained results, the following conclusions can be drawn. Mortar strengths increased with increasing spongilite content in the mortars. As the replacement level in lime mortars increased up to 20%, a decrease in the total open porosity of the samples associated with a reduction in the rate of capillary water absorption was observed. Despite the decreasing porosity of the samples with increasing spongilite content, spongilite mortars showed better ability to accumulate salts. The use of spongilite-modified lime mortars for desalination of chloride-contaminated masonry seems to be particularly suitable. Spongilite powder used as a pozzolanic additive has led to a considerable improvement in the frost resistance of lime mortars. Despite a slightly increased porosity after freezing cycles, spongilite mortars showed improved properties in all important aspects from the point of view of the utility of these mortars in restoration and conservation interventions on historic buildings, moreover, considering the use of unexploited waste material. Based on the achieved results, a 20% lime replacement can be recommended as an optimal level.

Acknowledgments. This research was funded by the Czech Science Foundation grant number 21-06582S "Experimental and computational analysis of salt transport, accumulation, and crystallization in non-hydrophobized rendering mortars".

References

1. Šrámek, J.: Relationships between mineralogy, physical-mechanical properties and durability of cretaceous calcitic spongilites. In: Rodrigues, J.D. et al. (eds.) Proceedings of the 7th International Congress on Deterioration and Conservation of Stone, pp. 57–66. Laboratório Nacional de Engenharia Civil, Lisbon, Portugal (1992)
2. Brotan, P., Gregerová, M.: Assessment of the spongilitic marlstone building blocks from the Slaný historical city walls. Geosci. Res. Rep. **50**, 9–16 (2017)
3. Táborský, T.: Omezení vzniku křemičitoalkalické reakce kameniva v betonu. In: Paříková, M. (ed.) Sborník semináře Vápno, cement, ekologie 2010. VUMO, Prague, Czech Republic (2010)
4. Abdulwahid, M.Y.: Influences of different stone powders on pervious concrete strength. Struct. Concr. **22**(1), E528–E534 (2021)
5. Vyšvařil, M., Žižlavský, T., Rovnaníková, P.: Fresh state properties of spongilite blended cement pastes. In: AIP Conference Proceedings, vol. 2322, p. 020008 (2021)
6. Vyšvařil, M., Bayer, P., Žižlavský, T.: Use of Spongilites as pozzolanic additives in cement mortars. Solid State Phenom. **325**, 65–70 (2021)
7. Perraki, T., Kakali, G., Kontori, E.: Characterization and pozzolanic activity of thermally treated zeolite. J. Therm. Anal. Calorim. **82**, 109–113 (2005)
8. Navrátilová, E., Rovnaníková, P.: Pozzolanic properties of brick powders and their effect on the properties of modified lime mortars. Constr. Build. Mater. **120**, 530–539 (2016)
9. Dvořák, K., Hájková, I.: The effect of high speed grinding technology on the properties of fly ash. Mater. Tehnol. **50**(5), 683–687 (2016)
10. Vyšvařil, M., Žižlavský, T., Bayer, P.: Foam glass dust as a supplementary material in lime mortars. J. Mater. Civ. Eng. **33**(4), 04021026 (2021)
11. Raverdy, M., Brivot, F., Paillére, A.M., Dron, R.: Appréciation de i'activité pouzzolanique des constituents secondaires. In: Vol. 3 of Proceedings of the 7th International Congress on the Chemistry of Cement, pp. 36–41. Éditions Septima, Paris, France (1980)
12. Lanas, J., Alvarez-Galindo, J.I.: Masonry repair lime-based mortars: factors affecting the mechanical behaviour. Cem. Concr. Res. **33**, 1867–1876 (2003)

13. Vyšvařil, M., Bayer, P., Žižlavský, T., Rovnaníková, P.: Use of natural zeolite aggregate in restoration lime renders. In: Álvarez, J.I., et al. (eds.) PRO 130: 5th Historic Mortars Conference HMC 2019, pp. 261–272. RILEM Publications S.A.R.L, Paris, France (2019)
14. Vyšvařil, M., Bayer, P.: Salt and ice crystallization resistance of lime mortars with natural lightweight aggregate. In: Serrat, C. Casas, J.R., Gibert, V. (eds.) XV International Conference on Durability of Building Materials and Components DBMC 2020. Scipedia, Barcelona, Spain (2020)
15. Brzyski, P.: The Effect of Pozzolan Addition on the Physical and Mechanical Properties of Lime Mortar. In: E3S Web Conference, vol. 49, p. 00009 (2018)
16. Aly, M., Pavía, S.: Properties of hydrated lime mortars with pozzolans. In: III International Congress on Construction and Building Research, pp. 54–56. Escuela Técnica Superior de Edificación, Universidad de Madrid, Madrid, Spain (2015)
17. Pavía, S., Aly, M.: Influence of aggregate and supplementary cementitious materials on the properties of hydrated lime (CL90s) mortars. Mater. de Construcción 66(324), e104 (2016)
18. Faria, P.: Resistance to salts of lime and pozzolan mortars. In: Groot, C. (ed.) PRO 067: International RILEM Workshop on Repair Mortars for Historic Masonry, pp. 99–110. RILEM Publications SARL, Paris, France (2009)
19. Ince, C., et al.: Factors affecting the water retaining characteristics of lime and cement mortars in the freshly-mixed state. Mater. Struct. 44(2), 509–516 (2011)
20. Gameiro, A., et al.: Physical and chemical assessment of lime-metakaolin mortars: influence of binder: aggregate ratio. Cem. Concr. Compos. 45, 264–271 (2014)
21. Vyšvařil, M., Žižlavský, T., Bayer, P.: Salt and ice crystallization resistance of lime-zeolite mortars. AIP Conf. Proc. 2022, in press
22. Veiga, M., Aguiar, J., Silva, S.A., Carvalho, F.: Methodologies for characterisation and repair of mortars of ancient buildings. In: Lourenço, P., Roca, P. (eds.) Proceedings of the 3rd International Seminar Historical Constructions, pp. 353–362. University of Minho, Braga, Portugal (2001)
23. Papayianni, I.: The longevity of old mortars. Appl. Phys. A 83, 685–688 (2006)

On the Effect of Poor-Quality Aggregates on the Physico-Mechanical Performance of Repair Lime-Based Mortars

Revecca Fournari, Loucas Kyriakou, and Ioannis Ioannou[✉]

Department of Civil and Environmental Engineering, University of Cyprus, Nicosia, Cyprus
ioannis@ucy.ac.cy

Abstract. The selection of building materials for any restoration project presupposes knowledge of their physico-mechanical properties, as compatible materials with the authentic ones need to be chosen at all times. In the case of composite building materials, such as lime-based mortars, the constituent raw materials, especially the aggregates which comprise the largest proportion of their volume, may affect the physico-mechanical performance of the end-product, both in the fresh and hardened states. It is therefore essential to identify the properties of the aggregates before these are incorporated in the mortar mixture. This paper reports on the effect of two different reef limestone crushed fine aggregates quarried in Cyprus on the physico-mechanical properties of repair lime-based mortars. The aggregates have been subjected to a series of standardized (i.e., soundness, Micro-Deval and water absorption) and non-standardized (powder X-Ray Diffraction) laboratory tests to identify their properties, before being used to produce lime-based mortars with fixed binder:aggregate ratio and workability. The results confirm the negative effect of poor-quality aggregates on the mechanical strength, the porosity and capillary absorption of the hardened end-products; this is corroborated through supplementary Mercury Intrusion Porosimetry (MIP), thermogravimetric (DTA/TG) analyses and Scanning Electron Microscopy.

Keywords: Lime Mortars · Aggregates · Strength · Porosity · Capillary Absorption

1 Introduction

Monuments are important cultural heritage assets, not only for their architectural value, but also because they provide evidence of traditional building techniques. Their conservation is, therefore, a responsibility of societies, in order to pass them on to future generations in good condition [1]. This relies very much on the use of compatible repair materials, including mortars. The physico-mechanical properties of the latter depend, among other parameters, on the nature and properties of the major constituent materials, i.e., the binder and the aggregates [2].

As a primary raw material, which comprises the largest proportion of the volume of lime-based mortars [3–5], aggregates have a major impact on the performance of these

V. Bokan Bosiljkov et al. (Eds.): HMC 2022, RILEM Bookseries 42, pp. 416–425, 2023.
https://doi.org/10.1007/978-3-031-31472-8_33

traditional composite materials. In fact, the quality of aggregates plays a fundamental role in the physico-mechanical properties of a mortar, and it should thus be given due attention when designing a repair mortar [6–9].

Historically, there was a large variation in the selection of aggregates for mortar production, depending on local availability. Typical natural aggregate sources were found in river basins and beaches. Nowadays, aggregates are mostly quarried from specific areas, following the crushing of parent rock. Although there may be fewer options commercially available, due to standards and regulations, these are generally of better quality than historically sourced aggregates.

Good aggregates for lime mortar production should consist of angular grains and should be free of contaminants, such as salt and organic matter. They should also be well-graded, with grain sizes between 0–4 mm. The composition of the aggregates is also important, as it can have an impact on the colour of the mortar and on its physical properties. A high clay or silt content in the aggregates is bound to affect the performance of the lime mortar and may lead to cracking and shrinkage. The sand used for lime mortar production should therefore be sieved and should be largely free from clay and silt [10–12]. It is worth noting that EN 13139 [13], which includes the specifications with which aggregates for the production of mortars must comply, recommends no more than 4% clay/silt in an aggregate.

The aforementioned highlight the importance of the quality of aggregates on the performance of lime-based mortars. This paper, therefore, investigates the influence of two different reef limestone crushed fine aggregates, quarried in Cyprus, on the physico-mechanical properties of repair lime-based mortars.

2 Materials and Methodology

Two different reef limestone crushed fine aggregates (0–4 mm) were used in this study. These aggregates originated from Pachna Formation (Terra and Koronia Members) quarries in Cyprus. The Terra Member (Lower Miocene) represents the first phase of reef growth on Cyprus [14]. Aggregate sample RL1, which comes from this Member, is generally hard and has a creamy-off-white colour. Its grains are mostly elongated and rough, whilst there is also a limited number of fossil skeletal fragments. Intraparticle porosity occurs within the grain pores. The distribution of grains in this sample seems to be uniform. The upper Koronia Member (Upper Miocene) contains mainly micritic calcite. Fossils are also abundant here [14]. The grains of sample RL2, which comes from this Member, appear to be more rounded and smooth. Although RL2 also has a creamy-off-white colour, there are some grains in it with reddish tones due to allochems. Aggregate sample RL2 further contains numerous fossil skeletal fragments. Significant primary intraparticle porosity occurs occasionally within partially unfilled fossils or between endoclasts. The distribution of grains in RL2 is not as uniform as in RL1, since the former has a higher percentage of fines.

A series of standardized tests, including magnesium sulphate soundness (EN 1367-2 [15]), Micro–Deval (ASTM D-7428 [16]) and water absorption (EN 1097-6 [17]) measurements, were first performed on both aforementioned aggregate samples to identify their properties. Their mineralogical composition was also determined through qualitative and semi-quantitative powder X-Ray Diffraction (PXRD) analyses, using a Bruker

D8 Advance system, with a Cu anode ($\lambda = 1.5406$ Å). Both samples were scanned with a scan rate of 0.5°/min, within the 2–100° 2θ angle range. The EVA (v.15.0) software and the International Centre for Diffraction Data (ICDD) PDF2 database were used to identify the crystalline phases. Semi-quantitative analysis was performed using the TOPAS software.

The aggregates were then used to produce lime mortars in the laboratory, in order to investigate the effect of aggregate quality on the performance of the end-product composite. Natural hydraulic lime 3.5 (Chaux Blanche Naturelle, supplied by Lafarge) was used as the binder in both mix designs. The binder:aggregate ratio was fixed at 1:3 w/w, in line with the prominent binder:aggregate ratio evidenced in prehistoric and historic lime composites in Cyprus [18] and in the Mediterranean basin [19–21], while the workability of the fresh mixtures was maintained within the range 170 ± 5 mm (Table 1).

Table 1. Mix design of laboratory composites. All quantities are measured by mass.

Mixture	Aggregate Code	Aggregates	Binder	Water/Binder	Workability (mm)
LM1	RL1	3	1	0.62	172
LM2	RL2	3	1	0.69	174

The fresh mortars were cast in standardized prismatic $40 \times 40 \times 160$ mm steel molds in two layers; each layer was compacted using a jolting table. The molds were covered with a glass surface to prevent evaporation and instant loss of humidity. The hardened specimens were demolded 3 days after casting and were stored in plastic containers covered in wet burlaps, in a room with constant temperature (23 ± 5) °C and humidity (50 ± 5%), throughout their curing period, in order to maintain high levels of humidity. Three specimens from each batch were tested at 28, 56 and 365 days of curing in order to determine their main mechanical (compressive and flexural strength, according to EN 1015-11 [22]) and physical (open porosity and capillary absorption, according to the methodologies described in Hall and Hoff [23]), properties, both in the short- and long-term. Mercury Intrusion Porosimetry (MIP) was also carried out on bulk samples, using a Micromeritics Autopore IV porosimeter, in order to provide additional information regarding the open porosity (p_o), apparent density (ρ_b) and pore size distribution. Differential Thermal and Thermogravimetric Analyses (DTA/TG), using a Shimadzu DTF-60H, were further carried out on powder samples taken from each specimen in order to investigate the presence of portlandite and subsequently the progress of carbonation/hydration reactions. The analyses were performed from 35 °C to 1200 °C, at a heating rate of 5 °C/min. Last but not least, Scanning Electron Microscopy (SEM), using a Jeol JSM-6610LV, assisted in the textural and morphological investigation of the microstructure of the mortars. The analyses were carried out on Au coated samples, under high vacuum conditions.

3 Results and Discussion

The results of the standardized tests, which were carried out to characterize the aggregates used in the production of the lime mortars in this study, are summarized in Table 2.

Table 2. Physico-mechanical properties of aggregates used in lime mortar production. MS: Magnesium Sulphate Soundness, MD: Micro-Deval, WA: Water Absorption

Samples	MS (%)	MD (%)	WA (%)
RL1	14	14.4	1.0
RL2	60	31.2	1.2

From Table 2, it is clear that aggregate sample RL1 may be classified as "good quality" aggregate, since it demonstrates lower values in all the tests it has been subjected to, compared to aggregate sample RL2, which may be classified as "poor quality" aggregate, since it demonstrates very high values in the magnesium sulphate soundness and Micro-Deval tests. It is worth mentioning that several researchers [24–26] have indicated a good correlation among the magnesium sulphate coefficient and the physico-mechanical properties of composite materials. The same stands for the Micro-Deval coefficient [26–28].

The very high magnesium sulphate soundness (MS) coefficient of aggregate sample RL2 may be attributed to its mineralogical composition and in particular to the presence of dolomite. From the results of the mineralogical analyses (Table 3), it is obvious that, whilst the predominant mineral in sample RL1 is calcite, sample RL2 contains almost equal amounts of dolomite and calcite. This high amount of dolomite in sample RL2 possibly affects its performance in use. In fact, there are researchers [29, 30] who have measured significant deformations in dolomitic limestones during drying and re-wetting experiments with the use of magnesium sulphate. Furthermore, EN 1367-2 [15] states that the soundness test with magnesium sulfate may not be suitable for all types of rocks; special reservations are expressed about some carbonate aggregates and some aggregates having a high proportion of magnesium bearing materials (e.g., dolomite aggregates).

Table 3. PXRD analyses of the aggregates used in lime mortar production.

Samples	PXRD Analysis
RL1	Calcite (98%), Dolomite (1%), Quartz (1%)
RL2	Dolomite (49%), Calcite (47%), Muscovite (2%), Halite (1%), Magnesite (1%)

Whilst the mineralogical composition of aggregates seems to affect their soundness, there are conflicting views in the literature regarding the correlation between the former and the wear resistance of aggregates. Some researchers [31] believe that there is no significant correlation between the mineral composition of aggregates and the Micro-Deval test, whilst others [32] claim that, among the key factors influencing the wear and fragmentation of aggregates are the geological properties of the deposit (i.e., its mineralogical composition, as well as the size, shape, and arrangement of grains). Thus, although it might have been expected that sample RL2 would have a lower Micro-Deval value, due to the fact that dolomite is slightly harder and denser than calcite [33], nevertheless, this factor seems not to have affected the result of the aforementioned standardized test. Hence, aggregate sample RL2 also appears to have a high Micro-Deval value (Table 2).

Table 4. Mechanical and physical properties of lime mortars. f_c: compressive strength, f_{cf}: flexural strength, C: capillary absorption, f_o: open porosity, p_b: apparent density (*standard deviation/COV*)

		Curing Period	LM1	LM2
f_c (MPa)		28d	3.3 (0.2/0.06)	1.9 (0.1/0.07)
		56d	6.2 (0.8/0.12)	4.2 (0.18/0.04)
		365d	14.5 (0.5/0.04)	11.2 (0.4/0.04)
f_{cf} (MPa)		28d	1.2 (0.2/0.15)	0.6 (0.02/0.03)
		56d	2.7(0.3/0.12)	1.7 (0.06/0.04)
		365d	4.6 (0.2/0.05)	3.7 (0.1/0.03)
C (g/m$^2 \cdot$s$^{0.5}$)		28d	129.4 (3.6/0.03)	217.0 (1.49/0.01)
		56d	123.4 (4.4/0.04)	192.6 (4.13/0.02)
		365d	116.0 (3.8/0.03)	170.4 (2.1/0.01)
f_o (%)		28d	27.6 (0.2/0.007)	33.5 (0.1/0.004)
		56d	26.1 (0.15/0.006)	29.8 (0.2/0.007)
		365d	21.6 (0.3/0.014)	24.7 (0.3/0.01)
MIP	f_o (%)	365d	24.3	31.6
	p_b (kg/m^3)		1828	1762

The results of the physico-mechanical properties of the hardened mortar specimens shown in Table 4 clearly demonstrate that mortar LM1, prepared with the better-quality aggregates RL1 (see Table 2 for aggregate characteristics), exhibits higher strengths than mortar LM2, prepared with the poor-quality aggregates RL2, at all curing ages tested. Similar results were obtained for the physical properties, since the aggregates with the higher soundness/Micro-Deval coefficients and water absorption (see Table 2) resulted in the mortar with the higher porosity and capillary absorption (LM2). Even though the aforementioned results may be attributed to the increased water/binder ratio used in mix design LM2 (see Table 1) [34], they are also certainly related to the poor quality of RL2 aggregates. As previously mentioned, several researchers [24–28] have indicated a good correlation among the magnesium sulphate and Micro-Deval coefficients and the mechanical and physical properties of mortars. Concerning water absorption, although there is a prevailing view among researchers [24–26] that this influences the performance of the end-product composite, no noticeable difference has been observed in the aggregate results hereby presented (see Table 2); hence, no safe conclusions can be drawn regarding the effect of this property on the properties of the mortars studied.

The Mercury Intrusion Porosimetry (MIP) results confirm that the open porosity of the mortar (LM2) prepared with the "poor quality" aggregates (RL2) is higher. At the same time, MIP measurements also reveal a modification in the pore structure of the composites hereby studied. The pore size distribution (Fig. 1) demonstrates a prominent increase in pore volume in the area between ca. 10–500 nm in mortar LM2, compared to mortar LM1. This may well be correlated to the poor quality of RL2 aggregate and suggests an increase in the volume of the composite that may be reached by atmospheric CO_2, since the pores involved in the carbonation process are those >0.1 µm in diameter [35, 36].

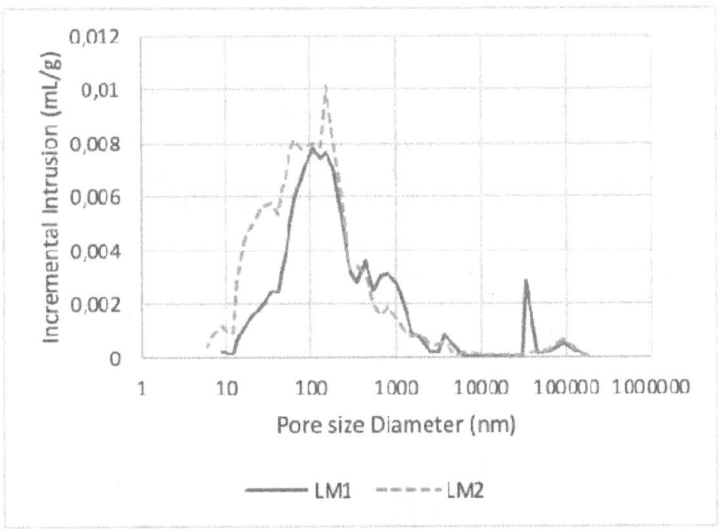

Fig. 1. MIP pore size distributions for both laboratory composites after 365 days of curing.

Thermal analyses carried out on the end-products after 365 days of curing show reduced portlandite content in mortar LM2. This is confirmed through the comparison of the endothermic reaction at ca. 400–500 °C, which corresponds to the dehydroxylation of portlandite (Fig. 2). The latter suggests the enhanced carbonation of LM2 composite, probably due to the increased volume of pores in the area between ca. 10–500 nm (Fig. 1).

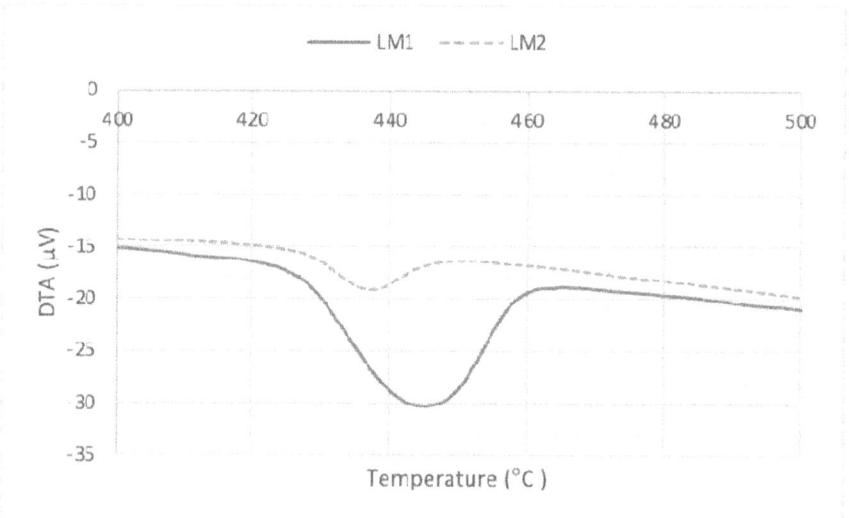

Fig. 2. DTA curves of both laboratory composites after 365 days of curing, showing the presence of portlandite (endothermic reaction at temperature ca. 400–500 °C).

Scanning Electron Microscopy (SEM) images of mortars LM1 and LM2 after 365 days of curing are shown in Fig. 3 and 4, accordingly. Figure 3 shows the absence of large pores in the matrix of lime mortar LM1, which appears to have a rather homogeneous microstructure. In contrast, several pores of varying diameter are identified in the microstructure of lime mortar LM2 (Fig. 4). This is in line with the higher porosity of this specific mix design (see Table 4). At the same time, no distinct microcracks have been observed at the interface between the binder and the aggregates in the microstructure of lime mortar LM1 (Fig. 3). In contrast, capillary cracks at the interfacial transition zone in LM2 are more evident (Fig. 4). These agree well with the much higher capillary absorption coefficients of LM2 (see Table 4).

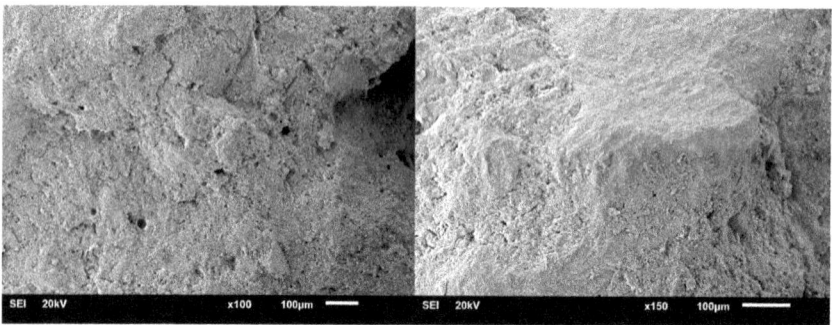

Fig. 3. Representative SEM images of mortar LM1.

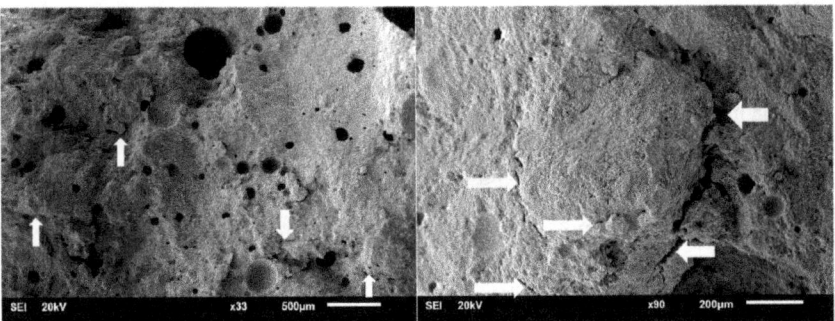

Fig. 4. Representative SEM images of mortar LM2. White arrows indicate the occurrence of capillary cracks.

4 Conclusions

This study aimed at highlighting the effect of the quality of aggregates on the performance of repair lime-based mortars. To this end, two different reef limestone crushed fine aggregates quarried in Cyprus have been thoroughly characterized and used in the production of lime mortars with fixed binder:aggregate ratio and workability. The end-products have been subjected to a number of tests, aiming at determining their physico-mechanical properties at different curing periods.

The results clearly indicate the negative effect of poor-quality reef limestone aggregates with high magnesium sulphate and Micro-Deval coefficients on the performance of hardened lime mortars. This is reflected both in the physical and the mechanical properties of the end-products, as well as in their microstructural characteristics. The mortar (LM2) prepared with the poor-quality aggregate (RL2) showed lower mechanical (compressive and flexural) strength and higher porosity and capillary absorption characteristics, at all curing periods tested. At the same time, its microstructure was evidently porous, with an increased volume of pores with diameters in the area between ca. 10–500 nm and cracks appearing in the interface between the aggregates and the binder.

From the above, it emerges that a careful selection of aggregates is essential prior to designing and preparing a repair lime-based composite. In the absence of strict aggregate quality control on construction sites during the restoration of monuments, the a priori selection of good quality aggregates is imperative in order to ensure the good quality and durability of the end-products. This may be facilitated by a combination of standardized aggregate characterization tests and mineralogical analyses.

Acknowledgements. The authors would like to thank Dr. Ioannis Rigopoulos for his help with the Scanning Electron Microscopy (SEM) analyses. LK would also like to acknowledge financial support from the University of Cyprus.

References

1. Vicente, R., Lagomarsino, S., Ferreira, T.M., Cattari, S., Mendes da Silva, J.A.R.: cultural heritage monuments and historical buildings: conservation works and structural retrofitting. In: Costa, A., Arêde, A., Varum, H. (eds.) Strengthening and Retrofitting of Existing Structures. BPR, vol. 9, pp. 25–57. Springer, Singapore (2018). https://doi.org/10.1007/978-981-10-5858-5_2

2. Pinto, A.F., da Fonseca, B.S., Silva, D.V.: The role of aggregate and binder content in the physical and mechanical properties of mortars from historical rubble stone masonry walls of the national palace of Sintra. Constr. Build. Mater. **268**, 121080 (2021)

3. Elsen, J.: Microscopy of historic mortars-a review. Cem. Concr. Res. **36**, 1416–1424 (2006)

4. Fournari, R., Ioannou, I., Vatyliotis, D.: A study of fine aggregate properties and their effect on the quality of cementitious composite materials. In: Lollino, G., Manconi, A., Guzzetti, F., Culshaw, M., Bobrowsky, P., Luino, F. (eds.) IAEG XII Congress Engineering Geology for Society and Territory-Volume 5: Urban Geology, Sustainable Planning and Landscape Exploitation, pp. 33–36. Springer, New York and London, Italy (2015)

5. Santos, A.R., Veiga, M.R., Santos Silva, A., de Brito, J., Alvarez, J.I.: Evolution of the microstructure of lime based mortars and influence on the mechanical behaviour: the role of the aggregates. Constr. Build. Mater. **187**, 907–922 (2018)

6. Ellis, S., Lawrence, M., Walker, P.: A critical review of the effect of calcitic aggregate on air lime. Spec. Issue Int. Congr. Mater. Struct. Stab.**5**, 97. Rabat, Morocco (2013)

7. Fournari, R., Ioannou, I.: Correlations between the properties of crushed fine aggregates. Minerals **9**, 86 (2019)

8. Carvalho, F.: Mortars from the palace of Knossos in Crete, Greece: a multi-analytical approach. Minerals **12**, 30 (2022)

9. Stefanidou, M., Koltsou, P.: The role of sand in mortar's properties. In: Hemeda S (ed.) Sand in Construction. IntechOpen, London (2022)

10. Smith, M.R., Collis, L.: Aggregates: Sand, Gravel and Crushed Rock Aggregates for Construction Purposes, 3rd edn. Geological Society of London, Engineering Geology Special Publications, London (2001)

11. Pavía, S., Toomey, B.: Influence of the aggregate quality on the physical properties of natural feebly-hydraulic lime mortars. Mater. Struct. **41**, 559–569 (2008)

12. Nayaju, A.B., Tamrakar, N.K.: Evaluation of fine aggregates from the budhi gandaki-narayani river, central Nepal for mortar and concrete. J. Nepal Geol. Soc. **58**, 69–81 (2019)

13. Aggregates for mortar, EN13139, European Committee for Standardization (2002)

14. Ioannou, I., Fournari, R., Petrou, M.: Testing the soundness of aggregates using different methodologies. Constr. Build. Mater. **40**, 604–610 (2013)

15. Tests for thermal and weathering properties of aggregates - Part 2: Magnesium sulphate test, EN1367–2, European Committee for Standardization (2009)

16. Standard Test Method for Resistance of Fine Aggregate to Degradation by Abrasion in the Micro-Deval Apparatus, ASTM D-7428, ASTM International (2015)

17. Tests for mechanical and physical properties of aggregates - Part 6: Determination of particle density and water absorption, EN1097–6, European Committee for Standardization (2013)

18. Theodoridou, M., Ioannou, I., Philokyprou, M.: New evidence of early use of artificial pozzolanic material in mortars. J. Archeological Sci. **40**, 3263–3269 (2013)

19. Maravelaki-Kalaitzaki, P., Bakolas, A., Moropoulou, A.: Physico-chemical study of Cretan ancient mortars. Cem. Concr. Res. **33**, 651–661 (2003)

20. Maravelaki-Kalaitzaki, P., Bakolas, A., Karatasios, I., Kilikoglou, V.: Hydraulic lime mortars for the restoration of historic masonry in Crete. Cem. Concr. Res. **3**, 1577–1586 (2005)

21. Stefanidou, M., Pachta, V., Konopissi, S., Karkadelidou, F., Papayianni, I.: Analysis and characterization of hydraulic mortars from ancient cisterns and baths in Greece. Mater. Struct. **47**(4), 571–580 (2013). https://doi.org/10.1617/s11527-013-0080-y

22. Methods of test for mortar for masonry. Determination of flexural and compressive strength of hardened mortar, EN 1015–11, European Committee for Standardization (1999)

23. Hall, C., Hoff, W.D.: Water Transport in Brick, Stone, Concrete. Spon Press, London (2012)

24. Brandes, H.G., Robinson, C.E.: Correlation of aggregate test parameters to hot mix asphalt pavement performance in Hawaii. Int. J. Transp. Eng. **132**, 86–95 (2006)

25. Yılmaz, M., Tugrul, A.: The effects of different sandstone aggregates on concrete strength. Constr. Build. Mater. **35**, 294–303 (2012)

26. Fournari, R., Ioannou, I., Rigopoulos, I.: The influence of ophiolitic crushed fine aggregate properties on the performance of cement mortars. Bull. Eng. Geol. Env. **80**(12), 8903–8920 (2021). https://doi.org/10.1007/s10064-021-02195-5

27. Rogers, C.A., Lane, B.C., Senior, S.A.: The micro-deval abrasion test for coarse and fine aggregate in asphalt pavement. In: International Center for Aggregates Research 11th Annual Symposium: Aggregates - Asphalt Concrete, Bases and Fines, Texas (2003)

28. Hossain, M.S., Lane, D.S., Schmidt, B.N.: Use of the Micro-Deval test for assessing the durability of Virginia aggregates. Final Report VTRC 07-R29. Virginia Transportation Research Council, Virginia (2007)

29. Lopez-Arce, P., Garcia-Guinea, J., Benavente, D., Tormo, L., Doehne, E.: Deterioration of dolostone by magnesium sulfate salt: an example of incompatible building materials at Bonaval Monastery Spain. Constr. Build. Mater. **23**, 846–855 (2008)

30. Balboni, E., Espinosa-Marzal, R.M., Doehne, E., Scherer, G.: Can drying and re-wetting of magnesium sulfate salts lead to damage of stone? Environmental Earth Sciences **63**, 1463–1473 (2011)

31. Wang, D., Wang, H., Bu, Y., Schulze, C., Oeser, M.: Evaluation of aggregate resistance to wear with Micro-Deval test in combination with aggregate imaging techniques. Wear **338**(339), 288–296 (2015)

32. Strzałkowski, P., Kazmierczak, U.: Wear and fragmentation resistance of mineral aggregates - a review of micro-deval and Los Angeles tests. Materials **14**, 5456 (2021)

33. Lamar, J.E.: Handbook on limestone and dolomite for Illinois quarry operators. Illinois State Geological Survey, Bulletin 91 (1967)

34. Winnefeld, F., Bottger, K.G.: How clayey fines in aggregates influence the properties of lime mortars. Mater. Struct. **39**, 433–443 (2006)

35. Lawrence, R.M., Mays, T.J., Rigby, S.P., Walker, P., D'Ayala, D.: Effects of carbonation on the pore structure of non-hydraulic lime mortars. Cem. Concr. Res. **7**, 1059–1069 (2007)

36. Rigopoulos, I., Kyriakou, L., Ioannou, I.: Improving the carbonation of air lime mortars at ambient conditions via the incorporation of ball-milled quarry waste. Constr. Build. Mater. **27**, 124073 (2021)

Fine Pumice as Pozzolanic Additive
in Restoration Lime Mortars

Tomáš Žižlavský[(✉)] ⓘ, Martin Vyšvařil ⓘ, and Patrik Bayer ⓘ

Faculty of Civil Engineering, Brno University of Technology, Brno, Czech Republic
zizlavsky.t@fce.vutbr.cz

Abstract. In order to improve the properties of fresh and hardened lime mortars, inorganic substances of hydraulic or pozzolanic character have been added to the air lime already in the ancient times. The most commonly used natural pozzolans included tuffs, tuffites, diatomaceous earth, zeolitic rocks, trass or pumice. Although the use of natural pumice in lime mortars is known from history, professional work on the influence of finely ground pumice on the properties of air lime mortars is almost non-existent. Rather, the effects of coarse natural pumice used as aggregate in lime and cement mortars are described in the literature. For this reason, the paper aims to describe the effects of partial lime replacement with finely ground natural pumice on the mechanical, microstructural, and durability properties of air lime mortars. The ground pumice showed similar pozzolanic activity to trass or natural zeolite predicting an improvement in mechanical properties and durability. As the replacement level in air lime mortars increased, the amount of mixing water needed for the same mortar consistency decreased, and the performance properties of the mortars improved. The increase in strengths of mortars was manifested mainly for the 40% lime replacement. This mortar reached at 28 days of age the compressive strength comparable with hydraulic lime-based mortars. The incorporation of finely ground pumice led to the formation of slightly denser, less water absorptive, and more frost resistant and salt crystallization resistant structure in air lime mortars. Lime-pumice mortars showed improved properties in all important aspects from the point of view of the utility of these mortars in restoration and conservation interventions on historic buildings. Based on the achieved results, the 40% lime replacement was found to be optimal.

Keywords: Lime mortar · Pozzolan · Fine Ground Pumice · Durability · Salt Crystallization · Microstructure

1 Introduction

Even though durability and longevity are one of the most desirable factors while referring to the creations of mankind, especially buildings, every product requires some form of maintenance or repair works to ensure further function. While addressing such works on a building considered built heritage, there are several more factors to be taken into account than while working on a common modern building. Thus durability and longevity of constructions in harsh, outer-world environment is a problem that concerns

V. Bokan Bosiljkov et al. (Eds.): HMC 2022, RILEM Bookseries 42, pp. 426–435, 2023.
https://doi.org/10.1007/978-3-031-31472-8_34

generations of builders thorough the history. Even though lime-based materials should have been considered rather strong and durable while compared to their alternatives in these days, they were still far from perfect as are modern materials today. Since ancient times, various admixtures and additives were used to improve the properties of the lime-based mortars [1]. However, since these times, a lot of knowledge had been lost or omitted, especially after the early 19th century invention of Portland cement (OPC) which replaced wide range of lime-pozzolana systems due to its superior strength and durability. With the evolution of OPC and spread of its use, the compatibility issues arose regarding the pre-industrial era buildings and modern materials [2]. When these problems were acknowledged in the restoration practice and addressed by the, now generally acknowledged, like-for-like approach [3, 4], the renaissance of pozzolanic additives begun.

The natural pozzolanic materials are of volcanic origin such as tuffs, tuffites, diatomaceous earth, zeolitic rocks, trass or pumice. Pumice is lightweight volcanic aggregate with porous structure created by entrapment of gasses during the rapid cooling of lava. The current literature mostly deals with the use of pumice as lightweight aggregate (as proposed by Vitruvius [1]) for common or self-compacting concrete, and less studies are addressing the use of finest fraction of pumice aggregate as supplementary cementitious material [5–8]. Relatively high percentage of binder can be replaced by pumice fines in lime-based materials to ensure improvement in the mechanical properties [9, 10].

The current study addresses the use of pumice as pozzolanic additive in lime-based mortars regarding their mechanical properties and durability to freeze-thaw and salt-crystallization cycles, the two most common phenomena behind the degradation of mortars/plasters used in buildings.

2 Materials and Methods

2.1 Materials and Sample Preparation

Commercial dry-hydrated lime of CL 90 S class according to EN 459-1 (Carmeuse a.s., Mokrá, CZ) was mixed with a pure quartz sand (QS; Filtrační písky a.s., Chlum u Doks, CZ) in 1:1 volumetric ratio to prepare the reference mortar. The pumice (WR; Vulkalit WR, Vulcatec Riebensahm GmbH, DE) was ground in a ball mill to achieve the fineness specified as passing through 0.125 mm sieve. The grain size distribution of the obtained material is presented in Fig. 1 along with the grain size distributions of used lime and QS. The chemical and mineralogical composition of lime and aggregate can be found in the previous work of the authors [11], that of pumice in Table 1. The fine ground pumice was used as a pozzolanic additive, replacing 10%, 20%, and 40% of lime by weight in the mixture. To achieve a sufficient precision and repeatability during mixture preparation, the 1:1 volumetric ratio (lime + pumice:sand) was converted to weight proportions presented in Table 2. The mortars were prepared using the amount of water necessary to achieve a workability predefined as 160 ± 5 mm using the flow table test (EN 1015-3).

The standardised prisms of $40 \times 40 \times 160$ mm were cast, demoulded after 72 h at laboratory conditions (20 ± 2 °C; $50 \pm 5\%$ RH), and placed in the highly humid

Table 1. Chemical and mineralogical composition of pumice.

Chemical composition (%)											Mineralogical composition (%)			
Al$_2$O$_3$	SiO$_2$	Fe$_2$O$_3$	CaO	MgO	K$_2$O	Na$_2$O	TiO$_2$	P$_2$O$_5$	MnO	SO$_3$	Amorphous	Quartz	Anorthite	Muscovite
21.04	55.43	2.25	0.72	0.15	4.65	11.52	0.29	0.08	3.81	0.07	88.88	2.38	6.13	2.75

(>90% RH) environment up till the age of 28 days, at which they were placed again in the laboratory conditions for further ageing.

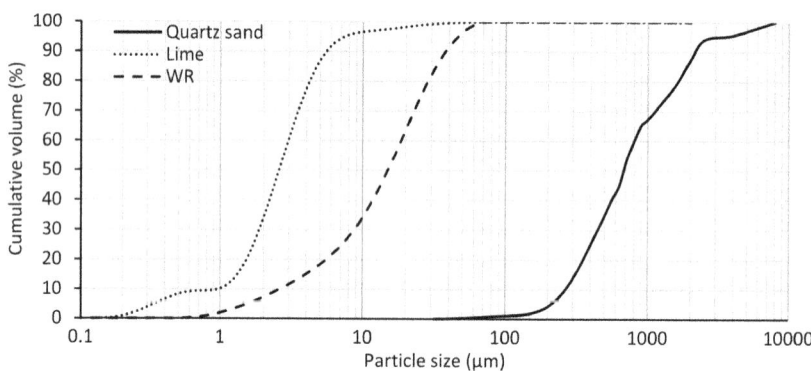

Fig. 1. Grain size distribution of input raw materials.

Table 2. Mix design and fresh state properties of the mortars.

	mix design (g)				fresh state properties		
	lime	WR	QS	water	Air. (%)	WRV (%)	D_{fm} (kg m^{-3})
REF	100	–	355	100	4.5	94.38	1918
L-WR 10	90	10	341	95	4.2	97.41	1937
L-WR 20	80	20	331	85	3.8	98.34	1983
L-WR 40	60	40	309	80	3.2	99.45	2010

2.2 Methods

The following tests were conducted on the fresh mortars: flow table test to determine the amount of water needed (EN 1015-3); test on air content in fresh mortar along with density of fresh mortar (D_{fm}) (EN 1015-7, resp. EN 1015-6); and determination of water retention value (WRV; EN 1015-8). The hardened samples were tested at the age of 28, 90, and 180 days for strength and dry bulk density in accordance with EN 1015-11 resp. EN 1015-10. Additionally, at the age of 90 days, the durability and microstructure of the mortars was studied using various tests. The water absorption under atmospheric pressure (EN 13755) was determined. The saturated samples were subjected to

15 temperature-varying (-20 °C to 20 °C with a 6 h hold at each extremity) cycles in closed, highly humid environment. After the cycling, the freeze-thaw coefficient (D_f) for flexural strength (ratio of strength of freeze-thawed (F) and reference (R) samples) was determined and the alterations in microstructure were observed: the pore size distribution using mercury intrusion porosimetry (MIP; Micrometrics PoreSizer 9310), and capillary water coefficient (C_m) according to EN 1015-18. The resistance of mortars to salt crystallization was tested, as reported elsewhere [11], in a similar way to EN 12370; the solutions were: 10% Na_2SO_4, 3% $NaCl$, and 3% NH_4NO_3.

3 Results and Discussion

3.1 Fresh State Properties

Fresh state properties are along with mix designs summarised in Table 2. The water required to prepare mortars with similar consistency decreases with growing amount of pumice. This is due to larger particle size for the ground pumice (Fig. 1), therefore also lower specific area of the particles [7]. The decrease in water/binder ratio to achieve the same workability corresponds with the results found in the literature for pumice and also other pozzolanic additives such as metakaolin, natural zeolite etc. [9, 12–15] and it correlates with the improvement of rheology of SCC with fine pumice as supplementary cementitious material as observed by several authors [8, 16, 17].

The variations in air content are most-likely related to various rheology of binder-paste which leads to a different level of air-entrapment during the mixing, as the presented values vary only slightly and are in a range of common values for non-air-entrained lime-based mortars. The D_{fm} changes accordingly to air content and water/binder ratio and is governed mainly by these two variables with possible small influence of pozzolana content. The already high WRV of lime-based mortar is further increased for the studied lime-pozzolana mortars growing alongside the pozzolana content. Consulting the work of C. Ince et al. [17] it can be concluded, that two main factors affecting the WRV would be: replacement of lime by pozzolana, thus lowering the mean specific area of binder phase which would lead to decrease in WRV; decrease in water/binder ratio which on the other hand leads to increase in WRV. As the WRV increases it can be assumed that the w/b ratio is the dominant of these two factors in this case. Even though higher WRV would not directly affect any of the studied properties, it can be beneficial for the durability of the material by slowing initial drying, thus diminishing the evolution of drying shrinkage cracks.

3.2 Strength and Hardened Bulk Density

The dry bulk densities of the mortars, stated in Fig. 2(c), are showing similar trends to fresh mortar densities (Table 2) regarding the ground pumice addition. Slightly time-wise growing trend in dry bulk densities of the mortars with lower amounts of WR added is a sign of growing carbonation, whereas the decrease in dry bulk density for L-WR 40 mortar is due to its higher humidity at the age of testing (28 days, right after removal from highly humid environment).

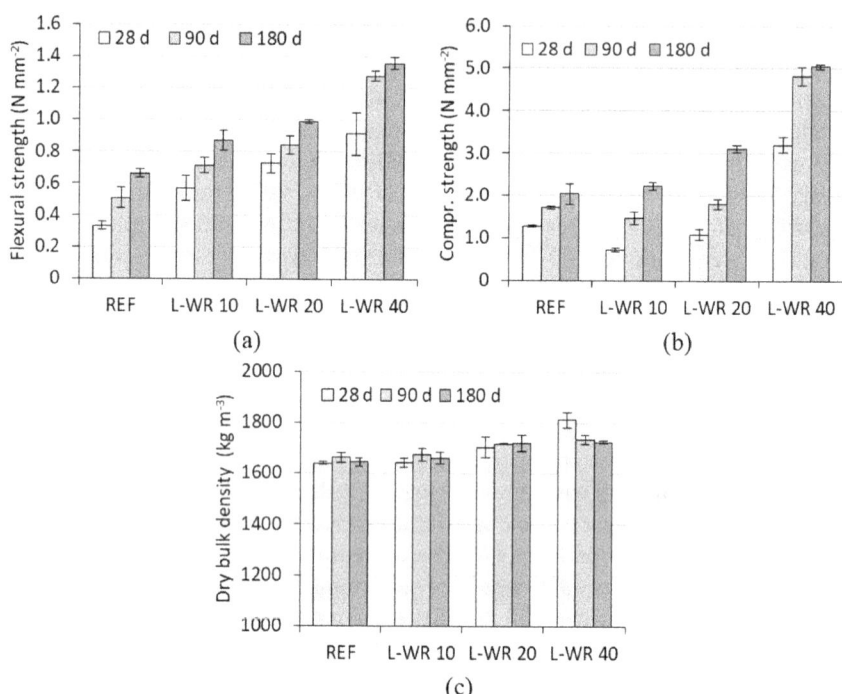

Fig. 2. (a) Flexural strength of the mortars; (b) Compressive strength of the mortars; (c) Dry bulk density of the mortars.

The strength increase with growing pumice dosage was expected due to its pozzolanic properties [7]. The flexural strength, as shown in Fig. 2(a), is increased almost three times at the age of 28 days for the highest pumice content. The ratio diminishes to twice as much at the age of 180 days as the carbonation of the lime-based mortar progresses leading to more significant improvement in strength of the reference mortar. The observed notable increase in flexural strength is comparable to other pozzolanic additives (metakaolin, natural zeolite, silica fume…) [12, 13, 15, 18].

The compressive strength (Fig. 2(b)) on the other hand is decreased at the age of 28 days for both, 10% and 20% lime replaced by pumice leaving the 40% lime replacement stand out more signifiably. However, it should be noted, that the 28 days compressive strength can be unfavourably affected by the increased humidity of specimens as they were tested right after the removal from highly humid environment and, as reported by e.g. Ramesh et al. [19, 20], the compressive strength is decreased due to sample humidity with the effect being more prominent for the non-hydraulic and feebly hydraulic binders. The environmental humidity may be also supported by increased WRV for the modified mortars (Table 1). Over time, the compressive strength of these samples develops, surpassing the strength of reference mortar in all cases by the age of 180 days.

3.3 Microstructural Characteristics

The microstructural characteristics of the mortars were studied at the age of 90 days. The water absorption under atmospheric pressure decreases with growing pumice content (Table 3). The values and degree of water absorption reduction based on pozzolana content are similar to other pozzolanic additives [12–14]. The decreased water absorption should be beneficial for the durability of the mortars.

The total open porosity (TOP R; Table 4) and pore size distribution (PSD; Fig. 3) are of typical values and curve-shapes for the lime-based materials with exception of PSD of mortar with the highest amount of pumice (L-WR 40). The TOP is in this case mainly governed by w/b ratio [21] and amount of pozzolanic additive, therefore increased occurrence of hydraulic reactions [13, 14]. The lower w/b the lower TOP, and the more hydraulic binder the lower TOP so as seen in Table 2, these two factors in this case appropriately complement each other. However, as w/b seems to be more dominant, this synergy led to a significant decrease in TOP only in the case of L-WR 40 mortar. For the mortars with 10% and 20% lime replaced by pumice, the pore structure remained almost the same with the slight reduction of volume of large pores (5–10 μm), which occurrence is driven by varied rheology of the mortar and air entrapment during mixing, slight reduction in volume of pores in the area of main pore volume (0.2–1 μm), and small increase in volume of sub 0.02 μm pores. The latter two changes being affected by the increased amount of pozzolanic additive, thus occurrence of hydraulic reactions. The L-WR 40 mortar presents distinct cumulative pore curve shape comprising of two separate intervals of diameters with significant pore volume. First interval is of pores of slightly larger diameter than the main pore volume of less modified mortars (1–3 μm vs 0.2–1 μm) and second one which contains pores of diameter about order of magnitude smaller than main pore volume (0.02–0.2 μm). The size range of second interval correlates with the common main pore volume of cementitious and highly hydraulic materials, indicating significant occurrence of hydraulic reactions and changes in the binder matrix structure.

The capillary water coefficient (C_m; Table 4) shows trends similar to TOP and water absorption. With decreasing TOP and water absorption, the C_m also decreases, with exception of L-WR 20 where it is slightly increased in correspondence with TOP. The decrease in moisture uptake speed with growing pozzolana addition is common phenomenon observed for various pozzolanic additives by several authors [12–14]. It is usually ascribed to evolution of more condensed structure with higher amount of smaller pores due to the presence of calcium-silicate hydrates formed during hydraulic reactions. This is in correlation with microstructure changes previously discussed in porosimetry study and is especially apparent in the case of L-WR 40 mortar.

3.4 Durability

The freeze-thaw durability, expressed as durability coefficient (D_f; Table 4) for flexural strength (ratio of flexural strengths of freeze-thawed and reference samples), significantly increases with incorporation of ground pumice into the mixture even in the lowest dosage. According to Czech standard ČSN 72 2452 the material is considered durable for the specified number of cycles if $D_f > 0.75$, thus all the mortars with lime partially replaced by ground pumice are durable for 15 cycles. The microstructural changes are expressed

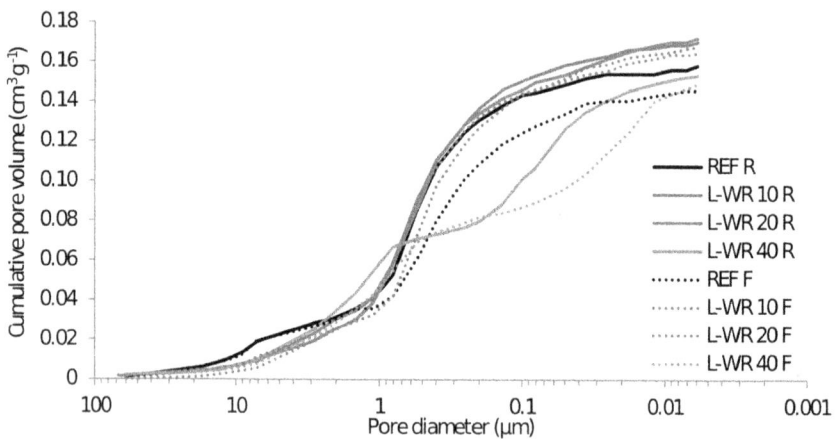

Fig. 3. Cumulative pore volume of mortars with pumice addition; F – mortars subjected to freeze-thaw cycles, R – reference mortars

as MIP results, and capillary water coefficient of freeze-thawed samples (denomination F in the variables). The TOP decreased in all cases, most significantly for the reference mortar, however, the PSD remained almost the same for most of the mortars with minor changes occurring in the area of main pore volume. Only L-WR 40 mortar showed notable variation in pore structure in decrease in volume of pores of size corresponding to main pore volume area for other mortars, and in decrease in pore diameter in the area of typical pore sizes for cementitious materials.

The C_m for the freeze-thawed mortars showed two distinctive behaviours. For the reference mortar and mortar with 10% of lime replaced by pumice, thus the less durable of the studied mortars, the water transport is slowed, which corresponds with decrease in TOP. On the other hand L-WR 20 and 40 mortars showed significant growth in C_m (almost double the value of non-frozen samples) despite the loss in TOP. Both these trends however point to the cracking of internal structure, where the cracks interrupted the capillary network in the reference and L-WR 10 mortars whereas they served as interconnections in the more condensed structure, especially of L-WR 40.

Table 3. Durability of studied mortars to salt crystallization and their water absorption.

	No. of cycles – salt crystallization			Water absorption (%)
	SO_4^{2-}	Cl–	NO_3^-	
REF	3	10	4	15.93
L-WR 10	3	10	5	14.63
L-WR 20	3	10	5	14.00
L-WR 40	5	10	5	13.87

Table 4. Microstructure and durability to freeze-thaw cycling of studied mortars.

	TOP R (%)	TOP F (%)	C_m R (kg m^{-2} s$^{-0.5}$)	C_m F (kg m^{-2} s$^{-0.5}$)	D_f (−)
REF	29.95	26.78	0.189	0.155	0.442
L-WR 10	29.93	29.27	0.152	0.136	0.758
L-WR 20	30.17	29.31	0.163	0.313	0.808
L-WR 40	27.72	27.08	0.080	0.158	0.819

The salt crystallization resistance of the mortars is presented in Table 3 as a number of cycles that the samples withstood before their disintegration. In the NaCl solution, all the mortars endured for the designed test duration of 10 cycles proving good resistance of lime-based materials to chloride corrosion, therefore several-times more cycles would be needed to reveal any differences between the mortars [2, 15]. The sulphate attack is more of chemical nature, resulting in a formation of gypsum through the mechanisms of wet or dry disposition of carbonate ions by sulphates. The gypsum formed is then wash out with the saline solution due to its relatively high water solubility (approximately 2.5 g l^{-1} vs 0.013 g l^{-1} for calcite), revealing underlying layers of calcitic structure and in this manner quickly consuming most of the binder matrix in the samples. Only L-WR 40 mortars with highest pozzolana addition showed improvement in the resistance as large portion of portlandite was used in the pozzolanic reactions with ground pumice and the structure of the samples is more likely dependent on the products of these reactions than the calcite formed during carbonation. Even the lowest pozzolana replacement led to slight improvement in the resistance of the samples subjected to ammonia nitrate solution however, further increase in pozzolana content had not affected the mortars resistance.

4 Conclusions

The study presents the use of ground pumice as a natural pozzolana partially replacing lime in the mortars and their fresh, mechanical, and microstructural properties and durability to salt and ice crystallization. The pozzolana addition leads to decrease in amount of kneading water required for the preparation of the mortar with the same workability. The density of fresh mortar along with water retention value are increased while air content in fresh mortar decreases with the degree of lime replacement. Growing ground pumice content promotes both, compressive and flexural strength however, in the case of compressive strength, the positive effects present themselves after longer ageing time for the mortars with lower pumice content. The microstructure of the mortars is only mildly affected in most cases as presented by slightly lower water absorption value and hindered water transport expressed as capillary water coefficient. Only the mortar with 40% of lime replaced by ground pumice showed notable variations in pore size distribution presented as two distinctive pore size intervals of main pore volumes, and significantly lower capillary water coefficient. Even the 10% lime replacement led to significant improvement in frost resistance of the mortars, further improving with growing replacement degree. The salt crystallization on the other hand was only mildly affected by the ground pumice leading to moderately enhanced durability in ammonium nitrate solution and increased resistance to sulphate attack for the mortar with 40% lime replacement.

The fine pumice was found to be suitable pozzolanic additive for the lime-based mortars, mainly while addressing freeze-thaw resistance where even 10% lime replacement led to significant improvement while still producing soft and permeable mortar. Where the final compressive strengths on the level of natural hydraulic lime are desired, there the higher percentage of lime should be replaced by ground pumice however, sacrificing notably capillary water transport in the mortar.

Acknowledgement. The work has been financially supported by Czech Science Foundation, grant number 21-06582S "Experimental and computational analysis of salt transport, accumulation, and crystallization in non-hydrophobized rendering mortars".

References

1. Vitruvius Polio, M.: Vitruvius: The Ten Books on Architecture. Dover Publications, New York, New Dover (1960)
2. Rodrigues, P.F., Henriques, F.M.A.: Current mortars in conservation: an overview. Restor. Build. Monum. **10**, 609–622 (2004). https://doi.org/10.1515/rbm-2004-5901
3. Forster, A.M.: Building conservation philosophy for masonry repair: part 1 – "ethics." Struct. Surv. **28**, 91–107 (2010). https://doi.org/10.1108/02630801011044208
4. Forster, A.M.: Building conservation philosophy for masonry repair: part 2 – "principles." Struct. Surv. **28**, 165–188 (2010). https://doi.org/10.1108/02630801011058906
5. Papanicolaou, C.G., Kaffetzakis, M.I.: Lightweight aggregate self-compacting concrete: state-of-the-art & pumice application. J. Adv. Concr. Technol. **9**, 15–29 (2011). https://doi.org/10.3151/jact.9.15
6. Herki, B.M.A: Lightweight concrete using local natural lightweight aggregate. J Crit Rev **7**, 490–497 (2020). https://doi.org/10.31838/jcr.07.04.93
7. Seraj, S., Cano, R., Ferron, R.D., Juenger, M.C.G.: The role of particle size on the performance of pumice as a supplementary cementitious material. Cem. Concr. Compos. **80**, 135–142 (2017). https://doi.org/10.1016/j.cemconcomp.2017.03.009
8. Hedayatinia, F., Delnavaz, M., Emamzadeh, S.S.: Rheological properties, compressive strength and life cycle assessment of self-compacting concrete containing natural pumice pozzolan. Constr. Build. Mater. **206**, 122–129 (2019). https://doi.org/10.1016/j.conbuildmat.2019.02.059
9. Allahverdi, A., Ghorbani, J.: Chemical activation and set acceleration of lime-natural pozzolan cement. Ceram. - Silikaty **50**, 193–199 (2006)
10. Bakis, A.: The usability of pumice powder as a binding additive in the aspect of selected mechanical parameters for concrete road pavement. Materials (Basel) **12**, 2743 (2019). https://doi.org/10.3390/ma12172743
11. Vyšvařil, M., Bayer, P.: Salt and ice crystallization resistance of lime mortars with natural lightweight aggregate. In: Serat C (ed) XV International Conference on Durability of Building Materials and Components. eBook of Proceedings. CIMNE, Barcelona (2020). https://doi.org/10.23967/dbmc.2020.121

12. Brzyski, P.: The effect of pozzolan addition on the physical and mechanical properties of lime mortar. In: E3S Web Conference, vol. 49, p.00009 (2018). https://doi.org/10.1051/E3S CONF/20184900009
13. Aly, M., Pavía, S.: Properties of hydrated lime mortars with pozzolans. In: III International Congress on Construction and Building Research (COINVEDI), pp. 54–56. Universidad de Madrid, Escuela Técnica Superior de Edificación (2015)
14. Pavía, S., Aly, M.: Influence of aggregate and supplementary cementitious materials on the properties of hydrated lime (CL90s) mortars. Mater. Construcción 66, 104 (2016). https://doi.org/10.3989/mc.2016.01716
15. Faria, P.: Resistance to salts of lime and pozzolan mortars. In: International RILEM Work Repair Mortars Hist Mason January, pp. 99–110 (2005)
16. Ghasemi, M., Rasekh, H., Berenjian, J., AzariJafari, H.: Dealing with workability loss challenge in SCC mixtures incorporating natural pozzolans: a study of natural zeolite and pumice. Constr. Build. Mater. 222, 424–436 (2019). https://doi.org/10.1016/j.conbuildmat.2019.06.174
17. Rashad, A.M.: An overview of pumice stone as a cementitious material – the best manual for civil engineer. Silicon 13, 551–572 (2021).https://doi.org/10.1007/s12633-020-00469-3
18. Ince, C., Carter, M.A., Wilson, M.A., et al.: Factors affecting the water retaining characteristics of lime and cement mortars in the freshly-mixed state. Mater. Struct. 44, 509–516 (2011). https://doi.org/10.1617/s11527-010-9645-1
19. Gameiro, A., Santos Silva, A., Faria, P., et al.: Physical and chemical assessment of lime–metakaolin mortars: Influence of binder: aggregate ratio. Cem. Concr. Compos. 45, 264–271 (2014). https://doi.org/10.1016/j.cemconcomp.2013.06.010
20. Ramesh, M., Azenha, M., Lourenço, P.B.: Mechanical properties of lime–cement masonry mortars in their early ages. Mater. Struct. 52(1), 1–14 (2019). https://doi.org/10.1617/s11527-019-1319-z
21. Ramesh, M., Azenha, M., Lourenço, P.B.: Impact of moisture curing conditions on mechanical properties of lime-cement mortars in early ages. In: 13th North American Masonry Conference Proceedings. The Masonry Society, Salt Lake City (2019)
22. Papayianni, I., Stefanidou, M.: Strength–porosity relationships in lime–pozzolan mortars. Constr. Build. Mater. 20, 700–705 (2006). https://doi.org/10.1016/j.conbuildmat.2005.02.012

The Relationship Between Natural Stone Joint Design, Surface Area and the Properties of Lime Mortar Joints

Matthew Cook[✉]

Cook Masonry Ltd., Sedbergh Cumbria, UK
info@cookmasonry.co.uk

Abstract. Historically, building stone was extracted and shaped by hand. To produce a flat surface using a mallet and chisel requires the time and energy of a skilled mason. As such, the highest level of workmanship was generally reserved only for the seen faces of stones. The joint surfaces were given less attention and would subsequently be "rougher".

In the modern era, diamond tipped gantry or wire saws are the standard equipment for stone processing. For the purposes of building conservation, the seen face of replacement stones are usually hand chiselled in keeping with the original design. However, it has become increasingly common for the joints of the new stones to be left as a clean diamond sawn surface.

This paper examines if and how the difference in surface area between various stone surface finishes changes the characteristics of the lime mortar joint. The paper includes a surface area comparison of modern and historic stone surface finishes. This is followed by practical testing to ascertain how lime mortar joint/adhesive bond strength changes in relation to the amount of stone surface area available for adhesion.

The results of the testing suggest a direct relationship between stone surface finish, joint surface area and lime mortar adhesion.

Keywords: Stone Conservation · Lime Mortar Adhesion · Stone Surface Area · Bond Wrench · Stone Joint Design · Stone Surface Roughness

1 Introduction

Using a variety of mortars, the act of removing excessively decayed stones from buildings and replacing them with newly cut stones replicating original designs has been taking place for many centuries [1] (p241–242).

However stone processing has undergone significant changes as a result of increased mechanisation. Replacement stones are increasingly being quarried and worked to shape using modern diamond tipped tools and are much smoother and more regularly cut than has been seen in the past [2] (79–84). It would seem logical to assume that this change in processing has reduced the surface area of the stone available for mortar adhesion.

V. Bokan Bosiljkov et al. (Eds.): HMC 2022, RILEM Bookseries 42, pp. 436–449, 2023.
https://doi.org/10.1007/978-3-031-31472-8_35

The majority of the sales in processed natural building stone is produced for use on new build projects [3]. Most modern stone is supplied in a sawn six sided format with the seen face(s) already worked as required. It is also common for quarries to supply sawn six sided stones with two polished sides and a polishing process is often incorporated into stone production procedures by default.

As a result of known incompatibilities between portland cement and natural stone, the use of lime mortar has become standard practice in stone building conservation [4]. Despite a number of well documented studies, the relationship between the lime mortar bond and natural stone is still not fully understood [5] (p13) [6] (p134).

Previous research [7] has highlighted the role of the interstitial mortar bond in governing moisture transfer between individual stones. Organisations specialising in stone conservation are also known to deliberately chisel the joints to "rough" the surface of new stones prior to the application of mortar. Roughing the stones is anecdotally thought to promote mortar adhesion.

A lack of official documentation regarding historic joint surface finishes and the finish of replacement stones makes comparative study problematic. It is therefore difficult to directly ascertain how stone surface finish/area may change the interaction between stone and lime mortar. Any possible effects upon the long term durability of the stone and mortar as a consequence of differences in surface area appear to be undocumented.

Tensile and/or adhesive strength measurements for lime mortar are conventionally calculated according to dimensional surface area [8]. Current testing standards do not make allowances for differences in surface "roughness" [8–10].

2 Sample Preparation

Seventy-Seven 150 mm × 30 mm × 150 mm (bed height last as per masonry convention/BS EN 1052-5) stones were diamond cut from a single slab of Dunhouse Buff Sandstone (17.7% porosity, 35.06 g/m · s^2 [3]). A further eleven stones were cut to the same specification from a single piece of Tadcaster magnesium limestone (20.50% porosity 56 g/m · s^2 [11]). This gave a total of eighty-eight stones (Figs. 1, 2, 3, 4 and 5).

The eleven magnesium limestone samples were polished to a grade of p3000. The Dunhouse stones were randomly divided up into seven groups of eleven stones each. Each of the eleven stones were then given one of seven possible stone finishes to the 150 mm × 150 mm face as per the following table:

Table 1. List of sample groups by surface finish.

Sample Group	Surface Finish
1	Diamond cut/sawn (p40)
2	Polished (p400)
3	Batted/tooled horizontally ($\approx$10 per inch/25 mm)
4	Batted/tooled vertically ($\approx$10 per inch/25 mm)
5	Batted/tooled diagonally ($\approx$10 per inch/25 mm)
6	"Dunted"/crudely point chiselled ($\approx$20 chisels per sample)
7	Rubble/needle gun
8	Magnesium lime – polished (p3000)

Fig. 1. Photograph of sample from group 1 being finished with a diamond wheel.

Fig. 2. Photograph of a sample group 2 with diamond polishing pads.

Fig. 3. Photograph of sample group 3 with tooling chisel.

Fig. 4. Photograph of sample group 6 with mallet and chisel.

Fig. 5. Photograph of sample 7 being worked with a needle gun.

3   Laser Scanning and 3D Modelling

A single sample of each group of stones from Table 1 (excluding group 4) was selected at random. The stone samples selected were sent for combined laser scanning and 3d modelling undertaken by an accredited laboratory [12]. Laser scanning at a resolution of 22 μm (0.022 mm) was undertaken using a Nikon LC15DX scanner. The results were inputted into the Nikon Focus Inspection programme. The compiled scanning results were used to create a point cloud model. The data from the 3d model was used to triangulate the total surface area of each design.

The results for group 4 were estimated using the data from group 3 as both stones use the same design rotated 90°.

Due to the complexity of the raw data obtained from the scanning, the following post processing was applied using Nikon Focus Inspection:

- Importing raw point clouds
- Merging the 6 individual paths into 1
- Filtering using a 0.1 mm grid
- Optimize and Mesh
- Align using detected planes and mid-lines
- Export STL file

Geomagic Wrap was then used to fix any potential issues by:

- Import STL file
- Open Manifold
- Convert to points removing nominals
- Re-mesh
- Fill holes (flat)
- Remove spikes
- Mesh Doctor
- Export STL file

Finally, the STL files were imported into Geomagic Design X and trimmed to 145 mm × 145 mm to remove the inconsistency found within the edges of the stones. The area/volume was then calculated by the Design X programme. The calculations were manually modified by multiplying the 145 mm × 145 mm surface area by 1.070155 (rounded to 6 decimal places) to simulate a continuous 150 mm × 150 mm surface with no edges.

Fig. 6. Computer generated 3D model of sample group 7.

4   Adhesive Bond Strength Test

Ten stones of each grouping prepared as outlined in Table 1 were separated and soaked in buckets of clean tap water for 36 hours. This was done to fully saturate the stones, ensuring that the moisture content was consistent/repeatable between samples.

Approximately twenty minutes prior to application, a mortar mix was made using 2.5 parts Nosterfield washed river sand to 1 part St Astier NHL 3.5 (ratio by volume). The mixing process was carried out in accordance with BS EN 1015-2 [13]. However, the consistency test outlined in BS EN 1015-3 [14] or BS EN 1015-4 [15] was substituted for the traditional method of cutting and suspending the mortar from the bottom of a trowel [16].

After 36 hours soaking in tap water, the stones were removed from the buckets in chronological order according to group and number. As the stones were removed, they were paired with another stone of the same group and number. Mortar was then applied to the worked face of one stone. The second stone was then placed face down on top of the combination of stone and mortar.

The top of the second stone was tapped down using a rubber mallet to create a 5 mm mortar joint. This process was repeated five times for each group of stones creating a total of forty test units. Once all eighty stones had been processed, the forty test units were left in a sheltered outdoor area to set/cure.

After 24 hours any excess mortar on the sides of the test units was struck off using a tuck trowel. The test units remained in the sheltered outdoor area to cure for a total of 28 days.

In anticipation of the 28 day curing period a 10 tonne hydraulic press [17] was prepared for testing. The base of the press frame was wedged in position using a block of wood.

The press was equipped with a 2.5 MPa digital pressure gauge. A recent calibration test [18] showed the gauge to be accurate to within 0.15% of the UKAS full scale benchmark. The gauge was set to zero prior to starting the experiment and the inertia of the ram measured at around 0.030 MPa.

A customised vice was positioned and secured 10 mm to the side of the mandrel of the hydraulic press head. This positioned the centre of the press head in the middle of the second stone of the test sample (see Fig. 7).

Following the 28 day curing period, the sample units were placed onto the 10 tonne hydraulic press for testing. The stone test units were placed into the custom vice with one stone (half the test unit) resting on the mid rail of the vice. The mortar joint and second stone of the test unit was left overhanging beyond the rail of the vice with no support underneath (see Fig. 7).

The top rail of the vice was then tightened to secure the top of the test unit to a force of 20 newton meters of torque. Care was taken to evenly distribute the force between the two sides of the vice by regularly alternating the side being tightened. The angle and position of the testing unit was confirmed to be square and correct using a sinking square tool aligned with the frame of the hydraulic press.

The digital pressure gauge was set to record the maximum force reading. The hydraulic press was then charged and the piston head brought to bear with the release valve partially open to prevent a sudden impact.

'Once positioned, the pressure was increased until such time as the test unit failed by breaking along the mortar joint. At this point the maximum pressure and nature of the failure was recorded. The procedure was repeated sequentially through the numbers and groupings until all 40 test units had been stressed to failure.

Fig. 7. Photograph of the hydraulic press during the bond strength testing for sample A1.

Fig. 8. Photograph of sample A1 after testing. The sample exhibits a typical failure pattern, splitting along a single stone to mortar interface.

5   Results, Analysis and Discussion

5.1   Surface Area Results

Table 2. Table showing the results of the 3d laser scanning surface area calculations.

Sample	145 mm × 145 mm post processing mm^2 (1dp)	150 mm × 150 mm estimate mm^2 (1dp)	Percentage increase from Dunhouse Polished (group 2)
1. Diamond Sawn	21495.5	23003.5	1.0
2. Polished	21288.1	22781.6	0.0
3. Batted with bed	21656.1	23175.4	1.7
4. Batted 90° to bed	21656.1*	23175.4*	1.7
5. Batted Diagonal	21546.4	23058.0	1.2
6. Point Chisel	22674.4	24265.1	6.5
7. Needled	21902.2	23438.7	2.9
8. Polished Lime	21072.1	22550.4	**−2.7**

* = estimate

5.2 Surface Area Analysis and Discussion

The test data in Table 2 indicates that the surface area of natural building stone changes according to the way the stone has been worked to shape or finished. In this experiment, roughly chiselling the surface of diamond cut stone using a point chisel offered the greatest impact at around a 6.5% increase above a diamond polished stone of the same type and dimensions.

Although the greatest increase seen in this test was 6.5%, it seems highly likely that hand working a stone using a point chisel in the traditional manner would result in a significantly greater increase in the surface area than seen in the testing. This is because test group 6 was only chiselled to around 30% coverage to ensure that the size of the joint and dimensions of the sample remained consistent to facilitate comparative testing. Similarly, the sample stones were carefully cut straight between fixed dimensions and it seems unlikely that unseen stonework would have been traditionally cut to such an exacting tolerance.

The testing also shows that surface area varies by stone type. This can be seen in the results of sample group 8 that measured 1.7% below the sandstone samples with an equivalent surface finish. It seems likely that this difference is due to the fine grained nature of the magnesium limestone vs the medium grained composition of the sandstone.

5.3 Surface Area to Adhesion Results

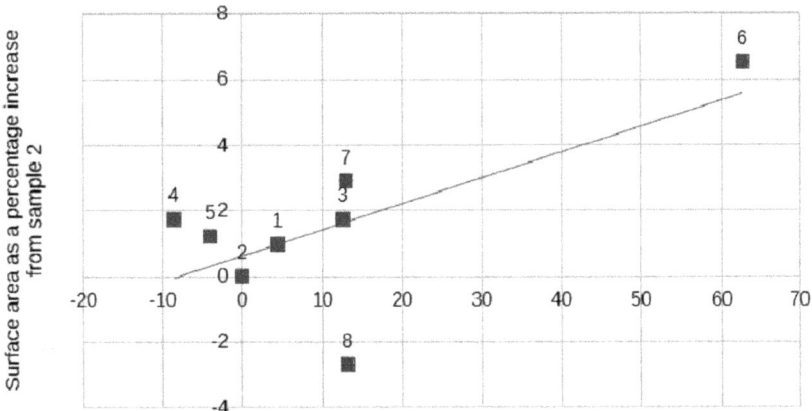

Highest failure force as a percentage increase from sample 2

Fig. 9. Graph comparing highest single failure force per group and surface area as a percentage of Dunhouse polished (sample group 2). Trend line applied.

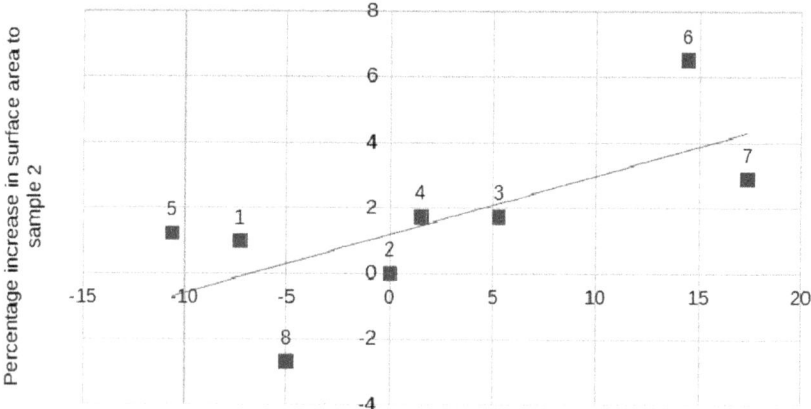

Fig. 10. Graph comparing group mean failure force (excluding suspected anomalies) and surface area as a percentage of Dunhouse polished (sample group 2). Trend line applied.

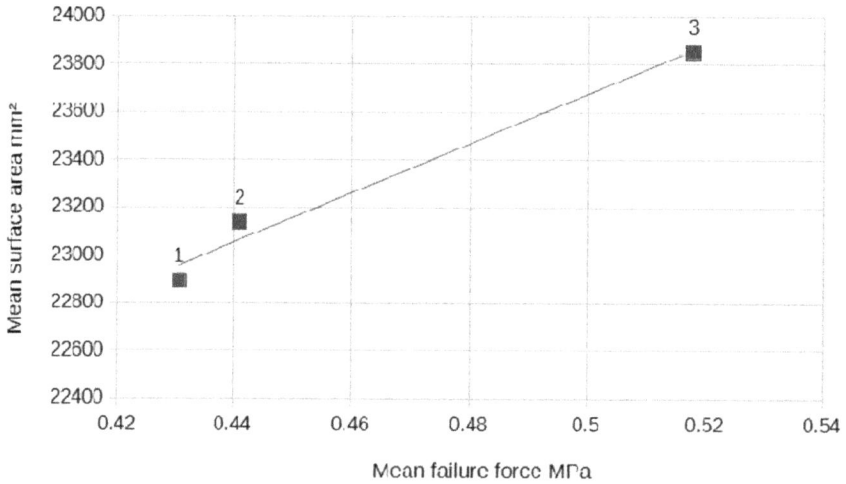

Fig. 11. Graph comparing mean surface area and mean failure force (excluding suspected anomalies) by workmanship type grouping. Trend line applied. Workmanship type group 1 (diamond cut) - Dunhouse diamond polished and Dunhouse diamond sawn S1/S2. Workmanship type group 2 (chiselled) - Dunhouse flat chiselled vertically, horizontally and diagonally S3/S4/S5. Workmanship type group 3 (rubble) - Dunhouse point chiselled and needled S6/S7.

5.4 Adhesive Bond Strength Analysis and Discussion

Figures 9 and 10 show that differences in both average (mean) adhesive bond strength and maximum achievable bond strength vary according to the surface finish applied to comparable stones. Group 6 produced both the highest and second highest individual failure force results within the testing.

It is notable that the results of the adhesive testing contained significant variation/inconsistency. Group 7 had the highest modified mean results (0.524 MPa) as well as the joint highest occurrence of possible anomalous results with two readings falling below the 0.300 MPa threshold for possible anomalous readings.

It is clear that the bond strength is influenced by factors unrelated to surface finish. This variation can be seen by the performance of group 8 in Figs. 9 and 10. The magnesium limestone samples demonstrated adhesive strength consistently above the sandstone samples with comparable surface finishes.

Group 8 also had an unusually high rate of atypical failures. Three of the five sample units failed along both stone to mortar interfaces with a break running through the mortar joint. All but three of the total combined Dunhouse samples failed in the typical single sided manner as shown in Fig. 8. No other sample group had more than a single atypical failure per type.

It would seem that natural variations within the stone, workmanship, sample or mortar preparation could be more relevant to the adhesive bond strength of individual samples than the surface finish type (such as in the case of discounted anomalies). However, Figs. 9, 10 and 11 all suggest that the average or potential maximum adhesive bond strength of similar materials is directly influenced by surface finish and can generally be grouped accordingly.

5.5   Surface Area to Bond Strength Relationship

Figure 9 shows a strong positive correlation between surface area and the maximum achievable stone adhesive bond strength. It should be noted that sample 8 is an outlier because the change in stone type introduces uncontrolled variables outside the scope of this investigation.

There is evidence of a moderate correlation between mean adhesive bond strength per group and surface area. This connection can be seen in Fig. 10.

When the samples are sorted and compared according to processing method, diamond cut/hand chiselled/rubble type as shown by Fig. 11, the sample sizes are increased to help remove anomalies. Figure 11 suggests a strong positive correlation between mean surface area and average adhesive bond strength.

The relationship between surface area and adhesive strength appears disproportionate. The linear rate of progression shown in Fig. 9 suggests that if the surface area of a fixed dimension is increased by 5% the maximum possible adhesive bond strength increases by approximately 55% (compared to polished samples of the same stone type).

Figure 11 suggests a similar relationship between surface area of a fixed dimension and mortar adhesion. As surface area increases by 1% the mean adhesive bond strength increases by approximately 5% (compared by numerical measurement data). Figure 10 demonstrates the same relationship at approximately 4% more adhesion per 1% increase in surface area.

6 Conclusion

It should be noted that the testing outlined in this paper is a new procedure, carried out on a small scale and outside laboratory conditions. However, it is clear that there is a strong link between stone surface finish, surface area and the adhesive bond strength of associated lime mortar.

Altering the surface finish of a stone directly impacts upon its surface area. This in turn changes average and/or maximum achievable adhesive bond strength between the stone surface and the lime mortar. Any increase in stone surface area causes a significant increase in adhesive bond strength. This difference could be as much as a ten fold percentage to percentage increase (when compared to polished samples of the same stone).

An explanation for the behaviour is available from the accepted current literature on the subject. The results suggest that an increase in adhesive strength occurs because the amount of pore space within the stone that is available for the lime to fill increases as surface area increases [5, 7]. This in turn enhances the bond between the two materials.

It therefore seems probable that other important joint characteristics such as stone to mortar moisture transfer and permeability, perceived/relative mortar permeability, lateral adhesive strength and weather resistance are similarly affected by joint surface area/surface tooling.

References

1. English Heritage. Practical Building Conservation Stone (2012)
2. Warland, G.E.: Modern Practical Masonry (1929)
3. Fine Grained Creamy Buff Sandstone. https://www.dunhouse.co.uk/stonefinder/dunhouse-buff/. Accessed 05 Dec 2020
4. BSI. BS EN 7913 Guide to the conservation of historic buildings (2013)
5. Wiggins, D.: Technical Paper 27 (2018)
6. Barr, S., McCarter, M., Suryanto, B.: Bond-strength performance of hydraulic lime and natural cement mortared sandstone masonry (2015)
7. De Freitas, V.P., Abrantes, V., Crausse, P.: Moisture migration in building walls analysis of the interface phenomena (1995)
8. BSI. BS EN 1052-5 Method of test for masonry - Determination of bond strength by the bond wrench method (2005)
9. BSI. BS EN 771-6 Methods of test for mortar for masonry - Part 12 Determination of adhesive strength of hardened rendering and plastering mortars on substrates (2016)
10. BSI. BS EN 772-16 Methods of test for masonry units Part 16: Determination of dimensions (2011)
11. Cadeby Technical Results. https://www.cadebystone.com/wp-content/uploads/2017/10/Cadeby-Bed-1-Tech-Details-26.01.17.pdf. Accessed 05 Dec 2020
12. BSI. BS EN ISO 9001 General requirements for the competence of testing and calibration laboratories (2017)
13. BSI. BS EN 1015-2 Methods of test for mortar for masonry - Part 2 Bulk sampling of mortars and preparation of test mortars (2002)
14. BSI. BS EN 1015-3 Methods of test for mortar for masonry - Part 3 Determination of consistence of fresh mortar (by flow table) (1999)

15. BSI. BS EN 1015-4 Methods of test for masonry mortar - Part 4 Determination of consistence of fresh mortar (by plunger penetration) (1999)
16. What you wanted to know about lime mortar and were afraid to ask (2018). https://www.you tube.com/watch?v=-0zycioctlg
17. Clarke CSA10BB 10 Tonne Hydraulic Bench Press. https://www.machinemart.co.uk/p/cla rke-csa10bb-10-tonne-bench-press/. Accessed 05 Dec 20
18. BSI. BS EN ISO 17025 General requirements for the competence of testing and calibration laboratories, 2

Development of a Gypsum-Based Grout for the Stabilisation of Gypsum-Based Plasters

Gvantsa Potskhishvili[1](✉), Chiara Pasian[2], and Francesca Piqué[3]

[1] Faculty of Restoration, Art History and Theory, Tbilisi State Academy of Art, Tbilisi, Georgia
gvantsa.potskhishvili@art.edu.ge
[2] Department of Conservation and Built Heritage, University of Malta, Msida, Malta
[3] Institute of Materials and Constructions (IMC), University of Applied Sciences and Arts of Southern Switzerland (SUPSI), Manno, Switzerland

Abstract. Research on injection grouting for the stabilisation of delaminated wall paintings and historic plasters mostly focuses on grouts based on aerial or hydraulic lime and on earthen-based grouting. Materials for the stabilisation of delaminated gypsum-based plasters have never been fully studied, even if gypsum has been used as a constituent material of wall paintings since ancient times. Grout mixtures developed during this research were specifically designed for the stabilisation of the 11th-century gypsum-based wall paintings in the Ateni Sioni church in Georgia. Due to the gypsum sensitivity to water, it was important to reduce the water content of the grout; this was achieved by partially substituting water with an alternative liquid, i.e. ethanol. The grout mixture design involved laboratory testing to assess the working properties and some performance characteristics of the grouts. In addition to the regular tests, grouts were injected into replicas simulating the challenging horizontal delamination present in the vaults of the church. The research proposes gypsum as the binder for a grout together with the use of a water-reduced dispersion medium, aiming to stabilise delaminated gypsum-based plasters.

Keywords: Injection grout · gypsum-based · gypsum plaster · site-specific · design · water-reduced

1 Introduction

1.1 Injection Grouting

Wall paintings, defined as paintings executed on the bearing structure, may be composed of different layers: from the primary support to plaster layer(s), paint layer(s), varnish layer(s) and attachment(s). Lack of adhesion between plaster layers resulting in delamination is a widespread deterioration phenomenon in wall paintings and is defined as a separation between coherent plaster layers or between plaster and primary support. This often results in the formation of an empty space, a void between the delaminating layers. This unstable situation which can result in the loss of painted plaster, can be fixed with the direct intervention of grouting. Grouting aims to introduce in the delamination an adhesive material with bulking properties to re-establish adhesion [1]. Such intervention is one of the most challenging in wall paintings conservation since the problem is concealed and grouting is irreversible [2] (p. 472). Injection grouts are mixtures composed of binder(s), aggregates, and a liquid (suspension medium), which is typically water; additives may be added to the formulation [3]. Research on injection grouting for the stabilisation of delaminated wall paintings and historic plasters mostly focuses on hydraulic lime/lime-based grouts (among others, [3]; [1]; [4–7]) and/or earthen-based grouts [8, 9]. In the literature, such binders may be also used in combination (ex. earth-lime, [10]). Studies on historic gypsum-based plasters rather focus on their characterisation [11, 12] and the design of compatible repair plasters [13–15]. To the authors' best knowledge, there are no publications (in English or Italian) regarding gypsum as a binder for grouting of gypsum-based plasters where working properties and performance characteristics are systematically assessed. The properties of gypsum-based grouts have not been studied, even if gypsum is widely spread as an original material, and it has been used as a constituent material of wall paintings since ancient times [16, 17]. Since the grout introduced during the intervention becomes a non-extractable part of the stratigraphy, it is important to ensure compatibility and stability of the grout, as well as retreatability. The physico-chemical compatibility with the original materials and a physico-mechanical behavior close to the one of the original plaster are paramount: therefore, especially for relatively unexplored binders and extensive interventions on site, a systematic study evaluating the physico-mechanical properties of the injection grout in comparison to the original is necessary before undertaking the treatment.

1.2   Case Study

The 7th-century Ateni Sioni church (Fig. 1), located in Georgia, is one of the most significant religious sites in the country and is listed as an immovable cultural monument of national significance. The church is built with various types of tuff stone, which can be found at the site around the building (vitroclastic tuff containing analzim, yellowish tuff and green coarse grained tuffs which are high in quartzite) [18]. Beside its fine architecture, the 10th-century stone reliefs on the church exterior facades, the 7th-century aniconic decorations and the 11th-century wall paintings on the interior walls of the church are extremely important. The 11th-century painting's phase depicts stories of the New and Old Testament, including a large cycle of the life of the Virgin as well as numerous figures of saints and ornamental decorations [19] (Fig. 1).

Various types of painted plasters have been identified in the Ateni Sioni church. Analyses of plaster fragments, thin and cross sections were performed at the University of Applied Sciences and Arts of Southern Switzerland (SUPSI) with Polarized Light Microscopy (PLM) and FT-IR (Fourier-Transform Infrared spectroscopy). Results showed that a thin layer (1 mm) of lime-based plaster containing particles of charcoal and probably red iron oxide was applied before the gypsum-based plaster, on the stone primary support. The most widely spread plaster type in the church is gypsum-based. It has been applied over the thin lime-based plaster, on the interior walls between the 7th–8th centuries [19]. The gypsum plaster does not contain aggregates and is usually applied in two layers, i.e. a thinner lower layer and a thicker upper layer reaching up to 1 cm of total thickness. The second most common plaster type in the church is a

(a)   (b)

Fig. 1. (a) Ateni Sioni church, Georgia; view from the southeast. © Gvantsa Potskhishvili, 2015; (b) The 11th-century decoration phase; south apse, Ateni Sioni church, Georgia. © Gvantsa Potskhishvili, 2015.

lime-based plaster with straw-like organic inclusions, usually present either in patches next to gypsum-based plaster or overlapping it. This plaster must have been applied between the 8th and 11th centuries [19]. The 11th-century wall paintings are painted using *a secco* technique on both the gypsum and the lime-based plasters. Another kind of lime-based plaster, mainly found in the pendentives, is characterised by high hair-like fiber content and organic inclusions, and most probably was applied between the 8th and 11th-centuries [19]. The church underwent several changes and restorations over the centuries and probably suffered from an early stage of problems of plaster adhesion. Starting from the middle of the 20th-century, a series of restoration works were undertaken on site due to the continuous water infiltration coming from the stone roof and due to the structural instability of the dome and the substructure. The stone roofing of the church was fully replaced between 1957 and 1970 with the same materials and construction technique. The roofing replacement seems to have continued every few years till 2005–2006, as the problems of water infiltration were still present [20]. Regarding the wall paintings, just brief notes can be found in the literature, which report restoration works [19]. The first note, dated to the 1930s, reports that the paintings were partially cleaned and "washed" by a Russian team of restorers. Later, in the 1950s, the paintings were cleaned and "safely fixed/stabilised" by the Georgian painter-restorers S. Abramishvili and K. Bakuradze [19]. Widespread delamination was found throughout the church and is documented in the condition assessment performed in 2012 by Prof. Francesca Piqué and Stefano Volta and in 2015 by Gvantsa Potskhishvili as part of her MA research work.

2 Definition of the Problem and of the Grouting Performance Criteria

The wall paintings condition assessment performed in the church showed the presence of large lacunas, of recent losses and of severe problems of lack of adhesion (i.e. delamination). In most of the cases, the delamination occurs between the stone and the plaster(s), i.e. the primary and the secondary supports. The most severely delaminated areas, being unstable, are located in the upper parts of the apses, on the horizontal and semi horizontal surfaces of the vaults and in the areas where there are joints between different plaster types (on horizontal, semi horizontal and vertical surfaces) (Fig. 2). When not clearly visible, the delamination was located in the stratigraphy by tapping, and its severity assessed according to the vibration and the sound produced.

A thorough graphic and photographic documentation was produced to illustrate the extent and gravity of this precarious situation and the need for a stabilisation intervention. One of the possible interventions to be designed is that of grouting.

Fig. 2. Delamination of the gypsum-based plaster; west apse, Ateni Sioni church, Georgia. © Francesca Piqué, 2012.

General performance criteria (such as physico-chemical stability, no introduction of soluble salts, etc.) are common to any conservation intervention [21]. Criteria which are specific to the type of intervention (e.g. grouting, consolidation, cleaning, etc.) are important for the design of the conservation treatment [21]. Working properties for grouting decorated surfaces should be sufficiently fluid to allow injection; in addition, minimal shrinkage upon hardening, mechanical strength similar or lower than that of the original materials, porosity and water vapour permeability similar to those of the original materials are necessary to ensure stability of the results over time [1]. Keeping in mind the criteria for any grouting intervention, *site-specific* performance criteria must then be defined case-by-case [21], and were indeed defined for the specific case at Ateni. The desired properties for the grouting intervention at Ateni were outlined, including working properties, i.e. short-term performance (during the intervention when the grout is in the fluid state), and performance characteristics, i.e. long-term performance (after the intervention and over time, when the grout is set and hardened) [21]. The site-specific working properties and performance characteristics of the grouts designed in this research were defined according to the composition of the original materials (gypsum plaster and stone support) and the specific condition on site, i.e. specifics of the delamination. Delamination between gypsum-based plaster occurred on horizontal, semi-horizontal or vertical surfaces. The void between the delaminated plaster and support in the vault could reach up to 5 cm. The severity of the delamination varied between unstable and dangerously unstable, with risk of loss of original material. Regarding injectability, since in the Ateni Sioni church access points need to be created in order to inject the grout, two size catheters to attach to the syringe for injection (one relatively narrow and the other larger in diameter, see Table 1) were tested, in order to design a versatile material, allowing flexibility on site and adapt the intervention to the different situations eventually found. The additional desired site-specific working properties and performance characteristics for the grout to be used in the Ateni Sioni church, including minimal water content because of the sensitivity to water of the original gypsum-based plaster, are described in Table 1.

3 Materials and Methods: Site-Specific Grout Formulation

As stated, the most common type of plaster present in the Ateni Sioni church is a gypsum-based plaster. Although this type of plaster has been known as a support for wall paintings around the world since ancient times, the chemical and physical characteristics of gypsum and in particular its solubility in water [22] and the rapid hardening time are well-known potential limitations in the use of this material for building and/or decorative purposes [23, 24]. In the case of injection grouting, these issues can raise complications in terms of choice of materials for grout formulations and/or treatment methodology. If gypsum is chosen as the binder of the injection grout, its fast setting may be a hindrance for the implementation of the intervention, since the grout needs to remain fluid during the intervention and cannot harden too fast. This may be one of the reasons why other binders such as lime are generally used for the stabilisation of such plasters. In addition to this, generally it is common practice to use water as the suspension medium for injection grouts, which is problematic due to the water-sensitive original materials present [4], such as gypsum-based plaster. Gypsum is slightly water-soluble, even if its solubility is much lower than other soluble salts commonly found in building materials [25]. Therefore, since original materials at the Ateni Sioni church are water-sensitive, it was decided it was important to reduce the water content of the grout, partially substituting it with an alternative liquid, i.e. ethanol [4]. On the other hand, being gypsum the binder of the injection grout, water is necessary in the hydration reaction occurring during setting [25]. Stoichiometrically the minimum amount of water is indicated by the setting chemical reaction:

$$CaSO_4 \cdot 0.5\ H_2O + 1.5\ H_2O \rightarrow CaSO_4 \cdot 2\ H_2O.$$

The research focuses on developing a gypsum-based grout with a reduced amount of water, both to: i) reduce setting time of the grout with the use of gypsum (compared to a

Table 1. Site-specific working properties (WP) and Performance characteristics (PC) of the grout to design for the Ateni SIoni Church.

Working Properties (WP)	Performance Characteristics (PC)
Good injectability: ability of the grout to easily pass through two catheters attached to the syringe (3.3 mm diameter and 4.7 mm diameter respectively)	Sufficient adhesion to both interfaces, primary stone support and gypsum plaster
Proper flow: ability of the grout to flow for a suitable distance in order to reach the edge of the delaminated area, vertically (on brick tiles and in replicas with gypsum plaster) and horizontally (in replicas with gypsum plaster)	Minimal dry density, especially for horizontal surfaces
Minimal water content: one of the most important requirements because of the sensitivity of gypsum-based plasters to water	Minimal hygral and thermal expansion and contraction
Good initial tackiness: to both interfaces, primary stone support and plaster, especially important for the delamination present on the horizontal surfaces	
Minimal wet density, especially for horizontal surfaces	
Reasonable hardening time: the grout should not start hardening during the intervention; however, relatively rapid hardening time is important to limit the period during which the wet grout is adding weight to the thin plaster layer, especially on horizontal surfaces and a support is necessary	

lime-based one) and use a material akin to the original ones, and ii) reduce solubility of the original material with the use of an alternative liquid beside water. The formulation of the grout proceeded in two stages: the first one focused on assessing the feasibility of the use of gypsum as the binder (*Stage 1*, Sect. 3.1 below), and the second one focused on refining the formulation (*Stage 2*, Sect. 3.2 below).

In the methodology for the formulation of a grout, the mix design and the testing of paramount properties must proceed in parallel to refine the formulation [5]. *Stage 1* aimed, through preliminary testing, to assess the feasibility of using grouts prepared with gypsum (on its own and in combination with lime), while lime was used as a control (see Sect. 3.1). Such preliminary testing involved the assessment of injectability, flow, shrinkage and adhesion, as these were judged as the fundamental working properties required, and helpful in guiding the grout design. Based on these results, the most promising three formulations were identified and then further refined during *Stage 2*, which led to the two final formulations (see Sect. 3.2). Such mixtures were further tested –beside injectability, flow and shrinkage– for expansion and bleeding, wet and dry density, hardening time, cohesion and adhesion into replicas. This multi-step formulation methodology is synthesized in Table 2; details of the different stages will be given in Sect. 3.1 and 3.2 below. Description of testing procedures and results will be fully reported in Sect. 4.

Table 2. Stages in the grout formulation.

Stage of formulation	Aim	Testing carried out	Mixture obtained
Stage 1	Feasibility of using gypsum or gypsum-lime as binder	**Preliminary testing of:** Injectability Flow Shrinkage Adhesion	Mix 1 (gypsum) Mix 2 (gypsum-lime) Mix 3 (lime)
Stage 2	Refining of the formulation to fulfil the site-specific performance criteria, particularly adhesion	**Full testing of:** Injectability Flow Expansion and bleeding Wet and dry density Shrinkage Hardening time Cohesion and adhesion (replicas)	Grout 1 (gypsum) Grout 2 (gypsum-lime)

3.1 *Stage 1* of Formulation and Testing

At the first stage of the grout mixture design, a wide range of materials (binder, fillers and suspension media) were considered according to their composition, properties and physico-chemical compatibility with original materials.

The following variables were considered in *Stage 1* of design and testing. Mixes which are the result of this first stage are then shown in Table 3.

- **Binder:** Gypsum was considered as the binder of the grout (Mix 1 in Table 3), due to the gypsum-based composition of the original plaster. It is important to underline that pure calcium sulfate hemihydrate ($CaSO_4 \cdot 0.5\ H_2O$) without the presence of any additive (from LAGES, Lavorazione Gessi Speciali SPA, Italy) was used for this research; this is crucial, since commercial reactive calcium sulfate hemihydrate (gypsum hemihydrate) typically contains retarders and/or other additives to modify its properties and improve workability. Since the feasibility of the use of solely gypsum as the binder was still to verify and the authors wanted to assess the difference between gypsum and lime, two further binders were considered as well, i.e. 50% lime - 50% gypsum (Mix 2 in Table 3) and just lime (control, Mix 3 in Table 3). The choice of the binder for Mix 2 (and later Grout 2, see Sect. 3.2) was based on the well-known historical use of these two binders mixed in different proportions to obtain intermediate properties, e.g. addition of gypsum to speed up the setting time of lime-based mixtures (ex. in stucco technology) [26]. In addition to this, the first very thin (1 mm) plaster layer applied on the stone at the Ateni Sioni church is lime-based, so both lime and gypsum are present in the stratigraphy. The lime used in both cases was slaked lime putty (Grassello di Calce Candor, 48-months-aged, Italy) (ca. 50% $Ca(OH)_2$ and 50% H_2O), drained of the excess water, i.e. just the paste was used.

- **Fillers:** Several fillers were selected for the preliminary testing, depending on their chemical composition (reactive with lime vs. inert), particle size and morphology. Aggregates reactive with lime (while no chemical reaction with gypsum occurs) included pumice (from CTS, Italy) in different granulometry (<90 µm, <140 µm, <240 µm, <280 µm), Pozzolana Romana Gialla Micronizzata (<63 µm) (from Opificio Bio Aedilitia, Italy), Scotchlite K1® (sodalime borosilicate glass, reactive with lime [4]; from 3M, <120 µm according to the technical data sheet). Quartz sand (from Taiana, Switzerland, 100–250 µm), inert, was also considered. Scotchlite K1®, round, hollow and non-porous, helps to improve injectability in grouts [9, 5, 27] and aids in obtaining mixtures with low density; the other aggregates tend to have an angular shape, helping packing geometry and therefore cohesion. Pumice 0–140 µm (1 pt/V) coupled with Scotchlite K1® (3 pt/V) was chosen as the best fillers combination (see Table 3), giving good injectability and minimal/no shrinkage. The other types of fillers coupled with Scotchlite K1® were discarded for the following reasons:

  1. Mixes tended to segregate when Quartz sand and/or Pozzolana were added (with all three binders);
  2. Flow got dramatically worse when Pumice 0–90 µm was added (with all three binders): mixes tended to segregate and material to accumulate at the bottom of the drip during the flow test, giving an inhomogeneous drip;
  3. The amount of suspension medium required to obtain a fluid grout with proper injectability and flow was higher when Pumice 0–240 µm or Pumice 0–280 µm were added to the mixes compared to mixes with Pumice 0–140 µm, and in case of Pumice 0–280 µm material tended to accumulate at the bottom of the drip during the flow test, giving an inhomogeneous drip.

  Furthermore, it was assessed that the amount of Scotchlite K1® has a significant impact on the properties of the grout: 2 pt/V was the minimum amount necessary in order to have sufficient flow and medium/low shrinkage of the mix. The mix with 2 pt/V Scotchlite K1® was compared with mixes with 2.5 pt/V and 3 pt/V Scotchlite K1®; the most satisfactory results were obtained for the mixes with 3 pt/V Scotchlite K1®, in terms of fluidity and minimal/no shrinkage.

- **Suspension medium:** Solubility bench tests were carried out on a fragment of original plaster from the Ateni Sioni church: water vs. solutions with a different proportions of water: ethyl alcohol. These were added drop by drop on the original plaster, and its hardness was verified gently pressing it with a metal spatula and assessing the indents left on the surface. Bench tests confirmed that the original gypsum-based plaster is sensitive to water, even the first few drops. However, results also showed that the gypsum plaster is very stable and seems not to soften/visibly solubilize (from a macroscopic assessment) with a solution of 50% water - 50% ethyl alcohol. In this case, pressure with the spatula did not leave any indents on the surface. In order to further reduce the water amount, also solutions with water-alcohol 40%–60% and 30%–70% were tested in the grout design, thinking that, considering the potential (high) amount of grout injected on site, lowering further the amount of water would have been advantageous. However, no promising results were obtained compared to the solution water-alcohol 50%–50%. Mixes with water-alcohol 40%–60% and 30%–70%, in fact, tended to significantly shrink and even collapse in the moulds. Therefore, the 50%–50% solution was chosen as the grout suspension medium. It was calculated that the amount of water in the gypsum grout was amply sufficient for the hydration reaction to fully occur. On the other hand, for the lime-based grouts, previous research has proved that lime carbonation and hydraulic reactions (lime + pumice) can occur also with a reduced water content (here amply in the minimum amount) and in presence of ethanol [4].

After this first formulation round (*Stage 1*) where injectability, flow and shrinkage were assessed, three preliminary formulations (see Table 3) were chosen, in which the only variable is the binder; the rest of the components remain constant for all the formulations (in terms of typology, proportion and amount).

Table 3. Preliminary formulations of the grouts.

Mix Components	Mix 1 (Gypsum)			Mix 2 (Gypsum & Slaked lime)			Mix 3 (Slaked lime)		
	pt/V	Vol. (mL)	Weight (g)	pt/V	Vol. (mL)	Weight (g)	pt/V	Vol. (mL)	Weight (g)
Gypsum hemihydrate	1	100	86	0.5	50	44	-	-	-
Slaked lime	-	-	-	0.5	50	76	1	100	150
Pumice 0–140 μ	1	100	81	1	100	81	1	100	81
Scotchlite K1®	3	300	20	3	300	20	3	300	20
Water–Alcohol 50%–50%	2.2	220	195	2.2	220	195	2.2	220	195

Adhesion of such selected grouts was then assessed in the laboratory onto the two materials lacking adhesion in the Church of Ateni: gypsum plaster and tuff stone. To assess the adhesion to the plaster, a gypsum plaster prepared in the laboratory was applied on a brick tile and a boundary was placed around the tile itself; fluid grout was poured on the plaster and allowed to harden for 10 days. 50 mm diameter circular incisions were carved from the surface down to the brick. Metal plugs were glued on such incisions and pulled off with a dynamometer to measure the force needed for the failure to occur. To assess the adhesion of the grout to the tuff stone support, cylindrical plastic moulds 50 mm diameter × 20 mm height leant on the stone were filled with fluid grout (Fig. 3). After 10 days of hardening, metal plugs were glued to the surface of the grout cylinders and then pulled off with a dynamometer. It was found that the adhesion properties of the gypsum-based grouts (Mix 1 and 2) were not satisfactory, as in both cases the failure occurred between the grout and the stone support/gypsum plaster with a minimal force applied. Therefore the formulations needed to be refined (*Stage 2*, Sect. 3.2), since a good adhesion to both interfaces (plaster and stone), in fact, is one the most important requirement for a grout, and crucial in the case of the Ateni Sioni Church.

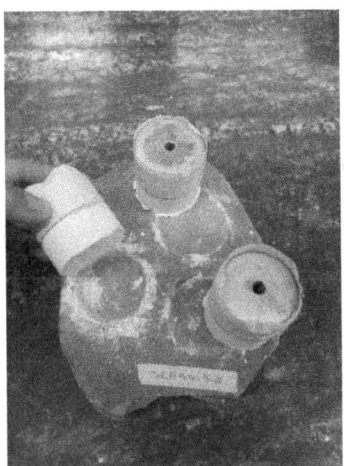

Fig. 3. Assessment of the adhesion of the grout to the tuff stone support. © Gvantsa Potskhishvili, 2015.

3.2 *Stage 2* of Formulation and Testing

Since in *Stage 1* of preliminary testing it was found that the adhesion of the gypsum grouts was not satisfactory, introduction of an additive was necessary to improve adhesion, particularly considering the delamination occurring on horizontal surfaces, where gravity has an important role and the grout needs to support the overlaying stratigraphy.

A literature review [3, 28–31] showed that two main categories of materials could be considered when looking at additives to improve adhesion for grouts: natural and/or synthetic. One of the most used synthetic material for this purpose is known to be an acrylic dispersion in water, PRIMAL® [30, 31], however the actual references detailing its use are limited. One of the earliest cited use of PRIMAL® AC-33 as an additive for grouts is for the *in situ* reattachment of earthen plasters in Peru, 1975–1977 [31] (p. 53). Nowadays, the most widely used acrylic dispersions for consolidation and adhesion interventions are PRIMAL® CM 330 (previously E330S) and PRIMAL® B60A ER. These products, according to the corresponding technical data sheets, have very similar properties, especially in terms of solid content (CM 330: 46.5–47.5%; B60A: 46–47%), pH (CM 330: 9.5–10.5; B60A: 9.4–9.9), appearance and minimal film formation temperature (CM 330: 10 °C; B60A: 9 °C). For the given research, it was decided to test PRIMAL® B60A, as its use seems to be better documented as an additive for grouts in previous studies [30, 31]. A minimum amount of acrylic was added in order to minimally alter the porosity and water transport properties of the hardened mixtures, yet obtaining adequate adhesion.

Table 4 shows the final formulations of the two grouts chosen for a possible implementation on site, subject to further testing (fully reported in Sect. 4). Mix 3 (just lime-based) was excluded from the further testing, since the objective of this research was to assess gypsum-based grouts to be potentially used on site and gypsum-based mixtures were preferred for a matter of chemical compatibility with the original plaster. Mix 3, in fact, is lime-based and it was originally included in the preliminary testing as a control rather than a grout developed for *in situ* use. The final Grout 1 and Grout 2 shown in Table 4 have different binder, i.e. gypsum in case of Grout 1 and gypsum-slaked lime in case of Grout 2. The amount of pumice is slightly higher in Grout 2 with gypsum-lime, because a higher amount of aggregate was necessary to obtain no shrinkage (Grout 1 with gypsum had already no shrinkage). On the other hand, the addition of Primal required a slight increase in solution in Grout 1 to obtain suitable injectability and flow: the reasons for this, the interaction acrylic-gypsum and its influence on the rheology of the mix should be further investigated and were beyond the scope of the present research. In Grout 2, on the other hand, slaked lime putty was used, which already contains 50% water and can aid in working properties [32, 33], including injectability and flow. The purpose was not to directly compare these two grouts to understand the influence of the single variables, since the change in more than one variable for practical purposes impedes such comparison. These modifications to the grouts were made to ensure their satisfactory working properties and performance characteristics, in order to design mixes to be used effectively on site.

Table 4. Final formulations of the grouts.

Mix	Grout 1 (Gypsum)			Grout 2 (Gypsum & Slaked lime)		
Components	pt/V	Vol. (mL)	Weight (g)	pt/V	Vol. (mL)	Weight (g)
Gypsum hemihydrate	1	100	86	0.5	50	44
Slaked lime	-	-	-	0.5	50	76
Pumice 0–140 μ	1	100	81	1.2	120	104
Scotchlite K1®	3	300	20	3	300	20
Primal B60A (47%)	0.02	2	2.05	0.02	2	2.05
Water-Alcohol 50%–50%	2.4	240	210	2.2	220	195

4   Testing Program and Results

As seen in Sect. 3, this research involved firstly testing of basic properties (such as injectability, flow and shrinkage) in parallel to the design of mixtures, to then move to a more thorough assessment of working properties (in the fluid state) and performance characteristics (in the hardened state) of the final grouts. This included injection into replicas simulating the horizontal delamination present in the ceilings and vaults of the church. The laboratory testing procedures adopted are described below and results are summarised in Table 5.

Injectability. Injectability of the grouts was assessed by evaluating the ability of a grout to pass through a syringe and/or through a syringe with a catheter attached to it, when approximately the same pressure is applied on the syphon by hand (adapted from [5]). A 60 mL syringe with two different catheters (diameters 3.3 mm and 4.7 mm) was used for the test. 30 mL grout was placed in the syringe and injected, firstly through the syringe and after through the syringe plus catheter. This procedure was performed at least three times for each syringe/catheter setting for each grout. According to how many mL of grout was passing through in a set amount of time (5 s), the grout was assessed as 'easy' (>20 mL), 'medium' (10–20 mL) or 'difficult' (<10 mL) to inject. Both grouts showed an 'easy' injectability through the tip of the syringe and the 4.7 mm catheter, and 'medium' injectability through the 3.3 mm catheter.

Flow. A brick tile (tuff would have replicated better site conditions, but it was not available) with vertical grooves was positioned vertically, and 10 mL grout was injected from the top to let it flow through the vertical channels present on the tile (adapted from [34] (p. 75)) (Fig. 4). Approximately the same pressure on the syphon was applied each time to have as similar testing conditions as possible. Distance and time of flow were recorded, and the homogeneity and body of the drip were observed and evaluated (Fig. 4). The test was repeated at least 10 times for each grout. The average flow distance (cm) and the average time taken for the grout to flow (s) were recorded, and the flow rate (cm/s) was calculated. Grout 2 (50% gypsum-50% lime) showed a longer flow distance and higher flow rate compared to Grout 1 (gypsum-based). Both produced a homogeneous drip with a good body.

Expansion and Bleeding. In this research, the standard test (ASTM C940–10) to measure expansion and bleeding was modified using a smaller amount of grout (30 mL instead of 800 mL). The test was carried out in plastic syringes without the syphon, and the tip of the syringe was blocked with a tape; the syringe was fixed in a vertical position so that it was stable. 30 mL of grout was placed in the syringe and the top of the syringe was covered with cling film to impede evaporation. The grout in the syringe was observed every 10 min for one hour, to assess if expansion and/or bleeding occurred. The two grouts showed neither expansion nor bleeding.

Wet and Dry Density. Density (wet and dry) of the grouts was calculated with the formula density = mass/volume (g/cm^3). Wet density is the density of the freshly prepared grout; dry density is the density of the hardened grout. To assess wet and dry density of the grouts, 30 mL of freshly prepared grout was placed in a Petri dish and weighed over time till the weight did not vary more than 0.01 g in three consecutive daily readings

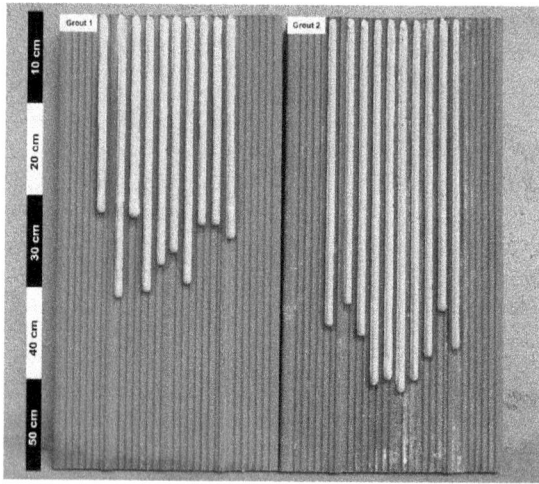

Fig. 4. Grout 1 and Grout 2 injected in the vertical channels incised in the brick tile. © Gvantsa Potskhishvili, 2016.

(adapted from [5]). Grout 1 and 2 showed similar wet and dry density, Grout 1 (gypsum-based) having a slightly lower wet density and Grout 2 (50% gypsum- 50% lime) having a slightly lower dry density.

Shrinkage. The shrinkage was qualitatively assessed with a porous support. 30 mL of freshly prepared grout was placed into a gypsum plaster cup (adapted from [34] (p. 83)), previously pre-wetted with 5 mL of water: alcohol 1:1 solution, and the system was covered with a gypsum plaster disc to reproduce a real case, in which the grout hardens in a pocket and it is not directly exposed to air [5]. After hardening, the surface was firstly observed to detect cracks and/or shrinkage close to the plaster cup walls, and the grout was then gently excavated with a spatula close to the walls, in order to observe the shrinkage in depth [5]. Both grouts showed no shrinkage, neither at the surface nor in depth.

Hardening Time. The hardening time of the grouts was measured by assessing it with the Vicat needle apparatus. The Vicat needle apparatus is typically used to measure the "setting time" of cement-based mortars (standard UNI-EN 196–3); just the bell-shaped needle (to assess the setting end) was here considered. Different definitions of hardening and setting are used in conservation and in materials technology [35] and in this research, the conservation definitions are considered. 30 mL of freshly prepared grout was placed in a gypsum plaster cup (previously pre-wetted with 5 mL of water: alcohol solution 1:1); measurements were taken approximately every 10 min, observing the depth of penetration of the Vicat bell-shaped needle and the mark left by it. Grouts were considered hardened when only the central point of the bell-shaped Vicat needle was marked on the surface of the specimen after the measurement. The time the grouts took to harden was recorded. Grout 1 (gypsum-based) hardened in 1 h, while Grout 2 (50% gypsum- 50% lime) hardened in 2 h and a half.

Cohesion and Adhesion. The specific system was designed to assess the cohesion and adhesion of grouts as well as their bulking property/filling capacity in replicas simulating a delamination between a tile (fired brick; tuff would have replicated better site conditions, but it was not available) and a gypsum-based plaster. For this purpose, replicas were prepared as follows: a layer of gypsum-based plaster was placed into a concave mould simulating the delamination profile (approximately 3 cm thick in the deepest point); holes were carved in the plaster as access points for the injection to be performed later. Once the plaster was dry, it was removed from the mould and attached to the brick tile using the same gypsum-based plaster, so that a pocket was left between the support and the concave layer of plaster. Injection of a freshly prepared grout into the replica was performed through the hole(s) with a syringe and a catheter attached to it (diameter of the catheter 4.7 mm).

The testing procedure included:

- Pre-wetting of the internal surface of the void with a solution of water and ethyl alcohol (1:1);
- Injection of the grout into replicas positioned vertically and horizontally respectively; replicas were positioned horizontally (support facing up and plaster facing down as in a ceiling) to imitate delamination of the plaster present on the semi-horizontal and horizontal surfaces in the church (Fig. 5);
- Monitoring of the intervention with an infrared thermography (IRT) imaging technique (AGEMA Thermovision 570 camera). Thermo Images were taken at the beginning, before the intervention, after the pre-wetting, immediately after the injection and 30, 60 and 120 min after the injection. The image taken after one hour from the injection showed that the surface was almost evenly homogenous in terms of temperature;
- Letting the grouts harden for two weeks in the laboratory with relative humidity ~70% and temperature ~23 °C;

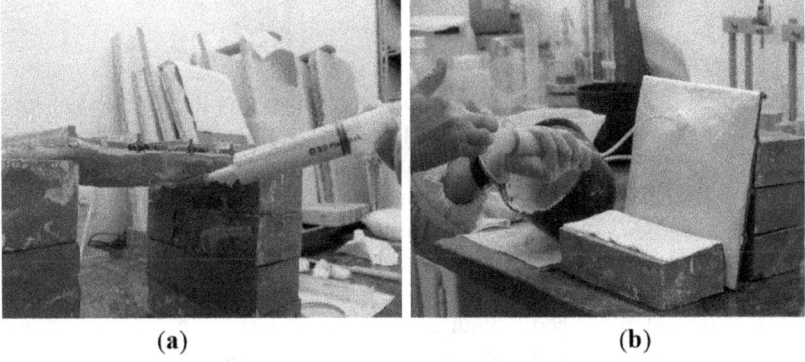

(a)   (b)

Fig. 5. Injecting the grout into the replica positioned horizontally (a) and vertically (b). © Gvantsa Potskhishvili, 2016.

- Cutting the replicas and observing their cross-sections to assess cohesion of the grout, adhesion and bulking properties.

Neither Grout 1 nor Grout 2 showed shrinkage: the adhesion to both the support and the plaster was very good, and no cracks were observed. The grouts resulted well cohesive and filled all the gap, showing excellent bulking properties (Fig. 6).

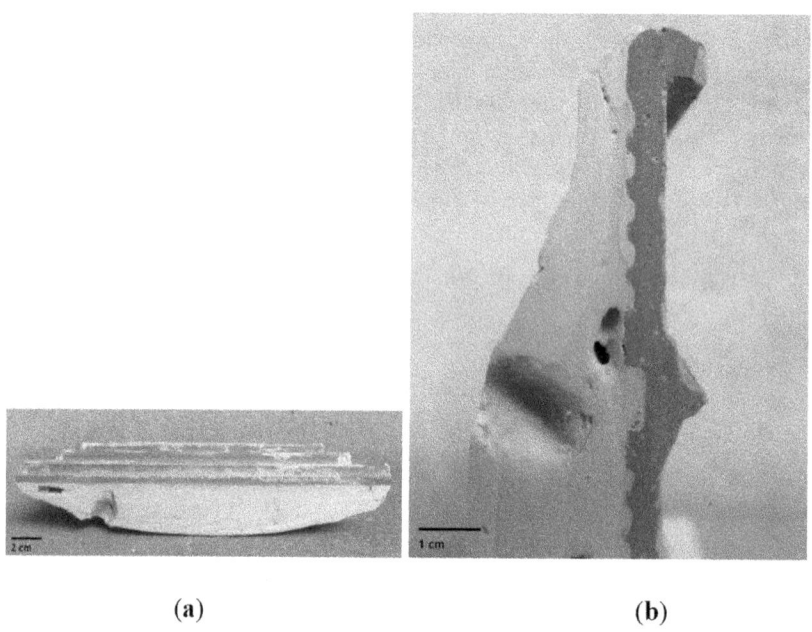

(a)   (b)

Fig. 6. (a) Cross-section of the replica; Grout 1 has been injected into the replica positioned horizontally. © Gvantsa Potskhishvili, 2016; (b) Cross-section of the replica; Grout 1 has been injected into the replica positioned vertically. © Gvantsa Potskhishvili, 2016.

5   Discussion and Conclusions

The objective of this research was to develop an injection grout(s) to address the severe delamination of the plaster afflicting the 11[th]century wall paintings in the Ateni Sioni church in Georgia. The focus was to produce a material compatible with the original gypsum- and lime-based plasters of the church and suitable for the type of deterioration encountered. Two grouts (one gypsum-based and one gypsum-lime-based) with a composition similar to that of the original materials (lime and gypsum-plasters and the tuff stone primary support) were developed following a thorough testing procedure and according to the set site-specific requirements. Bench tests revealed that the original gypsum-based plaster is susceptible to water and therefore grouts with a reduced water content were developed. To achieve this, water in the suspension medium of the grout was partially substituted by ethyl alcohol. Based on numerous bench tests it was concluded

Table 5. Summary of the working properties testing results.

Working properties										
Grout	Injectability			Flow			Density		Shrinkage	Hardening
	Syringe 60 mL	Syringe & catheter ø4.7 mm	Syringe & catheter ø3.3 mm	Average time (sec.)	Average distance (cm)	Flow rate(cm/s)	Wet (g/cm^3)	Dry (g/cm^3)	In replicas	(Hour)
Grout 1 (Gypsum hemihydrate)	Easy	Easy	Medium	12	26	2.1	0.86	0.44	No	~1 h
Grout 2 (Gypsum hemihydrate & Slaked lime)	Easy	Easy	Medium	9	35	3.8	0.89	0.42	No	~2 h 30 min

that a 50% water- 50% ethyl alcohol solution does not seem to damage the original plaster and gives the best results in terms of grout properties. At the same time, the amount of water present is sufficient for the chemical reactions to occur [4, 25] and produces a well-cohesive grout. Due to the unsatisfactory adhesion property of the initially formulated *Mix 1* and *Mix 2 (Stage 1)*, it was necessary to consider the addition of a synthetic adhesive *(Stage 2)*. Tests demonstrated that the addition of an acrylic water dispersion, Primal B60A, 0.4–0.5% in weight on the overall mixture, significantly improved the adhesion of both the formulations. Overall, *Grout 1*, gypsum-based, showed to have satisfactory injectability and flow. No expansion and no bleeding were detected. It showed similar wet and dry densities compared to the other grout, gypsum-lime-based, tested in this research. Such grout showed excellent ability to fill voids of different thicknesses without shrinking (results obtained through replica test) as well as good adhesion to both the interfaces in the replicas (gypsum and brick tile) and good internal cohesion. *Grout 1* also proved to have an appropriate hardening time (1 h) to be used for *in situ* implementation, long enough for the grout to be comfortably used, and quick enough to stabilise the delamination rapidly (not having the wet grout loading the stratigraphy of the ceiling for too long). *Grout 2*, gypsum-lime based, similarly to *Grout 1*, showed satisfactory injectability and flow properties. *Grout 2* has longer flow distance and higher flow rate compared to *Grout 1*, which could be advantageous in case of a ceiling/vault delamination where the grout needs to travel horizontally. No expansion and no bleeding were detected. In terms of density, *Grout 2* showed similar results to Grout 1. Also in this case, the grout had satisfactory results in filling the gaps of different thicknesses, as well as good adhesion to both the interfaces in the replicas, and good cohesion. The hardening time, as expected, was longer than the one of the gypsum-grout. The fact that hardening time is longer leads to the wet weight of the grout to be borne by the original materials in stratigraphy for longer, and this may be a problem for thick horizontal unstable delamination. The results obtained for Grout 1 (gypsum-based) proved that gypsum, in combination with fillers (Pumice and glass microspheres), an adhesive (Primal B60A) and a suspension medium (water: ethyl alcohol), has satisfactory working properties and performance characteristics for the *in situ* implementation of the grout. Furthermore, a second grout formulation (*Grout 2*) with 50% gypsum and 50%

slaked lime used as the binder (also in combination with fillers Pumice and glass microspheres, and small amounts of adhesive-Primal B60A- and a suspension medium-water: ethyl alcohol) showed equally satisfactory results, and can be also further implemented *in situ,* according to the need of the particular areas. For example, Grout 2 may be a better option for less critical areas which require, though, a better flow of the grout (ex. horizontal delamination not dangerously unstable where the grout needs to travel far), while, Grout 1 with a faster hardening time may be preferable for dangerously unstable areas requiring a faster stabilisation.

Acknowledgements. Undertaking this research definitely would not have been possible without the help of many professionals and colleagues. We would like to thank Professor Albert Jornet for his knowledge and experience shared during this research work. We would like to extend our thanks to Arch. Giacinta Jean, director of the BA and MA Conservation-Restoration program at the University of Applied Sciences and Arts of Southern Switzerland (SUPSI) as well as to Dr Christian Paglia, Director of the Institute of Materials and Constructions (IMC), SUPSI, for the access granted to the laboratories of the Institute. Many thanks to Samuel Antonietti and Cristina Mosca and to the technical staff, Guido Corredig, Massimo Mezzetti and Ezio Pesenti (IMC, SUPSI) for their invaluable help and assistance during the laboratory study and testing phases.

Enormous thanks to Giovanni Nicoli, Corinna Koch Dandolo and Andreas Küng for all the help and assistance in performing analyses and collection of the necessary materials. Special thanks to stone conservator Stefano Volta who has been extremely generous in sharing his knowledge and experience on site regarding injection grouting and consolidation treatments.

Many thanks to professor Nana Kuprashvili (Tbilisi State Academy of Arts), we are extremely grateful for all the help and support she has provided for this research. Thanks to all the colleagues in Georgia and especially to the ones working in the Ateni Sioni church, for their time and the enthusiasm during on site assessment.

References

1. Griffin, I.: Pozzolanas as additives for grouts: an investigation of their working properties and performance characteristics. Stud. Conserv. **49**(1), 23–34 (2004)
2. Rickerby, S., et al.: Development and testing of the grouting and soluble-salts reduction treatments of cave 85 wall paintings. In: Agnew, N. (ed.) Proceedings of the Second International Conference on the Conservation of Grotto Sites, Mogao Grottoes, Dunhuang, Conservation of Ancient Sites on the Silk Road, People's Republic of China, 28 June–3 July 2010, pp. 471–79. The Getty Conservation Institute, Los Angeles (2004)
3. Biçer, Ş.B., Griffin, I., Palazzo-Bertholon, B., Rainer, L.: Lime-based injection grouts for the conservation of architectural surfaces. Rev. Conserv. **10**, 3–17 (2009)
4. Pasian, C., Secco, M., Piqué, F., Artioli, G., Rickerby, S., Cather, S.: Lime-based injection grouts with reduced water content: an assessment of the effects of the water-reducing agents ovalbumin and ethanol on the mineralogical evolution and properties of grouts. J. Cult. Herit. **30**, 70–80 (2018)
5. Pasian, C., Martin de Fonjaudran, C., Rava, A.: Innovative water-reduced injection grouts for the stabilisation of wall paintings in the Hadi Rani Mahal, Nagaur, India: design, testing and implementation. Stud. Conserv. **65**(sup1), P244–P250 (2020)
6. Padovnik, A., Piqué, F., Jornet, A., Bokan-Bosiljkov, V.: Injection grouts for the re-attachment of architectural surfaces with historic value—measures to improve the properties of hydrated lime grouts in Slovenia. Int. J. Archit. Heritage **10**(8), 993–1007 (2016)

7. Papayianni, I., Pachta, V.: Experimental study on the performance of lime-based grouts used in consolidating historic masonries. Mater. Struct. **48**(7), 2111–2121 (2014). https://doi.org/10.1617/s11527-014-0296-5

8. Wong, L., Agnew, N. (eds.) The Conservation of Cave 85 at the Mogao Grottoes, Dunhuang: Development and Implementation of a Systematic Methodology to Conserve the Cave Wall Paintings and Sculpture. Getty Conservation Institute, Los Angeles (2011)

9. Simon, S., Geyer, D.: Comparative testing of earthen grouts for the conservation of historic earthen architectural surfaces. In: The 10th International Conference on the Study and Conservation of Earthen Architectural Heritage, Terra 2008. Getty Publications (2011)

10. Silva, R.A., Schueremans, L., Oliveira, D.V.: Grouting as a mean for repairing earth constructions. In: Terra em seminário 2010: 6º Seminario Arquitectura de terra em portugal: 9º Seminário Ibero-americano de arquitectura e construçao com Terra, pp. 82–86 (2010)

11. Freire, T., Silva, A.S., Rosário Veiga, M.D., Brito, J.D.: Characterisation of decorative Portuguese gypsum plasters from the nineteenth and twentieth centuries: the case of the Bolsa Palace in Oporto. In: Válek, J., Hughes, J.J., Groot, C.J.W.P. (eds.) Historic Mortars: Characterisation, Assessment and Repair, pp. 141–151. Springer, Dordrecht (2012). https://doi.org/10.1007/978-94-007-4635-0_11

12. Freire, M.T., Santos Silva, A., Veiga, M.D.R., Dias, C.B., Manhita, A.: Stucco marble in the Portuguese architecture: multi-analytical characterisation. Int. J. Archit. Heritage **14**(7), 977–993 (2020)

13. Freire, T., Veiga, M.R., Silva, A.S., Brito, J.D.: Improving the durability of Portuguese historical gypsum plasters using compatible restoration products. In: Proceedings of the 12th International Conference on Durability of Building Materials and Components XII DBMC, Porto, Portugalm, pp. 12–15 (2011)

14. Freire, T., Veiga, M.R., Silva, A.S., de Brito, J.: Restoration of decorative elements moulded on site: design and selection of gypsum-lime compatible products. In: Proceedings of the 3rd Historical Mortars Conference, Glasgow, Scotland, pp. 1–8 (2013)

15. Freire, M.T., Veiga, M.D.R., Santos Silva, A., de Brito, H.J.: Restoration of ancient gypsum-based plasters: design of compatible materials. In: Cement and Concrete Composites, vol. 120 (2021)

16. Moropoulou, A., Bakolas, A., Anagnostopoulou, S.: Composite materials in ancient structures. Cement Concr. Compos. **27**(2), 295–300 (2005)

17. Capitan-Vallvey, L.F., Manzano, E., Medina Florez, V.J.: A study of the materials in the mural paintings at the 'Corral del Carbon' in granada, spain. Stud. Conserv. **39**(2), 87–99 (1994). https://doi.org/10.1179/sic.1994.39.2.87

18. Kuprashvili, N., Kavsadze, M., Liluashvili, T.: Integrated study of the structural damage developed on the Western Façade of Ateni Sioni Church. J. ACADEMIA, Tbilisi State Academy of Art (2013)

19. Virsaladze, T.: From the History of Georgian Painting. G. Chubinashvili Centre, Tbilisi (2007)

20. Elizbarashvili, I., Suramelashvili, M., Chachkhunashvili, T., Tchurghulia, K.: Architectural restoration in Georgia; historiography, tradition and experience analysis. In: Tumanishvili, D. (ed.) Shota Rustaveli National Science Foundation; Society and Cultural Heritage Association, Tbilisi (2012)

21. Cather, S.: Trans-technological methodology: setting performance criteria for conserving wall paintings. In: Dipinti murali dell'estremo Oriente: diagnosi, conservazione e restauro: quando oriente e occidente s' incontrano e si confrontano, pp. 89–95. Longo editore (2006)

22. Lewry, A.J., Williamson, J.: The setting of gypsum plaster. J. Mater. Sci. **29**(21), 5524–5528 (1994). https://doi.org/10.1007/BF00349943

23. Rodríguez-Navarro, C.: Binders in historical buildings: traditional lime in conservation. In: International Seminar on Archaeometry and Cultural Heritage: the Contribution of Mineralogy. Sociedad Española de Mineralogía, Bilbao (2012)

24. Kühlenthal, M., Fisher, H. (eds.) Petra. Die Restaurierung der Grabfassaden [The Restoration of the Rockcut Tomb Façades]. Arbeitshefte des Bayerischen Landesamtes für Denkmalpflege, Munich (2000)

25. Charola, A.E., Pühringer, J., Steiger, M.: Gypsum: a review of its role in the deterioration of building materials. Environ. Geol. **52**(2), 339–352 (2007)

26. Caroselli, M., et al.: Study of materials and technique of late baroque stucco decorations: Baldassarre Fontana from Ticino to Czechia. Heritage **4**, 1737–1753 (2021). https://doi.org/10.3390/heritage4030097

27. Pachta, V.: The role of glass additives in the properties of lime-based grouts. Heritage **4**(2), 906–916 (2021)

28. Phillips, M.W.: Adhesives for the reattachment of loose plaster. Bull. Assoc. Preserv. Technol. **12**(2), 37–63 (1980)

29. Sickels, L.B.: Organic additives in mortars. Edinb. Archit. Res. **8**, 7–20 (1981)

30. Ferragni, D., Forti, M., Malliet, J., Mora, P., Teutonico, J.M., Torraca, G.: Injection grouting of mural paintings and mosaics. Stud. Conserv. **29**(sup1), 110–116 (1984)

31. Fong, K.L.: Design and Evaluation of Acrylic-Based Grouts for Earthen Plasters. Graduate Thesis in Historic Preservation. University Pennsylvania, Philadelphia (1999)

32. Hansen, E.F., Rodríguez-Navarro, C., Balen, K.: Lime putties and mortars. Stud. Conserv. **53**(1), 9–23 (2008)

33. Rodriguez-Navarro, C., Ruiz-Agudo, E., Ortega-Huertas, M., Hansen, E.: Nanostructure and irreversible colloidal behavior of Ca (OH) 2: implications in cultural heritage conservation. Langmuir **21**(24), 10948–10957 (2005)

34. Biçer Şimşir, B., Rainer, L.: Evaluation of lime-based hydraulic injection grouts for the conservation of architectural surfaces. A manual of laboratory and field test methods. The Getty Conservation Institute, Los Angeles (2013)

35. Pasian, C., Piqué, F., Riminesi, C., Jornet, A.: How not to bother salts while grouting. In: Laue, S. (ed.) Proceedings of the 4th International Conference on Salt Weathering of Buildings and Stone Sculptures, pp. 158–167. Fachhochschule Potsdam, Potsdam (2017)

Morphological Evolution of Calcium Carbonate Crystals in Dry Hydrated Lime Mortar

V. A. Anupama$^{(\boxtimes)}$ ⓘ and Manu Santhanam ⓘ

Indian Institute of Technology, Madras, India
vaanupama9@gmail.com

Abstract. Heritage air lime mortars are highly porous, flexible and allow the egress of moisture from the structure by the property of breathability. Commercial dry hydrated lime mortar is widely used to repair heritage structures. The binder is manufactured by adding water to the crushed quicklime; hydrated lime gets converted to calcium carbonate ($CaCO_3$) by absorbing atmospheric carbon dioxide. $CaCO_3$ exhibits different morphology such as calcite (most stable form), vaterite (least stable form) and aragonite, depending on the local conditions of temperature, pH, carbon dioxide concentration etc. Acicular aragonite can be present during the initial stages of carbonation. Amorphous Calcium Carbonate (ACC) could also exist initially as small spheres.

The present study focuses on carbonating dry hydrated lime mortar in accelerated carbonation condition ($3\% CO_2$) to investigate the effect of CO_2 concentration in the morphology of calcium carbonate formed. The study examines the extent of carbonation at various ages of the mortar specimens using the Phenolphthalein indicator test and X-Ray Diffraction (XRD). Morphology of the calcium carbonate crystals is examined using Scanning Electron Microscopy (SEM). The study is expected to provide insights into the relationship between $CaCO_3$ polymorphism and carbonation condition in dry hydrated lime mortar mixes.

Keywords: Morphology · Accelerated Carbonation · Calcium Carbonate · Lime Mortar · Scanning Electron Microscopy · Polymorph

1   Introduction

Lime-based mortars are extensively used in the construction of heritage masonry structures in India. However, the repair and restoration of these structures using lime mortar poses certain challenges, including increased time required for hardening by carbonation of the mortar. Complete carbonation of the mortar in a structure could even take centuries. Temperature, relative humidity, porosity and water to binder ratio are some of the factors that influence this process [1]. The carbonation reaction in a lime mortar system can be chemically represented as follows:

$$Ca(OH)_2 + H_2CO_3 \rightarrow CaCO_3 + 2H_2O \tag{1}$$

Initially, atmospheric carbon dioxide combines with moisture during the carbonation reaction to form carbonic acid. Subsequently, calcium hydroxide reacts with carbonic

V. Bokan Bosiljkov et al. (Eds.): HMC 2022, RILEM Bookseries 42, pp. 469–475, 2023.
https://doi.org/10.1007/978-3-031-31472-8_37

acid to form calcium carbonate. This exothermic reaction liberates about 74 kJ/mol of heat. As progressive carbonation modifies the mortar properties, laboratory experiments that investigate them require accelerated carbonation that could stabilize the properties at a faster rate [2]. Portlandite, present as hexagonal plate-like structures, carbonates to calcium carbonate, crystallizing as calcite, vaterite, or aragonite. At low relative humidity, Amorphous Calcium Carbonate (ACC) forms initially, which gets converted to metastable and unstable phases [3]. ACC dissolves and reprecipitates as polar $\{213^-4\}$ scalenohedral calcite, under excess $[Ca^{2+}]$ and high pH. These further interact with Ca^{2+} and get stabilized. At ambient temperature and pressure, calcite displays a rhombohedral habit, $\{101^-4\}$. The transition of scalenohedral to rhombohedral calcite is by dissolution and precipitation processes favoured by the drop in pH due to portlandite consumption and CO_2 dissolution in pore solution. Other crystal habits observed are $\{213^-4\}$ or $\{213^-1\}$ scalenohedra, $\{0112^-\}$ acute rhombohedron, $\{101^-0\}$ prism and $\{0001\}$ tabular. When the pH is above 10.5, calcite crystals are directly formed without forming vaterite [4]. Vaterites are metastable phases that exist as spherulites, whereas aragonites appear to be prismatic or needle-like crystals [2]. The higher CO_2 concentration and lower RH resulted in better-defined crystal morphology and a more integrated matric-aggregate appearance [3].

Carbon dioxide concentration is critical to the progress of carbonation [1]. The consequences of accelerating the carbonation process in the strength and durability of the lime mortar need to be investigated. Hence, the current paper investigates the evolution of calcium carbonate morphology when lime mortar is subjected to accelerated carbonation at 3% CO_2 concentration.

2   Materials and Methods

Dry hydrated lime powder of 90% purity was used to study the morphological evolution of calcium carbonate crystals in lime mortar. Binder to aggregate ratio of 1:3 by mass was followed, and standard sand of grades I, II, and III was mixed at equal proportions to cast cubic specimens of 50 mm size. The water to binder ratio of 0.8 was chosen such that a flow of 165 ± 5 mm is achieved in the flow table test conducted as per IS5512. Such a flow provides adequate workability on-site for mortar applications. The cubic specimens were cast and kept at room temperature and relative humidity for seven days after de-moulding. The specimens were stored in an accelerated carbonation chamber with 3% CO_2 concentration, 27 °C and 60–80% relative humidity till the age of testing.

The extent of carbonation at different ages was tested using the phenolphthalein indicator. The specimens were epoxy-coated at all sides except one to facilitate one-dimensional penetration of atmospheric CO_2 through the uncoated surface. They were broken at various ages along a plane parallel to the CO_2 penetration, ensuring minimal smudging of the sample powder, and the indicator was sprayed on the cut surface. The uncarbonated parts of the specimens with pH > 9 turned pink on the application of the indicator. Subsequently, the carbonated regions of the specimens were collected for mineralogical and morphological analysis. X-Ray Diffraction (XRD) was carried out for mineralogical characterization of the specimens at different ages. MiniFlex Rigaku benchtop powder X-Ray Diffraction equipment with CuKα radiation (1.5405Å) generated at 45 kV and 15 mA was used for X-Ray Diffraction. Peaks between the 2-theta (θ) values of 3° and 90° were captured with a step size of 0.01° and at a scanning rate of 10° per minute. The diffractograms were qualitatively analyzed using the XPert High-Score Plus software. The carbonated sample chunks were sputter-coated with Au for adequate conductivity prior to the microstructural analysis. Scanning Electron Microscope with a Cu detector and Smart lab Studio-II Miniflex software was used for the analysis. Secondary electron images of the specimens were obtained at 50 kX and 100 kX magnification using an accelerating voltage of 5 kV. The micrographs were analyzed to identify the morphological evolution of calcium carbonate crystals in the samples. The mechanical strength of the mortar cubes was determined using Universal Testing Machine.

3 Laboratory Experiments

3.1 The Extent of Carbonation and Mechanical Strength

The phenolphthalein indicator test was conducted to determine the extent of carbonation of specimens at different ages. The penetration depth is observed to be more than half the dimension of one side of the cube of side 50 mm by the end of 14 days of accelerated carbonation. The photographs taken after phenolphthalein indicator test at 0, 2, 6, 8, 12 and 14 days of accelerated carbonation are shown in Fig. 1 and the results are presented in Table 1.

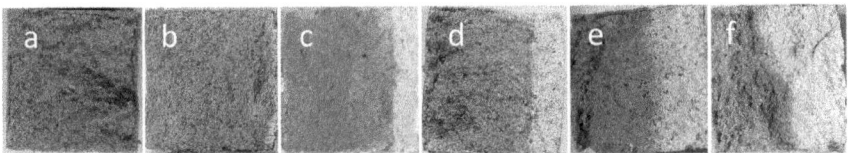

Fig. 1. Depth of carbonation at a) 0 days, b) 2 days, c) 6 days, d) 8 days, e) 12 days, and f) 14 days in the accelerated carbonation chamber

The results indicate that at 3% CO_2 concentration atmosphere, 50 mm cubic specimens get completely carbonated within two weeks.

The progress in carbonation is also reflected in the compressive strength results. Seven days of accelerated carbonation of 50 mm mortar cube in 3% CO_2 concentration atmosphere showed 2.15 MPa and 14 days of accelerated carbonation showed an increased compressive strength of 3.07 MPa. Increased carbonation and better crystal growth with increased duration of carbonation resulted in improved compressive strength.

Table 1. Depth of carbonation at different ages of specimens in accelerated carbonation chamber at 3% CO_2 concentration

Day	0	2	6	8	12	14
Depth of carbonation (mm)	0	0.2	10.4	11.2	24.1	25.5

3.2 Mineralogical Characterization

The carbonated regions of the specimens determined after the phenolphthalein indicator test were collected for mineralogical characterization. The results obtained from X-Ray diffraction of carbonated regions of samples at 7 days and 14 days in the accelerated carbonation atmosphere (AC-7 and AC-14, respectively) are shown in Fig. 2. From the qualitative analysis, it is observed that calcium carbonate is the major component in the sample. Traces of quartz observed in the diffractogram can be attributed to the sand in the mortar, which got powdered to particle size less than 75 μm during the sample preparation process.

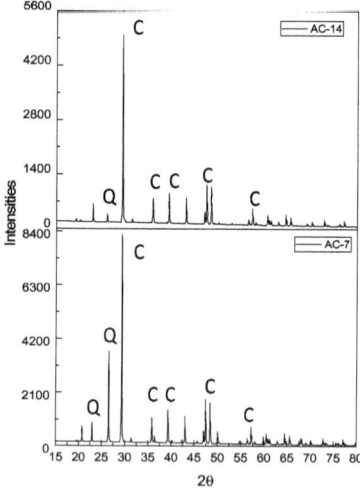

Fig. 2. Diffractograms of carbonated regions of samples in accelerated carbonation atmosphere for 7 days and 14 days in 3% CO_2 concentration

3.3 Morphological Characterization

Secondary electron images obtained from a scanning electron microscope were analyzed to understand the morphological changes in the calcium carbonate crystals. The carbonated parts at different ages of accelerated carbonation (4 days, 6 days, 8 days, 12 days, and 14 days) were used for the morphological analysis. The micrograph of the 4-day old sample in an accelerated carbonation chamber at 1 μm resolution indicates the presence of small spheres of size less than 1 μm, which could be amorphous calcium carbonate. Clusters of such spheres can be seen in Fig. 3. Four-day old samples also show carbonated regions with different morphology. Figure 3 also indicates the transformation of spherical amorphous calcium carbonate into scalenohedral calcite crystals. However, the micrograph shows poorly defined crystal structure regions, which could be inferred as a transition phase. As the carbonation duration increases, the crystal structure gets more defined, as shown in Fig. 4. After storage in the carbonation chamber for six days, scalenohedral calcite is abundant in the sample's carbonated region. As suggested in the literature investigating calcite morphology in an environment with high CO_2 concentration, rosette-shaped structures are observed in Fig. 4 [1]. After eight days of storage, the carbonated region shows a mixture of different morphologies of calcium carbonate. Scalenohedral calcite is prominent in a few regions, whereas needle-like and prismatic shapes are found in several other locations, which could be the aragonite polymorph of calcium carbonate. The micrographs of the 12-day old samples also show the presence of scalenohedral calcite crystals (Fig. 6). As the carbonation age increases, morphology changes from scalenohedra to plate-like. This transition can be observed in the micrographs of the carbonated regions of the 14-day old specimens (Fig. 7). The sample shows a variety of morphology for the calcite crystals. Scalenohedral crystals are present along with the plate-lite crystals. Small spheres of amorphous calcium carbonate are formed on the plate-like calcite crystals. These morphologies are different from the morphologies seen in naturally carbonated samples. The observations indicate that the carbonation atmosphere, mainly the concentration of carbon dioxide and duration of carbonation influences the morphology of calcium carbonate crystals in the lime mortar (Fig. 5).

Fig. 3. Micrograph of the carbonated region of the lime mortar specimen after four days of accelerated carbonation

Fig. 4. Micrograph of the carbonated region of the lime mortar specimen after six days of accelerated carbonation

Fig. 5. Micrograph of the carbonated region of the lime mortar specimen after eight days of accelerated carbonation

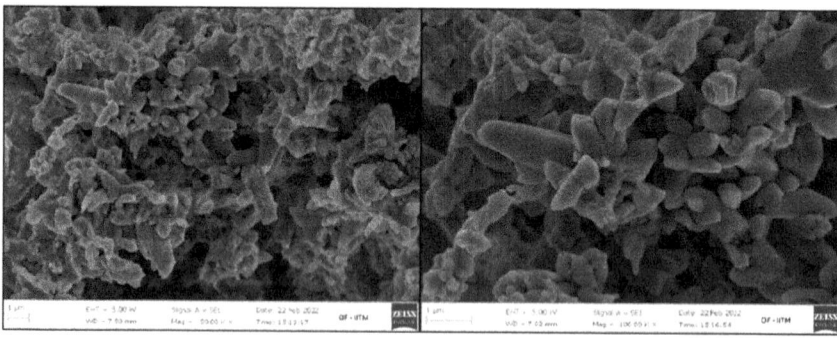

Fig. 6. Micrograph of the carbonated region of the lime mortar specimen after twelve days of accelerated carbonation

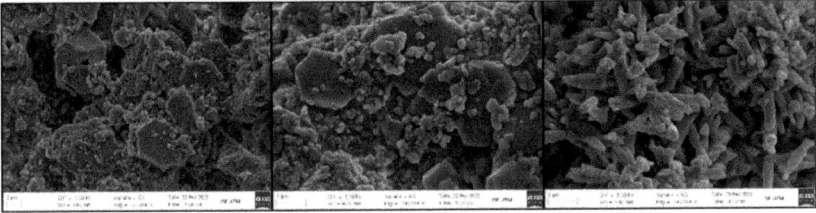

Fig. 7. Micrograph of the carbonated region of the lime mortar specimen after fourteen days of accelerated carbonation

4   Conclusions

The study focuses on observing the evolution of calcium carbonate morphology in an accelerated carbonation atmosphere of 3% CO_2. Phenolphthalein indicator test and X-Ray diffraction were used to determine the extent of carbonation of samples at different ages of accelerated carbonation. The 50 mm cube specimens were completely carbonated within 14 days of accelerated carbonation in a 3% CO_2 atmosphere. Improved mechanical strength of cubes at the age of 14 days compared to the strength at seven days indicates a better crystal development with increased duration of carbonation, which was confirmed through microscopic analysis.

Secondary electron images indicated that the morphology of calcium carbonate crystals changes with the duration of carbonation. The initial microstructure is prominently composed of spheres. From a poorly defined crystal structure, the shape of $CaCO_3$ crystals change to scalenohedral and then to plate-like as the carbonation duration increases. The microstructure is composed of calcium carbonate crystals of different morphologies even after complete carbonation. The presence of diverse morphology indicates that the evolution of morphology to the most stable form continues after the complete carbonation of the lime mortar. Analysis immediately after the complete carbonation of the specimens indicates that accelerated carbonation produces calcium carbonate crystals of morphology different than that developed through prolonged natural carbonation. The study revealed that the morphology of calcium carbonate crystals changes with the concentration of carbon dioxide and duration of carbonation.

References

1. Ergenç, D., Fort, R.: Accelerating carbonation in lime-based mortar in high CO_2 environments. Constr. Build. Mater. **188**, 314–325 (2018). https://doi.org/10.1016/j.conbuildmat.2018.08.125
2. Silva, B.A., Pinto, A.P.F., Gomes, A., Candeias, A.: Effects of natural and accelerated carbonation on the properties of lime-based materials. J. CO_2 Util. **49**(May), 101552 (2021). https://doi.org/10.1016/j.jcou.2021.101552
3. Ergenç, D., Gómez-Villalba, L.S., Fort, R.: Crystal development during carbonation of lime-based mortars in different environmental conditions. Mater. Charact. **142**(May), 276–288 (2018). https://doi.org/10.1016/j.matchar.2018.05.043
4. Cizer, Ö., Rodriguez-Navarro, C., Ruiz-Agudo, E., Elsen, J., Van Gemert, D., Van Balen, K.: Phase and morphology evolution of calcium carbonate precipitated by carbonation of hydrated lime. J. Mater. Sci. **47**(16), 6151–6165 (2012). https://doi.org/10.1007/s10853-012-6535-7

Design Rationale and Field Testing of a Gypsum-Based Grout for Wall Painting Stabilization in the Chapel of Niketas the Stylite, Cappadocia, Turkey

Jennifer Herrick Porter[1]([✉]) [ID], Yoko Taniguchi[2] [ID], and Hatice Temur Yıldız[3]

[1] Department of Conservation and Built Heritage, University of Malta, Msida 2080, Malta
jennifer.porter@um.edu.mt
[2] Graduate School of Humanities and Anthropology and Research Center for West Asian Civilization, Faculty of Humanities and Social Sciences, University of Tsukuba, 1-1-1 Tennodai, Tsukuba 305-8577, Ibaraki, Japan
[3] Nevşehir Directorate of Restoration and Conservation Regional Laboratory, Ministry of Culture and Tourism, 350 Evler Mah. Kültür Merkezi Binası No: 1, Nevşehir, Merkez, Turkey

Abstract. The rock-cut chapel of Niketas the Stylite (Cappadocia, Turkey) is decorated with 7–8th C CE painted gypsum plasters on a tuff support. A 2015 survey found that detached areas of plaster required stabilization through injection grouting. This paper presents the rationale behind the preliminary development of calcium sulfate hemihydrate (HH, $CaSO4\cdot0.5H20$)-based injection grouts for this purpose and their testing in the field. Research into gypsum-based conservation materials is limited compared to lime, and there are very few examples of the development of proprietary gypsum grouts for the stabilization of wall paintings on gypsum plaster. This study therefore reviews the existing literature, and explains the rationale behind the selection of materials for the conservation of the paintings, including a discussion of the potential risks and benefits associated with the use of lime and gypsum binders. Field testing focused on the development of HH-based grouts formulated with a range of aggregates and fluidizers. The grout mixes were tested against clearly-defined working properties and some of the performance characteristics for the intervention. Existing field-testing protocols were adapted as necessary to the working conditions, context and materials of the site and paintings.

Keywords: Gypsum grout · gypsum · plaster · wall painting conservation · field testing

1 Introduction

Between 2014–2016, the University of Tsukuba (Japan) and Turkish Ministry of Culture & Tourism carried out a collaborative project to conserve the rock-cut chapel of Saint Niketas (also known as Üzümlü church) in Kızıl Çukur, Ürgüp, Cappadocia, Turkey

V. Bokan Bosiljkov et al. (Eds.): HMC 2022, RILEM Bookseries 42, pp. 476–493, 2023.
https://doi.org/10.1007/978-3-031-31472-8_39

(Fig. 1). As part of this project, led by Prof. Yoko Taniguchi, a survey of the wall paintings within the Chapel highlighted the need to stabilize detached areas of painted plaster, mainly through non-structural injection grouting [1, 2].

The chapel's wall paintings are executed on a gypsum (calcium sulfate dihydrate, DH) plaster, and the selection of a compatible grout binder – primarily the choice between lime or gypsum – was the first major treatment design decision to be made. This issue has been confronted and widely researched in the context of lime-based plasters and mortars, and to a certain extent for earthen materials [3, 4], but is still comparatively under-researched for gypsum plasters [5, 6].

At the time of this study (2015), research into traditional gypsum wall painting and architectural plasters, and compatible gypsum-based repair materials, was limited [7–9]. Only three previous studies were found which mentioned the development of proprietary gypsum-based grouts for conservation purposes [10]. Since then, research into gypsum plasters has advanced, but mainly in the areas of repair plasters and the study of original technology; grout research still lags behind [5, 11].

This study therefore was an early attempt to address this issue, though the central aim was the development of a compatible grout for the wall paintings in the Chapel of Niketas. Despite the methodological limitations of field testing, this study highlights key questions surrounding the use of gypsum-based grouts for the conservation of gypsum-based original plasters.

2 The Site and Paintings

2.1 Site

The rock-cut Chapel of Niketas is carved into the base of a volcanic tuff cone formation in the Red Valley (Kızıl Çukur) near Ürgüp, Cappadocia, Turkey. The site takes its name from Niketas the Stylite, a Christian ascetic who lived at the site [12, 13]. The complex is one of only 2 surviving stylite sites in Cappadocia, and houses some of the earliest wall paintings in the region (late 7th–early 8th C CE) [14].

2.2 Painting Technique and Deterioration

The walls and ceilings of the chapel are decorated with painted scrolls and abstract foliate motifs, including a grapevine pattern (from which the site derives its informal name, Üzümlü, meaning 'grapes' in Turkish), saints and the Crucifixion, on a bright white plaster background [1]. The plaster is applied directly to the excavated walls of the tuff cone, composed primarily of feldspar with some quartz and small quantities of mica and clays. Fragments of pumice and other rock types are also bound within the tuff matrix [15]. The plaster layer is very thin (2–5 mm) and bright white, with no signs of aggregate. XRD analysis confirmed that plaster is composed of gypsum (calcium sulfate dihydrate, $CaSO_4 \cdot 2H_2O$) [1, 16].

Cracking of the plaster layer and detachment from the rock substrate was fairly extensive, and mainly associated with structural cracks within the rock fabric (Fig. 2), corresponding with geologic joint sets in the surrounding tuff formation [1, 15]. There were no signs of other sources of deterioration, such as water infiltration or rising damp, and no indications of salt activity [1].

Fig. 1. The chapel of Niketas (center) in the Red Valley, Cappadocia.

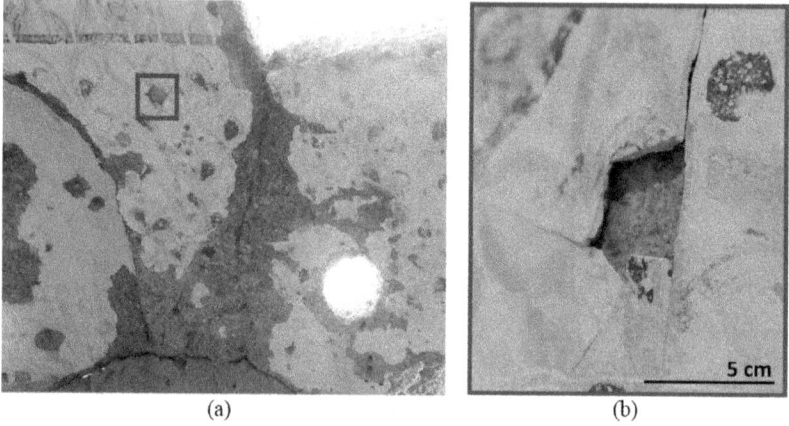

(a)   (b)

Fig. 2. (a) Structural cracks and associated cracking of painted plaster; (b) Detachment and lifting of plaster.

3 Conservation Needs

Stabilization of vulnerable areas of cracked and detached plaster was recommended. This required the use of an injection grout, a fluid mixture with bulking properties, which could flow within and fill the voids between plaster and substrate, thereby reintroducing contact between the surfaces [17]. A partial treatment was planned to focus only on detachments which appeared unstable and at risk of loss, rather than treating all detached or cracked areas. Similarly, the treatment did not aim to completely fill voids between plaster and substrate, but only to introduce sufficient grout material to stabilize selected areas. The grouting intervention was complemented by the application of edge repairs, to further stabilize and seal exposed plaster edges, using a gypsum-based plaster which was developed concurrently with the grout [1].

4   Choice of Grout Binder

Compatibility between original and intervention materials is a requisite for any conservation treatment, and it is generally acknowledged that using repair materials of similar chemical composition to the originals is one of the best ways of achieving this goal [17, 18]. Currently, research focusing on the development and testing of compatible gypsum-based materials for the repair of historic gypsum plasters is limited compared to lime and earthen materials, but growing rapidly [5, 8].

Inherent properties can make gypsum challenging or even risky for use as a conservation material, especially as a grout component (discussed in more detail in the next section). However, there are also potential risks associated with the use of lime for the conservation of gypsum plasters in systems where lime was not an original component:

- Lime and gypsum-based materials have significantly different set properties (e.g. mechanical strength, porosity, hygroscopicity, water transport properties) which may result in very different behavior over time [19–21];
- Introducing lime into a system containing gypsum and hydraulic materials (present in the tuff substrate) can lead to the formation of salts such as thaumasite/ettringite and deterioration over time [8, 19, 22, 23]; there are precedents for conservation treatment design with this in mind [8].

Based on these considerations, gypsum was chosen as the binder for the injection grout for the conservation of the wall paintings in the Chapel of Niketas.

4.1   Challenges Associated with the Use of Gypsum as a Conservation Material

Chemical and physical properties, the production process and even nomenclature can complicate the use of gypsum as a conservation material. Gypsum exists as a natural material and also as an artificial product, in a number of hydration states with widely varying properties and associated terminology [9, 24]. Calcium sulfate dihydrate (DH) ($CaSO_4 \cdot 2H_2O$; also referred to as gypsum and dihydrate) and calcium sulfate anhydrite (AH) ($CaSO_4$; insoluble or natural anhydrite) are both naturally-occurring, stable forms. DH can be heated to produce calcium sulfate hemihydrate (HH) ($CaSO_4 \cdot 0.5H_2O$; bassanite or Plaster of Paris), a metastable and reactive form [24]. When mixed with water, HH undergoes an exothermic hydration reaction to reform DH [25].

Different DH firing conditions produce HH variants (α-HH or β-HH) which are chemically identical but difficult to distinguish analytically [24, 26], and which have different setting times and set properties. Variations in traditional or industrial HH composition can also result from naturally-occurring impurities in raw DH materials (e.g. calcium carbonate, clays, AH) which affect working properties and long-term performance of gypsum-based repair materials [8, 20, 27, 28]. The type of HH is often not reported by suppliers; manufacturers of industrial gypsum will often combine α- and β-HH, and possibly other additives, to obtain specific properties [19]; impurities may go undetected or uncharacterized. Some of these issues may be counteracted by preparing one's own HH or sourcing directly from a manufacturer; analysis may also help characterize and screen HH products for impurities before use. However, these options may not be available to all conservators/in all working contexts.

Fast setting times/short working times may be one of the greatest challenges in employing HH as a conservation material. The HH hydration reaction proceeds in stages, with an induction period between initial water/HH contact and the point when the water/HH slurry first begins to thicken; this generally begins ~5–20 min after mixing with water [5, 6, 28, 29]. The mixture viscosity then increases to the point of solidification [25, 26, 28, 30]. Grout working time will be equivalent to the induction period, since the subsequent increase in viscosity will drastically alter grout properties. This may be particularly problematic if a large area of plaster or painting requires treatment and therefore a large batch of material needs to be prepared in order to work efficiently. Finally, gypsum expands slightly (0.15–0.3%) during exothermic hydration reaction to DH [31], a behavior which could help to ensure the conformance to surfaces and the bulking of voids, but could also be problematic in some circumstances.

4.2 Previous Studies

The aforementioned lack of relevant research is a further complication in the use of HH as a conservation material [5, 6], but especially for the design of grouting interventions.

At the time of this study (2015), only three previous studies were found which focused on the development of gypsum-based grouts for the conservation of historic gypsum plasters. Only one reported on the preparation of a proprietary HH-based injection grout for the conservation of a wall painting on gypsum plaster [10]. A 2014 conference paper briefly described the gypsum-grout stabilization of decorative architectural tiles, but the type of gypsum was not specified [32]. A third study described the injection of an estrich gypsum (a reactive form of $CaSO_4$ and CaO produced by high firing temperatures) and CaO slurry in methanol to stabilize detached areas of an estrich gypsum floor plaster [33], followed by injection of water to induce binder setting. While the creativity of the treatment design was noteworthy, the use of an estrich gypsum binder limited its relevance for the Niketas study.

Since 2015, only a few additional studies mention the preparation of proprietary HH-based injection grouts for the conservation of wall paintings on gypsum or architectural gypsum plasters [2, 34–37]. Similar to the previous examples, most do not provide details of the design, composition or testing of these grouts. An exception is the 2016 study by Potskhishvili [38] (undertaken shortly after the Niketas project but published for the first time in the current volume), who designed and tested HH-based injection grouts for the stabilization of wall paintings in Georgia.

Approaches and materials selection in the Niketas project therefore drew upon the findings from the single previous study which had developed a HH-based grout for wall painting treatment, and research into the development of CSH-based plasters for conservation and industrial applications, e.g. [7, 8, 21, 39, 40].

5 Treatment Design

5.1 Intervention Criteria

The required grout working properties and performance characteristics were clearly defined at the outset of treatment design, and guided the selection of materials and subsequent test methodologies (Table 1). Original materials, deterioration characteristics, the extent and aims of the intervention, and site working conditions were all considered when establishing intervention criteria.

Working properties were largely dictated by application requirements, such as the size and shape of voids to be filled (1–4 mm wide, extending 10–20 mm behind plaster surface) and access points (~2–4 mm at the widest) for delivering grout into these voids via syringe injection. Working time was also of critical concern, due to the rapid setting of HH, and many design choices aimed at modifying this property.

Mechanical strength criteria were also prioritized, since plaster detachment in the Chapel is associated with geologic activity, and therefore potentially ongoing movement. A grout with higher mechanical strength than the original plasters may induce further cracking and detachment of original materials under stress.

Table 1. Performance criteria established for the Niketas grouting intervention.

Working property	Description
Injectability	Must pass through an 18 G needle (inner diameter 0.838 mm)
Separation time	Slow enough to allow delivery and flow of homogenously distributed mixture
Working time	Sufficient to allow preparation and delivery of mixture before setting begins
Flow	Must be able to enter openings ~1–4 mm wide and flow 1–2 cm within voids
Performance characteristic	Description
Re-treatability	Partial filling of voids to allow future treatment
Bulking	Sufficient to fill spaces (1–4 mm wide, 10–20 mm deep) between plaster and stone
Volume stability	Minimal expansion; no slumping; no cracking
Separation	No fractionation or film formation
Adhesion	Should bond well with plaster and stone but fail under mechanical stress
Cohesion/mechanical strength	Mechanical strength lower than original plaster and stone
Hygro-thermal response	Similar to original materials

5.2 Component Selection

Injection grouts are composed of a binder, aggregates and suspension medium/a [17]. Additives, often organic, are sometimes used to further modify properties, though with potentially negative side effects [5, 28, 41]. Some of these ingredients can also act as HH set retardants which can help to achieve longer working times [5, 6, 8, 21, 28]. All components will of course affect grout working and set properties in various ways, and may play overlapping roles. The components for the Niketas test grouts were selected based on the pre-established intervention criteria.

A locally-available HH (purchased in the town of Nevşehir and therefore referred to as Nevşehir HH), was selected for economic and logistical reasons. FT-IR and XRD analysis confirmed its composition; preliminary tests indicated an initial setting time of 17 min.

Water and ethanol were selected as suspension media. Water is necessary for the HH hydration reaction to go forward, and an excess may also prolong working time [28, 29]. Ethanol was included for its potential to improve flow by reducing surface tension, and to minimize risk to water-sensitive original materials and potential soluble salt mobilization.

Aggregates were not a requisite grout ingredient since gypsum does not shrink during setting. Nonetheless, aggregates were included in the Niketas grout mixes to reduce weight, increase flow, and reduce density and therefore mechanical strength. Two silica-based aggregates, Poraver® expanded glass spheres [42] and Scotchlite™ K1 glass microspheres [43] were chosen for their morphology, particle size distribution and low density.

Additives were included to increase viscosity, water retention, and flow; and to promote the suspension of components in the fluid mix [7]. They may also increase working times [5, 6, 28]. Based on previous use of cellulose ethers as HH additives in industry and conservation[8, 39], two low-viscosity, well-characterized and commonly-used cellulose ethers were chosen: Tylose® MH 300 P2 [44–46] (in a 2% w/v solution in water) and Klucel™ G [47] (3% w/v in a 75% ethanol: 25% water solution). The concentrations were decided based on previous studies, and after initial trials indicated that these concentrations produced solutions with roughly similar viscosity, tack and dried adhesive strength.

The selected materials are summarized in Table 2 along with details of their composition, reasons for selection and predicted disadvantages.

Table 2. Materials selected for inclusion in grout formulations.

Material	Composition	Reasons for inclusion	Disadvantages

Binder: *Imparts cohesion between grout components and adhesion to substrate and detached plaster.*

Material	Composition	Reasons for inclusion	Disadvantages
Nevşehir gypsum	HH, $CaSO_4$ $0.5H_2O$	Locally-available, inexpensive	Potential batch variability

Suspension medium: *Provides flow, dispersion and suspension of grout components.*

Material	Composition	Reasons for inclusion	Disadvantages
Water	H_2O; local bottled drinking water	Water required for HH hydration reaction and as solvent for additives	Sensitivity of original materials, possible mobilization of salts
Ethanol	CH_3CH_2OH	Reduced risk of water-sensitive original material/salt mobilization. May improve flow by reducing surface tension.	Reduces water available for HH hydration reaction/may interfere with reaction[48]

Aggregates: *Influence flow; provide bulking; reduce weight; reduce density; reduce mechanical strength.*

Poraver® expanded glass	Porous soda-lime-silica glass spheres ; 40-250 µm[42]	Very fine particle size; spherical shape; lightweight; reduce density and mechanical strength set mixtures.	Alkaline (pH 8-12)[42, 43], aqueous slurry ; buoyancy can lead to separation.
Scotchlite™ K1 microballoons	Hollow soda-lime-borosilicate glass spheres; median diameter 65 µm[43]		

Additives: *Influence flow; improve suspension of mix components; increase water retention.*

Klucel™ G	Hydroxypropylcellulose; soluble in ethanol and water[47]	Weak adhesives; low viscosity[47]; commonly Used in conservation; remains soluble over time[44]	Film formation, reduced porosity, increase Hardness/adhesion[5, 28] Hygroscopic[44]
Tylose® MH 300 P	Methylhydroxyethylcellulose; soluble in water[46]		

6 Test Methodology

Because of the lack of previous studies into the development of proprietary HH-based injection grouts at the time of the Niketas project, there were no established HH-specific laboratory or field methods which could be drawn upon for testing and evaluation. Methods developed for the field testing of lime-and earth-based grouts were therefore adapted as necessary to the material and working context [18, 49]. The characterization of certain properties, such as expansion and hygro-thermal activity, could not be usefully assessed. The goal therefore was to develop a grout according to the properties which could be usefully evaluated in the field, while more rigorous laboratory testing would be carried out at a later stage, as needed.

Testing proceeded in three stages:

1. **Working properties:** Evaluated during mixing, after drawing into a syringe, and during injection through an 18 G needle.
2. **Set properties in cups:** Grout properties during and after setting were evaluated in isolation, by injecting into plastic cups and drying for 2 days before evaluation.
3. **Working and set properties *in situ*:** The final stages of testing were performed *in situ*. Grouts were applied to a tuff substrate in working and environmental conditions similar to those of the conservation intervention. Pre-cast gypsum lozenges (1:1 vol water:Nevşehir HH) were adhered to an external wall of the chapel by applying a gypsum plaster (also developed during the project) around lozenge edges (Fig. 4, Sect. 7). Voids between the preexisting irregularities of the tuff substrate and the flat lozenge surface were pre-wet with a 50% ethanol in water solution, followed by injection of the test grout. The grouts' set properties within the voids were assessed two days after application, by removing the lozenges from the wall by hand.

Table 3 summarizes the set properties which were assessed and the methods used. Two properties, adhesion and flow, could only be evaluated during the in-situ tests.

Most of the test methods were qualitative and somewhat subjective. Nonetheless, there are many precedents for the successful development of conservation grouts evaluated primary through field testing, using similar methodologies [4, 18, 49–51]. During the Niketas testing, procedures were established to minimize error (same operator and tools were used throughout each test) and the tests provided sufficient information about grout behavior to allow for the elimination of under-performing mixes.

Set grout mechanical strength and cohesion were evaluated manually, via pinch test or compression with the point of a metal spatula, and hardness was quantified using a Shore A Durometer. The durometer is a handheld device which measures material hardness through resistance to indentation by spring-loaded needle depression, based on a relative and dimensionless scale of 0 (extra soft) to 100 (extra hard) [52–54]. The Shore A device is normally used for measuring the hardness of a polymeric materials [53], but has also been used as a simple method for the field evaluation of construction materials [55]. Cup-cast samples were tested by placing the durometer perpendicular to the back of the demolded sample; the best-performing in situ grouts were measured directly on the wall. Three readings were taken and the results averaged. Hardness of the original plaster was not measured, due to risk of damage or destabilization, and tuff hardness was not measured either, so results could only be evaluated based on mix inter-comparison.

The testing program proceeded in an iterative manner: a first series of grouts was designed according to the expected performance of the selected components, and evaluated. Based on these results, the composition of the next set of test grouts was decided and tested. This process was repeated until a grout had was formulated which met all of the pre-established performance criteria. In all, 31 formulations were evaluated over 4 rounds of testing before an acceptable grout was produced. The process took 7 days to complete.

Table 3. Grout properties assessed and methods of evaluation

Working property	Test/assessment method		
Injectability	Assessed ability of grout to pass through syringe with 18 G needle		
Separation during use	Timed the separation of components in syringe		
Working time	Tactile assessment of time at which mixture viscosity first began to increase		
Flow	Assessed extent of travel and deposition behind gypsum lozenges		
Performance characteristic	**Test/assessment method**	**Cup test**	***In situ* test**
Bulking	Visual assessment of dried grouts	Y	Y
Volume stability	Visual assessment of dried grouts for slumping, cracking, deformation	Y	Y
Separation	Visual assessment of dried grouts for signs of component separation	Y	Y
Adhesion	Assessment of manual strength required to detach lozenge from substrate & examination after removal		Y
Cohesion/ mechanical strength	Shore A durometer readings; manual pinch test	Y	Y

7 Results

Most grout components were measured by volume, and ratios established as parts/volume. The volume of HH was kept constant throughout (10 mL) while non-binder quantities varied. All grout formulations contained gypsum, aggregates, and water. Initial mixes which contained only these components separated and released water too quickly, and therefore all subsequent mixes were formulated with a fluidizing additive (0.1–0.8% w/volume total mix). About half of the mixes contained ethanol, as medium and/or additive solvent (24–40% v/v); Poraver® and Scotchlite™ were used in combination or on their own. Liquid:solid ratios by volume ((water + ethanol):(gypsum + aggregate)) varied from 1:0.7–1:2. Examples of mix formulations and test results (including of the best-performing mix, #31) are provided Tables 4 and 5, Sects. 7.2 and 7.3.

7.1 Working Properties

The initial setting of the unmodified Nevşehir HH began at around 17 min after mixing with water. When mixed with other grout components, working times were generally extended to about 20–25 min, which was deemed acceptable given the limited extent of the planned treatment. Further increases in working times would have required significant increases in additive or the use of a chemical set retardant, which was undesirable.

Higher liquid:solid ratios (1:0.7–1:1.1 v/v) led to separation shortly after mixing, and the hollow, buoyant Scotchlite™ microspheres were particularly prone to separation. Most mixes showed at least some separation within 1–2 min of aspiration into the syringe, but this could usually be controlled by shaking the syringe to redistribute materials. Compaction in the syringe during injection was a problem with most mixes, possibly due to the porosity and restricted particle size distribution of the Poraver® microspheres.

Larger liquid proportions clearly increased fluidity and injectability; higher additive and ethanol concentrations were also linked to improved working properties but the correlation was not as clear.

7.2 Results of Cup Tests

Many of the set properties of the cup-cast mixes corresponded with the observed working properties (Table 4). Separation of grout components and significant volume instability were observed in set grouts with liquid:solid ratios between 1.07–1:1.1 range, and generally worse in mixes with higher quantities of ethanol. The separation of glass microballoons during setting/drying led to the formation of a white, friable surface skin, while higher additive concentrations (0.4–0.8% w/v total mix) resulted in dense, rubbery surface films. Ethanol and higher fluidizer concentrations were also often associated with volume instability during setting (slumping; cracking parallel and perpendicular to surface) (Fig. 3). However, this behavior may not accurately reflect how the grouts would perform in situ: mixtures cast in plastic do not rapidly desorb liquid as they would when applied to a porous tuff substrate, and therefore remain fluid longer, leaving more time for components to separate. Nonetheless, given the problems associated with ethanol content, and the lack of significant improvements in flow, ethanol and, by extension, the Klucel™ solution, were removed from subsequent testing formulations.

Table 4. Selection of working property and cup test results

Mix #	Formulation						Working properties		Set performance in cup
	HH (mL)	P (mL)	S (mL)	W (mL)	E (mL)	% A (w/v)	Inj	Sep	
2	10	10	5	9.5	12.5	0.4	++	+	2 mm surface skin; large horizontal crack; spongey
7	10	15		11.5	23	0.3	+++	++	Separation. Spongey surface skin; slumped 3 mm
8	10	15		12	10	0.5	−	++	Very thin white layer on surface but well-adhered
12	10	10		11.5	10.5	0.4	+	++	Fluffy, cracked surface; slumped 2 mm
13	10	10		9.5	6.5	0.5	−	+++	Good, but some fractionation at surface
16	10	10		13	0	0.3	+	+++	Very good; slight fractionation; no slumping

P: Poraver®; S: Scotchlite™; W: water (suspension medium + Tylose® solvent); E: ethanol (suspension medium + Klucel™ solvent); %A: % dry solid additive (Klucel™ or Tylose®) in total mix (g/mL). Inj: injection; Sep: separation

7.3 Results of In Situ Tests

About 1/3 of the total number of mixes was applied *in situ*. Ethanol and Klucel™ G were not included in these tests.

Most of the mixes had good bulking properties and volume stability. Adhesion and mechanical strength, which could only be evaluated at this point of the testing, revealed major differences between the mixes. In some cases, the adhesive bond between the tuff and/or gypsum lozenge was stronger than the cohesive/shear strength of the grout, resulting in removal of a layer of tuff while the grout remained adhered to the lozenge

Fig. 3. Separation, film formation, volume instability and cracking observed in grouts 2, 7, 12 (left) due to excess ethanol and/or additive, compared with 16 (right).

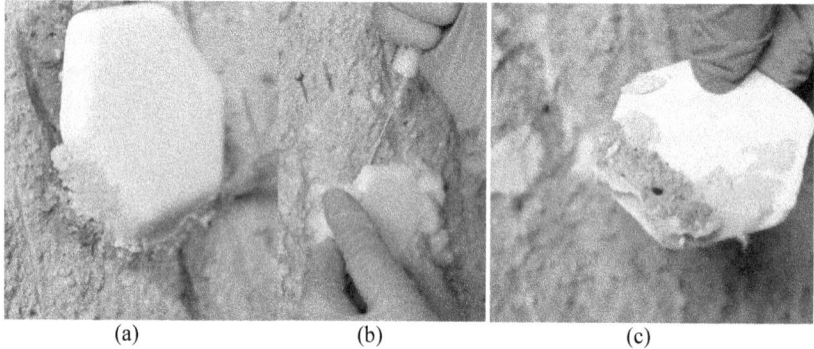

      (a)   (b)   (c)

Fig. 4. In situ tests. (a) Attachment of lozenge to tuff wall; (b) Grout injection; (c) Removal 2 days after injecting Mix #16, showing flow distribution; grout remained attached to the lozenge but removes a layer of tuff.

(Fig. 4). This may be due to factors such as greater surface roughness and greater porosity of the tuff substrate. However, exterior tuff surfaces are likely to be more deteriorated, and therefore less cohesive, than tuff surfaces within the Chapel.

There was a clear correlation between the quantity of Tylose® and the strength of adhesion to the substrate; mixtures with acceptable ranges of hardness and adhesion contained the lowest concentrations of fluidizer, between 0.1–0.2% Tylose® per total volume of mixture. The results of the hardness testing, when performed on the in situ mixes, roughly correlated with the adhesion tests and higher Tylose® concentrations. Table 5 provides some results of grout formulations tested in situ, including the best-performing mix, #31.

Table 5. Selection of in situ test results

Mix #	Formulation						Working properties			Performance characteristics	
	HH (mL)	P (mL)	S (mL)	W (mL)	T soln (mL)	%T (w/v)	Inj	Sep	Flow	Shore A Hardness (HA)	Adhesion/ cohesion
16	10	10		8	5	0.30	–	*NR*	*NR*	67.3	Very hard;
24	10	10		10	3	0.18	+	–	++	85.3	well-adhered, removed rock
23	10	10	5	10	3	0.16	+	++	++	58.2	Good: well adhered, but failed during removal, did not remove rock
26	10	10	2	11	2	0.11	–	*NR*	++	58.2	
27	10	15	3	14	3	0.13	–	*NR*	–	*NR*	
31	10	12	2	14	2	0.10	++	++	++	*NR*	

P: Poraver®; S: ScotchliteTM; W: water; T soln: 3% Tylose® in water. %T: %Tylose® dry solid in total mix (g/mL). Inj: injection; Sep: separation. NR: not recorded.

8 Discussion

Field-testing of a grout designed for the conservation of the Niketas chapel wall paintings demonstrates that it is possible to prepare gypsum-based injection grouts with good working times without the use of set retardant additives, and to test their performance in the field, following methods established for grouts based on other binders (e.g. lime and earth). The results of this field-testing campaign laid the groundwork for laboratory-based development of the Niketas grout, described in Taniguchi 2017 [2].

Grout performance was largely dependent on liquid:solid ratios (where liquid = suspension media + additive solvent volume; solid = HH + aggregate volume). It was found that grouts with L:S ranges:

- 1:1–1:2 were too thick (poor injectability, poor flow);
- 1:0.7–1:1.1 were too thin (separation during use; separation and volume instability during setting in cups);
- 1:0.7–1:1.5 performed best overall, particularly mixes in the 1:1.3–1:1.5 range.

Discrepancies in these patterns usually related to the type and ratio of aggregate used, but may also result from imprecisions in the measuring methodology (parts/volume rather than parts/weight). In some mixes, an increase in fluidity appeared partially related to ethanol content. Ethanol was excluded from testing due its connection with volume instability and separation in cup-cast samples, but this may also have been an effect of methodology. In fact, subsequent studies have reported the successful formulation of HH-based grouts containing ethanol [37, 38] and further research on the behavior of ethanol in HH mixes would be valuable, including potential implications of reducing the amount of water available for the HH hydration reaction [48].

Most examples of HH-based conservation grouts, including the Niketas test grouts, incorporate polymeric organic additives to modify rheology and working times [10, 34, 37, 38]. However, as has been shown in with lime-based grouts, organic additives may have negative impacts on set grout properties and long-term performance (chemical instability, film formation, reduction in porosity, and increases in mechanical strength) [28, 41, 44]. Future work might assess the possibility of controlling rheological and setting properties though the careful selection of aggregates and balancing with fluid proportions, to produce more compatible and stable HH-based conservation grouts [41].

Laboratory testing is required to fully assess the long-term performance characteristics of the proposed HH-based Niketas grout. To date, only Potskhishvili [38] has assessed set properties such as porosity and capillary absorption in a HH grout. Characteristic properties of gypsum systems, such as expansion and heat evolution upon setting, and instability in high/fluctuating RH environments [9, 56, 57], have yet to be addressed in any grout development studies.

9 Conclusions

The Chapel of Niketas project provides an example of relevant and representative *in situ* testing for the development of a gypsum-based injection grout for the stabilization of wall paintings on gypsum plaster. Despite a reliance on field testing, the study demonstrates the feasibility of preparing and working with gypsum-based injection grouts for the stabilization *in situ* wall paintings on gypsum plaster, and highlights some of the main challenges/open questions to address in future research into HH-based conservation grouts.

References

1. Taniguchi, Y.: Scientific Studies on Conservation for Üzümlü Church and its Wall Paintings in Cappadocia, Turkey, vol. 1: Report on the activities in 2014. Research Center for West Asian Civilization, University of Tsukuba, Tsukuba (2015)
2. Taniguchi, Y.: Scientific Studies on Conservation for Üzümlü Church and its Wall Paintings in Cappadocia, Turkey, vol. 2: Report on the activities in 2015–6. Research Center for West Asian Civilization, University of Tsukuba, Tsukuba (2017)
3. Rickerby, S., et al.: Development and testing of the grouting and soluble-salts reduction treatments of cave 85 wall paintings. In: Agnew, N. (ed.) Conservation of Ancient Sites on the Silk Road. Proceedings of the Second International Conference on the Conservation of Grotto Sites, pp. 471–479. Getty Conservation Institute, Los Angeles (2010)
4. Vernaza, C., Cancino, C., Rainer, L.: Pruebas e implementación. In: Informe sobre el análisis de condiciones, diagnóstico y pruebas de protección para las pinturas murales– Templo Santiago Apóstol de Kuñotambo, pp. 75–100. Getty Conservation Institute, Los Angeles (2018)
5. Freire, M.T., do Rosário Veiga, M., Silva, A.S., de Brito, J.: Restoration of ancient gypsum-based plasters: design of compatible materials. Cem. Concr. Compos. **120**, 104014 (2021). https://doi.org/10.1016/j.cemconcomp.2021.104014
6. Brunello, V., Bersani, D., Rampazzi, L., Sansonetti, A., Tedeschi, C.: Gypsum based mixes for conservation purposes: evaluation of microstructural and mechanical features. Mater. Constr. **70**, 207 (2020). https://doi.org/10.3989/mc.2020.05019

7. Freire, T., Veiga, M.R., Silva, A.S., de Brito, J.: Restoration of decorative elements moulded on site: design and selection of gypsum-lime compatible products. In: Hughes, J. (ed.) Proceedings of the 3rd Historical Mortars Conference, Glasgow, Scotland, pp. 1–8. RILEM Publications SARL, Paris (2013)

8. Igea Romera, J., Martínez-Ramírez, S., Lapuente, P., Blanco-Varela, M.T.: Assessment of the physico-mechanical behaviour of gypsum-lime repair mortars as a function of curing time. Environ. Earth Sci. **70**, 1605–1618 (2013). https://doi.org/10.1007/s12665-013-2245-y

9. Rodríguez-Navarro, C.: Binders in historical buildings: traditional lime in conservation. In: International Seminar on Archaeometry and Cultural Heritage: The Contribution of Mineralogy, pp. 91–112. Sociedad Española de Mineralogía, Bilbao (2012)

10. Sugihara, A., Fujisawa, A., Shimadzu, Y., Masuda, K., Yamauchi, K.: Conservation of the wall painting fragments excavated from the fortified city of Khulbuk in the collection of the National Museum of antiquities of Tajikistan. 保存科学 (Sci. Conserv.) **53**, 135–150 (2014)

11. Stoops, G., Tsatskin, A., Canti, M.G.: Gypsic mortars and plasters. In: Nicosia, C., Stoops, G. (eds.) Archaeological Soil and Sediment Micromorphology, pp. 201–204. Wiley, Chichester (2017)

12. Brennecke, H.C.: Stylite. In: Betz, H.D., Browning, D.S., Janowski, B., Jüngel, E. (eds.) Religion Past and Present Online. Brill, Leiden (2011). https://doi.org/10.1163/1877-5888_rpp_SIM_025228. Accessed 25 Mar 2022

13. Cooper, E., Decker, M.J.: Life and Society in Byzantine Cappadocia. Palgrave Macmillan, New York (2012)

14. Andaloro, M., Pogliani, P.: Materials and techniques of cappadocian wall paintings. In: Aoki, S., et al. (eds.) Conservation and Painting Techniques of Wall Paintings on the Ancient Silk Road. CHS, pp. 43–88. Springer, Singapore (2021). https://doi.org/10.1007/978-981-33-4161-6_4

15. Topal, T., Doyuran, V.: Analyses of deterioration of the Cappadocian tuff. Turkey. Environ. Geol. **34**, 5–20 (1998). https://doi.org/10.1007/s002540050252

16. Taniguchi, Y., Koizumi, K., Iba, C., Porter, J., Açıkgöz, F., Gülyaz, M.E.: Scientific research for conservation of the Rock Hewn Church of Üzümlü, Cappadocia. In: 37th International Symposium of Excavations, Surveys and Archaeometry, Erzurum, pp. 361–378 (2016)

17. Biçer-Şimşir, B., Griffin, I., Palazzo-Bertholon, B., Rainer, L.: Lime-based injection grouts for the conservation of architectural surfaces. Stud. Conserv. **54**, 3–17 (2009)

18. Agnew, N., Wong, L.: Preventive measures and treatment. In: The Conservation of Cave 85 at the Mogao Grottoes, Dunhuang: A Collaborative Project of the Getty Conservation Insti-tute and the Dunhuang Academy, pp. 259–295. Getty Publications, Los Angeles (2014)

19. Henry, A., Stewart, J.: English Heritage Practical Building Conservation Mortars Renders and Plasters. Ashgate Publishing Limited, Farnham (2011)

20. Middendorf, B.: Physico-mechanical and microstructural characteristics of historic and restoration mortars based on gypsum: current knowledge and perspective. Geol. Soc. London Spec. Publ. **205**, 165–176 (2002)

21. Salavessa, E., Jalali, S., Sousa, L.M., Fernandes, L., Duarte, A.M.: Historical plasterwork techniques inspire new formulations. Constr. Build. Mater. **48**, 858–867 (2013)

22. Collepardi, M.: Thaumasite formation and deterioration in historic buildings. Cem. Concr. Compos. **21**, 147–154 (1999)

23. Tesch, V., Middendorf, B.: Occurrence of thaumasite in gypsum lime mortars for restoration. Cem. Concr. Res. **36**, 1516–1522 (2006)

24. Prieto-Taboada, N., Gomez-Laserna, O., Martinez-Arkarazo, I., Olazabal, M.Á., Madariaga, J.M.: Raman spectra of the different phases in the $CaSO_4$–H_2O system. Anal. Chem. **86**, 10131–10137 (2014)

25. Singh, N.B., Middendorf, B.: Calcium sulphate hemihydrate hydration leading to gypsum crystallization. Prog. Cryst. Growth Charact. Mater. **53**, 57–77 (2007). https://doi.org/10.1016/j.pcrysgrow.2007.01.002

26. Lewry, A.J., Williamson, J.: The setting of gypsum plaster. 1. The hydration of calcium-sulfate hemihydrate. J. Mater. Sci. **29**, 5279–5284 (1994)

27. Lewry, A.J., Williamson, J.: The setting of gypsum plaster. 3. The effect of additives and impurities. J. Mater. Sci. **29**, 6085–6090 (1994)

28. Mróz, P., Mucha, M.: Hydroxyethyl methyl cellulose as a modifier of gypsum properties. J. Therm. Anal. Calorim. **134**(2), 1083–1089 (2018). https://doi.org/10.1007/s10973-018-7238-3

29. Yu, Q.L., Brouwers, H.J.H.: Microstructure and mechanical properties of β-hemihydrate produced gypsum: an insight from its hydration process. Constr. Build. Mater. **25**, 3149–3157 (2011). https://doi.org/10.1016/j.conbuildmat.2010.12.005

30. ASTM: ASTM C 472-20: Standard Test Methods for Physical Testing of Gypsum, Gypsum Plasters, and Gypsum Concrete. ASTM International, West Conshohocken (2020)

31. Gartner, E.M.: Cohesion and expansion in polycrystalline solids formed by hydration reactions—the case of gypsum plasters. Cem. Concr. Res. **39**, 289–295 (2009)

32. Motalebi, Z., Aslani, H., Vafaei, V.: Identification of damaging factors and optimizing the techniques used in tile work decorations of domes in restoration process. In: First Conference on Traditional Structures and Technology with Particular Reference to Domes, Tehran (2014)

33. Stec, M.: The restoration of the 12th century engraved estrich gypsum floor from the church in Wislica. In: 2nd Conference on Historic Mortars-HMC 2010 and RILEM TC 203-RHM final workshop, pp. 711–721. RILEM Publications SARL, Paris (2010)

34. Boostani, A., Tonietti, U.: History and crucial aspects of strengthening the arch system. In: Secco Suardo, L. (ed.) The Nine Domes of the Universe. The Ancient Noh Gonbad Mosque: The Study and Conservation of an Early Islamic Monument at Balkh, pp. 75–121. Aga Khan Trust for Culture, Kabul (2016)

35. Ethiopian Heritage Fund: Tigray conservation programme. Abuna Daniel, Qorqor. The Ethiopian Heritage Fund Newsletter (2017). http://us4.campaign-archive2.com/?e=%5BU NIQID%5D&u=141d4f13b89ea3a213eef5d6b&id=87e7afa087. Accessed 14 Nov 2021

36. Refaat, F., Mahmoud, H.M., Brania, A.: Uncovering nineteenth-century Rococo-style interior decorations at the National Military Museum of Cairo: the painting materials and restoration approach. J. Archit. Conserv. **26**, 87–104 (2020). https://doi.org/10.1080/13556207.2019.1695173

37. Taniguchi, Y., et al.: Conservation of the Ini-Sneferu-Ishetef wall paintings from the Old Kingdom: a joint project between Japan and Egypt. In: Bridgland, J. (ed.) Transcending Boundaries: Integrated Approaches to Conservation: ICOM-CC 19th Triennial Conference Preprints, Beijing, 17–21 May 2021. International Council of Museums, Paris (2021)

38. Potskhishvili, T.G.: Development of a site specific injection grout for gypsum based plaster in the Ateni Sioni church in Georgia. MA thesis, Scuola Universitaria Professionale della Svizzera Italiana, Lugano (2016)

39. Cardoso, F.A., Agopyan, A.K., Carbone, C., Pileggi, R.G., John, V.M.: Squeeze flow as a tool for developing optimized gypsum plasters. Constr. Build. Mater. **23**, 1349–1353 (2009)

40. Magallanes-Rivera, R.X., Escalante-Garcia, J.I., Gorokhovsky, A.: Hydration reactions and microstructural characteristics of hemihydrate with citric and malic acid. Constr. Build. Mater. **23**, 1298–1305 (2009)

41. Pasian, C., Porter, J.H., Gorodetska, M., Parisi, S.: Developing a lime-based injection grout with no additives for very thin delamination: the role of aggregates and particle size. In: Bosiljkov, V.B., Padovnik, A., Turk, T., Štukovnik, P. (eds.) 6th Historic Mortars Conference 21–23 September 2022, University of Ljubljana, Ljubljana, Slovenia. University of Ljubljana, Ljubljana (2022)

42. Poraver: Product safety information sheet (2019)
43. 3 M: 3 M^{TM} Scotchlite^{TM} Glass Bubbles. K Series, S Series: Product Information, 98-0212-3710-6 (2003)
44. Horie, V.: Materials for Conservation: Organic Consolidants, Adhesives and Coatings, 2nd edn. Butterworth-Heinemann, Burlington (2010)
45. SE Tylose GmbH & Co. KG: Tylose® for building materials (2019)
46. SE Tylose GmbH & Co. KG: Tylose® MH 300 P2 safety data sheet 10896-0128 (2021)
47. Ashland: Klucel^{TM} hydroxypropylcellulose: physical and chemical properties, PC-11229.3 (2017)
48. Lewry, A.J., Williamson, J.: The setting of gypsum plaster. 2: the development of microstructure and strength. J. Mater. Sci. **29**, 524–5528 (1994)
49. Biçer-Şimşir, B., Rainer, L.: Evaluation of lime-based hydraulic injection grouts for the conservation of architectural surfaces: a manual of laboratory and field test methods. Getty Conservation Institute, Los Angeles (2011)
50. Pasian, C., Martin de Fonjaudran, C., Rava, A.: Innovative water-reduced injection grouts for the stabilisation of wall paintings in the Hadi Rani Mahal, Nagaur, India: design, testing and implementation. Stud. Conserv. **65**, P244–P250 (2020). https://doi.org/10.1080/00393630.2020.1761179
51. Porter, J.H., Pasian, C., Secco, M., Salameh, M., Debono, N.: Diethyl oxalate-based microgrouts in calcium carbonate systems: formulation, field testing and mineralogical characterization. In: Ignacio Álvarez, J., Fernández, J.M., Navarro, Í., Durán, A., Sirera, R. (eds.) 5th Historic Mortars Conference (HMC 2019), pp. 1291–1305. RILEM Publications SARL, Paris (2019)
52. A Guide to Shore Hardness: https://www.polyglobal.co.uk/a-guide-to-shore-hardness/. Accessed 27 Mar 2022
53. ASTM: ASTM D2240-15: standard test method for rubber property—durometer hardness. ASTM International, West Conshohocken (2015)
54. Mix, A.W., Giacomin, A.J.: Dimensionless durometry. Polym. Plast. Technol. Eng. **50**, 288–296 (2011)
55. Wang, Z.-F., Chen, Y.: Strength of cement-stabilised clay by hardness testing. Proc. Inst. Civil Eng. – Constr. Mater. **170**, 250–257 (2017). https://doi.org/10.1680/jcoma.15.00057
56. Charola, A.E., Pühringer, J., Steiger, M.: Gypsum: a review of its role in the deterioration of building materials. Environ. Geol. **52**, 339–352 (2007). https://doi.org/10.1007/s00254-006-0566-9
57. Schug, B., et al.: A mechanism to explain the creep behavior of gypsum plaster. Cem. Concr. Res. **98**, 122–129 (2017). https://doi.org/10.1016/j.cemconres.2017.04.012

Comparative Evaluation of Properties of Laboratory Test Specimens for Masonry Mortars Prepared Using Different Compaction Methods

Vadim Grigorjev[1]($\boxtimes$), Miguel Azenha[2], and Nele De Belie[1]

[1] Ghent University, Ghent, Belgium
`vadim.grigorjev@ugent.be`
[2] University of Minho, ISISE, Guimarães, Portugal

Abstract. Development of new solutions for masonry mortars is heavily reliant on laboratory-based experimental procedures. This study provides insights into the properties of masonry mortars, prepared by four distinct compaction methods, based on existing standards: tamping, tapping, jolting and vibrating. The particular mortar mix under study has been designed in volumetric proportions of 1:1:6 with air lime, cement and sand, respectively. Evaluation of differences among compaction methods is based on bulk density, mechanical strength, porosity and water absorption measurements at 7 and 28 days. Density and strength testing results indicate statistically significant differences, where mechanically compacted mortars are denser and stronger than their manually compacted counterparts. Similar development is observed through assessment of mortar porosity. The variation is noticeable in gel and capillary pore range as shown by mercury intrusion, while open porosity evaluated by vacuum immersion also indicates some distinction between manual and mechanical compaction, with the latter producing less porous mortars. On the other hand, capillary water absorption results reveal higher coefficients for jolted and vibrated samples, hinting at different pore interconnectivity in mechanically and manually compacted mortar specimens.

Keywords: Lime-based mortar · compaction methods · fresh properties · hardened properties · analysis of variation

1 Introduction

Even though masonry construction has been paramount to humankind since around 8000 years ago, it still remains a crucial field for new developments and preservation of historic structures [1]. In both cases masonry is often composed of building blocks connected together by mortars, where lime-based materials have been and still are of primary importance [2, 3].

Robust experimental methods are needed to ensure that laboratory-developed lime-based products are representative of real-life structures. In the case of masonry mortar preparation in the laboratory, a specimen of $160 \times 40 \times 40 \text{ mm}^3$ is considered a standard

V. Bokan Bosiljkov et al. (Eds.): HMC 2022, RILEM Bookseries 42, pp. 494–506, 2023.
https://doi.org/10.1007/978-3-031-31472-8_40

mortar prism for strength testing, and this property is the only requirement based on Eurocode 6 [4]. However, there are more properties of a masonry mortar impacted by the compaction of the prismatic mortar specimens. Current standard for masonry mortars EN 1015-11 [5] describes two manual compaction procedures: the first involves compaction of the mortar by stroking it with a tamper, while the alternative suggests tilting the mould at a 30° angle and tapping it on the table. Even though these manual compaction methods are prescribed in the masonry mortar-specific standard, researchers sometimes [6–9] opt for machine-compaction similar to what is described in cement-specific standard EN 196-1 [10]. This standard also presents two compaction methods: the former involves compacting the mortar on a jolting table, while the latter suggests using a vibrating table. Inevitably, due to the nature of these compaction procedures, they introduce different levels of compaction energy into the mortar mixes, thus affecting properties such as hardened density, porosity, mechanical strength, water absorption capacity and susceptibility to potentially harmful substances. All these properties are important not only in characterizing laboratory-produced mortars, but also with regards to their real service life in modern masonry construction and conservation and repair of historic structures.

This study describes an experimental campaign designed to compare the four different mortar compaction methods mentioned above. Lime-cement mortar mixes are cast in prismatic moulds, compacted and cured for 7 and 28 days to evaluate the time-dependent change of properties. At the respective ages, differently compacted mortars are measured and weighted, allowing the calculation of their density. Various tests are performed, including mechanical strength, water absorption by capillarity and mercury intrusion porosimetry supplemented by open porosity test to evaluate the broader range of pore sizes. This setup could allow drawing conclusions not only regarding how different compaction methods fare in comparison with one another, but also which is the most representative compared to real-life masonry applications in new building, repair and conservation. However, the curing times used in this study are insufficient to assess any developments in mortar properties arising from the carbonation of air lime.

2 Materials and Methods

2.1 Mortar Mix and Preparation

A mix of 1:1:6 parts by volume of air lime, cement and sand was selected as the main mortar composition for this study. Lime is high-calcium hydrated lime CL90S, conforming to the requirements of EN 459-1 [11], provided by Lhoist [12]. It was paired with limestone cement CEM II/A-L 32,5 R as specified in EN 197-1 [13], supplied by Tarmac [14]. CEN standard sand, based on EN 196-1 [10], was chosen wittingly to limit the variability of aggregates when preparing different mortar batches. In addition to dry components, regular tap water was used; the amount was adjusted to achieve mortar consistency of 175 ± 10 mm flow table value as specified in EN 1015-11 [5]. Exact quantities, based on bulk densities of the above-mentioned materials, are presented in Table 1 below.

Table 1. Composition of lime-cement mortar.

Lime: Cement: Sand (by volume)	Lime (g)	Cement (g)	Sand (g)	Water (g)	w/b ratio (by mass)	w/b ratio (by volume)	Flow table value (mm)
1:1:6	57	141	1350	208	1.05	0.765	170 ($\pm$4)

2.2 Compaction Methods

Manual Compaction

- Tamping

Based on masonry-specific standard EN 1015-11 [5], the process of tamping involves stroking fresh mortar layers 25 times using a tamper rod. Compaction is achieved through impact loading, which is highly dependent on the operator.

In practice, compaction by tamping was performed by the same operator for all mortar batches, minimizing potential differences in tamper strokes. The tamper rod was a metal cylinder with a diameter of 20 mm. Extra detail was devoted to covering the full area of fresh mortar layer to limit the unevenness and formation of possible air pockets. This was practically achieved by consistently alternating strokes from one side of the imaginary centreline to the other, along the length of the prismatic mould while counting the strokes.

- Tapping

Proposed as an alternative to tamping, the action of tapping requires no extra tools. Fresh mortar layers are compacted by tilting the mould at an angle of approximately 30° and tapping it on the working surface (i.e. table or bench) 10 times.

Based on experience, successful execution of this method required tilting and tapping both sides of the mould alternately. Similarly, a simple guide marking the 30° angle was made to perform compaction in a more controlled pattern. The working surface was a wooden laboratory bench and the time required to complete compaction was approximately 30 s.

Mechanical Compaction

- Jolting

Contrary to EN 1015-11 [5], cement-specific standard EN 196-1 [10] proposes the use of mechanical compaction methods for cement-based laboratory mortars, still applied in the case of masonry, as mentioned previously. Of these compaction methods, jolting is most common, making use of a jolting table – a mechanical apparatus which secures the mould and shakes every mortar layer for a total of 60 times, one every second. This procedure is automated and the apparatus exerts the same force with every jolt.

- Vibrating

Another way of compacting fresh masonry mortar mechanically is by means of a vibrating table. In contrast to jolting, a vibrating table operates at a constant frequency with the mould placed on top for a total of 120 s. The operation starts immediately after filling the first mortar layer, while the second is added after 60 s.

The vibrating table used in this study was a portable vibrating table from Testing [15]. The technical specifications and geometry were different from those described in EN 196-1 [10].

2.3 Curing Conditions

After preparation, mortars were immediately placed into curing chamber with relative humidity at $95 \pm 5\%$ and temperature of 20 ± 2 °C [5]. Demoulding was performed at 2 days and the mortars were kept in the same conditions for 5 more days. After a week, mortars were either tested for respective properties, or transferred to 60% relative humidity and 20 °C room until the age of 28 days.

2.4 Mortar Properties

Density. Densities of mortar prisms were obtained at 7 and 28 days. The actual dimensions of prisms were measured with a Vernier scale, allowing calculation of volume; mass was measured on a laboratory balance (0.05 g precision). Division of mass by the calculated volume resulted in bulk densities of mortar samples. A total of 6 samples were prepared for density evaluation on two separate occasions.

Mechanical Strength. Flexural and compressive strengths were measured at 7 and 28 days with a walter + bai combined testing machine [16], following EN 196-1 [10] standard loading protocol. A total of 6 samples were prepared for strength testing on two separate occasions.

Porosity. Mercury intrusion porosimetry (MIP) is a non-standardized, yet widely applied method to evaluate pore volume, size and distribution in a mortar matrix [17–19]. For this experimental campaign, the samples for MIP were extracted from the intact mortar prism pieces, remaining after mechanical testing. Only mortars aged for 28 days had sufficient structural stability to extract the samples, indicating an adequately developed microstructure for pore detection. The microstructure was preserved by solvent-exchange method with isopropanol, designed to stop further hydration and carbonation of the samples [20, 21].

Method to measure porosity by immersion under vacuum, adopted from RILEM CPC 11.3 [22], was applied to estimate the total open porosity of mortar samples. However, the drying step was performed at the end of the experiment for the purpose of preserving the microstructure during testing. This test was chosen in order to complete the gel and capillary porosity by MIP with macro-porosity estimation. A total of 3 samples were prepared for the test.

Water Absorption. Water absorption by capillarity was tested according to EN 1015-18 [23]. The number of measurement times was increased in order to produce a typical water absorption curve. A total of 3 samples were prepared for the test. To aid the expression of results, water absorption coefficients were computed using the following equation:

$$C = \frac{M_{90} - M_{10}}{A\sqrt{t_{90} - t_{10}}} \tag{1}$$

where:
M_{10} – absorbed water after 10 min, kg;
M_{90} – absorbed water after 90 min, kg;
A – area of the submerged face - 0.0016 m^2;
t_{10} – 10 min;
t_{90} – 90 min;
C – coefficient of water absorption, kg/(m^{2}min$^{0.5}$).

Statistical Analysis. To aid the comparison between compaction methods, statistical analysis of variance (ANOVA) was performed on results of density and strength (based on 6 mortar samples) as well as water absorption and open porosity (based on 3 samples). Statistically significant differences were further processed with Tukey's honestly significant difference (HSD) test and where outliers were found and removed from the dataset by modified Thomson Tau test – with Tukey-Kramer test.

3 Results

3.1 Density

Bulk density is a useful property which aids the comparison of previously discussed compaction methods. Despite being the easiest parameter to calculate and assess, bulk density provides the first insights into the level of compaction attainable with different methods. It could help to indicate the variation of air voids and pores, hint at their size and predict the strength of mortars.

Recorded in Table 2, bulk density results at 7 days are based on calculations of average values from 6 mortar samples, prepared on two separate occasions for each method of compaction, further supplemented by the standard deviations and coefficients of variation. Based on the results, mechanically compacted samples have higher density than manually compacted ones, but the difference between jolting and vibrating is larger than that between tamping and tapping.

Similarly, density results at 28 days are presented in Table 3. These values are based on different samples than those recorded in Table 2, but equivalently, the variability is assessed from 6 samples prepared on two different occasions per compaction method. As expected, bulk densities are lower than those at 7 days due to water evaporation, but the overall trend is still observed – both manual compaction methods produce similar results, while higher bulk densities are achieved by mechanical compaction methods. The difference between jolting and vibrating is more prominent as well.

Table 2. Bulk densities of mortars at 7 days.

Compaction method	Manual		Mechanical	
	Tamping	Tapping	Jolting	Vibrating
Bulk density avg. (kg/m^3)	2180.6	2168	2237.1	2216.1
Standard deviation (kg/m^3)	36.6	29.4	17.5	39.4
COV (%)	1.7	1.4	0.8	1.8

Table 3. Bulk densities of mortars at 28 days.

Compaction method	Manual		Mechanical	
	Tamping	Tapping	Jolting	Vibrating
Bulk density avg. (kg/m^3)	1980.6	1989.3	2027.4	2053.4
Standard deviation (kg/m^3)	7.3	12.7	17.5	16.3
COV (%)	0.4	0.6	0.9	0.8

3.2 Mechanical Strength

Strength is arguably the most important property of a masonry mortar, at least from quality control point of view (according to existing regulations), with compressive strength being the only elective parameter in Eurocode 6 [4]. However, the experimental procedure involves both flexural and compressive testing of a mortar specimen at the same time, therefore these results are paired together.

Results of compressive strength are displayed in Fig. 1. At the age of 7 days, the strength ranges from 4.39 MPa to 5.21 MPa. When tested at 28 days, mortars show developed compressive strength of 7.25–8.4 MPa. Particularly striking in Fig. 1 is the peaking compressive strength of vibrated specimens at 28 days. Such a result indicates that vibrational compaction is completely unmatched for the compressive strength, when even jolted samples yield results closer to manual compaction methods.

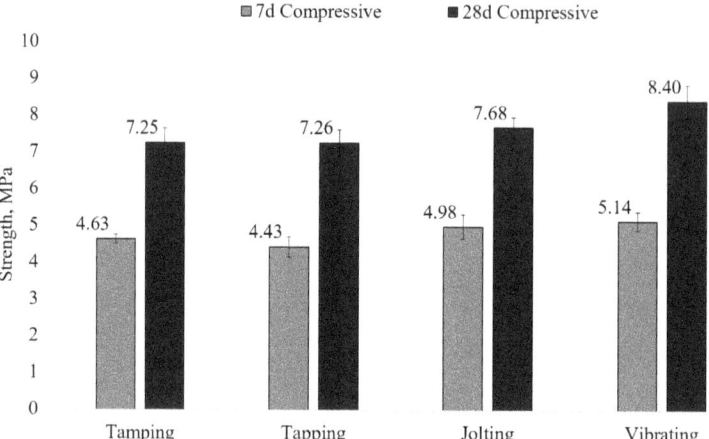

Fig. 1. Compressive strength of mortars at 7 and 28 days of age.

As in the case of compressive strength and Fig. 1, comparable trends can be observed in flexural strengths, where manual compaction methods produce samples with lower bending strengths than mechanical compaction methods. At the age of 7 days, as presented in Fig. 2, lime-cement mortars have flexural strength in the range of 0.92–1.12 MPa and at 28 days in the range of 1.66–1.83 MPa. Manual compaction methods are more comparable to one another than mechanical methods, although the difference between the latter two is not as pronounced as previously observed for compressive strengths in Fig. 1.

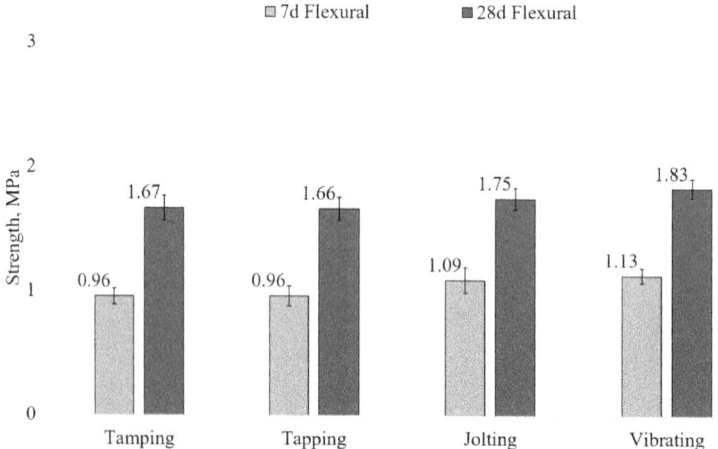

Fig. 2. Flexural strength of mortars at 7 and 28 days of age.

Results in Figs. 1 and 2 present standard deviation bars which are more noticeable when compared to standard deviation values in case of densities. Still, this variation is expected due to the destructive nature of the experiment and many different factors influencing the result.

Overall, mechanical strength values are within the range reported in literature for similar mortar mixes [7], with some sources reporting slightly lower values [6, 9], but these results would be considerably influenced by the choice of binder and aggregate materials and actual mix design quantities.

3.3 Porosity

Mercury Intrusion Porosimetry. Figure 3 presents the results of cumulative porosity as estimated based on the volume of intruded mercury. The graphs indicate rather insubstantial changes in the gel and capillary porosity for manually tamped and tapped, as well as mechanically jolted mortar samples. These compaction methods yield samples with porosity of ~18–19%.

Notable difference is observed in vibrated samples, where the amount of both gel and capillary pores is considerably lower, with ~16% in estimated total porosity.

The results shown in Fig. 3 indicate significantly lower porosity values by mercury intrusion than those reported in literature for 1:1:6 mortar mix designation [18].

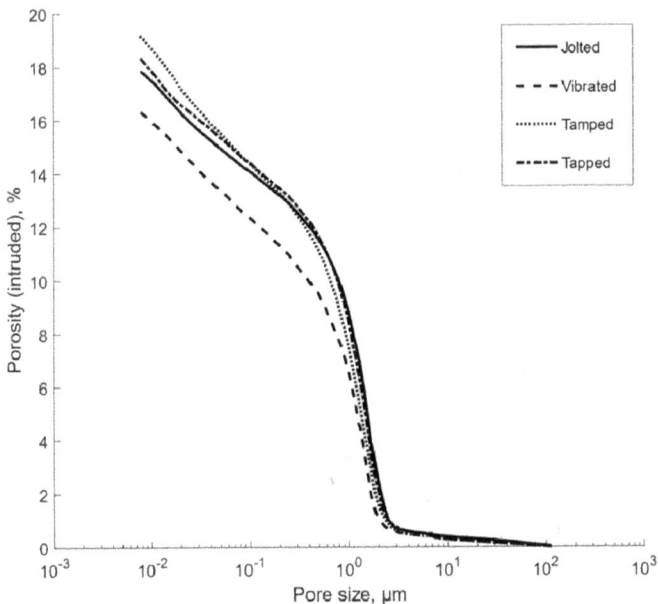

Fig. 3. Cumulative intrusive porosity of differently compacted mortars at 28 days.

Porosity by Immersion Under Vacuum. The results of open porosity in mortar samples are based on the percentage of water uptake under vacuum, as expressed in Fig. 4. It reveals the slightly higher values of total porosity of manually compacted samples, compared to mechanical methods. However, between the latter, vibrated samples are showing larger average result as well as broader standard deviation than jolted ones.

Contrary to porosity evaluation by mercury intrusion, the open porosity results are closer to those achieved by other researchers [6, 18].

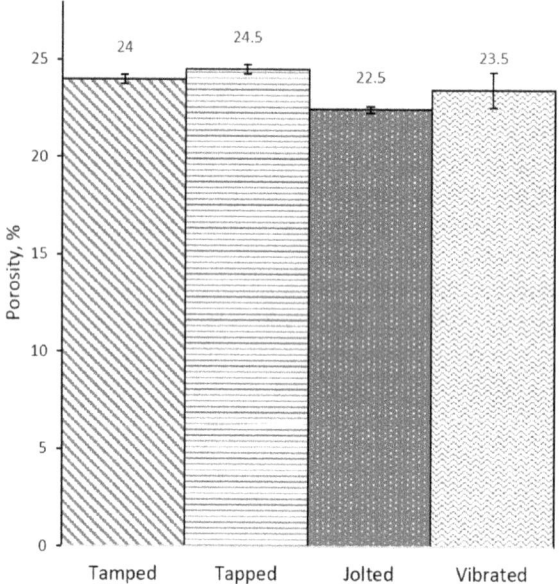

Fig. 4. Open porosity of differently compacted mortars at 28 days.

3.4 Water Absorption by Capillarity

Water absorption of differently compacted mortar samples is presented in Fig. 5 with the aid of curves, constructed from absorbed water mass measurements at 0, 1, 3, 5, 10, 30, 90, 180 and 1440 min. Furthermore, the water absorption coefficients, as specified in EN 1015-18 [23], are calculated based on Eq. 1.

Unexpectedly, the results presented in Fig. 5 suggest that mechanically compacted samples absorb more water more rapidly than manually compacted ones, despite having lower porosity. This is also confirmed by water absorption coefficients, where lowest values are produced by tamped samples, followed by tapped, vibrated and jolted specimens.

3.5 Statistical Analysis

ANOVA results with Tukey-Kramer HSD are presented in Table 4 below. Generally, in all measured properties the differences between compaction methods were significant, with specific pair-wise differences highlighted in the table. Evidently, the results of tapping

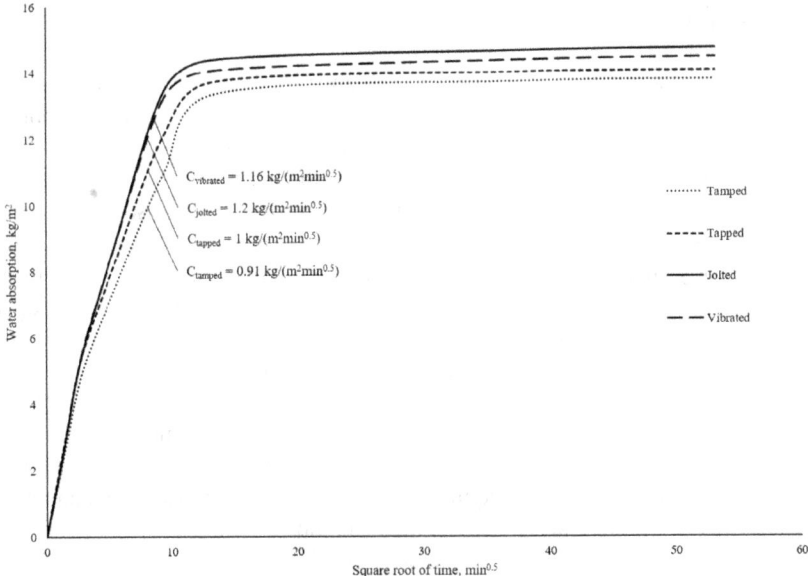

Fig. 5. Water absorption of mortars at 28 days.

and tamping are not significantly different, contrary to all the manual-mechanical method pairs, where 5 out of 8 measured properties indicate significant difference. Comparison of jolting and vibration also demonstrates significantly different results for density and compressive strength measurements at 28 days.

Table 4. ANOVA with Tukey-Kramer HSD results.

		Density		Compressive strength		Flexural strength		Porosity	Water absorption
	Age	7	28	7	28	7	28	28	28
	Pair	Statistically significant difference found?							
Tamping	Tapping	No	No	No	No	No	No	No	No
Tamping	Jolting	Yes	Yes	No	No	Yes	No	Yes	Yes
Tamping	Vibration	No	Yes	Yes	Yes	Yes	Yes	No	No
Tapping	Jolting	Yes	Yes	Yes	No	Yes	No	Yes	No
Tapping	Vibration	No	Yes	Yes	Yes	Yes	Yes	No	No
Jolting	Vibration	No	Yes	No	Yes	No	No	No	No

4 Conclusion

This study presents a comparative analysis of laboratory-produced masonry mortars, compacted using four different methods, namely manual tamping and tapping and mechanical jolting and vibration, all of which are specified in masonry and cement-specific standards. It should be noted that while all discussed compaction methods are used in laboratory practice, the choice is rarely justified. Even so, if the topic of debate concerns controlled and repeatable laboratory mortar production, the results of the present study indicate no particular superiority or inferiority of any compaction method with regards to these aspects. Within the scope of this work, all methods have led to similar variations in measurements of specific properties. This was unanticipated in view of the less controlled nature of manual compaction.

On the other hand, analysis of mortar properties elucidates the distinction between manual and mechanical compaction. Lower results for density and strength of mortars compacted manually present an expected outcome of the hand-operated process, which remotely mimics the bricklaying construction. By contrast, mechanical compaction methods introduce significantly higher compaction energy into fresh mortar mixes, resulting in denser and stronger mortar specimens. Nonetheless, the results of open porosity evaluation also suggest better performance by the jolted and vibrated specimens, albeit the difference being not particularly convincing when compared with tamped and tapped samples. In relation to micro-porosity, the trend breaks as jolted mortars produce similar pore size distribution to manually compacted mortars, while vibrated stand out by showing lower amounts of gel and capillary pores. Counterintuitively, water absorption by capillarity reveal larger rates and capacities of less porous mechanically compacted samples, than of those compacted manually. Even though mostly statistically insignificant, such phenomenon is not easily explicable, hindering the correlation with other observations in mortar properties.

Additionally, a pronounced difference of compressive strength results attainable by jolting and vibration raises potential concern regarding the suggestion to use these mechanical compaction procedures as alternatives. Especially noteworthy in this sense are the similar results in mortars produced by manual tamping and tapping, rationalizing the notion to use them interchangeably. Pairing this fact with representative properties and acceptable variability among different mortar batches could suggest that manual compaction methods are more suited for laboratory-prepared masonry mortar specimens.

However, it is important to note that the choice is heavily dependent on the intended application. Unsurprisingly, masonry-specific compaction methods produce results which are more comparable to real-life masonry structures due to mostly manual nature of brick working. Likewise, enhanced mortar properties achievable by mechanical compaction are inherent to cement mortars, directly linked to concrete research. Since concrete requires compaction in most applications, this practice is likely reflected in laboratory setting with mortars. Although in this study mechanical compaction methods appear to be less comparable to one another and produce similar variability as manual methods, such outcomes have to be treated carefully. A note of caution is due here since there are multiple sources of potential error: availability and choice of equipment, operator-dependent inputs, machine-dependent faults, calibration, environmental conditions, quality and storage of raw materials are only a few considerations. The present study has laid grounds for further investigations, which could consider these factors along with the variable mortar composition and comparative evaluations based on field mortars.

Acknowledgements. This project has received funding from the European Union's Horizon 2020 research and innovation programme under Marie Sklodowska-Curie project SUBLime [Grant Agreement n. 955986].

This work was partly financed by FCT / MCTES (Portugal) through national funds (PIDDAC) under the R&D Unit Institute for Sustainability and Innovation in Structural Engineering (ISISE), under reference UIDB / 04029/2020.

References

1. Como, M.: Statics of historic masonry constructions, p. 9 (2017). https://doi.org/10.1007/978-3-319-54738-1
2. Claisse, P.A.: Mortars and grouts. Civ. Eng. Mater. 303–311 (2016). https://doi.org/10.1016/B978-0-08-100275-9.00028-0
3. Elert, K., Rodriguez-Navarro, C., Pardo, E.S., Hansen, E., Cazalla, O: Lime mortars for the conservation of historic buildings. **47**(1), 62–75 (2013). https://doi.org/10.1179/SIC.2002.47.1.62
4. EN 1996-1-1+A1:2013 Eurocode 6 - Design of masonry structures - Part 1-1: General rules for reinforced and unreinforced masonry structures
5. EN 1015-11:2019 Methods of test for mortar for masonry - Part 11: Determination of flexural and compressive strength of hardened mortar
6. Ramesh, M., Azenha, M., Lourenço, P.B.: Quantification of impact of lime on mechanical behaviour of lime cement blended mortars for bedding joints in masonry systems. Constr. Build. Mater. **229**, 116884 (2019). https://doi.org/10.1016/J.CONBUILDMAT.2019.116884
7. Qadir, W., Ghafor, K., Mohammed, A.: Evaluation the effect of lime on the plastic and hardened properties of cement mortar and quantified using Vipulanandan model. Open Eng. **9**(1), 468–480 (2019). https://doi.org/10.1515/ENG-2019-0055
8. Yedra, E., Ferrández, D., Morón, C., Saiz, P.: New test methods to determine water absorption by capillarity. Experimental study in masonry mortars. Constr. Build. Mater. **319** (2022). https://doi.org/10.1016/J.CONBUILDMAT.2021.125988
9. Nalon, G.H., Alves, M.A., Pedroti, L.G., Lopes Ribeiro, J.C., Hilarino Fernandes, W.E., Silva de Oliveira, D.: Compressive strength, dynamic, and static modulus of cement-lime laying mortars obtained from samples of various geometries. J. Build. Eng. **44**, 102626 (2021). https://doi.org/10.1016/J.JOBE.2021.102626
10. EN 196-1:2016 Methods of testing cement - Part 1: Determination of strength
11. EN 459-1:2015 Building lime - Part 1: Definitions, specifications and conformity criteria
12. Lhoist - Minerals and lime producer (n.d.). https://www.lhoist.com/be_en. Accessed 11 Mar 2022
13. EN 197-1:2011 Cement - Part 1: Composition, specifications and conformity criteria for common cements
14. Tarmac | Sustainable construction solutions (n.d.). https://tarmac.com/. Accessed 11 Mar 2022
15. TESTING Bluhm & Feuerherdt GmbH. Operating Manual Vibrating Table (n.d.). https://testing.de/sites/default/files/productpdf_original/2.023x-operating-manual-vibrating-tables-en_0.pdf

16. walter+bai Testing Machines. Series D and DB (n.d.). https://www.walterbai.com/page/pro ducts/Building_Materials_Testing/Cement_Testing_Systems/Series_D_and_DB.php
17. Ma, H.: Mercury intrusion porosimetry in concrete technology: tips in measurement, pore structure parameter acquisition and application. J. Porous Mater. **21**(2), 207–215 (2014). https://doi.org/10.1007/S10934-013-9765-4/FIGURES/10
18. Marvila, M.T., de Azevedo, A.R.G., Ferreira, R.L.S., Vieira, C.M.F., de Brito, J., Adesina, A.: Validation of alternative methodologies by using capillarity in the determination of porosity parameters of cement-lime mortars. Mater. Struct. **55**(1), 1–15 (2022). https://doi.org/10. 1617/S11527-021-01877-6
19. Ghorbel, E., Wardeh, G., Gomart, H., Matar, P.: Formulation parameters effects on the per- formances of concrete equivalent mortars incorporating different ratios of recycled sand. J. Building Phys. **43**(6), 545–572 (2020). https://doi.org/10.1177/1744259119896093
20. Snoeck, D., et al.: The influence of different drying techniques on the water sorption properties of cement-based materials. Cem. Concr. Res. **64**, 54–62 (2014). https://doi.org/10.1016/J. CEMCONRES.2014.06.009
21. Snellings, R., et al.: RILEM TC-238 SCM recommendation on hydration stoppage by solvent exchange for the study of hydrate assemblages. Mater. Struct. **51**(6), 1–4 (2018). https://doi. org/10.1617/s11527-018-1298-5
22. RILEM TC: RILEM CPC 11.3 Absorption of water by immersion. RILEM Recommendations for the Testing and Use of Constructions Materials, p. 33 (1994)
23. EN 1015-18:2003 Methods of test for mortar for masonry - Part 18: Determination of water absorption coefficient due to capillary action of hardened mortar

The Challenge on Development of Repair Mortars for Historical Buildings in Severe Marine Environment: Paimogo Fort, A Case Study

Maria do Rosário Veiga👤🆔 and Ana Rita Santos👤(✉)🆔

National Laboratory for Civil Engineering, Buildings Department, Lisbon, Portugal
arsantos@lnec.pt

Abstract. The Paimogo Fort, listed as a public interest property in 1957, is one of the several Portuguese military fortifications built near the ocean, in the 17th century. Within the scope of the restoration project developed by the Municipality of Lourinhã and its partners, for the revitalization and safeguarding of the Paimogo's Fort, an extensive characterization of the original mortars was made and new repair mortars are being developed by National Laboratory for Civil Engineering.

In the present work the development of the durable and compatible repair mortars for this case study is described. Some compositions based on the original mortars' composition and characteristics are briefly described and their main physical and mechanical characteristics are analysed and compared in successive ages. Applications of the same mortars on porous composite substrates were subjected to real environmental conditions in the Fort external area to check their performance and durability.

The experimental results showed the importance of the *in situ* applications, since distinct behaviour was found. Moreover, it reveals that all the studied mortars absorb higher volume of water and have lower strength and higher deformability, when compared to the original ones. Thus, the choice of the formulations for application as a substitution render of the Fort walls is still in process of evaluation. Their performance and durability are being studied at longer ages and the formulations will be fine-tuned for further evaluation.

Keywords: Conservation · Durability · Lime · Mortar · Performance

1 Introduction

Portugal has an extensive coastal area where several military fortifications were built with the intention to protect the Portuguese coastline territory from the constant military threat from the ocean.

These historical constructions located near the ocean, have been subjected during centuries to a very aggressive environment of salty water spray, high humidity, intense sun radiation, thermal shock, strong wind and, sometimes, even to waves' strength.

Renders, whose main function is the protection of walls, are particularly exposed to such actions.

Samples collected from several forts located in Lisbon and coastal surroundings show clearly that the original mortars used for masonry and renders were mechanically resistant and durable lime mortars. Usually, they were based in calcitic air lime and additions or aggregates that promoted pozzolanic reactions [1, 2]. Lime lumps were generally found as well as more unusual materials, such as ceramic elements and coal. The aggregates were commonly composed by sand-to-gravel sized mineral grains and fossil fragments [2–4] and usually came from local beaches or fluvial sand deposits available nearby the fortifications [5, 6].

Paimogo Fort is one of these Portuguese fortifications, built in 1674 and located in the cliffs of Paimogo beach, near Lourinhã city (approximately 75 km from Lisbon). The Fort is classified of public interest since 1957 and, as other coastal forts from the 17th and 18th centuries, has proved to be very well constructed, with well-selected materials and craftsmanship.

Within the scope of the "Coast Memory Fort" EEA Grants project, promoted by Lourinhã Municipality, which aims to safeguard and revitalize the Paimogo Fort, an assessment of the conservation condition of the building and an analysis of the original materials was carried out.

The tests results showed that the old masonry walls were made with stone from local quarries and lime baked in the artisanal ovens of the region. The original mortars are of air lime and sand of siliceous and limestone nature, with varying proportions, depending on the type of mortar, between 1:1.5 and 1:4 (lime: aggregate mass proportions), which are much richer in lime than is usual in other types of old constructions. Moreover, it was also identified that the aggregate is free of clay and that all mortars present high mechanical strengths (ranging between 2.5 and 5 MPa in compression) [7, 8].

However, most of the original air lime renders and plasters have been removed in an intervention of 2006 and replaced by a hydraulic lime mortar, which presents nowadays some degradation, especially erosion and loss of cohesion (Fig. 1).

Thus, in the scope of this project, new repair mortars are being developed for the preservation of the Fort, based on the compositions and characteristics obtained for the original mortars and considering the raw materials of the region.

The mortars to use in this kind of interventions must have special properties to assure durability to the very aggressive environment, as well as keeping compatibility characteristics: The mortars must have high mechanical strength to resist erosion of the ocean wind and salt crystallization pressure, moderate elasticity modulus to accommodate deformations due to thermal variations, moderate capillary coefficient and high water vapour permeability to retard the entrance of water and allow a quick drying. Besides, they should not have high contents of salts, to avoid the increase of salt contamination into the walls [9, 10].

Therefore, to choose substitution renders for these constructions is a challenging situation. A Lime-white cement mortar applied in two thin coats of 1:1:6 and 1:2:9 (cement:lime:siliceous aggregate volume proportions), was used in a similar case, with acceptable results for the last thirty years [11] (Fig. 2a), however it is well known that cement must be avoided whenever possible, because it favours salt damage due to its content on alkaline ions [12]. The use of lime mortars with industrial pozzolanic additions were also an alternative, however the results presented in a similar case in a fortification in Lisbon, after twenty years, are not encouraging (Fig. 2b).

Fig. 1. State of conservation of the Paimogo Fort.

(a)   (b)

Fig. 2. Forts in the coast area of Lisbon subjected to rehabilitation, where different repair solutions were used: (a) repair mortars of lime-white cement mortar; (b) repair mortars of lime with industrial pozzolanic additions.

Since many lime lumps were found in the original mortars, the addition of quicklime to the hydrated lime mix was considered an option. According to the method proposed by *Antoine-Joseph Loriot*, from the late seventeenth century, the mortars made with this kind of mix would be "weatherproof and water resistant", providing volume stability, mechanical strength, and durability, avoiding cracks, as the drying of the excess water of the mix is consumed by the quicklime instead of evaporated, which should provide a balance in volume variation [13]. However, recent studies didn't confirm *Loriot's* assumptions and showed cracking and degradation problems in mortars with this type of mix [14].

In this study, lime-based mortars are proposed and characterised with the aim of selecting a durable and compatible mortar to be used in the restoration of the Fort. Three compositions, based on the original mortars' composition, are presented and their main physical characteristics – as capillary absorption and open porosity – and mechanical characteristics – as flexural and compressive strength, dynamic elastic modulus and surface hardness – are analysed and compared in successive ages. Applications of the same mortars on porous composite substrates (bricks with large joints of lime-based mortars) were also subjected to real environmental conditions in the Fort external area, to check their performance and durability.

2 Materials and Methods

2.1 Mortars Composition and Mixing

The mortars formulations selected for this study were chosen considering three main aspects:

- the compatibility with the ancient masonry and original renders from the physical, mechanical and chemical points of view [15];
- the physical performance concerning the severe exposure to ocean spray and salts deposition (porosimetry) [16–18];
- the materials available in the region and technology accessible to common construction workers.

The options selected were mortars composed by slaked air lime (**AL**) and composed by mixing slaked air lime and quicklime (**AL+Q30%**), being the optimum amount of quicklime previously tested in laboratory. Additionally, a slaked air lime and white cement mortar (**C+AL**) was also considered, having well known behaviour due to its application in a similar environment [11]. In this first phase of tests the aggregate used for all the formulations was natural siliceous sand, from Tagus River. The mortars compositions are summarized in Table 1.

The **AL** and **C+AL** mortars were prepared in accordance with European standard EN 1015-2 [20], using a mortar mixer *Controls 65-L0005*. However, with the use of quicklime in the formulation of **AL+Q30%** mortar, it was necessary to adapt the mixing process for this composition, as follows:

- Mix all the dry solid content of the **AL+Q30%** mortar, as given in Table 1;
- Add the predetermined amount of water to the dry solid mix into the mixer, over a period of 30 s, with the mixer running at low speed. Then continue mixing, at the same speed, for further 4 min and 30 s;
- Rest of the mix for 10 min;
- Complete the mixing, at low speed, for 5 min.

The consistence of each mortar was optimised by experimental application on a brick with the criteria of using the minimum mixing water to allow good workability, independently of the flow table value (Fig. 3). This method resulted in different kneading water ratios for the different mortars (Table 1).

Table 1. Mortars compositions.

Samples identification	Binder	Aggregate	b/a ratio (by volume)	w/b ratio (by weight)	Flow diameter[a] (mm)
AL	CL90-S (ρ = 360 kg/m^3)	Natural siliceous sand (0–2 mm) (ρ = 1450 km/m^3)	1:2	1.9	154 ± 1.9
AL + Q30%	CL90-S (ρ = 360 kg/m^3) + CL90-Q (R5, P2) (ρ = 730 kg/m^3)		(1 + 0.3):2.6	1.7	169 ± 5.0 145 ± 4.0*
C + AL	CEM II/B-L 32.5R (ρ = 930 kg/m^3) + CL90-S (ρ = 360 kg/m^3)		1:3:12	1.8	155 ± 2.7

ρ – loose bulk density; [a] Determined according to standard EN 1015-3 [19]; * flow diameter after 10 min rest.

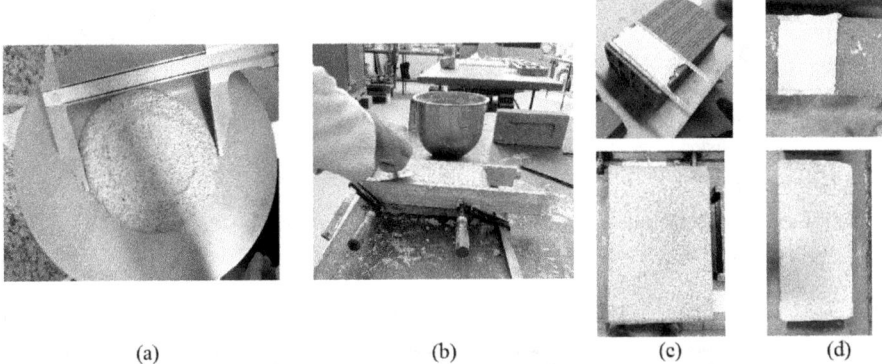

(a)   (b)   (c)   (d)

Fig. 3. Assessment of the workability: (a) flow table; (b) applicability; (c) behaviour on a common ceramic brick; (d) behaviour on a traditional porous brick.

The mortars were cast in 40 × 40 × 160 mm^3 normalized steel moulds and cured for 7 days in the moulds at 20 ± 2 °C and 65 ± 5% RH. In the first 3 days a sprinkling with water was performed on the mortar surface of each specimen every 24 h. After being removed from the moulds, the specimens were cured at the same conditions of temperature and relative humidity until the test date.

Additionally, the same mortars compositions were applied on common ceramic bricks and on traditional porous bricks (Fig. 4) and subjected to the same laboratory cure conditions of the prismatic specimens. The bricks used and their hygric characteristics are identified on Table 2.

Fig. 4. Applied mortars on normalized steel moulds, on a common ceramic brick and on a traditional porous brick.

Table 2. Characteristics of the bricks used (average values).

Test	Substrate	
	Common ceramic brick (30 cm x 20 cm x 11 cm)	**Traditional porous brick** (23 cm x 11 cm x 7 cm)
Water permeability under low pressure by karsten tubes[1] (ml/cm²)	0.2 after 30 min	0.7 after 30 min
Capillarity coefficient[2] (kg/m² min⁰·⁵)	0.6 ± 0.1	2.4 ± 0.4
Open Porosity[3] (%)	nd	34 ± 3.6

[1] based on a RILEM specification [21]; [2] determined according to EN 772-11 [22]; [3] determined according to EN 1936 [23].

Furthermore, the same mortars were applied, with a thickness of 20 ± 2 mm (based on the medium thickness usually applied in service), on a porous composite substrate (traditional porous bricks with large joints of lime-based mortars) - mini wallettes (Fig. 5a); afterwards, the mortars were also sprinkled with water for the first 3 days, and stored and cured under the same controlled conditions of temperature and relative humidity as the prismatic specimens (Fig. 5b). After 15 days, the applications were exposed to real environmental conditions at the Fort external area, with exposure to West, since this is the direction of the prevailing winds in the region and faced forward to the ocean (Fig. 5c). During the curing period on the Fort (september to december), the natural climatic conditions were characterized by temperature between 15 and 26 °C, quick showers and some periods with hot sun and strong wind (average of 20 km/h, with some wind gusts of 50 km/h). The relative humidity of the air ranged between 60 and 100%.

(a)   (b)   (c)

Fig. 5. Applications and cure conditions of the mini wallette: (a) application of the mortars on the experimental mini wallettes; (b) laboratory water sprinkling; (c) exposition to real environmental conditions of the Fort external area.

2.2 Methods

To monitor the physical-mechanical changes in the mortars owing to the hardening process, the mortar samples were analysed in laboratory after 28 and 90 days.

Flexural (Rt) and compressive (Rc) strengths were measured according to the EN 1015-11 standard [24] (Fig. 6a) and the dynamic elastic modulus (E) was determined by the resonance frequency method, following the EN 14146 standard [25] (Fig. 6b). The water absorption by capillary action test was performed based on the EN 15801 standard [26], where the prismatic specimens (40 × 40 × 160 mm^3) were placed in a container on non-absorbent small bars, with one of the 40 mm × 40 mm faces in contact with tap water (approximately 10 mm deep) (Fig. 6c). The drying process of the specimens based on EN 16322 [27], was conducted immediately after the absorption test (with the specimens completely saturated with water), by measuring the weight loss over time, until the samples reached equilibrium with the environment conditions (Fig. 6d). The porosity of a mortar is of great importance as it has a significant effect on the performance of the materials in relation to water, salt weathering and chemical weathering and, therefore, on the mortar's durability. Thus, the porosity changes were evaluated through the total open porosity (Po) by immersion and hydrostatic weighing, based on EN 1936 [23] (Fig. 6e).

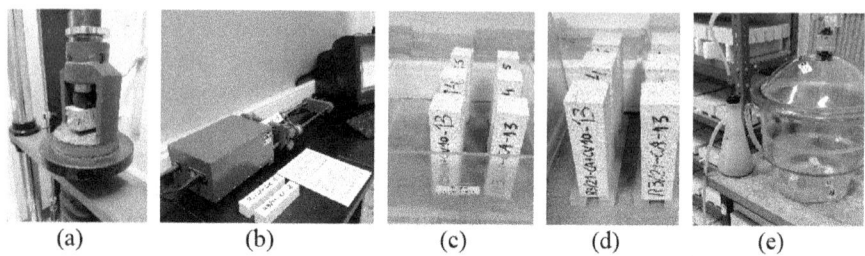

Fig. 6. Laboratory performed tests: (a) mechanical strengths; (b) elastic modulus; (c) water absorption by capillary; (d) water drying; (e) open porosity.

The tests performed on the bricks (common ceramic and traditional porous bricks) and on the mini wallettes to evaluate the behaviour and characteristics of the applied renders, were analysed after 28 and 90 days using the following techniques:

- Shore Hardness (with a Shore A Durometer): The test consists of the evaluation of the surface hardness, and indirectly of the cohesion, of an applied render using a Shore A durometer with a point which is pressed against the surface by the action of a spring with a standardized force (Fig. 7a). The penetration of the point gives a measure of the surface hardness. The test is based on the ISO 7619-1 [28] and ASTM D2240 [29]. This evaluation has been found in previous studies to have a good correlation with the surface cohesion [30].
- In order to evaluate the deterioration and the presence of anomalies (cracks, discontinuities and loss of cohesion) in the specimens, ultrasonic pulse velocity testing was used, based on EN 12504-4 [31]. This technique is based on measuring the speed of propagation of longitudinal ultrasonic waves (P-waves), through the specimens, using, in this case study, the transducers positioned on the same surface (indirect transmission), with a minimum distance of 20 mm between them and successively increasing the distance by displacing one of the transducers (Fig. 7b). The ultrasonic pulse velocity is determined as the slope of the distance/time line obtained with the measurements.

Additionally, after 90 days were also evaluated the compressive strength and the porosity through the total open porosity by immersion and hydrostatic weighing of the mortars applied on the bricks and on the mini wallettes.

3   Results and Discussion

3.1   Physical and Mechanical Behaviour

The results of the laboratory tests on the prismatic specimens are summarized in Table 3 and Fig. 8.

According to the results in Table 3, all mortars compositions show, in general, an improvement of the characteristics from 28 to 90 days, attributed to the progression of the carbonation process.

(a)   (b)   (c)

Fig. 7. Tests performed on the mortars applied on the bricks and on the mini wallettes: (a) Shore Hardness; (b) ultrasonic test; (c) compressive strength.

Table 3. Physical and mechanical properties obtained in laboratory on prismatic samples.

Characteristics		Mortar		
		AL	AL+Q30%	C+AL
Bulk density (kg/m^3)	28 days	1679 ± 2.8	1715 ± 4.6	1746 ± 6.9
	90 days	1704 ± 5.4	1742 ± 5.1	1766 ± 3.4
Modulus of elasticity (MPa)	28 days	1716 ± 30	2043 ± 52	2444 ± 94
	90 days	2367 ± 43	2908 ± 68	2853 ± 109
Flexural strength (MPa)	28 days	0.41 ± 0.04	0.36 ± 0.09	0.34 ± 0.01
	90 days	0.45 ± 0.01	0.54 ± 0.01	0.41 ± 0.03
Compressive strength (MPa)	28 days	0.32 ± 0.04	0.31 ± 0.03	0.69 ± 0.03
	90 days	0.62 ± 0.01	0.65 ± 0.05	0.85 ± 0.04
A Shore Hardness[a]	28 days	83 ± 3	80 ± 5	88 ± 3
	90 days	82 ± 6	88 ± 3	88 ± 2
Open Porosity (%)	28 days	28 ± 0.1	27 ± 0.3	28 ± 0.2
	90 days	28 ± 0.2	27 ± 0.3	28 ± 0.0
Capillarity coefficient (kg/m^2 min$^{0.5}$)	28 days	3.3	2.5	3.7
	90 days	3.0	2.3	3.6

[a] Unit measurement scale SHORE A from 0 to 100: 70 to 87: normal; >88: very stiff.

The bulk density (Table 3) is an easy to determine property able to give indirect comparative information about a range of other physical and mechanical characteristics. The results show that the bulk density of the **C+AL** mortar, as expected, present the highest value of the bulk density. However, despite the higher bulk density of **C+AL** mortar, the **AL+Q30%** mortar leads to lower open porosity values (Table 3), which can be related to the lower water content of this mix (Table 1) but also to adding quicklime to the mix, that could absorb the excess of mix water, thus making it less porous. The open porosity values between **AL** and **C+AL** mortars are similar.

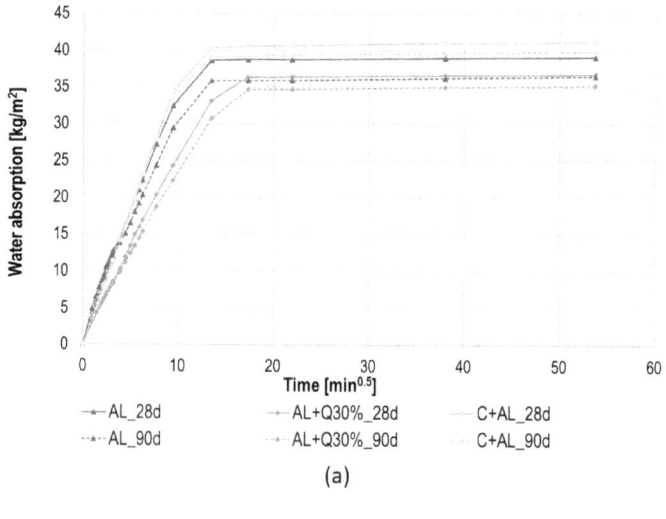

(a)

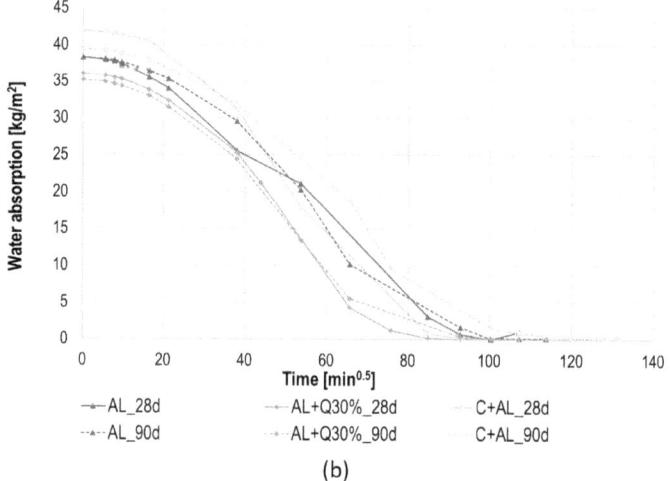

(b)

Fig. 8. Water absorption curves of the prismatic specimens: (a) capillary water absorption; (b) water desorption.

In what concerns the mechanical performances of the prismatic specimens (Table 3) it is observed that **AL** mortar is the most deformable, however with compressive strength values of the same order of magnitude as **AL+Q30%** mortar. Moreover, the highest mechanical behaviour was obtained with the **C+AL** mortar, however, at 90 days, the **AL+Q30%** mortar shows a significant evolution, comparing to **C+AL** mortar. This good behaviour of the **AL+Q30%** mortar may confirm the literature [13] argument that adding quicklime to slaked lime reduces the drying shrinkage and consequently the microcracking during the drying process, thus increasing the mechanical strength.

Concerning water behaviour (Fig. 8), it is possible to observe that all mortars compositions have high rate of water absorption considering the results obtained in the original mortars of the Fort (capillary water absorption coefficient between 0.35 and 1.70 kg/m^2.min$^{1/2}$ [7]) which can make them more vulnerable to degradation. Nevertheless, as expected, they tend to decrease with time (from 28 days to 90 days), due to the progression of carbonation and reactions of dissolution and re-carbonation of the matrix [2].

Despite the similar open porosity values (Table 3), the **C+AL** mortar showed the highest capillary water absorption coefficient (Table 3) and induces higher values of water absorption (Fig. 8a). Furthermore, the drying also occurs slower Fig. 8b) than the other mortars. The slower drying rate may be due to the reduced water vapour permeability, related with the hydraulic compounds, which slows down the evaporation of the water.

On the other hand, the **AL+Q30%** mortar showed a better behaviour from this point of view, with lower capillary water absorption coefficient and water content, both at 28 and at 90 days. Furthermore, the drying also occurs faster, which may be related to a pore structure of this mortar with low diameter capillary pores.

3.2   The Influence of Cure Conditions and Substrate Characteristics

The test results of the mortars applied on the bricks and on the mini wallettes (performed at laboratory and *in situ*, respectively) are synthesised in Table 4.

Table 4. Results of tests performed on the applications on a substrate (laboratory and *in situ*).

Characteristics		AL			AL+Q30%			C+AL		
		Laboratory		Fort	Laboratory		Fort	Laboratory		Fort
		C	T	T	C	T	T	C	T	T
A Shore Hardness[a]	28 days	88 ± 4	87 ± 6	86 ± 4	90 ± 2	88 ± 6	90 ± 5	87 ± 3	92 ± 3	86 ± 5
	90 days	86 ± 5	85 ± 5	89 ± 6	90 ± 3	88 ± 5	nd	88 ± 3	90 ± 4	88 ± 5
Modulus of elasticity - US (MPa)	28 days	1755	1359	1435	1917	1223	1124	2874	2511	2391
	90 days	1961	1570	2262	2216	1153	nd	2908	3688	2807
Compressive strength (MPa)	90 days	1.25	0.76	2.61	1.71	5.76	2.86	1.61	3.87	nd
Open Porosity (%)	90 days	28	27	26	28	27	28	27	26	27

[a] Unit measurement scale SHORE A from 0 to 100: 70 to 87: normal; >88: very stiff; **C** – Common ceramic brick (Cc = 0.6 kg/m^2 min$^{0.5}$); **T** - Traditional porous brick (the same bricks used in the mini wallettes at the Fort exposition - Cc = 2.4 kg/m^2 min$^{0.5}$); nd – not determined.

Considering the applications on different substrates at laboratory (Table 4), all mortars increase mechanical strengths when applied on more porous substrates (compared with prismatic samples - Table 3). However, the **AL** mortar decreases its strength from common to traditional brick, with very high rate of water absorption (Fig. 9). Moreover, the use of the most porous substrate (traditional porous bricks) also decreases the surface hardness and increases the deformability on the air lime mortars (**AL** and **AL+Q30%**)

(Table 4) compared with applications on the common brick. The **C+AL** mortar shows an opposite behaviour, with better performances when applied on the most porous and absorbent substrate.

The decrease of the air lime mortars' performance may be due to the quick reduction of the mix water available for the slaking and carbonation processes, which is absorbed at large scale by the porous substrate, accelerating the drying of the mortars and hindering the hardening lime reactions. This faster drying rate also generates high stress in the matrix and, as consequence, microcraking occurs (Fig. 10).

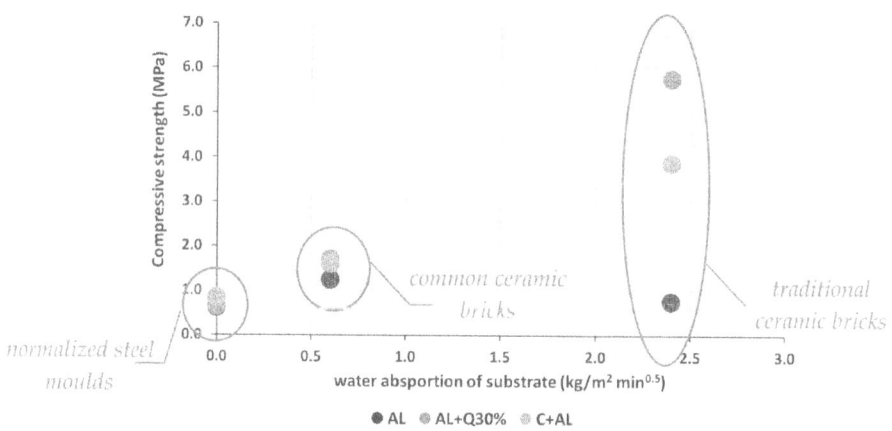

Fig. 9. Compressive strength vs water absorption of the substrates, at 90 days.

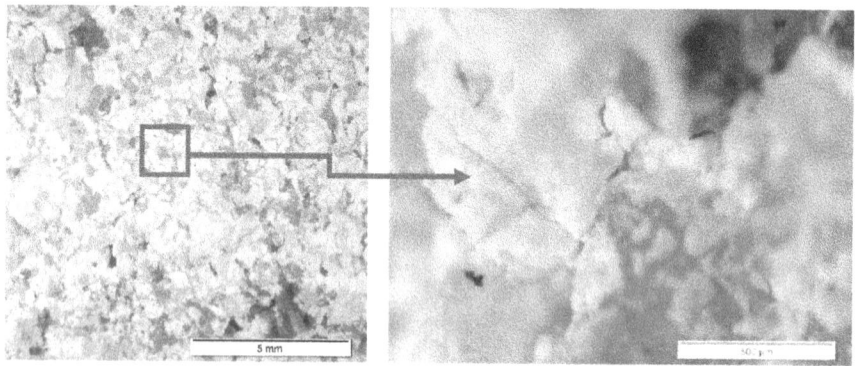

Fig. 10. General microcracking observed on matrix at **AL+Q30%** mortar.

In the characterization performed on the mini wallettes *in situ* (Table 4), only the **AL** mortar showed an improvement of its performance when exposed to real environment conditions of the Fort, when compared to the applications subjected to the laboratory conditions, which was attributed to a possible blockage of the original pores due to the presence of salts crystals inside them [2], since the pore structure of the **AL** mortars, with a large volume of macropores [32], and its deformable structure could accommodate the precipitation of salts inside the large pores without producing sufficient stress to damage the pore structure; consequently decreasing the porosity and enhancing their performance [2].

On the other hand, contrasting with the good behaviour presented by prismatic samples, the Fort exposition leaded to a reduction of the **AL+Q30%** mortar characteristics. The applied mortar was found detached from the substrate when measurements were to be made at 90 days (Fig. 11). This anomalous behaviour of the **AL+Q30%** mortar was attributed to its pore structure, with large volume of capillary porosity with low diameter, since there is a reduction of the active pores, due to the presence of salts crystals inside the pore structure, that reduce the strength bond between the mortars and the substrate and consequently reduce their performance when subjected to this very aggressive environment applied on very porous substrates.

Fig. 11. Lack of adherence of the AL+Q30% mortar exposed to real environmental conditions of the Fort external area with failure pattern mainly adhesive fracture.

4 Conclusions

Paimogo Fort is one of the several Portuguese coastal fortifications built in the 17th century to protect the country from military attacks. Located in the cliffs of Paimogo beach, near Lourinhã City, the Fort is classified of public interest since 1957.

In 2006, the original air lime renders and plasters of the Fort have been removed and new hydraulic lime mortars were applied, which present nowadays high degradation, especially erosion and loss of cohesion. Thus, new repair mortars are being developed for new applications in this severe marine environment.

The work carried out showed that the three mortars compositions analysed in this study (slaked air lime - **AL**, slaked air lime with quicklime – **AL+Q30%** and slaked air lime with white cement – **C+AL**) at 28 and 90 days showed, in general, good appearance, without cracking in laboratory conditions. In general, prismatic samples of the **AL+Q30%** mortar show the highest strengths and lowest water absorption rate, namely at 90 days. However, this mortar composition when applied on substrates with very high rate of absorption and in the real environmental conditions of the Fort decreases its performance: This mortar when applied on higher absorption bricks, by comparison with common bricks, lead to a significant reduction of the mixing water and, consequently, to a reduction of the volume and diameter of large capillary pores, that reduce the flow of water from the fresh mortars to the interface region of the substrate; this reduction weakens the effective bond, since the number of active pores, which allow the transport of fine particles that strengthen the effective bond, is reduced. Moreover, in the real environmental conditions, the salts crystals fill the pore structure of this mortar, making them more compact, however it produces stress in small pores which damage the pore structure and consequently reduces its performance, as well as reduces, even more, the few remaining active pores which weaken the mortar's adhesion.

Like **AL+Q30%**, also **C+AL** mortar present good behaviour in laboratory, however, in *in situ* conditions, a decrease of its performance is observed. On the other hand, mortars composed by air lime and low content of cement showed an increase of characteristics when applied on higher absorption bricks by comparison with common bricks, contrary to what was observed in the lime mortars. These facts may be due to the hydraulic reaction that occurs before the water is absorbed by the porous substrate.

On the other hand, in general, the slaked air lime mortar (**AL**), with coarser pores, shows some potential of improvement in the real environmental conditions of the Fort with salty water spray, high humidity, intense sun radiation, thermal shock and strong wind. The improvement of the **AL** mortar mechanical characteristics in this cure conditions can be attributed to the reactions of dissolution and re-carbonation of the matrix that partially fulfil pores. Another strengthening factor may be the possible blockage of the macropores (due to the presence of salts crystals inside the pore structure) without causing sufficient stress to produce damage, due to the large pores and to the low modulus of elasticity of the mortars. However, their low mechanical characteristics and high rate of water absorption can compromise the durability of these renders and of the masonry.

A new phase of tests is now being carried out, to enhance the mortar formulations based on air lime with a fine-tuning of the aggregate (using a different grain size distribution and substituting part of the siliceous sand by limestone aggregate from the region) in order to improve their performance and durability, taking into account the raw materials and the compatibility requirements for this case study.

Acknowledgements. The authors wish to thank LNEC technicians, Bento Sabala and Ana Maria Duarte, for their collaboration in carrying out the tests, the collaboration of *Lhoist Portugal* and *Secil Argamassas* in supplying the testing materials, as well as the support and collaboration of Lourinhã Municipality. This work is a contribution to the EEA Grants Culture Programme 2014–2021 "Coastal Memory Fort – Paimogo Fort".

References

1. Veiga, M.R., Santos Silva, A., Tavares, M., Santos, A.R., Lampreia, N.: Characterization of renders and plasters from a 16th century Portuguese military structure. Restorat. Build. Monuments **19**(4), 223–237 (2013)

2. Borges, C., Santos Silva, A., Veiga, R.: Durability of ancient lime mortars in humid environment. Constr. Build. Mater. **66**, 606–620 (2014)
3. Elsen, J.: Microscopy of ancient mortars: a review. Cem. Concr. Res. **36**, 1416–1424 (2006)
4. Santos Silva, A., et al.: Mineralogical and chemical characterization of historical mortars from military fortifications in Lisbon harbour (Portugal). Environ. Earth Sci. **63**, 1641–1650 (2011)
5. Santos Silva, A., et al.: Characterization of historical mortars from Alentejo's religious buildings. Int. J. Archit. Herit. **4**, 1–16 (2010)
6. Santos Silva, A., Santos, A.R., Veiga, R., Llera, F.: Characterization of mortars from the Fort of Nossa Senhora da Graça, Elvas (Portugal) to support the conservation of the monument. In: Proceedings of the 4th Historic Mortars Conference - HMC2016, Santorini, Greece, 10th–12th October 2016, pp. 42–49 (2016)
7. Veiga, M.R., Santos Silva, A., Marques, A.I., Sabala, B.: Characterization of existing mortars of the Fort of Nossa Senhora dos Anjos do Paimogo (Caracterização das argamassas existentes no Forte de Nossa Senhora dos Anjos do Paimogo). Report LNEC 67/2022 – DED/NRI, Lisboa: LNEC, March 2022, 114 p. (2022). (in Portuguese)
8. Traditional mortars of the Fort of Paimogo: composition, performance, replication. https://www.eeagrants.gov.pt. Accessed March 2022
9. Veiga, M.R., Fragata, A., Velosa, A., Magalhães, A.C, Margalha, G.: Substitution mortars for application in historical buildings exposed to sea environment. Analysis of the viability of several types of compositions. In: Proceeding of MEDACHS 08 - Construction Heritage in Coastal and Marine Environments: Damage, Diagnostic, Maintenance and Rehabilitation, Portugal, Lisbon, pp. 1–9 (2008)
10. Groot, C., Gunneweg, J.: Choosing mortars compositions for repoint mortars of historic masonry under severe environmental conditions. In: Proceeding of 3rd Historic Mortar Conference. Scotland, Glasgow (2013)
11. Veiga, M.R., Aguiar, J.: Rehabilitation of S. Bruno Fort. Notes on the results of the collaboration provided by LNEC to CMO. (Reabilitação do Forte de S. Bruno. Notas sobre os resultados da colaboração prestada pelo LNEC à CMO). Proceeding of IV Encontro Nacional de Municípios com Centro Histórico, Oeiras, november 1996 (in Portuguese) (1996)
12. Veiga, M.R.: Conservation of historic renders and plasters: from laboratory to site. In: Válek, J., Hughes, J., Groot, C. (eds.) Historic Mortars. RILEM Bookseries, vol. 7, pp. 207–225. Springer, Dordrecht (2012). https://doi.org/10.1007/978-94-007-4635-0_16. ISBN 978-94-007-4635-0
13. Loriot, A.-J.: Mémoire sur une découverte importante dans l'art de bâtir faite par le Sr. Loriot, mécanicien, dans lequel l'on rend publique, par ordre de Sa Majesté, la méthode de composer un ciment ou mortier propre à une infinité d'ouvrages, tant pour la construction que pour la décoration, p. 1774. L'Imprimierie Michel Lambert, Paris (1774)
14. Magalhães, A.C., Muñoz, R., Oliveira, M.M.: O uso da mistura de cal viva e cal extinta nas argamassas antigas: o método Loriot. In: Proceedings of the HCLB - I Congresso Internacional de História da Construção Luso-brasileiro, Espírito Santo, Brasil, 04th–06th September 2013 (2013) (in Portuguese)
15. Veiga, M.R.: Air lime mortars: what else do we need to know to apply them in conservation and rehabilitation interventions? A review. Constr. Build. Mater. **157**(30), 132–140 (2017)
16. Santos, A.R., Veiga, M.R., Santos Silva, A., de Brito, J.: As argamassas de revestimento como elementos de reabilitação: a influência da microestrutura. In: Menezes, M., et al. (eds.) Atas do 4º Encontro de Conservação e Reabilitação de Edifícios - ENCORE 2020. Lisboa, Portugal, 3–6th November 2020, pp. 503–514 (2020). (in Portuguese)
17. Santos, A.R., Veiga, M.R., Matias, L., Santos Silva, A., de Brito, J.: Durability and compatibility of lime-based mortars: the effect of aggregates. Infrastructures **3**(3), 34 (2018)

18. Arizzi, A., Viles, H., Cultrone, G.: Experimental testing of the durability of lime-based mortars used for rendering historic buildings. Constr. Build. Mater. **28**(1), 807–818 (2012)

19. European Committee for Standardization (CEN) - EN 1015-3, Methods of test for mortar for masonry; Part 3: Determination of consistence of fresh mortar (by flow table). Brussels, CEN (1999)

20. European Committee for Standardization (CEN) - EN 1015-2, Methods of test for mortar for masonry; Part 2: Bulk sampling of mortars and preparation of test mortars. Brussels, CEN (1998)-

21. RILEM – Water absorption under low pressure. Pipe method. Test n° II.4

22. European Committee for Standardization (CEN) - EN 772-11, Methods of test for masonry units; Part 11: Determination of water absorption of aggregate concrete, autoclaved aerated concrete, manufactured stone and natural stone masonry units due to capillary action and the initial rate of water absorption of clay masonry units. Brussels, CEN (2011)

23. European Committee for Standardization (CEN) - EN 1936, Natural stone test methods; Determination of real density and apparent density, and of total and open porosity. Brussels, CEN (2006)

24. European Committee for Standardization (CEN) - EN 1015-11, Methods of test for mortar for masonry; Part 11: Determination of flexural and compressive strength of hardened mortar. Brussels, CEN (1999)

25. European Committee for Standardization (CEN) - EN 14146, Natural stone test methods; Determination of the dynamic modulus of elasticity (by measuring the fundamental resonance frequency). Brussels, CEN (2004)

26. European Committee for Standardization (CEN) - EN 15801, Conservation of cultural property; Test methods: Determination of water absorption by capillarity. Brussels, CEN (2009)

27. European Committee for Standardization (CEN) - EN 16322, Conservation of cultural heritage; Test methods: Determination of drying properties. Brussels, CEN (2013)

28. International Organization for Standardization (ISO) - ISO 7619-1, Rubber, vulcanized or thermoplastic — Determination of indentation hardness—Part 1: Durometer method (Shore hardness). Geneva, ISO (2010)

29. American Society for Testing Materials (ASTM) - ASTM D2240-05, Standard Test Method for Rubber Property-Durometer Hardness. West Conshohocken, ASTM International (2010)

30. van Hees, R., Veiga, R., Slížková, Z.: Consolidation of renders and plasters. Mater. Struct. **50**(1), 1–16 (2016). https://doi.org/10.1617/s11527-016-0894-5

31. European Committee for Standardization (CEN) - EN 12504-4, Testing concrete in structures; Part 4: Determination of ultrasonic pulse velocity. Brussels, CEN (2004)

32. Santos, A.R., Veiga, M.R., Santos Silva, A., de Brito, J., Alvarez, J.I.: Evolution of the microstructure of lime-based mortars and influence on the mechanical behaviour: the role of the aggregates. Constr. Build. Mater. **187**, 907–922 (2018)

Practical Test for Pozzolanic Properties by A. D. Cowper: Implementation and Innovation

Marlene Sámano Chong[1]([✉]), Alberto Muciño Vélez[2], Ivonne Rosales Chávez[3], and Luis Fernando Guerrero Baca[4]

[1] National Autonomous University of Mexico and National Institute of Anthropology and History, Mexico City, Mexico
marlene_samano_c@encrym.edu.mx

[2] Materials and Structural Systems Laboratory, National Autonomous University of Mexico, Mexico City, Mexico

[3] Chemistry Faculty, National Autonomous University of Mexico, Mexico City, Mexico

[4] National Autonomous University of Mexico and Autonomous Metropolitan University, Mexico City, Mexico

Abstract. Various tests have been developed to evaluate the reactivity of some pozzolans to study the feasibility of using an accessible deposit, without the need of importing or transporting materials. Pozzolanicity tests mostly require sophisticated and expensive laboratories. For some conservation projects, this type of analysis is out of reach due to the complexity of analysis, their elevated costs, and the time it takes to carry them out. A. D. Cowper in 1927 describes a simple and *on field* test to determine if a material has pozzolanic activity, but it is not deeply explained why samples of materials with possible hydraulic action can be tested, after seven days in a qualitative indication of hydraulicity. In this research, pozzolanicity reaction and flocculation process are reviewed, the Cowper Test is implemented, and new simple procedures are added to corroborate the Cowper Test efficiency. The increase of volume due to flocculation, sedimentation velocity, observation in microscope, changes in pH and changes in refractive index are applied and discussed. This is a contribution to transit from a qualitative test into a semiquantitative, it is expected to be the fundament that can lead into a future *on field* quantitative method for determining pozzolanicity.

1 Introduction

1.1 The Air Lime Mortars with Pozzolanic Additions in the Conservation Field

In Mexican historic buildings restoration, an addition of cement (usually at least of 10%) in the air lime/aggregate mortar is a common recommendation [1] to modify the air lime mortar properties by inducing the setting in humid environments or in low presence of CO_2, reducing time or to increase compression strength resistance. This practice leads into a range of complications in the conservation field, because of the incompatibility between historic and repair mortars that inevitably causes damage to the original stone and/or lime mortar systems.

V. Bokan Bosiljkov et al. (Eds.): HMC 2022, RILEM Bookseries 42, pp. 523–541, 2023.
https://doi.org/10.1007/978-3-031-31472-8_42

Repair mortars should be a weaker material than masonry so that they can accommodate in differential movements and be a sacrificial material in weathering. Therefore, interventions made with cement are not weak or soft enough for historical materials, creating mechanical incompatibility [2] Cement mortars tend to have less permeability than lime mortars, which interferes with the evaporation and diffusion of moisture in historical materials. Cement mortar introduction results in humidity trapped inside the materials; normally the moisture outflow is given by the lime joints, however, with cement joints, evaporation occurs through the masonry pieces that weakens deeply due to the dissolution and recrystallization of soluble materials. The presence of soluble salts in Portland cement is another factor that significantly affects historical materials.

In the presence of moisture and in accordance with the point described above, the salts migrate to the masonry, recrystallizing and damaging the structure of the materials promoting disintegration, weakening and loss of the material. The rigidity of cement mortars inhibits the natural movements of historical structures. The historical material fails and cracks before such movements, due to hardness and stiffness differences, promoting greater deterioration [2].

To confer hydraulic properties to air lime mortars, pozzolans are added to develop *formulated aerial lime mortars*; term established by the *European Standards of Building Lime* [3] referring to lime with hydraulic properties mainly consisting of air lime and/or natural hydraulic lime with added pozzolans.

The property that made pozzolan precious in the classical age and still quite useful today is the ability of the vitreous material to react with lime and water [4] where the setting reaction does not require the presence of air, so it can occur in a humid environment, even under water, or in the centre of a thick wall. Such characteristics allow the use of a construction technique named by Vitruvius *opus caementicium* [5], something very similar to modern concrete technology. In the course of history and in various parts of the world, other volcanic materials were found to possess the same properties as the Italian pozzolans.

Pozzolans are aggregates for mortars [6] of siliceous or alumino-siliceous origin, which by themselves have little or no cementing value, but when they have been finely divided and in the presence of water, they react chemically with the hydroxide of calcium $Ca(OH)_2$ at room temperature to form compounds with cementitious properties.

In terms of physical and chemical properties, pozzolans confer to the air lime mortar the desired characteristics found by adding cement to them, like setting in humid environment or low presence of CO_2, reduced setting time and increased compression strength resistance. Beside the compatibility in material properties between historical and repair mortars, this practice also generates a sustainable practice by reducing environmental impact, promoting local consumption and safeguarding traditional constructive culture.

The challenge of developing formulated lime mortars in an *on field* conservation project is the lack of certainty that some material found regionally can be useful. Pozzolans, having different origins, have a variable chemical composition and their reactivity not only depends on this but also on their vitreous structure and their degree of fineness [7]. The degree to which a pozzolan reacts with lime is known as pozzolanicity. The pozzolanicity of a material can vary significantly, even among the same class of materials.

Various tests have been developed to evaluate the reactivity of some pozzolans to study the feasibility of using an accessible pozzolanic deposit. Many tests, specified in both the literature and international standards, require sophisticated and expensive laboratories. For the development of some projects, this type of analysis is out of reach due to the complexity of analysis, their elevated costs, and the time it takes to carry out the studies.

The *practical test for pozzolanic properties* proposed by A. D. Cowper in 1927 has been selected because of its simplicity, economy and efficiency, and because it has not been especially widespread, even though it presents results after seven days without the need of a specialized laboratory. This method analyses, in a qualitative way, samples of materials with possible hydraulic reaction [8].

However, in several references, the pozzolanicity test proposed by Cowper has been discarded because it releases only qualitative results and, even tough, it can be inferred that Cowper understood the chemical reaction and the physical changes in the materials, it is not explained deeply why the phenomenon occurs.

The objective of this research is the review, explanation, implementation and of the Cowper Test and the innovation of semiquantitative procedures to verify it, in a simple and *on field* perspective to complement the released information when determining if a material can be used as a pozzolanic addition due to its reactivity.

1.2 Pozzolanicity Reaction Tests

As it has been discussed previously, pozzolanicity depends not only on mineralogic composition but also in the molecular structure and the particle finesse. For developing pozzolanicity standardized tests, sophisticated and expensive equipment is required as well as time, because there are wide properties to look for, with special variables that unlock the pozzolanic process. In addition, natural pozzolans can vary in its properties even in the same deposit, so in a day-to-day basis, in a normal conservation project, determining pozzolanicity can be a very complex task. Tests are required to analyse the viability of a potential pozzolanic deposit and to provide quality control [9].

The *practical test for pozzolanic properties* designed by Cowper to determine if a material has pozzolanic activity can be carried out *in field* or in a laboratory with simple equipment in 7 days.

2 Methods

2.1 The Practical Test for Pozzolanic Properties

A.D. Cowper in his book "Lime and Lime Mortars" published in 1927 compiles a simple a qualitative test which he denominates *Practical Test for Pozzolanic Properties and* states the following [8]:

> ...*A simple practical test to determine whether a given sample of burnt clay or other material possesses pozzolanic properties, and (to some extent) to obtain a measure its relative activity as compared with other samples, is that utilized in connection with the investigation by the addition of pozzolanas to combine with*

free lime. This consists in adding a small quantity of the material, finely powered, to rather less its weight of slaked lime, in a tube or small narrow bottle, and covering it to a depth of a couple of inches with pure water; shaking up every 12 hours for a week; and then observing the appearance of the tube (shortly after shaking) and the bulk of the sediment, as compared with a fresh mixture or with another pozzolana similarly treated. The actual quantities recommended are 0.5 gm. powdered pozzolana, 0.3 gm. slaked lime, and 20 cc. distilled water, in medium sized test tubes.

The interaction between the solution of lime in water and the pozzolana will give rise to hydrated calcium alumino-silicates of complex or indefinite chemical structure; these hydrates will be formed in the colloidal condition, and are much more bulky than the lime, with the result that after seven days the tube will contain an increased volume of solid matter as measured by the height it extends up the tube, also the rate of settlement, owing to its flocculent nature, will be slower; the activity of the pozzolana can therefore be judged by inspecting the tubes at different times after shaking. In comparison, the volume of the solid matter in a similar tube immediately after mixing will be much smaller, and it will settle almost immediately.

There are several questions to be solved after reading what Cowper states, it explains mostly what happens but not why. Why do we need the pozzolans finely powdered? Why does the pozzolan reacts with lime? Why does it need shaking? Why does the hydrated calcium alumino- silicates are bulkier? What happens during the seven days? Why are flocs formed? Why does they settle on a slower rate?

It is also necessary to innovate the test by complementing from a qualitative observation to a semiquantitative measurement of volume increase, to verify the floc formation and to define the settlement speed. Transferring the test from qualitative to semiquantitative allows to manage the procedure, to eradicate the subjective appreciation, and it will allow to study patterns of results and to use measurable data to formulate facts. To understand why the Practical *Test for Pozzolanic Properties* performs it is necessary to review the chemical and physical aspects of the pozzolanicity reaction and relate it to the floc formation dynamic.

2.2 The Pozzolanicity Reaction

Pozzolans mainly consist of amorphous silica and probably also amorphous alumina Si-O and Al-O [4], the vitreous particles react with water and slaked lime - calcium hydroxide $Ca(OH)_2$ because the high alkalinity of the hydroxyl anions break the bonds of silicates and aluminates to form silicon hydroxide $[SiO(OH)_3]^-$ and aluminium hydroxide $[Al(OH)_4]^-$ which have a negative charge [10]. This hydroxides with negative charges, that repel between the two negatives, in contact with Ca^{2+} ions, of positive charge, react joining negative and positive particles to form a more stable form of hydrated silicates of C–S–H type, and C–A–H hydrates.

The hydrated calcium silicate C-S-H is a gel formed by a flocculation process. The small particles of the silicon and aluminium hydroxide have a negative surface charge, so they repel each other. This rejection prevents the particles from agglomerating, causing them to remain in suspension, as shown in Fig. 1a). Figures 1b present how Calcium Ca^{2+} ions can be adsorbed onto the particles and balance the charges. The introduction of this positive opposite charges allows the joining particles to form stable, suspended submicron flocs [11] by stabilizing the suspended particles, achieving a collision between them by shaking the suspension and then the growth of masses or flocs occurs. In Fig. 1 c, the ions can act as an agglutinating nucleus of the particles in suspension, coalescing the particles until they have a flocculated texture, the cations, in this case the hydroxides, attenuate the repulsions forces between particles allowing their union [12]. The approach of the particles creates the effective range of attractive Van Der Waals forces to lower the energy barrier to flocculation and loose floc formation. Aggregation, binding and strengthening of the flocs proceed until visibly suspended macro flocs are formed [11] as shown in Fig. 1d.

It is known that the silicate compounds tend to dissolve more rapidly as in comparison with the aluminate compounds (which need more calcium ions, too) therefore C-S-H gels precipitate firstly [10].

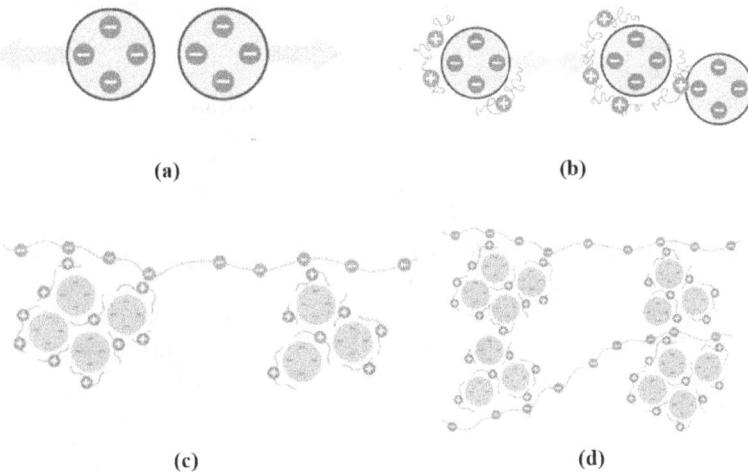

(a)   (b)

(c)   (d)

Fig. 1. C-S-H formation is a two-step particle aggregation process in which many small particles form a small number of large flocs: coagulation and flocculation. Coagulation process involves small particles which have negative surface charges. (a); The introduction of opposite charges allows the joining particles to form stable, suspended submicron flocs (b); Flocculation process requires mixing to promote particle collisions and the formation of submicron flocs (c); The approach of the particles by attractive Van Der Waals forces lower loose floc formation. Aggregation, binding, and strengthening of the flocs proceed until macro flocs are visibly suspended in suspension (d) [11].

Not only the siliceous composition is important for the pozzolan effect, but also other data must be considered, such as a high specific surface and an amorphous vitreous state:

The physical surface and finesse degree of the particles influences the speed reaction [7]. And it is necessary to notice that the presence of hydrated layers of water on the surface of interacting particles could lower the Van der Waals interaction energies, even by a factor from 5 to 50, depending on the thickness of the hydrated layers and the radius of the particle [13]. To comply with the specification "ASTM C432-59T", pozzolans must have finesse: retained in 30 mesh (ASTM) 2% max. (590μm mesh) and retained in 200 mesh (ASTM) 10% max. (74 μm mesh) [14].

About the amorphous vitreous state, the "ASTM c618" standard specifies that N natural pozzolans are derived from volcanic minerals and rocks, except for diatomaceous earth [6]. During volcanic eruptions and with rapid cooling of the magma, glassy phases with a disordered structure composed mainly of aluminosilicates are formed as an amorphous vitreous state. Due to the presence of gases, the solidified matter frequently acquires a porous texture and the aluminosilicates, with a disordered structure, will not remain stable when exposed to a lime solution [15].

The addition of pozzolans that have the mineralogical compounds, the vitreous state and the finesse degree can form C-S-H (hydrated calcium silicate) gel that promotes a setting process in air lime mortars in humid environments or low presence of CO_2. This performance is known as hydraulicity.

3 Methods to Support the Cowper Practical Test for Pozzolanic Properties

3.1 Refractive Index to Study the Degree of Flocculation

As stated before, the pozzolans combined with lime water tend to form flocs whose *particle size distribution* (PSD) can be used for its characterization. To determine particle size, various techniques can be used, from surface analysis techniques such as optical microscopy of transmitted or scattered light and scanning electronic microscopy in which the study of these deposited on a surface is carried out, obtaining two-dimensional images, or by means of atomic force microscopy with images that provide information not only on their morphology but also on their roughness, since it performs a three-dimensional analysis of the sample. Another of the techniques used is dynamic light scattering (DLS), in which the dynamics of particles immersed in a liquid are studied by means of the interaction of the solution with a monochromatic point light source such as a laser.

An alternative methodology proposed by this working group also consists of using a monochromatic light source, which, when passing from one medium to another, changes its speed and direction between two mediums of different density. This difference between the speed of light in a vacuum (c) and when it enters a medium (v) is known as the refractive index. Equation (1) defines refractive index.

$$n = \frac{c}{v} \tag{1}$$

The change of the path of light according to geometrical optics can be modelled as straight-path rays that change direction when passing from one medium to another [16]. This description of the path of light follows two laws: the law of reflection and the law of refraction.

The law of reflection implies the reflection of light when interacting with matter and states. Equation (2) defines where θi is the angle of incidence and θr is the angle of reflection, measured with respect to an axis normal to the surface from the point of incidence.

$$\theta_i = \theta_r \tag{2}$$

While the refraction of light is described by Snell's law. As shown in Eq. 3 which defines where ni and nR correspond the incidence and refractive index of media 1 and 2, respectively, and θi and θR are the angles between the media just at the point where they change direction [17].

$$n_i \sin \theta_i = n_R \sin \theta_R \tag{3}$$

With the above information as a preamble, the following experimental design, shown in Fig. 2, is established based on the work carried out by Ronald Newburgh and collaborators [18].

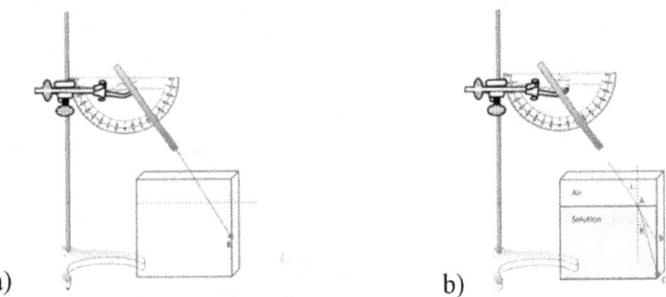

a)   b)

Fig. 2. Scheme of the system to study the degree of flocculation of mortars, in this the beam of light from the laser hits point A and hits point B when the container is empty a) once the solution is placed this beam hits point C b).

According to a trigonometric analysis, the ratio of heights of the incident light beam with and without solution nR = ni (h/h′) is determined, providing the refractive index of the solution, which will be related to the concentration of flocs in the solutions.

3.2 Digital Microscopy Observation to Confirm Floc Formation

A digital microscope is like a traditional optical microscope, only now it has a built-in docking and charging device in the camera that is used to view samples and specimens extensively through a monitor or computer screen [19].

It is necessary to recognize the limitations and the objective of this test. It is not intended to study this compound in greater depth through the application of new techniques, developing different structural models that try to explain the structure of hydrated calcium silicate, such as the electron microprobe (EMP), scanning electron microscopy (SEM) and transmission (TEM), infrared (FTIR) and Raman spectroscopy, gel permeation chromatography (GPC) and nuclear magnetic resonance (NMR) [20].

This optical microscope observation aims to corroborate how the pozzolana samples changes in contact with calcium hydroxide by confirming the formation of flocs by comparing samples at 7 days and freshly prepared samples. Optical microscopy allows to observe the shapes and textures of the samples. Thus, this paper presents the applicability of conventional optical microscopy in the study of floc formation as evidence of c-s-h formation.

The observation will be carried out with the means available in field. A YOMYM, 2MP resolution and 8 LEDs with USB 2.0 port digital portable optical microscope with a 1000X magnification, shown in Fig. 3a, will be used. The calibration will be made with the *Microscope Micrometre Calibration Ruler* in the section the measures the *round-shaped particle diameter measuring tape* in *um* as unit, as seen in Fig. 3b.

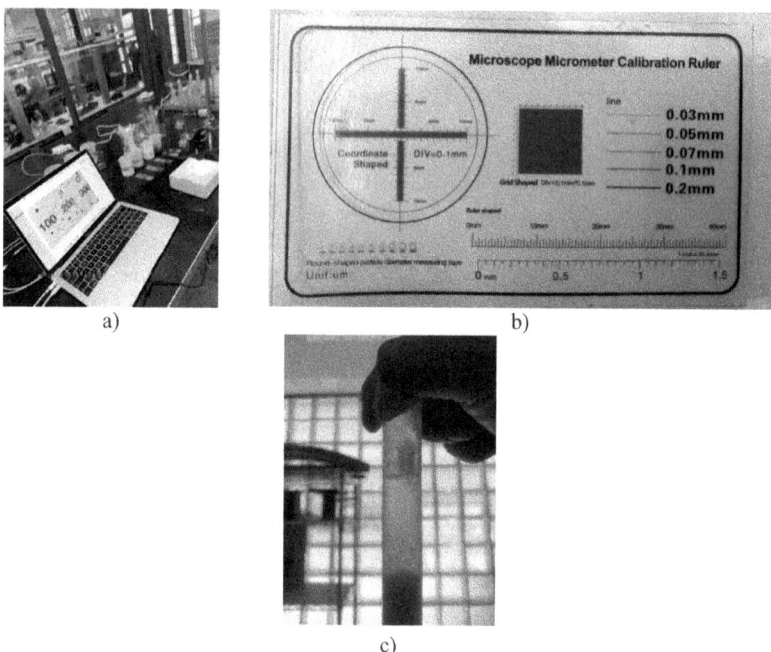

a)   b)

c)

Fig. 3. Observation with portable optical microscope (a); calibration with *Microscope Micrometre Calibration Ruler* (b); extraction of the floating products after the agitation

The tests will be done with a freshly made mixture and with another one of 7 days later to compare. The observation will be made with the extraction of the floating products after the agitation as seen in Fig. 3c.

3.3   pH Variation Qualitative Test

Boffey and Hirst in "The use of pozzolans in lime mortars" establish the *pH variation qualitative test* which states that reactive silica (in a pozzolan) will combine with lime in a saturated limewater solution and reduce the alkalinity of the solution. By testing the pH of the solution at regular intervals, the rate of pH reduction and thus the rate of pozzolanic activity can be monitored [21].

An Oakton® Waterproof PD 450 pH/DO Portable Meters will be used for this test, as shown in Fig. 4. It has a ±0.01 pH accuracy (Fig. 4) [22]. The measures will be taken in a fresh produced mix, at 7 and 14 days.

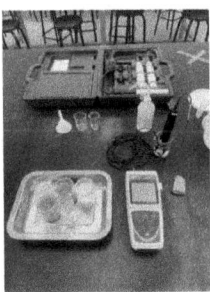

Fig. 4. Oakton® Waterproof PD 450 pH/DO Portable Meters form measuring pH of samples.

3.4   Sedimentation Velocity

Sedimentation velocity refers to the rate of deposition or descent at which a particle settles within a heterogeneous mixture. Mixtures of various phases can be mechanically separated by the effect of gravitational force (sedimentation); the requirement for this separation to occur is that the different components have different densities and that they do not dissolve in each other [23].

Starting from a simplified assumption where there is a laminar flow, the sedimentation rate could be determined with Stokes' Law based on the balance of forces, where information such as diameter of the particle, density differences between solid and liquid and dynamic viscosity of the mixture are needed [23].

The focus of this research group is to provide tools to confirm Cowper proposal with *on field* semiquantitative alternatives, so we need an alternative to use Strokes' Law. As Cowper establishes the qualitative observation that the mixture sediments more slowly after 7 days, this research group proposes to implement a semiquantitative measure of this test by determining the velocity in which the particles settle after agitation. The information will be by observation and with the measures of time and distance as shown in Eq. 4 which defines velocity.

$$\tilde{v} = \frac{\Delta x}{\Delta t} = \frac{displacement}{time\ change} \tag{4}$$

Sedimented samples will be needed and the measures of water level and sediment (Fig. 5a and b), samples will be agitated at equal intervals of time (Fig. 5c), the time needed for sedimentation since agitation stops (Fig. 5d), until the mixture reaches the initial point before agitation is monitored (Fig. 5e).

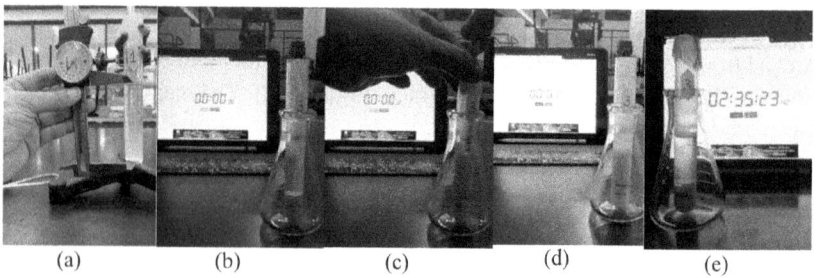

(a)   (b)   (c)   (d)   (e)

Fig. 5. Measures of water level and sediment (Figure a and b); samples will be agitated at equal intervals of time (figure c); time needed for sedimentation since agitation stops (Figure d); until the mixture reaches the initial point before agitation (Figure e).

Measurements will be performed with a *Vernier Mitutoyo dial calliper* to measure the length, and a chronometer to measure time. The test will be performed in fresh day 1 and at day 7 to verify the change of the settlement speed.

3.5 Statistical Treatment

The data obtained from laboratory analyses will be processed using the methods for obtaining the *Combined Uncertainty Uc*, defined in Eq. 5, which is an estimate of the standard deviation equal to the root of the total variance obtained by combining all the uncertainty components [24].

$$uc = \sqrt{u_A^2 + u_B^2} \tag{5}$$

In this equation u_A is the standard deviation by the root of the number of samples, as shown in Eq. 6 and U_B^2 as Measurement resolution.

$$u_A = \frac{s}{\sqrt{n}} \tag{6}$$

Equation   7   defines   standard   resolution   formula   where x_i is Data addition and $\bar{x}$ is data average.

$$S = \sqrt{\frac{\Sigma(x_i - \bar{x})^2}{N - 2}} \tag{7}$$

Results will be expressed as in the Eq. 8, data average ± combined uncertainty.

$$\bar{x} \pm uc \tag{8}$$

The tests will be executed 5 times.

4 Set Up of the Experiments

Natural pozzolans have mineralogical variants with diverse reactive constituents, it is possible, in a very generic way, to group them into volcanic glasses, volcanic tuffs and diatomaceous earths [15].

A diatomaceous earth consists of hydrated opaline or amorphous silicas that originate from the skeleton of small water plants, where their internal walls have a silica film [15].

To perform the *Practical Test for Pozzolanic Properties,* four different materials will be analysed: Limestone, and natural pozzolans as pumice, well known in Mexico as Tepojal, diatomaceous earth and volcanic slag, well known in Mexico as Tezontle. The limestone will be used as a control material that produces no reaction, because it is not a pozzolan. It is expected that the limestone sample won't have changes so it will be possible to compare what happens when there is no c-s-h formation (Table 1).

Table 1. Materials to be tested in the *Practical Test for Pozzolanic Properties*

Identification	Name	Characteristics
1	Limestone (Control)	It is a yellowish-white sedimentary rock mainly composed of Calcium Carbonate ($CaCO_3$), although it may contain contaminants such as Iron and Magnesium [26]
2	Tepojal. Pumice	Stone of volcanic origin, igneous extrusive and pyroclastic. Greyish white or yellowish in colour. Composed mainly of silica trioxide and aluminium trioxide [27]
3	Diatomaceous earth	Sedimentary rock formed by microfossils of algae characteristic of lacustrine areas, usually bright white although they may be coloured. Its chemical composition is mainly opaline or hydric silica and may contain small amounts of inorganic components such as alumina, iron, earth, and alkali metals [28]
4	Tezontle. Volcanic slag	Igneous volcanic rock, extrusive. Also known as volcanic slag or basalt lava. Porous texture and black or reddish-orange colour. It is rich in highly vesicles volcanic glass which gives it high porosity and low density [29]

The lime to be used in the test is paste *Lime Vitrubio NL 90* Reference L-240717–01.

4.1 Preparation

An amount of finely powdered pozzolan (0.5 g) will be added into slaked lime (0.3 g) in a test tube and it will be covered with 20 cc of distilled water. Over the course of a week, the tube is shaken every 12 h and its appearance after shaking is compared to that of a fresh mix just made [8] (Table 2).

Table 2. Materials, sieve, and measurement for preparing the samples

Pozzolana	Sieve	Pozzolana Weight	Slaked lime	Distilled water
1 Limestone (Control)	200	5 gr	3 gr	200 ml
2 Pumice	43% 200 57% 80	5 gr	3 gr	200 ml
3 Diatomaceous earth	200	5 gr	3 gr	200 ml
4 Tezontle	200	5 gr	3 gr	200 ml

5 Results

5.1 The Practical Test for Pozzolanic Properties

After seven days, the results about the volume increase are:

1. Limestone: Absence of volume increase (Fig. 6a, b)
2. Pumice: Absence of volume increase (Fig. 6c, d)
3. Diatomaceous earth: Volume increase (Fig. 6e, f)
4. Tezontle: Volume increase (Fig. 6g, h)

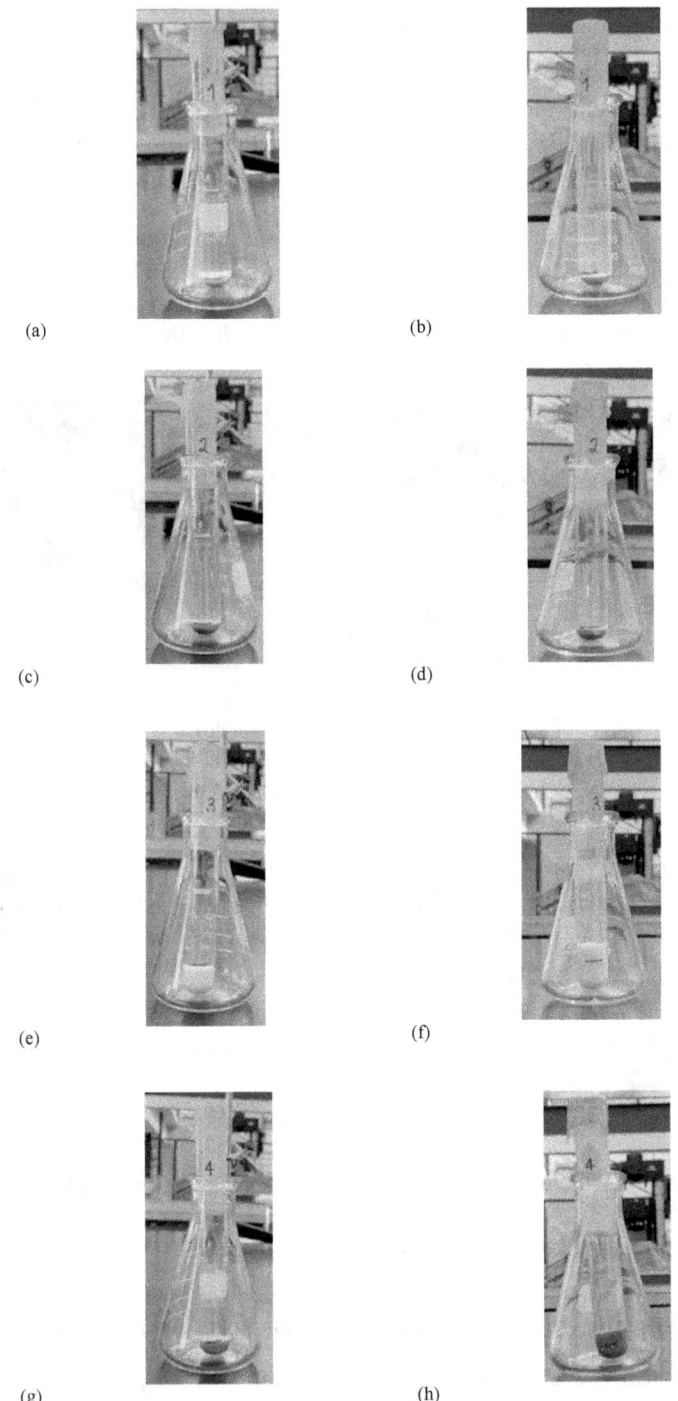

Fig. 6. Limestone day 1 (a); limestone day 7 Absence of volume increase (b). Tepojal day 1 (c); Tepojal day 7 Absence of volume increase (d). Diatomaceous earth day 1 (e); Diatomaceous earth day 7 Volume increase (f). Tezontle day 1 (g); Tezontle day 7 Volume increase (h)

5.2   Refractive Index to Study the Degree of Flocculation

The refractive index of vacuum, water, and agitated samples 1, 2, 3 and 4 were taken to compare between them (Fig. 7a). The lime-pozzolan-water mix was prepared in a 100 ml proportion and 7 days were allowed for flocculation to occur. It was stirred at equal intervals and left to settle for two minutes. The superior 50 ml of the suspension was used for this test.

Water refractive index will be the reference of how the suspension changes with the presence of flocs in the suspension. The measure of the refraction of the laser ray was taken in the inside part of the glass to avoid differences produced by the glass (Fig. 7b). A *Vernier Mitutoyo dial calliper* was used to measure the length differences between vacuum, water, and suspension 1, 2, 3, and 4 (Fig. 7c).

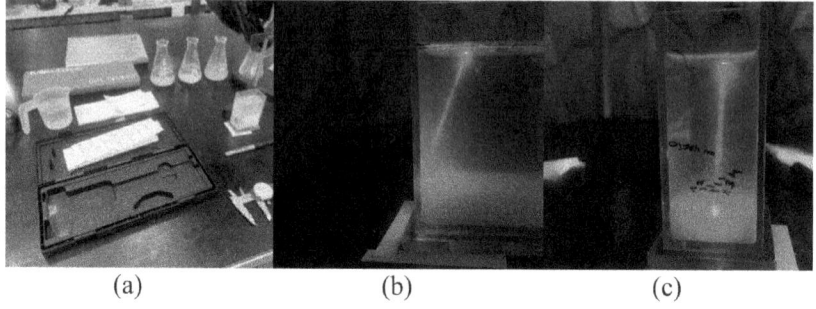

(a)   (b)   (c)

Fig. 7. The refractive index of vacuum, water, and agitated samples 1, 2, 3 and 4 were taken (Figure a). The measure of the refraction of the ray was taken in the inside part of the glass to avoid differences produced by the glass (Figure b). The length differences between vacuum, water, and suspension 1, 2, 3, and 4 were taken (Figure c).

The system was calibrated, at each round of measures, to position the ray at the vacuum height as a constant, which will be h and every suspension height will be h'. The data is reported with the average of five measures with combined uncertainty (Table 3).

Table 3. Refractive Index results

	Water	1 Limestone	2 Tepojal	3 Diatomaceous earth	4 Tezontle
Refractive Index	1.60 ± 0.05	1.56 ± 0.04	1.51 ± 0.04	1.29 ± 0.04	1.32 ± 0.04

The results show that there is a considerable variation in the refractive index between the suspensions. Where diatomaceous earth and tezontle show a difference compared to water, which confirms the presence of flocs.

5.3   Digital Microscopy Observation to Confirm Floc Formation

After seven days, the results about the microscope observation, it was possible to confirm that:

1. Limestone: Absence of flocs (Fig. 8a, b)

2. Pumice: Absence of flocs (Fig. 8c, d)
3. Diatomaceous earth: Present flocs (Fig. 8e, f)
4. Tezontle: Present flocs (Fig. 8g, h)

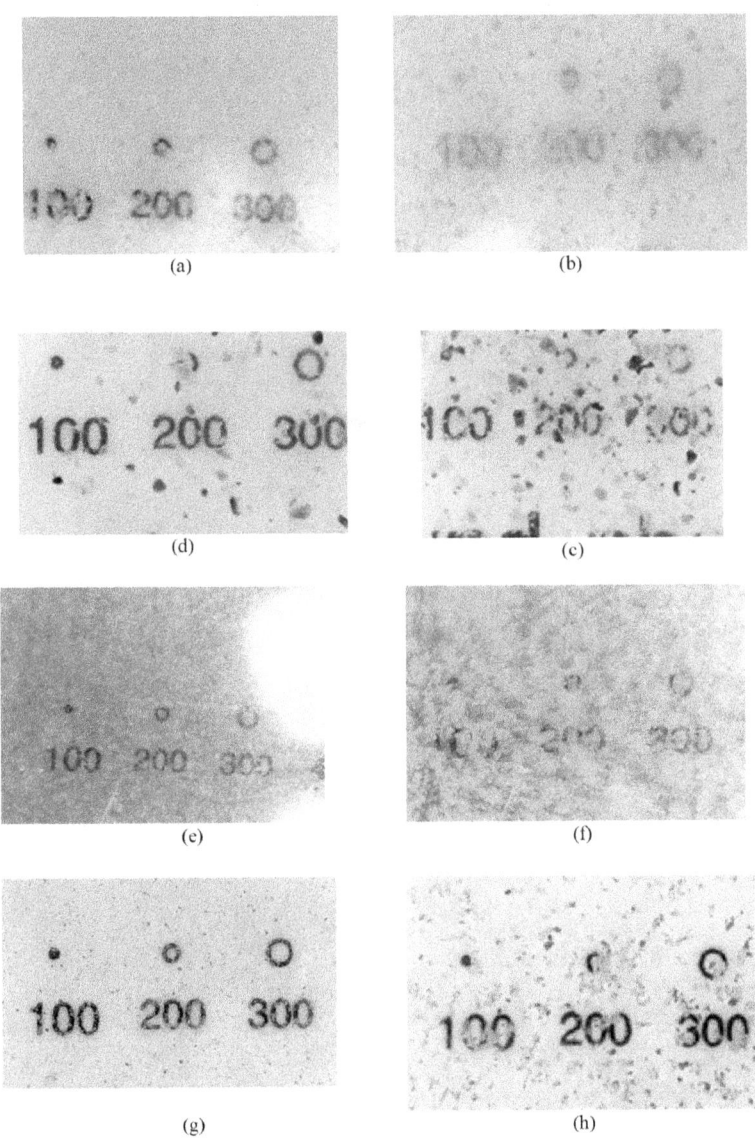

Fig. 8. Limestone day 1 (a); limestone day 7 (b). Tepojal day 1 (c); Tepojal day 7 (d). Diatomaceous earth day 1 (e); Diatomaceous earth day 7 (f). Tezontle day 1 (g); Tezontle day 7 (h)

5.4 pH Variation Qualitative Test

The pH measurement was done at day 1, at day 7 and at day 14. Five measures of each sample where taken. In the graphics the average is stated. The results show an increase of the pH in all the samples (Fig. 9), not as it was expected in the *pH variation qualitative test* proposed by Boffey and Hirst [21].

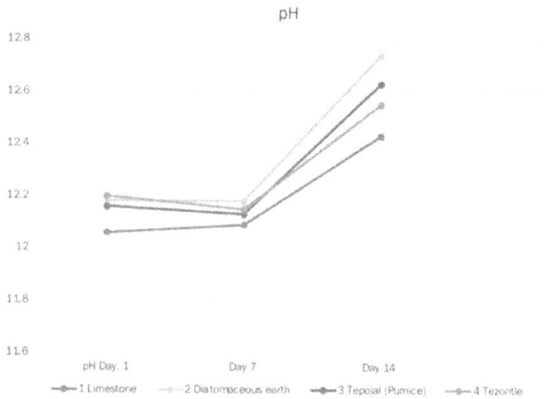

Fig. 9. Increase in the average value of pH in all the samples.

5.5 Sedimentation Velocity

The sedimentation velocity was done at day 1 and at day 7. Five measures of each sample where taken. The results show a decrease in diatomaceous earth and tezontle (Table 4) which denotes the presence of flocs.

Table 4. Sedimentation Velocity results

	DAY 1	DAY 7
1 Limestone	1.6 cm/min	1.3 cm/min
2 Tepojal	1.3 cm/min	1.4 cm/min
3 Diatomaceous earth	0.7 cm/min	0.02 cm/min
4 Tezontle	2.5 cm/min	0.14 cm/min

6 Discussion

In this study, the use of the pozzolana was obtained in Mexico's Valley. Any analysis or conclusion resulting from this research, on pozzolanic material without generalizing, refers only to this type of material.

With the qualitative observation of the *The Practical Test for Pozzolanic Properties* it was possible to determine that at seven days, as it was expected, the limestone didn't have a volume increase because it is not a pozzolan. It was useful to us it as a control sample. The pumicite (tepojal) didn't show a pozzolanic reaction. This confirms that pozzolanicity has different variables such as the origin, the mineralogic composition, the vitreous state, and the particle size. In this case, tepojal could be considered a pozzolan, and it could be inferred that there was no reaction because 57% of the sample was in an 80 mesh, and that probably interfere in the reaction. In the case of the diatomaceous earth a volume increase of the sample occurred at day seven, as well as it happened with the tezontle.

The complementary tests proposed in this study confirms in a semiquantitative way what initially was done by observation and inference by Cowper Test.

The refractive index corroborates the presence of a pozzolanic reaction by showing the presence of flocs. It is necessary to work in the precision of the measure because there can be variations according to the agitation of the mixture. It is important to keep adjusting the procedure because it is determinant to confirmation flocculation.

The observation in digital microscope, also confirmed the pozzolanic reaction in the Diatomaceous earth and the Tezontle and not in the limestone and tepojal, in a very simple and on field test. It will be important to confirm what it was observed with other more accurate methods to contrast them.

The pH variation qualitative test didn't show the results that were expected and proposed by Boffey and Hirst [21]. The possible answer to this phenomenon is that in the process of c-s-h formation there are also silicon and aluminium hydroxides that react with the calcium of the calcium hydroxide. The hydroxides of silicon, aluminium or calcium have a high pH. It could be possible that there are free -OH; there is a change in compounds but if some of them don't have enough calcium to convert into c-s-h they will remain hydroxides, so the levels of pH won't decrease. It necessary to continue studying this procedure to determine if an increase in the amount of calcium hydroxide in the mixture will promote the drop pH drop.

The sedimentation velocity test also confirmed the pozzolanic reaction in the Diatomaceous earth and the Tezontle and not in the limestone and tepojal. The speed in which the suspension sediments depend on the floc size. This is also a useful *on site* analysis that will be interesting to compare with other more accurate methods to validate what it has been proposed here.

7   Conclusion

The macroscopic physical changes of materials that produces a pozzolanic reaction can be observed in a qualitative way at seven days, and by measuring this transformation, we can determine a semiquantitative estimation of pozzolanicity. We can conclude that by this test, in 7 days and with limited resources, we can confirm the pozzolanic reactivity in the Diatomaceous earth and in the volcanic slag Tezontle.

This is a first contribution to transit from a qualitative test into a semiquantitative one. The research process allows to formulate deeper questions and the need of other analysis to confirm data and we expect to continue with an investigation that can lead into a *on field* quantitative method for determining pozzolanicity.

This is a first contribution to transit from a qualitative test into a semiquantitative one. The research process allows to formulate deeper questions and the need of other analysis to confirm data and we expect to continue with an investigation that can lead into a *on field* quantitative method for determining pozzolanicity.

For the research prospective Dynamic Light Scattering, Scanning Electron Microscopy, Sedimentation Velocity, Electrical Conductivity, as well as X Ray Diffraction, X Ray Fluorescence and Petrography are the analytical techniques to be used confirm the data of the *on site tests* on a semiquantitative or quantitative way.

The study and characterization of the materials used in the experimental model has the objective of understanding their behavior; complement and contrast observation, physical examination, imaging techniques and instrumental analysis to explain their shapes, structures and composition, will predict the potential performance of pozzolanic additions in aerial lime mortars. In the field of restoration, it is important to identify and understand the materials to be used in intervention processes, in this way it is possible to modify the variables that influence the behavior of the material and thus design materials under physical-chemical performance control and respond to the specific needs and demands for the conservation of the cultural heritage to be intervened. The choice of materials must be based on scientific support to justify the suitability of its use.

Acknowledgments. This research was designed and executed in the Materials and Structural Systems Laboratory of the Architecture Faculty of the National Autonomous University of Mexico; With the support of the Traditional Technology and Sustainability for Conservation Laboratory of the National School of Conservation and Restoration of Mexico.

References

1. Consejo Consultivo del Centro Histórico: Manual básico de intervención para inmuebles del centro histórico, pp. 1–78. Consejo Consultivo del Centro Histórico, San Luis Potosí (2008)
2. Brocklebank, I.: Building Limes in Conservation. Routledge (2012)
3. Foster, S.: BS EN 459-The new Standard for Building Lime (2009)
4. Torraca G.: Lectures on Materials Science for Architectural Conservation, 1st edn. The Getty Conservation Institute (2009). https://www.getty.edu/conservation/publications_resources/pdf_publications/pdf/torraca.pdf. Accessed 06 Mar 2022
5. Polión, V.: Los diez libros de arquitectura (No. 2)
6. ASTM International: ASTM C-618-17a Standard Specification for Coal Fly as and Raw or Calcined Natural Pozzolan for Use in Concrete. West Conshohocken (2017)
7. Sepulcre Aguilar, A.: Influencia de las adiciones puzolánicas en los morteros de restauración de fábricas de interés histórico-artístico, Escuela Técnica Superior de Arquitectura (2005)
8. Cowper, A.D.: Lime and Lime Mortars. Routledge, New York (2015)
9. Practical Action. Testing methods of pozzolanas. https://answers.practicalaction.org/pa-wp/our-resources/item/testing-methods-for-pozzolanas/. Accessed 06 Mar 2022
10. Alvarez, J.I., et al.: RILEM TC 277-LHS report: a review on the mechanisms of setting and hardening of lime-based binding systems. Mater. Struct./Materiaux Constr. **54** (2021)
11. Toledo M. Desarrollo de un proceso de floculación con control de la distribución de partículas. https://www.mt.com/mx/es/home/applications/L1_AutoChem_Applications/L2_ParticleProcessing/Formulation_Flocculation.html#:~:text=La%20floculaci%C3%B3n%20es%

20un%20proceso,peque%C3%B1a%20cantidad%20de%20grandes%20fl%C3%B3culos.&
text=Las%20part%C3%ADculas%20peque%C3%B1as%20suelen%20tener,y%20la%20e
stabilizaci%C3%B3n%20(1a). Accessed 20 Feb 2022

12. Arcillas 1./Arcillas e Ingeniería Civil Coruña. http://caminos.udc.es/info/asignaturas/grado_
    tecic/211/algloki/pdfs/ARCILLAS.pdf. Accessed 23 Feb 2022
13. Romero Shirai, C.P.: Floculación y viscosidad de suspensiones de sílice coloidal en presencia
    de sales de agua de mar, Tesis de Magister en Ciencias de la Ingeniería, Universidad de
    Concepción (2018)
14. Costafreda Mustelier, J.L.: Granulometría y reacción puzolánica. In: IV Congreso cubano de
    minería. Sociedad Cubana de Geología (2011)
15. Méndez Mariano, R.R.: Determinación de la reactividad puzolánica de adiciones minerales
    de origen natural con el cemento Portland. Oaxaca (2008)
16. Ling, S.J., Sanny, J., Moebs, W.: University physics, vol. 3. Houston (2018). https://openstax.
    org/books/university-physics-volume-3/pages/1-introduction. Accessed 28 Mar 2022
17. Halliday, D., Resnick, R., Krane, K.: Física vol. 2. 6th ed. Vol. 2. D.F.: Compañia editorial
    Continental S.A. de C.V. (1999). http://www.fulviofrisone.com/attachments/article/485/Res
    nick-Fisica%20Vol%202.pdf. Accessed 29 Mar 2022
18. Newburgh, R., Rueckner, W., Peidle, J., Goodale, D.: Using the small-angle approximation
    to measure the index of refraction of water. Phys. Teach. (2000)
19. Microscopio Digital. https://usbmicroscopiodigital.com.mx/microscopio-digital/. Accessed
    13 Apr 2022
20. Martín Garrido, M.: Modificaciones estructurales del gel C-S-H irradiado con láser continuo
    de CO_2, Tesis para el grado de Doctor. Universidad Complutense de Madrid, Madrid (2019)
21. Boffey, G., Hirst, E.: The use of pozzolans in lime mortars. J. Archit. Conserv. (1999)
22. Oakton. Oakton PD 450 Waterproof pH/DO Portable Meter with Probes. https://www.
    coleparmer.com/i/oakton-pd-450-waterproof-ph-do-portable-meter-with-probes/3563230.
    Accessed 13 Apr 2022
23. Flottweg. Técnica de separación. Velocidad de sedimentación. https://www.flottweg.com/es/
    wiki/tecnica-de-separacion/velocidad-de-sedimentacion/. Accessed 13 Apr 2022
24. Instituto de Hidrología M y EA: Instructivo para estimación de la incertidumbre de medición
    (2020)
25. Mehta, P.K., Monteiro, P.: Concreto: estructura, propiedades y materiales. Instituto Mexicano
    del Cemento y del Concreto (1998)
26. Barba Pingarrón, L., Villaseñor Alonso, I.: La cal. Historia, propiedades y usos. Universidad
    Nacional Autónoma de México, Ciudad de México (2013)
27. Plataforma Educativa Virtual Primaria. Tepojal. http://pep.ieepo.oaxaca.gob.mx/recursos/mul
    timedia/rocas_minerales/publi_rocas/pomez.htm. Accessed 13 Apr 2022
28. Coordinación General de Minería: Perfil de Mercado de la Diatomita (2013)
29. Plataforma Educativa Virtual Primaria. Tezontle. http://pep.ieepo.oaxaca.gob.mx/recursos/
    multimedia/rocas_minerales/publi_rocas/tezontle.htm. Accessed 13 Apr 2022

Determination of the Salt Distribution in the Lime-Based Mortar Samples Using XRF and SEM-EDX Characterization

Marina Aškrabić[1(✉)] [iD], Dimitrije Zakić[1] [iD], Aleksandar Savić[1] [iD], Ljiljana Miličić[2], Ivana Delić-Nikolić[2], and Martin Vyšvařil[3] [iD]

[1] Faculty of Civil Engineering, University of Belgrade, Belgrade, Serbia
amarina@imk.grf.bg.ac.rs
[2] Institute IMS, Belgrade, Serbia
[3] Faculty of Civil Engineering, Brno University of Technology, Brno, Czech Republic

Abstract. Although, the salt crystallization is one of the most common causes of the deterioration of lime-based mortars, testing of their resistance to the soluble salt action has not yet been standardized. The problems following the development of the globally accepted testing method are, among others: defining the type of mortar samples, ways of samples' contamination, the type and the concentration of the salt solutions used, environmental conditions during testing, determination of the damage development and the durability assessment. Another task of the testing method is to explain and connect the processes developing in the materials when they are applied in laboratory and real conditions. In this paper, soluble salt resistance testing of lime mortars on the composite samples is presented. The main focus of the paper is on the determination of the salt distribution in this type of samples after the five wetting and drying cycles. Samples consisted of two lime rendering layers (inner – 1/3 and outer – 1/1), both prepared according to the experiences found in the literature for these types of lime mortars when applied on historical structures, placed on the natural stone bases. They were cured in laboratory conditions for 90 days, before drying and exposing to soluble salts action. Two types of 10% salt solutions were used for the test: sodium-chloride and sodium-sulfate. Salt contamination was performed by capillary action only in the first cycle, while in the other cycles samples were wetted by deionized water. After the finalization of the cycles, the detached pieces of mortar and efflorescence were removed from the samples. One of the samples from both groups were then cut in two halves, from which one was used for X-Ray Fluorescence (XRF) and another for Scanning Electron Microscopy with Energy Dispersive X-Ray Analysis (SEM – EDX) characterization. For the XRF analysis samples were divided into four layers, and then crushed and sieved through 0.5 mm sieve before testing. For the SEM-EDX analysis the polished thick cross sections were prepared. The paper presents the results of these two analyses, and discusses the advantages and disadvantages of their application for this purpose. Mineralogical analysis of the samples was performed using XRD analysis. It was shown that XRF analysis provides more precise quantification of the elements within one sample, while SEM-EDX analysis gives possibilities for testing of layers with smaller depth within one cross – section.

© The Author(s), under exclusive license to Springer Nature Switzerland AG 2023
V. Bokan Bosiljkov et al. (Eds.): HMC 2022, RILEM Bookseries 42, pp. 542–553, 2023.
https://doi.org/10.1007/978-3-031-31472-8_43

Keywords: lime-based mortars · soluble salt resistance · salt distribution · XRF · SEM-EDX

1 Introduction

Salt crystallization is considered to be one of the most common damage mechanisms in porous materials [1, 2]. Effective durability test for the materials, in contact with soluble salts, should determine the longevity of the material, and in the case of restoration or conservation works on historical monuments, also compatibility between existing and newly placed materials. The choice of the most effective test, depends on several important parameters: the size, type and the number of samples, type of salt and the solution concentration, ways of the salt accumulation in the sample, environmental conditions during wetting and drying cycles, and the ways of following the damage development and determination [3].

The assessment of the material condition after the finalization of the test is the main issue presented in this paper. It can be divided in two directions of interest. The first one and the most common one is evaluation of the surface degradation of the material. The second one is determination of the salt distribution through the sample.

Evaluation of the material and surface degradation is performed through different methods: number of cycles until total degradation of the samples [4], adhesion of the material [5], visual determination of damage [6], mass changes [6], changes of ultrasonic pulse velocity [7] and changes of mechanical properties [8]. However, the amount of salt ions, and the salt distribution are performed through microscopy, XRD, FTIR, TG/DTA evaluation, ion and chloride distribution and SEM microscopy [6, 8–10].

In this paper the possibilities of XRF and SEM-EDX characterization of different depths of samples after the finalization of the testing of the resistance to soluble salts action are presented and discussed.

2 Materials and Specimens Preparation

Testing of the soluble salt resistance was performed on composite specimens consisted of stone base (siga stone), and two layers of lime mortars. Lime mortars were designed following the recommendations and experiences presented in literature for historical lime-based renders [11]. Inner render layer contained natural river aggregate, sized 0–4 mm, while binder to aggregate ratio was 1:3 by volume. Outer render layer contained aggregate sized 0–0,5 mm, with binder to aggregate ratio 1:1 by volume. Lime – putty, produced by Javor, Veternik, with approximately 50% of water, was used in all mixtures. Chemical composition of the component materials, obtained through chemical analysis is presented in Table 1.

Detailed mortar design is presented in Table 2. Physical, mechanical and micro-structural properties of these mortars at the ages of 28, 60, 90 and 180 days, are published previously in [12]. Thickness of the inner rendering layer was 2 cm, and of the outer rendering layer – 1 cm.

Preparation and curing of the composite specimens (presented in Fig. 1), as well as the testing procedure are described in detail in [13]. Samples were exposed to the action of 10% solutions of both sodium chloride and sodium sulfate (5 specimen for each test). The solution was introduced in specimens through capillary action, during the 20 min of exposure of the stone base. After the drying period, in the 4 consequent cycles, only deionized water was introduced in the specimens keeping the same period of exposure. The drying periods consisted of repeating following cycle until 80% of absorbed water would evaporate: temperature 40 °C and relative humidity 20% during 16 h and temperature 20 °C and relative humidity of 50% during 8 h. Top view of the tested specimens at the end of the final cycle are presented in Fig. 2.

Table 1. Chemical composition of component materials (%)

Element	Lime - putty	Coarse aggregate (0/4 mm)	Fine aggregate (0/0.5 mm)
CaO	73.25	4.21	8.02
MgO	1.05	1.38	2.15
SiO_2	0.37	80.49	69.57
Al_2O_3	0.65	3.90	4.20
Fe_2O_3	0.35	4.32	5.17
CO_2	1.65	–	–
Na_2O	–	1.09	1.30
K_2O	–	1.10	1.18
SO_3	–	0.04	0.05
L.O.I	22.67	2.78	7.34

L.O.I. – loss on ignition

Table 2. Composition of the tested mortars (kg/m^3)

Mortar type	Lime	Aggregate	Water
Inner layer	190	1530	342
Outer layer	393	916	491

In total, five specimens were tested on the soluble salt action of sodium chloride and sodium sulfate solutions. After the end of the test, specimens were dried to the constant mass at the temperature of 40 °C. Then the layer of efflorescence and the loose parts of mortars were removed from the specimens, and the mass of the material collected was measured. Specimens preserved from the first cycle, and one chosen specimen from the final cycle were then cut in two halves, by vertical cut, as shown in Fig. 3. One half was used for the XRF analysis, while the other half was used for the SEM-EDX analysis.

Fig. 1. Composite specimens after the preparation

Sodium cloride   Sodium sulfate

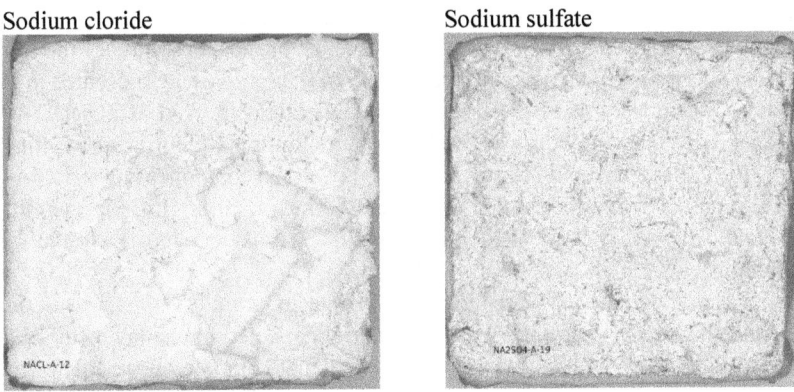

Fig. 2. Top view of the samples exposed to the soluble salt solutions after 5th cycle

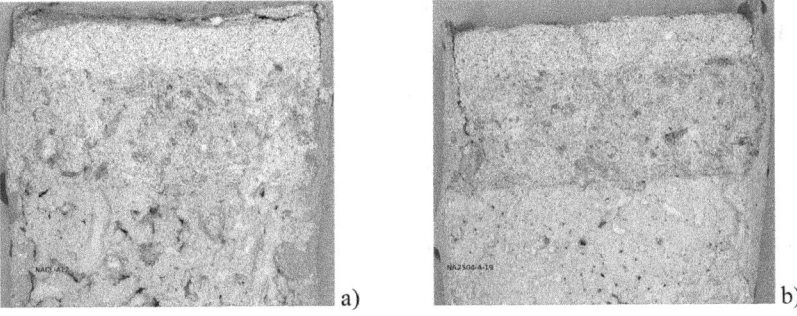

Fig. 3. Cross section of samples after cutting: a) specimen exposed to the sodium chloride solution, b) specimen exposed to sodium sulfate solution

3 Methods

3.1 X-ray Fluorescence

X-ray fluorescence (XRF) is a non-destructive analytical technique used for the determination of the elemental composition of the material. XRF analyzer determines the chemical composition of the sample by measuring the fluorescence (or secondary) X-rays, emitted from the sample, when it is excited by the primary X-ray. Each of the elements present within the sample produces several characteristic fluorescence X-rays that are unique for the element in question.

XRF analysis was performed on crushed samples, taken from different layers of the specimen. Both types of mortar were divided in two layers, and grinded. Samples collected from the outer layer mortar were taken in the depth of 5 mm, while the depth for the samples collected for the inner layer mortar was 10 mm. This was the consequence of the different maximum grain size of the aggregates used in both of the layers. Samples collected from the outer render layer were used as is in the analysis, while inner layer samples were additionally sieved through the sieve sized 0.5 mm, in order to eliminate coarse aggregates from the sample. In each of the steps mass of the samples was measured. Described steps in the preparation of the samples are presented in Fig. 4.

For the element analysis in question, Thermo Scientific Niton XL3t GOLDD + XRF energy-dispersive spectrometer was used. Detection limits for chlorine and sulfur were 60 mg/kg (ppm) and 70 mg/kg (ppm), respectively. Since component materials used in the preparation of the specimens contained no chlorine or sulfur, it is expected that the total detected amount of these elements is consequence of the exposure to the soluble salts solutions.

The measured parameters were the mass of the tested sample (m_{ts}), mass of the fine fraction in on layer (m_f) and total mass of material collected in one layer (m_t). For the layers 1 and 2, mass of the fine fraction and the total mass of material are equal, while in layers 3 and 4 there are differences between these values due to the presence of coarser aggregate in the total mass of material. The measured amounts of chlorine ($m_{Cl, ts}$) and sulfur ($m_{S, ts}$) is expressed in ppm (parts per million) in the mass of the tested sample. In order to calculate the total amount of these ions in the total mass of the sample following calculations have been performed.

$$m_{Cl,f} = m_{Cl,ts} \cdot m_f / 1000 \tag{1}$$

$$m_{S,f} = m_{S,ts} \cdot m_f / 1000 \tag{2}$$

Since through the chosen calibration of XRF measurements it was not possible to obtain the amounts of sodium and oxygen, the amounts of sodium chloride and sodium sulfate in different layers of the sample were calculated using stoichiometry, as shown in Eqs. 3, and 4.

$$m_{NaCl} = m_{Cl} \cdot 58.443 / 35.453 \tag{3}$$

$$m_{Na_2SO_4} = m_S \cdot 142.04 / 32.065 \tag{4}$$

3.2 Scanning Electron Microscopy

The second half of the sections used for the XRF measurements, were used for preparation of thick sections, taking into account the full diameter of the specimens. They were cut at the some distance below the cutting plane, in order to avoid irregularities of surface in the cutting plane. The sections were about 0.5 cm below the initial sawing plane, i.e. half a centimeter towards the lateral surface.

Scanning electron microscopy (SEM) uses electrons in order to form an image. It can be performed in two ways, using secondary or back-scattered electrons. The second method was used in presented research, on the thick polished section, coupled with EDX analysis that relies on an interaction of source of X-ray excitation and a sample. SEM was performed on prepared sections by use of a FEI Quanta FEG 250. The acceleration voltage was 20 resp. 25 kV. Both samples were investigated without sputtering at low-vacuum.

Micrographs with the back-scattered (BSE) detector were taken at 100× magnification (about 2.6 × 2.2 mm per photo) for transversal section areas of approx. 28 mm depth and 6.5 mm width, these centrally located areas included the surface of each sample in the cross-section and the full thickness of the mortar layers, reaching down to the upper layer of inner rendering mortar. Some specific details presenting salt positions were presented with 600× magnification and 1200× magnification.

Quantitative analysis by energy-dispersive X-ray spectroscopy (EDX) was performed for each of the micrographs, i.e. for sample areas of approx. 2.6 × 2.2 mm each. The results for the elements Na, Mg, Al, Si, S, K, Ca, Fe were normalized to 100%. The EDX-detector used was of the type EDAX Octane Elect, the software for quantification was Genesisâ. Acquisition time for each spectrum was in the range of 2 min.

4 Results and Analysis

Elemental analysis obtained through described methods is presented in Tables 3, 4, 5, and 6. Results of XRF analysis are presented together with the measurement error (in brackets). Both of the methods have it's limitations in detecting all the elements found in the samples. According to the chemical analysis of component materials presented in Table 1, it was expected to detect calcium (Ca) and silicon (Si) in greater extent, together with aluminium (Al) and iron (Fe) found in aggregates, with magnesium (Mg), sodium (Na) and potassium (K) in minor quantities. Most importantly for the research in question were the possibilities of detecting the chlorine (Cl) and sulfur (S), that were imported in the samples through sodium chloride and sulfate solutions.

Through the chosen calibration of XRF analysis, it was possible to quantify amounts of Fe, Al, Si, Mg, Cl and S.

The highest amount of both chlorine and sulfur were found in the top layer (layer 1). The highest overall values were detected for silicone, with tendency of increase towards the layers where higher amounts of aggregates were present.

As it was explained in Sect. 3, amounts of sodium chloride and sodium sulfate were calculated for each tested layer, and presented in in Fig. 5 in mg per the total mass of material collected in each layer (m_t).

SEM-EDX   **XRF**

Fig. 4. Preparation of the samples for the XRF and SEM – EDX measurements

Table 3. Elemental composition of specimen treated with sodium - chloride obtained through XRF analysis

Element (ppm)	Layer 1	Layer 2	Layer 3	Layer 4
Fe	4103.4 (58.5)	3698.6 (56.4)	13086.8 (134.4)	4618.0 (59.7)
Al	3538.3 (984.5)	3848.0 (978.0)	6073.7 (835.8)	6221.1 (729.7)
Si	72663.4 (1164.2)	67041.8 (1244.7)	133709.5 (1411.4)	136620.0 (1321.0)
Cl	92062.0 (502.6)	23544.8 (226.7)	10856.6 (131.8)	3936.6 (78.0)
Mg	10212.3 (5193.1)	<LOD	<LOD	<LOD

<LOD – less than the level of detection

Through SEM-EDX analysis almost all elements present in the sample could be detected. Still, because carbon and oxygen were not detected, and they were expected to be present in higher amounts in lime mortars, it was not possible to quantify the results in the same way as presented in XRF analysis. In the case of sodium chloride, since the instrument detected both sodium and chlorine, it was possible to determine the percentage of sodium chloride in the total amount of detected elements. These results are shown in Fig. 6.

Similarly to the results obtained through XRF analysis the highest amount of chlorine and sulfur, as well as sodium, were found in the top layer. Since the observed layers in SEM-EDX analysis were thinner, it is shown that most of the salts are preserved in the

Table 4. Elemental composition of specimen treated with sodium - sulfate obtained through XRF analysis

Element (ppm)	Layer 1	Layer 2	Layer 3	Layer 4
Fe	4781.3 (62.4)		10516.0 (119.4)	12645.2 (131.8)
Al	6364.3 (960.0)		7081.7 (832.7)	7732.8 (822.6)
Si	114712.3 (1440.9)		129847.6 (1382.8)	139890 (1383.3)
Cl	1861.2 (79.4)		1759.5 (66.8)	1232.1 (58.8)
S	21861.5 (338.9)		11509.6 (230.3)	12655.4 (231.6)
Mg	<LOD		<LOD	<LOD

<LOD – less than the level of detection

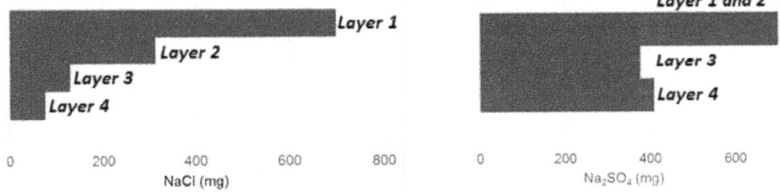

Fig. 5. Calculated amounts of sodium chloride and sodium sulfate

first 2.5 mm of the sample. It should be here repeated that the salts found on the surface of the samples (efflorescence layer), were removed before the samples for the described analysis were cut and prepared.

Table 5. Elemental composition of specimen treated with sodium - chloride obtained through SEM-EDX analysis

Element	Layer a	Layer b	Layer c	Layer d	Layer e	Layer f	Layer g
Fe (%)	2.59	2.46	2.09	1.56	1.79	1.89	2.68
Al (%)	3.35	3.74	4.15	3.96	4.08	3.53	3.49
Si (%)	43.29	48.27	48.88	48.11	52.58	52.78	62.17
Cl (%)	4.08	1.93	1.50	1.46	1.18	0.64	1.48
Mg (%)	0.96	1.33	1.03	1.68	1.28	0.83	1.52
Na (%)	1.98	1.20	0.94	1.68	1.51	1.25	1.12
K (%)	1.12	1.10	1.64	1.31	1.46	1.06	1.37
Ca (%)	35.05	32.31	32.11	32.48	28.72	30.52	20.20

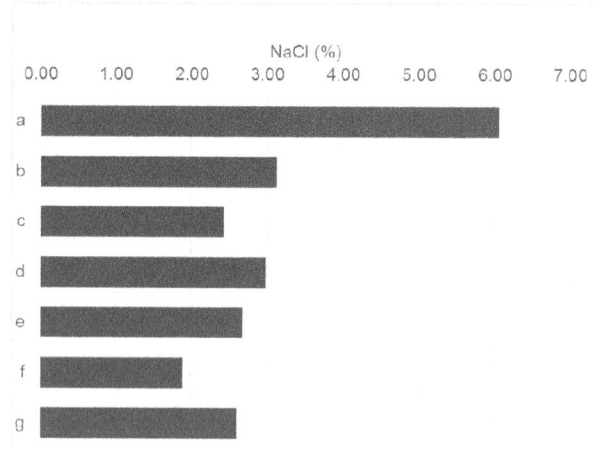

Fig. 6. Calculated amounts of sodium chloride from SEM-EDX analysis

Table 6. Elemental composition of specimen treated with sodium sulfate obtained through SEM-EDX analysis

Element	Layer a	Layer b	Layer c	Layer d	Layer e	Layer f	Layer g
Fe (%)	2.47	2.02	2.13	2.36	5.34	4.04	1.91
Al (%)	3.92	3.87	3.39	3.61	4.20	2.78	3.05
Si (%)	50.03	47.24	51.23	52.39	56.56	71.99	73.69
S (%)	1.85	1.08	0.97	0.62	1.18	0.84	0.60
Mg (%)	1.05	1.24	1.15	1.33	3.00	1.71	1.01
Na (%)	2.96	2.31	1.35	1.18	1.33	0.75	0.80
K (%)	1.84	1.52	1.40	1.59	1.15	1.06	1.66
Ca (%)	35.87	40.71	38.38	36.91	27.23	16.84	17.27

Another benefit from the SEM analysis is the possibility to obtain both visual data on the position of the salts in the layer. As it can be seen in the Fig. 7, first line presents top images of top layers (layer a) of both tested samples. Sodium chloride is preserved in the pores around sand grains, while sodium sulfate is intermixed with the binder, and can be observed only through slight differences in colour. Going deeper into the sample, it can be seen that both types of salt are preserved near the coarse grains of aggregate, which is especially visible in the layer that detects contact between two types of mortar (layer e). In this layer additional details, with higher magnification were observed and presented in the third line of pictures, where previously presented comments can be confirmed.

The differences between the salt depositions in lower layers of the mortar can be noticed in the magnified figures. Sodium sulfate creates needle-like products in the contact zone with coarser aggregate grains that could be explained by formation of gypsum ($CaSO_4$). These products induce narrowing of the pores and partially block the movement of the salts towards surface layers. In this way amounts of salts remaining in lower layers are increased, as confirmed by the XRF analysis (Fig. 5), when compared to the sodium chloride, where no similar behaviour was noticed.

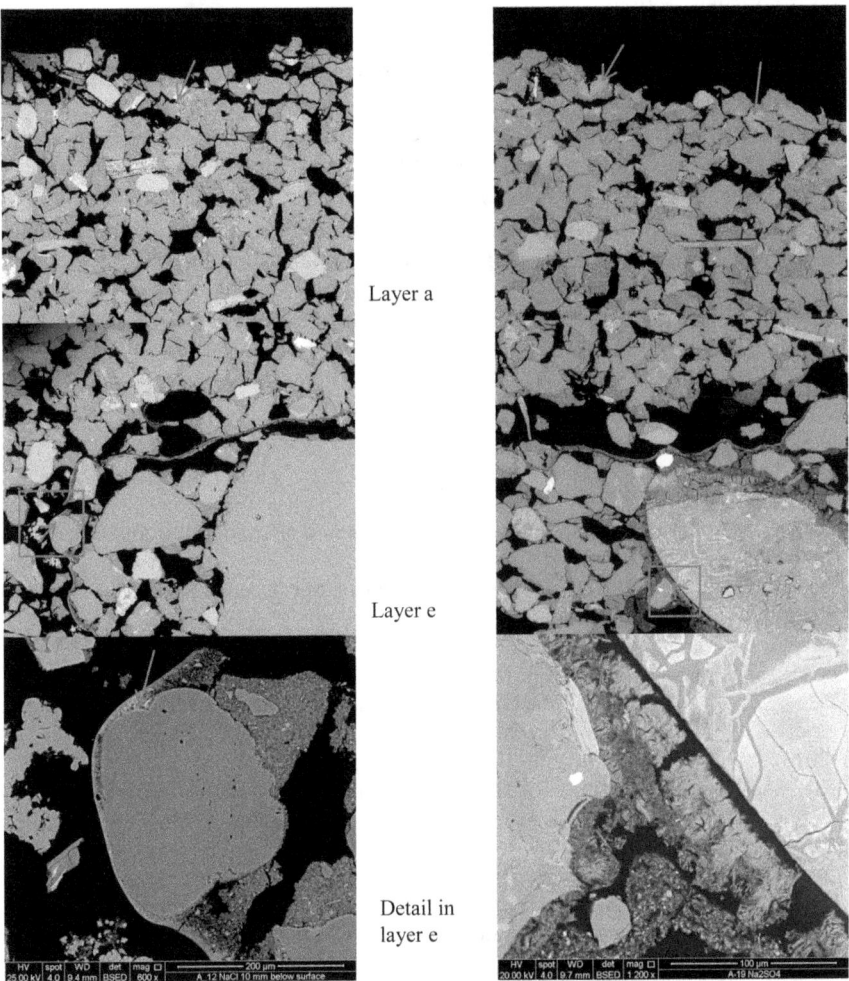

Layer a

Layer e

Detail in layer e

Fig. 7. Selected figures from SEM analysis (from layers a and e) for both types of samples

5   Conclusion

Two ways of determining the position and distribution of salts in the composite samples after the testing of the resistance to the soluble salts action were presented within the paper. Both of the methods demand adequate preparation of the samples, either through grinding and sieving, or through preparation of thick cross sections by the trained personnel. In the case of grinding and sieving part of the material could be lost, if it is not properly detached from the coarser grains of aggregate, since it was shown that both sodium sulfate and sodium chloride were found attached to the coarser aggregates. Also, some part of material and salts could be lost from the sample during the preparation of the thick polished sections.

The advantage of the XRF analysis is direct quantification of the presence of detected elements. The advantage of SEM-EDX analysis is the possibility to perform visual and chemical analysis at the same time.

Both analysis showed similar trends, when the distributions of the salts, or chlorine and sulfur are in question. It was shown that the highest amount of salt was preserved in the top layer (layer 1 for XRF analysis, and layer a for SEM-EDX analysis).

Both techniques show good potential for determining the damage and the behaviour of the samples after the finalization of the soluble salts resistance testing.

Acknowledgement. The work reported in this paper is a part of the investigation within the research project 200092, supported by the Ministry of Education, Science and Technological Development, Republic of Serbia. This support is gratefully acknowledged.

References

1. Charola, A.: Salts in the deterioration of porous materials: an overview. J. Am. Inst. Conserv. **39**(3), 327–343 (2000)
2. Doehne, E.: Salt weathering: a selective review. Geol. Soc. London Spec. Publ. **205**, 51–64 (2002)
3. Lubelli, B., et al.: Towards a more effective and reliable salt crystallization test for porous building materials: state of the art. Mater. Struct. **51**(55), 1–21 (2018)
4. Peterová, A., Doubravová, K., Machovič, V., Jiroušek, J.: mortars from roman cement and their properties. In: Proceedings of the 2nd Historic Mortars Conference, Prague, pp. 1151–1157 (2010)
5. Klisińska-Kopacz, A., Tišlova, R.: The effect of composition of roman cement repair mortars on their salt crzstallization resistance and adhesion. Procedia Eng. **57**, 565–571 (2013)
6. Faria, P., Martins, A.: Influence of curing conditions on lime and lime-metakaolin mortars. In: Proceedings of XII Durability of Building Materials and Components, pp. 1–8 (2011)
7. Bianco, N., Calia, A., Denotarpietro, G., Negro, P.: Laboratory assessment of the performance of new hydraulic mortars for restoration. Procedia Chem. Youth Conser. Cult. Heritage **8**, 20–27 (2013)
8. Andrejkovičova, S., Alves, C., Velosa, A., Rocha, F.: Bentonite as a natural additive for lime and lime-metakaolin mortars used for restoration of adobe buildings. Cem. Concr. Compos. **60**, 99–110 (2015)
9. Cerulli, T., Pistolesi, C., Maltese, C., Salvioni, D.: Durability of traditional plasters with respect to blast furnace slag-based plaster. Cem. Concr. Res. **33**, 1375–1383 (2003)
10. Gulotta, D., Goidanich, S., Tedeschi, C., Toniolo, L.: Commercial NHL-containing mortars for the preservation of historical architecture, part 2: durability to salt decay. Constr. Build. Mater. **96**, 198–208 (2015)
11. del Mar Barbero-Barrera, M., Maldonado-Ramos, L., Van Balen, K., García-Santos, A., Neila-González, F.J.: Lime render layers: an overview of their properties. J. Cult. Heritage **15**, 326–330 (2015)

12. Aškrabić, M., Vyšvařil, M., Zakić, D., Savić, A., Stevanović, B.: Effects of natural zeolite addition on the properties of lime putty-based rendering mortars. Constr. Build. Mater. **270**, 121363 (2021)
13. Aškrabić, M., Zakić, D., Savić, A., Miličić, L., Delić-Nikolić, I., Ilić, Z.: Comparison between damage development on composite and standardized mortar specimens exposed to soluble salts. In: Proceedings of Fifth International Conference on Salt Weathering of Buildings and Stone Sculptures, pp. 129–139 (2021). ISBN 978-94-6366-439-4

Developing a Lime-Based Injection Grout with no Additives for Very Thin Delamination: The Role of Aggregates and Particle Size/Morphology

Chiara Pasian[✉], Jennifer H. Porter, Mariia Gorodetska, and Stephanie Parisi

Department of Conservation and Built Heritage, University of Malta, Msida, Malta
chiara.pasian@um.edu.mt

Abstract. A thin grout was designed and tested for the stabilisation of the 16th c. Perez d'Aleccio's wall painting cycle of the Great Siege, in the Grandmaster's Palace in Valletta, Malta. The painting, on a single layer of lime-based plaster applied on Globigerina Limestone, presented delaminated pockets up to 5 mm thick, with narrow access points ≤ 1.5 mm wide (cracks) to inject the grout. For this reason, the grout could just be injected through very fine needles, a 18G (0.84 mm inner diameter) and/or 19G needle (0.69 mm inner diameter). Considering the porous original plaster, to avoid a reduction in porosity of the grout observed in previous studies, the site-specific grout was designed without the use of additives such as superplasticisers, and included just binder (slaked lime putty), aggregates, water. Different aggregates were selected and tested in different proportion, the variables for their selection being: particle shape, particle size and water absorption. Testing of the grouts included: injectability, flow on plastered tile, bleeding, shrinkage and adhesion in Globigerina Limestone cups, cohesion. A very thin grout, passing through a needle 0.69 mm wide, was obtained, without the use of superplasticisers, relying only on the role of aggregates and their particle size and morphology.

Keywords: Injection Grout · Thin Delamination · Site-specific · Injectability · Aggregate · Particle Size/Morphology

1 Introduction

Injection grouting to stabilise delaminated wall paintings and historic plasters is a challenging intervention, addressing wall paintings at potential risk of loss. Since both problem and treatment are concealed behind the surface, their assessment is typically difficult, and often lack of accuracy and control are unavoidable [1] (p. 472). When thinly delaminated areas (<5 mm gap) need to be stabilised, the intervention can be particularly complex, especially when the grout needs to be injected through very narrow access points. Particularly in the case of wall paintings, it is not advisable to enlarge access points or to open new ones, unless this is done in areas of existing loss and is strictly necessary. When such areas are not present, existing access points such as cracks should be used to avoid damage to and further destabilisation of original materials.

© The Author(s), under exclusive license to Springer Nature Switzerland AG 2023
V. Bokan Bosiljkov et al. (Eds.): HMC 2022, RILEM Bookseries 42, pp. 554–566, 2023.
https://doi.org/10.1007/978-3-031-31472-8_44

Starting from the knowledge of original materials and technology and the definition of the problem encountered, grouts should be designed to fulfil performance criteria established for the particular case under consideration (site-specific, [2]). In the case of grouts for thin (<5 mm) delamination, beside minimal shrinkage and good adhesion and cohesion, very good injectability and flow are crucial. These are typically enhanced with the use of additives improving such properties [3] (see Sect. 1.2).

1.1 Perez d'Aleccio's Great Siege Wall Painting Cycle, Grandmaster's Palace in Valletta, Malta

The monumental 16th c. wall painting cycle of the Great Siege by Matteo Perez d'Aleccio, in the Throne Room of the Grandmaster's Palace in Valletta, Malta, portrays in chronological order the historical account of the 1565 Great Siege of Malta led by Suleiman the Magnificent, sultan of the Ottoman Empire. Commissioned by Grand Master de la Cassière, and composed of twelve scenes interspaced by allegorical figures, the cycle was painted by d'Aleccio between 1572 and 1581 [4] (p. 55), very shortly after the historic events it depicts.

The single layer of lime-based plaster, just 5–8 mm ca. thick, has been applied on the stone support, i.e. ashlars of the very porous local Globigerina Limestone [5]. The plaster application in patches suggests a possible a fresco technique; the original technique though is far more complex than a pure fresco [6] and studies on the matter are ongoing. Analysis of plaster thin sections reveals that the plaster composition varies across the scenes of the cycle: all plasters are lime-based, while the aggregates type and sorting can significantly vary [6, 7]. The plaster of the scenes for which the thin grout was developed contains carbonatic aggregates 40–70% ca., *cocciopesto* (ground brick) 25–60% ca. and low amount of quartz [6, 7], and its porosity is up to 18% [7].

1.1.1 Definition of the Problem

As mentioned, the − relatively thin− plaster is applied directly to a very porous and absorbent stone, and this must have made its adhesion challenging per se, already at the time of the plastering. Acoustic tapping [8] was systematically carried out, and results showed that hollow sounding areas are widespread throughout the painted cycle (hollow sounding areas respond to tapping with a deeper sound, while adhered areas give off a higher tone or no sound at all), possibly suggesting areas of poor adhesion, which on the other hand IR thermography could not identify. Nonetheless, the vast majority of the hollow sounding areas is actually stable. Unstable delamination was relatively rare considering the size of the whole painted cycle, and mostly concentrated in specific areas, where surface manifestation such as cracking was present.

The thin grout object of this paper was designed for the delamination present on the allegorical figure of *Prudentia* and the second scene of the cycle ('*La smontata dell'armata a Marsascirocco...*') (Fig. 1), on the NE wall of the Throne Room. Here a cracking system was present, with physical evidence suggesting it may have been active; delaminated unstable areas were often located across the two sides of the crack (Fig. 1). Cracks depth ranged between <4 and 20 mm, running perpendicularly to the surface, and, when wider and deeper, cutting through plaster and stone. The void of the delaminated pockets, estimated to be ca. 2–5 mm thick, was not visible and its extent could not be fully assessed; access points to grout were often just ≤1.5 mm wide, corresponding to the width of the crack. No other access points were available to grout or could be created in already damaged areas such as losses. It is possible that some of these areas had been already grouted in the previous conservation campaign (by the Hochschule für Bildende Künste Dresden in 2000–2004).

Wide crack (3-6 mm width)

Medium crack (1-3 mm width)

Narrow crack (0.5-1 mm width)

Delaminated unstable area

Delaminated possibly unstable area

Fig. 1. Allegory 2 (*Prudentia*) and Scene II of the Great Siege wall paintings by d'Aleccio, on the NE wall of the Throne Room (Grand Master's Palace, Valletta, Malta), with mapping of cracks and unstable delaminated areas. Basemap obtained from a picture by ©Cilia 2009, commissioned by Heritage Malta. Mapping ©DCBH, UM 2021.

1.1.2 Definition of the Grout Performance Criteria

General performance criteria for grouting (valid for any grouting intervention) include among working properties (WP; short-term properties): good injectability, no liquid-solid separation, minimal shrinkage after setting; among performance characteristics (PC, long-term properties): good cohesion, good adhesion, porosity and hygrothermal behaviour similar to those of the original plasters, mechanical strength similar to or lower than that of the original plasters [9]. Building on the definition of the specific problem encountered (Sect. 1.1.1), site-specific performance criteria were set, and they are listed in Table 1.

Table 1. Site-specific performance criteria for the thin grout

Intervention criteria	Property	Further details
Working Properties	Good injectability	Ability to pass through a 19G or bigger needle applied on a syringe (19G: outer Ø 1.07 mm, inner Ø 0.69 mm) [10]
	Good flow	Ability to easily travel horizontally in the pocket, e.g. just from one access point such a crack
Performance characteristics	Mechanical strength lower than that of the original plaster	Mechanical strength lower than that of the original plaster
	Good adhesion	Good adhesion
Additional comments about the intervention	–	Avoid to bridge/fill cracks between plaster fragments isolated by cracks, since movement (the cause of cracking) in the area may be active
	–	Not aiming to fill cracks running perpendicular to the surface

1.2 Enhancing Injectability and Flow in Injection Grouts: Superplasticisers, and the Role of Aggregates in the Mixture

It is common practice to include additives in injection grouts — 'in small quantity' [11] — to modify the properties of the mixture [3]. Among such additives, fluidisers/plasticisers/superplasticisers are employed to modify the injectability and flow properties of the grouts [3]. Such properties are crucial when grouts need to address a thin delamination (<5 mm gap) and/or to be injected through narrow spaces/access points, where a needle is applied to the syringe. Natural additives have been used to enhance injectability [3, 12], or — very commonly and more widely — synthetic additives such as plasticisers/superplasticisers [13–17]. These are low-molecular-weight polymers reducing the water necessary to achieve the desired injectability and flow, or improving injectability and flow at a constant water/binder ratio [18]. The influence of different types of superplasticisers was studied for air lime-based [17] and air lime-pozzolan grouts [13, 14]. While the use of the superplasticiser enhanced injectability, the porosity of the hardened grouts was reduced, in percentage [17] and in terms of pore size distribution [13, 14]. A lower porosity determined — often significantly — higher values in mechanical strength [13, 14, 17]. This is not desirable in the case of highly porous wall paintings and historic plasters with a relatively low mechanical strength, where a non-structural intervention is needed, and it was indeed not desirable for d'Aleccio's Great Siege wall painting cycle.

A different approach to improve injectability and flow of injection grouts can revert on the role of aggregates (and particles, in general) in the mixture. Lime-based grouts are suspensions, i.e. particles (including those of the binder) are suspended in a liquid (typically water). Hence the behaviour of the overall mixture will depend on how such particles are sorted and interact in the liquid (with an influence on WP) and then in the hardened material (influence on PC). In the set grout, packing density and geometry play a key role in cohesion and mechanical strength, as well as in porosity, which in turn will have an influence on a number of other properties [19]. In the fluid grout, the following factors will have an influence on the rheology of the grout:

- Particle shape: aggregates with a round shape generally improve WP such as injectability and flow [12, 15], while angular ones generally reduce them;
- Particle size distribution: a wide range of particle size (called *polydispersity*) decreases the viscosity of the mixture, hence improving flow [20];
- Water absorption of the particles: the higher the absorption capacity of the aggregates the poorer the injectability and flow of the grout (and the higher its viscosity), since the water of the mixture is absorbed by its own components.

Such factors were taken into account in the design of the d'Aleccio thin grout to balance WP and PC and fulfil the performance criteria set in Sect. 1.1.2.

2 Grout Design and Testing

2.1 Materials Selected

Lime was chosen as the binder of the thin grout, since the original plaster to stabilise is lime-based; specifically, slaked lime putty was selected (drained of all excess water, just paste). Portlandite crystals [$Ca(OH)_2$] in a slaked lime putty, particularly if aged, have been reported to be smaller compared to those in a putty made with hydrated lime [21, 22], and to have the capacity to adsorb more liquid water, which acts as a lubricating film between $Ca(OH)_2$ particles: this can improve working properties such as plasticity and water retention ([21], studied for plasters).

Different aggregates were selected, according to their composition, shape, particle size distribution, water absorption capacity (depending on their porosity) (Tables 2, 3, and Fig. 2). While the carbonatic aggregates and quartz sand are inert, glass bubbles (soda-lime borosilicate glass) are reactive with lime [23]. This needed to be kept in mind in order to design a grout which was not too strong compared to the original plaster.

Table 2. Materials tested for the formulation of the grout

Component		Particle shape (+ notes)	Particle size considered	H$_2$O absorption capacity (qualitative)
Slaked lime putty (Grassello di Calce, Calchera San Giorgio®, IT, 8 months-aged)	Ca(OH)$_2$ + water	hexagonal platelets [21]	N/A	N/A
Xaħx (crushed Globigerina Limestone) (Esrom, MT)	CaCO$_3$	equant, very angular and angular, medium sphericity (Fig. 2a)	<90 μm	very high[1]
Ramel (crushed Upper Coralline Limestone) (Esrom, MT)	CaCO$_3$	equant, angular and sub-angular, low sphericity (Fig. 2b)	<90 μm	high[2]
Carrara Marble powder (Halmann Vella Marble, MT)	CaCO$_3$	equant, angular and sub-angular, medium sphericity (Fig. 2c)	<90 μm	very low[3]
Quartz sand (JCR imports, MT)	SiO2	equant, sub-angular/sub-rounded, high sphericity (Fig. 2d)	<90 μm	negligible/very low[4]
Poraver® (PoraverNorth America Inc.)	'soda-lime glass' [24]	round (rough surface, full sphere) [24]	40–90 μm	medium-high[5]
Scotchlite™ *K1* (3M)	'soda-lime borosilicate glass' [25]	round (smooth surface, empty sphere) [25]	<120 μm [25]	none[6]
Scotchlite™ *S22* (3M)	'soda-lime borosilicate glass' [25]	round (smooth surface, empty sphere) [25]	<75 μm [25]	none[6]

[1] Porosity of the stone from which it derives up to ca. 40% [5]
[2] Porosity of the stone from which it derives up to ca. 18% [26]
[3] Porosity of the stone from which it derives ca. 0.45–2% [27]
[4] Although the composition of quartz sand may vary, e.g. 94.5 to 99.7% SiO$_2$ content [28], and so its porosity, columns filled with quartz sand were used to test gas permeability [29] and grouts injectability [30], and quartz sand was added to grouts for its negligible porosity minimising liquid absorption [23]
[5] Porous [24]
[6] Non-porous [25]

Table 3. Particle size distribution, in percentage, of the angular aggregates considered[1]

Aggregate	90–63 μm	<63 μm
Xaħx	7.3%	92.7%
Ramel	13.8%	86.2%
Carrara Marble powder	30.4%	69.6%

[1] Quartz sand had no fraction <63 μm

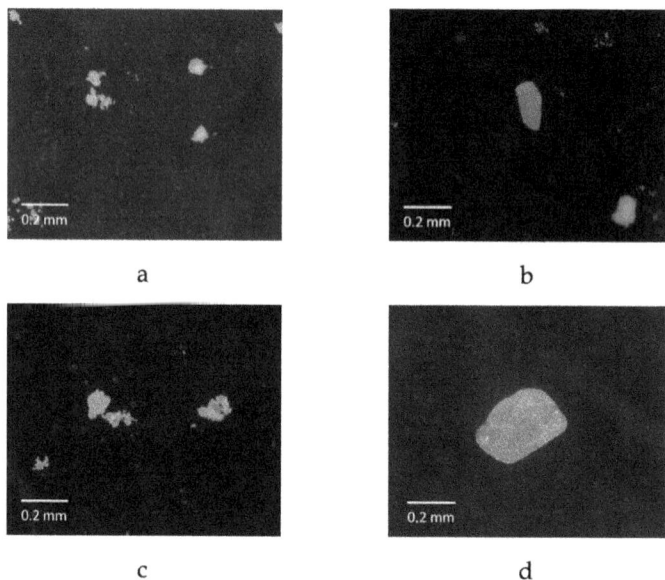

Fig. 2. Photomicrographs of angular aggregates taken with a portable microscope Dino-Lite AM4515ZT-EDGE: (a) Xaħx; (b) Ramel; (c) Carrara Marble powder; (d) Quartz sand. ©DCBH, UM 2022.

2.2 Formulation and Testing

In the design process, the formulation and testing of injection grouts went hand-in-hand: the formulation was incrementally modified and/or refined according to the results of the testing, in an iterative process [12]. A simple but worthwhile testing programme was established according to the performance criteria set and to properties that are always essential for a (thin) grout to fulfil. Properties tested and testing procedures are listed in Table 4.

In the formulation, keeping in mind that round particles tend to improve WP such as injectability and flow [12, 15], a balance of different particle shapes, both round and angular was selected to also ensure good cohesion. On the other hand, aggregates with a predominant fine fraction (<63 μm) overall (ex. ScotchliteTM [25]) were coupled with aggregates with a higher coarse fraction (>63 μm) overall (ex. Ramel): such polydispersity also improves WP [20] while ensuring good cohesion.

Table 4. Testing programme

Property	Testing procedure
Injectability	Up to 8 mL grout was injected through a 18G (inner diameter 0.84 mm) and 19G (inner diameter 0.69 mm) needle (using a 10 mL syringe). The amount of grout smoothly passing through without clogging was recorded
Flow	A Globigerina Limestone slab had been previously plastered with a 1: 2 lime-ramel <1 mm plaster, and vertical channels ca. 5 mm wide and 25 cm long had been carved into it [31] (pp. 75–76). The channels were pre-wet, and 2 mL grout was injected from the top of the channels and let flow. The length, homogeneity and body of the grout drip were observed, compared and evaluated
Bleeding	5 mL grout was injected in a 10 mL syringe placed vertically (without the syphon and with its exit sealed with tape), and evaporation was prevented covering it at the top with tape. The separation of the liquid from the solid component of the grout, if any, was recorded
Shrinkage and Adhesion	5 mL grout was injected in pre-wet cups [31] (pp. 83–85) made of Globigerina Limestone, simulating the support. The cups were covered with plaster slabs to simulate a semi-closed pocket [12]. Shrinkage was assessed at the surface [31] (p. 85), and the set grout was excavated with a spatula to assess shrinkage in depth [12]. In this way also adhesion to the cup walls could be assessed
Hardness	As much as possible, the same pressure was applied with a spatula on the different grouts set in the stone cups, and the indents were compared. Pieces of set grout were crushed with a spatula to assess its strength

After preliminary injectability testing, the top particle size for angular aggregates was set to be 90 μm (see Table 2), since larger particle sizes greatly reduced injectability, especially through the narrower 19G needle.

For ease of reasoning in the formulation design, components have been considered in pt/V, and will be reported in this paper as such (pt/V were translated into weight for each component to ensure consistency, repeatability and reproducibility, but are not reported here).

Table 5 shows the final trials leading to the refinement of the formulation. The table also reports the results obtained in the injectability, flow and shrinkage tests: these were the main tests which guided changes in the formulation. No grout showed bleeding. The hardness results, qualitative, did not show such a variation in the different formulations, and therefore they are not reported in the paper: in all grouts indents could be created with a spatula, and pieces of the grout could be crushed applying some pressure. Although the grouts are likely to develop some more strength over time (carbonation, and pozzolanic reactions due to the presence of Scotchlite™ [23]), no grout was judged too hard to fit the purpose (Table 1). Over the course of the design process, the following were assessed:

- **Binder to aggregate ratio:**

With a constant amount of liquid at 1.5 pt/V, minimal/low shrinkage in the stone cups was generally observed with a binder: aggregate ratio of 1: 3.5 pt/V; therefore this ratio was kept for most of the testing in order to understand the influence of the aggregates in the mixes, varying their type and size to improve working properties. The amount of water was slightly modified just for promising mixes (G44 from G42; G48b from G46, see Table 5), i.e. decreased to adjust shrinkage or increased to adjust injectability;

- **Water absorption of aggregates:**

As expected, non-porous/very poorly porous aggregates (Scotchlite™, quartz sand, marble powder) generally had a better impact on WP compared to porous aggregates (Poraver®, xaħx, ramel), because they did not absorb/absorbed to a much lesser extent the mixing liquid — of course the impact on WP also depends on their particle size distribution (see below in the list).

Mixes containing a relatively high proportion of porous aggregates with a large fraction <63 μm (Table 3) had worse injectability and flow, because the fine fraction (with a greater surface area) absorbed mixing water to a higher extent compared to the coarser fraction. This was particularly visible with xaħx (<63 μm ca. 93%), but also to a lesser extent with ramel (<63 μm ca. ca. 86%). In addition to this, the porosity of xaħx particles can be up to 40% ([5] for Globigerina limestone) versus the porosity of ramel particles up to ca. 18% ([25] for Upper Coralline Limestone). Therefore, in the design refinement, among the porous angular aggregates, xaħx was excluded; on the other hand, ramel was set at 63–90 μm (cutting out the < 63 μm fraction) in the formulation refinement, i.e. from G42 to G48b (see below next point and Table 5);

- **Particle size distribution:**

Since a wide particle size distribution enhances flow (and good cohesion), overall the full range < 90 μm was employed. However, not to overload the granulometric curve with very fine material (<63 μm) generally decreasing flow and increasing absorption, in different trials the curve of one of the angular aggregates (either marble or ramel) was cut, retaining just its 63–90 μm fraction, while the other aggregate was kept at < 90 μm. The curve cutting was mostly effective for WP with ramel (see Table 5), since of the two angular aggregates shortlisted, it is the most porous and has the highest fraction < 63 μm (see Table 3);

- **Shape - Round aggregates:**

Although round, Poraver® (porous, rough surface) did not provide good injectability and flow, while Scotchlite™ (non-porous, smooth surface), as expected, greatly enhanced them. Over testing, Scotchlite™ S22 proved to provide better injectability and flow compared to Scotchlite™ K1 (or compared to a mix of S22 and K1), and therefore S22 was selected. An amount of Scotchlite™ S22 above 1.5 pt/V in the mix caused clogging of the needle and a very poor flow, most probably due to its high very fine fraction;

- **Shape - Angular aggregates:**

While angular aggregates are necessary for good packing geometry and cohesion, they may make injectability difficult through small needles. Comparing quartz sand vs. marble powder (both very poorly porous aggregates, Table 2), quartz sand, having a more roundish shape (Fig. 2d), improved flow, but it reduced injectability (see Table 5, G39 vs. G40); marble was therefore selected in the formulation refinement (see Table 5).

Table 5. Grout formulation refinement

Grout #	S22 < 75 μm (pt/V)	Ramel < 90 μm (pt/V)	Ramel 63-90 μm (pt/V)	Marble < 90 μm (pt/V)	Marble 63-90 μm (pt/V)	Quartz sand < 90 μm (pt/V)	Water (pt/V)	Inject (mL)	Flow1 (cm)	Shrinkage
39	1.5	1.5	–	–	–	0.5	1.5	18G:1 19G:1	21	cracking and sinking
40	1.5	1.5	–	–	0.5	–	1.5	18G:4 19G:3	18	deep crack
41	1.5	1	–	–	1	–	1.5	18G:4 19G:4	15	deep crack
42	1.5	–	1	1	–	–	1.5	18G:7 19G:5	14	deep crack
44	1.5	–	1	1	–	–	1.4	18G:2 19G:1	12	narrow cracks
46	1.5	–	0.8	1.2	–	–	1.5	18G:6 19G:2	18.5	no cracks; no shrink. in depth
48b	1.5	–	0.8	1.2	–	–	1.65	18G:6 19G:5	18.5	no cracks; no shrink. in depth

[1] All the grouts in the table gave a homogeneous drip, not flat.

2.3 Overall Assessment and Discussion

While focusing on fundamental WP in order to achieve a mixture fulfilling the performance criteria and being usable on site (very narrow access points), a close look was necessary to another crucial property: shrinkage, having an influence on adhesion. No grout can be used if cracking and shrinking, since that can cause failure of the overall intervention. For this reason, it is advisable to reproduce as accurately as possible site conditions, including support (Globigerina Limestone in this case) and pre-wetting: if the grout already cracks under controlled laboratory conditions, it is more likely to do so on site, where a number of variables are not controllable and where the conservator cannot see how the grout actually performs (problem and intervention concealed).

Therefore, while injectability and flow were promising already in grout G40 (ramel < 90 μm and marble 63–90 μm, Table 5), adjustments were needed to reduce shrinkage. Since the adjustment of the proportion of these two aggregates (G41, Table 5) was ineffective (deep cracks were still observed), a different particle size distribution was adopted, with ramel 63–90 μm and marble < 90 μm (from G42 onwards, Table 5): the predominant (ca. 86%, Table 3) very fine and absorbent fraction of ramel < 63 μm was cut, keeping instead the fine fraction of marble, having much lower absorption. In this way, on the one hand WP were improved (G42 vs. G41, Table 5), while shrinkage was also improved: with a lower absorption of the mixing liquid by the aggregates, cracking was reduced. Since G42 was still cracking, two options were explored to address the problem: water reduction (G44, Table 5) and aggregates proportion adjustment (G46, Table 5). Water reduction in G44 impaired WP to a non-acceptable extent, while a slight increase in marble (poorly porous and poorly absorbent, all curve < 90 μm present, helping packing geometry and shrinkage) and decrease in ramel led to no cracking and no shrinkage, with good flow and injectability through 18G with 0.84 mm diameter (G46, Table 5). To improve injectability through the smallest needle 19G (0.69 mm diameter), the water content of the mixture was increased (G48b, Table 5): such grout still showed no shrinkage, while retaining good flow and improving in injectability.

Grout 48b was therefore selected as the final mix.

3   Conclusions

Specific sections of the single layer of lime-based plaster of the 16th c. wall paintings cycle of the Great Siege by Perez d'Aleccio (Grand Master's Palace, Valletta, Malta) showed delamination from the Globigerina Limestone support, with pockets estimated to be 2–5 mm thick. Access points for injection grouting were limited to narrow cracks, just ≤ 1.5 mm wide, and could not be enlarged due to the high significance of the painting and the risk of inducing further destabilisation.

Starting from the knowledge of the original materials and the definition of the problem, site-specific performance criteria were set, including injectability through 18G and 19G needles and good flow. Grout design and testing of fundamental properties proceeded hand-in-hand, incrementally modifying the formulation according to the results of the tests. A range of angular $CaCO_3$-based aggregates was employed, and soda-lime borosilicate glass round aggregates, with different porosity and therefore water absorption, and different particle size distribution. These factors were carefully considered and balanced in order to achieve the necessary injectability and flow while obtaining good cohesion and no shrinkage. A thin lime-based grout was designed, injectable through a 0.69 mm inner diameter needle (19G) and showing no shrinkage in a pre-wet porous support (Globigerina Limestone cup) simulating conditions on site. This was achieved just relying on the role of aggregates in the mixture, carefully considering particle morphology, particle size distribution, and particle water absorption. Such factors, which proved to have an impact on the overall grout formulation, were varied and deliberately manipulated to obtain a thin grout without the use of additives such as superplasticisers.

Acknowledgements. The Department of Conservation and Built Heritage, University of Malta, is conserving the wall paintings cycle of the Great Siege by Perez d'Aleccio, in the Grand Master's Palace in Valletta, Malta, in partnership with Heritage Malta, and under the auspices of the Office of the President of the Republic of Malta. The project is funded by Gasan Foundation, Planning Authority and Melita Foundation through RIDT (Research Innovation & Development Trust) University of Malta. Thanks to Heritage Malta and Daniel Cilia for the pictures of the NE wall used as basemaps.

References

1. Rickerby, S., Shekede, L., Wei, T., Hai, Q., Jinjian, Y., Piqué, F.: Development and testing of the grouting and soluble-salts reduction treatments of Cave 85 wall paintings. In: Agnew, N. (ed.) Conservation of Ancient Sites on the Silk Road. Proceedings of the Second International Conference on the Conservation of Grotto Sites, pp. 471–479. Getty Conservation Institute, Los Angeles (2010)

2. Cather, S.: Trans-technological methodology: setting performance criteria for conserving wall paintings. In: Dipinti murali dell'estremo Oriente: diagnosi, conservazione e restauro: quando Oriente e Occidente si incontrano e si confrontano, pp. 89–95. Longo Editore, Ravenna (2006)

3. Biçer-Şimşir, B., Griffin, I., Palazzo-Bertholon, B., Rainer, L.: Lime-based injection grouts for the conservation of architectural surfaces. Rev. Conserv. **10**, 3–17 (2009)

4. Espinosa Rodriguez, A.: The Great Siege fresco by Perez d'Aleccio. In: Ganado, A. (ed.) Palace of the Grandmasters in Valletta, pp. 55-71. Patrimonju Publishing, Valletta (2001)

5. Cassar, J.: Deterioration of the globigerina limestone of the Maltese islands. Geol. Soc. Lond. Spec. Publ. **205**(1), 33–49 (2002)

6. Gorodetska, M.: The plastering technique of Matteo Perez d'Aleccio's Great Siege wall painting cycle in the Grand Master's Palace, Valletta, Malta. MSc dissertation, University of Malta, Malta (2022)

7. Pro Arte s.n.c: Indagine mineralogico-petrografica. Impasti artificiali decorati - Malta. Rapporto di prova n° 44/1. Unpublished report for the Department of Conservation and Built Heritage, University of Malta, Noventa Vicentina (2021)

8. Skłodowski, R., Drdácký, M., Skłodowski, M.: Identifying subsurface detachment defects by acoustic tracing. NDT E Int. Indep. Nondestr. Test. Eval. **56**, 56–64 (2013)

9. Griffin, I.: Pozzolanas as additives for grouts: an investigation of their working properties and performance characteristics. Stud. Conserv. **49**(1), 23–34 (2004)

10. Needle Gauge Table. https://darwin-microfluidics.com/blogs/tools/syringe-needle-gauge-table. Accessed 15 Mar 2022

11. CEN, EN 16572:2015(Main). Conservation of cultural heritage - Glossary of technical terms concerning mortars for masonry, renders and plasters used in cultural heritage (2015)

12. Pasian, C., Martin de Fonjaudran, C., Rava, A.: Innovative water-reduced injection grouts for the stabilisation of wall paintings in the Hadi Rani Mahal, Nagaur, India: design, testing and implementation. Stud. Conserv. **65**(sup1), 244–250 (2020)

13. González-Sánchez, J.F., Taşcı, B., Fernández, J.M., Navarro-Blasco, Í., Alvarez, J.I.: Combination of polymeric superplasticizers, water repellents and pozzolanic agents to improve air lime-based grouts for historic masonry repair. Polymers **12**(4), 887 (2020)

14. Duran, A., González-Sánchez, J.F., Fernández, J.M., Sirera, R., Navarro-Blasco, Í., Alvarez, J.I.: Influence of two polymer-based superplasticizers (poly-naphthalene sulfonate, PNS, and lignosulfonate, LS) on compressive and flexural strength, freeze-thaw, and sulphate attack resistance of lime-metakaolin grouts. Polymers **10**(8), 824 (2018)

15. Pachta, V.: The role of glass additives in the properties of lime-based grouts. Heritage **4**(2), 906–916 (2021)
16. Papayianni, I., Pachta, V.: Experimental study on the performance of lime-based grouts used in consolidating historic masonries. Mater. Struct. **48**(7), 2111–2121 (2014). https://doi.org/10.1617/s11527-014-0296-5
17. Padovnik, A., Piqué, F., Jornet, A., Bokan-Bosiljkov, V.: Injection grouts for the re-attachment of architectural surfaces with historic value—measures to improve the properties of hydrated lime grouts in Slovenia. Int. J. Archit. Heritage **10**(8), 993–1007 (2016)
18. Flatt, R.J., Schober, I.: Superplasticizers and the rheology of concrete. In: Roussel, N. (ed.) Understanding the Rheology of Concrete, pp. 144–208. Woodhead Publishing Limited, Sawston (2012)
19. Pasian, C., Secco, M., Piqué, F., Rickerby, S., Artioli, G., Cather, S.: Performance of grout with reduced water content: the importance of porosity and related properties. In: Papayianni, I., Stefanidou, M., Pachta, V. (eds.) Proceedings of the 4th Historic Mortars Conference (HMC 2016), Santorini, Greece, pp. 639–46. Laboratory of Building Materials, Department of Civil Engineering, Aristotle University of Thessaloniki, Greece (2016)
20. Duffy, J.: Controlling suspension rheology: the physical characteristics of dispersed particles have a large impact on overall rheological properties. Chem. Eng. **122**(1), 34–39 (2005)
21. Hansen, E.F., Rodriguez-Navarro, C., Balen, K.: Lime putties and mortars, insights into fundamental properties. Stud. Conserv. **53**, 9–23 (2008)
22. Rodriguez-Navarro, C., Ruiz-Agudo, E., Ortega-Huertas, M., Hansen, E.: Nanostructure and irreversible colloidal behavior of $Ca(OH)_2$: implications in cultural heritage conservation. Langmuir **21**(24), 10948–10957 (2005)
23. Pasian, C., Secco, M., Piqué, F., Artioli, G., Rickerby, S., Cather, S.: Lime-based injection grouts with reduced water content: an assessment of the effects of the water-reducing agents ovalbumin and ethanol on the mineralogical evolution and properties of grouts. J. Cult. Herit. **30**, 70–80 (2018)
24. Poraver®, Technical data sheet. http://5.imimg.com/data5/XV/AR/HW/SELLER-1247927/poraver-light-weight-aggregates.pdf. Accessed 29 Mar 2022
25. Scotchlite[TM], Technical data sheet. https://multimedia.3m.com/mws/media/619093O/3m-glass-bubbles-types-k-and-s-uk-data-sheet.pdf. Accessed 29 Mar 2022
26. Cassar, J., Vannucci, S.: Petrographical and chemical research on the stone of the megalithic temples. Malta Archaeol. Rev. **5**, 40–45 (2003)
27. Siegesmund, S., Ruedrich, J., Koch, A.: Marble bowing: comparative studies of three different public building facades. Environ. Geol. **56**(3), 473–494 (2008)
28. Platias, S., Vatalis, K.I., Charalampides, G.: Suitability of quartz sands for different industrial applications. Procedia Econ. Finan. **14**, 491–498 (2014)
29. Li, G., Li, X.S., Lv, Q.N., Zhang, Y.: Permeability measurements of quartz sands with methane hydrate. Chem. Eng. Sci. **193**, 1–5 (2019)
30. Xue, B.: A study on chemical grouting of quartz sand. MSc dissertation, The University of Western Ontario, Canada (2018)
31. Biçer-Simsir, B., Rainer, L.: Evaluation of lime-based hydraulic injection grouts for the conservation of architectural surfaces. A manual of laboratory and field test methods. The Getty Conservation Institute, Los Angeles, USA (2013)

Enhancement of Latent Heat Storage Capacity of Lime Rendering Mortars

Andrea Rubio-Aguinaga⑩, José María Fernández⑩, Íñigo Navarro-Blasco⑩, and José Ignacio Álvarez(✉) ⑩

University of Navarra, Pamplona, Spain
jalvarez@unav.es

Abstract. Microencapsulated Phase Change Materials (PCMs) were included in air lime rendering mortars in order to improve the thermal comfort of the inhabitants and the energy efficiency of buildings of the Architectural Heritage under the premises of minimum intervention and maximum compatibility. Three different PCMs were tested and directly added during the mixing process to fresh air lime mortars in three different percentages: 5, 10 and 20 wt. %. Some chemical additives were also incorporated to improve the final performance of the renders: a starch derivative as an adhesion booster; metakaolin as pozzolanic addition to shorten the setting time and to increase the final strength; and a polycarboxylated ether as a superplasticizer to adjust the fluidity of the fresh renders avoiding an excess of mixing water. The specific heat C_p, the enthalpy ΔH ascribed to the phase change and the melting temperature of the PCMs were determined by Differential Scanning Calorimetry (DSC). The capacity of the renders to store/release heat was demonstrated at a laboratory scale. The favourable results proved the effect of these PCMs with respect to the thermal performance of these rendering mortars, offering a promising way of enhancement of the thermal efficiency of building materials of the Cultural Heritage.

Keywords: Air Lime Mortars · Phase Change Materials (PCM) · Thermal Energy Storage Materials · Thermal Efficiency · Renders

1 Introduction

One of the goals of today's society, with greater concern for the environment and sustainability, is to reduce global energy consumption. According to the International Energy Agency (IEA), the construction sector is one of the largest consumers of energy due to the energy costs associated with heating and cooling buildings [1]. For this reason, in recent years, thermal energy storage (TES) systems have been extensively studied in order to improve the thermal efficiency of buildings, with Phase Change Materials (PCM) standing out in particular [1].

Phase change materials are systems capable of storing latent thermal energy through phase changes. Generally, these phase changes are solidification and melting. Thus, during the melting process, the heat gain is stored in the form of latent heat of fusion and,

V. Bokan Bosiljkov et al. (Eds.): HMC 2022, RILEM Bookseries 42, pp. 567–581, 2023.
https://doi.org/10.1007/978-3-031-31472-8_45

during the solidification process; this latent heat is released [2]. Therefore, depending on the external temperature of the system, heat is released or absorbed regulating moderately the temperature of the medium [1].

Phase change materials are a family of materials with high heat values of melting and solidification with the ability to absorb or release large amounts of thermal energy at constant temperature (latent heat) when subjected to a phase change [3]. The characteristics required for a material to be considered a PCM are: non-corrosive, non-toxic, inert, high latent heat, high thermal conductivity, low-priced, congruent melting and not undergo (or at least minimally) subcooling [4]. Phase change materials can be classified according to different criteria. A distinction is made between liquid-liquid PCMs and solid-liquid PCMs according to the state transition. The latter are the most commonly used because of the higher enthalpy associated with the phase change due to the greater freedom of movement of the molecules in the liquid state with respect to those in the solid state [3]. At the same time, solid-liquid PCMs are classified according to their composition as: inorganic (hydrated salts, molten salts and eutectic salts), organic (fatty acids, sugars, crystalline polymers and paraffin waxes) and metallic (metals and alloys) [3, 4].

Ancient monuments or traditional residential buildings show low levels of thermal efficiency and when repair works are addressed the enhancement of the thermal comfort conditions and the reduction of energy consumption is required. There are numerous strategies for the reduction of the thermal demand; however, most of the available envelope upgrades cannot be applied in repair works of Cultural Heritage buildings. This is because their application, their high price, and their incompatibility with some architectural heritage structures do not meet the general requirements of recognition of the tangible and intangible values of Cultural Heritage as well as the performance standards for protection and restoration, such as minimal modification of the aesthetic and history of the property, minimal intervention, highest compatibility of the materials to be introduced and use of new materials and methods for restoration and conservation. For this reason, the incorporation of PCMs in lime renders is, apparently, a suitable option for enhancement the thermal efficiency of historic buildings and monuments as it respects all the above mentioned requirements. However, there is little literature available on the incorporation of phase change materials into lime-based mortars.

There are different methods for the incorporation of PCMs in building materials, the most important are: direct incorporation (mixed directly with the building material), direct immersion (immersion of the porous building material into the molten PCMs), macroencapsulation (panels, tubes and spheres filled with large amounts of PCMs), microencapsulation (PCMs with sizes between 0.1 μm and 1 mm that are enclosed in microcapsules), shape stabilization (melting and mixing of PCM with a polymeric support material), and form-stable method (PCM entrapped in porous polymeric matrix) [1]. In the present work, solid-liquid PCMs are used, whose main drawback is the volume variations they undergo during phase changes that can lead to diffusion of the PCMs in the composite and cause microstructural changes and leakage [3–5]. Therefore, microencapsulated PCMs are used so the macroscopic shape of the PCM is maintained, the heat transfer area is increased and undesired movements in the matrix are avoided, in addition to meeting the restoration and conservation criteria [3].

Paraffin is one of the most studied and widely used solid-liquid PCMs [4]. Paraffin waxes are mainly composed of mixtures of linear alkanes (around 75–100%) together with a small percentage of branched alkanes (isoalkanes, cycloalkanes, alkylbenzenes…) [2]. Commercial paraffin waxes are inexpensive, have a wide range of melting temperatures, have a high latent heat, are thermally and chemically stable, are non-toxic and do not undergo subcooling (4,5). However, the application of paraffin waxes as PCMs is limited by their low thermal conductivity (~0.2 W/m K) which causes a reduction of their rates of heat storage and release during melting and solidification [4, 5]. Consequently, microencapsulation is normally used in order to enhance thermal conductivity by maximizing their heat transfer area.

The aim of this work is the incorporation of microencapsulated PCMs in lime mortars in order to improve the energy efficiency and thermal comfort of the buildings of the architectural heritage. The selection of PCMs for its application as thermal energy storage systems was carried out according to the melting temperature. Specifically, in this work microencapsulated paraffin and bio-based PCMs were assayed. Melting temperatures were selected in order to storage/release heat under rather cold climate conditions ($T_{melting}$ ~ 18 °C) or hot climates ($T_{melting}$ ~ 25 °C). This way, the incorporation of these PCMs in lime mortars will be able to reduce the differences between peak and off-peak thermal loads, reduction of energy consumption, cut in electricity demand, indoor thermal comfort improvement and, ultimately, reduction of CO_2 release to the atmosphere.

In this work, percentages of 5, 10 and 20 wt. % of three different microencapsulated PCMs were added during the mixing process to fresh air lime mortars. The PCMs used were two paraffin-based PCMs with melamine microcapsule with melting points of 18 °C and 24 °C and a bio-based microencapsulated PCM with melting temperature of 29 °C. Other chemical additives were also incorporated to improve the final performance of the renders: a starch derivative as an adhesion booster, metakaolin as pozzolanic addition to shorten the setting time and to increase the final strength; and a polycarboxylate ether as a superplasticizer to adjust the fluidity of the fresh renders avoiding an excess of mixing water.

2 Materials and Methods

2.1 Materials

Rendering air lime mortars were prepared by mixing calcitic air lime supplied by Cal Industrial S.A. (Calinsa Navarra), classified as CL-90-S by European regulations and calcitic sand (supplied by CTH Navarra). Percentages of binder/aggregate weight ratio were 21.7/78.3, whereas the percentage of mixing water was fixed at 25 wt. % of the total weight of the mortar.

To obtain easily workable renders with good performance, different additives and mineral admixtures were also used to optimize the mix composition. A superplasticizer (polycarboxylated ether derivative, MasterCast GT 205) was added to adjust the fluidity of the fresh renders avoiding an excess of mixing water. Different percentages of a starch derivative (Casaplast) were added as an adhesion booster. Metakaolin (MK, supplied by

METAVER) was added in some of the mortars (20 wt. % with respect to the weight of lime, bwol) in order to increase the final strength and durability of the rendering mortars.

Three different solid-liquid microencapsulated PCMs supplied by Microtek, were used: two paraffin-based PCMs with melamine microcapsule with melting points of 18 °C and 24 °C (denoted as 18PCM and 24PCM) and a bio-based microencapsulated PCM with melting temperature of 29 °C (29PCM). Percentages bwol of 5%, 10% and 20% of PCM were directly added to fresh air lime mortars during the mixing process.

As control group two PCM-free mortars (CTRL-1, MK-free, and CTRL-2, with 20 wt.% of MK) were prepared in order to compare the PCM performance (Table 1 and Table 2 gather the composition of the mixes).

The consistency of all the samples was assessed by the flow table test and seen to be within the range of 160–215 mm of slump.

Table 1. Composition of PCM-free renders and renders containing 24PCM.

Render	PCM-free		24PCM					
	CTRL-1	CTRL-2	24PCM-1	24PCM-2	24PCM-3	24PCM-4	24PCM-5	24PCM-6
PCM (wt. %)	0	0	5	5	10	10	20	20
SP (wt. %)	0.6	0.75	0.75	0.75	0.75	0.75	0.75	0.75
MK (wt. %)	0	20	0	20	0	20	0	20
Starch (wt. %)	0.50	0.50	0.50	0.50	0.50	0.50	0.50	0.50

Table 2. Composition of renders containing 18PCM.

Render	18PCM					
	18PCM-1	18PCM-2	18PCM-3	18PCM-4	18PCM-5	18PCM-6
PCM (wt. %)	5	5	10	10	20	20
SP (wt. %)	0.75	0.75	0.75	0.75	0.75	0.75
MK (wt. %)	0	20	0	20	0	20
Starch (wt. %)	0.50	0.50	0.50	0.50	0.50	0.50

Table 3. Composition of renders containing 29PCM.

Render	29PCM					
	29PCM-1	29PCM-2	29PCM-3	29PCM-4	29PCM-5	29PCM-6
PCM (wt. %)	5	5	10	10	20	20
SP (wt. %)	0.75	0.75	0.75	0.75	0.75	0.75
MK (wt. %)	0	20	0	20	0	20
Starch (wt. %)	0.50	0.50	0.50	0.50	0.50	0.50

2.2 Experimental Methods

For the preparation of the fresh grouts, air lime, sand, metakaolin, adhesion booster, the corresponding PCM and an initial percentage of superplasticizer (0.25% with respect to the weight of lime, bwol) were blended for 5 min using a solid additives mixer BL-8-CA (Lleal, S.A., Granollers, Spain) to achieve a homogenous mix.

A fixed percentage of mixing water (25 wt. %) was then added at low speed for 270 s in a Proeti ETI 26.0072 (Proeti, Madrid, Spain) mixer. Accumulative additions of 0.25% bwol of superplasticizer were added until adequate fluidity (as measured by the flow table test). For each render, the percentages of SP (along with the adhesion booster) were modified until applicable, adherent and low-cracking renders were achieved, as detailed in a further work.

Regarding the hardened state study, fresh mixtures were moulded into cylindrical specimens with dimensions of 33 mm diameter and 39 mm height, and then cured at lab conditions (20 °C $\pm$ 0.5 °C and 45% $\pm$ 5% RH).

2.3 Characterization Methods

The characterization and study of the thermal properties of the mortars was carried out through different techniques, such as Thermogravimetric Analysis (TGA), Differential Scanning Calorimetry (DSC) and thermal conductivity measurements.

Differential Scanning Calorimetry (DSC). Thermal performance of the enhanced lime-based mortars was studied by Differential Scanning Calorimetry (DSC). The parameters studied were specific heat capacity (C_p), the enthalpy (ΔH_m) ascribed to the phase change and the melting temperature (T_m) of the PCMs, I.E., the latent heat storage. The equipment used to perform the DSC analysis was the DSC25 TA Instruments and the data evaluation was carried out with TRIOS from TA Instruments. C_p was determined using modulated DSC in a range of temperatures of -25 °C to 60 °C. ΔH_m and T_m were determined in powdered representative samples of ca. 7 mg, in 40 ML aluminum pans with hermetic lids, applying 3 cycles of heating-cooling between 0 °C and 50 °C with an increment of 5 °C/min. At the beginning and end of each cycle an isotherm is applied for 5 min. All analyses are performed under nitrogen atmosphere (flow of 50 mL/min).

Thermogravimetric Analysis (TGA). Thermogravimetric analysis was used to study the thermal stability of the renders. The equipment used to perform the thermogravimetric analysis was the SDTA650 TA Instruments and the data evaluation was carried out with TRIOS from TA Instruments. Samples of ca. 10 mg in 90 µL alumina pans were analysed. The heating program applied consisted of a heating ramp (20 °C/min) from 35 °C to 1000 °C under a nitrogen atmosphere (flow of 100 mL/min).

Thermal Conductivity Measurements. The equipment used to perform the thermal conductivity measurements was the FOX50 Heat Flow Meter from TA Instruments. This instrument consists of two parallel plates controlled by two independent Peltier cells. Each plate is subjected to a specific temperature with a temperature difference between plates of 10 °C, thus generating a thermal gradient and obtaining the thermal conductivity at the average temperature between the two plates. Strict thermal equilibrium criteria were applied to the measurements consisting of 6 consecutive blocks of data acquisition. These criteria included, for each plate, a temperature deviation of less than 1 °C from the setpoint temperature and a variation of less than 200 µV of the transducer signals from those obtained in the previous block. The thermal conductivity of 55 mm diameter and 20 mm thick disks of the rendering mortars was measured at 5 °C, 15 °C, 25 °C and 35 °C. In this way, the conductivity of the macroscopic material is measured at temperatures where the PCM is in the solid state and in the liquid state. In all cases, renders were, at least, 28 days cured before the thermal conductivity measurement. In addition, in order to obtain more representative values three disks of each render were measured and the average was calculated along with its standard deviation.

3   Results and Discussion

3.1   Evaluation of Latent Heat Thermal Energy Storage

Firstly, pure microencapsulated PCMs thermal stabilities were analyzed by TGA (Fig. 1). The degradation of the additive is observed in all cases at around 400 °C typical of melamine [8]. It is notable in the case of 24PCM that it starts to degrade at lower temperatures with a smaller mass loss at 155 °C.

The thermal stabilities of PCM-bearing renders were also studied. Figure 2 includes TGA curves of some PCM-bearing renders in comparison with the PCM-free render (CTRL-1). It is shown in all cases the degradation of the PCM around 400 °C and the calcite decarbonation at 800 °C [9].

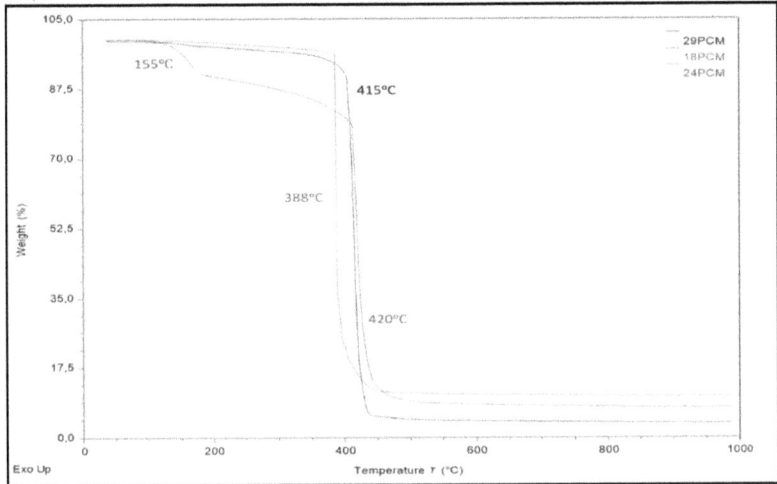

Fig. 1. TGA curves of pure PCMs.

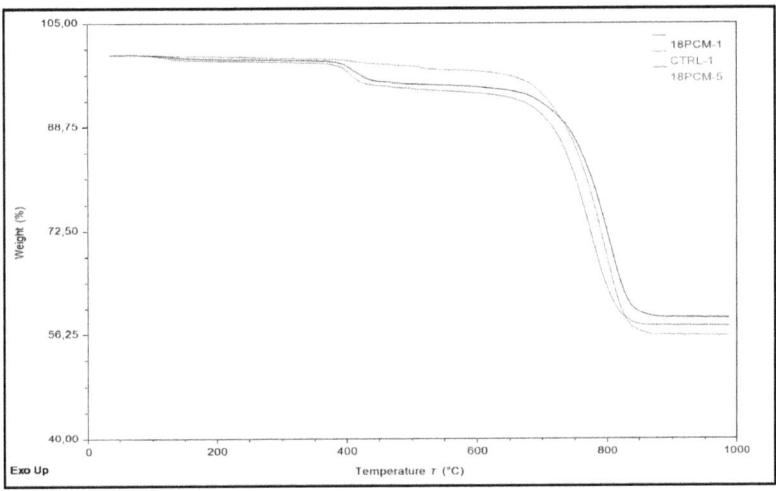

Fig. 2. TGA curves of 18PCM-1, 18PCM-5 and CTRL-1.

The parameters obtained with the DSC measurements were the melting temperature (T_m), the enthalpy ascribed to the phase change (ΔH_m) (that is, the latent heat thermal energy storage capacity) and the specific heat capacity (C_p). These parameters are useful for the study of the thermal behaviour of mortars after the addition of different percentages of PCMs.

Pure PCMs were studied. Figure 3 depicts the heat flows during the melting processes of the pure PCMs. It can be seen how all of them present similar enthalpies around 150 J/g, the highest (158.97 J/g) corresponding to the 24PCM additive. In addition, it is possible to check that their melting temperatures are in accordance with their data sheets.

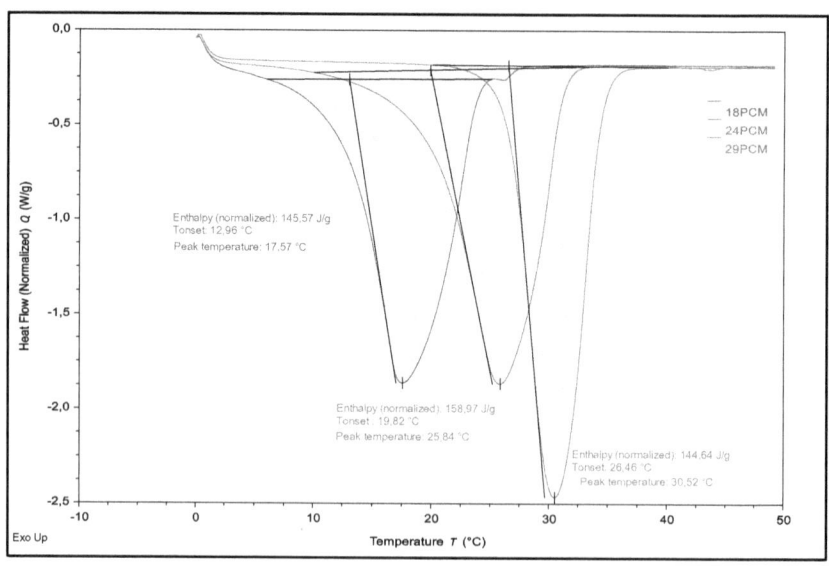

Fig. 3. DSC curves of pure PCMs.

Figure 4 includes DSC curves of the MK-free renders with 10% of each type of PCM compared to their control render (CTRL-1). It is possible to clearly distinguish the melting process of each PCM after its addition to the lime-based mortars.

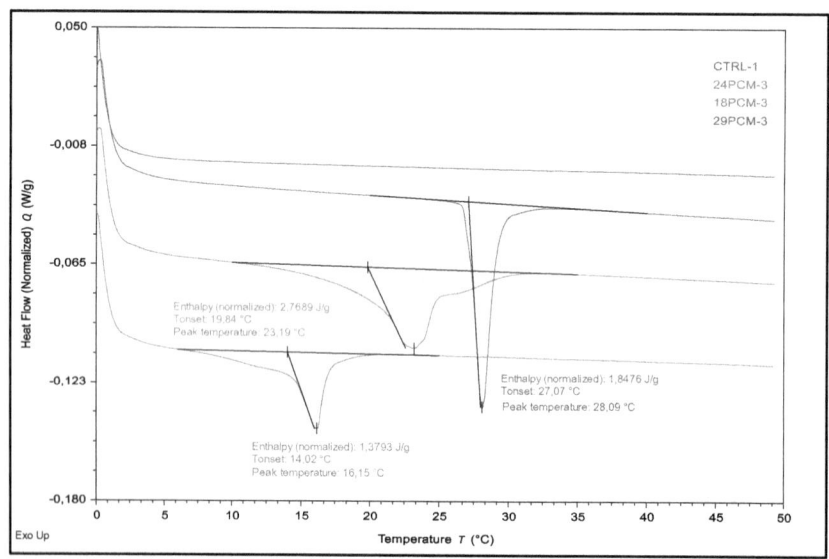

Fig. 4. DSC curves of MK-free renders compared to CTRL-1.

The latent heat thermal energy storage capacity (measured as enthalpy) along with the onset temperatures and peak temperatures of PCM-bearing renders are included in Tables 4 and 5.

Table 4. ΔH_m, T_{peak} and T_{onset} values of PCM-free renders and 24PCM-renders.

Render	PCM (%)	Curing days	ΔH_m (J/g)	T_{peak} (°C)	T_{onset} (°C)
CTRL-1	0	28	0.0	–	–
	0	91	0.0	–	–
CTRL-2	0	28	0.0	–	–
	0	91	0.0	–	–
24PCM-1	5	28	0.74 ± 0.01	24.20 ± 0.01	19.80 ± 0.09
	5	91	0.83 ± 0.03	21.79 ± 0.09	19.68 ± 0.06
24PCM-2	5	28	0.84 ± 0.01	24.16 ± 0.01	16.43 ± 0.11
	5	91	0.92 ± 0.01	24.36 ± 0.01	16.73 ± 0.04
24PCM-3	10	28	2.78 ± 0.02	23.19 ± 0.01	19.80 ± 0.09
	10	91	2.13 ± 0.03	24.05 ± 0.02	19.68 ± 0.06
24PCM-4	10	28	2.08 ± 0.01	23.51 ± 0.01	19.84 ± 0.08
	10	91	2.12 ± 0.02	24.23 ± 0.01	19.08 ± 0.04
24PCM-5	20	28	5.81 ± 0.01	23.87 ± 0.01	19.88 ± 0.09
	20	91	6.15 ± 0.01	24.18 ± 0.01	20.32 ± 0.08
24PCM-6	20	28	4.03 ± 0.02	23.89 ± 0.01	19.82 ± 0.09
	20	91	5.49 ± 0.02	24.18 ± 0.01	20.60 ± 0.07

As expected, the capacity of the PCM-bearing mortars to store (or release) heat, as monitored by the enthalpy values, show an increasing pattern as the amount of PCM incorporated in the mortar increases (Fig. 5). This pattern is maintained in both curing ages and after the addition of MK. In general, MK-free renders show higher enthalpy values and therefore heat storage capacity, this is particularly true for high percentages of PCM.

Regarding the C_p measurements (that is, the sensible heat capacity), it was observed that the addition of the PCMs does not affect severely the C_p of the render either when the PCM is in the solid phase or when it is in the liquid phase (Fig. 6 and 7) (that is, excluding the temperature interval in which a phase change – melting or solidification - takes place). In the case of PCM-bearing mortars, the thermodynamic melting process is clearly observed in where the C_p values vary drastically (Fig. 6 and 7). The C_p values obtained in all cases are within the usual range for these materials (0.73–1.26 J/g °C) [6, 7]. However, in this work, there was not a clear trend to increase the heat capacity of the material when PCMs were incorporated [6, 7].

Table 5. ΔH_m, T_{peak} and T_{onset} values of 18PCM-renders and 29PCM-renders.

Render	PCM (%)	Curing days	ΔH_m (J/g)	T_{peak} (°C)	T_{onset} (°C)
18PCM-1	5	28	0.83 ± 0.01	16.16 ± 0.01	14.18 ± 0.04
	5	91	0.42 ± 0.01	16.36 ± 0.01	13.80 ± 0.05
18PCM-2	5	28	0.62 ± 0.02	16.27 ± 0.01	14.31 ± 0.06
	5	91	0.41 ± 0.01	16.52 ± 0.01	14.54 ± 0.05
18PCM-3	10	28	1.38 ± 0.01	16.16 ± 0.01	13.97 ± 0.07
	10	91	1.52 ± 0.01	16.12 ± 0.01	13.67 ± 0.04
18PCM-4	10	28	1.33 ± 0.01	16.15 ± 0.01	13.92 ± 0.01
	10	91	1.63 ± 0.02	16.27 ± 0.01	14.06 ± 0.01
18PCM-5	20	28	4.41 ± 0.01	16.10 ± 0.01	13.66 ± 0.01
	20	91	3.18 ± 0.01	16.15 ± 0.01	13.76 ± 0.01
18PCM-6	20	28	2.53 ± 0.01	16.10 ± 0.01	13.72 ± 0.08
29PCM-1	5	28	0.92 ± 0.01	28.07 ± 0.03	27.08 ± 0.01
29PCM-2	5	28	1.17 ± 0.02	28.08 ± 0.01	27.05 ± 0.01
29PCM-3	10	28	1.84 ± 0.01	28.10 ± 0.01	27.07 ± 0.01
29PCM-4	10	28	1.24 ± 0.02	28.04 ± 0.01	27.05 ± 0.01
29PCM-5	20	28	4.80 ± 0.01	28.30 ± 0.01	27.15 ± 0.02
29PCM-6	20	28	4.99 ± 0.02	28.23 ± 0.01	26.93 ± 0.01

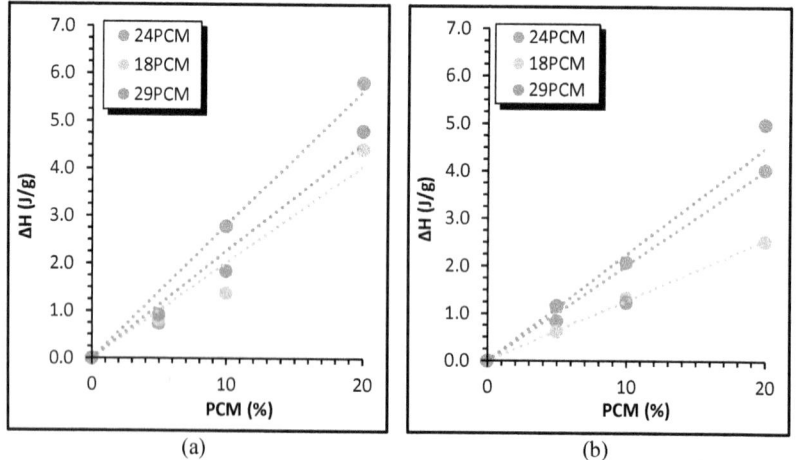

Fig. 5. Enthalpy values of renders at 28 curing days: (a) MK-free; (b) 20% MK.

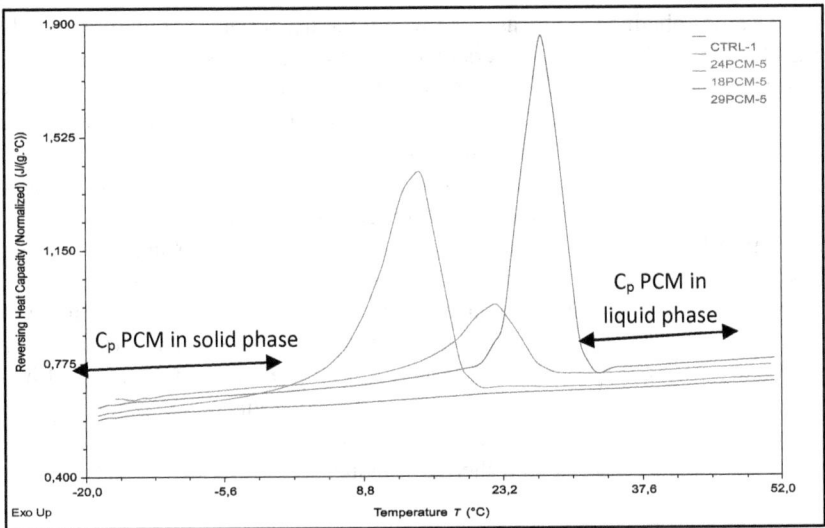

Fig. 6. C_p values of MK free renders at 28 curing days.

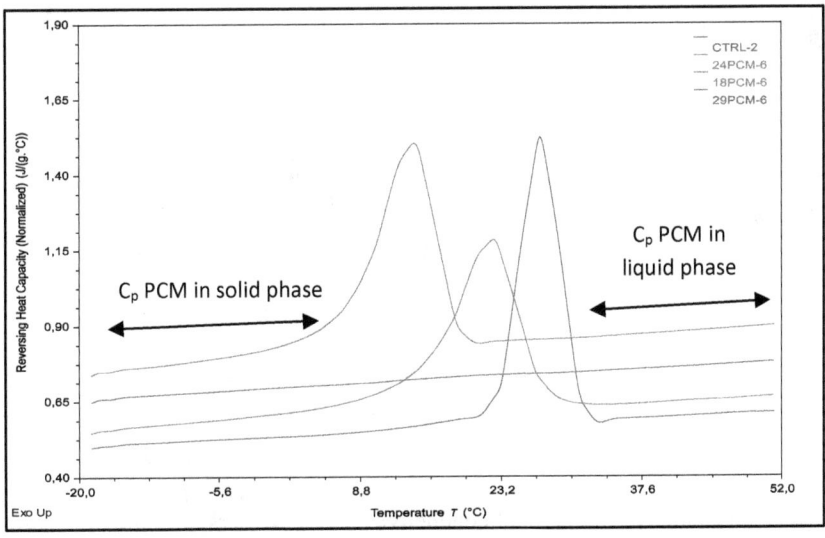

Fig. 7. C_p values of 20% MK renders at 28 curing days.

3.2 Thermal Conductivity Measurements

Tables 3, 4 and 5 include thermal conductivity coefficient (λ) values of the lime renders measured at 5 °C, 15 °C, 25 °C and 35 °C. Thermal conductivity is the rate at which heat flows through a material. Therefore, the higher the thermal conductivity coefficient (λ), the higher the thermal conductivity of the material.

The Λ values obtained in all cases are within the usual range for these materials [6, 7]. It is observed in PCM-free renders that the presence of metakaolin does not affect the thermal conductivity of the samples. In addition, PCMs apparently do not have a strong influence on the thermal conductivity either. These values are consistent with the pore size distributions as the addition of PCMs had little effect on the pore size distributions (Fig. 8 and 9) due to the previous optimization of the mix compositions by using appropriate chemical additives. As it can be seen, main pore size and critical pore size (threshold for the main Hg intrusion) are kept almost constant for the different renders. However, some previous works [6, 7] reported a decrease in thermal conductivity after the addition of PCM ascribing it to the lower conductivity of this additive. This fact might be ascribed to the fact that in those works, conversely to the current one, the percentage of mixing water to achieve adequate workability was adjusted having probably strong influence in the macroscopic structure of the material. In this work the water ratio remained constant in all renders resulting in tiny differences in the structure of the renders and therefore in their thermal conductivity (Tables 6, 7 and 8).

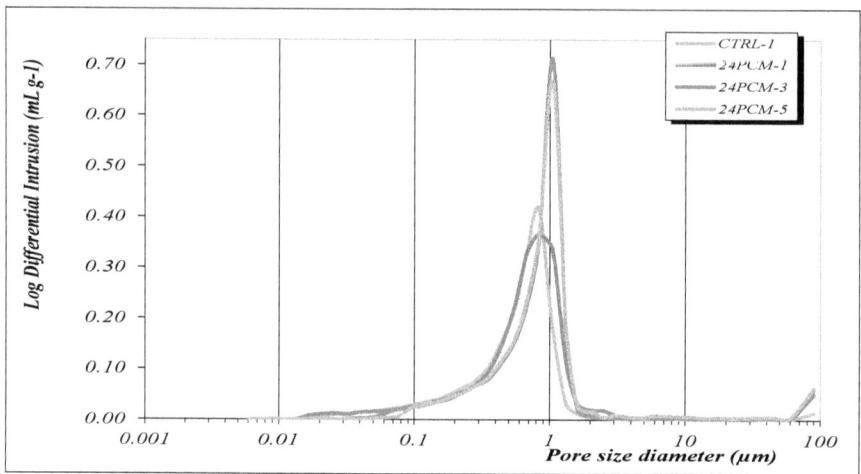

Fig. 8. Comparison of pore size distribution of MK-free renders containing 24PCM at 91 curing days.

The maintenance of the thermal conductivity of the renders after the incorporation of the additives could be advantageous to achieve maximum heat storage. In this sense, the matrix of the rendering mortar should be able to allow the heat to reach the inner part of the microencapsulated PCMs, to undergo the phase change and thus be able to regulate moderately the temperature.

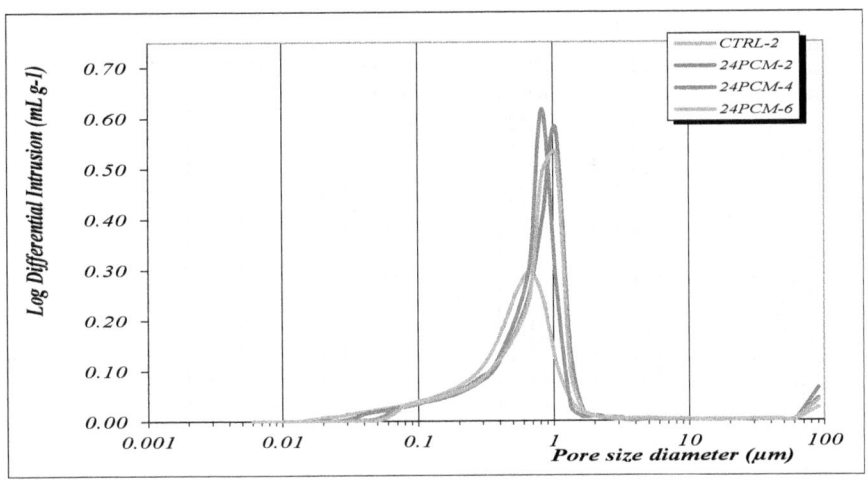

Fig. 9. Comparison of pore size distribution of renders containing 24PCM and 20% MK at 91 curing days.

Table 6. Thermal conductivity coefficients of PCM-free renders at different temperatures.

PCM-free		
	0% MK	20% MK
T (°C)	λ (W/m · K)	λ (W/m · K)
5	0.37 ± 0.05	0.38 ± 0.03
15	0.37 ± 0.05	0.38 ± 0.03
25	0.37 ± 0.05	0.38 ± 0.03
35	0.37 ± 0.05	0.38 ± 0.03

Table 7. Thermal conductivity coefficients of 24PCM renders at different temperatures.

	24PCM					
	5% PCM		10% PCM		20% PCM	
	0% MK	20% MK	0% MK	20% MK	0% MK	20% MK
T (°C)	λ (W/m · K)	λ (W/m · K)	λ (W/m · K)	λ (W/m · K)	λ (W/m · K)	λ (W/m · K)
5	0.44 ± 0.04	0.36 ± 0.03	0.40 ± 0.03	0.39 ± 0.05	0.36 ± 0.06	0.38 ± 0.01
15	0.45 ± 0.04	0.36 ± 0.03	0.40 ± 0.03	0.39 ± 0.04	0.36 ± 0.06	0.38 ± 0.01
25	0.45 ± 0.05	0.35 ± 0.03	0.40 ± 0.03	0.40 ± 0.03	0.37 ± 0.06	0.38 ± 0.01
35	0.45 ± 0.05	0.35 ± 0.04	0.40 ± 0.04	0.40 ± 0.03	0.37 ± 0.06	0.38 ± 0.01

Table 8. Thermal conductivity coefficients of 18PCM renders at different temperatures.

	18PCM					
	5% PCM		10% PCM		20% PCM	
	0% MK	20%MK	0% MK	20%MK	0% MK	20%MK
T (°C)	λ (W/m · K)	λ (W/m · K)	λ (W/m · K)	λ (W/m · K)	λ (W/m · K)	λ (W/m · K)
5	0.39 ± 0.02	0.40 ± 0.02	0.40 ± 0.08	0.41 ± 0.01	0.46 ± 0.02	0.37 ± 0.08
15	0.39 ± 0.02	0.40 ± 0.02	0.41 ± 0.08	0.41 ± 0.01	0.46 ± 0.02	0.38 ± 0.08
25	0.39 ± 0.02	0.40 ± 0.02	0.41 ± 0.08	0.40 ± 0.01	0.46 ± 0.02	0.38 ± 0.08
35	0.39 ± 0.02	0.40 ± 0.02	0.40 ± 0.08	0.40 ± 0.01	0.45 ± 0.02	0.38 ± 0.08

4 Conclusions

Thermal performance of PCM-enhanced air lime renders was studied. The specific heat C_p, the enthalpy ΔH ascribed to the phase change and the melting temperature of the PCMs were determined by Differential Scanning Calorimetry (DSC). Results showed that C_p values barcly vary with the percentage of PCM. As expected, the latent heat thermal energy storage capacity (measured as enthalpy) values ascribed to the melting process of the PCMs added to the renders show an almost linear trend, the higher the amount of this additive, the higher the enthalpy.

Thermal conductivity measurements have shown that the addition of PCMs does not weaken the thermal conductivity of the renders. This could be due to the optimization of the mix composition during the mortar preparation, in which the percentage of mixing water is remained constant resulting in little variation in the macrostructure of the material.

Favourable results have proven the effect of these PCMs on the thermal performance of the renders offering promise for improved thermal efficiency of Cultural Heritage building materials.

The capacity of the renders to store/release heat demonstrated at a laboratory scale and the use of models that imitate building envelopes might be also suggested as future achievements.

Acknowledgements. Funded by Spanish Ministerio de Ciencia e Innovación, grant number PID2020-119975RB-I00 LIMORTHER and by Gobierno de Navarra, MULTIFICON PC143.

References

1. Frigione, M., Lettieri, M., Sarcinella, A., Barroso de Aguiar, J.: Sustainable polymer-based phase change materials for energy efficiency in buildings and their application in aerial lime mortars. Constr. Build. Mater. **231**, 1–11 (2020)
2. Himran, S., Suwono, A., Mansoor, G.A.: Characterization of alkanes and paraffin waxes for application as phase change energy storage medium. Energy Sources **16**(1), 117–128 (1994)

3. Liu, H., Wang, X., Wu, D.: Innovative design of microencapsulated phase change materials for thermal energy storage and versatile applications: a review. Sustain. Energy Fuels **3**(5), 1091–1149 (2019)
4. Farid, M.M., Khudhair, A.M., Razack, S.A.K., Al-Hallaj, S.: A review on phase change energy storage: Materials and applications. Energy Convers. Manag. **45**(9–10), 1597–1615 (2004)
5. Ukrainczyk, N., Kurajica, S., Šipušić, J.: Thermophysical comparison of five commercial paraffin waxes as latent heat storage materials. Chem. Biochem. Eng. Q. **24**(2), 129–137 (2010)
6. Theodoridou, M., Kyriakou, L., Ioannou, I.: PCM-enhanced lime plasters for vernacular and contemporary architecture. Energy Procedia **97**, 539–545 (2016)
7. Haurie, L., Serrano, S., Bosch, M., Fernandez, A.I., Cabeza, L.F.: Single layer mortars with microencapsulated PCM: Study of physical and thermal properties and fire behaviour. Energy Build. **111**, 393–400 (2016)
8. García-Viñuales, S., et al.: Study of melamine-formaldehyde/phase change material microcapsules for the preparation of polymer films by extrusion. Membranes **12**(3), 266 (2022)
9. Lanas, J., Sirera, R., Alvarez, J.I.: Compositional changes in lime-based mortars exposed to different environments. Thermochim. Acta **429**(2), 219–226 (2005)

Obtaining of Repair Lime Renders with Microencapsulated Phase Change Materials: Optimization of the Composition, Application, Mechanical and Microstructural Studies

Andrea Rubio-Aguinaga , José María Fernández , Íñigo Navarro-Blasco ,
and José Ignacio Álvarez(✉)

University of Navarra, Pamplona, Spain
jalvarez@unav.es

Abstract. Different batches of repair lime rendering mortars were designed by mixing microencapsulated Phase Change Materials (PCMs) and other additives. The final aim of these renders is to improve the thermal efficiency of the envelope of the Built Heritage, while allowing the practitioners to apply a render with positive final performance. The combinations of the PCMs in different weight percentages, a superplasticizer (to increase the fluidity of the render keeping constant the mixing water), an adhesion improver and a pozzolanic additive were studied. The adhesion of these renders onto bricks and limestone specimens and the shrinkage and cracking of the mortars were studied in detail. X-ray diffraction technique was used to study the composition and evolution of the carbonation process. Compressive strength measurements were studied in hardened specimens. In addition, the porous structure of the rendering mortars was studied by mercury intrusion porosimetry to assess the effect of the PCMs' addition. Results have shown that these thermally enhanced mortars are feasible materials for real-life application in the context of architectural heritage restoration and conservation.

Keywords: Air Lime Mortars · Phase Change Materials (PCM) · Durability · Compressive Strength · Shrinkage · Adhesion · Architectural Heritage · Restoration

1 Introduction

A widely recognized good practice in rehabilitation is the use of lime-based mortars for restoration works of the Architectural Heritage. It has positive environmental benefits compared to cement, due to their lower energy consumption in the production phase, lower ingredient contributions and potential global warming processes. Furthermore, their atmospheric CO_2 absorption properties are also useful minimizing the carbon footprint [1]. Lime mortars have been widely used both in the interior and exterior of buildings, some of its most frequent applications are: renders and plasters, masonry bedding mortars, ornamental pieces, finishing mortars, flooring mortars… [1].

V. Bokan Bosiljkov et al. (Eds.): HMC 2022, RILEM Bookseries 42, pp. 582–598, 2023.
https://doi.org/10.1007/978-3-031-31472-8_46

However, traditional residential buildings, ancient monuments and historic buildings commonly show low levels of thermal efficiency. Therefore, when repair works are addressed, the enhancement of the thermal performance of the building is required. There are many thermal consumption reduction strategies; nevertheless, most envelope upgrades available cannot be applied to cultural heritage repair work. This is because their application, their high price, and their incompatibility with some architectural heritage structures do not meet the general requirements of recognition of the tangible and intangible values of Cultural Heritage as well as the performance standards for protection and restoration, such as minimal modification of the aesthetic and history of the property, minimal intervention, highest compatibility of the materials to be introduced and use of new materials and methods for restoration and conservation.

In this context of reduction of the thermal demand, phase change materials are becoming increasingly important. Phase Change Materials, also known as PCMs, are systems capable of storing latent thermal energy through phase changes. Usually, these phase changes are solidification and melting. Thus, during the melting process, the heat gain is stored in the form of latent heat of fusion and, during the solidification process, this latent heat is released [2]. Therefore, depending on the external temperature of the system, heat is released or absorbed regulating moderately the temperature of the medium [3].

This work focuses on the "in bulk" incorporation of microencapsulated PCMs in lime rendering mortars as an interesting option for improving the thermal properties of historic buildings and monuments. Specifically, three different solid-liquid microencapsulated PCMs were directly incorporated during the mixing process to fresh air lime mortars. Solid-liquid PCMs were selected because of the higher enthalpy associated with the phase change. The PCMs were also microencapsulated with the aim of maintaining the macroscopic shape of the PCM, increasing the heat transfer area and avoiding undesired movements in the matrix of the mortar [4].

The aim of this work is the optimization of the composition of air lime-based renders containing PCMs, studying their physical-mechanical performance after their application. Since the incorporation of PCMs in bulk would lead to alterations of the fresh and hardened mortar's properties, this research work addresses the use of different additives to overcome practical problems. Therefore, different percentages of an adhesion booster were added along with different proportions of a superplasticizer to adjust the fluidity of the fresh renders avoiding an excess of mixing water.

In addition, metakaolin was added as a pozzolanic agent. Metakaolin is generally processed by calcining high purity clay and it contains active forms of silica and alumina which react with portlandite ($Ca(OH)_2$) yielding hydrated calcium silicate phases (C-S-H). The formation of new hydrated phases provides the enhancement of several properties of the lime-based mortars such as high values of mechanical strength and durability, low water permeability and good cohesion between binders and aggregates [5, 6].

In this way, several batches of mortars were prepared in order to optimize and obtain the best composition of PCM-bearing rendering mortar, according to the final application of the material.

2 Materials and Methods

2.1 Materials

Calcitic air lime mortars were prepared by mixing calcitic air lime supplied by Cal Industrial S.A. (Calinsa Navarra), classified as CL-90-S by European regulations and calcitic sand (supplied by CTH Navarra). Percentages of binder/aggregate weight ratio were 21.7/78.3, whereas the percentage of mixing water was fixed at 25 wt. % of the total weight of the mortar.

Different additives and mineral admixtures were also used to optimize the mix composition of the renders. As an adhesion booster, different percentages of a starch derivative (Casaplast) were added along with various percentages of a polycarboxylated ether derivative (MasterCast GT 205) as a superplasticizer to adjust the fluidity of the fresh renders avoiding an excess of mixing water.

Metakaolin (MK, supplied by METAVER) was added in some of the mortars (20 wt. % with respect to the weight of lime, bwol) in order to increase the final strength and durability of the rendering mortars.

Three different solid-liquid microencapsulated PCMs supplied by Microtek, were used: two paraffin-based PCMs with melamine microcapsule with melting points of 18 °C and 24 °C (denoted as 18PCM and 24PCM) and a bio-based microencapsulated PCM with melting temperature of 29 °C (29PCM). Percentages bwol of 5%, 10% and 20% of PCM were directly added to fresh air lime mortars during the mixing process.

As control group two PCM-free mortars (CTRL-1, MK-free, and CTRL-2, with 20 wt.% of MK) were prepared in order to compare the PCM performance (Table 1 and Table 2 gather the composition of the mixes).

2.2 Experimental Methods

For the preparation of the fresh grouts, air lime, sand, metakaolin, adhesion booster, the corresponding PCM and an initial percentage of superplasticizer (0.25% bwol) were blended for 5 min using a solid additives mixer BL-8-CA (Lleal, S.A., Granollers, Spain) to achieve a homogenous mix.

A fixed percentage of mixing water (25 wt. %) was then added at low speed for 270 s in a Proeti ETI 26.0072 (Proeti, Madrid, Spain) mixer. Accumulative additions of 0.25% bwol of superplasticizer were added until adequate fluidity (as measured by the flow table test) and adhesion on a brick surface for render application were achieved at the discretion of a technician specialized in the production of mortars' mixtures and their application in the form of a single layer.

Regarding the hardened state study, fresh mixtures were moulded into cylindrical specimens with dimensions of 33 mm diameter and 39 mm height, and then cured at lab conditions (20 °C ± 0.5 °C and 45% ± 5% RH).

2.3 Characterization Methods

Fresh State Tests. Several fresh state tests were carried out in order to characterize the fresh mortars. The fluidity of the fresh mortars was measured using the flow table test

(UNE-EN 1015-3 standard [7]). The density of the paste and the percentage of entrained air were determined according to the UNE-EN 1015-6 standard [8] and UNE-EN 1015-7 standard [9] respectively. In addition, the water retentivity was measured using the UNE-EN 83-816-93 [10]. Lastly, the setting time was measured according to the UNE-EN 1015-9 standard [11].

Single Coat Mortars: Adhesion, Shrinkage, Cracks Formation. As these mortars are intended to be used as renders, the performance of a single coat applied on a water-saturated brick was studied. Renders of 0.5 cm thickness of the different mortars were layered on saturated bricks. In order to evaluate the degree of adhesion and the formation or not of either cracks or fissures, a qualitative evaluation based on visual appearance after 1 month of the application was carried out. The criterion was the following: degree 0 (complete adhesion to the substrate and no evidence of cracks), degree 1 (complete adhesion and presence of very shallow and few cracks), degree 2 (complete adhesion and presence of numerous and shallow cracks) and degree 3 (poor adhesion and presence of numerous and deep cracks).

XRD. X-ray diffraction technique was used to study the composition and evolution of the carbonation process. X-Ray diffraction measurements were carried out in a Bruker D8 Advance diffractometer with Cu Kα1 radiation, from 5° to 80° (2θ), 1 s per step and a step size of 0.03°. The evaluation of the data was executed with DIFFRACplusEVA ® from Bruker.

Compressive Strength. The determination of the compressive strength of the hardened mortars at the ages of 28 days and 91 days was carried out with a Frank/Controls 81565 press with a Proeti ETI 26.0052 compressive breaking device and a breaking speed of 20–50 N/s with a time interval between 30 and 90 s. Three specimens of 33 mm diameter and 39 mm height were used for each case in order to achieve representative values.

Mercury Intrusion Porosimetry. Mercury intrusion porosimetry was used to study the porous structure of the rendering mortars with the aim of assessing the effect of the PCMs' addition. Mercury intrusion porosimetry was carried out in a Micromeritics AutoPore IV 9500 apparatus. Measurements of cubic fragments of the mortars of ca. 1 cm edge were performed with a pressure range between 0.0015 and 207 MPa.

3   Results and Discussion

3.1   Optimization of the Mix Composition: Assessment of the Adherence

Firstly, some mortars were prepared without the use of an adhesion booster (Fig. 1), which showed a poor performance. It was observed that the fresh mixture needed an additive capable, on one hand, of improving adhesion and, on the other hand, of avoiding/minimizing the crack formation. A viscosity enhancer with a sticky nature such as the starch and the starch derivatives might be useful to this aim. Starch derivatives have been shown to be able to increase the adherence of a plain lime mortar and to reduce the number of cracks [12]. Their performance as water retainers prevents a quick drying off when the renders were applied onto absorbent substrates. For this reason, percentages by weight (with respect to the weight of lime) of 0.25% and 0.50% of a starch derivative as an adhesion booster were added to the admixture.

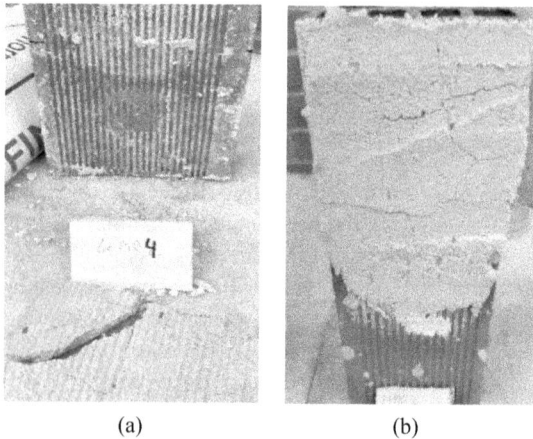

(a)   (b)

Fig. 1. Rendering on saturated brick of mortar: (a) containing 10% PCM, 20% MK and 0.45% SP; (b) containing 20% MK and 0.35% SP.

As it is shown in the comparison between Fig. 2(a–b), with 0.25 wt. % of the additive, and Fig. 3(a–b), 0.50 wt.% of the additive, the low dosage (0.25%) was not optimal (no really good adherence, some cracks). However, 0.50% (Fig. 3(a–b)) significantly improved adhesion and prevented crack formation. For this reason, this ratio was selected for the preparation of the mortars.

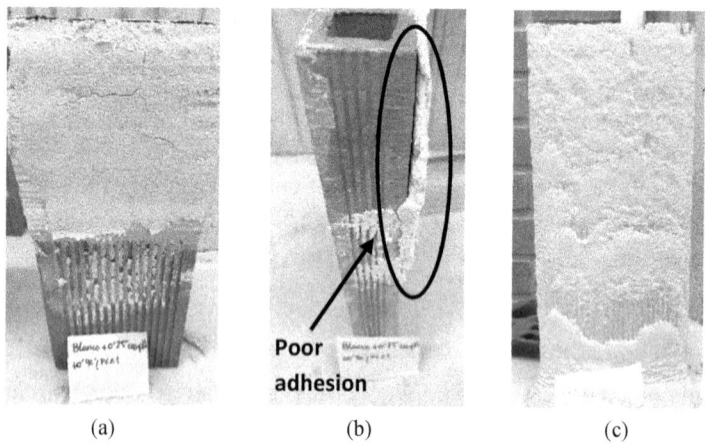

(a)   (b)   (c)

Fig. 2. Renderings on saturated brick of mortar containing: (a) 0.25% starch and 0.40% SP (frontal); (b) 0.25% starch and 0.40% SP (side face); (c) 1.25% bwol of SP.

Simultaneously, for each render composition, the percentages of the superplasticizing additive were adjusted in order to achieve a workable consistency. Percentages from 0.1% to 1.25% bwol were tested, obtaining at the respective ends either unworkable dry renders (yielding a non-applicable render) or, on the other hand, highly fluid and slippery renders, which came unstuck from the brick, dropping off (Fig. 2(c)).

The effectiveness of the polycarboxylate-based SP for lime mortars had been pointed out in the literature [13] and in the current work was seen to be really useful to mitigate the impact in fluidity ascribed to the MK and PCM addition. Low dosages of the tested SP were effective in providing suitable consistency to the renders without extra mixing water requirements.

In this way, controlling fluidity, adhesion and prevention of cracks formation optimal percentages of adhesion booster and superplasticizer were selected for each render composition with PCMs (Table 1). Some examples of the optimal single coat mortars are included in Fig. 3 and 4.

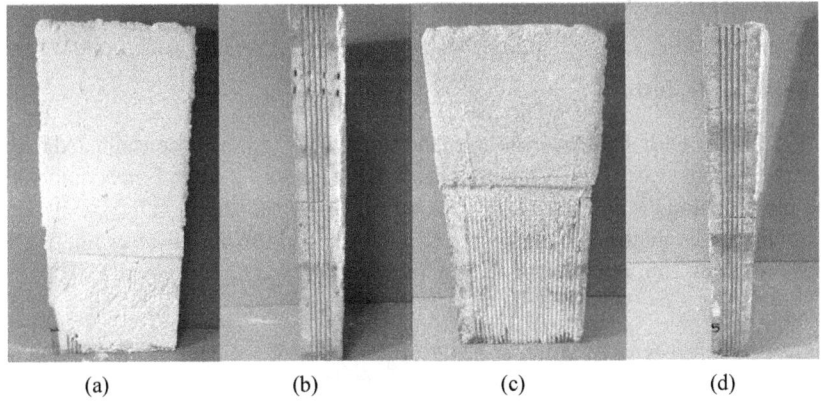

| (a) | (b) | (c) | (d) |

Fig. 3. Rendering on saturated brick of mortar containing (a) 0.50% starch and 0.60% SP (frontal); (b) 0.50% starch and 0.60% SP (side face); (c) 0.50% starch, 0.75% SP and 5% 24PCM (frontal); (d) 0.50% starch, 0.75% SP and 5% 24PCM (side face).

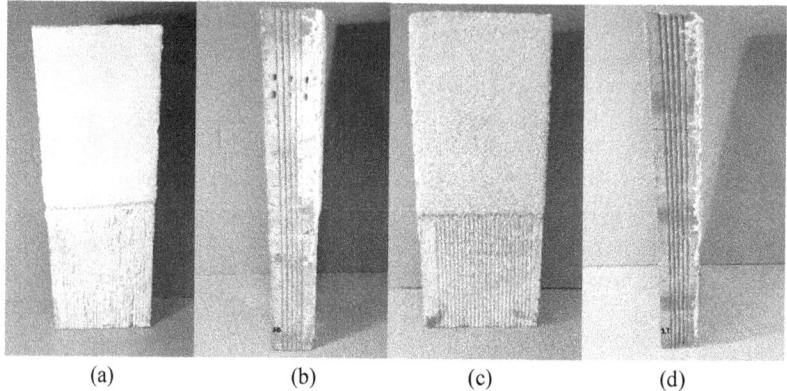

(a)   (b)   (c)   (d)

Fig. 4. Rendering on saturated brick of mortar containing (a) 0.50% starch, 0.75% SP, 20% MK and 10% 18PCM (frontal); (b) 0.50% starch, 0.75% SP, 20% MK and 10% 18PCM (side face); (c) 0.50% starch, 0.75% SP, 20% MK and 20% 29PCM (frontal); (d) 0.50% starch, 0.75% SP, 20% MK and 20% 29PCM (side face).

3.2 Fresh State Tests

Once the percentages of adhesion booster and superplasticizer had been adjusted for each render, fresh state tests were carried out on all the batches of mortar. Final compositions of the mortars along with their fluidity, as measured by the slump values (Fig. 5), and their qualitative evaluation on the degree of adhesion and the formation or not of cracks and fissures after 1 month are included in Tables 1, 2 and 3.

Table 1. Compositions for PCM-free and 24PCM-bearing renders and their fluidity and adhesion test.

Render	PCM-free		24PCM					
	CTRL-1	CTRL-2	24PCM-1	24PCM-2	24PCM-3	24PCM-4	24PCM-5	24PCM-6
PCM (wt. %)	0	0	5	5	10	10	20	20
SP (wt. %)	0.6	0.75	0.75	0.75	0.75	0.75	0.75	0.75
MK (wt. %)	0	20	0	20	0	20	0	20
Starch (wt. %)	0.50	0.50	0.50	0.50	0.50	0.50	0.50	0.50
Air lime (wt. %)	21.7	21.7	21.7	21.7	21.7	21.7	21.7	21.7
Calcitic sand (wt. %)	78.3	78.3	78.3	78.3	78.3	78.3	78.3	78.3

(continued)

Table 1. (*continued*)

Render	PCM-free		24PCM					
	CTRL-1	CTRL-2	24PCM-1	24PCM-2	24PCM-3	24PCM-4	24PCM-5	24PCM-6
Water (wt. %)	25	25	25	25	25	25	25	25
Slump (mm)	182	180	177	177	175	171	161	164
Qualitative evaluation of rendering (0–3)	0	0	0	0	0	0	0	1

Table 2. Compositions for 18PCM-bearing renders and their fluidity and adhesion test.

Render	18PCM					
	18PCM-1	18PCM-2	18PCM-3	18PCM-4	18PCM-5	18PCM-6
PCM (wt. %)	5	5	10	10	20	20
SP (wt. %)	0.75	0.75	0.75	0.75	0.75	0.75
MK (wt. %)	0	20	0	20	0	20
Starch (wt. %)	0.50	0.50	0.50	0.50	0.50	0.50
Air lime (wt. %)	21.7	21.7	21.7	21.7	21.7	21.7
Calcitic sand (wt. %)	78.3	78.3	78.3	78.3	78.3	78.3
Water (wt. %)	25	25	25	25	25	25
Slump (mm)	178	173	189	175	183	195
Qualitative evaluation of rendering (0–3)	0	0	0	0	1	1

All mortars showed complete adhesion to the substrate and no (or at most minimal) cracks were formed. In addition, all consistencies were suitable for their application as a render (no slippage).

Table 3. Compositions for 29PCM-bearing renders and their fluidity and adhesion test.

Render	29PCM					
	29PCM-1	29PCM-2	29PCM-3	29PCM-4	29PCM-5	29PCM-6
PCM (wt. %)	5	5	10	10	20	20
SP (wt. %)	0.75	0.75	0.75	0.75	0.75	0.75
MK (wt. %)	0	20	0	20	0	20
Starch (wt. %)	0.50	0.50	0.50	0.50	0.50	0.50
Air lime (wt. %)	21.7	21.7	21.7	21.7	21.7	21.7
Calcitic sand (wt. %)	78.3	78.3	78.3	78.3	78.3	78.3
Water (wt. %)	25	25	25	25	25	25
Slump (mm)	179	195	190	211	214	194
Qualitative evaluation of rendering (0–3)	0	0	0	0	0	0

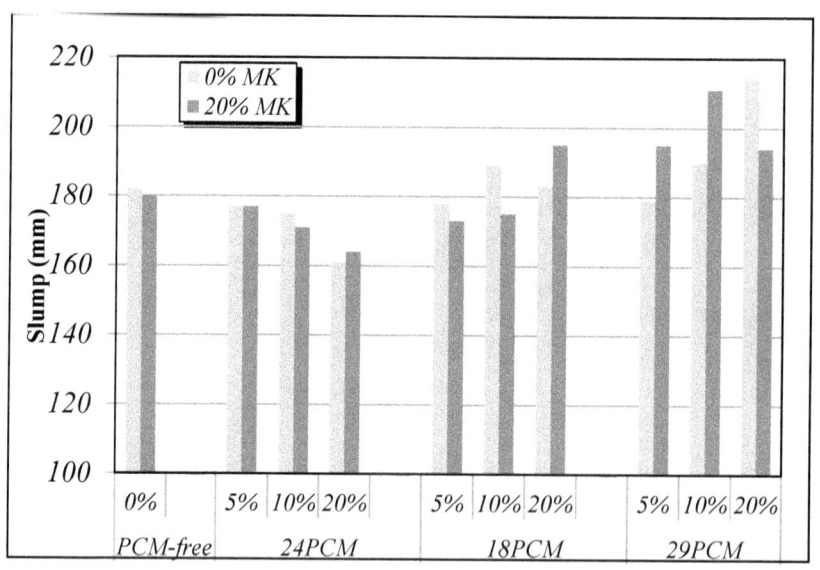

Fig. 5. Fluidity values (slump measured in the flow table test) of the different renders.

As it is shown in Fig. 5, the bio-based 29PCM increases in general the fluidity of the render, even with the highest percentages assayed (20%). This could be due to the PCM particle size, spherical shape and a subsequent ball-bearing effect.

Fresh state tests results are included in Tables 4, 5 and 6.

Table 4. Fresh state tests results of PCM-free renders and renders containing 24PCM.

Render	PCM-free		24PCM					
	CTRL-1	CTRL-2	24PCM-1	24PCM-2	24PCM-3	24PCM-4	24PCM-5	24PCM-8
Stiffening time (min)	1157	1305	1211	1168	1392	1540	1432	1295
Paste density (kg/L)	1.94	1.94	1.83	1.83	1.85	1.83	1.79	1.78
Entrained air (%)	4.3	2.4	6.3	7.2	7.0	6.4	7.0	5.6
Water retentivity (%)	95.9	95.6	95.2	93.1	93.1	93.7	92.2	94.8

Table 5. Fresh state tests results of mortars containing 18PCM.

Render	18PCM					
	18PCM-1	18PCM-2	18PCM-3	18PCM-4	18PCM-5	18PCM-6
Stiffening time (min)	1119	1306	1580	1341	1769	1377
Paste density (kg/L)	1.95	1.93	1.93	1.94	1.89	1.89
Entrained air (%)	2.2	2.8	1.7	1.7	1.1	3.0
Water retentivity (%)	93.5	94.3	92.3	95.0	93.1	93.4

Table 6. Fresh state results of mortars containing 29PCM.

Render	29PCM					
	29PCM-1	29PCM-2	29PCM-3	29PCM-4	29PCM-5	29PCM-6
Stiffening time (min)	1397	1179	1047	1212	1277	1185
Paste density (kg/L)	1.90	1.89	1.89	1.91	1.87	1.85
Entrained air (%)	4.3	6.0	5.2	4.5	6.0	5.6
Water retentivity (%)	96.9	93.7	93.7	94.9	92.7	94.3

Generally, as it is shown in Tables 4, 5 and 6, due to the use of different additives, stiffening time, entrained air, paste density and water retentivity are not dramatically affected by the addition of any of the PCMs in any percentage.

3.3 Mineralogical Characterization (XRD)

X-ray diffraction patterns of the PCM-free renders at 28 and at 91 days of curing are shown in Fig. 6. The progress of the carbonation process is observed since the intensities of the characteristic peaks of portlandite (PDF 44-1481) decay while the intensities of the characteristic peaks of calcite (PDF 05-0586) increase. These variations are subtle since the curing days are not far apart. In addition, it is possible to observe quartz (PDF 33-1161) coming from the pozzolanic agent. The C-S-H phases resulting from the pozzolanic reaction are hardly distinguished due to their amorphous nature [6].

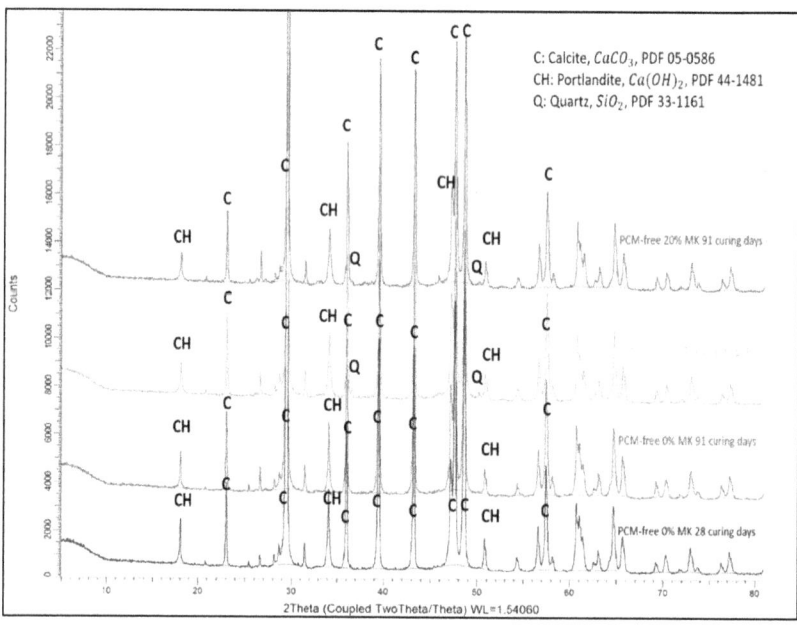

Fig. 6. XRD patterns of the PCM-free renders.

The XRD patterns of the Phase Change Materials are displayed in Fig. 7. The halo between 10° and 30° 2θ show in 18PCM' diffractogram is due to its amorphous condition. The sharp diffraction peaks shown in 24PCM pattern correspond to a paraffin wax (PDF 53-1532) and to the melamine-based microcapsule (PDF 02-0164). 29PCM datasheet does not include composition information. However, bio-based PCMs are usually organic fatty acid esters made from underutilized and renewable raw materials, mainly animal fats and vegetable oils [14].

Percentages of portlandite ($Ca(OH)_2$, PDF 44-1481) present for each render at each curing day are included in Table 7. It is observed that the amount of portlandite decreases with the age of curing, evidencing the progress of carbonation.

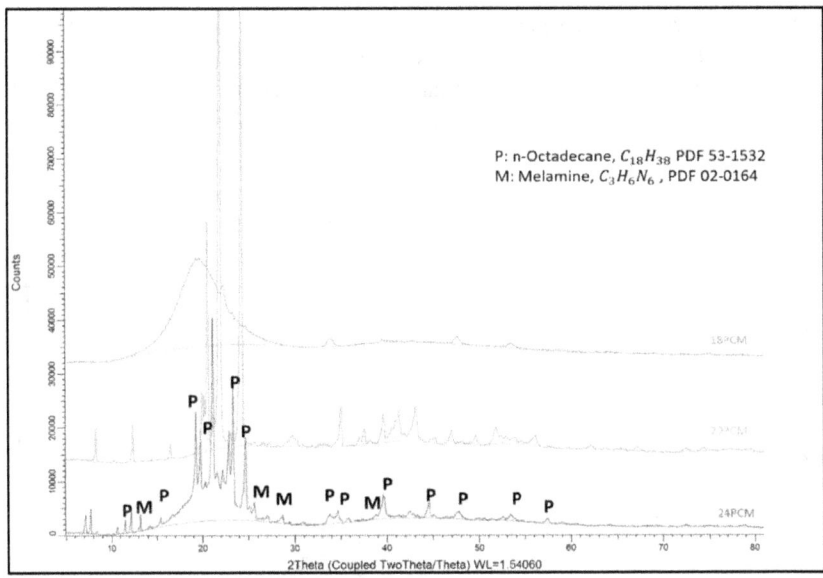

Fig. 7. XRD patterns of the PCMs.

Table 7. Percentages of portlandite of the different renders calculated with XRD.

Render	Curing days	Ca(OH)$_2$ (%)	Render	Curing days	Ca(OH)$_2$ (%)	Render	Curing days	Ca(OH)$_2$ (%)
CTRL-1	28	3.0	24PCM-6	28	3.9	29PCM-1	28	4.9
	91	2.7		91	3.3		91	3.5
CTRL-2	28	2.6	18PCM-1	28	3.0	29PCM-2	28	4.5
	91	2.5		91	2.6		91	3.3
24PCM-1	28	8.6	18PCM-2	28	4.4	29PCM-3	28	6.3
	91	4.6		91	3.4		91	4.8
24PCM-2	28	2.4	18PCM-3	28	3.3	29PCM-4	28	3.2
	91	2.3		91	2.1		91	2.4
24PCM-3	28	4.7	18PCM-4	28	3.2	29PCM-5	28	4.3
	91	3.1		91	2.2		91	2.5
24PCM-4	28	2.6	18PCM-5	28	3.8	29PCM-6	28	4.2
	91	2.3		91	3.5		91	3.1
24PCM-5	28	5.7	18PCM-6	28	3.7			
	91	5.0		91	3.5			

3.4 Compressive Strength

Compressive strength values of the mortars at 28 and 91 curing days are included in Fig. 8. In general, the addition of the pozzolanic agent increases the compressive strength of the material. It is observed that the PCM addition does not severely affect the mechanical strength of the mortars. This is particularly true for the lowest percentage of PCM (5%). The 18PCM yielded, on average, the highest compressive strength values. For MK-free renders, the values were in general similar, or higher, than that of the control mortar. However, the presence of PCM was detrimental for the strength of MK-bearing renders compared to control mortar.

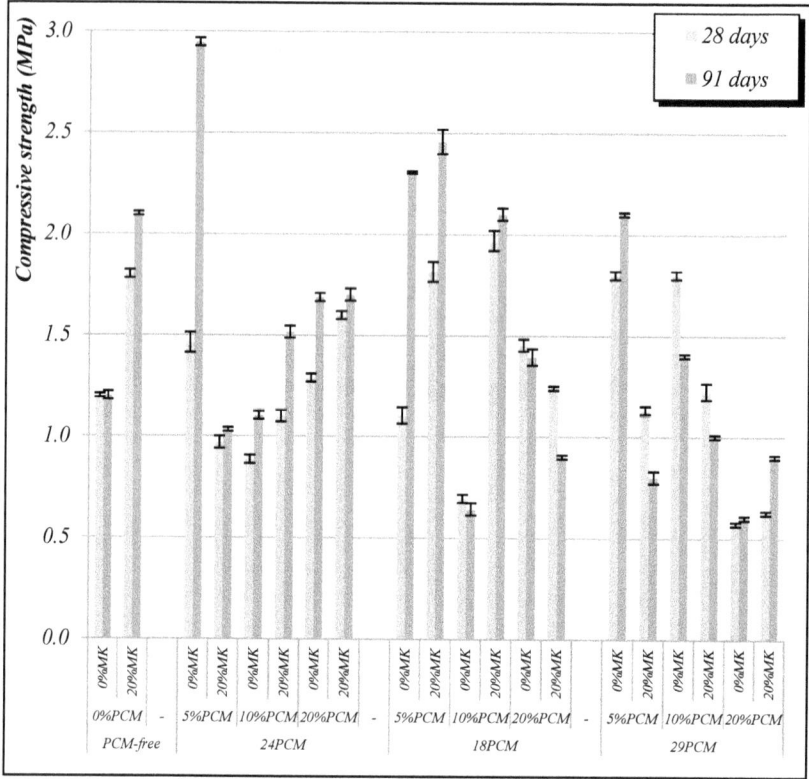

Fig. 8. Compressive strength of renders at different curing times.

3.5 Pore Size Distribution

The effect of the PCM addition on the pore size distribution has been studied (Table 8). PCM-free renders have shown a reduction in porosity and average pore size as the curing process progresses. This porosity reduction was sharper for MK-free control, in which the absence of pozzolanic reaction leads to increased carbonation.

For PCM-bearing renders, an increase in the total porosity and in the average pore size as the percentage of PCM increases was generally observed. This increase could be due to the inherent porosity of the PCM and/or the possible generation of discontinuity with the conglomerate matrix (to be checked with SEM studies).

In addition, the filling effect of metakaolin has been verified by comparing the PCM-free renders with and without 20% MK CTRL-1 and CTRL-2 [6]. This pattern was also observed with some mortars containing PCM, for example, 24PCM-3 (MK-free) and 24PCM-4 (20% MK) and 24PCM-5 (MK-free) and 24PCM-6 (20% MK).

Table 8. Porosity percentages and average pore diameter.

Render	PCM (%)	Curing days	Porosity (%)	Average pore diameter (μm)
CTRL-1	0	28	36.7	0.3543
	0	91	26.0	0.3335
CTRL-2	0	28	36.4	0.3222
	0	91	35.0	0.3043
24PCM-1	5	28	41.6	0.3276
	5	91	39.0	0.4176
24PCM-2	5	28	40.3	0.4530
	5	91	41.5	0.5434
24PCM-3	10	28	42.0	0.7033
	10	91	40.1	0.6936
24PCM-4	10	28	39.7	0.5687
	10	91	38.6	0.5352
24PCM-5	20	28	44.1	0.9492
	20	91	39.5	0.7840
24PCM-6	20	28	35.9	0.5742
	20	91	38.9	0.5882
18PCM-5	20	28	36.4	0.4400
	20	91	34.1	0.5229

Comparison of pore size distributions are included in Fig. 9 and 10. It is shown how the unimodal distribution is maintained after the addition of the PCMs in every percentage. However, a shift of the main peak of the distribution together with changes in the area under the curve corresponding to the increase in mean pore size and total porosity are observed.

In general, the variations in porosity are consistent with the compressive strength results (please, see Sect. 3.4). However, in some cases, an increase in mechanical strength is observed along with higher porosity (e.g. 24PCM-1). Further work will be necessary to study the microstructural interaction between the different rendering mortar's components by SEM. The increase in porosity might be related to the inherent porosity of the PCM, while at the same time increasing by filling effect the compactness of the binding matrix, thus explaining the higher compressive strength.

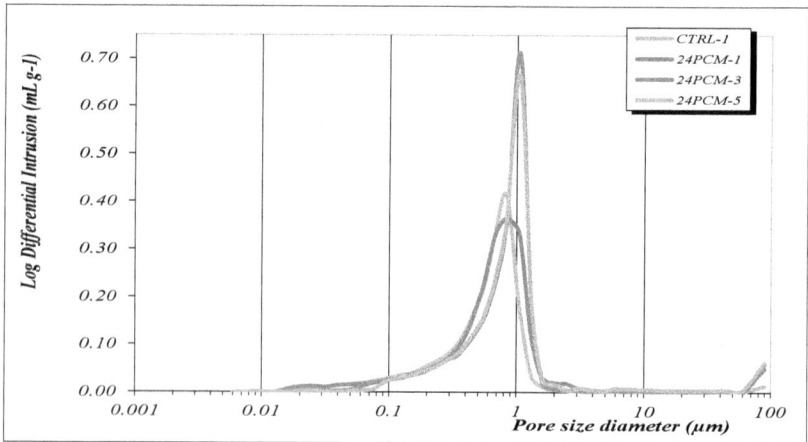

Fig. 9. Comparison of pore size distribution of MK-free renders containing 24PCM at 91 curing days.

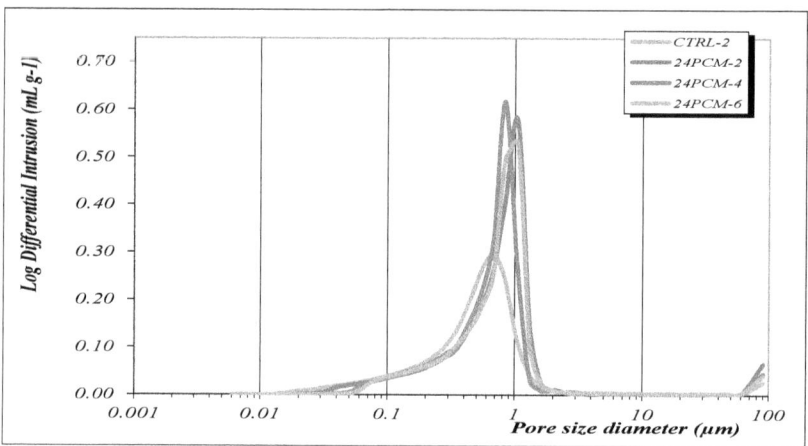

Fig. 10. Comparison of pore size distribution of renders containing 24PCM and 20% MK at 91 curing days.

4 Conclusions

Microencapsulated PCMs have been successfully incorporated in bulk into air lime-based rendering mortars. Furthermore, the composition of these mortars has been optimized considering their final application as building envelopes. Different chemical additives have been used to obtain workable, easily spreadable mortars with good adhesion and low cracking. Different quantities of a polycarboxylated-based superplasticizer have been adjusted to control the flowability of the paste. An adhesion booster (starch derivative) has been added to improve adhesiveness and to reduce crack formation.

Metakaolin as pozzolanic agent has also been included to improve the compressive strength performance.

The fresh state results have shown that the addition of these additives can mainly affect the fluidity of the mortar, while the water retention, setting time, entrained air and paste density have not been greatly affected. The compressive strengths of the mortars have not been dramatically modified for the lowest dosage of PCM. The 18PCM yielded on average the highest strengths. Finally, the addition of PCMs resulted in an increase in porosity and in the average pore diameter compared to PCM-free renders.

After the optimization of the composition of the mixes and due to the use of chemical additives, the incorporation of PCMs has been feasible and not detrimental to the properties of the render. Specific mix compositions have been ascertained for each one the renders as a function of the microencapsulated PCM type and its percentage. These renders have been demonstrated to be easily applicable.

In a further work, the assessment of the thermal efficiency has been carried out to study the thermal behaviour of the PCMs. The use of models that imitate building envelopes might be also suggested as future achievement.

Acknowledgements. Funded by Spanish Ministerio de Ciencia e Innovación, grant number PID2020-119975RB-I00 LIMORTHER and by Gobierno de Navarra, MULTIFICON PC143.

References

1. Alvarez, J.I., et al.: RILEM TC 277-LHS report: a review on the mechanisms of setting and hardening of lime-based binding systems. Mater. Struct. **54**, 63 (2021)
2. Himran, S., Suwono, A., Mansoor, G.A.: Characterization of alkanes and paraffin waxes for application as phase change energy storage medium. Energy Sour. **16**(1), 117–128 (1994)
3. Frigione, M., Lettieri, M., Sarcinella, A., Barroso de Aguiar, J.: Sustainable polymer-based Phase Change Materials for energy efficiency in buildings and their application in aerial lime mortars. Constr. Build. Mater. **231**, 1–11 (2020)
4. Liu, H., Wang, X., Wu, D.: Innovative design of microencapsulated phase change materials for thermal energy storage and versatile applications: a review. Sustain. Energy Fuels **3**(5), 1091–1149 (2019)
5. Arizzi, A., Cultrone, G.: Aerial lime-based mortars blended with a pozzolanic additive and different admixtures: a mineralogical, textural and physical-mechanical study. Constr. Build. Mater. **31**, 135–143 (2012)
6. Duran, A., González-Sánchez, J.F., Fernández, J.M., Sirera, R., Navarro-Blasco, I., Alvarez, J.I.: Influence of two polymer-based superplasticizers (poly-naphthalene sulfonate, PNS, and lignosulfonate, LS) on compressive and flexural strength, freeze-thaw, and sulphate attack resistance of lime-metakaolin grouts. Polymers **10**(8) (2018)
7. European Committee for Standarization. UNE-EN 1015-3 Methods of Test for Mortar for Masonry. Part 3: Determination of consistence of fresh mortar (by flow table). EN 2006
8. European Committee for Standarization. UNE-EN 1015-6 Métodos de ensayo de los morteros para albañilería. Parte 6: Determinación de la densidad aparente del mortero fresco. EN 1999
9. European Committee for Standarization. UNE-EN 1015-7:1999 Methods of Test for Mortar for Masonry. Part 7: Determination of Air Content of Fresh Mortar. EN 1999
10. European Committee for Standarization. UNE-EN 83-816-93 Test methods. Fresh Mortars. Determination of water retentivity. EN 1993

11. European Committee for Standarization. UNE-EN 1015-9:2000 Methods of Test for Mortar for Masonry. Part 9: Determination of Workable Life and correction Time of Fresh Mortar. EN 2000

12. Izaguirre, A., Lanas, J., Álvarez, J.I.: Behaviour of a starch as a viscosity modifier for aerial lime-based mortars. Carbohydr. Polym. **80**(1), 222–228 (2010)

13. Fernández, J.M., Duran, A., Navarro-Blasco, I., Lanas, J., Sirera, R., Alvarez, J.I.: Influence of nanosilica and a polycarboxylate ether superplasticizer on the performance of lime mortars. Cem. Concr. Res. **43**(1), 12–24 (2013)

14. Naresh, R., Parameshwaran, R., Vinayaka, V.: Bio-Based Materials and Biotechnologies for Eco-Efficient Construction. Woodhead Publishing, Cambridge (2020)

Time-Dependent Deformations of Lime-Based Mortars and Masonry Specimens Prepared with Them

Ioanna Papayianni[1](✉) and Emmanuella Berberidou[2]

[1] Professor Emeritus, Aristotle University of Thessaloniki, Thessaloniki, Greece
papayian@civil.auth.gr
[2] Civil Engineer MSc in Preservation of Historic Structures, Aristotle University of Thessaloniki, Thessaloniki, Greece

Abstract. Lime-based mortars are preferred in repairing interventions of Historic Masonries (HM), which are the bearing elements of old structures, since they are considered compatible with the existing mortars. In historic castles, towers and particularly in cases of thick mortar joints reconstruction, long term deformations are of great interest for the structural stability of the repaired monuments. The exposure of lime-based mortars to drying and sustained loading influences both phenomena of shrinkage and creep. In this experimental work, the behaviour of different composition mortars (based on hydrated lime, lime-pozzolan, hydraulic lime and lime+pozzolan+cement) have been studied under sustained load by using spring-loaded creep frames.

The same types of mortars were also used for the masonry specimens with traditional roman type bricks in successive layers. All specimens, mortar prisms $(4 \times 4 \times 16)$ cm and masonry specimens were placed in a moisture-controlled room of RH 55–65% and 20 °C temperature. Bricks and mortars' compressive strength have been separately estimated by crushing proper samples. Deformations were recorded for more than 6 months because of drying and sustained loading. It seems that the strain measured is higher than in the case of cement-based mortars/concretes. Furthermore, for thick joints the composition of mortar and its quality, expressed as compressive strength play a crucial role in the total deformation measured apart from the sustained load value. In addition, the rate of creep evolution and the order of creep magnitude are indicated.

Keywords: lime-based mortars · masonry specimens · creep deformations

1 Introduction

Creep deformations of Historic Masonry (HM) have been studied since the 70s [1]. During the decades 80s and 90s, many research papers were published on this phenomenon since it was associated with the collapse mechanism of high and heavy HM of monumental structures such as Pavia Civic Tower (1989), Noto Cathedral (1996) [2–6]. Other researchers proceeded to predict the failure of HM by modelling creep deformations contributing to structural Analysis and behaviour of Historic Structures [7].

© The Author(s), under exclusive license to Springer Nature Switzerland AG 2023
V. Bokan Bosiljkov et al. (Eds.): HMC 2022, RILEM Bookseries 42, pp. 599–607, 2023.
https://doi.org/10.1007/978-3-031-31472-8_47

In [8, 9], researchers have shown that the nature of creep phenomenon is related to internal movement of adsorbed or intercrystalline water to internal seepage to the outside of mortars since mortars/concrete from which all evaporable water has been removed exhibit practically no creep.

Pohle et al. [10] mentioned in their very systematic work concerning Frauenkirche of Dresden "creep of historic masonry is determined exclusively by the mortar. The most significant creep comes from the joint mortar and depends on its thickness and type of mortar used".

Apart from external factors influencing creep, such as a direct proportionality between creep and applied stress, ambient relative humidity, age of loading and exposed area of the total area of the structure [9], in the case of brick masonry of historic structures the mortar's characteristics play an important role to the long-term behaviour of masonry under sustained loading. The type of binder (i.e., hydrated lime, lime+pozzolan, hydraulic lime, lime+pozzolan+cement) and the water/binder ratio, by which the porosity of the matrix is mainly affected, are closely related to the strength development of the mortar and seepage of moisture available out of the binders' matrix. In general lime-based mortars are "soft" and (relatively to cement-based ones) of high porosity. It is very vulnerable to the drying process, which leads to shrinkage and creep that co-occur under constant load and are measured as long-term deformations.

The creep is influenced by the origin, total content, gradation and in particular the elastic modulus of elasticity of aggregates. The higher the modulus of elasticity of aggregates, the greater the restraint of creep offered to the mortar. As was shown in research [11], the presence of coarse aggregates in the lime-based mortar mixtures contributes to the considerable (up to 70%) reduction of the long-term deformations. This seems to justify why pebbles or coarse aggregates were used in the composition of joint mortars in castes and massive structures.

Furthermore, by comparing creep deformations of mortars' composition, which only differ in the presence of crushed brick as powder replacing pozzolan or as part of aggregates [11], it seems that brick powder or fine brick aggregates restrain both shrinkage and creep deformations.

In reference [12], two mortar compositions of the same binding system, but in one of them half of the aggregate content was replaced by crushed brick, have shown considerably different time-dependant deformations. The mortar with crushed brick showed about half of the long-term deformation compared to the corresponding one without crushed brick.

Considering, the high water retentivity of fine crushed brick or brick powder, it could be said that this component of mortars contributes to the restraint of long-term deformations.

Reconstruction of missing parts, lacunae, or deep pointing with lime-based mortars is common in repairing old structures. In this paper, an effort was made to study the long-time behaviour of lime-based mortar (without crushed brick powder) in which the water demand for a workability of 15 ± 1cm (EN 1053–3) and water/binder ratio was reduced by using 1% by mass of binders superplasticizers of carboxylic basis.

2   Experimental Part

The raw materials used for the experiments were:
  Binders:

- Hydrated lime (powder) CL90
- Hydraulic lime, NHL 3.5
- Natural pozzolan from Milos Island ground to a fineness, 10% by mass retained on 45μm sieve. The reactivity of it with lime was checked according to ASTM C311–83.

  Aggregates:

- A mix of river siliceous sand containing 60% sand (0–4) mm and 40% of coarse sand (4–8) mm. The granulometric curve of the mix was even.

  Superplasticizer: 1% by mass of binders, superplasticizer of carboxylic – acid basis.
  A traditional type of fired bricks of dimensions (30x15x3) cm were produced by a manufacturer and used for the making the masonry specimens after cutting them properly. The basic characteristics were tested, and the mean values of samples were:

- Compressive strength 12.7 MPa
- Flexural strength 5.42 MPa
- Dynamic modulus of elasticity 6.18 MPa

The preparation of the mortar mixtures followed the relevant EN Standards (EN 1015 series). A flow table of 15 ± 1 cm was adopted for the mortars that will be used as bedding mortars. The use of superplasticizer (1% by mass of binders for all compositions) allowed the achievement of water/binder ratios below 0.6 apart from 1(L) composition for which the w/b ratio was 0.93.

A slight vibration of molded specimens $(4 \times 4 \times 16)$ cm was used to remove enclosed air during mixing. After demolding (3–4 days after casting), the mortar specimens were placed in proper curing conditions as described in EN 459.

The proportions of the mortar compositions are given in Table 1. Many mortar specimens were prepared to be adequate for testing their physical and mechanical characteristics such as porosity, specific gravity, flexural and compressive strength at 28d, 90d and 180d age. In addition, for long-time deformations testing two mortar specimens from each composition were placed after demolding in a climate chamber (RH $65 \pm 5\%$, temp 20 ± 2 °C) and their shrinkage was measured as dV/V (volume change) for 6 months.

Two mortar specimens from each composition were placed in a specific metallic spring-type apparatus (following the description of ASTM C512–83) see Fig. 1, for measuring creep deformations under a sustained load equal to 30% of the mean value of their 28d compressive strength.

In parallel, the traditional type bricks were properly cut into pieces, and four masonry specimens for each mortar compositions were made, layed as shown in Fig. 2.

It must be noted that the brick pieces were embedded in water for some minutes and used surface dried for building the specimens. To estimate the sustained load for masonry specimens of each mortar composition, two of them at 90d age were crushed in compression and the mean compressive strength was defined.

At 90d age, the masonry specimens were loaded with a sustained load equal to 40% of their mean value compressive strength for each mortar type, following ASTM C593 (2000) method.

Table 1. Mortar mixture compositions

Materials	Proportions Parts by mass			
	1 (L)	2 (L+P)	3 (NHL)	4 (L+P+C)
Hydrated Lime CL90 (L)	1	1	–	1
Pozzolan milled retained 10% on 45μm sieve (P)	–	1	–	0.7
Hydraulic Lime NHL 3.5 (NHL)	–	–	1	–
River sand (0-8mm)	2.5	5.0	2.5	5.0
Superplasticizer 1% by mass of binders	√	√	√	√
White cement CEM 52.5N (C)	–	–	–	0.3
Workability (cm) EN 1015–3	14	16	15	15
Water/binder	0.93	0.58	0.53	0.59

Fig. 1. Apparatus for testing creep deformations

Fig. 2. Masonry specimen with traditional type bricks and lime-based mortars

3 Results

The basic properties of mortar compositions are shown in Table 2 as well as the ultimate volume change due to drying shrinkage. The changes in the volume of mortar specimens were measured by using digital calipers. Creep deformations of mortars were measured continuously by micrometers of high precision (0.001 mm) and shown in Fig. 3. The compressive strength of masonry specimens at 90 d age is given in Fig. 4, while the creep deformations of them up to 180 days have been plotted in Fig. 5.

Table 2. Basic Properties of Mortar Compositions

Mortar compositions	90d Porosity %	90d Specific gravity	Compressive strength (MPa)			Flexural strength (MPa)			Ultimate volume change after 180d (dV/V)
			28d	90d	180d	28d	90d	180d	
1 (L)	29.0	1.74	0.7	0.8	0.9	0.035	–	–	0.030
2 (L+P)	23.2	1.82	3.6	7.1	7.8	1.04	1.60	1.64	0.027
3 (NHL)	22.7	1.90	4.1	8.6	8.7	1.05	3.00	3.13	0.022
4 (L+P+C)	25.2	1.80	5.6	10.6	12.6	2.40	3.10	3.30	0.034

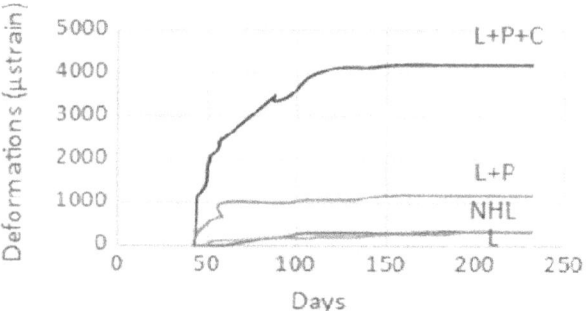

Fig. 3. Creep deformations of mortar compositions

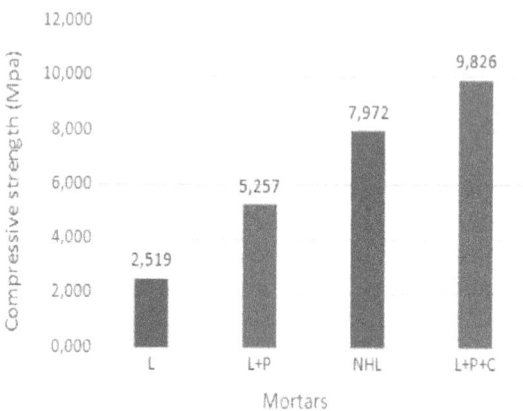

Fig. 4. Compressive strength of masonry specimens at 90-d age

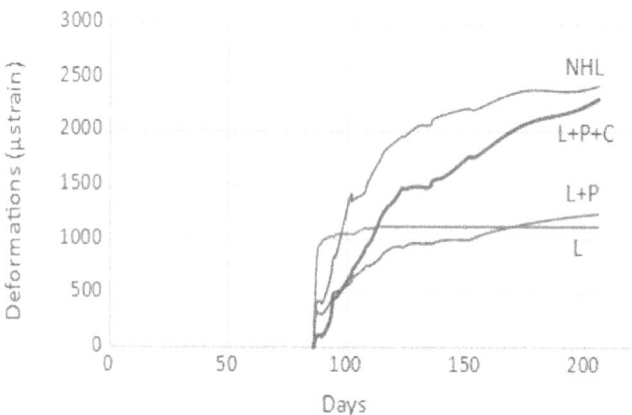

Fig. 5. Creep deformations of masonry specimens built with mortar mixtures

4   Discussion

Regarding properties of lime-based mortars in Table 2, it could be said that apart from pure lime mortars, the compressive strength of which remains below 1 MPa (90d 0.85 MPa), the other compositions developed at 28d adequate strength above 3.5 MPa while at 90d their strength reaches a value near the maximum measured at 180 days. As expected the stronger the hydraulic component of the binder, the higher the developed strength. The 90d porosity of mortars containing hydraulic component ranges from 22.7% to 25.2%, while the porosity for pure lime mortars is 29.0%.

Regarding the ultimate volume changes, it seems that drying and the resultant volume changes that occur after the demoulding of mortars and up to 180d drying range from 2.2% to 3.4%, the higher value shown by composition of the ternary binding system (L+P+C). The evolution of changes is higher within the first 30–35 days after the exposure to climatic chamber conditions and then continue to increase slowly up to 150 days reaching the maximum of volume change. The highest volume change due to shrinkage was recorded by the mortar composition No 4 with L+P+C ternary binder.

The creep deformations of mortar compositions were measured at 28d age by applying 30% of their 28d compressive strength. During loading of the two pure lime mortar specimens, one of them was crushed, and then a lower load was applied to about 15% of mortars' strength (instead of 30%). Furthermore, for the hydraulic lime mortars the deformations up to 60 days after loading were lost due to the detachment of the micrometer. Comparing the curves, it could be said that creep deformations of lime-based mortars without crushed brick but with superplasticizer are higher than that of cement-based mortars/concretes (300–800μstr) and range from 1200 to 4200 mstr.

Regarding the masonry specimens 90d strength, the highest value was exhibited by specimens built with mortar of composition 4 (L+P+C) and was 9.82 MPa. The lowest strength of 2.51 MPa was that of masonry specimens with pure lime mortars. The applied load was defined to be 40% of the mean compressive strength for each pair of masonry specimens with mortars 1 (L), 2(L+P), 3(NHL), 4 (L+P+C). However, checking with load cells the sustained load it was found that the actual loading in the case of 1 (L) and 2 (L+P) series was slightly lower and equal to the 30% of corresponding strength. Therefore, the comparison of creep deformations plotted in Fig. 5 was not possible since the proportionality of stress/strength was not kept the same for all specimens. However, the creep deformations developed in masonry specimens ranged from 1200–2400 μstrain. Problems with masonry specimens with pure lime mortars concerning their mechanical characteristics are often mentioned by researchers [13], as well as the disparity of the experimental results concerning creep deformations of masonry wallets with lime-based mortars [14]. However, parameters involved in relevant prediction models show a significant scatter to be trusted more than experimental results.

5   Conclusions

The creep deformations of mortars are higher than those of masonry specimens made with these mortars. This could be related to the exposure area of mortars to drying in masonry specimens and stress/strength ratio.

The issue that the lower strength lime mortars exhibit the higher creep deformation reported in literature was not confirmed in this research. The stress/strain ratio seemed to be an important influencial factor.

The creep deformations of lime-based mortar modified with the addition of super-plasticizer were lower than those referred to in literature and ranged from 1200–4200 μstrain while the masonry specimens with these mortars developed creep deformations up to 2400 μstrain. The order of creep magnitude is higher than that of cement-based mortars but lower than deformations mentioned in literature for lime-based mortars without the addition of superplasticizers.

The rate of creep evolution is different in pure lime mortars compared with lime+pozzolan or hydraulic lime mortars. In the former, the evolution occurs in a short time, while in the latter much longer time is needed to reach final creep.

The decrease of water demand for a required level of workability by using super-plasticizers seems to result in the reduction of the long-time deformations of lime-based mortars.

References

1. Lenczner, D., Salahuddin, J.: Creep and moisture in piers and walls. In: Jessop, E.L., Ward, M.A. (eds.) Proceedings of the 1st Canadian Masonry Symposium, pp. 72–86
2. Shrive, N.G., England, G.L.: Elastic, creep and shrinkage behaviour of masonry. Int. J. Masonry Constr. $1(3)$, 103 (1981)
3. Schubert, P.: Deformation of masonry due to shrinkage and creep. In: Wintz, J.A., Yorkdale, A.H. (eds.) Proceedings of the 5th International Brick Masonry Conference (VIBMaC) held in Washington, DC, 5–10 October 1979. McLean, Virginia (1982)
4. Binda, L., et al.: Collapse of the civic tower of Pavia: a survey of the materials and structure. Masonry Int. $6(1)$, 11–20 (1992)
5. Binda, L., Anzani, A., Saisi, A.: Failures due to long-term behaviour of heavy structures: the Pavia civic tower and the noto cathedral. In: Brebbia, C.A. (ed.) Proceedings of Structural Studies Repairs and Maintenance of Heritage Architecture VII, WIT Transactions on The Built Environment, vol. 66 (2003)
6. Anzani, A., Roberti, G.M., Binda, L.: Time dependent behaviour of masonry: experimental results and numerical analysis. In: Brebbia, C.A., Frewer, R.I.B. (eds.) Proceedings of the 3rd International Conference on STREMA III, Computational Mechanics Publications, Southampton-Boston, pp. 415–422 (1993)
7. Papa, E., Taliercio, A.: Creep modelling of masonry historic towers. WIT Trans. Built Environ. **66**, 10 (2003)
8. Glucklich, J., Ishai, O.: Creep mechanism in cement mortar. In: Journal Proceedings, vol. 59. No. 7 (1962)
9. Neville, A.M.: Theories of creep in concrete. In: Journal Proceedings, vol. 52. No. 9 (1955)
10. Pohle, F., Jäger, W.: Material properties of historical masonry of the Frauenkirche and the masonry guideline for reconstruction. Constr. Build. Mater. $17(8)$, 651–667 (2003)
11. Papayianni, I.: Creep deformation of lime-based repair mortars. The effect of aggregate size. In: Proceedings of the 13th international brick/block masonry conference. Vermeltfoort Amsterdam (2004)
12. Karaveziroglou, M., et al.: Creep behaviour of mortars used in restoration. In: Brebbia, C.A., Lefteris , B. (eds.) Proceedings of the Computational Mechanics Publications, pp. 231–239 (1995)

13. Costigan, A., Pavía, S., Kinnane, O.: An experimental evaluation of prediction models for the mechanical behavior of unreinforced, lime-mortar masonry under compression. J. Build. Eng. **4**, 283–294 (2015)
14. Ignoul, S., et al.: Creep behavior of masonry structures–failure prediction based on a rheological model and laboratory tests. Struct. Anal. His. Construct. 913–920 (2006)

Influence of Methyl Cellulose in Injection Grout on Mould Growth on Mural Paintings – Preliminary Results

Andreja Padovnik[1(✉)], Violeta Bokan Bosiljkov[1], Polonca Ropret[2], and Janez Kosel[2]

[1] Faculty of Civil and Geodetic Engineering, University of Ljubljana, Ljubljana, Slovenia
andreja.padovnik@fgg.uni-lj.si

[2] Institute for the Protection of Cultural Heritage of Slovenia, Conservation Centre, Research Institute, Ljubljana, Slovenia
janez.kosel@zvkds.si

Abstract. Commercially non-structural injection grouts often contain cellulose ether as an additive, mainly to modify the viscosity and stability of the grout. Nevertheless, in our previous field studies we observed mould growth after injection of commercial grout that contained methyl cellulose (MC) as an additive. Therefore, our objective was to determine the effect of MC on mould development after grouting of painted plaster layers. For this purpose, a painted panel sandwich model, simulating delaminated plaster layers, was inoculated with fungal strains isolated from various cultural heritage sites. After incubation (27 days), mould growth was assessed using Calcofluor white fluorescent staining. Our preliminary results show that during the incubation the moisture content on the surface of the dry part of the model increased steadily due to the high humidity in the chamber and the absorption capabilities of the porous plaster. Moreover, grouting increased the moisture content on the surface of the paint layer by ~10%. As a result, fungal stains EXF-15333 and EXF-15047 grew exclusively on the dry injected part of the surface (growth of 30%) and no growth was observed on the dry non-injected part. Alarmingly, when the mortar was removed, it was revealed that the injected grout in the air pockets had not cured after 27 days.

Keywords: mural painting · casein binder · methyl cellulose · injection grout · mould

1 Introduction

Cellulose ethers (CEs), namely ethyl cellulose (EC), hydroxyethyl cellulose (HEC), hydroxypropyl cellulose (HPC), carboxy methyl cellulose (CMC), methyl cellulose (MC) and ether mixtures: hydroxypropyl methyl cellulose (HPMC) and hydroxyethyl methyl cellulose (HEMC); are produced though chemical processing (sodium hydroxide) of ground cellulose derived from various plants [1]. They are mainly used as viscosity modifiers, plasticity modifiers and water retention agents, and they improve bonding and thickening [2, 3]. They play an important role in the technology of lime mortars, improving their compressive strength and adhesion, and extending the setting and hardening processes. In the mortar, they slow down the growth rate and crystallinity of CaCO3 crystals, producing a more compact lime mortar structure [4]. In addition, commercially available injection grouts often also contain cellulose ethers to provide some required properties to the grout, in particular, to change the viscosity and stability of the mass. Lastly, cellulose ethers are widely used in conservation and restoration, as a medium for pigments [5–7], as a consolidant [8] and for basketry [9], for relining canvas [10], as an adhesive for textiles [11, 12] and wallpapers [13–15].

Steger et al. [16] reported that 33% of the examined CEs released harmful volatile organic compounds (VOC), which caused slight corrosion of at least one of the metal coupons in the Oddy test. The corrosion compounds included oxides such as massicot, litharge, cuprite, and tenorite among carbonates (hydrocerussite, plumbonacrite), and acetates such as basic lead acetate, lead acetate trihydrate as well as lead format. During field surveys, we observed rapid indoor mould development on the painted ceilings and walls that had recently been consolidated with injection grout containing methyl cellulose (MC). To date, no published studies could be found that explain the observed biodegradation. Therefore, the aim of this work was to determine the effect of methyl cellulose in injection grout used to consolidate delaminated painted plaster layers on the development of surface mould. For this purpose, special panel sandwich models simulating delaminated plaster layers were used.

2 Materials and methods

2.1 Panel Sandwich Model and Injection Grout

For the panel sandwich model, rough and fine lime plasters were used. The rough plaster was prepared from 1:3 slaked lime putty: coarse sand (0/4 mm) lime mortar, and the fine plaster from 1:2 slaked lime putty: fine sand (0/1 mm) lime mortar. Both mortars were mixed with a Hobart mixer for 180 seconds. The slaked lime putty was 2 years old and contained about 51% water and 49% $Ca(OH)_2$, and this was the only water used to prepare rough and fine plaster.

The panel sandwich models were prepared following the instructions of Padovnik et al. [17] and Azeiteiro et al. [18], to simulate 5 mm high detachment of the fine plaster from the rough plaster by air pockets.

Casein tempera with yellow ochre pigment was applied to the fine plaster of a three-year-old panel sandwich model to simulate a painted mural surface. The casein tempera was made from two parts fresh casein obtained from skimmed cow's milk and one part slaked lime putty. The casein was mixed and dissolved in lime putty to obtain a sticky and dense emulsion, which was diluted with three parts of lime water. The yellow ochre pigment was mixed with the liquid binder at a ratio of 60:40. The paint was then applied with a brush to a fine plaster layer (Fig. 1).

The grout mixture consisted of commercial hydrated lime of class CL 70-S (IAK, Kresnice, Slovenia) according to EN 459-1 [19], finely ground limestone (CALCIT, Stahovica, Slovenia) as a filler and two chemical admixtures. The admixtures used were polycarboxylate ether-based super plasticiser (PCE) and hydroxyethyl methyl cellulose

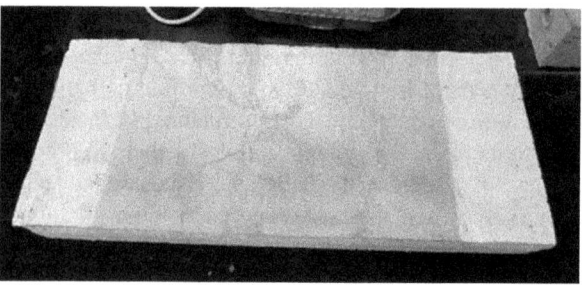

Fig. 1. Panel sandwich model with casein paint layer.

(HEMC) from Kremer Pigmente GmbH & Co KG, Germany (product name: Tylose MH). The properties of the HEMC used are reported in [20].

The grout mixture in this study is based on the volume ratio of 1:3 (lime: filler) [21], which was converted into a weight ratio to obtain identical compositions of the tested grout (Table 1). The grout mixture was composed of 290 g lime, 1030 g filler, and 540 g water (water/binder ratio of 1.86). The content of superplasticiser and HEMC was equal to 1 % of the total mass of solids (lime + filler).

First, the 1% HEMC solution was prepared. A certain amount of HEMC powder was slowly added to the water with constant stirring (water temperature $\leq 50°C$) so that the HEMC dissolved completely. Then the solution was settled and cooled.

The grout mixture was prepared using a KitchenAid mixer with a power of 300 W and a stainless-steel gate anchor blade. First, the lime and the filler were mixed. Then, 70% of the water with HEMC was added and mixed for 2 min at low speed (540 rpm). In the last 15 s of the low-speed mixing, the PCE-SP and 30% of the water were added. Finally, each grout was mixed at high speed (1200 rpm) for 3 min.

The average values of wet density, mini-slump flow, bleeding after 3 hr and water retention capacity of the grout are shown in Table 1 [21]. The water retention capacity of the grout is high (98%), which means that the water in the grout is kept at the same level for an extended period, which is especially important when consolidating delaminated plaster layers under changing temperature and humidity conditions [22].

Table 1. Wet density (g/cm^3); mini-slump flow (mm); bleeding after 3 hr (%) and water retention capacity (%) of the fresh injection grout with HEMC [21].

	Wet density (g/cm3)	Mini slump flow (mm)	Bleeding after 3 h (%)	Water retention capacity (%)
Injection grout with HEMC	1.32	203	0.4	98

After the casein paint had dried for one month, the dry air pockets were injected with a syringe. The injected air pockets were sealed with tape to prevent air infiltration (Fig. 3a).

2.2 Fungal Isolates

Strains isolated from the following artefacts and environments were used to inoculate the surfaces of the prepared model samples: air in the depot of the restoration centre, oil paintings on canvas, an African wooden sculpture and an old baptism book (Table 2). All isolates are designated as EXF and were supplied by the Infrastructural Centre Mycosmo-Culture-Collection, Slovenia. All were grown to sporulation on malt extract

Table 2. Five fungal strains isolated mostly from the interiors of cultural heritage institutions.

Isolate designation	Identified species	Source	Isolation procedures	Identification method	Collected by/ Reference
EXF-9717	*Penicillium canescens*	Baptism book of the parish of Šmartena: Die ll:ma Baptizatus est Josephus (from around 1400 ad)	Dilution plates (MEA) after swabbing as described by Jurjević et al. [23], and DNA isolation from colonies	ITS (primers ITS1, NL4 and ITS4) amplification and sequencing described in Sklenář et al. [24]	P. Zalar[A], M. Matul[A] (May 2014)
EXF-15047	*Aureobasidium melanogenum*	Air in the depot of the RC of IPCHS	Bio-aerosol impaction sampler as described by Peterson & Jurjević [25], plating on MEA and DNA isolation from colonies	ITS (primers ITS1, NL4 and ITS4) amplification and sequencing described in Sklenář et al. [24]	P. Zalar[A], M. Matul[A] (May 2014)
EXF-7651	*Aspergillus destruens*	Oil painting on canvas of the NGS	Dilution plates (MEA) after swabbing as described by Jurjević et al. [23], and DNA isolation from colonies	ITS (primers ITS1, NL4 and ITS4) and benA (primers Bt2a, T10 and Bt2b) amplification and sequencing described in Sklenář et al. [24]	[24]
EXF-15333	*Cladosporium halotolerans*	Oil painting on canvas (Vittore Carpaccio) from the RC of IPCHS	Dilution plates (MEA) after swabbing as described by Jurjević et al. [23], and DNA isolation from colonies	ITS (primers ITS1, NL4 and ITS4) and Act (primers ACT-512F and ACT-783R)	P. Zalar[A], M. Matul[A] (May 2014)

(*continued*)

Table 2. (*continued*)

Isolate designation	Identified species	Source	Isolation procedures	Identification method	Collected by/ Reference
EXF-15148	*Aspergillus creber*	Religious wooden sculpture from Mali, (first half of the 20th cent., SEM)	Dilution plates (MEA) after swabbing as described by Jurjević et al. [23], and DNA isolation from colonies	ITS (primers ITS1, NL4 and ITS4) and benA (primers Bt2a, T10 and Bt2b) amplification and sequencing described in Sklenář et al. [24]	P. Zalar[A], M. Matul[A] (May 2014)

Restoration Centre (RC) of the Institute for the protection of Cultural Heritage of Slovenia (IPCHS); Slovene Ethnographic Museum (SEM); Internal transcribed spacer (ITS); β-tubulin (benA); Actin (Act); and Infrastructural Centre Mycosmo-Culture-Collection, Slovenia (EXF, https://www.ex-genebank.com/). [A]Microbiology, Department of Biology; Chair of Molecular Genetics and Microbiology, Večna pot 111, Ljubljana, Slovenia.

agar (MEA, 30 g/L of malt extract (Sigma Aldrich) and 15 g/L of agar (Sigma Aldrich)) solid medium at 26 °C.

2.3   Inoculation of the Painted Injected Mortar Plate

Spores of fungal monocultures on MEA plates were collected and stored at 4 °C in skimmed milk (1 mL) which was instantly dehydrated by mixing it with 10 mL of anhydrous granular silica gel. Prior to mixing, the latter was dry-heat sterilised at 175 °C for 2 h in 15 mL glass vials [26], turning it into yellow granules. To prepare for the inoculation, 0.5 mL of granules containing spores were suspended in 0.5 ml of Milli-Q water, and spore count was determined under the microscope Zeiss LSM 800 using a standard Bürker-Türk counting chamber (hemocytometer) [27]. The final spore concentration was adjusted to around 1×10^7 spores/mL by additional dilutions in Milli-Q water.

For final inoculation, the painted mural surface was divided into a grid of 1.5 cm x 1 cm squares (Fig. 2). Then, the spore suspension was applied to the entire surface of an individual square using a sterile brush. Each fungal strain was inoculated on 3 separate squares and two surface (Fig. 2) – on the dry injected and dry non-injected parts.

After that, the sandwich model was placed in a container with the paint layer facing up. Prior to that, the bottom surface of the container was covered by a 3 cm layer of distilled water; the distance between the water level and the paint layer was 20 cm. The container was subsequently covered by a glass panel (Fig. 3b). Constant temperature (23.7 ± 2 °C) and relative humidity (95 ± 3%) were maintained inside the container (Fig. 4), which was placed in a dark room. The entire inoculation period of the samples spanned over 27 days.

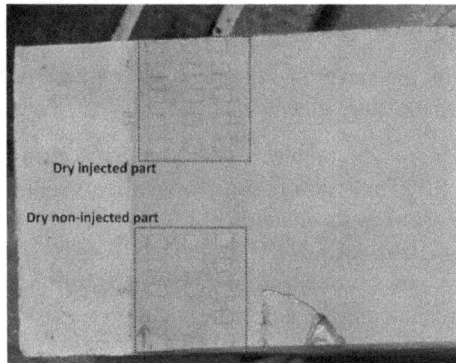

Fig. 2. Divided area for the application of spore suspension.

Fig. 3. (a) Injected air pockets are sealed with tape to prevent air infiltration. (b) The panel sandwich model in a special container.

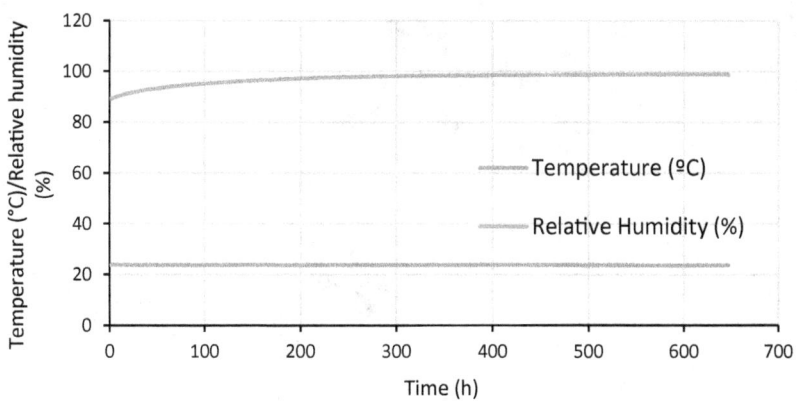

Fig. 4. Temperature and relative humidity in the container.

2.4   Measuring the Surface Moisture Content

The moisture content on the surface of the paint layer of the dry parts was measured after grouting using two moisture meters (Table 4):

- Extech Moisture Hygro-Thermometer MO300 (Sensor 1) with sensor Ahlborn Almemo FH A696- MF. The results of the measurements are given in % and represent the mass fraction of water in the material.
- Humidity meter Humidcheck CONTACT 30.5503 (Sensor 2); as a result of the measurement, relative values are obtained and converted into a descriptive estimate or water content in percent using the correlation table.

2.5   Mould Growth Analysis Using Fluorescent Microscopy

The exact percentage of inoculated mould formation on the surface of the painted mortar plate (within a particular grid square) was determined using the fluorescent microscope analysis. Firstly, each grid square had to be separated/isolated, using a hammer and a scalpel. The resulting pieces were then fixed onto a microscope glass slide using a Fimo polymer clay (Staedtler Mars GmbH & Co. KG, Germany) so that their painted surfaces faced upwards and were completely aligned with the ground surface.

Then the pieces were stained using a fluorescent dye Calcofluor White, which specifically binds to the chitin cell wall of the mould (Fig. 2). For this purpose, 50 µL of dye Calcofluor White, previously mixed in a 1:1 ratio with 10% KOH, was pipetted onto

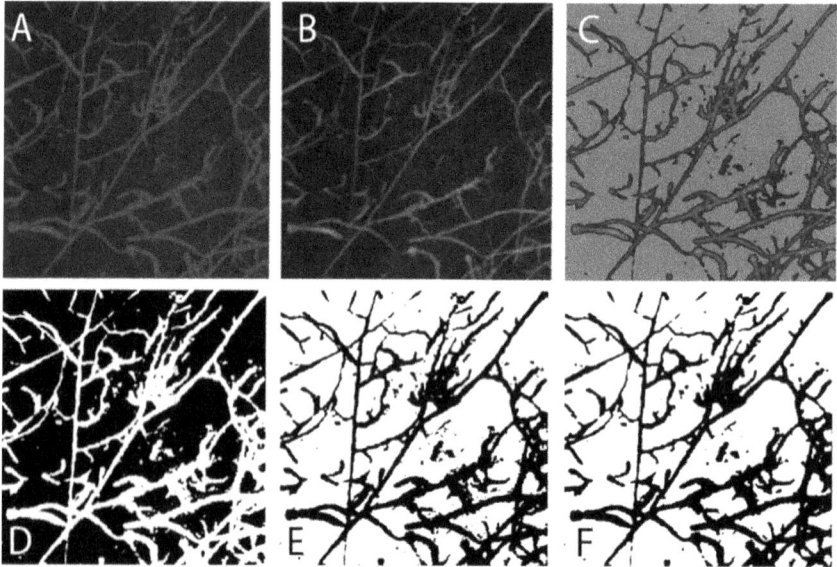

Fig. 5. Fluorescent microscopic photographs used to determine the percentage of mould surface coverage. A: original photograph; B: black and white 8 bit photograph; C: boundary focus determination between the mould and the background; D: binary format with black background and white mould; E: inverted image; and F: noise removal

the pained surface and cover glass was applied on top. Staining was analysed using the Zeiss LSM 800 confocal fluorescence microscope with a fluorescence filter with a maximum emission of 460 nm and a magnification of 50x. The individual images for each slide were analysed with an opensource image processing package of ImageJ called Fiji (version 1.52c) [28], to obtain the mould surface coverage using a triangle method and an ImageJ measuring function [29]. The images were converted to binary format with a black background and white biofilm, with an automatically determined threshold value separating the biofilm from the background. Visual noise was removed from all the images in the set, using the "despeckle" function. Considering the entire field of view of the image, we then calculated the mould surface coverage percentage using the "measure" function, and from 3 images the average percentage of coverage was calculated (Fig. 5).

3   Results and discussion

Regardless of the experimental conditions, the optical and fluorescent microscopic investigations revealed no surface growth for fungal isolates EXF-9717, EXF-7651 and EXF-15148. However, isolates EXF-15333 and EXF-15047 grew well on the dry injected part of the painted mural surface (Fig. 6, 7, 8, and 9), reaching 28.6 % and 30.4 % of surface mould growth coverage, respectively (Table 3). No mould growth was observed on the dry non-injected part of the painted mural surface for these two isolates. From the obtained results, it can be concluded that HEMC injection has a profound effect on surface mould development, since it elevates the moisture content of the dry part from 4.2 % to 13.5 % 10 min after grouting (Table 4). Since a high constant moisture content in a substrate is a major growth-limiting factor for fungi, this elevation directly sustains mould development throughout the 1-month incubation period, which ranges from 16.2 to 22.5 % (Table 4). The positive effect of HEMC on mould development was observed in the field when grout injecting was performed with commercial mix grout containing methyl cellulose as an additive (Fig. 10).

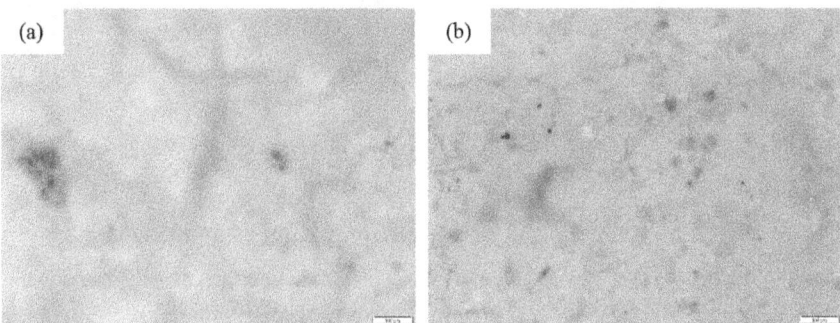

Fig. 6. Optical microscopy of the surface area within each individual grid square pre-inoculated with EXF-15047: a) dry injected part; and b) dry non-injected part. The scale bar is 100 μm.

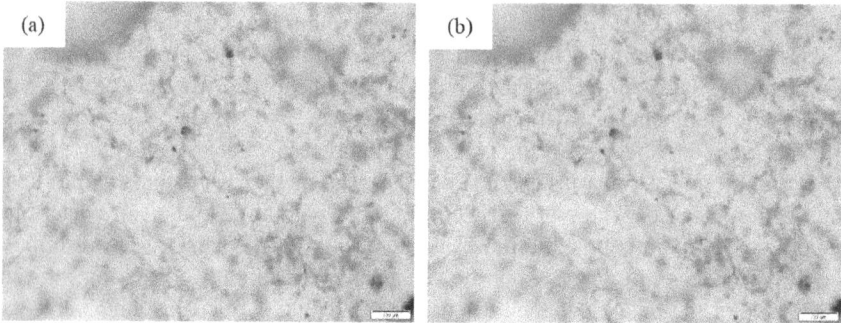

Fig. 7. Optical microscopy of the surface area within each individual grid square pre-inoculated with EXF-15333: a) dry injected part; and b) dry non-injected part. The scale bar is 100 μm.

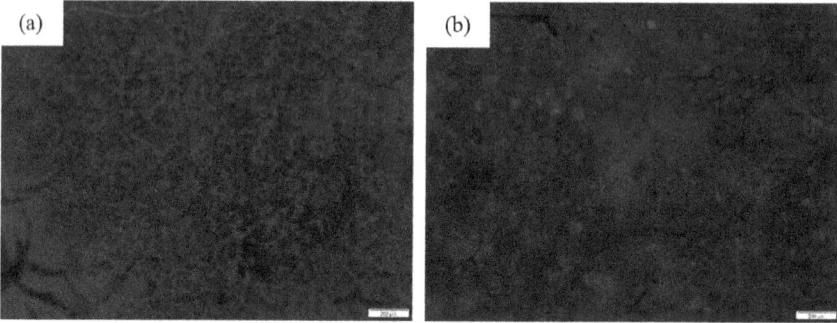

Fig. 8. Fluorescent microscopy of Calcofluor White stained surface area of each individual grid square pre-inoculated with EXF-15047: a) dry injected part; and b) dry non-injected part. The scale bar is 200 μm.

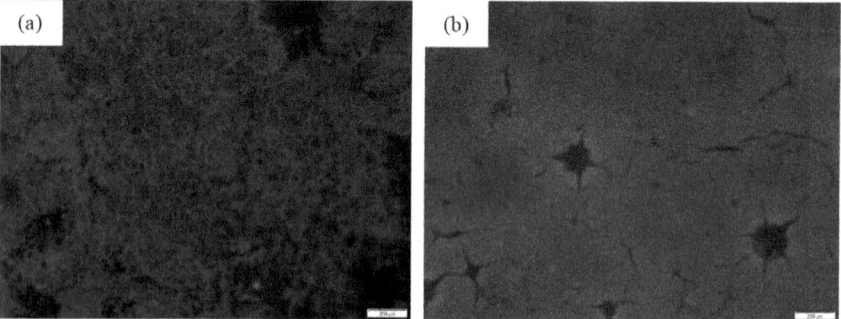

Fig. 9. Fluorescent microscopy of Calcofluor White stained surface area of each individual grid square pre-inoculated with EXF-15333: a) dry injected part; and b) dry non-injected part. The scale bar is 200 μm.

Table 3. Area of mould surface growth (%) within each individual grid square pre-inoculated with 5 different fungal isolates. Surface growth was calculated from fluorescent images captured after Calcofluor White staining.

	Area of mould surface growth (%)	
	Dry injected part	Dry non-injected part
EXF-15333	28.6 ± 2.6	0.0
EXF-15047	30.4 ± 1.3	0.0
EXF-9717	0.0	0.0
EXF-7651	0.0	0.0
EXF-15148	0.0	0.0

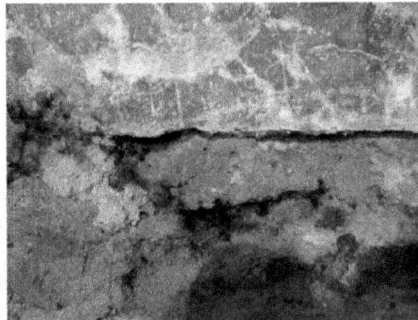

Fig. 10. Mould growth after injection of commercial grout containing methyl cellulose as an additive.

Table 4 shows that the injected part had a much higher moisture content on the painted surface than the non-injected part 10 minutes and 27 days after grouting. It is, therefore, clear that grouting increases the moisture on the surface of the paint layer.

Table 4. Moisture content on the surface of the paint layer 10 min and 27 days after grouting.

	Sensor 1 (%)	Sensor 2	Sensor 1 (%)	Sensor 2
	10 min after grouting		After 27 days	
Dry injected part	13.5	Wet (2.9 %)	16.2	Wet (3.7%)
Dry non-injected part	4.2	Dry (0.8 %)	11.5	Half-dry (2.1 %)

The comparison of moisture content values 10 minutes and 27 days after grouting shows that after 27 days, the values of the moisture content on the surface of the dry non-injected part increase due to the high humidity in the container (Figure 3) and the moisture absorption by the porous lime plaster. However, high humidity in the container leads to much higher moisture on the paint layer for the injected parts. In addition, when the paint layer with the fine mortar was removed for the analysis of mould growth, we could see that the injected grout in the air pockets was still wet after 27 days. The main reason for this is probably the high water retention of the grout (Table 1), which means that the water remains in the grout for a long time in an environment of high humidity.

4   Conclusions

The preliminary results presented allow the following conclusions:

- Grout with HEMC has a high-water retention capacity, which means that water is retained in the grout for an extended period, and the grout cannot harden, especially when the humidity of the environment is high; thus porous plaster can constantly absorb water from the air.
- Grouting increases the moisture content on the surface of the paint layer by ~10%.
- The fungal stains EXF-15333 and EXF-15047 grew only on the dry injected part of the surface (growth of 30 %).
- No growth was observed on the dry non-injected part.

However, a more extensive testing campaign with different grout compositions and moisture conditions is required for the future.

Acknowledgment. The authors would like to thank the Slovenian Research Agency for the following grants: BI-RS/20-21-013, P2-0185 and J7-3147. Moreover, we are grateful to Mojca Mlakar, Erik Putar and to Jan Bregar for their technical assistance and overall help and to Prof. Dr. Polona Zalar (Department of Biology, Biotechnical Faculty, University of Ljubljana) for her donation of fungal strains designated as EXF. These strains are maintained in the Ex Culture Collection of the Department of Biology, Biotechnical Faculty, University of Ljubljana (Infrastructural Centre Mycosmo, MRIC UL, Slovenia).

This work was financially supported by the Slovenian Research Agency, through Programme Group P2-0185.

References

1. Lilienfeld, L.: Alkyl ethers of cellulose and process of making the same. US, **188**(1), 376 (1916)
2. Franco, A.P., Recio, M.A.L., Szpoganicz, B., Delgado, A.L., Felcman, J., Mercê, A.L.R: Complexes of carboxymethylcellulose in water. Part 2. Co2+ and Al3+ remediation studies of wastewaters with Co2+, Al3+, Cu2+, VO2+ and Mo6+. Hydrometallurgy, **87**(3–4), 178–189 (2007). https://doi.org/10.1016/J.HYDROMET.2006.08.013.
3. Mierczynska-Vasilev, A., Beattie, D.A.: Adsorption of tailored carboxymethyl cellulose polymers on talc and chalcopyrite: correlation between coverage, wettability, and flotation. Miner. Eng. **23**(11–13), 985–993 (2010). https://doi.org/10.1016/J.MINENG.2010.03.025.
4. Liu, H., Zhao, Y., Peng, C., Song, S., López-Valdivieso, A.: Lime mortars – The role of carboxymethyl cellulose on the crystallization of calcium carbonate. Constr. Build. Mater. **168**, 169–177 (2018). https://doi.org/10.1016/J.CONBUILDMAT.2018.02.119

5. Ranacher, M.: Painted Lenten veils and wall coverings in Austria: technique and conservation **25**, 142–148 (2013). https://doi.org/10.1179/sic.1980.25.Supplement-1.142

6. O'Donoghue, E., Johnson, A.M., Mazurek, J., Preusser, F., Schilling, M., Walton, M.S.: Dictated by media: conservation and technical analysis of a 1938 joan miró canvas painting, **51**(2), 62–68 (2013). https://doi.org/10.1179/sic.2006.51.Supplement-2.62

7. Dignard, C., Douglas, R., Guild, S., Maheux, A., McWilliams, W.: Ultrasonic misting. Part 2, treatment applications. J. Am. Inst. Conserv. 127–141 (1997)

8. Rosenqvist, A.M.: The stabilizing of wood found in the Viking ship of oseberg—Part I, **4**(1), 13–22 (2014). https://doi.org/10.1179/sic.1959.004

9. Thomsen, F.G: Repair of a Tlingit basket using molded cotton fibers. In: ICOM Committee for Conservation 6th Triennial Meeting, 21–25 September 1981, pp. 3. Ottawa (1981)

10. Wilt, M.: Evaluation of cellulose ethers for conservation. J. Paul Getty Trust (1990)

11. Hillyer, L., Tinker, Z., Singer, P.: Evaluating the use of adhesives in textile conservation: Part I: an overview and surveys of current use **21**, 37–47 (2010). https://doi.org/10.1080/01410096.1997.9995114

12. Masschelein-Kleiner, L., Bergiers, E.: Influence of adhesives on the conservation of textiles **29**, 70–73 (2013). https://doi.org/10.1179/sic.1984.29.Supplement-1.70

13. Thomson, R.: Paper leather wallpapers: a contradiction in terms. Beiträge zur Erhaltung von Kunst-und Kulturgut **2**, 16–19 (2004)

14. Karnes, C., Ream, J.D., Wendelin, E.C.: Wallpapers at Winterthur: Seeing Them in a 'New Light' (2000)

15. Schulte, E.K.: Wallpaper conservation at the longfellow national historic site: parlor and dining room **20**(2), 100–110 (2013). https://doi.org/10.1179/019713681806028668

16. Steger, S., Eggert, G., Horn, W., Krekel, C.: Are cellulose ethers safe for the conservation of artwork? New insights in their VOC activity by means of Oddy testing. Herit. Sci. **10**(1), 1–12 (2022). https://doi.org/10.1186/S40494-022-00688-4/FIGURES/2

17. Padovnik, A., Piqué, F., Jornet, A., Bokan-Bosiljkov, V.: Injection grouts for the re-attachment of architectural surfaces with historic value—measures to improve the properties of hydrated lime grouts in Slovenia. Int. J. Archit. Heritage: Conserv. Anal. Restor. **10**(8), 993–1007 (2016). https://doi.org/10.1080/15583058.2016.1177747

18. Azeiteiro, L.C., Velosa, A., Paiva, H., Mantas, P.Q., Ferreira, V.M., Veiga, R.: Development of grouts for consolidation of old renders. Constr. Build. Mater. **50**, 352–360 (2014). https://doi.org/10.1016/J.CONBUILDMAT.2013.09.006

19. EN 459-1. Building Lime - Part 1: Definitions, Specifications and Conformity Criteria. CEN, Brussels, Belgium (2010)

20. Tylose® MH 300. https://www.kremer-pigmente.com/en/shop/mediums-binders-glues/63600-tylose-mh-300-p2.html. Accessed 05 May 2022

21. Odić, M., Padovnik, A., Šeme, B.: The influence of methyl cellulose on the fresh properties of non-structural injection grouts. In: Summaries of the International Meeting of Conservators-Restorers 2022, pp. 114. Posavski muzej Brežice, Slovenia (2022)

22. Spychał, E.: The effect of lime and cellulose ether on selected properties of plastering mortar. Procedia Eng. **108**, 324–331 (2015). https://doi.org/10.1016/j.proeng.2015.06.154

23. Jurjević, Željko, Kubátová, A., Kolařík, M., Hubka, V.: Taxonomy of aspergillus section petersonii sect. nov. encompassing indoor and soil-borne species with predominant tropical distribution. Plant Syst. Evol. **301**(10), 2441–2462 (2015). https://doi.org/10.1007/s00606-015-1248-4

24. Sklenář, F., et al.: Phylogeny of xerophilic aspergilli (subgenus aspergillus) and taxonomic revision of section restricti. Stud. Mycol. **88**, 161–236 (2017). https://doi.org/10.1016/j.simyco.2017.09.002

25. Peterson. S.W., Jurjević, Z.: Talaromyces columbinus sp. nov., and genealogical concordance analysis in Talaromyces clade 2a. PLoS One, **8**(10), 78084 (2013). https://doi.org/10.1371/journal.pone.0078084

26. Perkins, D.D.: Preservation of Neurospora stock cultures with anhydrous silica gel. Can. J. Microbiol. **8**(4), 591–594 (1962)

27. Cadena-Herrera, D., et al.: Validation of three viable-cell counting methods: manual, semi-automated, and automated. Biotechnol. Rep. **7**, 9–16 (2015). https://doi.org/10.1016/j.btre.2015.04.004

28. Abràmoff, M.D., Magalhães, P.J., Ram, S.J.: Image processing with ImageJ. Biophotonics Int. **11**(7), 36–42 (2004)

29. Zack, G.W., Rogers, W.E., Latt, S.A.: Automatic measurement of sister chromatid exchange frequency. J. Histochem. Cytochem. **25**(7), 741–753 (1997). https://doi.org/10.1177/25.7.70454

Author Index

V. Bokan Bosiljkov et al. (Eds.): HMC 2022, RILEM Bookseries 42, pp. 621–623, 2023.
https://doi.org/10.1007/978-3-031-31472-8

Printed in the USA
CPSIA information can be obtained
at www.ICGtesting.com
LVHW020813170924
791295LV00003B/250